Contents

TECHNICAL DRAWING

Books by the Authors

Basic Technical Drawing, rev. ed., by H. C. Spencer and J. T. Dygdon (Macmillan Publishing Company, 1980)

Basic Technical Drawing Problems by H. C. Spencer and J. T. Dygdon (Macmillan Publishing Company, 1972)

Descriptive Geometry, 8th ed., by E. G. Paré, R. O. Loving, I. L. Hill, and R. C. Paré (Macmillan Publishing Company, 1991)

Descriptive Geometry Worksheets with Computer Graphics, Series A, 8th ed., by E. G. Paré, R. O. Loving, I. L. Hill, and R. C. Paré (Macmillan Publishing Company, 1991)

Descriptive Geometry Worksheets with Computer Graphics, Series B, 8th ed., by E. G. Paré, R. O. Loving, I. L. Hill, and R. C. Paré (Macmillan Publishing Company, 1991)

Engineering Graphics, 4th ed., by F. E. Giesecke, A. Mitchell, H. C. Spencer, I. L. Hill, R. O. Loving, and J. T. Dygdon (Macmillan Publishing Company, 1987)

Engineering Graphics Problems, Series 1, 4th ed., by H. C. Spencer, I. L. Hill, R. O. Loving, and J. T. Dygdon (Macmillan Publishing Company, 1987)

Principles of Engineering Graphics by F. E. Giesecke, A. Mitchell, H. C. Spencer, I. L. Hill, R. O. Loving, and J. T. Dygdon (Macmillan Publishing Company, 1990)

Principles of Engineering Graphics Problems by H. C. Spencer, I. L. Hill, R. O. Loving, and J. T. Dygdon (Macmillan Publishing Company, 1990)

Technical Drawing, 9th ed., by F. E. Giesecke, A. Mitchell, H. C. Spencer, I. L. Hill, J. T. Dygdon, and J. E. Novak (Macmillan Publishing Company, 1991)

Technical Drawing Problems, Series 1, 9th ed., by F. E. Giesecke, A. Mitchell, H. C. Spencer, I. L. Hill, J. T. Dygdon, and J. E. Novak (Macmillan Publishing Company, 1991)

Technical Drawing Problems, Series 2, 9th ed., by H. C. Spencer, I. L. Hill, J. T. Dygdon, and J. E. Novak (Macmillan Publishing Company, 1991)

Technical Drawing Problems, Series 3, 9th ed., by H. C. Spencer, I. L. Hill, J. T. Dygdon, and J. E. Novak (Macmillan Publishing Company, 1991)

NINTH EDITION

TECHNICAL DRAWING

FREDERICK E. GIESECKE
Late Professor Emeritus of Drawing
Texas A & M University

ALVA MITCHELL
Late Professor Emeritus of Engineering Drawing
Texas A & M University

HENRY CECIL SPENCER
Late Professor Emeritus of Technical Drawing;
Formerly Director of Department
Illinois Institute of Technology

IVAN LEROY HILL
Professor Emeritus of Engineering Graphics;
Formerly Chairman of Department
Illinois Institute of Technology

JOHN THOMAS DYGDON
Professor of Engineering Graphics,
Chairman of the Department,
and Director of the Division of Academic Services
and Office of Educational Services
Illinois Institute of Technology

JAMES E. NOVAK
Associate Director/Executive Officer of Educational Services
Illinois Institute of Technology

Macmillan Publishing Company
NEW YORK

Collier Macmillan Canada
TORONTO

Maxwell Macmillan International
NEW YORK   OXFORD   SINGAPORE   SYDNEY

Editor: John Griffin
Production Supervisor: Elisabeth Belfer
Production Manager: Nicholas Sklitsis
Text Designer: Patrice Fodero

This book was set in 10/12 Caslon by Waldman Graphics,
printed and bound by Von Hoffmann Press, Inc.
The cover was printed by Lehigh Press, Inc.

Macmillan Publishing Company
866 Third Avenue, New York, New York 10022

Collier Macmillan Canada, Inc.
Suite 200
1200 Eglinton Avenue, E.
Don Mills, Ontario, M3C 3N1

LIBRARY OF CONGRESS CATALOGING-IN-PUBLICATION DATA

Technical drawing / Frederick E. Giesecke . . . [et al.].—9th ed.
     p.   cm.
   Includes index.
   ISBN 0-02-342605-5
   1. Mechanical drawing.   I. Giesecke, Frederick Ernest, 1869–
T353.T28   1991
604.2—dc20   90-23040
                                                     CIP

Printing: 1 2 3 4 5 6 7 8   Year: 1 2 3 4 5 6 7 8 9 0

Preface

This book is intended to provide a thorough coverage of technical drawing for use as a classroom text and/or reference manual. It contains a great number of problems covering every phase of the subject and constitutes a complete teaching unit in itself. The original aim of the authors was and still is to prepare a book that teaches the language of the engineer. This goal has prompted the authors to illustrate and explain the basic principles from the standpoint of the student— that is, to present each principle so clearly that the student is certain to understand it, and to make the text interesting enough to encourage all students to read and study on their own initiative. By this means the authors hope to free the instructor from the repetitive labor of teaching each student individually the subject matter that the textbook can teach. Thus more attention can be given to those students having real difficulties.

History

The extensive use of this text since 1933 in college classes, technical schools, and industrial drafting and design departments has encouraged the authors to continue with the original purpose of the book. Each succeeding edition of the book reflects the latest trends and practices in technical education, the newest developments in technology and industry, and especially the various current sections of ANSI Y14 *American National Standard Drafting Manual.*

The high quality of drafting in the illustrations and problems has been maintained throughout the years and continues in this ninth edition. It is logical that the drawings and illustrations in a drawing book are very important.

A long-standing feature is the emphasis on technical sketching throughout the text as well as in an early chapter devoted specifically to sketching. This chapter is unique in integrating the basic concepts of views with freehand rendering so that the subject of multiview drawing can be introduced through the medium of sketches.

Features of This Text

The increased use of computer technology for drafting, design work, and manufacturing processes is reflected in many chapters. Two chapters are specifically devoted to this new technology. Chapter 3 presents a generic introduction to computer-aided design and drafting and a survey of computer equipment, or hardware, of current CAD systems. Chapter 8 includes a general discussion of the use and operation of a CAD system focusing on computer graphics programs, or software. Rather than describing one particular software package, examples are given showing how several popular programs can be applied by the user. In addition, relevant material has been added to the other chapters with examples of how computer graphics may be used in particular applications. Many illustrations of computer-generated drawings and the equipment used to make them have been included. These discussions emphasize the relationship between traditional drafting techniques and computer graphics. A comprehensive glossary of CAD/CAM terms and concepts is given in the Appendix.

In addition to the new material on computer graphics, the chapters on manufacturing processes, dimensioning, tolerancing, reproduction and control of drawings, electronic diagrams, welding representation, graphs and diagrams, alignment charts, empirical equations, and graphical mathematics have all been revised to reflect the latest ANSI standards and industrial practice. The Appendix has similarly been updated.

With the cooperation of many leading manufacturers, engineers, and software developers, the authors have been able to include in the text many new photographs and drawings to further develop and present the subject material.

The growing importance of the engineer's design function is emphasized, especially in the chapter on design and working drawings. The chapter is designed to give the student an understanding of the fundamentals of the design process.

The book consistently reflects the latest trends and practices in education, industry, and especially the various current sections of the ANSI Y14 *American National Standard Drafting Manual* and other relevant ANSI standards.

Many problems and illustrations have been updated. A large number of drawings include the approved system of metric dimensions, now that the metric system is more fully used by international industries. The current editions of ANSI standards also indicate a preference for the use of metric units. Many problems, especially those in Chapter 16, present an opportunity for the student to convert dimensions to either the decimal-inch system or the metric system.

Supplements

In addition to the numerous problems in this edition of the text, three complete workbooks have been prepared especially for use with this text:

- *Technical Drawing Problems,* Series 1, by Giesecke, Mitchell, Spencer, Hill, Dygdon, and Novak.
- *Technical Drawing Problems,* Series 2, by Spencer, Hill, Dygdon, and Novak.
- *Technical Drawing Problems,* Series 3, by Spencer, Hill, Dygdon, and Novak.

Thus there are four alternate sources of problems, and problem assignments may be varied easily from year to year. It is expected that the instructor who uses this text together with one of the workbooks will supplement the problem sheets with assignments from the text, to be drawn on blank paper. Many of the text problems are designed for Size A4 or Size A sheets, the same size as the easily filed problem sheets.

Acknowledgments

The authors wish to express their thanks to the many individuals and companies who have so generously contributed their services and materials to the production of this book. In particular, we are especially indebted to Mr. E. J. Mysiak, Engineering Manager, Phoenix Company of Chicago; Mr. Stephen A. Smith, Manager, Product Design and Development, Packaging Corporation of America; Professor Jamshid Mohammadi, Department of Civil Engineering at Illinois Institute of Technology; Mr. Byron Urbanick, Vice President, Paneltech LTD, and Mr. James W. Zagorski, Kelly High School, Chicago Public Schools. The authors also express their thanks to Mr. Gary W. Rybicki and Mr. William T. Briggs Jr., Department of Engineering Graphics, Illinois Institute of Technology, for their helpful suggestions and cooperation.

Special thanks are due to Mrs. Frances A. Koller for her assistance in typing the manuscript for this book, to our editor John Griffin, and to our Production Supervisor Elisabeth Belfer.

Students, teachers, and engineers, designers, and drafters are invited to write concerning any questions that may arise. All comments and suggestions will be welcome.

Ivan Leroy Hill
Clearwater, FL

John Thomas Dygdon
Illinois Institute
   of Technology
Chicago, IL

James E. Novak
Illinois Institute
   of Technology
Chicago, IL

Contents

TECHNICAL
DRAWING

CHAPTER 1

The Graphic Language and Design

The old saying "necessity is the mother of invention" continues to hold, and a new machine, structure, system, or device is the result of that need. If the new device, machine, system, or gadget is really needed or desired, people will buy it, providing it does not cost too much. Then, naturally, these questions may arise: Is there a wide potential market? Can this device or system be made available at a price that people are willing to pay? If these questions can be answered satisfactorily, then the inventor, designer, or officials of a company may elect to go ahead with the development of production and marketing plans for the new project or system.

A new machine, structure, or system, or an improvement thereof, must exist in the mind of the engineer or designer before it can become a reality. This original concept or idea is usually placed on paper, or as an image on a computer screen, and communicated to others by way of the *graphic language* in the form of freehand *idea sketches,* Figs. 1.1 and 6.1. These idea or design sketches are then followed by other sketches, such as *computation sketches,* for developing the idea more fully.

1.1 The Young Engineer*

The engineer or designer must be able to create idea sketches, calculate stresses, analyze motions, size the parts, specify materials and production methods, make design layouts, and supervise the preparation of drawings and specifications that will control the numerous details of production, assembly, and maintenance of the product. In order to perform or supervise these many tasks, the engineer makes liberal use of freehand sketches. He or she must be able to record and communicate ideas quickly to associate and support personnel. Facility in freehand sketching (Chapter 6) or the ability to work with computer-controlled drawing techniques, §16.19, requires a thorough knowledge of the graphic language. The engineer or designer who uses a computer for drawing and design work must be proficient in drafting, designing, and conceptualizing.

*Henceforth in this text, all conventional titles such as student, drafter, designer, engineer, engineering technician, engineering technologist, and so on are intended to refer to all persons, male and female.

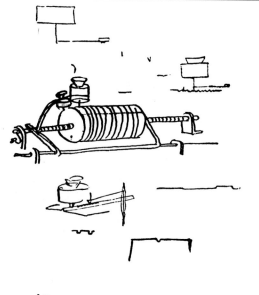

Typical engineering and design departments are shown in Figs. 1.2 and 1.3. Many of the staff have considerable training and experience; others are recent graduates who are gaining experience. There is much to be learned on the job, and it is necessary for the inexperienced person to start at a low level and advance to more responsibility as experience is gained.

1.2   The Graphic Language

Although people around the world speak different languages, a universal graphic language has existed since the earliest of times. The earliest forms of writing were through picture forms, such as the Egyptian hieroglyphics, Fig. 1.4. Later these forms were simplified and became the abstract symbols used in our writing today.

A drawing is a *graphic representation* of a real thing, an idea, or a proposed design for later

Fig. 1.1   Edison's Phonograph. *Original sketch of Thomas A. Edison's first conception of the phonograph; reproduced by special permission of Mrs. Edison.*

Fig. 1.2   Computer-Aided Design and Drafting Section of an Engineering Department. *Courtesy of Jervis B. Webb Co.*

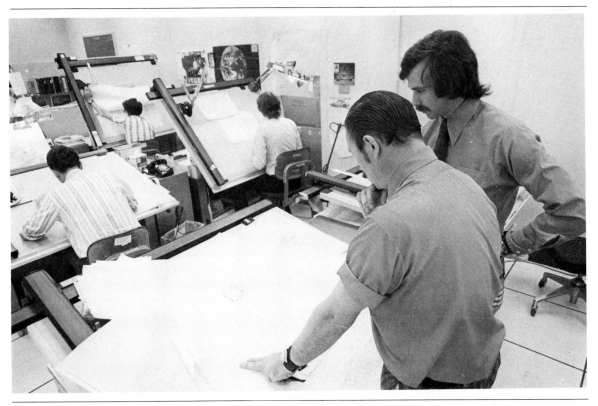

Fig. 1.3   Engineering Drafting Department. *Courtesy of AT&T Bell Laboratories.*

Fig. 1.4   Egyptian Hieroglyphics.

manufacture or construction. Drawings may take many forms, but the graphic method of representation is a basic natural form of communication of ideas that is universal and timeless in character.

1.3 Two Types of Drawings

Graphic representation has been developed along two distinct lines, according to the purpose: (1) artistic and (2) technical.

From the beginning of time, artists have used drawings to express aesthetic, philosophic, or other abstract ideas. People learned by listening to their elders and by looking at sculptures, pictures, or drawings in public places. Everybody could understand pictures, and they were a principal source of information. The artist was not just an artist in the aesthetic sense, but also a teacher or philosopher, a means of expression and communication.

The other line along which drawing has developed has been the technical. From the beginning of recorded history, people have used drawings to represent the design of objects to be built or constructed. Of these earliest drawings no trace remains, but we definitely know that drawings were used, for people could not have designed and built as they did without using fairly accurate drawings.

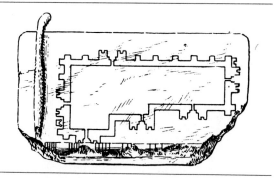

Fig. 1.5 Plan of a Fortress. This stone tablet is part of a statue now in the Louvre, in Paris, and is classified in the earliest period of Chaldean art, about 4000 B.C. *From Transactions ASCE, May 1891.*

1.4   Earliest Technical Drawings

Perhaps the earliest known technical drawing in existence is the plan view for a design of a fortress drawn by the Chaldean engineer Gudea and engraved upon a stone tablet, Fig. 1.5. It is remarkable how similar this plan is to those made by modern architects, although "drawn" thousands of years before paper was invented.

In museums we can see actual specimens of early drawing instruments. Compasses were made of bronze and were about the same size as those in current use. As shown in Fig. 1.6, the old compass resembled the dividers of today. Pens were cut from reeds.

The theory of projections of objects on imaginary planes of projection (to obtain *views*, Chapter 7) apparently was not developed until the early part of the fifteenth century—by the Italian architects Alberti, Brunelleschi, and others. It is well known that Leonardo da Vinci used drawings to record and transmit to others his ideas and designs for mechanical constructions, and many of these drawings are still in existence, Fig. 1.7. It is not clear whether Leonardo ever made mechanical drawings showing orthographic views as we now know them, but it is probable that he did. Leonardo's treatise on painting, published in 1651, is regarded as the first book ever printed on the theory of projection drawing; however, its subject was perspective and not orthographic projection.

The scriber-type compass gave way to the compass with a graphite lead shortly after graphite pencils were developed. At Mount Vernon we can see the drawing instruments used by the great civil engineer George Washington, bearing the date 1749. This set, Fig. 1.8, is very similar to the conventional drawing instruments used today, consisting of a divider and a compass with pencil and pen attachments plus a ruling pen with parallel blades similar to the modern pens.

1.5   Early Descriptive Geometry

The beginnings of descriptive geometry are associated with the problems encountered in designs for building construction and military fortifications of France in the eighteenth century. Gaspard Monge (1746–1818) is considered the "inventor" of descriptive geometry, although his efforts were preceded by publications on stereotomy, architecture, and perspective in which many of the principles were used. It was while he was a professor at the Polytechnic School in France near the close of the eighteenth century

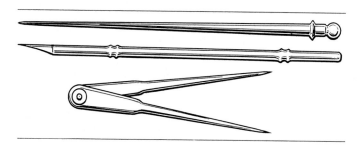

Fig. 1.6 Roman Stylus, Pen, and Compass. *From Historical Note on Drawing Instruments, published by V & E Manufacturing Co.*

Fig. **1.7**   An Arsenal, by Leonardo da Vinci. *The Bettmann Archive.*

Fig. **1.8**   George Washington's Drawing Instruments. *From Historical Note on Drawing Instruments, published by V & E Manufacturing Co.*

that Monge developed the principles of projection that are now the basis of our technical drawing. These principles of descriptive geometry were soon recognized to be of such military importance that Monge was compelled to keep his principles secret until 1795, following which they became an important part of technical education in France and Germany and later in the United States. His book, *La Géométrie Descriptive,* is still regarded as the first text to expound the basic principles of projection drawing.

Monge's principles were brought to the United States from France in 1816 by Claude Crozet, an alumnus of the Polytechnic School and a professor at the United States Military Academy at West Point. He published the first text on the subject of descriptive geometry in the English language in 1821. In the years immediately following, these principles became a regular part of early engineering curricula at Rensselaer Polytechnic Institute, Harvard University, Yale University, and others. During the same period, the idea of manufacturing interchangeable parts in the early arms industries was being developed, and the principles of projection drawing were applied to these problems.

1.6   Modern Technical Drawing

Perhaps the first text on technical drawing in this country was *Geometrical Drawing,* published in 1849 by William Minifie, a high school teacher in Baltimore. In 1850 the Alteneder family organized the first drawing instrument manufacturing company in the United States (Theo. Alteneder & Sons, Philadelphia). In 1876 the blueprint process was introduced at the Philadelphia Centennial Exposition. Up to this time the graphic language was more or less an art, characterized by fine-line drawings made to resemble copper-plate engraving, by the use of shade lines, and by the use of water color "washes." These techniques became unnecessary after the introduction of blueprinting, and drawings gradually were made less ornate to obtain the best results from this method of repro-

duction. This was the beginning of modern technical drawing. The graphic language now became a relatively exact method of representation, and the building of a working model as a regular preliminary to construction became unnecessary.

Up to about 1900, drawings everywhere were generally made in what is called first-angle projection, §7.38, in which the top view was placed under the front view, the left-side view was placed at the right of the front view, and so on. At this time in the United States, after a considerable period of argument pro and con, practice gradually settled on the present *third-angle projection* in which the views are situated in what we regard as their more logical or natural positions. Today, third-angle projection is standard in the United States, but first-angle projection is still used throughout much of the world.

During the early part of the twentieth century, many books were published in which the graphic language was analyzed and explained in connection with its rapidly changing engineering design and industrial applications. Many of these writers were not satisfied with the term "mechanical drawing" because they recognized that technical drawing was really a graphic language. Anthony's *An Introduction to the Graphic Language,* French's *Engineering Drawing,* and Giesecke et al., *Technical Drawing* were all written with this point of view.

1.7   Drafting Standards

In all the previously mentioned books there has been a definite tendency to standardize the characters of the graphic language, to eliminate its provincialisms and dialects, and to give industry, engineering, and science a uniform, effective graphic language. Of prime importance in this movement in the United States has been the work of the American National Standards Institute (ANSI) with the American Society for Engineering Education, the Society of Automotive Engineers, and the American Society of Mechanical Engineers. As sponsors they have prepared the *American National Standard Drafting Man-*

ual—Y14, which is comprised of several separate sections that were published as approved standards as they were completed over a period of years. See Appendix 1.

These sections outline the most important idioms and usages in a form that is acceptable to the majority and are considered the most authoritative guide to uniform drafting practices in this country today. The Y14 Standard gives the characters of the graphic language, and it remains for the textbooks to explain the grammar and the penmanship.

1.8 Definitions

After this brief survey of the historical development of the graphic language, and before we begin a serious study of theory and applications, a few terms need to be defined.

Descriptive geometry The grammar of the graphic language; it is the three-dimensional geometry forming the background for the practical applications of the language and through which many of its problems may be solved graphically.

Instrumental or **mechanical drawing** Properly applies only to a drawing made with drawing instruments. The use of "mechanical drawing" to denote all industrial drawings is unfortunate not only because such drawings are not always mechanically drawn, but also because that usage tends to belittle the broad scope of the graphic language by naming it superficially for its principal mode of execution.

Computer graphics The application of conventional computer techniques with the aid of one of many graphic data processing systems available to the analysis, modification, and the finalizing of a graphical solution. The use of computers to produce technical drawings is called computer-aided design or computer-aided drafting (CAD) and also computer-aided design and drafting (CADD). A typical CAD workstation is shown in Fig. 1.9.

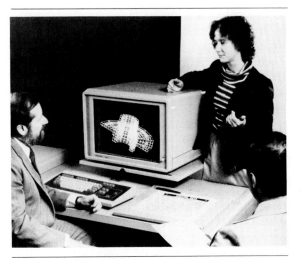

Fig. 1.9   CAD Workstation. *Courtesy of Control Data Corporation.*

Engineering drawing and **engineering drafting** Broad terms widely used to denote the graphic language. However, since the language is used not only by engineers but also by a much larger group of people in diverse fields who are concerned with technical work or with industrial production, these terms are still not broad enough.

Technical drawing A broad term that adequately suggests the scope of the graphic language. It is rightly applied to any drawing used to express technical ideas. This term has been used by various writers since Monge's time at least and is still widely used, mostly in Europe.

Engineering graphics or **engineering design graphics** Generally applied to drawings for technical use and has come to mean that part of technical drawing that is concerned with the graphical representation of designs and specifications for physical objects and data relationships as used in engineering and science.

Technical sketching   The freehand expression of the graphic language, whereas ***mechanical drawing*** is the instrumental expression of it. Technical sketching is a most valuable tool for the engineer and others engaged in technical work because through it most technical ideas can be expressed quickly and effectively without the use of special equipment.

Blueprint reading   The term applied to the "reading" of the language from drawings made by others. Actually, the blueprint process is only one of many forms by which drawings are reproduced today, but the term "blueprint reading" has been accepted through usage to mean the interpretation of all ideas expressed on technical drawings, whether or not the drawings are blueprints.

1.9   What Engineering, Science, and Technology Students Should Know

From the dawn of history the development of technical knowledge has been accompanied, and to a large extent made possible, by a corresponding graphic language. Today the intimate connection between engineering and science and the universal graphic language is more vital than ever before, and the engineer, scientist, or technician who is ignorant of or deficient in the principal mode of expression in his or her technical field is professionally illiterate. Thus, training in the application of technical drawing is required in virtually every engineering school in the world.

The old days of fine-line drawings and of shading and "washes" are gone forever; artistic talent is no longer a prerequisite to learning the fundamentals of the graphic language. Instead, today's student of graphics needs precisely the aptitudes, abilities, and computer skills that will be needed in the science and engineering courses that are studied concurrently and later.

The well-trained engineer, scientist, or technician must be able to make and read correct graphical representations of engineering structures, designs, and data relationships. This means that the individual must understand the fundamental principles, or the *grammar,* of the language and be able to execute the work with reasonable skill, which is *penmanship.*

Graphics students often try to excuse themselves for inferior results (usually caused by lack of application) by arguing that after graduation they do not expect to do any drafting at all; they expect to have others make any needed drawings under their direction. Such a student presumptuously expects, immediately after graduation, to be the accomplished engineer concerned with bigger things and forgets that a first assignment may involve working with drawings and possibly revising drawings, either on the board or with computerized aids, under the direction of an experienced engineer. Entering the engineering profession via graphics provides an excellent opportunity to learn about the product, the company operations, and the supervision of others.

Even a young engineer who has not been successful in developing a skillful penmanship in the graphic language will have use for its grammar, since the ability to *read* a drawing will be of utmost importance. See Chapter 16.

Furthermore, the engineering student is apt to overlook the fact that, in practically all the subsequent courses taken in college, technical drawings will be encountered in most textbooks. The student is often called upon by instructors to supplement calculations with mechanical drawings or sketches. Thus, a mastery of a course in technical drawing utilizing both traditional methods and computer systems (CAD) will aid materially not only in professional practice after graduation but more immediately in other technical courses, and it will have a definite bearing on scholastic progress.

Besides the direct values to be obtained from a serious study of the graphic language, there are a number of very important training values that, though they may be considered by-products, are as essential as the language itself. Many students learn the meaning of neatness, speed, and accuracy for the first time in a drawing course. These are basic habits that every successful engineer, scientist, and technician must have or acquire.

All authorities agree that the ability to *think in three dimensions* is one of the most important requisites of the successful scientist and engineer. This training to visualize objects in space, to use the constructive imagination, is one of the principal values to be obtained from a study of the graphic language. The ability to *visualize* is possessed to an outstanding degree by persons of extraordinary creative ability. It is difficult to think of Edison, De Forest, or Einstein as being deficient in constructive imagination.

With the increase in technological development and the consequent crowding of drawing courses by the other engineering and science courses in our colleges, it is doubly necessary for students to make the most of the limited time devoted to the language of the profession, to the end that they will not be professionally illiterate, but will possess an ability to express ideas quickly and accurately through the correct use of the graphic language.

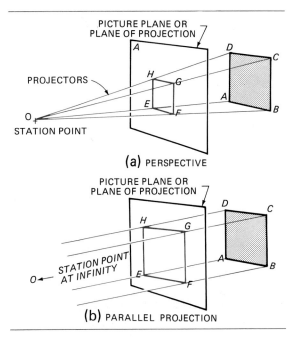

Fig. 1.10   Projections.

1.10   Projections

Behind every drawing of an object is a space relationship involving four imaginary things.

1. The *observer's eye,* or the *station point.*
2. The *object.*
3. The *plane* or *planes of projection.*
4. The *projectors,* also called visual rays and lines of sight.

For example, in Fig. 1.10 (a) the drawing EFGH is the projection, on the plane of projection A, of the square ABCD as viewed by an observer whose eye is at the point O. The projection or drawing on the plane is produced by the points where the projectors pierce the plane of projection (piercing points). In this case, where the observer is relatively close to the object and the projectors form a "cone" of projectors, the resulting projection is known as a perspective.

If the observer's eye is imagined as infinitely distant from the object and the plane of projection, the projectors will be parallel, as shown in Fig. 1.10 (b); hence, this type of projection is known as a *parallel* projection. If the projectors, in addition to being parallel to each other, are perpendicular (normal) to the plane of projection, the result is an *orthographic,* or right-angle, projection. If they are parallel to each other but oblique to the plane of projection, the result is an *oblique* projection.

These two main types of projection—perspective and central or parallel projection—are further broken down into many subtypes, as shown in Fig. 1.11, and will be treated at length in the various chapters that follow.

A classification of the main types of projection according to their projectors is shown in Table 1.1.

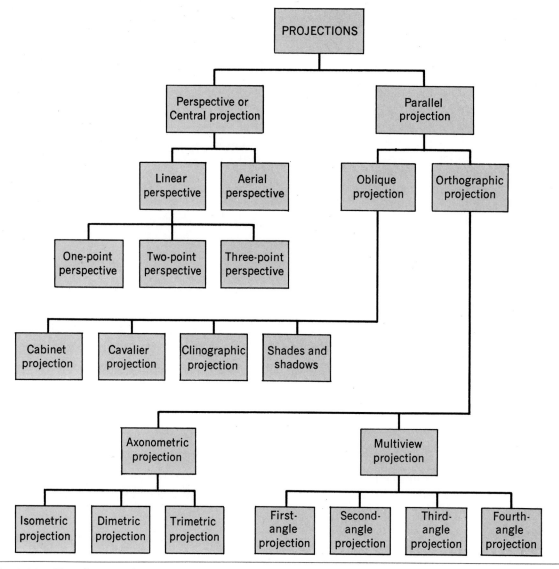

Fig. 1.11   Classification of Projections.

Table 1.1   *Classification by Projectors*

Classes of Projection	Distance from Observer to Plane of Projection	Direction of Projectors
Perspective	Finite	Radiating from station point
Parallel	Infinite	Parallel to each other
Oblique	Infinite	Parallel to each other and oblique to plane of projection
Orthographic	Infinite	Perpendicular to plane of projection
Axonometric	Infinite	Perpendicular to plane of projection
Multiview	Infinite	Perpendicular to planes of projection

CHAPTER 2

Instrumental Drawing

For many years the items of equipment essential to students in technical schools, and to engineers and designers in professional practice, remained unchanged. One needed a drawing board, T-square, triangles, an architects' or engineers' scale, and a professional quality set of drawing instruments. Recently, however, there has been a shift toward greater use of the drafting machine, the parallel-ruling straightedge, the technical fountain pen, and other modern equipment— not to mention the significant increase in the use of the computer as a drafting tool.

The basic items of equipment are shown in Fig. 2.1. To secure the most satisfactory results, the drawing equipment should be of high grade. When drawing instruments (item 3) are to be purchased, the advice of an experienced drafter or designer, or a reliable dealer,* should be sought because it is difficult for beginners to distinguish high-grade instruments from those that are inferior.

2.1 Typical Equipment

A complete list of equipment for students of technical drawing follows. The numbers refer to the equipment illustrated in Fig. 2.1.

1. Drawing board (approx. 20″ × 24″), drafting table, or desk, §2.4.
2. T-square (24″, transparent edge), drafting machine, or parallel ruling edge, §§2.5, 2.56, and 2.57.
3. Set of instruments, §§2.33 and 2.34.
4. 45° triangle (8″ sides), §2.16.
5. 30° × 60° triangle (10″ long side), §2.16.
6. Ames Lettering Guide or lettering triangle, §4.16.
7. Architects' triangular scale, §2.28.
8. Engineers' triangular scale, §2.27.
9. Metric triangular scale, §2.25.
10. Irregular curve, §2.53.
11. Protractor, §2.18.
12. Mechanical pencils and/or thin-lead mechanical pencils and HB, F, 2H, and 4H to 6H leads, or drawing pencils. See §§2.8 and 2.9.

*Keuffel & Esser Co., Morristown, NJ; Eugene Dietzgen Co., Des Plaines, IL; Charles Bruning Co., Mt. Prospect, IL; Teledyne Post, Des Plaines, IL; Vemco Corp., Pasadena, CA; Staedtler Mars, Elk Grove Village, IL; Koh-I-Noor Rapidograph, Inc., Bloomsbury, NJ; and Tacro, Brooklyn, NY, are some of the larger distributors of this equipment; their products are available through local dealers.

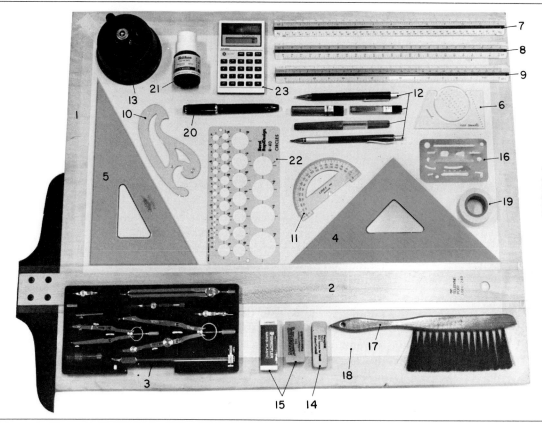

Fig. 2.1   Principal Items of Equipment.

13. Lead pointer and sandpaper pad, §2.10.

14. Pencil eraser, §2.12.

15. Plastic, drafting eraser or Artgum cleaning eraser, §2.12.

16. Erasing shield, §2.12.

17. Dusting brush, §2.12.

18. Drawing paper, tracing paper, tracing cloth, or films as required. Backing sheet (drawing paper—white, cream, or light green) to be used under drawings and tracings.

19. Drafting tape, §2.7.

20. Technical fountain pens, §2.48.

21. Drawing ink, §2.50.

22. Templates, §2.55.

23. Calculator.

24. Cleansing tissue or dust cloth (not shown).

2.2   Objectives in Drafting

On the following pages, the correct methods to be used in instrumental drawing are explained. The student who practices and learns correct manipulation of the drawing instruments will eventually be able to draw correctly by habit, thus giving his or her full attention to the problems at hand.

The following are the important objectives the student should strive to attain:

Fig. 2.2 Orderliness Promotes Efficiency and Accuracy.

1. *Accuracy.* No drawing is of maximum usefulness if it is not accurate. The student cannot achieve success in a college career or later in professional employment if the habit of accuracy is not acquired.

2. *Speed.* "Time is money" in industry, and there is no demand for the slow drafter, technician, or engineer. However, speed is not attained by hurrying; it is an unsought by-product of *intelligent and continuous work.* It comes with study and practice.

3. *Legibility.* The drafter, technician, or engineer should remember that the drawing is a means of communication to others, and that it must be clear and legible in order to serve its purpose well. Care should be given to details, especially to lettering, Chapter 4.

4. *Neatness.* If a drawing is to be accurate and legible, it must also be clean; therefore, the student should constantly strive to acquire the habit of neatness. Untidy drawings are the result of sloppy and careless methods, §2.13, and will be unacceptable to an instructor or employer.

2.3 Drafting at Home or School

In the school drafting room, as in the industrial drafting room, the student is expected to give thoughtful and continuous attention to the problems at hand.

Technical drawing requires headwork and must be done in quiet surroundings without distractions. The efficient drafting student sees to it that the correct equipment is available and refrains from borrowing—a nuisance to everyone. While the student is drawing, the textbook, the chief source of information, should be available and in a convenient position, Fig. 2.2.

When questions arise, first use the index of the text and endeavor to find the answer for yourself. Try to develop self-reliance and initia-

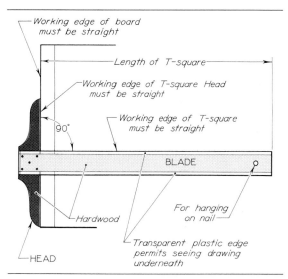

Fig. 2.3   The T-square.

tive, but when you really need help, ask your instructor. Students who go about their work intelligently, with a minimum waste of time, first study the assignment carefully to be sure that they understand the principles involved; second, they make sure that the correct equipment is in proper condition (such as sharp pencils); and third, they make an effort to dig out answers for themselves (the only true education).

One of the principal means of promoting efficiency in drafting is orderliness. Efficiency, in turn, will produce accuracy in drawing. All needed equipment and materials should be placed in an orderly manner so that everything is in a convenient place and can readily be found when needed, Fig. 2.2 The drawing area should be kept clear of equipment not in direct use. Form the habit of placing each item in a regular place outside the drawing area when it is not being used.

When drawing at home, if possible work in a room by yourself. Place a book under the upper portion of the drawing board to give the board a convenient inclination, or you can pull out the study table drawer and use it to support the drawing board at a slant.

It is best to work in natural north light coming from the left and slightly from the front. Never work on a drawing in direct sunlight or in dim light, as either may be injurious to the eyes. If artificial light is needed, the light source should be such that shadows are not cast where lines are being drawn and that there will be as little reflected glare from the paper as possible. Special fluorescent drafting lamps are available with adjustable arms so that the light source may be moved to any desired position. Drafting will not hurt eyes that are in normal condition, but the exacting work will often disclose deficiencies not previously suspected.

Left-handers   Place the head of the T-square on the right, and have the light come from the right and slightly from the front.

2.4   Drawing Boards

If the left edge of the drafting table top has a true straightedge and if the surface is hard and smooth (such as masonite), a drawing board is unnecessary, provided that drafting tape is used to fasten the drawings. It is recommended that a backing sheet of heavy drawing paper be placed between the drawing and the table top.

However, in most cases a drawing board will be needed. These vary from 9″ × 12″ (for sketching and field work) up to 48″ × 72″ or larger. The recommended size for students is 20″ × 24″, Fig. 2.1, which will accommodate the largest sheet likely to be used.

Drafters use drafting tape, which in turn permits surfaces such as hardwood, masonite, or other materials to be used for drawing boards.

The left-hand edge of the board is called the *working edge* because the T-square head slides against it, Fig. 2.3. This edge must be straight, and you should test the edge with a T-square blade that has been tested and found straight, Fig. 2.4. If the edge of the board is not true, it should be replaced.

2.5   T-square

Fig. 2.3   The T-square is composed of a long strip, called the *blade*, fastened rigidly at right angles to a shorter piece called the *head*. The upper edge of the blade and the inner edge of

14

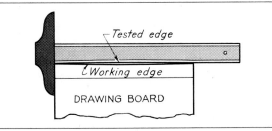

Fig. 2.4   Testing the Working Edge of the Drawing Board.

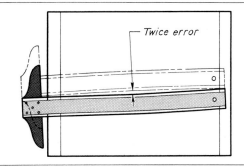

Fig. 2.5   Testing the T-square.

the head are *working edges* and must be straight. The working edge of the head must not be convex or the T-square will rock when the head is placed against the board. The blade should have transparent plastic edges and should be free of nicks along the working edge. Transparent edges are recommended, since they permit the draftsman to see the drawing in the vicinity of the lines being drawn.

Do not use the T-square for any rough purpose. Never cut paper along its working edge, as the plastic is easily cut and even a slight nick will ruin the T-square.

2.6   Testing and Correcting the T-square

To test the working edge of the head, see if the T-square rocks when the head is placed against a straightedge, such as a drawing board working edge that has already been tested and found true. If the working edge of the head is not straight, the T-square should be replaced.

Fig. 2.5   To test the working edge of the blade, draw a sharp line very carefully with a hard pencil along the entire length of the working edge; then turn the T-square over and draw the line again along the same edge. If the edge is straight, the two lines will coincide; otherwise, the space between the lines will be twice the error of the blade.

It is difficult to correct a crooked T-square blade, and if the error is considerable, it may be necessary to discard the T-square and obtain another.

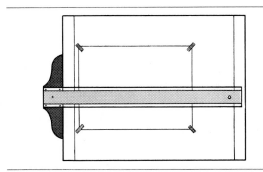

Fig. 2.6   Placing Paper on Drawing Board.

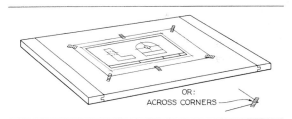

Fig. 2.7   Positions of Drafting Tape.

2.7   Fastening Paper to the Board

The drawing paper should be placed close enough to the working edge of the board to reduce to a minimum any error resulting from a slight "give," or bending, of the blade of the T-square, and close enough to the upper edge of the board to permit space at the bottom of the sheet for using the T-square and supporting the arm while drawing, Fig. 2.6.

Drafting tape is preferred for fastening the drawing to the board, Fig. 2.7, because it does

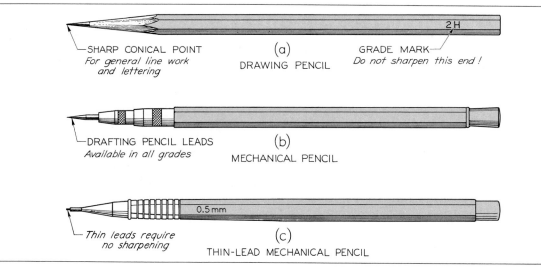

Fig. 2.8   Drawing Pencils.

not damage the board and it will not damage the paper if it is removed by *pulling it off slowly toward the edge of the paper.*

To fasten the paper in place, press the T-square head firmly against the working edge of the drawing board with the left hand, while the paper is adjusted with the right hand until the top edge coincides with the upper edge of the T-square. Then move the T-square to the position shown and fasten the upper left corner, then the lower right corner, and finally the remaining corners. Large sheets may require additional fastening, whereas small sheets may require fastening only at the two upper corners.

Tracing paper should not be fastened directly to the board because small imperfections in the surface of the board will interfere with the line work. Always fasten a larger backing sheet of heavy drawing paper on the board first; then fasten the tracing paper over this sheet.

2.8   Drawing Pencils
Fig. 2.8 (a)   High-quality drawing pencils should be used in technical drawing—never ordinary writing pencils.

Fig. 2.8 (b)   Many makes of mechanical pencils are available, together with refill drafting leads of conventional size in all grades. Choose the holder that feels comfortable in the hand and that grips the lead firmly without slipping. Mechanical pencils have the advantages of maintaining a constant length of lead while permitting the use of a lead practically to the end, of being easily refilled with new leads, of affording a ready source for compass leads, of having no wood to be sharpened, and of easy sharpening of the lead by various mechanical pencil pointers now available.

Fig. 2.8 (c)   Thin-lead mechanical pencils are available with 0.3, 0.5, 0.7, or 0.9 mm diameter drafting leads in several grades. These thin leads produce uniform width lines without sharpening, providing both a time savings and a cost benefit.

Mechanical pencils are recommended as they are less expensive in the long run.

2.9   Choices of Grade of Pencil
Drawing pencil leads are made of graphite with the addition of a polymer binder or of kaolin

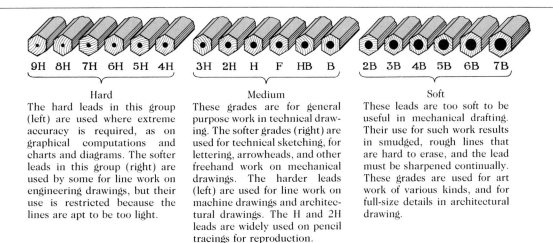

9H 8H 7H 6H 5H 4H	3H 2H H F HB B	2B 3B 4B 5B 6B 7B
Hard	Medium	Soft

Hard
The hard leads in this group (left) are used where extreme accuracy is required, as on graphical computations and charts and diagrams. The softer leads in this group (right) are used by some for line work on engineering drawings, but their use is restricted because the lines are apt to be too light.

Medium
These grades are for general purpose work in technical drawing. The softer grades (right) are used for technical sketching, for lettering, arrowheads, and other freehand work on mechanical drawings. The harder leads (left) are used for line work on machine drawings and architectural drawings. The H and 2H leads are widely used on pencil tracings for reproduction.

Soft
These leads are too soft to be useful in mechanical drafting. Their use for such work results in smudged, rough lines that are hard to erase, and the lead must be sharpened continually. These grades are used for art work of various kinds, and for full-size details in architectural drawing.

Fig. 2.9   Lead Grade Chart.

(clay) in varying amounts to make 18 grades from 9H, the hardest, down to 7B, the softest. The uses of these different grades are shown in Fig. 2.9. Note that small-diameter leads are used for the harder grades, whereas large-diameter leads are used to give more strength to the softer grades. Hence, the degree of hardness in the wood pencil can be roughly judged by a comparison of the diameters.

Specifically formulated leads of carbon black particles in a polymer binder are also available in several grades for use on the polyester films now found quite extensively in industry. See §2.62.

To select the grade of lead, first take into consideration the type of line work required. For light construction lines, guide lines for lettering, and for accurate geometrical constructions or work where accuracy is of prime importance, use a hard lead, such as 4H to 6H.

For mechanical drawings on drawing paper or tracing paper, the lines should be **black,** particularly for drawings to be reproduced. The lead chosen must be soft enough to produce jet black lines, but hard enough not to smudge too easily or permit the point to crumble under normal pressure. The same comparatively soft lead is preferred for lettering and arrowheads.

This lead will vary from F to 2H, roughly, depending on the paper and weather conditions. If the paper is hard, it will be necessary generally to use harder leads. For softer surfaces, softer leads can be used. The weather factor to consider is the humidity. On humid days the paper absorbs moisture from the atmosphere and becomes soft. This can be recognized because the paper expands and becomes wrinkled. It is necessary to select softer leads to offset the softening of the paper. If you have been using a 2H lead, for example, change to an F until the weather clears up.

2.10   Sharpening the Pencil

Keep your lead sharp! This is certainly the instruction needed most frequently by the beginning student. A dull lead produces fuzzy, sloppy, indefinite lines. Only a sharp lead is capable of producing clean-cut black lines that sparkle with clarity.

If a good mechanical pencil, Fig. 2.8 (b), is used, much time may be saved in sharpening,

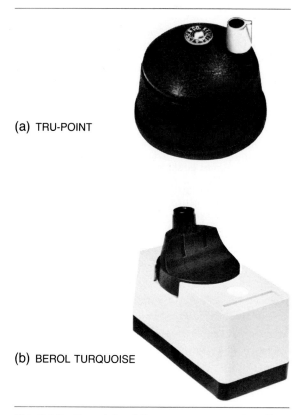

(a) TRU-POINT

(b) BEROL TURQUOISE

Fig. 2.10   Pencil Lead Pointers. *Courtesy of Keuffel & Esser Co.*

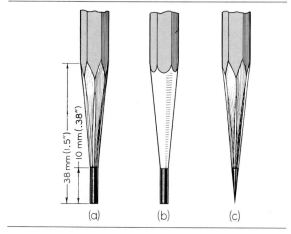

Fig. 2.11   Pencil Points.

since the lead can be fed from the pencil as needed. Two excellent lead pointers for mechanical pencils are shown in Fig. 2.10. Each has the advantage of one-hand manipulation, and of collecting the loose graphite particles inside, where they cannot soil the hands, the drawing, or other equipment.

If thin-lead mechanical pencils are used, Fig. 2.8 (c), no sharpening is required since the lead diameter determines the line width. Hence, several thin-lead mechanical pencils are required for the various line widths used in technical drawing. Each thin-lead mechanical pencil will accommodate only one diameter of lead.

If a wood drawing pencil is used, Fig. 2.8 (a), sharpen the unlettered end in order to preserve

the identifying grade mark. First, the wood is removed with a knife or a special drafting pencil sharpener, starting about 38 mm (1.5") from the end, and about 10 mm (.38") of uncut lead is exposed, Fig. 2.11 (a) or (b). Next the lead is shaped to a sharp conical point, and then the point is wiped clean with cloth or paper tissue to remove loose particles of graphite.

Mechanical devices for removal of the wood are shown in Fig. 2.12 (a) and (b). The procedures for shaping the lead are illustrated in Fig. 2.13.

Never sharpen your pencil over the drawing or any of your equipment.

Keep the pencil pointer close by, as frequent pointing of the pencil will be necessary.

When the sandpaper pad is not in use, it should be kept in a container, such as an envelope, to prevent the particles of graphite from falling on the drawing board or drawing equipment, Fig. 2.12 (c).

Many drafters burnish the point on a piece of hard paper to obtain a smoother, sharper point. However, for drawing visible lines the point should not be needle-sharp, but very slightly rounded. First sharpen the lead to a needle point, then stand the pencil vertically, and with

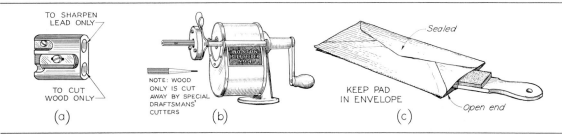

Fig. 2.12 Pencil Sharpeners.

a few rotary motions on the paper, wear the point down slightly to the desired shape.

2.11 Alphabet of Lines

Each line on a technical drawing has a definite meaning and is drawn in a certain way. The line conventions endorsed by the American National Standards Institute, ANSI Y14.2M–1979 (R1987), are presented in Fig. 2.14, together with illustrations of various applications.

Two widths of lines are recommended for use on drawings. All lines should be clean-cut, dark, uniform throughout the drawing, and properly spaced for legible reproduction by all commonly used methods. Minimum spacing of 1.5 mm (.06″) between parallel lines is usually satisfactory for all reduction and/or reproduction processes. The size and style of the drawing and the smallest size to which it is to be reduced govern the actual width of each line. The contrast between the two widths of lines should be distinct. Pencil leads should be hard enough to prevent smudging, but soft enough to produce the dense black lines so necessary for quality reproduction.

When photoreduction and blowback are not necessary, as is the case for most drafting laboratory assignments, three weights of lines may improve the appearance and legibility of the drawing. The "thin lines" may be made in two widths—regular thin lines for hidden lines and stitch lines and a somewhat thinner version for the other secondary lines such as center lines, extension lines, dimension lines, leaders, section lines, phantom lines, and long-break lines.

For the "thick lines"—visible, cutting plane, and short break—use a relatively soft lead such as F or H. All thin lines should be made with a sharp medium-grade lead such as H or 2H. All lines (except construction lines) must be *sharp* and *dark*. Make construction lines with a sharp 4H or 6H lead so thin that they barely can be seen at arm's length and need not be erased.

The high-quality photoreduction and reproduction processes used in the production of this book permitted the use of three weights of lines in many illustrations and drawings for increased legibility.

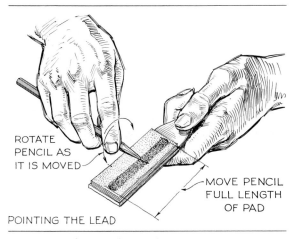

Fig. 2.13 Shaping the Lead.

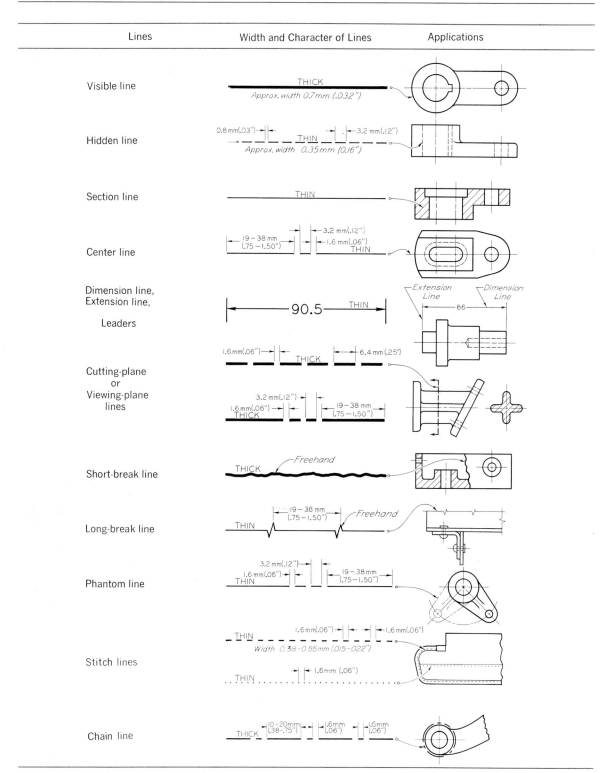

Fig. 2.14 Alphabet of Lines (Full Size).

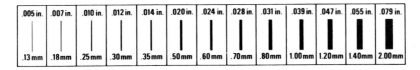

.005 in.	.007 in.	.010 in.	.012 in.	.014 in.	.020 in.	.024 in.	.028 in.	.031 in.	.039 in.	.047 in.	.055 in.	.079 in.
.13 mm	.18 mm	.25 mm	.30 mm	.35 mm	.50 mm	.60 mm	.70 mm	.80 mm	1.00 mm	1.20 mm	1.40 mm	2.00 mm

Fig. 2.15   Line Gage. *Courtesy of Koh-I-Noor Rapidograph, Inc.*

In Fig. 2.14, the ideal lengths of all dashes are indicated. It would be well to measure the first few hidden dashes and center-line dashes you make and thereafter to estimate the lengths carefully by eye.

Fig. 2.15   The line gage is a convenient reference for lines of various widths.

2.12   Erasing

Erasers are available in many degrees of hardness and abrasiveness. For general drafting the Pink Pearl or the Mars-Plastic is suggested, Fig. 2.16. These erasers are suitable for erasing pencil or ink line work. Best results are obtained if a hard surface, such as a triangle, is placed under the area being erased. If the surface has become badly grooved by the lines, the surface can be improved by burnishing the back side with a hard smooth object or with the back of the fingernail.

Fig. 2.17   The erasing shield is used to protect the lines near those being erased.

Fig. 2.18   The electric erasing machine saves time and is essential if much drafting is being done.

Fig. 2.19   A dusting brush is useful for removing eraser crumbs without smearing the drawing.

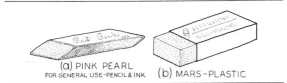

(a) PINK PEARL
FOR GENERAL USE-PENCIL & INK   (b) MARS-PLASTIC

Fig. 2.16   Erasers.

Fig. 2.17   Using the Erasing Shield.

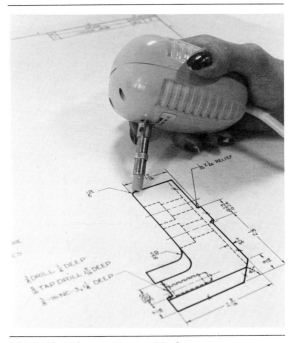

Fig. 2.18   Electric Erasing Machine.

Fig. 2.19   Dusting Brush.

2.13   Keeping Drawings Clean

Cleanliness in drafting is very important and should become a habit. Cleanliness does not just happen; it results only from a conscious effort to observe correct procedures.

First, the drafter's hands should be clean at all times. Oily or perspiring hands should be frequently washed with soap and water. Talcum powder on the hands tends to absorb excessive perspiration.

Second, all drafting equipment, such as drawing board, T-square, triangles, and scale, should be wiped frequently with a clean cloth. Water should be used sparingly and dried off immediately. A soft eraser may also be used for cleaning drawing equipment.

Third, the largest contributing factor to dirty drawings is *not dirt, but graphite* from the pencil; hence, the drafter should observe the following precautions:

1. Never sharpen a lead over the drawing or any equipment.

2. Always wipe the lead point with a clean cloth or cleansing tissue, after sharpening or pointing, to remove small particles of loose graphite.

3. Never place the sandpaper pad or file in contact with any other drawing equipment unless it is completely enclosed in an envelope or similar cover, Fig. 2.12 (c).

4. Never work with the sleeves or hands resting on a penciled area. Keep such parts of the drawing covered with clean paper (not a cloth). In lettering a drawing, always place a piece of paper under the hand.

5. Avoid unnecessary sliding of the T-square or triangles across the drawing. Pick up the tri-angles by their tips and tilt the T-square blade upward slightly before moving. A very light sprinkling of powdered Artgum or drafting powder on the drawing helps to keep the drawing clean by picking up the loose graphite particles as you work. It should be brushed off and replaced occasionally.

6. Never rub across the drawing with the palm of the hand to remove eraser particles; use a dust brush, Fig. 2.19, or flick—don't rub—the particles off with a clean cloth.

If the foregoing rules are observed, a completed drawing will not need to be cleaned. The practice of making a pencil drawing, scrubbing it with a soft eraser, and then retracing the lines is poor technique. It is also a waste of time and a habit that should not be acquired.

At the end of the period or of the day's work, the drawing should be covered with paper or cloth to protect it from dust.

If the drawing must be removed from the board before it is complete, it can be carried flat in a drawing portfolio or gently rolled and carried in a cardboard or plastic tube, preferably one with ends that can be closed.

2.14   Horizontal Lines

Fig. 2.20   To draw a horizontal line, press the head of the T-square firmly against the working edge of the board with the left hand; then slide the left hand to the position shown at (a) so as to press the blade tightly against the paper. Lean the pencil in the direction of the line at an angle of approximately 60° with the paper, (b), and draw the line from left to right. Keep the pencil in a vertical plane, (b) and (c); otherwise, the line may not be straight. While drawing the line, let the little finger of the hand holding the pencil glide lightly on the blade of the T-square, and rotate the pencil slowly, except for the thin-lead pencils, between the thumb and forefinger so as to distribute the wear uniformly on the lead and maintain a symmetrical point.

When great accuracy is required, the pencil may be "toed in" as shown at (d) to produce a perfectly straight line.

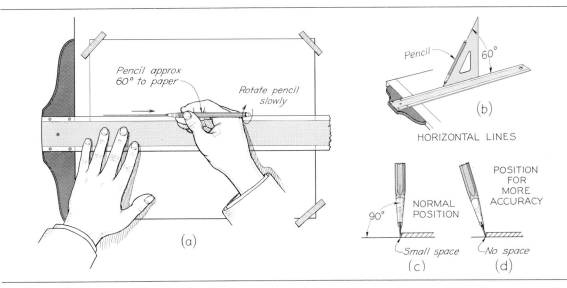

Fig. 2.20   Drawing a Horizontal Line.

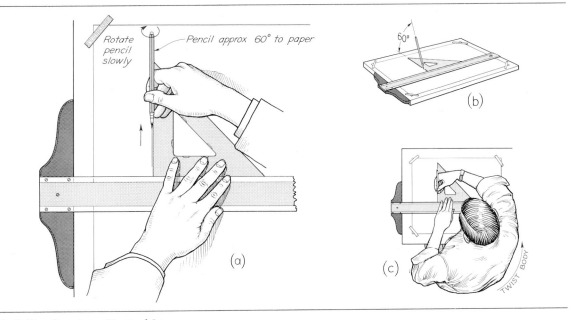

Fig. 2.21   Drawing a Vertical Line.

Thin-lead pencils should be held nearly vertical to the paper and not rotated. Also, pushing the thin-lead pencil from left to right, rather than pulling it, tends to minimize lead breakage.

Left-handers   In general, reverse the procedure just outlined. Place the T-square head against the right edge of the board, and with the pencil in the left hand, draw the line from right to left.

2.15   Vertical Lines
Fig. 2.21   Use either the 45° triangle or the 30° × 60° triangle to draw vertical lines. Place

Fig. 2.22 Triangles.

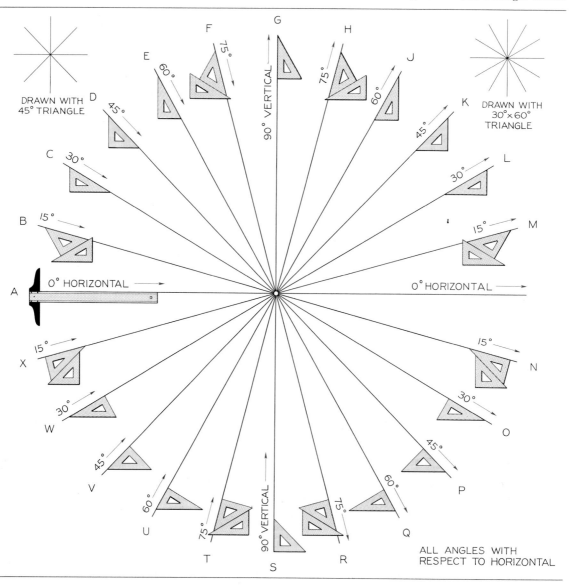

(a) 45° TRIANGLE (b) 30°×60° TRIANGLE

Fig. 2.23 The Triangle Wheel.

DRAWN WITH 45° TRIANGLE

DRAWN WITH 30°×60° TRIANGLE

ALL ANGLES WITH RESPECT TO HORIZONTAL

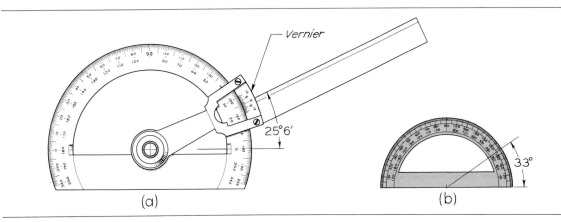

Fig. 2.24   Protractors.

the triangle on the T-square with the *vertical edge on the left* as shown in (a). With the left hand, press the head of the T-square against the board; then slide the hand to the position shown where it holds both the T-square and the triangle firmly in position. Then draw the line upward, rotating the pencil slowly between the thumb and forefinger.

Lean the pencil in the direction of the line at an angle of approximately 60° with the paper and in a vertical plane, (b). Meanwhile, the upper part of the body should be twisted to the right as shown at (c).

See §2.14 regarding the use of thin-lead pencils.

Left-handers   In general, reverse the above procedure. Place the T-square head on the right and the vertical edge of the triangle on the right; then, with the right hand, hold the T-square and triangle firmly together, and with the left hand draw the line upward.

The only time it is advisable for right-handers to turn the triangle so that the vertical edge is on the right is when drawing a vertical line near the right end of the T-square. In this case, the line would be drawn downward.

2.16   The Triangles
Fig. 2.22   Most inclined lines in mechanical drawing are drawn at standard angles with the

45° *triangle* and the *30° × 60° triangle.* The triangles are made of transparent plastic so that lines of the drawing can be seen through them. A good combination of triangles is the 30° × 60° triangle with a long side of 10″ and a 45° triangle with each side 8″ long.

2.17   Inclined Lines
The positions of the triangles for drawing lines at all of the possible angles are shown in Fig. 2.23. In the figure it is understood that the triangles in each case are resting on the blade of the T-square. Thus, it is possible to divide 360° into twenty-four 15° sectors with the triangles used singly or in combination. Note carefully the directions for drawing the lines, as indicated by the arrows, and that all lines in the left half are drawn *toward the center,* while those in the right half are drawn *away from the center.*

2.18   Protractors
For measuring or setting off angles other than those obtainable with the triangles, the *protractor* is used. The best protractors are made of nickel silver and are capable of most accurate work, Fig. 2.24 (a). For ordinary work the plastic protractor is satisfactory and much cheaper, (b). To set off angles with greater accuracy, use one of the methods presented in §5.20.

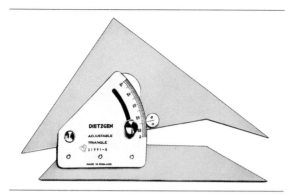

Fig. 2.25   Adjustable Triangle.

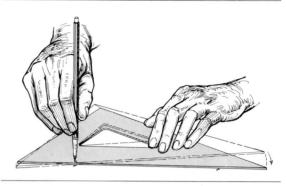

Fig. 2.26   To Draw a Pencil Line Through Two Points.

2.19   Drafting Angles

A variety of devices combining the protractor with triangles to produce great versatility of use are available, one type of which is shown in Fig. 2.25.

2.20   To Draw a Line Through Two Points

Fig. 2.26   To draw a line through two points, place the pencil vertically at one of the points, and move the straightedge about the pencil point as a pivot until it lines up with the other point; then draw the line along the edge.

2.21   Parallel Lines

Fig. 2.27   To draw a line parallel to a given line, move the triangle and T-square as a unit until the hypotenuse of the triangle lines up with the given line, (a); then, holding the T-square firmly in position, slide the triangle away from the line, (b), and draw the required line along the hypotenuse, (c).

Obviously any straightedge, such as one of the triangles, may be substituted for the T-square in this operation, as shown at (a).

To draw parallel lines at 15° with horizontal, arrange the triangles as shown in Fig. 2.28.

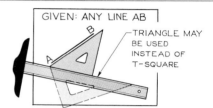

GIVEN: ANY LINE AB

TRIANGLE MAY BE USED INSTEAD OF T-SQUARE

(a) MOVE T-SQUARE AND TRIANGLE TO LINE UP WITH AB

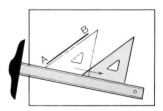

(b) SLIDE TRIANGLE ALONG T-SQUARE

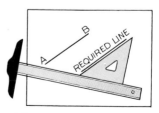

REQUIRED LINE

(c) DRAW REQUIRED LINE PARALLEL TO AB

Fig. 2.27   To Draw a Line Parallel to a Given Line.

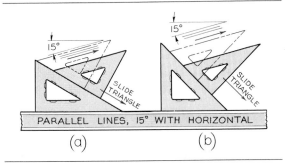

Fig. 2.28 Parallel Lines.

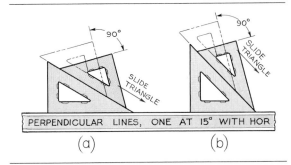

Fig. 2.30 Perpendicular Lines.

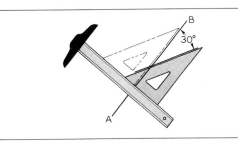

Fig. 2.31 Line at 30° with Given Line.

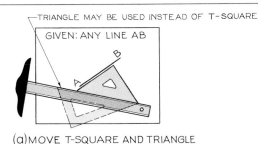

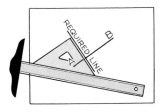

Fig. 2.29 To Draw a Line Perpendicular to a Given Line.

2.22 Perpendicular Lines

Fig. 2.29 To draw a line perpendicular to a given line, move the T-square and triangle as a unit until one edge of the triangle lines up with the given line, (a); then slide the triangle across the line, (b), and draw the required line, (c).

To draw perpendicular lines when one of the lines makes 15° with horizontal, arrange the triangles as shown in Fig. 2.30.

2.23 Lines at 30°, 60°, or 45° with Given Line

To draw a line making 30° with a given line, arrange the triangle as shown in Fig. 2.31. Angles of 60° and 45° may be drawn in a similar manner.

2.24 Scales

A drawing of an object may be the same size as the object (full size), or it may be larger or smaller than the object.

27

Fig. 2.32   Printed Circuit Board. *United Nations/ Guthrie.*

The ratio of reduction or enlargement depends on the relative sizes of the object and of the sheet of paper on which the drawing is to be made. For example, a machine part may be half size; a building may be drawn $\frac{1}{48}$ size; a map may be drawn $\frac{1}{1200}$ size; or a printed circuit board, Fig. 2.32, may be drawn four times size.

Fig. 2.33   Scales are classified as the *metric scale* (a), the *engineers' scale* (b), the *decimal scale* (c), the *mechanical engineers' scale* (d), and the *architects' scale* (e). A full-divided scale is one in which the basic units are subdivided throughout the length of the scale. Only the lower scale at (e) is an open divided scale, one in which only the end unit is subdivided.

Scales are usually made of plastic or boxwood. The better wood scales have white plastic edges. Scales are either triangular, Fig. 2.34 (a) and (b), or flat, (c) to (f). The triangular scales have the advantage of combining many scales on one stick, but the user will waste much time looking for the required scale if a *scale guard*, (g), is not used. The flat scale is almost univer-

sally used by professional drafters because of its convenience, but several flat scales are necessary to replace one triangular scale, and the total cost is greater.

2.25 Metric Scales

Fig. 2.33 (a)   The metric system is an international language of measurement that, despite modifications over the past 200 years, has been the foundation of science and industry and is clearly defined. The modern form of the metric system is the International System of Units, commonly referred to as SI (from the French name, Le Système International d'Unités). It is important to remember that SI differs in several respects from former metric systems. For example, cc was previously an accepted abbreviation for cubic centimeter; in SI the symbol used is cm^3. In the past, degree centigrade was accepted as an alternative name for degree Celsius; in SI only degree Celsius is used. Probably the most important characteristic of SI is that it is a unique system; each quantity has only one unit. The SI system was established in 1960 by international agreement and is now considered the standard international language of measurement.

The metric scale is used when the meter is the standard for linear measurement. The meter was established by the French in 1791 with a length of one ten-millionth of the distance from the Earth's equator to the pole. The meter is equal to 39.37 inches or approximately 1.1 yards.

The metric system for linear measurement is a decimal system similar to our system of counting money. For example,

$$
\begin{aligned}
1 \text{ mm} &= 1 \text{ millimeter } (\tfrac{1}{1000} \text{ of a meter}) \\
1 \text{ cm} &= 1 \text{ centimeter } (\tfrac{1}{100} \text{ of a meter}) \\
&= 10 \text{ mm} \\
1 \text{ dm} &= 1 \text{ decimeter } (\tfrac{1}{10} \text{ of a meter}) \\
&= 10 \text{ cm} = 100 \text{ mm} \\
1 \text{ m} &= 1 \text{ meter} \\
&= 100 \text{ cm} = 1000 \text{ mm} \\
1 \text{ km} &= 1 \text{ kilometer} = 1000 \text{ m} \\
&= 100\,000 \text{ cm} = 1\,000\,000 \text{ mm}
\end{aligned}
$$

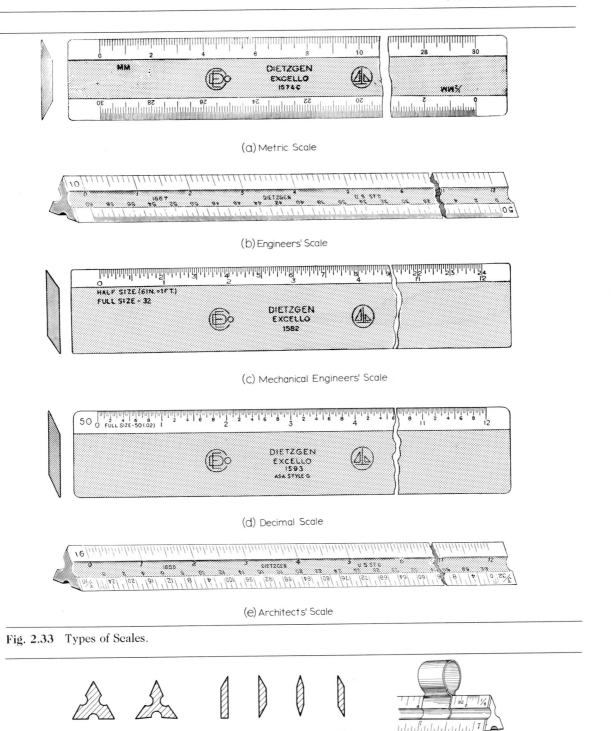

(a) Metric Scale

(b) Engineers' Scale

(c) Mechanical Engineers' Scale

(d) Decimal Scale

(e) Architects' Scale

Fig. 2.33 Types of Scales.

(a)   (b)   (c)   (d)   (e)   (f)   (g) SCALE GUARD

Fig. 2.34 Sections of Scales and Scale Guard.

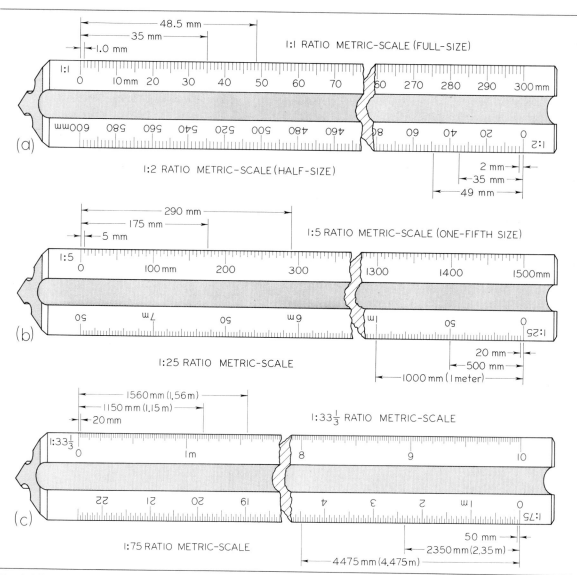

Fig. 2.35 Metric Scales.

The primary unit of measurement for engineering drawings and design in the mechanical industries is the millimeter (mm). Secondary units of measure are the meter (m) and the kilometer (km). The centimeter (cm) and the decimeter (dm) are rarely used.

In recent years, the auto and other industries have used a dual dimensioning system of millimeters and inches; see Fig. 16.22. The large agricultural machinery manufacturers have elected to use all metric dimensions with the inch equivalents given in a table on the drawing, Fig. 16.22.

Many of the dimensions in the illustrations and the problems in this text are given in metric units. Dimensions that are given in the customary units (inches and feet, either decimal or fractional) may be converted easily to metric values. In accordance with standard practice, the ratio

1 in. = 25.4 mm is used. Decimal equivalents tables can be found inside the front cover, and conversion tables are given in Appendix 31.

Metric scales are available in flat and triangular styles with a variety of scale graduations. The triangular scale illustrated in Fig. 2.35 has one full-size scale and five reduced-size scales, all full divided. By means of these scales a drawing can be made full size, enlarged size, or reduced size. To specify the scale on a drawing, see §2.31.

FULL SIZE   *Fig. 2.35 (a)* The 1 : 1 scale is full size, and each division is actually 1 mm in width with the numbering of the calibrations at 10-mm intervals. The same scale is also convenient for the ratios of 1 : 10, 1 : 100, 1 : 1000, and so on.

HALF SIZE   *Fig. 2.35 (a)* The 1 : 2 scale is one-half size, and each division equals 2 mm with the calibration numbering at 20-unit intervals. In addition, this scale is convenient for ratios of 1 : 20, 1 : 200, 1 : 2000, and so on.

The remaining four scales on this triangular metric scale include the typical scale ratios of 1 : 5, 1 : 25, 1 : $33\frac{1}{3}$, and 1 : 75, as illustrated at (b) and (c). These ratios may also be enlarged or reduced as desired by multiplying or dividing by a factor of 10. Metric scales are also available with other scale ratios for specific drawing purposes.

The metric scale is used in map drawing and in drawing force diagrams or other graphical constructions that involve such scales as 1 mm = 1 kg and 1 mm = 500 kg.

2.26   Inch-Foot Scales

Several scales that are based on the inch-foot system of measurement continue in domestic use today along with the metric system of measurement that is accepted worldwide for science, technology, and international trade.

2.27   Engineers' Scale

Fig. 2.33 (b) The *engineers' scale* is graduated in the decimal system. It is also frequently called

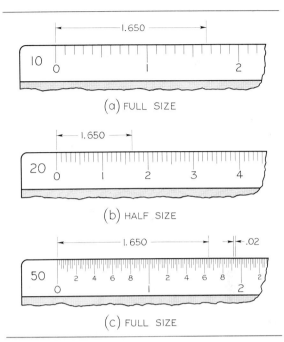

Fig. 2.36   Decimal Dimensions.

the *civil engineers' scale* because it was originally used mainly in civil engineering. The name *chain scale* also persists because it was derived from the surveyors' chain composed of 100 links, used for land measurements. The name "engineers' scale" is perhaps best because the scale is generally used by engineers of all kinds.

The engineers' scale is graduated in units of one inch divided into 10, 20, 30, 40, 50, and 60 parts. Thus, the engineers' scale is convenient in machine drawing to set off dimensions expressed in decimals. For example, to set off 1.650″ full size, Fig. 2.36 (a), use the 10-scale and simply set off one main division plus $6\frac{1}{2}$ subdivisions. To set off the same dimension half size, use the 20-scale, (b), since the 20-scale is exactly half the size of the 10-scale. Similarly, to set off a dimension quarter size, use the 40-scale.

The engineers' scale is also used in drawing maps to scales of 1″ = 50′, 1″ = 500′, 1″ = 5 miles, and so on and in drawing stress diagrams or other graphical constructions to such scales as 1″ = 20 lb and 1″ = 4000 lb.

2.28   Architects' Scale

Fig. 2.33 (e)   The *architects' scale* is intended primarily for drawings of buildings, piping systems, and other large structures that must be drawn to a reduced scale to fit on a sheet of paper. The full-size scale is also useful in drawing relatively small objects, and for that reason the architects' scale has rather general usage.

The architects' scale has 1 full-size scale and 10 overlapping reduced-size scales. By means of these scales a drawing may be made to various sizes from full size to $\frac{1}{128}$ size. *Note particularly, in all the reduced scales the major divisions represent feet, and their subdivisions represent inches and fractions thereof.* Thus, the scale marked $\frac{3}{4}$ means $\frac{3}{4}$ inch = 1 foot, not $\frac{3}{4}$ inch = 1 inch; that is, one-sixteenth size, not three-fourths size. And the scale marked $\frac{1}{2}$ means $\frac{1}{2}$ inch = 1 foot, not $\frac{1}{2}$ inch = 1 inch, that is, one-twenty-fourth size, not half-size.

All the scales, from full size to $\frac{1}{128}$ size, are shown in Fig. 2.37. Some are upside down, just as they may occur in use. These scales are described as follows.

FULL SIZE   *Fig. 2.37 (a)*   Each division in the full-size scale is $\frac{1}{16}$″. Each inch is divided first into halves, then quarters, eighths, and finally sixteenths, the division lines diminishing in length with each division. To set off $\frac{1}{32}$″, estimate visually one half of $\frac{1}{16}$″; to set off $\frac{1}{64}$″, estimate one-fourth of $\frac{1}{16}$″.

HALF SIZE   *Fig. 2.37 (a)*   Use the full-size scale, and divide every dimension mentally by two. (Do not use the $\frac{1}{2}$″ scale, which is intended for drawing to a scale of $\frac{1}{2}$″ = 1′, or one-twenty-fourth size.) To set off 1″, measure $\frac{1}{2}$″; to set off 2″, measure 1″; to set off $3\frac{1}{4}$″, measure $1\frac{1}{2}$″ (half of 3″), then $\frac{1}{8}$″ (half of $\frac{1}{4}$″); $6\frac{1}{2}$ to set off $2\frac{13}{16}$″ (see figure), measure 1″, then $\frac{13}{32}$″ ($\frac{6\frac{1}{2}}{16}$″ or half of $\frac{13}{16}$″).

QUARTER SIZE   *Fig. 2.37 (b)*   Use the 3″ scale in which 3″ = 1′. The subdivided portion to the left of zero represents 1 foot compressed to actually 3″ in length and is divided into inches, then half inches, quarter inches, and finally eighth inches. Thus, the entire portion repre-

senting 1 foot would actually measure 3 inches; therefore, 3″ = 1′. To set off anything less than 12″, start at zero and measure to the left.

To set off $10\frac{1}{8}$″, read off 9″ from zero to the left, and add $1\frac{1}{8}$″ and set off the total $10\frac{1}{8}$″, as shown. To set off more than 12″, for example, $1′–9\frac{3}{8}$″ (see your scale), find the 1′ mark to the right of zero and the $9\frac{3}{8}$″ mark to the left of zero; the required distance is the distance between these marks and represents $1′–9\frac{3}{8}$″.

EIGHTH SIZE   *Fig. 2.37 (b)*   Use the $1\frac{1}{2}$″ scale in which $1\frac{1}{2}$″ = 1′. The subdivided portion to the right of zero represents 1′ and is divided into inches, then half inches, and finally quarter inches. The entire portion, representing 1′, actually is $1\frac{1}{2}$″; therefore, $1\frac{1}{2}$″ = 1′. To set off anything less than 12″, start at zero and measure to the right.

DOUBLE SIZE   Use the full-size scale, and multiply every dimension mentally by 2. To set off 1″, measure 2″; to set off $3\frac{1}{4}$″, measure $6\frac{1}{2}$″; and so on. The double-size scale is occasionally used to represent small objects. In such cases, a small actual-size outline view should be shown near the bottom of the sheet to help the shop worker visualize the actual size of the object.

OTHER SIZES   *Fig. 2.37*   The other scales besides those just described are used chiefly by architects. Machine drawings are customarily made only double size, full size, half size, one-fourth size, or one-eighth size.

2.29   Decimal Scale

Fig. 2.33 (d)   The increasing use of decimal dimensions has brought about the development of a scale specifically for that use. On the full-size scale, each inch is divided into fiftieths of an inch, or .02″, as shown in Fig. 2.36 (c), and on the half- and quarter-size scales, the inches are compressed to half size or quarter size, and then are divided into 10 parts, so that each subdivision stands for .1″.

The complete decimal system of dimensioning, in which this scale is used, is described in §13.10.

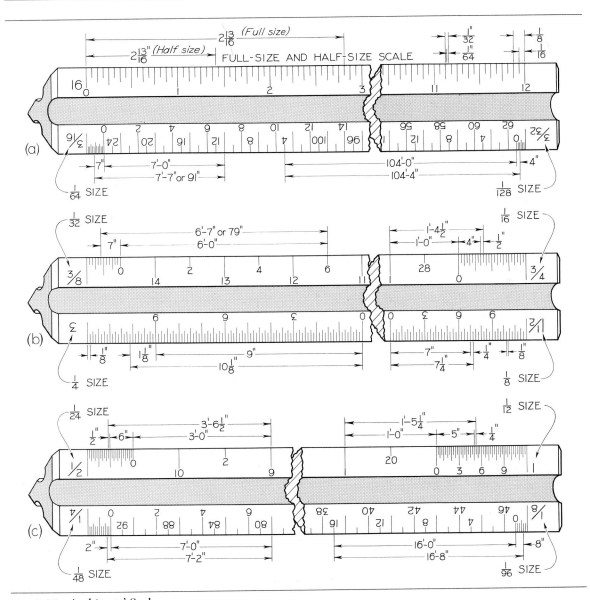

Fig. 2.37 Architects' Scales.

2.30 Mechanical Engineers' Scale

Fig. 2.33 (c) The objects represented in machine drawing vary in size from small parts, an inch or smaller in size, to machines of large dimensions. By drawing these objects full size, half size, quarter size, or eighth size, the drawings will readily come within the limits of the standard-size sheets. For this reason the mechanical engineers' scales are divided into units representing inches to full size, half size, quarter size, or eighth size. To make a drawing of an object to a scale of one-half size, for example, use the mechanical drafter's scale marked half size, which is graduated so that every $\frac{1}{2}''$ represents $1''$. Thus, the half-size scale is simply a full-size scale compressed to one-half size.

33

These scales are also very useful in dividing dimensions. For example, to draw a $3\frac{11}{16}''$ diameter circle full size, we need half of $3\frac{11}{16}''$ to use as radius. Instead of using arithmetic to find half of $3\frac{11}{16}''$, it is easier to set off $3\frac{11}{16}''$ on the half-size scale.

Triangular combination scales are available that include the full- and half-size mechanical engineers' scales, several architects' scales, and an engineers' scale.

2.31   To Specify the Scale on a Drawing

For machine drawings the scale indicates the ratio of the size of the drawing of the part or machine to its actual size irrespective of the unit of measurement used. The recommended practice is to letter FULL SIZE or 1:1; HALF SIZE or 1:2; and similarly for other reductions. Expansion or enlargement scales are given as $2:1$ or $2\times$; $3:1$ or $3\times$; $5:1$ or $5\times$; $10:1$ or $10\times$; and so on.

The various scale calibrations available on the metric scales and the engineers' scale provide almost unlimited scale ratios. The preferred metric scale ratios appear to be $1:1, 1:2, 1:5,$ $1:10, 1:20, 1:50, 1:100,$ and $1:200$. For examples of how scales may be shown on machine drawings, see Figs. 16.24 and 16.25.

Map scales are indicated in terms of fractions, such as   Scale $\frac{1}{62500}$, or graphically, such as 400 0 400 800 Ft . See also Fig. 25.2.

2.32   Accurate Measurements

Fig. 2.38   Accurate drafting depends considerably on the correct use of the scale in setting off distances. Do not take measurements directly off the scale with the dividers or compass, as damage will result to the scale. Place the scale on the drawing with the edge parallel to the line on which the measurement is to be made and, with a sharp pencil having a conical point, make a short dash at right angles to the scale and opposite the correct graduation mark, as shown at (a). If extreme accuracy is required, a tiny prick mark may be made at the required point with the needle point or stylus, as shown at (b), or with one leg of the dividers.

Avoid cumulative errors in the use of the scale. If a number of distances are to be set off end-to-end, all should be set off at one setting of the scale by adding each successive measurement to the preceding one, if possible. Avoid setting off the distances individually by moving the scale to a new position each time, since slight errors in the measurements may accumulate and give rise to a large error.

2.33   Drawing Instruments

Drawing instruments are generally sold in sets, in cases, but they may be purchased separately. The principal parts of high-grade instruments are usually made of nickel silver, which has a silvery luster, is corrosion-resistant, and can be

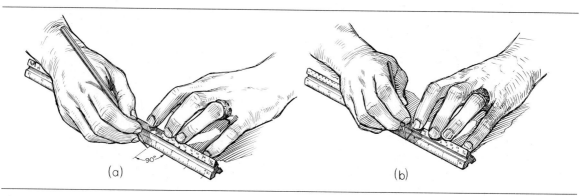

Fig. 2.38   Accurate Measurements.

readily machined into desired shapes. Tool steel is used for the blades of ruling pens, for spring parts, for divider points, and for the various screws.

In technical drawing, accuracy, neatness, and speed are essential, §2.2. These objectives are not likely to be obtained with cheap or inferior drawing instruments. For the student or the professional drafter, it is advisable, and in the end more economical, to purchase the best instruments that can be afforded. Good instruments will satisfy the most rigid requirements, and the satisfaction, saving in time, and improved quality of work that good instruments can produce will more than justify the higher price.

Unfortunately, the qualities of high-grade instruments are not likely to be recognized by the beginner, who is not familiar with the performance characteristics required and who is apt to be attracted by elaborate sets containing a large number of shiny, low-quality instruments. Therefore, the student should obtain the advice of the drafting instructor, an experienced drafter, or a reliable dealer.

2.34   Giant Bow Set

It was once general practice to make pencil drawings on detail paper and then to make an inked tracing from it on tracing cloth. As reproduction methods and transparent tracing papers were improved, it was found that a great deal of time could be saved by making drawings directly in pencil with dense black lines on the tracing paper and making prints or photocopies therefrom, thus doing away with the preliminary pencil drawing on detail paper. Today, though inked tracings are still made when a fine appearance is necessary and where the greater cost is justified, the overwhelming proportion of drawings is made directly in pencil on tracing paper, vellum, polyester films, or pencil tracing cloth.

Fig. 2.39   These developments have brought about the giant bow sets that are now offered by all the major manufacturers. The sets contain various combinations of instruments, but all feature a large bow compass in place of the traditional large compass. The large bow instrument is much sturdier and is capable of taking the

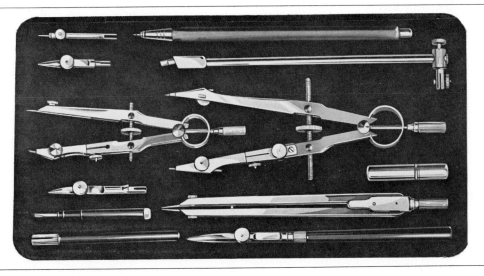

Fig. 2.39   Giant Bow Set. *Courtesy of Frank Oppenheimer.*

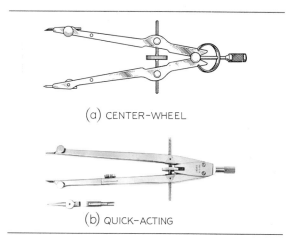

(a) CENTER-WHEEL

(b) QUICK-ACTING

Fig. 2.40   Giant Bow Compass. *(b) Courtesy of Frank Oppenheimer.*

heavy pressure necessary to produce dense black lines without losing the setting.

Most of the large bows are of the center-wheel type, Fig. 2.40 (a). Several manufacturers now offer different varieties of quick-acting bows. The large bow compass shown at (b) can be adjusted to the approximate setting by simply opening or closing the legs in the same manner as for the other bow-style compass.

2.35   The Compass

The giant bow compass, Figs. 2.39–2.41, has a socket joint in one leg that permits the insertion of either pencil or pen attachments. A *lengthening bar* or a *beam attachment* is often provided to increase the radius. For production drafting, in which it is necessary to make dense black lines to secure clear legible reproductions, the giant bow, or an appropriate template, Figs. 2.69 and 2.70, is preferred.

2.36   Using the Compass

Fig. 2.41   These instructions apply generally both to the old style and the giant bow compasses. The compass, with pencil and inking attachments, is used for drawing circles of approximately 25 mm (1″) radius or larger. Most compass needle points have a plain end for use

when the compass is converted into dividers and a shoulder end for use as a compass. Adjust the needle point with the shoulder end out and so that the small point extends *slightly* farther than the pencil lead or pen nib, Fig. 2.43 (d).

To draw a penciled circle, (1) set off the required radius on one of the center lines, (2) place the needle point at the exact intersection of the center lines, (3) adjust the compass to the required radius (25 mm or more), and (4) lean the compass forward and draw the circle clockwise while rotating the handle between the thumb and forefinger. To obtain sufficient weight of line, it may be necessary to repeat the movement several times.

Any error in radius will result in a doubled error in diameter; so, it is best to draw a trial circle first on scrap paper or on the backing sheet and then check the diameter with the scale.

On drawings having circular arcs and tangent straight lines, draw the arcs first, whether in pencil or in ink, as it is much easier to connect a straight line to an arc than the reverse.

For very large circles, a beam compass, §2.38, is preferred, or use the lengthening bar to increase the compass radius. Use both hands, as

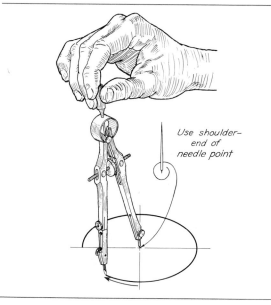

Use shoulder-end of needle point

Fig. 2.41   Using the Giant Bow Compass.

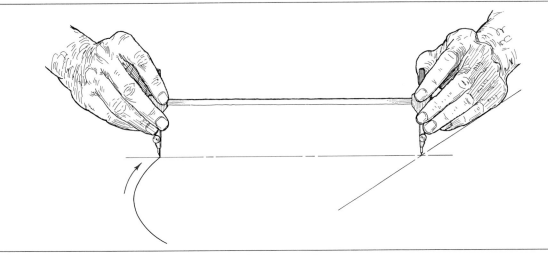

Fig. 2.42 Drawing a Circle of Large Radius with the Beam Compass.

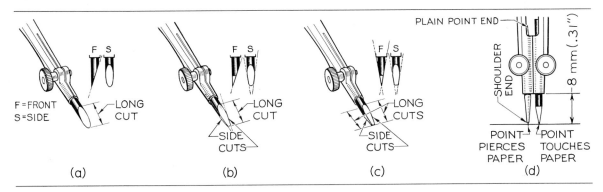

Fig. 2.43 Compass Lead Points.

shown in Fig. 2.42, but be careful not to jar the instrument and thus change the adjustment.

When using the compass to draw construction lines, use a 4H to 6H lead so that the lines will be very dim. For required lines, the arcs and circles must be black, and softer leads must be used. However, since heavy pressure cannot be exerted on the compass as it can on a pencil, it is usually necessary to use a compass lead that is one or two grades softer than the pencil used for the corresponding line work. For example, if a 2H lead is used for visible lines drawn with the pencil, then an F lead might be found suitable for the compass work. The hard leads supplied with the compass are usually unsatisfactory for most line work except construction lines. In summary, use leads in the compass that will pro-

duce arcs and circles that *match* the straight pencil lines.

It is necessary to exert pressure on the compass to produce heavy "reproducible" circles, and this tends to enlarge the compass center hole in the paper, especially if there are a number of concentric circles. In such cases, use a horn center, or center tack, in the hole, and place the needle point of the compass in the center of the tack.

2.37 Sharpening the Compass Lead

Fig. 2.43 Various forms of compass lead points are illustrated. At (a), a single elliptical face has been formed by rubbing on the sandpaper pad, as shown in Fig. 2.44. At (b), the point is nar-

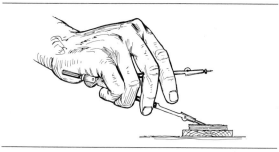

Fig. 2.44   Sharpening Compass Lead.

rowed by small side cuts. At (c), two long cuts and two small side cuts have been made so as to produce a point similar to that on a screwdriver. At (d), the cone point is prepared by chucking the lead in a mechanical pencil and shaping it in a pencil pointer. Avoid using leads that are too short to be exposed as shown.

In using the compass, *never use the plain end of the needle point.* Instead, use the shoulder end, as shown in Fig. 2.43 (d), adjusted so that the tiny needle point extends about halfway into the paper when the compass lead just touches the paper.

2.38   Beam Compass

Fig. 2.45   The beam compass, or trammel, is used for drawing arcs or circles larger than can be drawn with the regular compass and for transferring distances too great for the regular dividers. Besides steel points, pencil and pen attachments are provided. The beams may be made of nickel silver, steel, aluminum, or wood and are procurable in various lengths. A square nickel silver beam compass set is shown at (a), and a set with the beam graduated in millimeters and inches at (b).

2.39   Dividers

The dividers are similar to the compass in construction and are made in square, flat, and round forms.

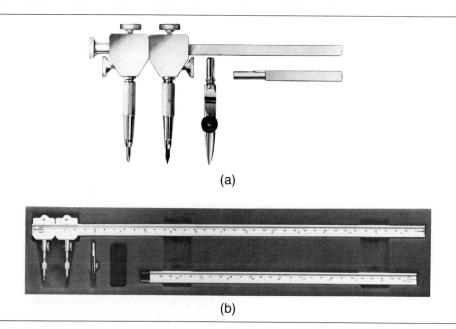

(a)

(b)

Fig. 2.45   Beam Compass Sets. *(a) Courtesy of Frank Oppenheimer; (b) Courtesy of Tacro, Div. of A&T Importers, Inc.*

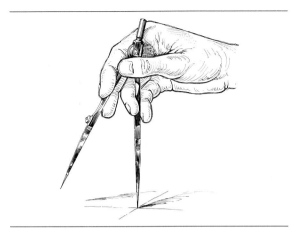

Fig. 2.46 Adjusting the Dividers.

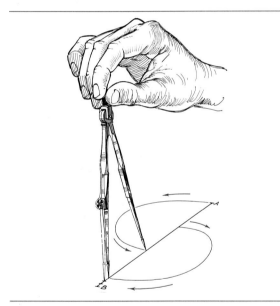

Fig. 2.47 Using the Dividers.

The friction adjustment for the pivot joint should be loose enough to permit easy manipulation with one hand, as shown in Fig. 2.46. If the pivot joint is too tight, the legs of the compass tend to spring back instead of stopping at the desired point when the pressure of the fingers is released. To adjust tension, use a small screwdriver.

Many dividers are made with a spring and thumbscrew in one leg so that minute adjustments in the setting can be made by turning the small thumbscrew.

2.40 Using the Dividers

The dividers, as the name implies, are used for *dividing* distances into a number of equal parts. They are also used for *transferring distances* or for *setting off* a series of equal distances. The dividers are used for spaces of approximately 25 mm (1″) or more. For less than 25 mm spaces, use the bow dividers, Fig. 2.50 (a). *Never use the large dividers for small spaces when the bow dividers can be used; the latter are more accurate.*

Fig. 2.47   To divide a given distance into a number of equal parts, the method is one of trial and error. Adjust the dividers with the fingers of the hand that holds them, to the approximate unit of division, estimated by eye. Rotate the dividers counterclockwise through 180°, and so on, until the desired number of units has been stepped off. If the last prick of the dividers falls short of the end of the line to be divided, increase the distance between the divider points proportionately. For example, to divide the line AB into three equal parts, the dividers are set by eye to approximately one-third the length AB. When it is found that the trial radius is too small, the distance between the divider points is increased by one-third the remaining distance. If the last prick of the dividers is beyond the end of the line, a similar decreasing adjustment is made.

The student should avoid *cumulative errors*, which may result when the dividers are used to set off a series of distances end to end. To set off a large number of equal divisions, say, 15, first set off 3 equal large divisions and then divide each into 5 equal parts. Wherever possible in such cases, use the scale instead of the dividers, as described in §2.32, or set off the total and then divide into the parts by means of the parallel-line method, §§5.14 and 5.15.

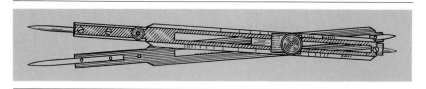

Fig. 2.48   Proportional Dividers.

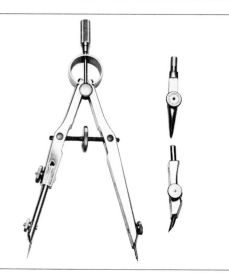

Fig. 2.49   Combination Pen and Pencil Bow. *Courtesy of Frank Oppenheimer.*

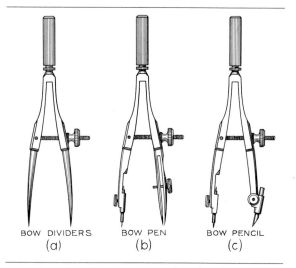

BOW DIVIDERS   BOW PEN   BOW PENCIL
(a)   (b)   (c)

Fig. 2.50   Bow Instruments with Side Wheel.

2.41   Proportional Dividers

Fig. 2.48   For enlarging or reducing a drawing, proportional dividers are convenient. They may also be used for dividing distances into a number of equal parts, or for obtaining a percentage reduction of a distance. For this purpose, points of division are marked on the instrument so as to secure the required subdivisions readily. Some instruments are calibrated to obtain special ratios, such as $1 : \sqrt{2}$, the diameter of a circle to the side of an equal square, and feet to meters.

2.42   The Bow Instruments

The bow instruments are classified as the *bow dividers, bow pen,* and *bow pencil.* A combination pen and pencil bow, usually with center-wheel adjustment, Fig. 2.49, and separate instruments with either side-wheel or center-wheel adjustment, Fig. 2.50, are available. The choice is a matter of personal preference.

2.43   Using the Bow Instruments

The bow pencil and bow pen are used for drawing circles of approximately 25 mm (1″) radius or smaller. The bow dividers are used for the same purpose as the large dividers, but for smaller (approximately 25 mm or less) spaces and more accurate work.

Whether the center-wheel or side-wheel instrument is used, the adjustment should be made with the thumb and finger of the hand that holds the instrument, Fig. 2.51 (a). The instrument is manipulated by twirling the head between the thumb and fingers, (b).

The lead is sharpened in the same manner as for the large compass, §2.37, except that for small radii, the inclined cut may be turned *inside* if preferred, Fig. 2.52 (a). For general use, the lead should be turned to the outside, as shown at (b). In either case, always keep the compass lead sharpened. *Avoid stubby compass leads,* which cannot be properly sharpened. At

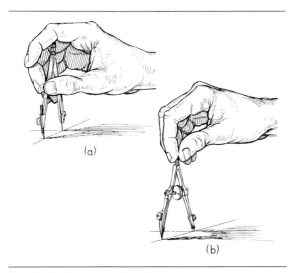

Fig. 2.51   Using the Bow Instruments.

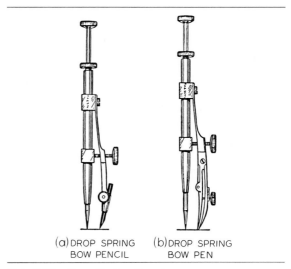

(a) DROP SPRING
BOW PENCIL

(b) DROP SPRING
BOW PEN

Fig. 2.53   Drop Spring Bow Instruments.

least 6 mm ($\frac{1}{4}''$) of lead should extend from the compass at all times.

In adjusting the needle point of the bow pencil or bow pen, be sure to have the needle extending slightly longer than the pen or the lead, Fig. 2.43 (d), the same as for the large compass.

In drawing small circles, greater care is necessary in sharpening and adjusting the lead and the needle point, and especially in accurately setting the desired radius. If a 6.35 mm ($\frac{1}{4}''$) diameter circle is to be drawn, and if the radius is "off" only 0.8 mm ($\frac{1}{32}''$) the total error on diameter is approximately 25 percent, which is far too much error.

Appropriate templates may also be used for drawing small circles. See Figs. 2.69 and 2.70.

2.44   Drop Spring Bow Pencil and Pen

Fig. 2.53   These compasses are designed for drawing multiple identical small circles, such as drill holes or rivet heads. A central pin is made to move easily up and down through a tube to which the pen or pencil unit is attached. To use the instrument, hold the knurled head of the tube between the thumb and second finger, plac-

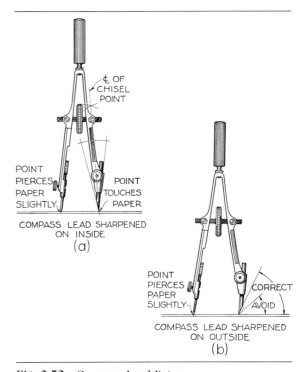

¢ OF
CHISEL
POINT

POINT
PIERCES
PAPER
SLIGHTLY

POINT
TOUCHES
PAPER

COMPASS LEAD SHARPENED
ON INSIDE
(a)

POINT
PIERCES
PAPER
SLIGHTLY

CORRECT

AVOID

COMPASS LEAD SHARPENED
ON OUTSIDE
(b)

Fig. 2.52   Compass-Lead Points.

ing the first finger on top of the knurled head of the pin. Place the point of the pin at the desired center, lower the pen or pencil until it touches the paper, and twirl the instrument clockwise with the thumb and second finger. Then lift the

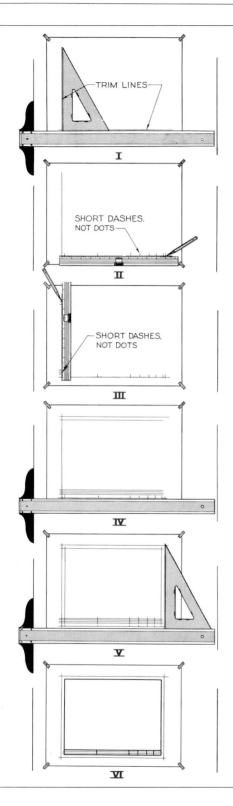

tube independently of the pin, and finally lift the entire instrument.

2.45   To Lay Out a Sheet

Fig. 2.54   After the sheet has been attached to the board, as explained in §2.7, proceed as follows (see also Layout A–2, inside back cover).

I.   Using the T-square, draw a horizontal *trim line* near the lower edge of the paper, and then using the triangle, draw a vertical trim line near the left edge of the paper. Both should be *light construction lines.*

II.   Place the scale along the lower trim line with the full-size scale up. Draw short and light dashes *perpendicular* to the scale at the required distances. See Fig. 2.38 (a).

III.   Place the scale along the left trim line with the full-size scale to the left, and mark the required distances with short and light dashes perpendicular to the scale.

IV.   Draw horizontal construction lines with the aid of the T-square through the marks at the left of the sheet.

V.   Draw vertical construction lines, *from the bottom upward,* along the edge of the triangle through the marks at the bottom of the sheet.

VI.   Retrace the border and the title strip to make them heavier. Notice that the layout is made independently of the edges of the paper.*

2.46   Technique of Pencil Drawing

By far the greater part of commercial drafting is executed in pencil. Most prints or photocopies are made from pencil tracings, and all ink tracings must be preceded by pencil drawings. It should therefore be evident that skill in drafting chiefly implies skill in pencil drawing.

Technique is a style or quality of drawing imparted by the individual drafter to the work. It

---

*In industrial drafting rooms the sheets are available, cut to standard sizes, with border and title strips already printed. Drafting supply houses can supply such papers, printed to order, to schools for little extra cost.

Fig. 2.54   To Lay Out a Sheet.
Layout A–2; see inside back cover.

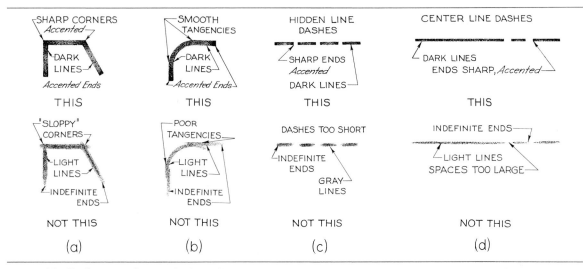

Fig. 2.55 Technique of Lines (Enlarged).

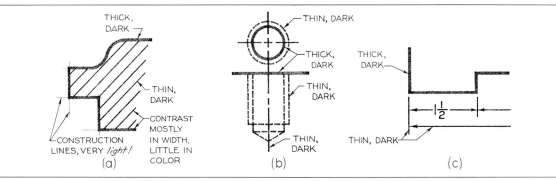

Fig. 2.56 Contrast of Lines (Enlarged).

is characterized by crisp black line work and lettering. Technique in lettering is discussed in §4.11.

DARK ACCENTED LINES *Fig. 2.55* The pencil lines of a finished pencil drawing or tracing should be very dark. Dark crisp lines are necessary to give punch or snap to the drawing. Ends of lines should be accented by a little extra pressure on the pencil, (a). Curves should be as dark as other lines, (b). Hidden-line dashes and center-line dashes should be carefully estimated as to length and spacing and should be of uniform width throughout their length, (c) and (d).

Dimension lines, extension lines, section lines, and center lines also should be dark. The

difference between these lines and visible lines is mostly in width—there is very little difference, if any, in blackness.

A simple way to determine whether your lines on tracing paper or cloth are dense black is to hold the tracing up to the light. Lines that are not opaque black will not print clearly by most reproduction processes.

Construction lines should be made with a sharp, hard lead and *should be so light that they need not be erased* when the drawing is completed.

CONTRAST IN LINES *Fig. 2.56* Contrast in pencil lines should be similar to that of ink lines; that is, the difference beween the various lines

43

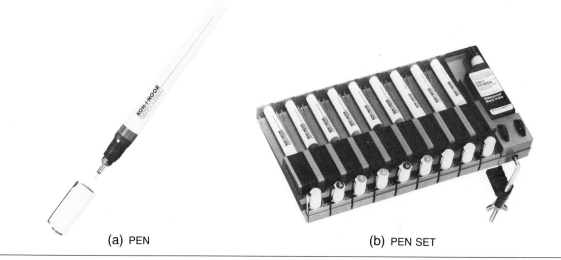

(a) PEN   (b) PEN SET

Fig. 2.57   Technical Fountain Pen and Pen Set. *Courtesy of Koh-I-Nor Rapidograph, Inc.*

should be mostly in the *widths* of the lines, with little if any difference in the degree of darkness. The visible lines should contrast strongly with the thin lines of the drawing. If necessary, draw over a visible line several times to get the desired thickness and darkness. A short retracing stroke backward (to the left), producing a jabbing action, results in a darker line.

2.47   Pencil Tracing

While some pencil tracings are made of a drawing placed underneath the tracing paper (usually when a great deal of erasing and changing is necessary on the original drawing), most drawings today are made directly in pencil on tracing paper, pencil tracing cloth, films, or vellum. These are not tracings but pencil drawings, and the methods and technique are the same as previously described for pencil drawing.

In making a drawing directly on a tracing medium, a smooth sheet of heavy white drawing paper, a backing sheet, should be placed underneath. The whiteness of the backing sheet improves the visibility of the lines, and the hardness of the surface makes it possible to exert

pressure on the pencil and produce dense black lines without excessive grooving of the paper.

All lines must be dark and cleanly drawn when drawings are intended to be reproduced.

2.48   Technical Fountain Pens

Fig. 2.57   The technical fountain pen, with the tube and needle point, is available in several line widths. Many prefer this type of pen, for the line widths are fixed and it is suitable for freehand or mechanical lettering and line work. The pen requires an occasional filling and a minimum of skill to use. For uniform line work, the pen should be used perpendicular to the paper, Fig. 2.58. For best results, follow the manufacturer's recommendations for operation and cleaning.

2.49   Ruling Pens

Fig. 2.59   The ruling pen should be of the highest quality, with blades of high-grade tempered steel sharpened properly at the factory. The nibs should be sharp, but not sharp enough to cut the paper.

Fig. 2.60   Filling the Ruling Pen.

Fig. 2.58   Using the Technical Fountain Pen.

The *detail pen,* capable of holding a considerable quantity of ink, is extremely useful for drawing long heavy lines, (b). This type of pen is preferred for small amounts of ink work as the pen is easily adjusted for the desired line width and is readily cleaned after use.

For methods of using the ruling pen, see §2.51.

Special drawing ink is available for use on acetate and polyester films. Such inks should not be used in technical fountain pens unless the pens are specifically made for acetate-based inks.

For removing dried waterproof drawing ink from pens or instruments, pen-cleaning fluids are available at dealers.

2.50   Drawing Ink

Drawing ink is composed chiefly of carbon in colloidal suspension and gum. The fine particles of carbon give the deep, black luster to the ink, and the gum makes it waterproof and quick to dry. The ink bottle should not be left uncovered, as evaporation will cause the ink to thicken.

2.51   Use of the Ruling Pen

The ruling pen, Fig. 2.59, is used to ink lines drawn with instruments, never to ink freehand lines or freehand lettering. The proper method of filling the pen is shown in Fig. 2.60.

The pen should lean at an angle of about 60° with the paper in the direction in which the line

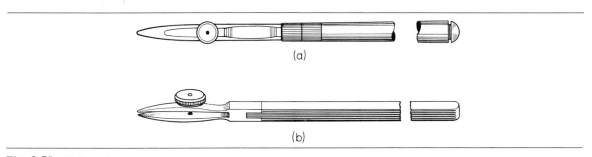

(a)

(b)

Fig. 2.59   Ruling Pens.

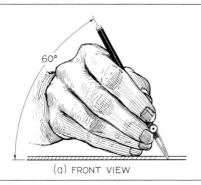
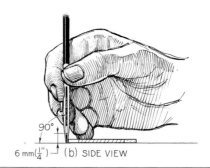

(a) FRONT VIEW

(b) SIDE VIEW

6 mm(¼")

Fig. 2.61   Position of Hand in Using the Ruling Pen.

is being drawn and in a vertical plane containing the line, Fig. 2.61.

The ruling pen is used in inking irregular curves, as well as straight lines, as shown in Figs. 2.66 and 2.67. The pen should be held more nearly vertical when used with an irregular curve than when used with the T-square or a triangle. The ruling pen should lean only slightly in the direction in which the line is being drawn.

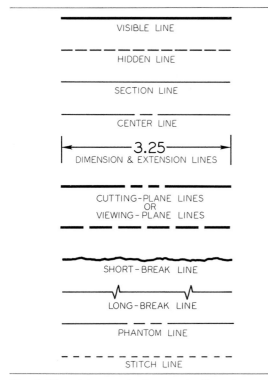

Fig. 2.62   Alphabet of Ink Lines (Full Size).

2.52   Technique of Inking

The various widths of lines used for inked drawings or tracings are shown in Fig. 2.62. A definite order should be followed in inking a drawing or tracing, Fig. 2.63, as follows.

1. (a) Mark all tangent points in pencil directly on drawing or tracing.
   (b) Indent all compass centers (with pricker or divider point).
   (c) Ink visible circles and arcs.
   (d) Ink hidden circles and arcs.
   (e) Ink irregular curves, if any.

2. (1st: horizontal; 2nd: vertical; 3rd: inclined)
   (a) Ink visible straight lines.
   (b) Ink hidden straight lines.

3. (1st: horizontal; 2nd: vertical; 3rd: inclined)
   Ink center lines, extension lines, dimension lines, leader lines, and section lines (if any).

4. (a) Ink arrowheads and dimension figures.
   (b) Ink notes, titles, etc. (Pencil guide lines directly on drawing or tracing.)

Some drafters prefer to ink center lines before indenting the compass centers because ink can go through the holes and cause blots on the back of the sheet.

When an ink blot is made, the excess ink should be taken up with a blotter, or smeared with the finger if a blotter is not available, and not allowed to soak into the paper. When the spot is thoroughly dry, the remaining ink can be erased easily.

46

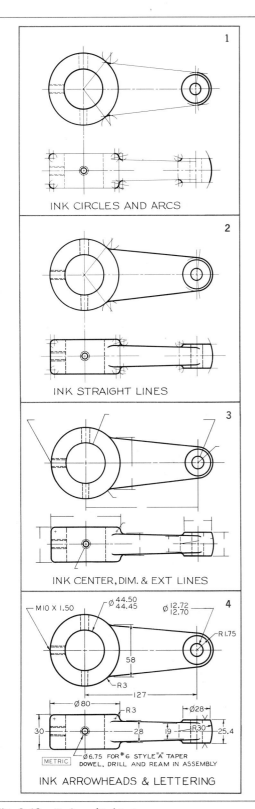

Fig. 2.63   Order of Inking.

For cleaning untidy drawings or for removing the original pencil lines from an inked drawing, the Pink Pearl or the Mars-Plastic eraser is suitable if used lightly.

When erasure on cloth damages the surface, it may be restored by rubbing the spot with soapstone and then applying pounce or chalk dust. If the damage is not too great, an application of the powder will be sufficient.

When a gap in a thick ink line is made by erasing, the gap should be filled in with a series of fine lines that are allowed to run together. A single heavy line is difficult to match and is more likely to run and cause a blot.

In commercial drafting rooms, the timesaving electric erasing machine, Fig. 2.18, is usually available.

2.53   Irregular Curves

When it is required to draw mechanical curves other than circles or circular arcs, an irregular or French curve is generally employed. Many different forms and sizes of curves are manufactured, as suggested by the more common forms illustrated in Fig. 2.64.

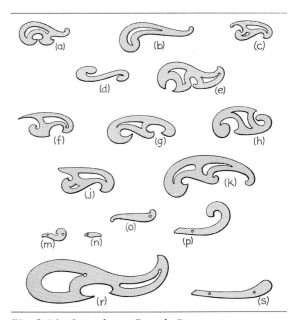

Fig. 2.64   Irregular or French Curves.

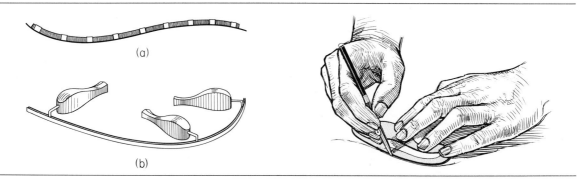

Fig. 2.65   Adjustable Curves.   **Fig. 2.66**   Using the Irregular Curve.

The curves are composed largely of successive segments of the geometric curves, such as the ellipse, parabola, hyperbola, and involute. The best curves are made of transparent plastic. Among the many special types of curves that are available are hyperbolas, parabolas, ellipses, logarithmic spirals, ship curves, and railroad curves.

Fig. 2.65   Adjustable curves are also available. The curve shown at (a) consists of a core of lead, enclosed by a coil spring attached to a flexible strip. The one at (b) consists of a spline, to which "ducks" (weights) are attached. The spline can be bent to form any desired curve, limited only by the elasticity of the material. An ordinary piece of solder wire can be used very successfully by bending the wire to the desired curve.

2.54   Using the Irregular Curve

The irregular curve is a device for the *mechanical drawing of curved lines and should not be applied directly to the points* or used for purposes of producing an initial curve. The proper use of the irregular curve requires skill, especially when the lines are to be drawn in ink. After points have been plotted through which the curve is to pass, a light pencil line should be sketched freehand smoothly through the points.

Fig. 2.66   To draw a mechanical line over the freehand line with the aid of the irregular curve, it is only necessary to match the various segments of the irregular curve with successive portions of the freehand curve and to draw the line with pencil or ruling pen along the edge of the curve. It is very important that the irregular curve match the curve to be drawn for some distance at each end beyond the segment to be drawn for any one setting of the curve, as shown in Fig. 2.67. When this rule is observed, the successive sections of the curve will be tangent to each other, without any abrupt change in the curvature of the line. In placing the irregular curve, the short-radius end of the curve should be turned toward the short-radius part of the curve to be drawn; that is, the portion of the irregular curve used should have the same curvilinear tendency as the portion of the curve to be drawn. This will prevent abrupt changes in direction.

The drafter should change position with the drawing when necessary, to avoid working on the lower side of the curve.

When plotting points to establish the path of a curve, it is desirable to plot more points, and closer together, where sharp turns in the curve occur.

Free curves may also be drawn with the compass, as shown in Fig. 5.42.

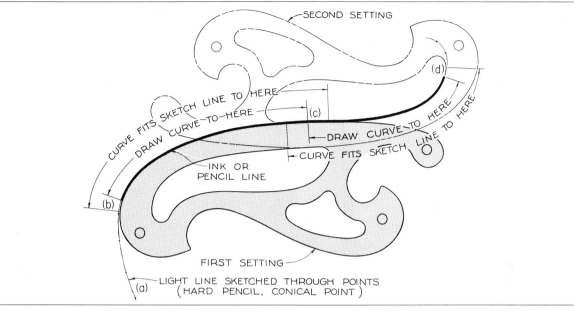

Fig. 2.67   Settings of Irregular Curve.

Fig. 2.68   For symmetrical curves, such as an ellipse, use the same segment of the irregular curve in two or more opposite places. For example, at (a) the irregular curve is matched to the curve and the line drawn from 1 to 2. Light pencil dashes are then drawn directly on the irregular curve at these points. (The curve will take pencil marks well if it is lightly "frosted" by rubbing with a hard pencil eraser.) At (b) the irregular curve is turned over and matched so that the line may be drawn from 2 to 1. In similar manner, the same segment is used again at (c) and (d). The ellipse is completed by filling in the gaps at the ends by using the irregular curve or, if desired, the compass.

2.55   Templates

Fig. 2.69   Templates are available for a great variety of specialized needs. A template may be found for drawing almost any ordinary drafting symbols or repetitive features. The engineers'

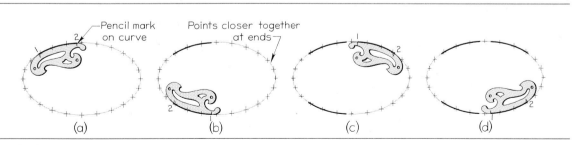

Fig. 2.68   Symmetrical Figures.

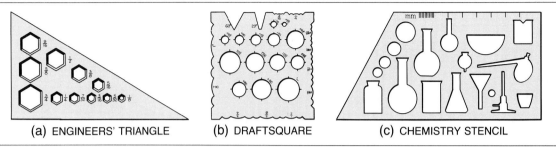

(a) ENGINEERS' TRIANGLE   (b) DRAFTSQUARE   (c) CHEMISTRY STENCIL

Fig. 2.69   Drafting Devices.

triangle, (a), is useful for drawing hexagons or for bolt heads and nuts; the draftsquare, (b), is convenient for drawing the curves on bolt heads and nuts, for drawing circles, thread forms, and so forth; and the chemistry stencil, (c), is useful for drawing chemical apparatus in schematic form.

Ellipse templates, §5.56, are perhaps more widely used than any other type. Circle templates are useful for drawing small circles quickly and for drawing fillets and rounds, and are used extensively in tool and die drawings. Some of the more commonly used templates are shown in Fig. 2.70.

2.56   Drafting Machines

Fig. 2.71   The drafting machine is an ingenious device that replaces the T-square, triangles, scales, and protractor. The links, or bands, are arranged so that the controlling head is always in any desired fixed position regardless of where it is placed on the board; thus, the horizontal straightedge will remain horizontal if so set. The

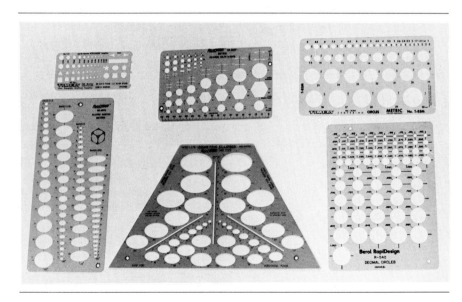

Fig. 2.70   Templates.

Fig. 2.71   Drafting Machine. *Courtesy of Keuffel & Esser Co.*

Fig. 2.72   Adjustable Drafting Table with Track Drafting Machine. *Courtesy of Keuffel & Esser Co.*

controlling head is graduated in degrees (including a vernier on certain machines), which allows the straightedges, or scales, to be set and locked at any angle. There are automatic stops at the more frequently used angles, such as 15°, 30°, 45°, 60°, 75°, and 90°.

Fig. 2.72   Drafting machines and drafting tables have been greatly improved in recent years. The chief advantage of the drafting machine is that it speeds up drafting. Since its parts are made of metal, their accurate relationships are not subject to change, whereas T-squares, triangles, and working edges of drawing boards must be checked and corrected frequently. Drafting machines for left-handers are available from the manufacturers.

2.57   Parallel-Ruling Straightedge
For large drawings, the long T-square becomes unwieldy, and considerable inaccuracy may result from the "give" or swing of the blade. In such a case the parallel-ruling straightedge, Fig. 2.73, is recommended. The ends of the straightedge are controlled by a system of cords and pulleys that permit the straightedge to be moved up or down on the board while maintaining a horizontal position.

2.58   Drawing Papers
Drawing paper, or *detail paper,* is used whenever a drawing is to be made in pencil but not for reproduction. For working drawings and for general use, the preferred paper is light cream or buff in color, and it is available in rolls of widths 24″ and 36″ and in cut sheets of standard sizes such as 8.5″ × 11″, 11″ × 17″, 17″ × 22″, and so on. Most industrial drafting rooms use standard sheets, §2.63, with printed borders and title strips, and since the cost for printing adds so little to the price per sheet, many schools have also adopted printed sheets.

The best drawing papers have up to 100 percent pure rag stock; have strong fibers that afford superior erasing qualities, folding strength, and toughness; and will not discolor or grow brittle

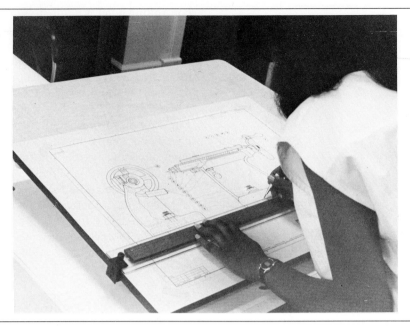

Fig. 2.73   Parallel-Ruling Straightedge.

with age. The paper should have a fine grain or tooth that will pick up the graphite and produce clean, dense black lines. However, if the paper is too rough, it will wear down the pencil excessively and will produce ragged, grainy lines. The paper should have a hard surface so that it will not groove too easily when pressure is applied to the pencil.

For ink work, as for catalog and book illustrations, white papers are used. The better papers, such as Bristol Board and Strathmore, come in several thicknesses such as 2-ply, 3-ply, and 4-ply.

2.59   Tracing Papers

Tracing paper is a thin transparent paper on which drawings are made for the purpose of reproducing by blueprinting or by other similar processes. Tracings are usually made in pencil but may be made in ink. Most tracing papers will take pencil or ink, but some are especially suited to one or to the other.

Tracing papers are of two kinds: (1) those treated with oils, waxes, or similar substances to render them more transparent, called *vellums;* (2) those not so treated, but which may be quite transparent, owing to the high quality of the raw materials and the methods of manufacture. Some treated papers deteriorate rapidly with age, becoming brittle in many cases within a few months, but some excellent vellums are available. Untreated papers made entirely of good rag stock will last indefinitely and will remain tough.

2.60   Tracing Cloth

Tracing cloth is a thin transparent muslin fabric (cotton, not "linen" as commonly supposed) sized with a starch compound or plastic to provide a good working surface for pencil or ink. It is much more expensive than tracing paper. Tracing cloth is available in rolls of standard widths, such as 30", 36", and 42", and also in sheets of standard sizes, with or without printed borders and title forms.

For pencil tracings, special pencil tracing cloths are available. Many concerns make their drawings in pencil directly on this cloth, dispensing entirely with the preliminary pencil drawing on detail paper, thus saving a great deal of time. These cloths generally have a surface that will produce dense black lines when hard pencils are used. Hence, these drawings do not easily smudge and will stand up well with handling.

2.61   Polyester Films and Coated Sheets

The polyester film is a superior drafting material available in rolls and sheets of standard size. It is made by bonding a mat surface to one or both sides of a clear polyester sheet. The transparency and printing qualities are very good, the mat drawing surface is excellent for pencil or ink, erasures leave no ghost marks, and the film has high dimensional stability. Its resistance to cracking, bending, or tearing makes it virtually indestructible, if given reasonable care. The film has rapidly replaced cloth and is competing with vellum in some applications. Some companies have found it more economical to make their drawings directly in ink on the film.

Large coated sheets of aluminum, which provides a good dimensional stability, are often used in the aircraft and auto industry for full-scale layouts that are scribed into the coating with a steel point rather than a pencil. The layouts are reproduced from the sheets photographically.

2.62   Standard Sheets

Two systems of sheet sizes together with length, width, and letter designations are listed by ANSI, as shown in the accompanying table.

The use of the basic sheet size, 8.5″ × 11.0″ or 210 mm × 297 mm and multiples thereof, permits filing of small tracings and of folded prints in standard files with or without correspondence. These sizes can be cut without waste from the standard rolls of paper, cloth, or film.

For layout designations, title blocks, revision

Nearest International Size[a] (millimeter)	Standard USA Size[a] (inch)
A4 210 × 297	A 8.5 × 11.0
A3 297 × 420	B 11.0 × 17.0
A2 420 × 594	C 17.0 × 22.0
A1 594 × 841	D 22.0 × 34.0
A0 841 × 1189	E 34.0 × 44.0

[a]ANSI Y14.1–1980 (R1987).

blocks, and list of materials blocks, see inside the back cover of this book. See also §16.5.

2.63   The Computer as a Drafting Tool

The development of CAD workstations and the availability of drafting software have made the computer a valuable drafting instrument. The use of the computer in drafting began in 1963 when Ivan Sutherland, of M.I.T., developed "Sketchpad," a program for engineers and drafters. The use of the computer continued to grow and by the late 1970s was widespread. All drafters and engineers today need to become familiar with computer systems in drafting and design.

Most CAD systems, regardless of their size, Fig. 2.74, consist of an input device such as a

Fig. 2.74   Computer-Aided Drafting System. *Courtesy Bausch & Lomb*

joystick or digitizer pad, a cathode-ray tube (CRT—much like a television screen) on which to "build" the drawing, and a plotter that prints the final results. These devices are shown and discussed in Chapters 3 and 8. It is appropriate to point out that this new approach is developing rapidly and that much routine drafting can be done on CAD systems.

It should be noted that a student will need to learn the basic concepts of orthographic and pictorial projections as well as dimensioning before a CAD system can be effectively used.

INSTRUMENTAL DRAWING PROBLEMS

All the following constructions, Figs. 2.75–2.85, are to be drawn in pencil on Layout A2 (see inside the back cover of this book). The steps in drawing this layout are shown in Fig. 2.54. Draw all construction lines *lightly,* using a hard lead (4H to 6H), and all required lines dense black with a softer lead (F to H). If construction lines are drawn properly—that is, *lightly*—they need not be erased.

If the layout is to be made on the A4 size sheet, width dimensions for title-strip forms will need to be adjusted to fit the available space.

The drawings in Figs. 2.80–2.85 are to be drawn in pencil, preferably on tracing paper or vellum; then prints should be made to show the effectiveness of the student's technique. If ink tracings are required, the originals may be drawn on film or on detail paper and then traced on vellum or tracing cloth. For any assigned problem, the instructor may require that all dimensions and notes be lettered in order to afford further lettering practice.

The problems at the end of Chapter 5, "Geometric Constructions," provide excellent additional practice to develop skill in the use of drawing instruments.

Since many of the problems in this chapter are of a general nature, they can also be solved on most computer graphics systems. If a system is available, the instructor may choose to assign specific problems to be completed by this method.

Problems in convenient form for solution may be found in *Technical Drawing Problems,* Series 1, by Giesecke, Mitchell, Spencer, Hill, Dygdon, and Novak; *Technical Drawing Problems,* Series 2, by Spencer, Hill, Dygdon, and Novak; and *Technical Drawing Problems,* Series 3, by Spencer, Hill, Dygdon, and Novak; all designed to accompany this text and published by Macmillan Publishing Company.

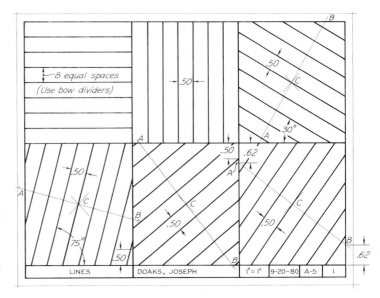

Fig. 2.75 Using Layout A–2 or A4–2 (adjusted), divide working space into six equal rectangles, and draw visible lines, as shown. Draw construction lines AB through centers C at right angles to required lines; then along each construction line, set off 0.50″ spaces and draw required visible lines. Omit dimensions and instructional notes.

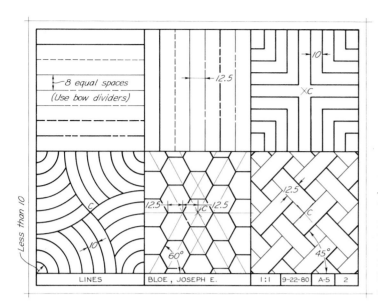

Fig. 2.76 Using Layout A–2 or A4–2 (adjusted), divide working space into six equal rectangles, and draw lines as shown. In first two spaces, draw conventional lines to match those in Fig. 2.14. In remaining spaces, locate centers C by diagonals, and then work constructions out from them. Omit the metric dimensions and instructional notes.

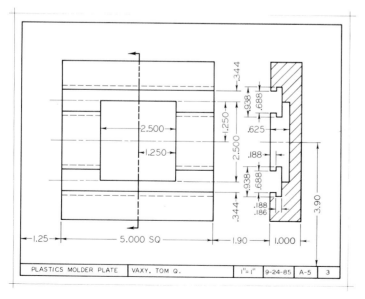

Fig. 2.77 Using Layout A–2 or A4–2 (adjusted), draw views in pencil, as shown. Omit all dimensions.

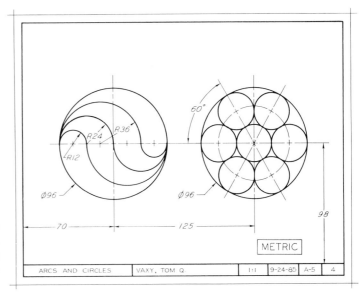

Fig. 2.78 Using Layout A–2 or A4–2 (adjusted), draw figures in pencil, as shown. Use bow pencil for all arcs and circles within its radius range. Omit all dimensions.

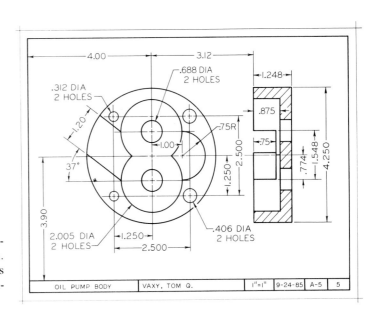

Fig. 2.79 Using Layout A–2 or A4–2 (adjusted), draw views in pencil, as shown. Use bow pencil for all arcs and circles within its radius range. Omit all dimensions.

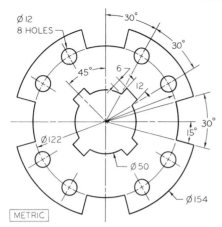

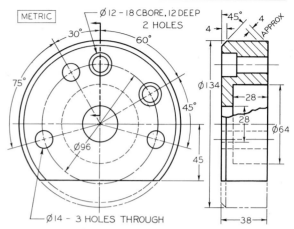

Fig. 2.80   Friction Plate.   Using Layout A–2 or A4–2 (adjusted), draw in pencil. Omit dimensions and notes.

Fig. 2.81   Seal Cover.   Using Layout A–2 or A4–2 (adjusted), draw views in pencil. Omit dimensions and notes. See §9.8.

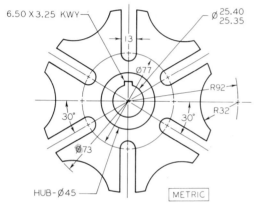

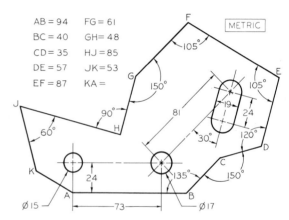

Fig. 2.82   Geneva Cam.   Using Layout A–2 or A4–2 (adjusted), draw in pencil. Omit dimensions and notes.

AB = 94   FG = 61
BC = 40   GH = 48
CD = 35   HJ = 85
DE = 57   JK = 53
EF = 87   KA =

Fig. 2.83   Shear Plate.   Using Layout A–2 or A4–2 (adjusted), draw accurately in pencil. Give length of KA. Omit other dimensions and notes.

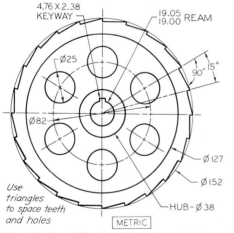

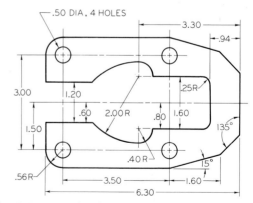

Fig. 2.84   Ratchet Wheel.   Using Layout A–2 or A4–2 (adjusted), draw in pencil. Omit dimensions and notes.

Fig. 2.85   Latch Plate.   Using Layout A–2 or A4–2 (adjusted), draw in pencil. Omit dimensions and notes.

Introduction to CAD

BY JAMES W. ZAGORSKI*

The use of electronic computers today in nearly every phase of engineering, science, business, and industry is well known. The computer has altered accounting and manufacturing procedures, as well as engineering concepts. The integration of computers into the manufacturing process, from design to marketing, is changing the methods used in the education and training of technicians, drafters, designers, and engineers.

Engineering, in particular, is a constantly changing field. As new theories and practices evolve, more powerful tools must be developed and perfected to allow the engineer and designer to keep pace with the expanding body of technical knowledge. Modern computers are capable of carrying out long and complex sequences of operations, manipulating and storing information, and making logical comparisons and routine decisions, if properly programmed, all at tremendous speed and with very little human intervention. The computer has become an indispensable and effective tool for design and practical problem solving. New methods for analysis and design, the creation of technical drawings, and the solving of engineering problems, as well as the development of new concepts in automation and robotics, are the result of the influence of the computer on current engineering and industrial practices.

One simple fact that is frequently overlooked is that computers are not new. Charles Babbage, an English mathematician, developed the idea of a mechanical digital computer in the 1830s, and many of the principles used in Babbage's design are the basis of today's computers. The computer appears to be a mysterious, uncompromising, often sinister, machine, but it is nothing more than a tool that just happens to be a highly sophisticated electronic device. It is capable of data storage, basic logical functions, and mathematical calculations, but not without a human interface to give it instructions. Because of the phenomenal technological advancements during the last decade, the term *computer* has become a part of everyone's vocabulary. Computer applications have expanded human capabilities to such an extent that virtually every type of business and industry utilizes a computer directly or indirectly.

---

*Instructor, Kelly High School, Chicago Public Schools.

3.1 Computer Systems and Components

Engineers and drafters have used computers for many years to perform the mathematical calculations required in their work. Only recently, however, has the computer been accepted as a valuable tool in the preparation of technical drawings. Traditionally, drawings were made by using drafting instruments and applying ink, or graphite, to paper or film. Revisions and reproductions of these drawings were time consuming and often costly. Now the computer is used to produce, revise, store, and transmit original drawings. This method of producing drawings is called *computer-aided design* or *computer-aided drafting* (CAD) and also *computer-aided design and drafting* (CADD). Since these and other comparable terms are used synonymously, and since industry and software creators are beginning to standardize, they will be referred to throughout this chapter simply as CAD.

Other terms such as *computer-aided manufacturing* (CAM) or *computer integrated manufacturing* (CIM) are often used in conjunction with the term CAD. CAD/CAM refers to the integration of computers into the design and production process. See Fig. 3.1. CAD/CAM/CIM is used to describe the use of computers in the total manufacturing process, from design to production, publishing of technical material, marketing, to cost accounting. The single concept that these processes refer to is the use of a computer and software to aid the designer or drafter in the preparation and completion of a task.

Computer graphics is a very broad field and is another term that is frequently used—and often misused. It covers the creation and manipulation of computer-generated images and may include areas in photography, business, cartography, animation, and publication, as well as drafting and design.

A complete computer system consists of *hardware* and *software*. The various pieces of physical equipment that comprise a computer system are known as hardware. The programs, instructions, and other documentation that permit the computer system to operate are classified as software. Computer programs are categorized as either application programs or operating system programs. Operating system programs are sets of instructions that control the operation of the computer and peripheral devices as well as the execution of specific programs and data flow to and from peripheral de-

Fig. 3.1 Typical CAD/CAM Application.   Chrysler CAD/CAM (computer-aided design, computer-aided manufacturing) locates the positions where robots spot weld body sections in the company's U.S. assembly plants. Weld positions on Chrysler LeBaron side body section shown on the CAD/CAM computer screen (*left*) becomes a reality in the St. Louis assembly plant (*right*). *Courtesy of Control Data Corporation.*

Fig. 3.2 Typical CAD Workstation. *Courtesy of Hewlett-Packard Company.*

Digitizer is the most widely used input device because of its versatility in many applications. Cursor cross-hairs on the screen follow the movement of a stylus or a puck on an electronic tablet.

Mouse can be used on ordinary surfaces such as a desktop, but slippage often causes inaccuracies.

Trackball gives good tactile feedback and provides for precise cursor positioning on the screen through fairly large ball displacements.

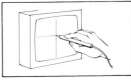

Lightpen is inexpensive and interacts directly with the screen for a natural drawing interface. But operation for long periods can produce arm fatigue.

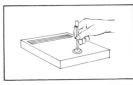

Joystick performs graphics operations quickly by providing rapid cursor movement for relatively small device displacement.

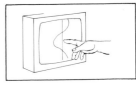

Touch-sensitive screen allows a direct interface with the screen and requires no external pointing device. But arm fatigue is a problem, and accuracy is limited.

Voice-data entry can provide for rapid entry of commands. But software must be refined before this technology gains widespread use.

Fig. 3.3 Graphic Input Devices. *Courtesy of* Machine Design.

vices. This type of program may also provide support for activities and programs such as input/output control, editing, storage assignment, data management, diagnostics, assign drives for I/O devices, in addition to providing support for standard system commands and networking. Application programs are the link between specific system use and its related tasks—design, drafting, desktop publishing, etc.—and the general operating system program. All computer-aided drafting systems consist of similar hardware components, Fig. 3.2, such as input devices, a central processing unit, data storage devices, and output devices. For input devices, Fig. 3.3, the system may have one or more of the following: a keyboard terminal, mouse, digitizer/graphics tablet, and light pen.

Output devices the CAD system may include would be a plotter, an alphanumeric/graphics character printer, and some type of *cathode ray tube* (CRT) or monitor, Fig. 3.4.

Frequently, some devices are combined. For example, a terminal or workstation can contain the typewriter keyboard, CRT, data storage device, and possibly a processing unit all in the same cabinet. Such a combined device is known

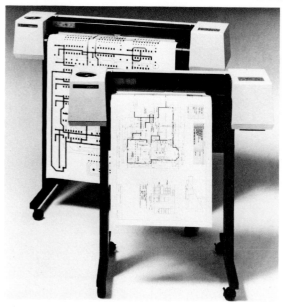

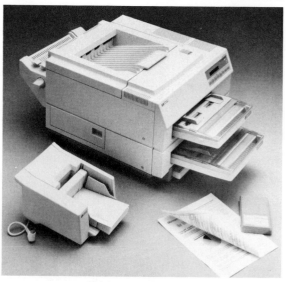

(b) LASER PRINTER

(a) PLOTTER

Fig. 3.4   Output Devices. *(a), (b) courtesy of Hewlett-Packard Company; (c) courtesy of California Computer Products, Inc.*

(c) CATHODE RAY TUBE (CRT)

as an I/O device because its input and output functions are no longer considered separately and is often called a *workstation*, Fig. 3.5.

The system must have a data storage device, Fig. 3.6, such as a tape drive, a hard (fixed) or soft (floppy) disk drive, or an optical disk drive (CD-ROM).

Finally, a computer or central processing unit (CPU), Fig. 3.7, is needed to do all the numerical manipulations and to control all the other devices connected to the system.

The system itself may be interactive or passive. An interactive system allows for data input to be viewed and modified as the commands are

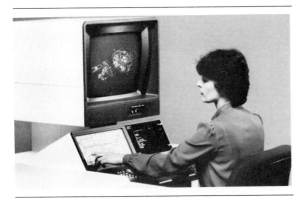

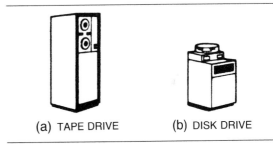

(a) TAPE DRIVE   **(b)** DISK DRIVE

Fig. 3.6   Data Storage Devices.

Fig. 3.5   CAD Workstation. *Courtesy of Auto-trol Technology Corporation.*

Fig. 3.7   Central Processing Unit with CAD Workstation. *Courtesy of Computervision Corporation, a subsidiary of Prime Computer, Inc.*

entered. In a passive system, the program, must be run to completion before the results can be viewed.

A CAD system configuration could include a single terminal, called a standalone workstation, or a group of terminals sharing the same storage and processing capabilities, known as a network. It is not uncommon to have several persons working on different tasks sharing the same system of storage and processing. The end product of any CAD system is still a set of specifications, working drawings, and, in some cases, numerical control tapes.

3.2   Computer Types

Computers may be classified as one of two distinct types, *analog* or *digital*. An analog computer measures continuously without steps, whereas the digital computer counts by digits, going from one to two, three, and so on, in distinct steps. A slide rule, an electric wall clock with minute and hour hands, and the radial speedometer on a car are all examples of analog devices. An abacus and a digital watch are examples of digital devices. Digital computers are more widely used than analog computers because they are more flexible and can do a greater variety of jobs.

Analog computers are generally used for mathematical problem solving and usually are not found outside of an educational institution or research facility. This type of computer, which measures continuous physical properties, is often used to monitor and control electronic, hydraulic, or mechanical equipment. Digital computers have extensive applications in business and finance, engineering, numerical control, and computer graphics.

Both types of computers have undergone great changes in appearance and in operation. Equipment that once filled the greater part of a large room has now been replaced by machines that occupy small desktop areas. This new generation of computers can do more operations in less time with less human input than did their ancestors. The single most important advancement in computer technology has been the development of the *integrated circuit* (IC). The IC chip has replaced thousands of components on the *printed circuit* (PC) board and made possible the development of microprocessors. The microprocessor is the processing unit of a computer. The difference in size between a PC board with individual components and an IC chip is shown in Fig. 3.8. The term *microminiaturization* is applied to advanced integrated circuit

Fig. 3.8   Size Comparison of a PC Board and an IC Chip.

chip technology. The evolution of IC chip technology has led to the increased production of low-cost microcomputers. Microcomputers are largely responsible for the increase in use of computer-aided drafting systems in industry. Low-cost microcomputer CAD systems can now be cost-justified by industrial users.

3.3   Computer-Aided Drafting

The traditional tools of the designer or engineer have been the drawing board, an array of drawing instruments, and the slide rule. In recent years templates have been developed for certain symbols that have to be drawn frequently and the slide rule has been replaced by an electronic calculator. Conventional drafting equipment is now being replaced by another tool called a computer.

The first demonstration of the computer as a design and drafting tool was given at the Massachusetts Institute of Technology, in 1963, by Dr. Ivan Sutherland, with his system called "Sketchpad," which used a cathode ray tube and a light pen for graphic input to a computer. The first commercial computer-aided drafting system was introduced in 1964 by International Business Machines. Many changes in CAD systems have taken place since the introduction of the first system. The changes in the systems are due to the advent of the microprocessor, more sophis-

Fig. 3.9 Computer-Controlled and -Programmed Robot Welding Line at Chrysler Corporation's Jefferson Assembly Plant. *Courtesy of Control Data Corporation.*

ticated software (programs), and new industrial applications. In most cases, the drafter/engineer can create, revise, obtain prints (hard copy) and store drawings with relative ease, utilizing less space. CAD was originally developed to aid in creating production drawings. The advent of three-dimensional CAD software made it apparent that a 3D computer model could assist not only in the manufacture of the part but also, together with the three-dimensional database, in testing the design with finite element analysis programs, in developing technical manuals and other documentation that combine illustrations of the design with text from word processing programs, and in marketing where the 3D solid models can be used with a rendering and animation program. Increases in productivity and cost effectiveness are two advantages constantly stressed by CAD advocates. In addition, CAD stations can be linked directly or through a *local area network* (LAN) to the manufacturing or production equipment, or with *numerical control* (NC) equipment to automatically program a tape that can be used on NC machines in manufacturing operations or in robotics, Fig. 3.9.

The primary users of CAD are in mechanical engineering and electronic design, civil engineering, and cartography. The design and layout of printed circuits are a principal application of CAD in the electronics industry, which, prior to 1976, was the largest CAD user. Mechanical engineering has since overtaken electronics and continues to expand its CAD applications and use. Continued expansion in mechanical design applications is expected because the design, analysis, and numerical control capabilities of CAD can be applied to such a varied range of products and processes. Cartography, seismic data display, demographic analysis, urban planning, piping layouts, and especially architectural design also show growth in CAD use. A relatively new area in computer graphics is *image processing*. Image processing includes animation, 35-mm slide preparation, photocolor enhancement, and font and character generation used in television broadcasting and the graphic arts industry.

It should not be assumed that all drawings in the future will be made by computers and that engineers, designers, and drafters will no longer be needed. A computer is capable of doing a great many things, but in reality it is an electronic device with no brain. It cannot think and will not do anything more, or anything less, than the human operator instructs it to do. Drafters will continue to be the backbone of the design or drafting department, although now their work will be more creative as many of the tedious and repetitive tasks are automated. A CAD system is not creative, but it can assist the individual to be more productive. The creation of a drawing will continue to be of prime importance; CAD simply provides the means for replacing technique in the preparation of a drawing to save time. Although CAD will continue to be an important part of engineering work in the future, a knowledge of the principles of technical drawing will still be a prime requisite for engineers, designers, and drafters engaged in this activity; if anything, additional training in this subject will be necessary.

3.4   CAD System Configurations

A CAD system, Fig. 3.10, will contain all of the components of a traditional computer system, with additional components for graphical data

Fig. 3.10   CAD Workstation with Keyboard and Keypad. *Courtesy of Hewlett-Packard Company.*

input, display generation, and graphical output. As stated previously, a complete CAD system will include various pieces of physical equipment called hardware and CAD software. The computer, or central processing unit (CPU), controls all functions of the system and processes all data from input devices. Since graphics programs on the computer are so mathematics intensive, the CPU usually requires the help of an additional IC chip called a math coprocessor to speed up the drawing process.

There are many devices that can feed data into the CPU. Entering data into the computer is called input. Input devices commonly used in a CAD system are the keyboard, mouse, graphics tablet, digitizing table, or menu pad. With the exception of the keyboard and menu pad, most graphical input devices are pointing devices, in that they use electronic signals to transfer movements of the device by the operator's hand to a cursor that moves about the display screen in corresponding fashion to a specific location.

After the computer has processed the data, it will generate the results through an output device. An output device may be a cathode ray tube (CRT or monitor), a printer, or a plotter. While the monitor will only display the design on the screen, printers and plotters will generate original drawings on paper or film. These drawings are called hard copy.

Internal and auxiliary storage devices are also considered part of the system hardware. These devices may be a tape drive, hard disk drive, soft or floppy disk drive, or an optical disk drive. Each of these can store the necessary data for later retrieval.

Programs, computer language codes, and other documents are considered part of the system software. Hardware and software together make up the complete CAD system.

3.5   Central Processing Unit

The central processing unit (CPU), or computer, receives all data and manages, manipulates, and controls all functions of the CAD system. CAD systems use digital computers, which receive input in the form of numbers (digits) and letters or special characters that represent predetermined combinations of numbers or digits. All data must be converted into a binary form or code for the computer to understand and accept it. This is known as digitizing the information. This code is called *binary coded decimal instructions* (BCDI). This binary code uses a two-digit system, 1 and 0, to transmit all data through the circuits. The number 1 is the "on" signal; the 0, the "off" signal. A *bit* is a binary digit (*binary digit*). Bits are grouped, or organized, into larger units called words to be accessed by computer instructions. Word length, which is expressed as bits, differs with various computers. The most common word lengths are 8, 16, and 32 bits. The word length indicates the maximum word size that can be processed, as a unit, during an instruction cycle. Often, computers are categorized by their word length, such as 8-bit, 16-bit, or 32-bit computers. The number of bits in the word length indicates the processing power of the computer (the larger the word length, the greater the processing power). A sequential group of adjacent bits in a computer is called a *byte*. The current industry standard is 8 bits equal 1 byte. A byte represents a character that is operated on as a unit by the CPU. The length of a word on a majority of computer systems currently is 4 bytes. This means that each word in any one of these systems occupies a 32-bit storage location. The memory capacity of a computer is therefore expressed as a number of bytes rather than bits.

Computers use two types of memory. Permanent memory, called ROM (read only memory), is encoded into a microprocessor chip and is not lost when the computer is turned off. ROM, coordinated with the disk operating system (DOS), performs a diagnostic check of the system to insure all circuits are operational when the computer is turned on. Random access memory (RAM) is considered temporary memory. Any program or file stored in RAM is lost when the computer is turned off. A computer's memory capacity is based on its RAM, or temporary memory.

Fig. 3.11 Mainframe Computer System. *Courtesy of Control Data Corporation.*

Since CAD systems utilize digital computers, we will restrict our discussion of computer types to digital computers. Digital computers are commonly divided into three classes: *mainframe, mini,* and *micro.* The difference between the three classes is essentially word length (capacity). The largest CPUs are called mainframes, Fig. 3.11. A mainframe computer will have internal memory measured in megabytes (1 MB = 1 million bytes), usually word lengths of 32 bits or more, and multimegabyte or gigabyte (GB, 1000 million bytes) disks, and high-speed tape drives for external data storage. Instruction times for these machines are measured and expressed in *nanoseconds* (thousandths of a microsecond) or in *mips* (millions of instructions per second). Mainframes are usually general-purpose computers; they can be used for a variety of functions within a company, such as accounting and inventory control, and as the core of a CAD system.

Minicomputers, Fig. 3.12, began as smaller size mainframes. The development of the microprocessor chip has lessened the distinction between the mainframe computer and the minicomputer. The minicomputer's internal memory is smaller, its speed of operation is somewhat slower, and it is somewhat limited in flexibility of use. Two desirable features of a minicomputer are real-time operation (immediate response to an alarm or command) and lower cost systems and peripherals. It is generally accepted that the development of the minicomputer was the beginning of the great demand for CAD systems.

Microcomputers, Fig. 3.13, basically composed of a CPU combined with a memory and control features, were originally single-purpose machines. Microcomputers are currently lower cost equivalents of full-scale minicomputers, often equalling the computational power and speed of many minicomputers. New microcomputer system architecture allows for internal memory expansion of up to 32 MB and internal data storage devices in excess of 100 MB. Microcomputers in use today are 8-bit, 16-bit, or 32-bit systems. Microcomputers currently used for CAD tend to require more internal memory, 640 kilobytes (1 kB = 1000 bytes) to 10 MB, because of the mathematical computations involved with CAD programs. Originally, CAD programs developed primarily for microcomputers were two-dimensional and used for production drawings only. CAD programs today can be used to create complete three-dimensional models, analyze the design, optimize the design, and generate all necessary data used in the production of the model.

Fig. 3.12   Minicomputer Workstation. *Courtesy of Hewlett-Packard Company.*

Fig. 3.13   Four Microcomputers Linked to Control Data's CYBERNET Data Services Network to Take Advantage of Large-System Computing Power and Applications Software. *Courtesy of Control Data Corporation*

The wide variety of peripherals and CAD software now available makes these microsystems cost effective for smaller firms.

3.6 Display Devices

Another major reason for the rapid growth in CAD systems is the further development of display devices. These display devices, commonly referred to as monitors, utilize a wide variety of imaging principles. Each device has definite characteristics with regard to brightness, clarity, resolution, response time, and color. The purpose of any graphics display is to project an image on a screen. The image that is displayed may be alphanumeric (text symbols, letters, and/or numerals) or graphical (pictorial symbols and/or lines). Users of interactive CAD systems communicate directly or indirectly through graphics terminals. The information requested by the user may be displayed as animated figures, graphs, color-coded diagrams, or simply a series of lines. Most interactive CAD systems use one of three main types of display devices—the storage tube, Fig. 3.14 (a), the vector refresh, (b), and the raster scan, (c)—all of which produce images through the basic cathode ray tube (CRT) principle.

A storage tube CAD monitor, also called a direct view storage tube (DVST), was the most popular computer display device for many years. In a storage tube, flood guns constantly blanket the entire CRT screen with electrons that alone do not have the energy to illuminate the phosphor on the screen. A focused writing beam produces a visible track or line by electron bombardment. The screen grid becomes positively charged in the area where the tightly focused beam strikes it, allowing electrons from the flood gun to be accelerated onto the screen and cause the phosphor to glow. The image is constantly displayed without the need for any retracing of the writing beam, and the image is stored on the screen, thus making very little demand on the computer. The major advantage of storage tubes is that large amounts of graphics data can be stored on the screen without the need to be con-

(a) STORAGE TUBE

(b) VECTOR REFRESH

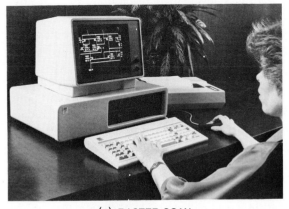

(c) RASTER SCAN

Fig. 3.14 Display Devices. *(b) courtesy of Adage, Inc.; (c) courtesy of Control Data Corporation.*

stantly refreshed from the computer memory. Other advantages of the storage tube are its excellent flicker-free resolution, high display capacity, and moderate cost. The limitation of storage tube devices is lack of dynamic capability. Once an image is created and stored, lines cannot be selectively erased, and the image cannot be changed without *repainting* (redrawing) the entire image, so that interactive display manipulation is not possible. In addition, storage tubes have low brightness and contrast. They are also monochromatic (single color), usually green. Despite their limitations, the high resolution and high data storage capacity of DVSTs make them well suited for presenting analysis results in the form of contour plots and other drawing formats.

Vector refresh monitors (also called stroke writing, random scan, or calligraphic system) use an electron writing beam to directly trace out the image lines on the CRT screen in a continuous sweep similar to oscilloscopes. After the beam has generated a complete image or line, it constantly retraces, or refreshes, the image before the phosphor fades. Because the displayed image is stored in memory, the need for additional hardware and memory for image storage and refreshing increases system complexity, making this monitor one of the most expensive. The advantages of vector refresh monitors are the ability to display motion or animation, high brightness level, high resolution, and selective erasing. These features make this display device an excellent monitor to use for complex geometric modeling and finite element analysis. The major disadvantages of vector refresh technology are the limited color capabilities and the flicker encountered when there is a large amount of information on the screen; complex images reduce the refresh rate below a frequency of 30 hertz, which appears to the human eye as a flicker.

Raster scan devices are similar to conventional television screens. These devices produce an image with a matrix of picture-element dots within a grid called *pixels*. Each pixel is either a light or dark image that falls within a square area on the grid and appears on the screen as a dot. As in conventional television, an electron beam is swept across the entire screen, line by line, top to bottom. This process is called raster scanning. A signal turns on or illuminates a pixel according to a pattern stored in memory, which is referred to as a bit map. The screen is scanned 60 times a second to update the image before the phosphor dims. Recent advances in raster scan technology have increased the resolution of these devices and decreased system cost through "intelligent" graphics monitors. The majority of new graphics systems being marketed contain raster scan devices. The advantages of this type of device are the wide range of colors that can be simultaneously displayed on the screen, the ability to show motion, high brightness level, and selective erasing. The major disadvantage of this technology is the high memory and processing requirements that result in higher cost.

There are several other types of graphic display devices in use, including the gas plasma, electroluminescent, and liquid crystal displays, that fall into the flat screen technology area. Each has greater limitations and associated problems than the devices mentioned, so that they are much less extensively used for graphics than other display devices.

3.7 Input Devices

The graphical display of data was available on first generation computers, but the process was time consuming. Programs had to be written and data entered by means of batch mode processing. A system developed in the 1950s for the Air Defense Command called SAGE used the light pen for data input. Further development of an interactive method of data input by Dr. Ivan Sutherland with his "Sketchpad" system increased the usage and popularity of computer-generated graphics. Since then, many diverse types of input devices have been developed. A CAD system may use one or a combination of input devices to create images on the display screen. Graphic input devices may be grouped into three categories: (1) *keyboard* and *touch-sensitive*, (2) *time-dependent*, and (3) *coordinate-dependent* devices.

(a) MENU PAD, LIGHT PEN, AND KEYBOARD

(b) KEYBOARD AND MOUSE

(c) KEYBOARD, THUMBWHEEL, AND GRAPHICS TABLET

(d) KEYBOARD, JOYSTICK, AND GRAPHICS TABLET

Fig. 3.15 Graphic Input Devices. *(a) courtesy of Jervis B. Webb Co.; (b) courtesy of Hewlett-Packard Company; (c) courtesy of Bausch & Lomb; (d) courtesy of California Computer Products, Inc.*

The alphanumeric keyboard is the universal input device by which data and commands are entered. A typical keyboard consists of alphanumeric character keys for keying in letters, numbers, and common symbols such as #, &, and %; cursor control keys with words or arrows printed on them indicating directional movement of the screen cursor; and special function keys that are used by some software programs for entering commands with a single keystroke. Many large mainframe-based CAD systems have used a special function keypad, or menu pad, that allows access to a command with a single keystroke, Fig. 3.15 (a). Single stroke command selection was considered so essential for cost effectiveness and ease of use that developers of mini- and microcomputer-based CAD systems included this feature of single and double keystroke command access into their program utilizing the CTRL, ALT, SHIFT, and function keys. Typically, a CAD system will use a keyboard for inputting commands and text, and another input device for cursor control.

Digitizing tablets are one of the most commonly used input devices. They can be used to create an original CAD drawing or to convert an existing pen or pencil and paper drawing into a CAD drawing. Digitizing tablets range in size from $8\frac{1}{2}'' \times 11''$ to $36'' \times 48''$. Tablets larger than $36'' \times 48''$, called digitizing tables, are used primarily for converting existing drawings to a CAD system format. The *resolution* of digitizing tablets is important. This determines how small a movement the input device can detect, usually expressed in thousandths of an inch, and depends on the number of wires per inch in the tablet's grid system. The working area on a tablet can be divided up into smaller sections. One section could then contain system control features; another, command functions; another, screen control commands; and so on. A much smaller area on the tablet then remains to represent the screen's working area. Attached to the tablet will be either a puck or a stylus. A *puck* is a small, hand-held, box-like device with a clear plastic extension, or window, containing crosshairs, that transfers the location of the puck on the tablet grid to the relative location on the screen. Single or multiple buttons on top of the puck are used to select points and/or commands for input to the system. A *stylus* appears to be a ball-point pen with an electronic cable attached to it. The tip of the stylus senses the position on the tablet grid and relays these coordinates to the computer. When the stylus is moved across the tablet, the screen cursor moves correspondingly across the screen. The stylus also contains a pressure-sensitive tip that enables the user to select points or commands by pressing down on the stylus.

An input device that has gained in popularity and use with both large and small CAD systems is called a *mouse*, Fig. 3.15 (b). A mouse may be of the mechanical type or the optical type. Both are similar in appearance to a puck without the plastic extension and crosshairs. A mechanical mouse uses a roller, or ball, on the underside of the device to detect movement. An optical mouse senses movement and position by bouncing a light off a special reflective surface. Both types of these devices will have from one to three buttons on top of them to select positions or commands. The advantages of this type of device are that it is easy to use, requires a very small working area, and is relatively inexpensive. A mouse cannot, however, be used to digitize existing drawings into a CAD format.

One of the oldest input devices used on CAD systems is the *trackball*, Fig. 3.15 (c). Trackballs were used on many large mainframe-based CAD systems and were often incorporated into the keyboard. A trackball consists of a ball nested in a holder, or cup, much like the underside of a mechical mouse, and from one to three buttons for entering coordinate data into the system. Within the holder are sensors that pick up the movement of the ball. The ball is moved in any direction with the fingers or hand to control cursor movement on the CRT screen. Cursor speed and button function can be set by the user.

A *joystick*, Fig. 3.15 (d), is an input device more commonly used with video games today than with CAD systems. This device looks like a small rectangular box with a lever extending vertically from the top surface. This hand-controlled lever is used to manipulate the screen cursor and manually enter coordinate data. Buttons on the device can be used to enter coordinate location or specific commands. This device is inexpensive and requires a very small working area, but sometimes presents a problem when very precise cursor positioning is important.

The *light pen*, next to the keyboard, is the oldest type of CAD input device currently in use, Fig 3.15 (a). It looks much like a ballpoint pen or the stylus on a digitizing tablet. A light pen is a hand-held, photosensitive device, which works only with raster scan or vector refresh monitors, that is used to identify displayed elements of a design or to specify a location on the screen where an action is to take place. The pen senses light created by the electron beam as it scans the surface of the CRT. When the pen is held close to or touches the CRT screen, the computer can determine its location and position the cursor under the pen. Because this input device more closely emulates the traditional drafter's pencil or pen than other devices, it quickly gained popularity. Its decline in popularity resulted from studies that showed it was tiring to use and therefore cut the productivity of the user.

Touch-panel displays, though not widely used with CAD systems, are another type of input device. These devices allow the user to touch an area on the screen with a finger to activate commands or functions. This area on the screen contains a pressure-sensitive mesh that picks up the location of the finger and transfers it to the computer, which activates the command associated with that specific location. These devices work well with function or command entry but are inaccurate for CAD coordinate entry.

One of the most recent developments in data input is *voice recognition* technology. This technology utilizes a combination of specialized integrated circuits and software to recognize spoken words. The system itself must first be "trained" by repeating commands into a microphone. The computer converts the operator's oral commands to digital form and then stores the characteristics of the operator's voice. Once trained, when the operator gives an oral command, the system will check the sound against the words stored in its memory and then execute the command. Since this technology is new, it has several distinct disadvantages for CAD users. The vocabulary supported by the system is limited. The memory required for storing complex sound or voice patterns is extremely large. Access time, or time between the spoken word and command activation, may be several seconds. Finally, if the operator's voice changes in some manner or words have similar voice patterns, the system may not recognize the oral input at all.

3.8 Output Devices

In most instances, the user of a CAD system will need a record of images that are stored on database files or displayed on the CRT. When an image is placed on paper, film, or other media, it is then referred to as *hard copy*. This hard copy can be produced by one of several types of output devices.

The device most commonly used for the reproduction of computerized drawings is the *pen plotter*, Fig. 3.16. A pen plotter is an electromechanical graphics output device. Pen plotters may be classified as drum, flatbed, or microgrip. The *drum plotter*, (a), utilizes a long narrow cylinder in combination with a movable pen carriage. The medium to be drawn on (paper, vellum, or film) is mounted curved rather than flat and conforms to the shape of the cylinder drum. The drum rotates, moving the drawing surface, and provides one axis of movement, while the pen carriage moves the pens parallel to the axis of the cylinder and provides the other axis of movement. The combined movement of drum and pen allows circles, curves, and inclined lines to be drawn. The pen carriage on a drum plotter typically holds more than one pen so that varied line weights or multicolor plots can be drawn. These plotters accept up to E size paper in single sheets or in a roll that is then cut after the drawing is plotted.

Flatbed plotters, Fig. 3.16 (c), differ from drum plotters in that the medium to be drawn on is mounted flat and held stationary by electrostatic or vacuum attraction while the pen carriage controls movement in box axes. The area of these plotters may be as small as A size ($8\frac{1}{2}'' \times 11''$) or larger than E size ($34'' \times 44''$), and from one to eight pens may be used for varied line weights and multicolor plots.

Microgrip plotters, Fig. 3.16 (b) and (d), have become one of the most widely used types of output devices. Their popularity is due to their adaptability to all types of computers, size ranges, low maintenance, and relatively inexpensive pricing. Microgrip plotters are similar to drum plotters in that the medium to be drawn on is moved in one axis while the pen moves along the other axis. These plotters get their name from the small rollers that grip the edges of the medium and move it back and forth under the pen carriage. These plotters range from A to

(a) DRUM

(b) MICROGRIP

(c) FLATBED

(d) MICROGRIP

Fig. 3.16   Plotters. *(a) courtesy of California Computer Products, Inc.; (b) courtesy of Control Data Corporation; (c) courtesy of Versatec, Inc.; (d) courtesy of Hewlett-Packard Company.*

E size, with single or multiple pen carriages, and may accept cut sheets or rolls.

All pen plotters are rated according to specified standards for accuracy, acceleration, repeatability, and speed. Accuracy is the amount of deviation in the geometry it is supposed to draw, usually ranging from .001″ to .005″. Acceleration is the rate at which the pen attains plotting speed and is expressed in Gs (for gravitational force). Pen speed is important because slower speeds usually produce darker lines. The faster the pen attains a constant speed, the more consistent the linework will be. The ability of a plotter to retrace the same drawing over and over again is called repeatability. The deviation of the pen in redrawing the same line is the measure of repeatability and usually varies from .001″ to .005″. Plotter pen speed determines how fast the pen moves across the drawing medium. Pen speed may range up to 22″ per second on some plotters. Most CAD software also allows the operator to set the speed at which drawings will be plotted to achieve maximum line quality and consistency. Slow pen speeds normally produce better quality plots than high pen speeds.

Other factors help determine the quality of a pen-plotted drawing. The variety of pens, inks, and drawing media available allows the operator to coordinate pen, ink, and paper to produce the most desirable hard copy.

Dot matrix plotting is another method by which hard copy can be produced. This technology, in use for some time, has typically been associated with printing devices that were not suitable for producing quality hard copy. Recent technology, however, has permitted the development of devices that can produce accurate, industrial quality dot matrix plots. These devices use a process called rasterization to convert images to a series of dots. The image is transferred optically (or sometimes by a laser) to the surface of the medium on a selenium drum that is electrostatically charged. Sometimes the image may be created by an array of nibs that electrically charge small dots on the medium. The drum, with the addition of a toner, usually a very fine carbon powder that adheres to the surface of the drum only in places where there is an image, then transfers that image to the surface of the paper. A high-quality raster scan plotter can produce an image so fine and of such quality that it is not obvious how the image was produced unless examined under a magnifying glass.

Electrostatic plotters, Fig. 3.17, produce hard copy by placing an electrostatic charge on specially coated paper and having a toner, or ink, adhere to the charged area. Drawing geometry is converted through rasterization into a series of dots. These dots represent the charged area.

Fig. 3.17 Electrostatic Plotter. *Courtesy of Hewlett-Packard Company.*

Resolution of these plotters is determined by the number of dots per inch (dpi), usually ranging from 300 to 600 dpi. This type of plotter produces hard-copy drawings in single or multicolors much faster than pen plotters, but the cost, power, and environmental requirements are also much greater.

Dot matrix technology has produced machines that are now dual purpose, although not originally designed as such. These machines were originally designed as printers to handle word processing and spreadsheet programs using alphanumeric characters. With the increasing popularity and use of CAD and other graphics software, manufacturers have improved the technology of these devices so that they can now create graphic hard copy close or equivalent to true plotter quality. The device can now be classified as a printer/plotter.

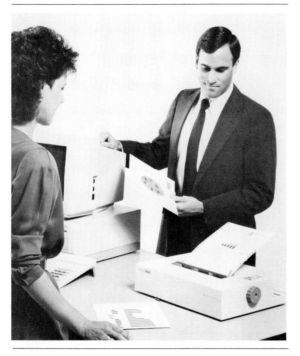

Fig. 3.18   Ink Jet Printer. *Courtesy of Hewlett-Packard Company.*

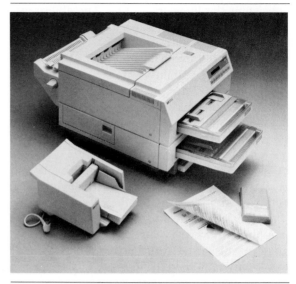

Fig. 3.19   Laser Printer. *Courtesy of Hewlett-Packard Company.*

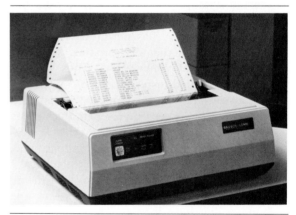

Fig. 3.20   Dot Matrix Printer. *Courtesy of Bausch & Lomb.*

An *ink jet printer/plotter*, Fig. 3.18, produces images by depositing droplets of ink on paper. These droplets correspond to the dots created by the rasterization process. This device places a charge on the ink rather than the paper, as in the electrostatic process. Ink jet plotters can produce good quality color-rendered images in addition to standard technical drawings.

Laser technology represents the newest evolution in plotter technology. A *laser printer/plotter*, Fig. 3.19, uses a beam of light to create images. This device utilizes electrostatic charging and raster scanning to produce a plotted image.

Dot matrix printers, Fig. 3.20, produce images by one of two processes. They can produce images through impact on carbon or ink ribbon, or they may use a heat (thermal) process. Each method uses a number of pins in a specific configuration, such as 5 × 7, 7 × 9, or 9 × 9, set in the printer head. Machine commands control the sequence in which the pins strike the medium to produce an image.

3.9   Data Storage Devices

Since all data kept in RAM (random access memory) will be lost when the computer is turned off, it must be saved, or stored, before the power is

(a) MAGNETIC TAPE   (b) HARD DISK

Fig. 3.21   Storage Devices.

off. Data storage devices provide a place to save information permanently for later use. CAD programs, for example, are stored on a disk or tape and when loaded (or activated), portions of the program go into RAM, which is temporary memory. While a drawing is being worked on, all data associated with that drawing is kept in the same temporary memory. When the drawing is completed, or the operator must stop, that drawing and all the associated data must be saved, or directed to a storage device, before the program is exited or the power shut off. Otherwise all accumulated data from that work session will be lost. These storage devices can be considered electronic file cabinets.

Magnetic tape drives and disk drives, Fig. 3.21, are two distinct categories of storage devices. *Magnetic tape* storage, (a), uses plastic tape coated with magnetic particles. When data is sent from the computer for storage on the tape, these particles are charged by a read/write head in the tape drive. The data being sent is recorded as a series of charges along the tape. Once these particles are charged, they will remain charged until the head writes over them or they are demagnetized. Tape drives file and read data in sequential order. This means they must look through data in the order that the tape is wound and unwound. This is similar to forwarding or rewinding a video tape to look for a spe-

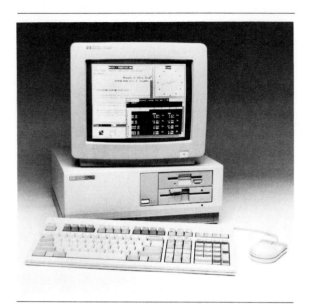

Fig. 3.22 Microcomputer with Two Flexible (Floppy) Disk Drives. *Courtesy of Hewlett-Packard Company.*

cific scene, or an audio tape to play a particular song. Magnetic tape drives may be either reel to reel or cassette type. Reel to reel tape drives consist of two reels and a drive controller. One reel is used for tape storage and the second reel as a take-up reel. These tape reels may vary in size from 6″ to 12″ in diameter. The longer the tape, the more data that can be stored on it. Over a gigabyte of data may be stored on some reel to reel tape systems. Cassette tape storage devices were originally created to back-up microcomputer systems. These cassettes were slow and did not hold the amount of data that can be found on current microcomputers. Recent technology, however, has produced tape cassettes that are high speed and can store over a gigabyte of data. These cassettes resemble audio cassettes in both size and appearance. Tape storage is essentially used for backing up data from a hard drive, or for archival purposes, since the tapes can be removed and stored for later use.

Disk storage devices are the most commonly used method of data storage. *Disk drives* may be of the fixed (hard disk) variety, flexible (floppy) variety, or optical type. Disk drives file and read data in random order. This means that the device writes data to any portion of the disk that is empty, and is able to locate data almost instantly, because it has access to the whole disk at once. Disk drives are rated according to their type, access time, capacity, and transfer rate.

The *fixed disk drive*, or hard disk, Fig. 3.21 (b), is the most common method of data storage. This type of drive uses an aluminum disk as the medium for storage. These drives may be internal, attached inside the computer case, or external, in a separate case or cabinet. A disk controller or controller card must be installed in the computer to allow the computer and drive to communicate or interface with each other. The storage capacity of these drives will range from 10 MB to several hundred megabytes. Access time is expressed in milliseconds (ms) and will range from 18 to 80 ms. The lower the number in milliseconds, the faster the access time. Transfer rate is the speed at which the drive can send data to the CPU and is expressed in bytes per second (bps). The higher the number in bytes per second, the faster the data can be sent.

Floppy disk drives, Fig. 3.22, derive their name from the removable flexible plastic disks used in this device. The disks used in this drive are available in three sizes, $3\frac{1}{2}″$, $5\frac{1}{4}″$, and 8″ in diameter. All three sizes come in either high density or low density formats. The density of a disk refers to the amount of magnetic particles on the disk. This helps to determine the amount of data the disk will hold. Disks will also be classified as single sided or double sided, depending on whether data can be stored on one or both surfaces of the disk. Typically, a $3\frac{1}{2}″$ disk, called micro-diskette, will hold 720 kB of data in double-sided low-density format or 1.44 MB of data in double-sided high-density format. A $5\frac{1}{4}″$ disk will hold 360 kB in double-sided low-density format, and 1.2 MB in double-sided high-density format. The floppy disk is inexpensive and convenient to use but holds less data and is slower than fixed disk drives.

The ever-increasing need for larger storage capacity has spurred the development of new technologies. The newest technology in data storage is *optical disk drives*. Optical disks are removable, like floppy disks, and capable of holding many gigabytes of data. These drives use a laser to read and write data to a chemically coated aluminum disk, Fig. 3.23. The original optical disk drive was a read only device and was called a *compact disk read only memory* (CD-ROM) drive. Each disk normally contained a program that could be read or accessed, but no data could be written on it after installation.

Further developments produced a *write once read many* (WORM) drive. This device allows data to be written on it, but the data becomes permanent on the disk and cannot be erased. This storage device is especially suited for archival purposes. The data is "burned" into the disk surface by a laser so the information becomes more permanent than magnetic storage, and the disk is removable. A third generation optical disk drive, write many read many, currently in its infancy, allows data to be erased and written over, but poor performance in access and transfer rates has curtailed its acceptance.

Fig. 3.23   Optical Disk. *Courtesy of Hewlett-Packard Company.*

3.10   Operating System Programs
As previously stated, the term *software* refers to all programs that are written to be performed on a computer. Operating system programs are the code (the written set of instructions) that the system follows to handle input, operate the system, and output data. They translate user programs into machine language programs and may be written in a variety of standard computer languages. Regardless of which language is used, drafters and designers need not be proficient in programming, for they will be users rather than developers of the programs. A computer *operating system* (OS) is the program that manages the operation of the physical equipment and controls the execution of other programs on the system. The speed at which a CAD system can interact with an application program depends on how efficiently the operating system manages the various tasks. Operating systems may be designated as *multiprocessing* or *multiprogramming*. Some operating systems can do both multiprocessing and multiprogramming but require large memory. Real-time multiprocessing operating systems are usually preferred over multi-

programming systems. Operating system software is completely hardware dependent.

3.11   Application Programs
The computer programs designed to perform specific tasks in response to a particular user's requirements are called *application programs*. When a user enters variables into a CAD system, for example, the application program is responsible for producing images on the display and completing any required calculations. Application programs are expensive to develop, and it is considered more economical to purchase existing software than to develop new programs. Application software is written for specific operating system software and can be used with any computer system that has the same operating system and therefore can be considered hardware independent.

A complete computer-aided drafting system is capable of a variety of operations and can produce many different types of drawings. Examples of some of the operations performed and drawings created are shown in Figs. 3.24–3.35.

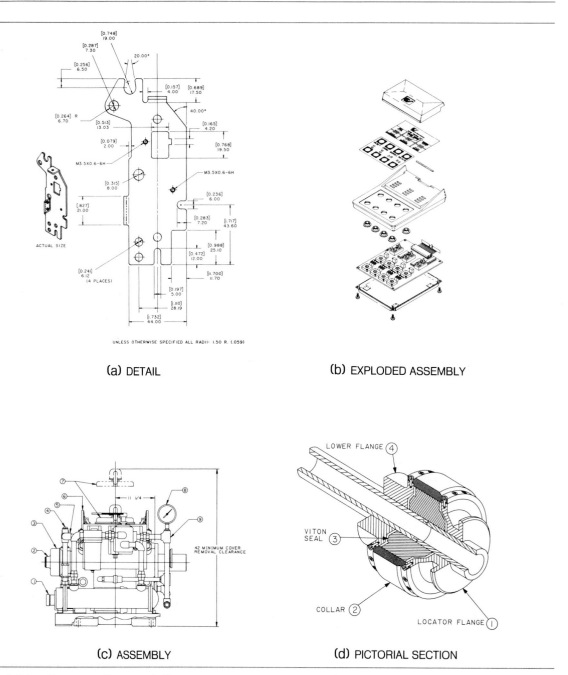

(a) DETAIL

(b) EXPLODED ASSEMBLY

(c) ASSEMBLY

(d) PICTORIAL SECTION

Fig. 3.24 Computer-Generated Drawings. *Courtesy of Computervision Corporation, a subsidiary of Prime Computer, Inc.*

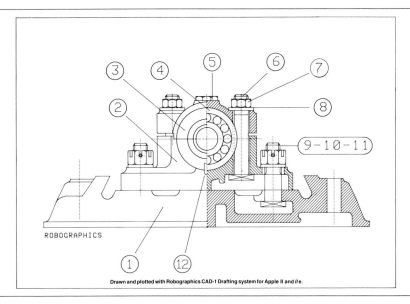

Fig. 3.25   Computer-Generated Assembly Drawing in Half Section. *Courtesy of Chessel-Robocom Corporation.*

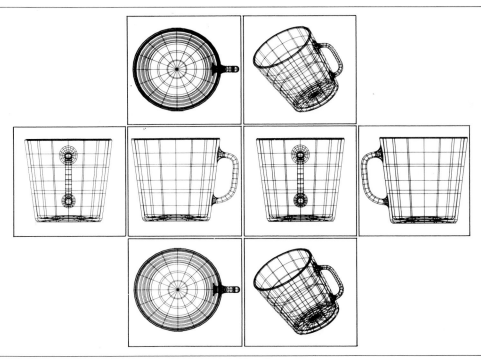

Fig. 3.26   Mold Design and Production. With the Auto-trol AD/380 automated design and drafting system, a part can be displayed in multiple views so the mold designer can see the part from all angles. *Courtesy of Auto-trol Technology Corporation.*

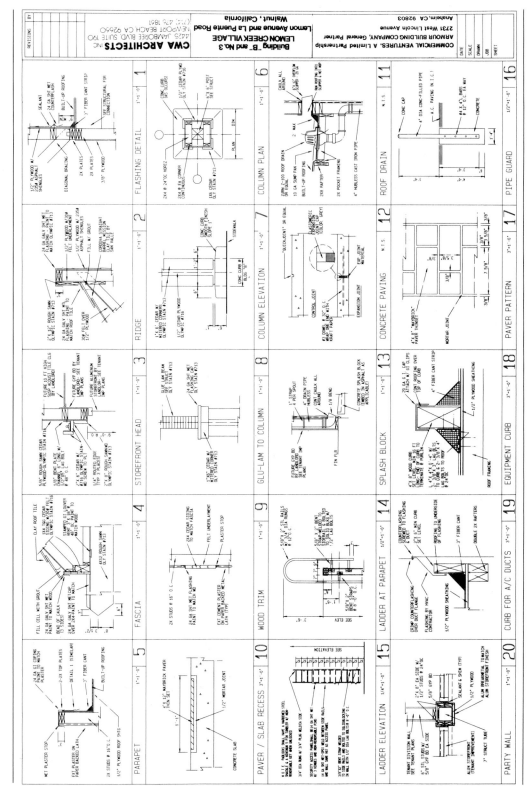

Fig. 3.27 Architectural Details Produced by Using the VersaCAD System. *Courtesy of VersaCAD, Inc.*

82

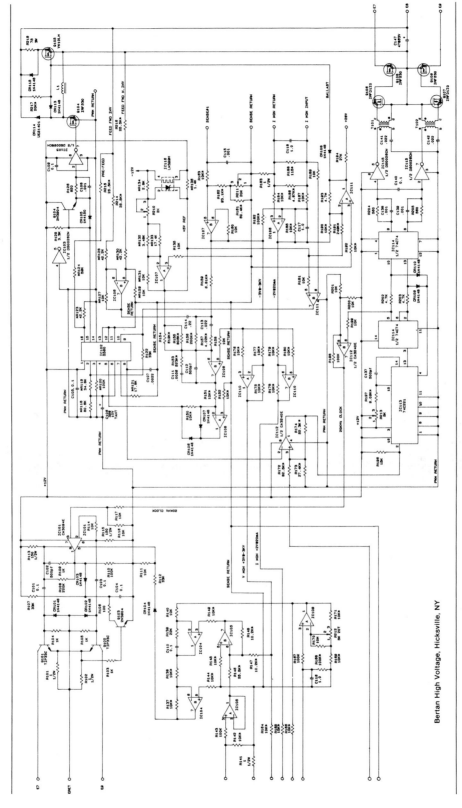

Fig. 3.28 Electronic Diagram Produced by Using the VersaCAD System. *Courtesy of VersaCAD, Inc.*

Bertan High Voltage, Hicksville, NY

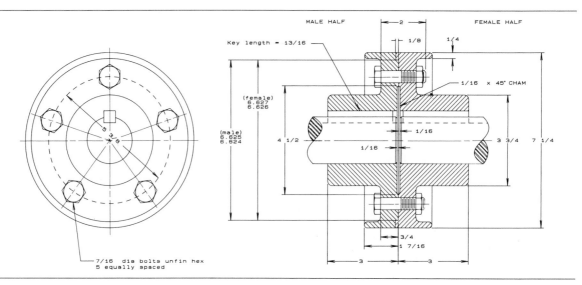

Fig. 3.29   Assembly Drawing in Section Produced by Using the VersaCAD Advanced System. *Courtesy of VersaCAD, Inc.*

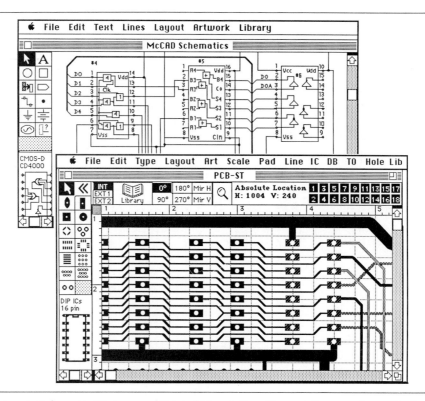

Fig. 3.30   Electronic Schematic Diagram and Printed Circuit Board Layout Generated on a Macintosh Computer with McCAD Software. *Courtesy of VAMP, Inc.*

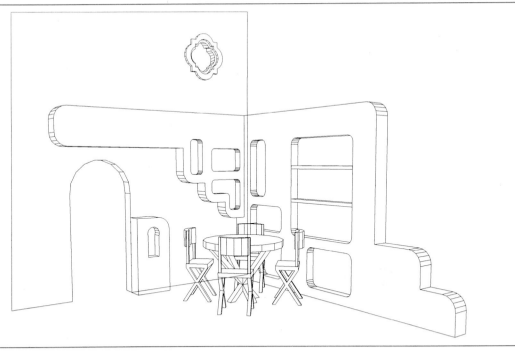

Fig. 3.31   Computer-Generated Perspective of Interior Room Detail Produced by Using SilverScreen CAD Program. *Courtesy of Schroff Development Corp.*

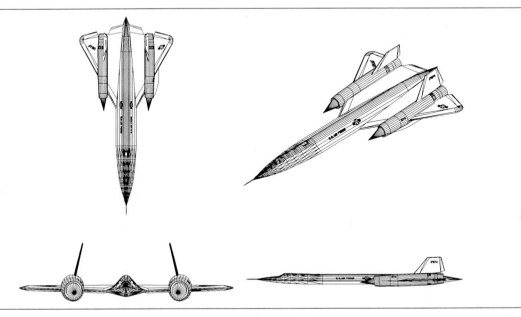

Fig. 3.32   Computer-Generated Multiview and Pictorial Drawing of U.S. Air Force SR-71 Reconnaisance Plane Produced by Using SilverScreen CAD Program. *Courtesy of Schroff Development Corp.*

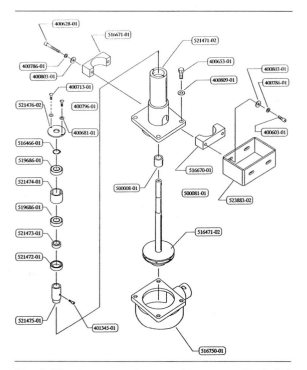

Fig. 3.33   Computer-Generated Isometric Exploded Assembly Drawing Produced by Using TRI-CAD Software. *Courtesy of Cadgrafix, Inc.*

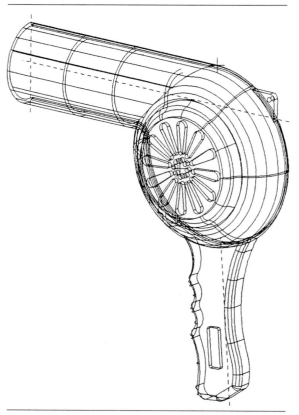

Fig. 3.34   Computer-Generated Wireframe Pictorial of Hair Dryer Produced by Using CADKEY Software. *Courtesy of CADKEY, Inc.*

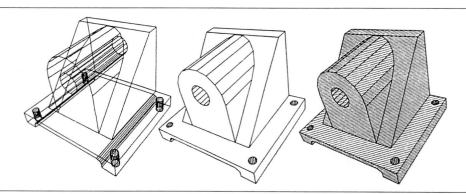

Fig. 3.35   Example of Wireframe, Solid, and Shaded Solid Pictorials Produced Using SilverScreen CAD Program. *Courtesy of Schroff Development Corp.*

3.12   Using a CAD System

After a CAD system has been installed, the beginning user must become thoroughly familiar with it and learn how to use it effectively. This will require learning some new skills as well as a different vocabulary. All CAD systems create drawings using basic geometric entities that are selected and placed on a screen by the CAD operator. The method and number of operations that are required to activate the various commands differ from one system to another.

Most CAD manufacturers offer training programs and tutorials that will make the learning process much easier. They will provide instruction and training manuals that give information and details about the operation of the system. These manuals can be used not only during the initial training period but also for reference purposes during later operation of the system.

All basic geometric entities are drawn in the same manner; for example, certain parameters must be set prior to the construction of an entity, such as line style, density, and position on

the drawing. Some systems require that other parameters be established prior to drawing, such as conventional English or metric (ISO) paper size and pen type.

Most experienced drafters have developed shorter or simpler methods for creating a drawing, such as using overlays or templates. CAD systems also have simplified methods for drawing. Some systems have symbol libraries that contain many of the frequently used symbols, such as electrical relays, switches, transformers, resistors, bolts, nuts, keys, piping, and architectural symbols. These symbols may be in a symbol library, Fig. 3.36, or the symbol may be located on one of the templates in the library of templates. Some CAD systems allow the users to customize their symbol libraries. The desired symbols must first be drawn by the user on the CAD system the same as they would appear on a drawing board. This process may initially take as much time as it would manually, but once the image has been entered in the computer database, it need never be drawn again. The symbol can easily be retrieved from the symbol library whenever required.

Any CAD system function can be classified as entity creation, manipulation, or storage. Once a drawing has been created, it can be manipulated on the screen by the CAD operator. Some commonly used commands to manipulate an image are MOVE, ROTATE, COPY, PAN, ZOOM, SCALE, DELETE, and LAYER.

After the CAD operator completes a drawing, it can then be placed in the computer's storage on a disk or tape. The command, and even the method, to perform this function may differ from one system to another. The terms SAVE and QUIT are frequently used for data storage, but the computer receiving this command may automatically transfer the data from its main memory to a soft disk in one system and to a hard disk on another system.

It is not possible to list all commands for every CAD system in a single chapter of a textbook. Every CAD system manufacturer does, however, provide a user's manual for its system that contains this information. To become pro-

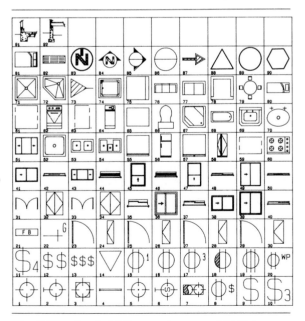

Fig. 3.36   Symbol Library. *Courtesy of T&W Systems, Inc.*

ficient in the use of any system, it is imperative that the user be aware of, and frequently refer to, the user's manual. See Appendix 3 for a glossary of CAD terms and definitions.

3.13   Selecting a CAD System

As the number of manufacturers of computer system equipment has increased, there has been a corresponding decrease in the cost of these systems. Some CAD systems may contain only those features necessary to produce simple two-dimensional entities, while others have the capability to create true three-dimensional views automatically. The manufacturers of some of these systems are eager to promote new or improved functions that may not be found on other systems. Almost daily the computer industry announces amazing advances.

Since the newer CAD systems are generally easier to operate, the words *user friendly* now are frequently used by manufacturers to promote their systems. There has, however, been a tendency by some to exaggerate what their particular systems can do. A prospective user should therefore view all claims with some skepticism, until proved, for they can be misleading and often lead to disappointment. Unfortunately, many firms have purchased a system only to discover later that it did not perform as expected.

Before purchasing a CAD system, it is highly recommended that a careful, well-thought-out plan for selecting a system be developed and followed. This plan can be divided into five phases:

I.   Establish the need for a CAD system.

II.   Survey and select system features.

III.   Request CAD system demonstrations.

IV.   Review selected systems.

V.   Select, purchase, and install a CAD system.

Let us now discuss each of these phases in greater detail.

I. ESTABLISH THE NEED FOR A CAD SYSTEM   The first consideration in the selection process is to determine whether a CAD system is needed at all. All potential users of the CAD system should be consulted as to how, when, and where a system would be used and whether it would be cost effective in their particular operations. Never purchase a system simply because it may be considered a first step into the future and your firm wants to project a progressive image. It is important to investigate and evaluate the time- and cost-saving claims of manufacturers by contacting firms that have CAD systems in operation. Contact as many firms as possible, especially those with a wide range of experience. Prepare a brief questionnaire to survey these firms asking questions regarding costs, training periods, system operation, and so on. Also ask at what point after installation the system became cost effective. Evaluate the responses to these questions and compare them with your specific requirements. This information will assist you in deciding if a CAD system can be beneficial to your firm at the present time.

Undoubtedly, you will hear many spectacular claims about the productivity of CAD systems. The productivity, however, will depend on the type of engineering operation. For example, among the first companies to introduce and use CAD systems were the electronic industry manufacturers. They found that drafters were constantly redrawing many symbols (resistor, transistor, etc.) and standard hardware parts (nuts, screws, etc.). Although they used plastic templates, common in all drafting departments, a considerable amount of time was spent doing tedious, repetitive drawing tasks. Since much of the time was devoted to tracing or redrawing something that was previously drawn, it was determined that a CAD system would save as much as 50 percent of the drafting time compared to using traditional methods. It is important to remember that with CAD it is only necessary to draw something once, even though initially it may take as long, or perhaps longer, to create it on the system as on the drawing board. From that point on, however, you need only to recall this information from data storage, which a CAD system can do with amazing speed. In this example, the benefits of CAD are obvious.

The process of determining the need for a CAD system can be time consuming and frustrating. Nevertheless, speed should be sacrificed to careful deliberation in this phase.

II. Survey and Select System Features It is generally agreed that software should be selected before hardware. However, some CAD systems are *turnkey* systems, that is, a total system with software and hardware combined and inseparable. Therefore, you should examine all the features of any given system very carefully before being attracted by spectacular hardware. For example, some systems have two cathode ray display screens. The chance of being impressed by this feature may overshadow the question of whether one really needs the two CRT screens.

Consider whether the system will be multipurpose or used strictly for CAD. Will other office operations such as word processing or accounting be done on this machine? The answer to this question may add or eliminate CAD programs based on their operating system software.

Investigate how well a system will exchange information or interface with other CAD or CAM systems. Many systems do not have this capability. One mechanism the industry is trying to standardize on for exchanging information with other CAD systems is known as initial graphics exchange specification (IGES), and if a system has it, this can be a definite advantage.

It is suggested that you survey the people who will use the system to determine what they think is desirable in a CAD system. From this survey, develop a checklist of hardware and software features that your future system should have.

III. Request CAD System Demonstrations After the checklist has been created and approved by all parties concerned, make arrangements to see the various CAD systems in operation. A list of vendors can be compiled from advertisements in trade journals, magazines, and so on, or from the various directories of computer graphics manufacturers that are published. Contact the vendors to arrange demonstrations. Explain to them exactly what you expect the system to do. If the vendors are completely aware of your requirements, they will be able to give a more realistic presentation. Most of all, be prepared to ask questions. With each succeeding demonstration your questions will also be more effective. Moreover, the answers that you receive will be more meaningful.

Some of the questions that you should ask are

1. What are the brand names of the equipment used?

2. Are service contracts available? If so, what are the details and the cost of the contract? In most instances, it is highly unlikely that your own staff will be able to repair the complicated equipment; therefore, a service contract is advisable. It is also possible in some situations to obtain service contracts from third parties that specialize in this type of business. The service provided by these firms for preventive and downtime maintenance may even be less costly than that offered by the original manufacturer.

3. What is the warranty period, or periods, if parts of the system are furnished by different vendors? Also, is the warranty paid for by the original vendor or is the cost shared by the user and vendor?

4. What types of CAD software are provided, and what operations can be performed?

5. What kind and length of training will be provided to the user staff? How long after installation will it take for the staff to be proficient enough for the system to be cost effective?

6. Does the vendor issue software updates? If so, how often is this done and what is the cost?

7. What is the reputation of the vendor for providing user support in the form of toll-free telephone assistance or other methods of communication? Is the vendor readily available when the user encounters software and/or hardware problems and requires assistance?

8. Who currently uses the particular system? Will the vendor provide a list of current

users? Users that have several months experience with a CAD system are best qualified to answer question 7.

A CAD system analysis worksheet can be very helpful when demonstrations are presented. Items to be included on the worksheet may be arranged according to hardware and software specifications. The hardware features that should be listed will generally be included in the following five categories: (1) central processor, (2) data input devices, (3) display monitors, (4) data storage, and (5) output devices.

1. Central Processor. When gathering information about a central processor, five areas that are common to computers should be investigated. These are (a) word size, (b) memory, (c) storage, (d) execution rate, and (e) operating system. Each of these areas should be carefully considered prior to the purchase of any software separate from the hardware. If, however, a turnkey system is being considered, then you have no choice, since the manufacturer has built the system centered on a central processor. A brief definition of each of these terms follows.

(a) Word size. A term used to indicate the microprocessor capability, usually expressed as 8 bit, 16 bit, dual 16 bit, and 32 bit. Computers process information in words; the larger the word, the faster the processing.

(b) Memory (or RAM, random access memory). RAM indicates how many bytes (1 byte = 8 bits) of temporary memory can be utilized by the CPU. For example, 64 kB means 64,000 bytes, and 10 MB means 10 million bytes, and so on. Most CPUs have expansion slots where additional memory cards can be inserted. This fact is significant when you wish to expand the system or create a network by adding workstations at a later date. If you add more workstations to any system without expanding the memory, you will notice a definite degrading of the system's performance. The system will be slow in movement and response.

(c) Execution rate. This is the rate at which the computer acts on information entered, and

it is usually expressed in nanoseconds (millionth of a second) per event, or operation. Frequently, this rate is a function of word size and main memory size. Or to explain it in terms of operation: a 32-bit word size CPU with 10 MB of main memory will produce a system capable of operating at an incredible speed.

(d) Operating system programs. These programs control the computer's method of processing data (different from applications software) and are usually supplied by the computer manufacturer. If the software is purchased separately, it is important to know whether or not it will function properly with the CPU's operating system and other hardware peripherals connected to the system.

2. Data Input Devices. A variety of the devices available for data entry also allow the user to communicate with the CAD system. A keyboard is normally the standard device included with a system. Other devices may be a graphics tablet, mouse, light pen, thumbwheel(s), trackball, and so on. One or more of these devices may be included with a turnkey system or may be offered as an optional feature. The higher quality graphics devices have finer resolution (expressed in thousandths of an inch) and greater accuracy. The higher the quality, the greater the cost of the device.

3. Display Monitors. All interactive CAD systems have display monitors: vector refresh, raster scan, or direct view storage tube (DVST). In turnkey systems the monitor is included as part of the workstation. Raster scan monitors are categorized as monochrome or color, and according to their resolution and number of colors they can display. The industry has now standardized classifications of monitors as

1. Monochrome.
2. Color graphics array (CGA).
3. Enhanced color graphics array (EGA).
4. Video graphics array (VGA).
5. Super VGA.
6. Professional graphics array (PGA).

The following features should be considered in selecting a monitor.

(a) Single or dual display monitors. Does the system use one or two screens? If two screens are used, how are they formatted? Is one screen for graphics and the other for text and menu display, or are both screens for graphics display? Although two screens may be an impressive attraction, many CAD systems operate very well with one graphic display screen and are far less expensive.

(b) Resolution. Resolution is the smallest spacing between two display elements that permits the elements to be distinguished on the terminal. It indicates the clarity of display and usually is expressed as pixels (picture elements), or lines, such as 1024×1024. The higher the numbers, the better the definition of lines, circles, arcs, and so on.

(c) Monochrome or color. Monochrome means single color only, such as green, amber, or white lines on a dark background. Color refers to a display containing a full spectrum of colors such as that on the screen of a color television set. Although a color display may be an attractive optional feature, a CAD system will function very well with a monochrome display and will be less costly.

(d) Screen size. The size of the viewing area (12″, 14″, or 19″) is measured diagonally across the screen. If the resolution is coarse, not much will be gained by using a large screen, but with fine resolution a larger screen will produce a clearer display.

4. Storage. The manner in which the computer system accumulates and retains information for future use is called storage. When reviewing these specifications carefully consider these features.

(a) Storage type. Storage is typically in the form of hard or soft disk, magnetic tape, and so on.

(b) Capacity. The amount of system storage is called capacity and is expressed (measured) in bytes, such as 712 kB or 40 MB. A larger number indicates a greater capacity.

(c) Access time. The speed at which the computer retrieves information is called access time. It is one measure of system response.

(d) Removable or nonremovable. Ascertain if data can be removed from the drive and stored elsewhere, such as on tape, disk packs, or floppy disks.

(e) Expandable. Determine if the storage capacity of the system can be expanded. If it can, to what extent?

5. Output Devices. The end result of any computer system is output. In a CAD system the output is usually in the form of hard-copy drawings, although mathematical calculations may also be part of the output. Two common devices that are used to create hard copy on a CAD system are plotters and printers. Several issues that should be considered regarding output devices are

(a) Will the device be included in the system price or do you purchase it separately?

(b) Is the device manufactured by a recognized company with an established reputation for quality, service, and so on?

(c) What is the maximum size of drawing that the device can generate?

(d) How fast does the device operate?

(e) What is the resolution and accuracy of the completed drawing, usually expressed in thousandths of an inch?

(f) If a plotter is used, what different types and sizes of pens and colors of ink can be used?

(g) If a printer is used, what type is it—ink jet, dot matrix, laser, and so on?

The questions and information provided in each of the preceding five categories pertaining to hardware features are offered as a guide to assist the prospective buyer and user in the comparison and evaluation of CAD systems. The worksheet may be expanded, or modified, to reflect particular requirements. A similar worksheet should also be developed listing CAD system software feature requirements.

With the completion of this phase you will have achieved several important goals. Your knowledge and understanding of CAD systems will have increased, and you will be in a position to identify those system features that you require and can afford. In addition, you will have assembled a list of current users of CAD systems for future reference or exchange of information.

IV. REVIEW SELECTED SYSTEMS   It is important at this stage that all of the collected information be organized and carefully reviewed. A list should be made now of only those systems that merit further serious consideration. You may wish to request another demonstration of the particular systems that are on your revised list and to request additional information from current users regarding equipment performance, staff training, vendor support, and so on. Remember, if you select a system made by a reputable and financially stable manufacturer, and purchase it from a dependable retailer, you will sacrifice nothing in terms of service and support. Throughout this chapter we have made every effort to feature or illustrate systems and equipment made by such firms.

V. FINAL SELECTION, PURCHASE, AND INSTALLATION OF A CAD SYSTEM.   This last phase occurs when the final decision will be made as to whether or not to purchase a CAD system. The decision invariably will depend on how you plan to use the system and how much you can afford to pay for it. If the decision is made to acquire a system, determine costs, choose a delivery date, and arrange for installation and training.

3.14   Summary

The information presented in this chapter is intended to familiarize the student with the basic "generic" concepts, hardware, peripherals, and systems in CAD. It is not possible, nor was it intended, to present programs, subroutines, or all the commands used on CAD systems.

When possible, the instructor should arrange for students to visit nearby engineering and drafting departments that have CAD systems in operation. Those students who wish to obtain additional information on this subject should consult their school or local library.

INTRODUCTION TO CAD PROBLEMS

The following problems are given to examine your retention and understanding of the subject matter presented in this chapter. When necessary, refer to the appropriate sections of the chapter to check your answers.

Prob. 3.1   Define the following terms: computer system, hardware, software, analog, digital, computer graphics, CAD, CADD, and CAM.

Prob. 3.2   What are the principal components of a computer system? A CAD system? Draw a systems flowchart that illustrates the sequence of operations for each of the systems.

Prob. 3.3   Prepare a list of CAD system hardware components and give examples of each.

Prob. 3.4   List and describe the main types of graphics monitors.

Prob. 3.5   Arrange a visit to the computer center at your school, or to a local engineering design office, and prepare a written report on the use of computers in design and drafting at these facilities.

CHAPTER 4

Lettering

The designs of modern alphabets had their origin in Egyptian hieroglyphics, Fig. 1.4, which evolved over time into a cursive hieroglyphic or hieratic writing. This writing was adopted by the Phoenicians who developed it into an alphabet of 22 letters. The Greeks later adopted the Phoenician alphabet but it evolved into two distinct types in different sections of Greece: an Eastern Greek type, used also in Asia Minor, and a Western Greek type, used in the Greek colonies in and near Italy, which became the Latin alphabet about 700 B.C. The Latin alphabet came into general use throughout the Old World.

Originally, the Roman capital alphabet consisted of 22 characters, and these have remained practically unchanged to this day. The numerous modern styles of letters were derived from the design of the original Roman capitals.

4.1 Lettering* Styles

Before the invention of printing by Gutenberg in the fifteenth century, all letters were made by hand, and their designs were modified and decorated according to the individual writer's taste. In England these letters became known as Old English. The early German printers adopted the Old English letters, and they are still in limited use.

The early Italian printers used Roman letters, which were later introduced into England and gradually replaced the Old English letters. The Roman capitals have come down to us virtually in their original form. A general classification of letter styles is shown in Fig. 4.1.

*Lettering, not "printing," is the correct term for making letters by hand. Printing means the production of printed material on a printing press.

4.2 Roman Letters

The term *Roman* refers to any letter that has wide downward strokes and thin connecting strokes, as would result from the use of a wide pen, and the ends of the strokes are terminated with spurs called *serifs*. Roman letters include the Old Roman and Modern Roman and may be vertical or inclined. Inclined letters are also referred to as *italic*, regardless of the letter style; those shown in Fig. 4.1 are inclined Modern Roman.

OLD ROMAN LETTERS *Fig. 4.2* The Old Roman letter is the basis of all our letters and is still regarded as the most beautiful. This letter is employed mostly by architects. Because of its great beauty, it is used almost exclusively on buildings and for inscriptions on bronze or stone.

ABCDEFGH abcdefghij Roman *All letters having elementary strokes "accented" or consisting of heavy and light lines, are classified as Roman.*

ABCDEFGHI abcdefghijklm *Italic- All slanting letters are classified as Italics~ These may be further designated as Roman-Italics. Gothic Italics or Text Italic.*

ABCDEFG abcdefghijkl *Text—This term includes all styles of Old English. German text. Bradley text or others of various trade names ~ Text styles are too illegible for commercial purposes.*

ABCDEFGH abcdefgh GOTHIC *All letters having the elementary strokes of even width are classified as Gothic*

Fig. 4.1   Classification of Letter Styles.

ABCDEFGHIJKLM
NOPQRSTUVW
XYZ1234567890abcd
efghijklmnopqrstuvwxyz

Fig. 4.2   Old Roman Capitals, with Numerals and Lowercase of Similar Design.

MODERN ROMAN LETTERS   *Figs. 4.3 and 4.4*
The Modern Roman, or simply "Roman," letters were developed during the eighteenth century by the type founders; the letters used in most modern newspapers, magazines, and books are of this style. The text of this book is set in Modern Roman capital and lowercase letters. These letters are often used on maps, especially for titles. They may be drawn in outline and then filled in, as shown in Fig. 4.3, or they may be produced with one of the broad-nib pens shown in Fig. 4.13.

Fig. 4.3   Modern Roman Capitals and Numerals.

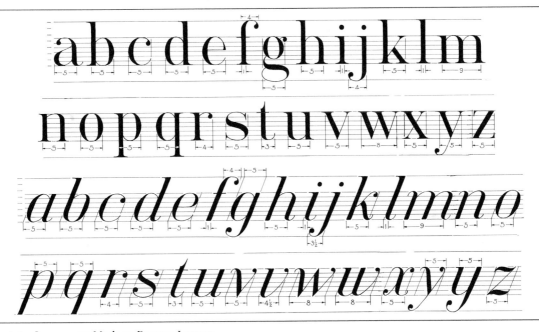

Fig. 4.4   Lowercase Modern Roman Letters.

A typical example of the use of Modern Roman in titles is shown in Fig. 4.43. Their use on maps is discussed in §4.27.

4.3   Gothic Letters

The *Text* letters shown in Fig. 4.1 are often loosely referred to as Old English, although these and other similar letters, such as German Text, are actually Gothic. German Text is the only form of medieval Gothic in commercial use today.

Commercial Gothic is a relatively modern development that originated from the earlier Gothic forms. Also called sans-serif Gothic, this letter is the only one of interest to engineers, Fig. 4.5. It is the plainest and most legible style and is the one from which our single-stroke engineering letters are derived. While admittedly not as beautiful as many other styles, sans-serif let-

97

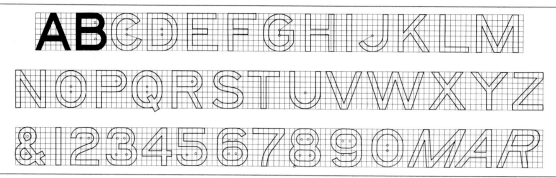

Fig. 4.5 Gothic Capital Letters.

CONDENSED LETTERS
EXTENDED LETTERS
Condensed Letters
Extended Letters

Fig. 4.6 Condensed and Extended Letters.

ters are very legible and comparatively easy to make. They may also be drawn in outline and filled in, Fig. 4.5.

EXTENDED AND CONDENSED LETTERS *Fig. 4.6*
To meet design or space requirements, letters may be narrower and spaced closer together, in which case they are called compressed or condensed letters. If the letters are wider than normal, they are referred to as extended letters.

LIGHTFACE AND BOLDFACE LETTERS *Fig. 4.7*
Letters also vary as to the thickness of the stems or strokes. Letters having very thin stems are

LIGHTFACE
BOLDFACE

Fig. 4.7 Lightface and Boldface Letters.

called LIGHTFACE, while those having heavy stems are called **BOLDFACE**. As the preceding sentence demonstrates, Modern Roman also exists as lightface and boldface.

4.4   Greek Alphabet
Greek letters are often used as symbols in both mathematics and technical drawing by the engineer. A Greek alphabet, showing both uppercase and lowercase letters, is given for reference purposes in Fig. 4.8.

4.5   Single-Stroke Gothic Letters
During the latter part of the nineteenth century the development of industry and of technical drawing in the United States made evident a need for a simple legible letter that could be executed with single strokes of an ordinary pen. To meet this need, C. W. Reinhardt, formerly chief draftsman for *Engineering News,* developed al-

A	α	alpha	I	ι	iota	P	ρ	rho

A α alpha   I ι iota   P ρ rho
B β beta   K κ kappa   Σ ς sigma
Γ γ gamma   Λ λ lambda   T τ tau
Δ δ delta   M μ mu   Υ υ upsilon
E ε epsilon   N ν nu   Φ φ phi
Z ζ zeta   Ξ ξ xi   X χ chi
H η eta   O ο omicron   Ψ ψ psi
Θ θ theta   Π π pi   Ω ω omega

Fig. 4.8   Greek Alphabet.

phabets of capital and lowercase inclined and "upright" letters,* based on the old Gothic letters. For each letter he worked out a systematic series of strokes. The single-stroke Gothic letters used on technical drawings today are based on Reinhardt's work.

4.6   Standardization of Lettering

Reinhardt's development of single-stroke letters was the first step toward standardization of technical lettering. Since that time, however, there has been an unnecessary and confusing diversity of lettering styles and forms. In 1935 the American National Standards Institute suggested letter forms that are now generally considered as standard. The lettering forms given in the present standard [ANSI Y14.2M–1979 (R1987)] are practically the same as those given in 1935

*Published in *Engineering News* in about 1893 and in book form in 1895.

except that lowercase forms have since been added.

The letters in this chapter and throughout this text conform to the American National Standard. Vertical letters are perhaps slightly more legible than inclined letters, but they are more difficult to execute. Both vertical and inclined letters are standard, and the engineer or drafter may be called on to use either.

Lettering on drawings must be legible and suitable for easy and rapid execution. The single-stroke Gothic letters shown in Figs. 4.25 and 4.26 meet these requirements. Either vertical or inclined letters may be used, but only one style should appear on any one drawing. Background areas between letters in words should appear approximately equal, and words should be clearly separated by a space equal to the height of the lettering. Only when special emphasis is necessary should the lettering be underlined.

It is not desirable to vary the size of the lettering according to the size of the drawing except when a drawing is to be reduced in reproduction.

Drawings for microfilm reproduction require well-spaced lettering to prevent "fill-ins." The microfont alphabet, Fig. 4.9, is an adaptation of the single-stroke Gothic characters developed by the National Microfilm Association. Only the vertical style is shown.

4.7   Uniformity

In any style of lettering, uniformity is essential. Uniformity in height, proportion, inclination, strength of lines, spacing of letters, and spacing

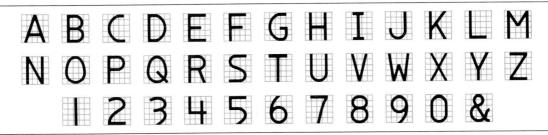

Fig. 4.9   Microfont Alphabet.

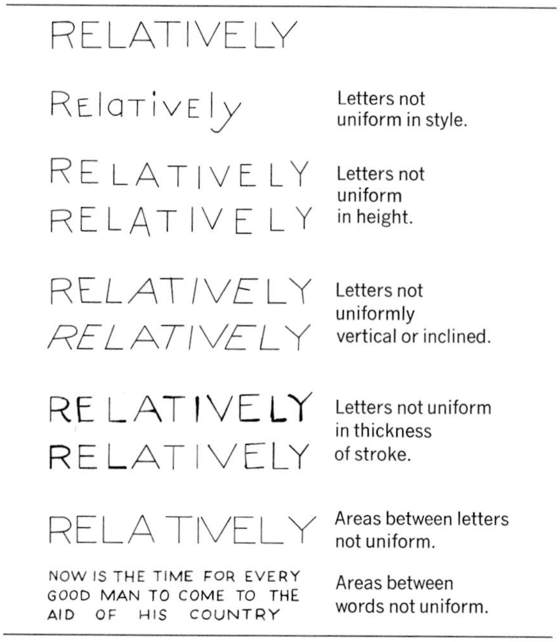

RELATIVELY

Relatively · Letters not uniform in style.

RELATIVELY
RELATIVELY · Letters not uniform in height.

RELATIVELY
RELATIVELY · Letters not uniformly vertical or inclined.

RELATIVELY
RELATIVELY · Letters not uniform in thickness of stroke.

RELATIVELY · Areas between letters not uniform.

NOW IS THE TIME FOR EVERY GOOD MAN TO COME TO THE AID OF HIS COUNTRY · Areas between words not uniform.

Fig. 4.10 Uniformity in Lettering.

of words insures a pleasing appearance, Fig. 4.10.

Uniformity in height and inclination is promoted by the use of light guide lines, §4.14. Uniformity in strength of lines can be obtained only by the skilled use of properly selected pencils and pens, §§4.10 and 4.11.

4.8 Optical Illusions

Good lettering involves artistic design, in which the white and black areas are carefully balanced to produce a pleasing effect. Letters are designed to *look* well, and some allowances must be made for errors in perception. Note that in Fig. 4.25 the width of the standard H is less than its height to eliminate a square appearance, the numeral 8 is narrower at the top to give it stability, and the width of the letter W is greater than its height, for the acute angles in the W give it a compressed appearance. Such acute angles should be avoided in good letter design.

4.9 Stability

If the upper portions of certain letters and numerals are equal in width to the lower portions, the characters appear top-heavy. To correct this, the upper portions are reduced in size where possible, thereby producing the effect of stability and a more pleasing appearance, Fig. 4.11.

If the central horizontal strokes of the letters B, E, F, and H are placed at midheight, they will appear to be below center. To overcome this optical illusion, these strokes should be drawn slightly above the center.

4.10 Lettering Pencils

Pencil letters are best made with a medium-soft lead with a conical point, Fig. 2.11 (c), or with a suitable thin-lead pencil, Fig. 2.8 (c).

Today the majority of drawings are finished in pencil and reproduced. To reproduce well by any process, the pencil lettering must be dense black, as should all other final lines on the drawing. The right lead to use depends largely on the amount of tooth or grain in the paper, the rougher papers requiring the harder pencils. The lead should be soft enough to produce jet black

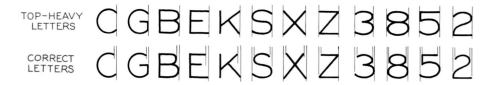

TOP-HEAVY LETTERS   CGBEKSXZ3852

CORRECT LETTERS   CGBEKSXZ3852

Fig. 4.11 Stability of Letters.

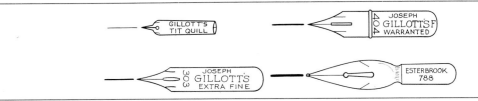

Fig. 4.12 Pen Points (Full Size).

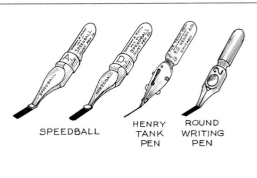

SPEEDBALL   HENRY TANK PEN   ROUND WRITING PEN

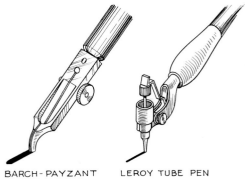

BARCH-PAYZANT   LEROY TUBE PEN

Fig. 4.13 Special Pens for Freehand Lettering.

lettering, yet hard enough to prevent excessive wearing down of the point, crumbling of the point, and smearing of the graphite.

4.11 Lettering Pens

The choice of a pen for lettering is determined by the size and style of the letters, the thickness of stroke desired, and the personal preference of the drafter. Fig. 4.12 shows a variety of the best pen points in a range from the *tit quill,* the finest, to the *ball-pointed,* the coarsest. The widths of the lines made by the several pens are shown full size. Letters more than $\frac{1}{2}''$ (12.7 mm) in height generally require a special pen, Fig. 4.13.

The technical fountain pen, Fig. 4.14, is a newer instrument that drafters now use for lettering and line work. The point is a small tube in which an automatic plunger rod keeps the ink flowing, and the pen has a reservoir cartridge for the storage of ink. The pen point produces a uniform thickness of line and makes the task of inking much simpler. These pens are available in sets of different point sizes and may be used with lettering instruments and templates.

Any lettering pen must be kept clean. All these pens should be frequently cleaned with cleaning fluid to keep them in service.

4.12 Technique of Lettering

Any normal person can learn to letter if a persistent and intelligent effort is made. Although it is true that "practice makes perfect," it must be understood that practice alone is not enough; it must be accompanied by *continuous effort to improve.*

Fig. 4.14 Technical Fountain Pen. *Courtesy of Keuffel & Esser Co.*

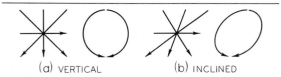

Fig. 4.15 Basic Lettering Strokes.

Fig. 4.16 Position of Hand in Lettering.

Lettering is freehand drawing and not writing. Therefore, the six fundamental strokes and their direction for freehand drawing are basic to lettering, Fig. 4.15. The horizontal strokes are drawn to the right, and all vertical, inclined, and curved strokes are drawn downward.

Good lettering is always accomplished by conscious effort and is never done well otherwise, though good muscular coordination is of great assistance. Ability to letter has little relationship to writing ability; excellent letterers are often poor writers.

There are three necessary aspects of learning to letter.

1. Knowledge of the proportions and forms of the letters and the order of the strokes. No one can make a good letter who does not have a clear mental image of the correct form of the letter.

2. Knowledge of composition—the spacing of letters and words. Rules governing composition should be thoroughly mastered, §4.24.

3. Persistent practice, with *continuous effort to improve.*

First, sharpen the pencil to a needle point; then dull the point *very slightly* by marking on paper while holding the pencil vertically and rotating the pencil to round off the point.

Pencil lettering should be executed with a fairly soft pencil, such as an F or H for ordinary paper; the strokes should be *dark* and *sharp,* not gray and blurred. In order to wear the lead down uniformly and thereby keep the lettering sharp, turn the pencil frequently to a new position.

The correct position of the hand in lettering is shown in Fig. 4.16. In general, draw vertical strokes downward or toward you with a finger movement, and draw horizontal strokes from left to right with a wrist movement without turning the paper.

Since practically all pencil lettering will be reproduced, the letters should be dense black. Avoid hard pencils that, even with considerable pressure, produce gray lines. Use a fairly soft pencil and keep it sharp by frequent dressing of the point on the sandpaper pad or file. An example (full size) of pencil lettering exhibiting correct technique is shown in Fig. 4.17.

4.13 Left-handers

All evidence indicates that the left-handed drafter is just as skillful as the right-hander, and this includes skill in lettering. The most important step in learning to letter is learning the correct shapes and proportions of letters, and these can be learned as well by the left-hander as by anyone else. The left-hander does have a problem of developing a system of strokes that seems personally most suitable. The strokes shown in Figs. 4.25 and 4.26 are for right-handers. The left-hander should experiment with each letter to find out which strokes are best. The habits of left-handers vary so much that it is futile to suggest a standard system of strokes for all left-handers.

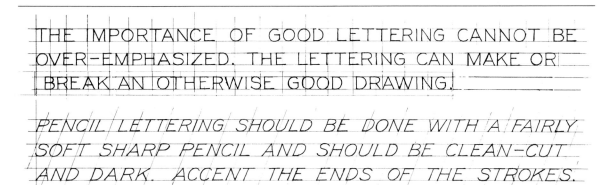

Fig. 4.17   Pencil Lettering (Full Size).

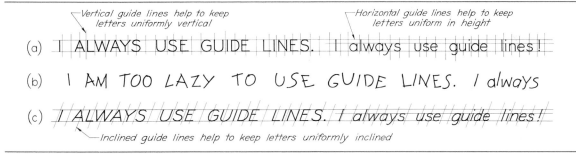

Fig. 4.18   Guide Lines.

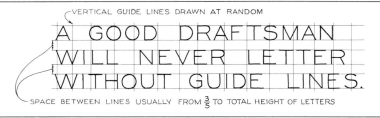

Fig. 4.19   Guide Lines for Vertical Capital Letters.

4.14 Guide Lines

Fig. 4.18 Extremely light horizontal guide lines are necessary to regulate the height of letters. In addition, light vertical or inclined guide lines are needed to keep the letters uniformly vertical or inclined. Guide lines are absolutely essential for good lettering and should be regarded as a welcome aid, not as an unnecessary requirement.

4.15 Guide Lines for Capital Letters

Guide lines for vertical capital letters are shown in Fig. 4.19. On working drawings, capital letters are commonly made $\frac{1}{8}''$ (3.2 mm) high, with the space between lines of lettering from three-fifths to the full height of the letters. See Table 16.1 for ANSI-recommended minimum letter heights on drawings. The vertical guide lines are not

103

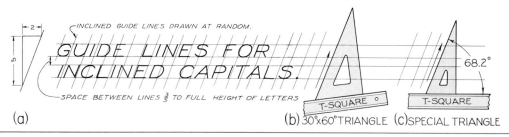

Fig. 4.20   Guide Lines for Inclined Capital Letters.

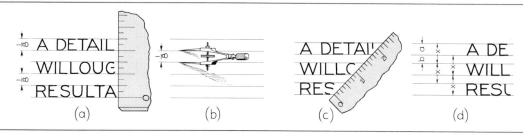

Fig. 4.21   Spacing of Guide Lines.

used to space the letters—this should always be done by eye while lettering—but only to keep the letters uniformly vertical. Accordingly, they should be drawn at random. Where several lines of letters are to be made, these vertical guide lines should be continuous from top to bottom of the lettered area, as shown.

Guide lines for inclined capital letters are shown in Fig. 4.20. The spacing of horizontal guide lines is the same as for vertical capital lettering. The American National Standard slope of 2 in 5 (or 68.2° with horizontal) may be established by drawing a "slope triangle," as shown at (a), and drawing the guide lines at random with the T-square and triangle, as shown at (b). Special triangles for the purpose may be used, as shown at (c), or the lines may be drawn with the Braddock–Rowe Lettering Triangle, Fig. 4.23, or the Ames Lettering Guide, Fig. 4.24.

A simple method of spacing horizontal guide lines is to use the scale, as shown in Fig. 4.21 (a), and merely set off a series of $\frac{1}{8}''$ spaces, making both the letters and the spaces between lines of letters $\frac{1}{8}''$ high. Another method of setting off equal spaces, $\frac{1}{8}''$ or otherwise, is to use the bow dividers, as shown at (b).

If it is desired to make the spaces between lines of letters less than the height of the letters, the methods shown at (c) and (d) will be convenient. At (c) the scale is placed diagonally, the letters in this case being four units high and the spaces between lines of lettering being three units. If the scale is rotated clockwise about the zero mark as a pivot, the height of the letters and the spaces between lines of letters diminish but remain proportional. If the scale is moved counterclockwise, the spaces are increased. The same unequal spacing may be accomplished with the bow dividers, as shown at (d). Let distance $x = a + b$, and set off x-distances, as shown.

When large and small capitals are used in combination, the small capitals should be three-fifths to two-thirds as high as the large capitals, Fig. 4.22. This is in conformity with the guide-line devices described in §§4.16 and 4.17.

Fig. 4.22   Large and Small Capital Letters.

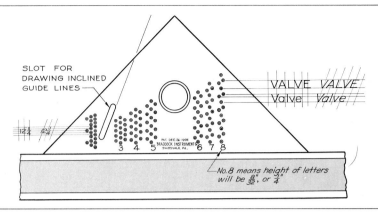

Fig. 4.23   Braddock-Rowe Lettering Triangle.

4.16   Lettering Triangles

Lettering triangles, which are available in a variety of shapes and sizes, are provided with sets of holes in which the pencil is inserted and the guide lines are produced by moving the triangle with the pencil point along the T-square. The Braddock-Rowe Lettering Triangle, Fig. 4.23, is convenient for drawing guide lines for lettering and dimension figures as well as for drawing section lines. In addition, the triangle is used as a utility 45° triangle. The numbers at the bottom of the triangle indicate heights of letters in thirty-seconds of an inch. Thus, to draw guide lines for $\frac{1}{8}''$ (3.2 mm) capitals, use the No. 4 set of holes. For lowercase letters, draw guide lines from every hole; for capitals, omit the second

hole in each group. The spacing of holes is such that the lower portions of lowercase letters are two-thirds as high as the capitals, and the spacing between lines of lettering is also two-thirds as high as the capitals.

The column of holes at the extreme left is used to draw guide lines for dimension figures $\frac{1}{8}''$ (3.2 mm) high and fractions $\frac{1}{4}''$ (6.4 mm) high and also for section lines $\frac{1}{16}''$ (1.6 mm) apart.

4.17   Ames Lettering Guide

Fig. 4.24   The Ames Lettering Guide is an ingenious transparent plastic device composed of a frame holding a disk with three columns of holes. The vertical distances between the holes

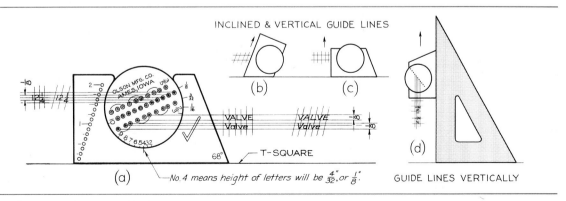

Fig. 4.24   Ames Lettering Guide.

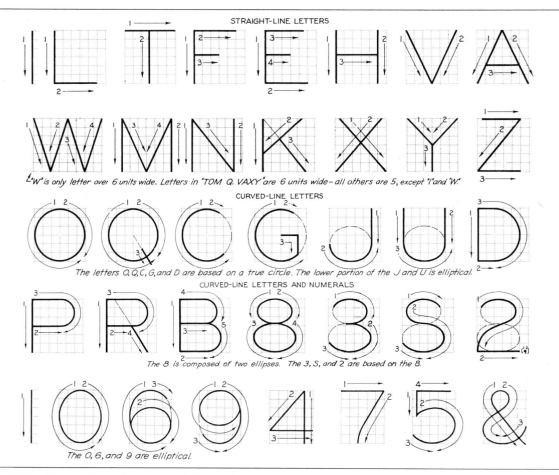

Fig. 4.25   Vertical Capital Letters and Numerals.

may be adjusted quickly to the desired spacing for guide lines or section lines by simply turning the disk to one of the settings indicated at the bottom of the disk. These numbers indicate heights of letters in thirty-seconds of an inch. Thus, for $\frac{1}{8}''$ high letters, the No. 4 setting would be used. The center column of holes is used primarily to draw guide lines for numerals and fractions, the height of the whole number being two units and the height of the fraction four units. The No. 4 setting of the disk will provide guide lines for $\frac{1}{8}''$ whole numbers, with fractions twice as high, or $\frac{1}{4}''$, as shown at (a). Since the spaces are equal, these holes can also be used to draw equally spaced guide lines for lettering or to draw section lines. The Ames Lettering Guide is also available with metric graduations for desired metric spacing.

The two outer columns of holes are used to draw guide lines for capitals or lowercase letters, the column marked three-fifths being used where it is desired to make the lower portions of lowercase letters three-fifths the total height of the letters and the column marked two-thirds being used where the lower portion is to be two-thirds the total height of the letters. In each case, for capitals, the middle hole of each set is not used. The two-thirds and three-fifths also indicate the spaces between lines of letters.

The sides of the guide are used to draw inclined or vertical guide lines, as shown at (b) and (c).

Fig. 4.26 Inclined Capital Letters and Numerals.

4.18 Vertical Capital Letters and Numerals

Fig. 4.25 For convenience in learning the proportions of the letters and numerals, each character is shown in a grid 6 units high. Numbered arrows indicate the order and direction of strokes. The widths of the letters can be easily remembered. The letter 1 or the numeral 1 has no width. The W is 8 units wide ($1\frac{1}{3}$ times the height) and is the widest letter in the alphabet. All the other letters or numerals are either 5 or 6 units wide, and it is easy to remember the 6-unit letters because when assembled they spell TOM Q. VAXY. All numerals except the 1 are 5 units wide.

All horizontal strokes are drawn to the right, and all vertical, inclined, and curved strokes are drawn downward, Fig. 4.15.

As shown in Fig. 4.25, the letters are classified as *straight-line letters* or *curved-line letters*. On the third row the letters O, Q, C, and G are all based on the circle. The lower portions of the J and U are semiellipses, and the right sides of the D, P, R, and B are semicircular. The 8, 3, S, and 2 are all based on the figure 8, which is composed of a small ellipse over a larger ellipse. The 6 and 9 are based on the elliptical zero. The lower part of the 5 is also elliptical in shape.

4.19 Inclined Capital Letters and Numerals

Fig. 4.26 The order and direction of the strokes and the proportions of the inclined capital letters and numerals are the same as those for the vertical characters. The methods of draw-

107

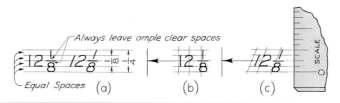

Fig. 4.27   Guide Lines for Dimension Figures.

ing guide lines for inclined capital letters are given in §4.15, and for numerals in §4.20. Inclined letters are also classified as straight-line or curved-line letters, most of the curves being elliptical in shape.

4.20   Guide Lines for Whole Numbers and Fractions

Complete guide lines should be drawn for whole numbers and fractions, especially by beginners. This means that both horizontal and vertical guide lines, or horizontal and inclined guide lines, should be drawn.

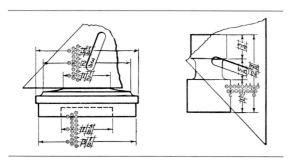

Fig. 4.28   Use of Braddock-Rowe Triangle.

Fig. 4.27   Draw five equally spaced guide lines for whole numbers and fractions. Thus, fractions are twice the height of the corresponding whole numbers. Make the numerator and the denominator each about three-fourths as high as the whole number to allow ample clear space between them and the fraction bar. For dimensioning, the most commonly used height for whole numbers is $\frac{1}{8}''$ (3.2 mm), and for fractions $\frac{1}{4}''$ (6.4 mm), as shown.

If the Braddock-Rowe Triangle is used, the column of holes at the left produces five guide lines $\frac{1}{16}''$ (1.6 mm) apart, Fig. 4.28.

If the Ames Lettering Guide, Fig. 4.24, is used with the No. 4 setting of the disk, the same five guide lines, $\frac{1}{16}''$ (1.6 mm) apart, may be drawn from the central column of holes.

Some of the most common errors in lettering fractions are illustrated in Fig. 4.29. Never let numerals touch the fraction bar, (a). Center the denominator under the numerator, (b). Never use an inclined fraction bar, (c), except when lettering in a narrow space, as in a parts list. Make the fraction bar slightly longer than the widest part of the fraction, (d).

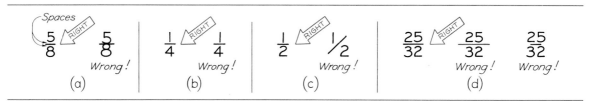

Fig. 4.29   Common Errors.

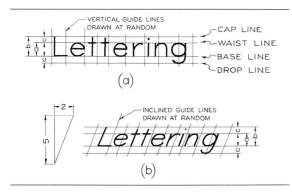

(a)

(b)

Fig. 4.30   Guide Lines for Lowercase Letters.

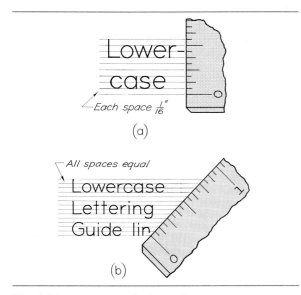

(a)

(b)

Fig. 4.31   Spacing with the Scale.

4.21 Guide Lines for Lowercase Letters

Fig. 4.30   Lowercase letters have four horizontal guide lines, called the cap line, waist line, base line, and drop line. Strokes of letters that extend up to the cap line are called ascenders, and those that extend down to the drop line, descenders. Since only five letters have descenders, the drop line is little needed and is usually omitted. In spacing horizontal guide lines, space a may vary from three-fifths to two-thirds of space b. Spaces c are equal, as shown.

If it is desired to set off guide lines for letters $\frac{3}{16}''$ (4.8 mm) high with the scale (using two-thirds ratio), it is only necessary to set off equal spaces each $\frac{1}{16}''$ (1.6 mm), Fig. 4.31 (a). The lower portion of the letter thus would be $\frac{1}{8}''$ (3.2 mm), and the space between lines of letters would also be $\frac{1}{8}''$ (3.2 mm). If the scale is placed at an angle, the spaces will diminish but remain equal, (b). Thus, this method may be easily used for various heights of lettering.

The Braddock-Rowe Triangle, Fig. 4.23, and the Ames Lettering Guide, Fig. 4.24, produce guide lines for lowercase letters as described here and are highly recommended.

4.22 Vertical Lowercase Letters

Fig. 4.32   Vertical lowercase letters are used largely on map drawings and very seldom on machine drawings. The shapes are based on a repetition of the circle or circular arc and the straight

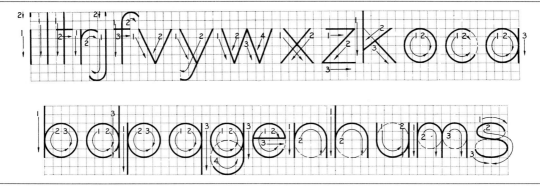

Fig. 4.32   Vertical Lowercase Letters.

Fig. 4.33   Inclined Lowercase Letters.

line, with some variations. The lower part of the letter is usually two-thirds the height of the capital letter.

4.23   Inclined Lowercase Letters
Fig. 4.33   The order and direction of the strokes and the proportions of inclined lowercase letters are the same as those of vertical lowercase letters. The slope of the letters is the same as for inclined capitals, or 68.2° with horizontal. The slope may be determined by drawing a "slope triangle" of 2 in 5, as shown in Fig. 4.30 (b), or with the aid of the inclined slot in the Braddock-Rowe Triangle, Fig. 4.23, or with the Ames Lettering Guide, Fig. 4.24 (b).

4.24   Spacing of Letters and Words
Fig. 4.34   Uniformity in spacing of letters is a matter of equalizing spaces by eye. *The background areas between letters, not the distances between them, should be approximately equal.* In (a) the actual distances are equal, but the letters do not appear equally spaced. At (b) the distances are intentionally unequal, but the background areas between letters are approximately equal, and the result is an even and pleasing spacing.

Some combinations, such as LT and VA, may even have to be slightly overlapped to secure good spacing. In some cases the width of a letter may be decreased. For example, the lower stroke of the L may be shortened when followed by A.

EQUAL DISTANCES – INCORRECT

EQUAL BACKGROUND AREAS – CORRECT

Fig. 4.34   Spacing Between Letters.

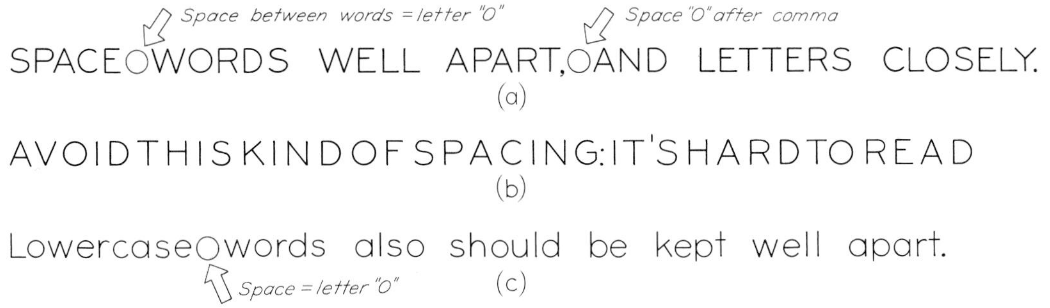

SPACE○WORDS　WELL　APART,○AND　LETTERS　CLOSELY.

Space between words = letter "O"　　*Space "O" after comma*

(a)

AVOIDTHISKINDOFSPACING:IT'SHARDTOREAD

(b)

Lowercase○words　also　should　be　kept　well　apart.

Space = letter "O"　　(c)

Fig. 4.35　Spacing Words.

***Fig. 4.35** Space words well apart, but space letters closely within words.* Make each word a compact unit well separated from adjacent words. For either uppercase or lowercase lettering, make the spaces between words approximately equal to a capital O. Avoid spacing letters too far apart and words too close together, as shown at (b). Samples of good spacing are also shown in Fig. 4.17.

When it is necessary to letter to a stop line as in Fig. 4.36 (a), space each letter from *right to left,* as shown in step II, estimating the widths of the letters by eye. Then letter from *left to right,* as shown at III, and finally erase the spacing marks.

When it is necessary to space letters symmetrically about a center line, Fig. 4.36 (b), which is frequently the case in titles, Figs. 4.43–4.45, number the letters as shown, with the space between words considered as one letter. Then place the middle letter on center, making allowance for narrow letters (Is) or wide letters (Ws) on either side. The X in Fig. 4.36 (b) is placed slightly to the left of center to compensate for the letter I, which has no width. Check with the dividers to make sure that distances *a* are exactly equal.

Another method is to letter roughly a trial line of lettering along the bottom edge of a scrap of paper, place it in position immediately above, as shown at (c), and then letter the line in place. Be sure to use guide lines for the trial lettering.

4.25　Lettering Devices

The Leroy Standard Lettering Instrument, Fig. 4.37, is perhaps the most widely used lettering device. A guide pin follows grooved letters in a

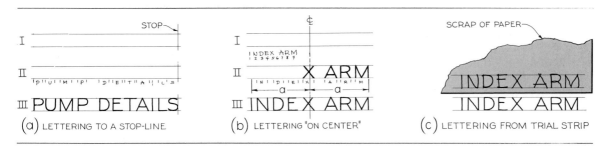

Fig. 4.36　Spacing to a Stop Line and "on Center."

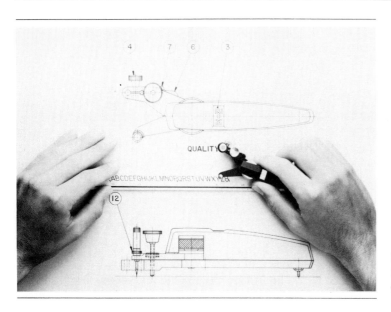

Fig. 4.37   Leroy Standard Lettering Instrument. *Courtesy of Keuffel & Esser Co.*

template, and the inking point moves on the paper. By adjusting the arm on the scriber, the letters may be made vertical or inclined. A number of templates and pen sizes for letters and symbols are available, including templates for a wide variety of "builtup" letters similar to those made by the Varigraph and Letterguide, described shortly. Inside each pen is a cleaning pin used to keep the small tube open. These pins are easily broken, especially the small ones, when the pen is not promptly cleaned. To clean a pen, draw it across a blotter until all ink has been absorbed; then insert the pin and remove it and wipe it with a cloth. Repeat this until the pin remains clean. If the ink has dried, the pens may be cleaned with Leroy pen-cleaning fluid, available at dealers. Leroy lettering sets, Fig. 4.38, are available as standard or metric sets. Both have the same style and features except for pen size designations. Leroy pens are also available in various sizes in standard or reservoir types.

The Wrico Lettering Guide, Fig. 4.39, consists of a scriber and templates similar to the Leroy system. Wrico letters more closely resemble American National Standard letters than do those of other sets.

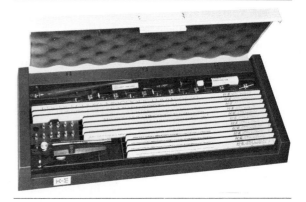

Fig. 4.38   Leroy Standard Lettering Set. *Courtesy of Keuffel & Esser Co.*

Fig. 4.39   Wrico Lettering Guide. *Courtesy of Wood-Regan Instrument Co., Inc.*

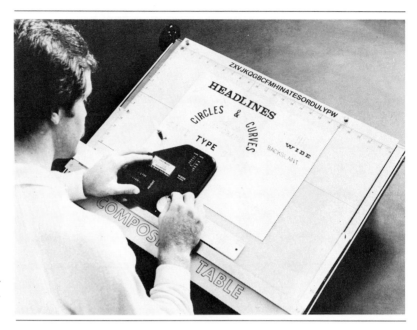

Fig. 4.40 The Varigraph Machine and Table. *Courtesy of Varigraph, Inc.*

The Varigraph is a more elaborate device for making a wide variety of either single-stroke letters or "builtup" letters. As shown in Fig. 4.40, a guide pin is moved along the grooves in a template, and the pen forms the letters.

The Letterguide scriber, Fig. 4.41, is a much simpler instrument, which also makes a large variety of styles and sizes of letters when used with the various templates available. It also operates with a guide pin moving in the grooved letters of the template, while the pen, which is mounted on an adjustable arm, makes the letters in outline.

The Kroy Lettering Machine, Fig. 4.42, is a unique lettering machine that creates type on tape which can then be applied on drawings, artwork, posters, and the like. Lettering is produced by dialing the type disk to the desired character and pressing the print button. Spacing is automatic and adjustable. The machine is available in electric or manual models that use either 61- or 80-character type disks. Several type styles and sizes are available.

Various forms of press-on lettering and special lettering devices (typewriters, etc.) are available. In addition, the various computer-aided drafting systems have the capability to produce letters of different heights and styles and to make changes as required. In whatever way the lettering is applied to the drawing and whatever style of lettering is used, the lettering must meet the requirements for legibility and microfilm reproduction.

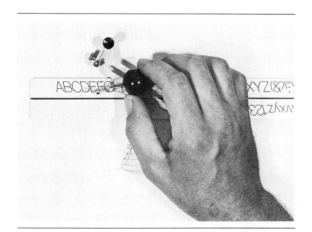

Fig. 4.41 Letterguide. *Courtesy of Letterguide Co.*

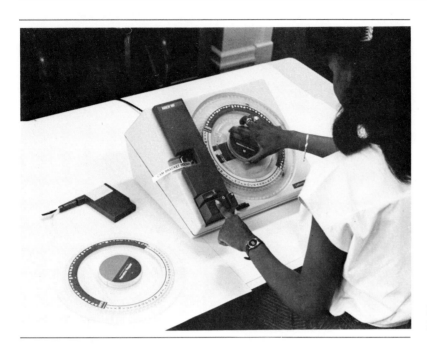

Fig. 4.42 Kroy Lettering Machine.

4.26 Titles

The composition of titles on machine drawings is relatively simple. In most cases, the title and related information are lettered in "title boxes" or "title strips," which are printed directly on the drawing paper, tracing paper, tracing cloth or polyester film. See, for example, Figs. 16.24–16.26. The main drawing title is usually centered in a rectangular space. This may be done by the method shown in Fig. 4.36 (b); or, if the lettering is being done on tracing paper or cloth, the title may be lettered first on scrap paper and then placed underneath the tracing, as shown in Fig. 4.43, and then lettered directly over.

If a title box is not used, the title of a machine drawing may be lettered in the lower right corner of the sheet as a "balanced title," Fig. 4.44. A balanced title is simply one that is arranged symmetrically about an imaginary center line. These titles take such forms as the rectangle, the oval, the inverted pyramid, or any other simple symmetrical form.

On display drawings, or on highly finished maps or architectural drawings, titles may be composed of filled-in letters, usually Gothic or Roman, Fig. 4.45.

Equal
DRAWING TITLE
Scrap of paper underneath

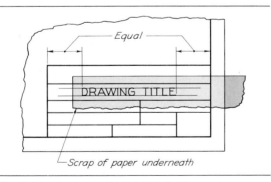

Fig. 4.43 Centering Title in Title Box.

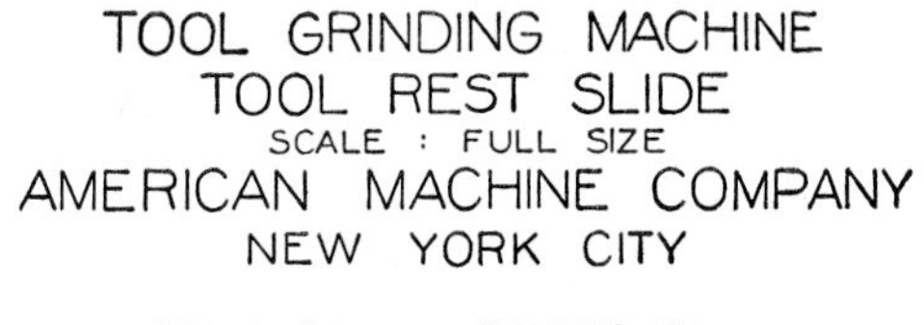

TOOL GRINDING MACHINE
TOOL REST SLIDE
SCALE : FULL SIZE
AMERICAN MACHINE COMPANY
NEW YORK CITY
DRAWN BY ____ CHECKED BY ____

Fig. 4.44 Balanced Machine-Drawing Title.

In any kind of title, the most important words are given most prominence by making the lettering larger, heavier, or both. Other data, such as scale and date, may be displayed smaller.

4.27 Lettering on Maps

Modern Roman letters are generally used on maps, as follows.

1. *Vertical capitals.* Names of states, countries, townships, capital cities, large cities, and titles of maps.

2. *Vertical lowercase* (first letter of each word a capital). Names of small towns, villages, post offices, etc.

3. *Inclined capitals.* Names of oceans, bays, gulfs, sounds, large lakes, and rivers.

4. *Inclined lowercase* or "stump" letters (first letter of each word a capital). Names of rivers, creeks, small lakes, ponds, marshes, brooks, and springs.

Gothic letters are used as follows.

1. *Vertical capitals.* Names of prominent land features such as mountains, plateaus, and canyons.

MAP OF
BRAZOS COUNTY
TEXAS

SCALE : 1 = 20,000

0   1   2   3   4000 FEET

Fig. 4.45 Balanced Map Title.

2. *Vertical lowercase.* Names of small land features such as small valleys, islands, and ridges.

3. *Inclined capitals.* Names of railroads, tunnels, highways, bridges, and other public structures.

4.28 Computer Graphics

Lettering is a standard feature available in computer graphics programs. Using CAD software, the drafter or designer can add titles, notes, and dimensioning information to a computer-generated drawing. Several fonts, Figs. 4.46 and 4.47,

SHAHEEN TALIB
ILLINOIS INSTITUTE OF TECHNOLOGY
ENGINEERING GRAPHICS 419 - 01
Date: 9/15/88   Scale:

Shaheen Talib
Illinois Institute of Technology
Engineering Graphics 419 - 01
Date: 9/15/88   Scale:

Shaheen Talib
Illinois Institute of Technology
Engineering Graphics 419 - 01
Date: 9/15/88   Scale:

SHAHEEN TALIB
ILLINOIS INSTITUTE OF TECHNOLOGY
ENGINEERING GRAPHICS 419 - 01
Date: 9/15/88   Scale:

Fig. 4.46 CAD Lettering Examples. *Courtesy of Department of Engineering Graphics, Illinois Institute of Technology.*

115

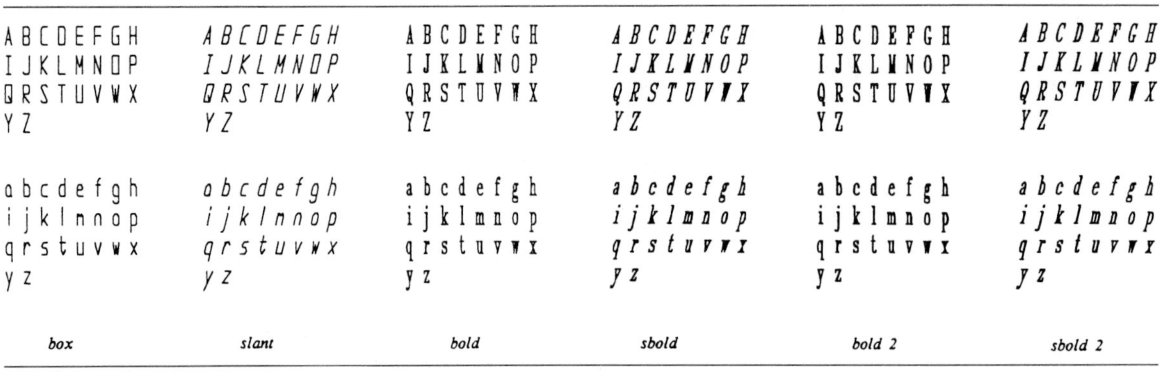

ABCDEFGH IJKLMNOP QRSTUVWX YZ	ABCDEFGH IJKLMNOP QRSTUVWX YZ	ABCDEFGH IJKLMNOP QRSTUVWX YZ	ABCDEFGH IJKLMNOP QRSTUVWX YZ	ABCDEFGH IJKLMNOP QRSTUVWX YZ	ABCDEFGH IJKLMNOP QRSTUVWX YZ
abcdefgh ijklnnop qrstuvwx yz	abcdefgh ijklnnop qrstuvwx yz	abcdefgh ijklmnop qrstuvwx yz	abcdefgh ijklmnop qrstuvwx yz	abcdefgh ijklmnop qrstuvwx yz	abcdefgh ijklmnop qrstuvwx yz
box	*slant*	*bold*	*sbold*	*bold 2*	*sbold 2*

Fig. 4.47   Typical CAD Type Fonts. *Courtesy of CADKEY.*

and a variety of sizes may be selected. When modifications are required, it is easy for the CAD operator to make appropriate lettering changes on the drawing.

LETTERING EXERCISES

Layouts for lettering practice are given in Figs. 4.48–4.51. Draw complete horizontal and vertical or inclined guide lines *very lightly*. Draw the vertical or inclined guide lines through the full height of the lettered area of the sheet. For practice in ink lettering, the last two lines and the title strip on each sheet may be lettered in ink, if assigned by the instructor. Omit all dimensions.

Lettering sheets in convenient form for lettering practice may be found in *Technical Drawing Problems,* Series 1, by Giesecke, Mitchell, Spencer, Hill, Dygdon, and Novak; *Technical Drawing Problems,* Series 2, by Spencer, Hill, Dygdon, and Novak; and Technical Drawing Problems, Series 3, by Spencer, Hill, Dygdon, and Novak; all designed to accompany this text and published by Macmillan Publishing Company.

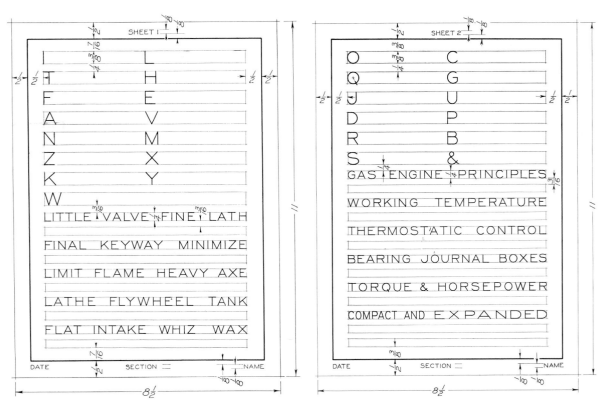

Fig. 4.48 Lay out sheet, add vertical or inclined guide lines, and fill in vertical or inclined capital letters as assigned. For decimal-inch and millimeter equivalents of given dimensions, see table inside of front cover.

Fig. 4.49 Lay out sheet, add vertical or inclined guide lines, and fill in vertical or inclined capital letters as assigned. For decimal-inch and millimeter equivalents of given dimensions, see table inside of front cover.

117

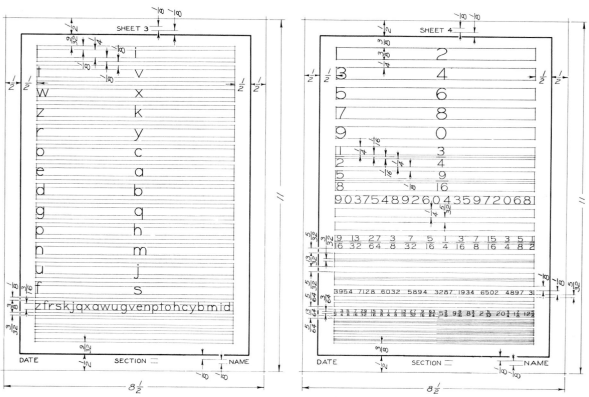

Fig. 4.50 Lay out sheet, add vertical or inclined guide lines, and fill in vertical or inclined lower-case letters as assigned. For decimal-inch and millimeter equivalents of given dimensions, see table inside of front cover.

Fig. 4.51 Lay out sheet, add vertical or inclined guide lines, and fill in vertical or inclined numerals as assigned. For decimal-inch and millimeter equivalents of given dimensions, see table inside of front cover.

CHAPTER 5

Geometric Constructions

Many of the constructions used in technical design drawings are based on plane geometry, and every drafter, technician, or engineer should be sufficiently familiar with them to be able to apply them to the solutions of problems. Pure geometry problems may be solved only with the compass and a straightedge, and in some cases these methods may be used to advantage in technical drawing. However, the drafter or designer has available the T-square,* triangles, dividers, and other equipment, such as drafting machines, that in many cases can yield accurate results more quickly by what we may term "preferred methods." Therefore, many of the solutions in this chapter are practical adaptations of the principles of pure geometry.

This chapter is designed to present definitions of terms and geometric constructions of importance in technical drawing, suggest simplified methods of construction, point out practical applications, and afford opportunity for practice in accurate instrumental drawing. The problems at the end of this chapter may be regarded as a continuation of those at the end of Chapter 2.

In drawing these constructions, accuracy is most important. Use a sharp medium-hard lead (H to 3H) in your pencil and compasses. Draw construction lines extremely light—so light that they can hardly be seen when your drawing is held at arm's length. Draw all final and required lines medium to thin but dark.

5.1 Points and Lines

Fig. 5.1 A *point* represents a location in space or on a drawing, and has no width, height, or depth. A point is represented by the intersection of two lines, (a), by a short crossbar on a line, (b), or by a small cross, (c). Never represent a point by a simple dot on the paper.

A line is defined by Euclid as "that which has length without breadth." A *straight line* is the shortest distance between two points and is commonly referred to simply as a "line." If the line is indefinite in extent, the length is a matter of convenience, and the endpoints are not fixed, (d). If the endpoints of the line are significant, they must be marked by means of small mechanically drawn crossbars, (e). Other common terms are illustrated from (f) to (h). Either straight lines or curved lines are parallel if the

---

*Hereafter, reference to the T-square could also refer to the parallel straightedge or drafting machine.

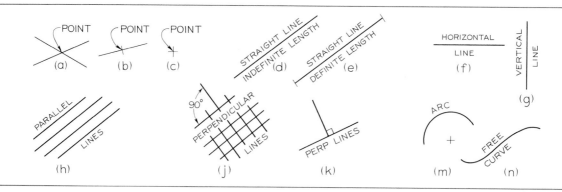

Fig. 5.1   Points and Lines.

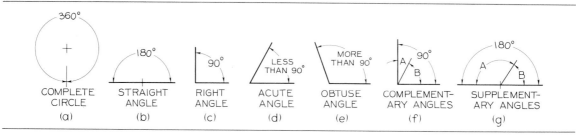

Fig. 5.2   Angles.

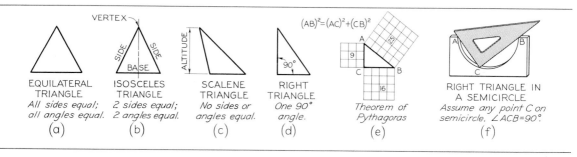

Fig. 5.3   Triangles.

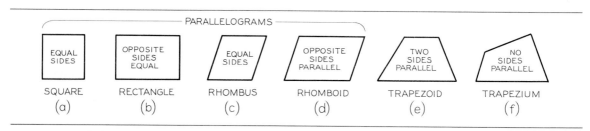

Fig. 5.4   Quadrilaterals.

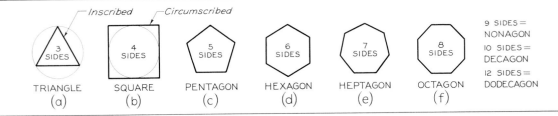

Fig. 5.5   Regular Polygons.

shortest distance between them remains constant. The common symbol for parallel lines is ∥, and for perpendicular lines it is ⊥ (singular) or ⊥s (plural). Two perpendicular lines may be marked with a "box" to indicate perpendicularity, as shown at (k). Such symbols may be used on sketches, but not in production drawings.

5.2   Angles
Fig. 5.2   An angle is formed by two intersecting lines. A common symbol for angle is ∠ (singular) or ∠s (plural). There are 360 degrees (360°) in a full circle, as shown at (a). A degree is divided into 60 minutes (60′), and a minute is divided into 60 seconds (60″). Thus, 37° 26′ 10″ is read: 37 degrees, 26 minutes, and 10 seconds. When minutes alone are indicated, the number of minutes should be preceded by 0°, as 0° 20′.

The different kinds of angles are illustrated in (b) to (e). Two angles are *complementary,* (f), if they total 90°, and are *supplementary,* (g), if they total 180°. Most angles used in technical drawing can be drawn easily with the T-square or straightedge and triangles, Fig. 2.23. To draw odd angles, use the protractor, Fig. 2.24. For considerable accuracy, use a *vernier protractor,* or the tangent, sine, or chord methods, §5.20.

5.3   Triangles
Fig. 5.3   A triangle is a plane figure bounded by three straight sides, and the sum of the interior angles is always 180°. A right triangle, (d), has one 90° angle, and the square of the hypotenuse is equal to the sum of the squares of the two sides, (e). As shown at (f), any triangle inscribed in a semicircle is a right triangle if the hypotenuse coincides with the diameter.

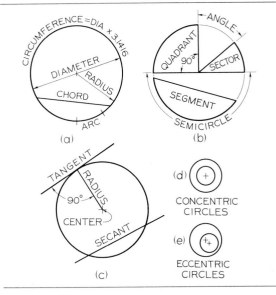

Fig. 5.6   The Circle.

5.4   Quadrilaterals
Fig. 5.4   A quadrilateral is a plane figure bounded by four straight sides. If the opposite sides are parallel, the quadrilateral is also a parallelogram.

5.5   Polygons
Fig. 5.5   A polygon is any plane figure bounded by straight lines. If the polygon has equal angles and equal sides, it can be inscribed in or circumscribed around a circle and is called a *regular polygon.*

5.6   Circles and Arcs
Fig. 5.6   A circle, (a), is a closed curve all points of which are the same distance from a point called the center. *Circumference* refers to the

121

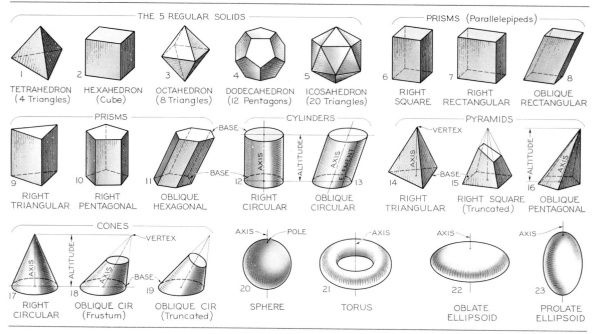

Fig. 5.7   Solids.

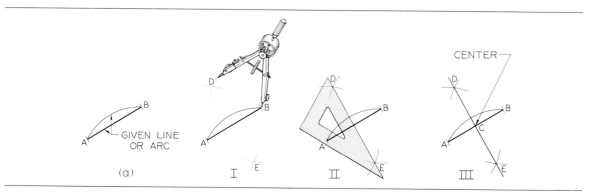

Fig. 5.8   Bisecting a Line or a Circular Arc (§5.8).

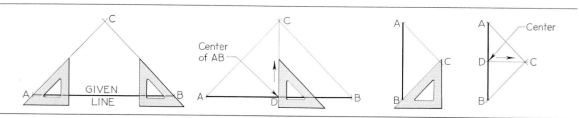

Fig. 5.9   Bisecting a Line with Triangle and T-square (§5.9).

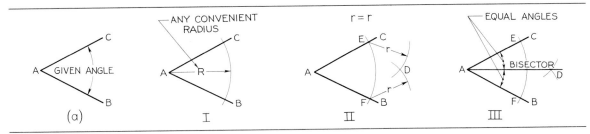

Fig. 5.10   Bisecting an Angle (§5.10).

circle or to the distance around the circle. This distance equals the diameter multiplied by π (called pi) or 3.1416. Other definitions are illustrated in the figure.

5.7   Solids

Fig. 5.7   Solids bounded by plane surfaces are *polyhedra*. The surfaces are called faces, and if these are equal regular polygons, the solids are regular polyhedra.

A *prism* has two bases, which are parallel equal polygons, and three or more lateral faces, which are parallelograms. A triangular prism has a triangular base; a rectangular prism has rectangular bases; and so on. If the bases are parallelograms, the prism is a parallelepiped. A right prism has faces and lateral edges perpendicular to the bases; an oblique prism has faces and lateral edges oblique to the bases. If one end is cut off to form an end not parallel to the bases, the prism is said to be truncated.

A *pyramid* has a polygon for a base and triangular lateral faces intersecting at a common point called the vertex. The center line from the center of the base to the vertex is the axis. If the axis is perpendicular to the base, the pyramid is a right pyramid; otherwise it is an oblique pyramid. A triangular pyramid has a triangular base; a square pyramid has a square base; and so on. If a portion near the vertex has been cut off, the pyramid is truncated, or referred to as a frustum.

A *cylinder* is generated by a straight line, called the generatrix, moving in contact with a curved line and always remaining parallel to its previous position or to the axis. Each position

of the generatrix is called an element of the cylinder.

A *cone* is generated by a straight line moving in contact with a curved line and passing through a fixed point, the vertex of the cone. Each position of the generatrix is an element of the cone.

A *sphere* is generated by a circle revolving about one of its diameters. This diameter becomes the axis of the sphere, and the ends of the axis are poles of the sphere.

A *torus* is generated by a circle (or other curve) revolving about an axis that is eccentric to the curve.

5.8   To Bisect a Line or a Circular Arc

Fig. 5.8   Given line or arc AB, as shown at (a), to be bisected.

I.   From A and B draw equal arcs with radius greater than half AB.

II.   and   III.   Join intersections D and E with a straight line to locate center C.

5.9   To Bisect a Line with Triangle and T-square

Fig. 5.9   From endpoints A and B, draw construction lines at 30°, 45°, or 60° with the given line; then through their intersection, C, draw a line perpendicular to the given line to locate the center D, as shown.

To divide a line with the dividers, see §2.40.

5.10   To Bisect an Angle

Fig. 5.10   Given angle BAC, as shown at (a), to be bisected.

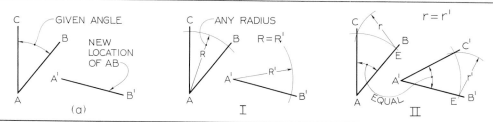

Fig. 5.11   Transferring an Angle (§5.11).

I.   Strike large arc R.

II.   Strike equal arcs r with radius slightly larger than half BC, to intersect at D.

III.   Draw line AD, which bisects angle.

5.11   To Transfer an Angle

Fig. 5.11   Given angle BAC, as shown at (a), to be transferred to the new position at A′B′.

I.   Use any convenient radius R, and strike arcs from centers A and A′.

II.   Strike equal arcs r, and draw side A′C′.

5.12   To Draw a Line Through a Point and Parallel to a Line

Fig. 5.12 (a)   With given point P as center, and any convenient radius R, strike arc CD to intersect the given line AB at E. With E as center and the same radius, strike arc R′ to intersect the given line at G. With PG as radius and E as center, strike arc r to locate point H. The line PH is the required line.

Fig. 5.12 (b)   PREFERRED METHOD   Move the triangle and T-square as a unit until the triangle lines up with given line AB; then slide the triangle until its edge passes through the given point P. Draw CD, the required parallel line. See also §2.21.

5.13   To Draw a Line Parallel to a Line and at a Given Distance

Let AB be the line and CD the given distance.

Fig. 5.13 (a)   With points E and F near A and B, respectively, as centers, and CD as radius,

draw two arcs. The line GH, tangent to the arcs, is the required line.

Fig. 5.13 (b)   PREFERRED METHOD   With any point E of the line as center and CD as radius, strike an arc JK. Move the triangle and T-square as a unit until the triangle lines up with the given line AB; then slide the triangle until its edge is tangent to the arc JK, and draw the required line GH.

Fig. 5.13 (c)   With centers selected at random on the curved line AB, and with CD as radius, draw a series of arcs; then draw the required line tangent to these arcs as explained in §2.54.

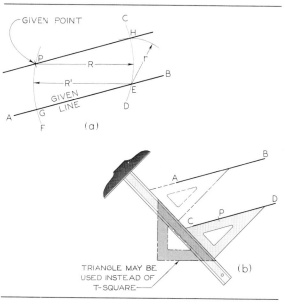

Fig. 5.12   Drawing a Line Through a Point Parallel to a Line (§5.12).

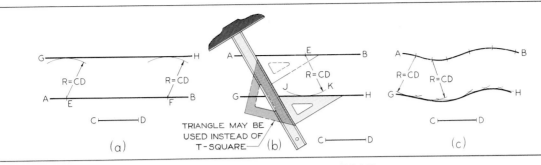

Fig. 5.13   Drawing a Line Parallel to a Line at a Given Distance (§5.13).

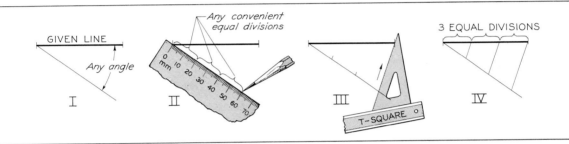

Fig. 5.14   Dividing a Line into Equal Parts (§5.14).

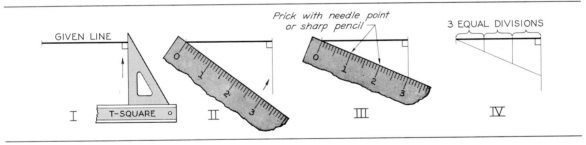

Fig. 5.15   Dividing a Line into Equal Parts (§5.15).

5.14   To Divide a Line into Equal Parts
Fig. 5.14

I.   Draw a light construction line at any convenient angle from one end of line.

II.   With dividers or scale, set off from intersection of lines as many equal divisions as needed, in this case, three.

III.   Connect last division point to other end of line, using triangle and T-square, as shown.

IV.   Slide triangle along T-square and draw parallel lines through other division points, as shown.

5.15   To Divide a Line into Equal Parts
Fig. 5.15

I.   Draw vertical construction line at one end of given line.

II.   Set zero of scale at other end of line.

125

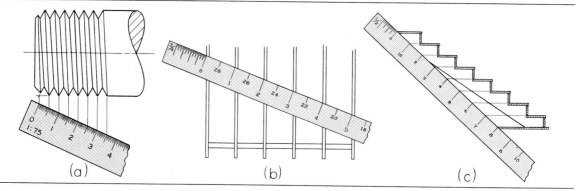

Fig. 5.16   Practical Applications of Dividing a Line into Equal Parts (§5.15).

III.   Swing scale up until third unit falls on vertical line, and make tiny dots at each point, or prick points with dividers.

IV.   Draw vertical construction lines through each point.

Some practical applications of this method are shown in Fig. 5.16.

5.16   To Divide a Line into Proportional Parts

Fig. 5.17   Let it be required to divide the line AB into three parts proportional to 2, 3, and 4.

Fig. 5.17 (a)   PREFERRED METHOD   Draw a vertical line from point B. Select a scale of convenient size for a total of nine units and set the zero of the scale at A. Swing the scale up until the ninth unit falls on the vertical line. Along the scale, set off points for 2, 3, and 4 units, as shown. Draw vertical lines through these points.

Fig. 5.17 (b)   Draw a line CD parallel to AB and at any convenient distance. On this line, set off 2, 3, and 4 units, as shown. Draw lines through the ends of the two lines to intersect at the point O. Draw lines through O and the points 2 and 5 to divide AB into the required proportional parts.

Constructions of this type are useful in the preparation of graphs (Chapter 28).

Fig. 5.17 (c)   Given AB, to divide into proportional parts, in this case proportional to the square of x, where $x = 1, 2, 3, \ldots$. Set zero of scale at end of line and set off divisions 4, 9, 16, $\ldots$. Join the last division to the other end of the line, and draw parallel lines as shown. This method may be used for any power of x.

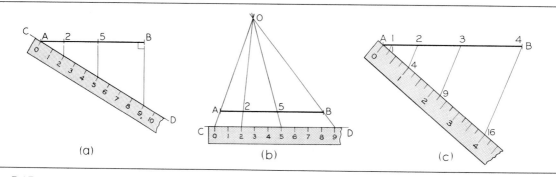

Fig. 5.17   Dividing a Line into Proportional Parts (§5.16).

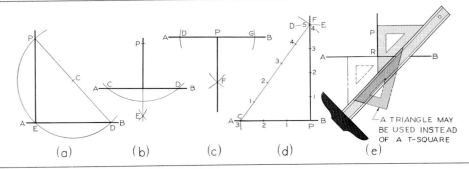

Fig. 5.18   Drawing a Line Through a Point and Perpendicular to a Line (§5.17).

5.17   To Draw a Line Through a Point and Perpendicular to a Line
Fig. 5.18   Given the line AB and a point P.

WHEN THE POINT IS NOT ON THE LINE   *Fig. 5.18 (a)*   From P draw any convenient inclined line, as PD. Find center C of line PD, and draw arc with radius CP. The line EP is the required perpendicular.

Fig. 5.18 (b)   With P as center, strike an arc to intersect AB at C and D. With C and D as centers, and radius slightly greater than half CD, strike arcs to intersect at E. The line PE is the required perpendicular.

WHEN THE POINT IS ON THE LINE   *Fig. 5.18 (c)*   With P as center and any radius, strike arcs to intersect AB at D and G. With D and G as centers, and radius slightly greater than half DG, strike equal arcs to intersect at F. The line PF is the required perpendicular.

Fig. 5.18 (d)   Select any convenient unit of length, for example, 6 mm or $\frac{1}{4}''$. With P as center, and 3 units as radius, strike an arc to intersect given line at C. With P as center, and 4 units as radius, strike arc DE. With C as center, and 5 units as radius, strike an arc to intersect DE at F. The line PF is the required perpendicular.

This method makes use of the 3–4–5 right triangle and is frequently used in laying off rectangular foundations of large machines, buildings, or other structures. For this purpose a steel tape may be used and distances of 30, 40, and 50 feet measured as the three sides of the right triangle.

Fig. 5.18 (e)   PREFERRED METHOD   Move the triangle and T-square as a unit until the triangle lines up with AB; then slide the triangle until its edge passes through the point P (whether P is on or off the line), and draw the required perpendicular.

5.18   To Draw a Triangle with Sides Given
Fig. 5.19   Given the sides A, B, and C, as shown at (a).

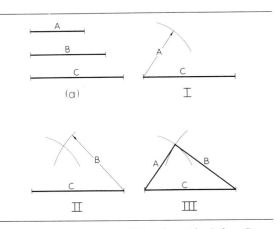

Fig. 5.19   Drawing a Triangle with Sides Given (§5.18).

127

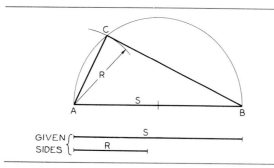

Fig. 5.20   Drawing a Right Triangle (§5.19).

I.   Draw one side, as C, in desired position, and strike arc with radius equal to side A.

II.   Strike arc with radius equal to side B.

III.   Draw sides A and B from intersection of arcs, as shown.

5.19   To Draw a Right Triangle with Hypotenuse and One Side Given

Fig. 5.20   Given sides S and R. With AB as a diameter equal to S, draw semicircle. With A as center, and R as radius, draw an arc intersecting the semicircle at C. Draw AC and CB to complete the right triangle.

5.20   To Lay Out an Angle

Fig. 5.21   Many angles can be laid out directly with the triangle, Fig. 2.23, or they may be laid out with the protractor, Fig. 2.24. Other methods, where considerable accuracy is required, are as follows.

TANGENT METHOD   *Fig. 5.21 (a)*   The tangent of angle θ is $\dfrac{y}{x}$, and $y = x \tan \theta$. To construct the angle, assume a convenient value for x, preferably 10 units of convenient length, as shown. (The larger the unit, the more accurate will be the construction.) Find the tangent of angle θ in a table of natural tangents, multiply by 10, and set off $y = 10 \tan \theta$.

EXAMPLE   To set off $31\frac{1}{2}°$, find the natural tangent of $31\frac{1}{2}°$, which is 0.6128. Then

$$y = 10 \text{ units} \times 0.6128 = 6.128 \text{ units}$$

SINE METHOD   *Fig. 5.21 (b)*   Draw line x to any convenient length, preferably 10 units as shown. Find the sine of angle θ in a table of natural sines, multiply by 10, and strike arc $R = 10 \sin \theta$. Draw the other side of the angle tangent to the arc, as shown.

EXAMPLE   To set off $25\frac{1}{2}°$, find the natural sine of $25\frac{1}{2}°$, which is 0.4305. Then

$$R = 10 \text{ units} \times 0.4305 = 4.305 \text{ units}$$

CHORD METHOD   *Fig. 5.21 (c)*   Draw line x to any convenient length, draw arc with any convenient radius R, say, 10 units. Find the chordal length C in a table of chords (see a machinists' handbook), and multiply the value by 10, since the table is made for a radius of 1 unit.

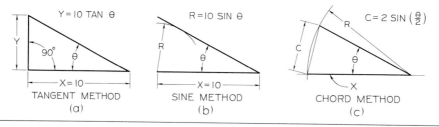

Fig. 5.21   Laying Out Angles (§5.20).

EXAMPLE   To set off 43° 20′, the chordal length C for 1 unit radius, as given in a table of chords = 0.7384, and if R = 10 units, then C = 7.384 units.

If a table is not available, the chord C may be calculated by the formula $C = 2 \sin \frac{\theta}{2}$.

EXAMPLE   Half of 43° 20′ = 21° 40′. The sine of 21° 40′ = 0.3692. C = 2 × 0.3692 = 0.7384 for a 1 unit radius. For a 10 unit radius, C = 7.384 units.

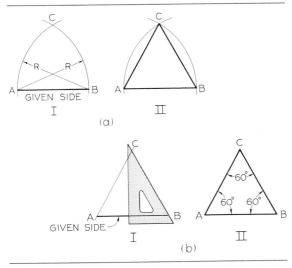

Fig. 5.22   Drawing an Equilateral Triangle (§5.21).

5.21   To Draw an Equilateral Triangle
Given side AB.

Fig. 5.22 (a)   With A and B as centers and AB as radius, strike arcs to intersect at C. Draw lines AC and BC to complete the triangle.

Fig. 5.22 (b)   PREFERRED METHOD   Draw lines through points A and B making angles of 60° with the given line and intersecting at C, as shown.

5.22   To Draw a Square
Fig. 5.23 (a)   Given one side AB. Through point A, draw a perpendicular, Fig. 5.18 (c). With A as center, and AB as radius, draw the arc to intersect the perpendicular at C. With B and C as centers, and AB as radius, strike arcs to intersect at D. Draw lines CD and BD.

Fig. 5.23 (b)   PREFERRED METHOD   Given one side AB. Using the T-square or parallel straightedge and 45° triangle, draw lines AC and BD perpendicular to AB and the lines AD and BC at 45° with AB. Draw line CD.

Fig. 5.23 (c)   PREFERRED METHOD   Given the circumscribed circle (distance "across corners"), draw two diameters at right angles to each other. The intersections of these diameters with the circle are vertexes of an inscribed square.

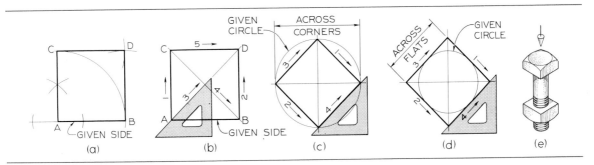

Fig. 5.23   Drawing a Square (§5.22).

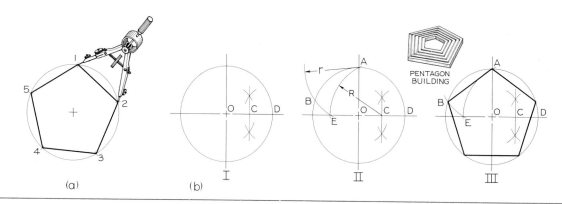

Fig. 5.24   Drawing a Pentagon (§5.23).

Fig. 5.23 (d) PREFERRED METHOD   Given the inscribed circle (distance "across flats," as in drawing bolt heads), use the T-square (or parallel straightedge) and 45° triangle and draw the four sides tangent to the circle.

5.23   To Draw a Regular Pentagon
Given the circumscribed circle.

Fig. 5.24 (a) PREFERRED METHOD   Divide the circumference of the circle into five equal parts with the dividers, and join the points with straight lines.

GEOMETRICAL METHOD   *Fig. 5.24 (b)*
 I.   Bisect radius OD at C.

 II.   With C as center, and CA as radius, strike arc AE. With A as center, and AE as radius, strike arc EB.

 III.   Draw line AB; then set off distances AB around the circumference of the circle, and draw the sides through these points.

5.24   To Draw a Hexagon
Given the circumscribed circle.

Fig. 5.25 (a)   Each side of a hexagon is equal to the radius of the circumscribed circle. There-fore, using the compass or dividers and the radius of the circle, set off the six sides of the hexagon around the circle, and connect the points with straight lines. As a check on the accuracy of the construction, make sure that opposite sides of the hexagon are parallel.

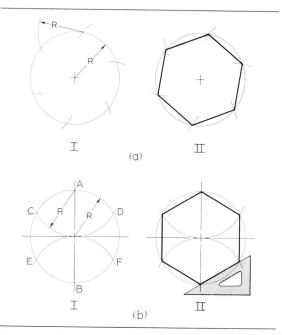

Fig. 5.25   Drawing a Hexagon (§5.24).

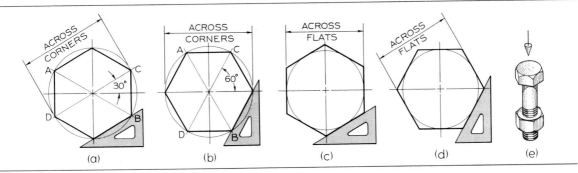

Fig. 5.26   Drawing a Hexagon (§5.25).

Fig. 5.25 (b) PREFERRED METHOD This construction is a variation of the one shown at (a). Draw vertical and horizontal center lines. With A and B as centers and radius equal to that of the circle, draw arcs to intersect the circle at C, D, E, and F, and complete the hexagon as shown.

5.25   To Draw a Hexagon

Given the circumscribed or inscribed circle. *Both Acceptable Methods.*

Fig. 5.26 (a) and (b) Given the circumscribed circle (distance "across corners"). Draw vertical and horizontal center lines, and then diagonals AB and CD at 30° or 60° with horizontal; then with the 30° × 60° triangle and the T-square, draw the six sides as shown.

Fig. 5.26 (c) and (d) Given the inscribed circle (distance "across flats"). Draw vertical and horizontal center lines; then with the 30° × 60° triangle and the T-square or straightedge draw the six sides tangent to the circle. This method is used in drawing bolt heads and nuts. For maximum accuracy, diagonals may be added as at (a) and (b).

5.26   To Draw a Hexagon

Fig. 5.27 Using the 30° × 60° triangle and the T-square or straightedge, draw lines in the order shown at (a) where the distance AB ("across corners") is given, or as shown at (b) where a side CD is given.

5.27   To Draw an Octagon

Fig. 5.28 (a) PREFERRED METHOD Given inscribed circle, or distance "across flats." Using the T-square or straightedge and 45° triangle, draw the eight sides tangent to the circle, as shown.

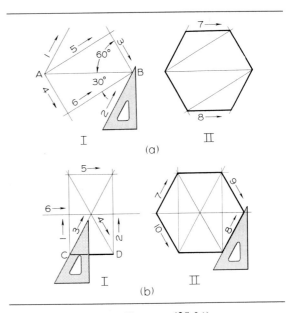

Fig. 5.27   Drawing a Hexagon (§5.26).

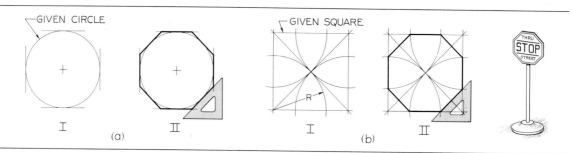

Fig. 5.28   Drawing an Octagon (§5.27).

Fig. 5.28 (b)   Given circumscribed square, or distance "across flats." Draw diagonals of square; then with the corners of the given square as centers, and with half the diagonal as radius, draw arcs cutting the sides as shown at I. Using the T-square and 45° triangle, draw the eight sides as shown at II.

5.28   To Transfer Plane Figures by Geometric Methods

TO TRANSFER A TRIANGLE TO A NEW LOCATION **Fig. 5.29 (a) and (b)**   Set off any side, as AB, in the new location, (b). With the ends of the line as centers and the lengths of the other sides of the given triangle, (a), as radii, strike two arcs to intersect at C. Join C to A and B to complete the triangle.

TO TRANSFER A POLYGON BY THE TRIANGLE METHOD   **Fig. 5.29 (c)**   Divide the polygon into triangles as shown, and transfer each triangle as explained previously.

TO TRANSFER A POLYGON BY THE RECTANGLE METHOD   **Fig. 5.29 (d)**   Circumscribe a rectangle about the given polygon. Draw a congruent rectangle in the new location and locate the vertexes of the polygon by transferring location measurements a, b, c, and so on, along the sides of the rectangle to the new rectangle. Join the points thus found to complete the figure.

TO TRANSFER IRREGULAR FIGURES   **Fig. 5.29 (e)**   Figures composed of rectangular and circular forms are readily transferred by enclosing the elementary features in rectangles and determin-

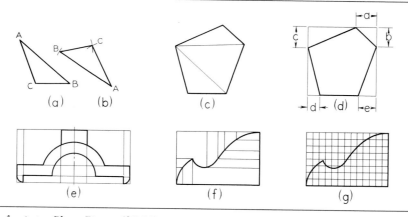

Fig. 5.29   Transferring a Plane Figure (§5.28).

ing centers of arcs and circles. These may then be transferred to the new location.

TO TRANSFER FIGURES BY OFFSET MEASUREMENTS *Fig. 5.29 (f)* *Offset location measurements* are frequently useful in transferring figures composed of free curves. When the figure has been enclosed by a rectangle, the sides of the rectangle are used as reference lines for the location of points along the curve.

TO TRANSFER FIGURES BY A SYSTEM OF SQUARES *Fig. 5.29 (g)* Figures involving free curves are easily copied, enlarged, or reduced by the use of a system of squares. For example, to enlarge a figure to double size, draw the containing rectangle and all small squares double their original size. Then draw the lines through the corresponding points in the new set of squares. See also Fig. 6.18.

5.29 To Transfer Drawings by Tracing-Paper Methods

To transfer a drawing to an opaque sheet, the following procedures may be used.

PRICKED-POINT METHOD Lay tracing paper over the drawing to be transferred. With a sharp pencil, make a small dot directly over each important point on the drawing. Encircle each dot so as not to lose it. Remove the tracing paper, place it over the paper to receive the transferred drawing, and maneuver the tracing paper into the desired position. With a needle point (such as a point of the dividers), prick through each dot. Remove the tracing paper and connect the pricked points to reproduce the lines of the original drawing.

To reproduce arcs or circles, it is only necessary to transfer the center and one point on the circumference. To duplicate a free curve, transfer as many pricked points on the curve as desired.

TRACING METHOD Lay tracing paper over the drawing to be transferred, and make a pencil tracing of it. Turn the tracing paper over and mark over the lines with short strokes of a soft pencil so as to provide a coating of graphite over every line. Turn tracing face up and fasten in position where drawing is to be transferred. Trace over all lines of the tracing, using a hard pencil. The graphite on the back acts as a carbon paper and will produce dim but definite lines. Heavy in the dim lines to complete the transfer.

Fig. 5.30 If one-half of a symmetrical object has been drawn, as for the ink bottle at I, the other half may be easily drawn with the aid of tracing paper as follows.

I. Trace the half already drawn.

II. Turn tracing paper over and maneuver to

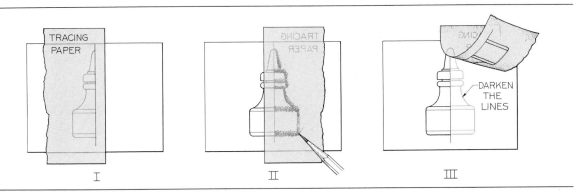

Fig. 5.30 Transferring a Symmetrical Half (§5.29).

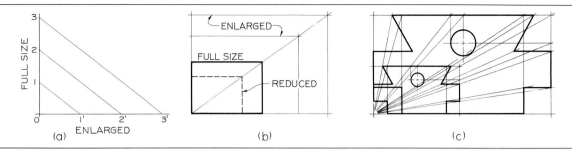

Fig. 5.31   Enlarging or Reducing (§5.30).

the position for the right half. Then trace over the lines freehand or mark over the lines with short strokes as shown.

III.   Remove the tracing paper, revealing the dim imprinted lines for the right half. Heavy in these lines to complete the drawing.

5.30   To Enlarge or Reduce a Drawing

Fig. 5.31 (a)   The construction shown is an adaptation of the parallel-line method, Figs. 5.14 and 5.15, and may be used whenever it is desired to enlarge or reduce any group of dimensions to the same ratio. Thus, if full-size dimensions are laid off along the vertical line, the enlarged dimensions will appear along the horizontal line, as shown.

Fig. 5.31 (b)   To enlarge or reduce a rectangle (say, a photograph), a simple method is to use the diagonal, as shown.

Fig. 5.31 (c)   A simple method of enlarging or reducing a drawing is to make use of radial lines, as shown. The original drawing is placed underneath a sheet of tracing paper, and the enlarged or reduced drawing is made directly on the tracing paper.

5.31   To Draw a Circle Through Three Points

Fig. 5.32 (a)

I.   Let A, B, and C be the three given points not in a straight line. Draw lines AB and BC, which will be chords of the circle. Draw perpendicular

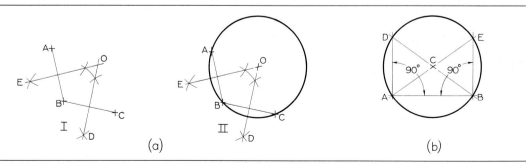

Fig. 5.32   Finding Center of Circle (§§5.31 and 5.32).

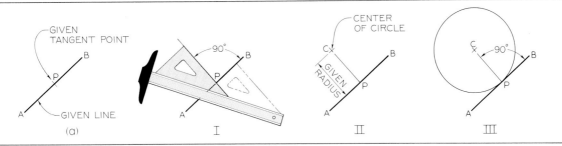

Fig. 5.33   Drawing a Circle Tangent to a Line (§5.33).

bisectors EO and DO, Fig. 5.8, intersecting at O.

II.   With center at O, draw required circle through the points.

5.32   To Find the Center of a Circle

Fig. 5.32 (b)   Draw any chord AB, preferably horizontal as shown. Draw perpendiculars from A and B, cutting circle at D and E. Draw diagonals DB and EA whose intersection C will be the center of the circle. This method uses the principle that any right triangle inscribed in a circle cuts off a semicircle, as was shown earlier, in Fig. 5.3 (f).

Another method, slightly longer, is to reverse the procedure of Fig. 5.32 (a). Draw any two nonparallel chords and draw perpendicular bisectors. The intersection of the bisectors will be the center of the circle.

5.33   To Draw a Circle Tangent to a Line at a Given Point

Fig. 5.33   Given a line AB and a point P on the line, as shown at (a).

I.   At P erect a perpendicular to the line.

II.   Set off the radius of the required circle on the perpendicular.

III. Draw circle with radius CP.

5.34   To Draw a Tangent to a Circle Through a Point

Fig. 5.34 (a)   PREFERRED METHOD   Given point P on the circle. Move the T-square and triangle as a unit until one side of the triangle passes through the point P and the center of the

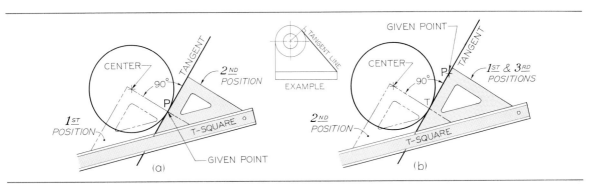

Fig. 5.34   Drawing a Tangent to a Circle Through a Point (§5.34).

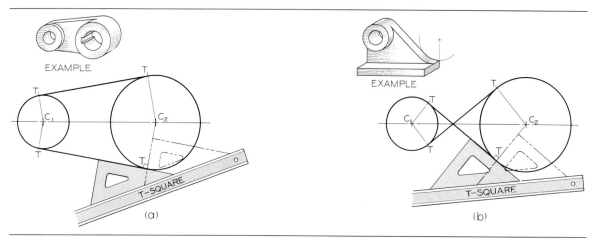

Fig. 5.35   Drawing Tangents to Two Circles (§5.35).

circle; then slide the triangle until the other side passes through point P, and draw the required tangent.

Fig. 5.34 (b)   Given point P outside the circle. Move the T-square and triangle as a unit until one side of the triangle passes through point P and, by inspection, is tangent to the circle; then slide the triangle until the other side passes through the center of the circle, and lightly mark the point of tangency T. Finally, move the triangle back to its starting position, and draw the required tangent.

In both constructions either triangle may be used. Also, a second triangle may be used in place of the T-square.

5.35   To Draw Tangents to Two Circles
Fig. 5.35 (a) and (b)   Move the triangle and T-square as a unit until one side of the triangle is tangent, by inspection, to the two circles; then slide the triangle until the other side passes through the center of one circle, and lightly mark the point of tangency. Then slide the tri-

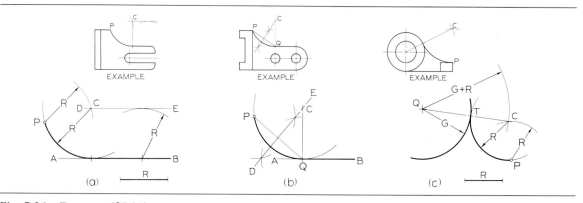

Fig. 5.36   Tangents (§5.36).

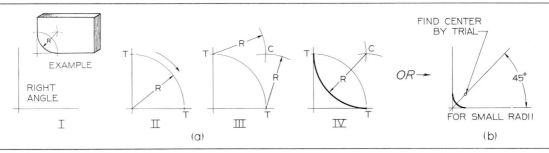

Fig. 5.37   Drawing a Tangent Arc in a Right Angle (§5.37).

angle until the side passes through the center of the other circle, and mark the point of tangency. Finally, slide the triangle back to the tangent position, and draw the tangent lines between the two points of tangency. Draw the second tangent line in a similar manner.

5.36   To Draw an Arc Tangent to a Line or Arc and Through a Point

Fig. 5.36 (a)   Given line AB, point P, and radius R. Draw line DE parallel to given line and distance R from it. From P draw arc with radius R, cutting line DE at C, the center of the required tangent arc.

Fig. 5.36 (b)   Given line AB, with tangent point Q on the line, and point P. Draw PQ, which will be a chord of the required arc. Draw perpendicular bisector DE, and at Q erect a perpendicular to the line to intersect DE at C, the center of the required tangent arc.

Fig. 5.36 (c)   Given arc with center Q, point P, and radius R. From P strike arc with radius R. From Q strike arc with radius equal to that of the given arc plus R. The intersection C of the arcs is the center of the required tangent arc.

5.37   To Draw an Arc Tangent to Two Lines at Right Angles

Fig. 5.37 (a)

I.   Given two lines at right angles to each other.

II.   With given radius R, strike arc intersecting given lines at tangent points T.

III.   With given radius R again, and with points T as centers, strike arcs intersecting at C.

IV.   With C as center and given radius R, draw required tangent arc.

FOR SMALL RADII   ***Fig. 5.37 (b)***   For small radii, such as $\frac{1}{8}$R for fillets and rounds, it is not practicable to draw complete tangency constructions. Instead, draw a 45° bisector of the angle and locate the center of the arc by trial along this line, as shown.

Note that the center C can be located by intersecting lines parallel to the given lines, as shown in Fig. 5.13 (b). However, the circle template can also be used to draw the arcs R for the parallel line method of Fig. 5.13 (b). While the circle template is very convenient to use for small radii up to about $\frac{5}{8}''$ or 16 mm, it is necessary that the diameter of a circle on the template precisely equals twice the required radius.

5.38   To Draw an Arc Tangent to Two Lines at Acute or Obtuse Angles

Fig. 5.38 (a) or (b)

I.   Given two lines not making 90° with each other.

II.   Draw lines parallel to given lines, at distance R from them, to intersect at C, the required center.

III.   From C drop perpendiculars to the given lines respectively to locate tangent points T.

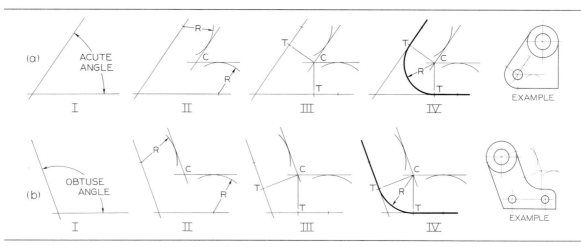

Fig. 5.38   Drawing Tangent Arcs (§5.38).

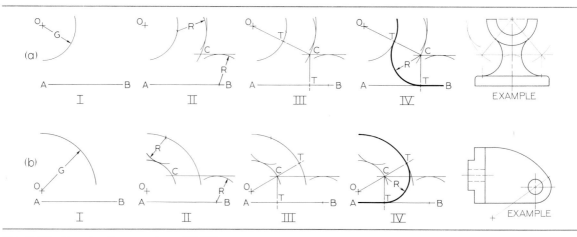

Fig. 5.39   Drawing Tangent Arcs (§5.39).

IV.   With C as center and with given radius R, draw required tangent arc between the points of tangency.

5.39   To Draw an Arc Tangent to an Arc and a Straight Line
Fig. 5.39 (a) or (b)

I.   Given arc with radius G and straight line AB.

II.   Draw a straight line and an arc parallel, respectively, to the given straight line and arc at the required radius distance R from them, to intersect at C, the required center.

III.   From C drop a perpendicular to the given straight line to obtain one point of tangency T. Join the centers C and O with a straight line to locate the other point of tangency T.

IV.   With center C and given radius R, draw required tangent arc between the points of tangency.

5.40   To Draw an Arc Tangent to Two Arcs
Fig. 5.40 (a) or (b)

I.   Given arcs with centers A and B, and required radius R.

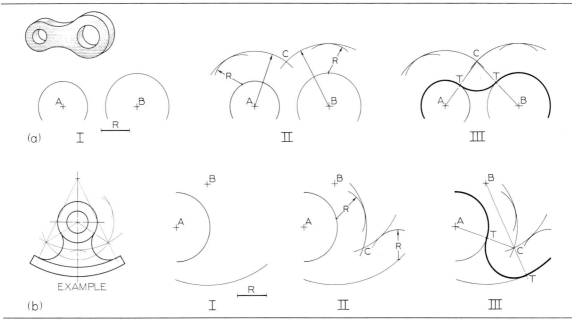

Fig. 5.40   Drawing an Arc Tangent to Two Arcs (§5.39).

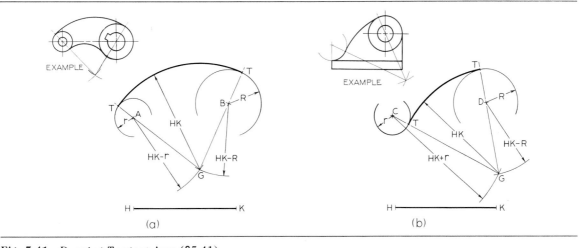

Fig. 5.41   Drawing Tangent Arcs (§5.41).

II.   With A and B as centers, draw arcs parallel to the given arcs and at a distance R from them; their intersection C is the center of the required tangent arc.

III.   Draw lines of centers AC and BC to locate points of tangency T, and draw required tangent arc between the points of tangency, as shown.

5.41   To Draw an Arc Tangent to Two Arcs and Enclosing One or Both
THE REQUIRED ARC ENCLOSES BOTH GIVEN ARCS
Fig. 5.41 (a)   With A and B as centers, strike arcs HK − r (given radius minus radius of small circle) and HK − R (given radius minus radius of large circle) intersecting at G, the center of the

139

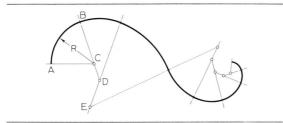

Fig. 5.42   A Series of Tangent Arcs (§5.42).

required tangent arc. Lines of centers GA and GB (extended) determine points of tangency T.

THE REQUIRED ARC ENCLOSES ONE GIVEN ARC Fig. 5.41 (b)   With C and D as centers, strike arcs HK + r (given radius plus radius of small circle) and HK − R (given radius minus radius of large circle) intersecting at G, the center of the required tangent arc. Lines of centers GC and GD (extended) determine points of tangency T.

5.42   To Draw a Series of Tangent Arcs Conforming to a Curve

Fig. 5.42   First sketch lightly a smooth curve as desired. By trial, find a radius R and a center C, producing an arc AB that closely follows that portion of the curve. The successive centers D, E, and so on, will be on lines joining the centers with the points of tangency, as shown.

5.43   To Draw an Ogee Curve

CONNECTING TWO PARALLEL LINES   *Fig. 5.43 (a)*   Let NA and BM be the two parallel lines. Draw AB, and assume inflection point T (at midpoint if two equal arcs are desired). At A and B erect perpendiculars AF and BC. Draw perpendicular bisectors of AT and BT. The intersections F and C of these bisectors and the perpendiculars, respectively, are the centers of the required tangent arcs.

Fig. 5.43 (b)   Let AB and CD be the two parallel lines, with point B as one end of the curve and R the given radii. At B erect perpendicular to AB, make BG = R, and draw arc as shown. Draw line SP parallel to CD at distance R from CD. With center G, draw arc of radius 2R, intersecting line SP at O. Draw perpendicular OJ to locate tangent point J, and join centers G and O to locate point of tangency T. Using centers G and O and radius R, draw the two tangent arcs as shown.

CONNECTING TWO NONPARALLEL LINES   *Fig. 5.43 (c)*   Let AB and CD be the two nonparallel lines. Erect perpendicular to AB at B. Select point G on the perpendicular so that BG equals any desired radius, and draw arc as shown. Erect perpendicular to CD at C and make CE = BG. Join G to E and bisect it. The intersection F of the bisector and the perpendicular CE, extended, is the center of the second arc. Join centers of the two arcs to locate tangent point T, the inflection point of the curve.

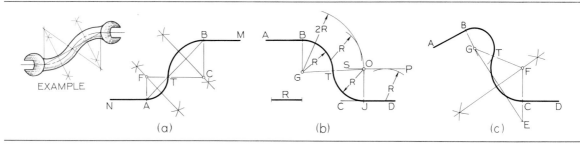

Fig. 5.43   Drawing an Ogee Curve (§5.43).

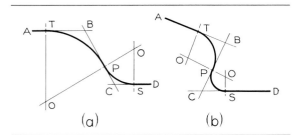

Fig. 5.44   Tangent Curves (§5.44).

5.44   To Draw a Curve Tangent to Three Intersecting Lines

Fig. 5.44 (a) and (b) Let AB, BC, and CD be the given lines. Select point of tangency P at any point on line BC. Make BT equal to BP, and CS equal to CP, and erect perpendiculars at the points P, T, and S. Their intersections O and Q are the centers of the required tangent arcs.

5.45   To Rectify a Circular Arc

To *rectify* an arc is to lay out its true length along a straight line. The constructions are approximate, but well within the range of accuracy of drawing instruments.

TO RECTIFY A QUADRANT OF A CIRCLE, AB *Fig. 5.45 (a)* Draw AC tangent to the circle and BC at 60° to AC, as shown. The line AC is almost equal to the arc AB, the difference in length being about 1 in 240.

TO RECTIFY ARC AB *Fig. 5.45 (b)* Draw tangent at B. Draw chord AB and extend it to C, making BC equal to half AB. With C as center and radius CA, strike the arc AD. The tangent BD is slightly shorter than the given arc AB. For an angle of 45° the difference in length is about 1 in 2866.

Fig. 5.45 (c) Use the bow dividers, and beginning at A, set off equal distances until the division point nearest to B is reached. At this point, reverse the direction and set off an equal number of distances along the tangent to determine point C. The tangent BC is slightly shorter than the given arc AB. If the angle subtended by each division is 10°, the error is approximately 1 in 830.

NOTE   If the angle θ subtending an arc of radius R is known, the length of the arc is

$$2\pi R \frac{\theta}{360°} = 0.01745R\theta.$$

5.46   To Set Off a Given Length Along a Given Arc

Fig. 5.45 (c) Reverse the preceding method so as to transfer distances from the tangent line to the arc.

Fig. 5.45 (d) To set off the length BC along the arc BA, draw BC tangent to the arc at B. Divide BC into four equal parts. With center at 1, the first division point, and radius 1–C, draw the arc CA. The arc BA is practically equal to BC for

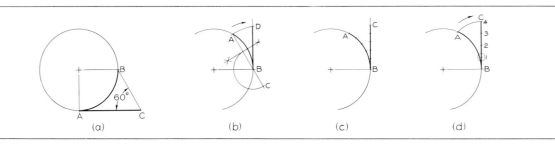

Fig. 5.45   Rectifying Circular Arcs (§§5.45 and 5.46).

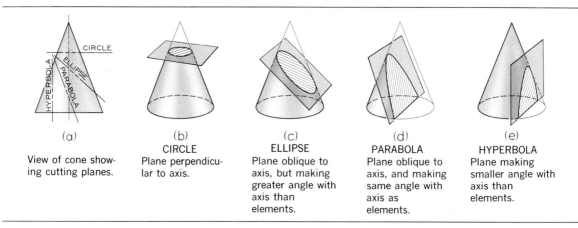

Fig. 5.46   Conic Sections (§5.47).

angles less than 30°. For 45° the difference is approximately 1 in 3232, and for 60° it is about 1 in 835.

5.47   The Conic Sections

Fig. 5.46   The conic sections are curves produced by planes intersecting a right circular cone. Four types of curves are produced: the *circle, ellipse, parabola,* and *hyperbola,* according to the position of the planes, as shown. These curves were studied in detail by the ancient Greeks, and are of great interest in mathematics, as well as in technical drawing. For equations, see any text on analytic geometry.

5.48   Ellipse Construction

The long axis of an ellipse is the major axis, and the short axis is the minor axis, Fig. 5.47 (a). The foci E and F are found by striking arcs with radius equal to half the major axis and with center at the end of the minor axis. Another method is to draw a semicircle with the major axis as diameter, then to draw GH parallel to the major axis and GE and HF parallel to the minor axis, as shown.

An ellipse is generated by a point moving so that the sum of its distances from two points (the foci) is constant and equal to the major axis. For example, Fig. 5.47 (b), an ellipse may be constructed by placing a looped string around

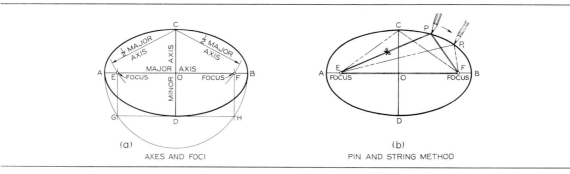

Fig. 5.47   Ellipse Constructions (§5.48).

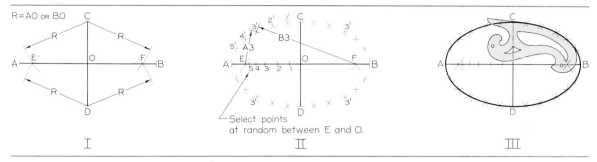

Fig. **5.48**   Drawing a Foci Ellipse (§5.49).

the foci E and F, and around C, one end of the minor axis, and moving the pencil point P along its maximum orbit while the string is kept taut.

5.49   To Draw a Foci Ellipse

Fig. 5.48   Let AB be the major axis and CD the minor axis. This method is the geometrical counterpart of the pin-and-string method. Keep the construction very light, as follows.

I.   To find foci E and F, strike arcs R with radius equal to half the major axis and with centers at the ends of the minor axis.

II.   Between E and O on the major axis, mark at random a number of points (spacing those on the left more closely), equal to the number of points desired in each quadrant of the ellipse. In

this figure, five points were deemed sufficient. For large ellipses, more points should be used—enough to insure a smooth, accurate curve. Begin construction with any one of these points, such as 3. With E and F as centers and radii A–3 and B–3, respectively (from the ends of the major axis to point 3), strike arcs to intersect at four points 3', as shown. Using the remaining points 1, 2, 4, and 5, for each find four additional points on the ellipse in the same manner.

III.   Sketch the ellipse lightly through the points; then heavy in the final ellipse with the aid of the irregular curve, Fig. 2.66.

5.50   To Draw a Trammel Ellipse

Fig. 5.49   A "long trammel" or a "short trammel" may be prepared from a small strip of stiff paper or thin cardboard, as shown. In both cases, set off on the edge of the trammel distances equal to the semimajor and semiminor axes. In one case these distances overlap; in the other they are end to end. To use either method, place the trammel so that two of the points are on the respective axes, as shown; the third point will then be on the curve and can be marked with a small dot. Find additional points by moving the trammel to other positions, always keeping the two points exactly on the respective axes. Extend the axes to use the long trammel. Find enough points to insure a smooth and symmetrical ellipse. Sketch the ellipse lightly through the points; then heavy in the ellipse with the aid of the irregular curve, Fig. 2.66.

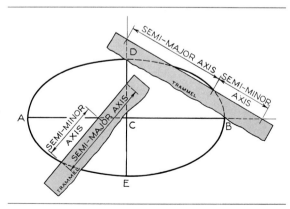

Fig. **5.49**   Drawing a Trammel Ellipse (§5.50).

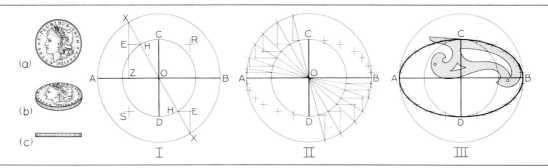

Fig. 5.50 Drawing a Concentric-Circle Ellipse (§5.51).

5.51 To Draw a Concentric-Circle Ellipse

Fig. 5.50 If a circle is viewed so that the line of sight is perpendicular to the plane of the circle, as shown for the silver dollar at (a), the circle will appear as a circle, in true size and shape. If the circle is viewed at an angle, as shown at (b), it will appear as an ellipse. If the circle is viewed edgewise, it appears as a straight line, as shown at (c). The case shown at (b) is the basis for the construction of an ellipse by the concentric-circle method, as follows (keep the construction very light).

I. Draw circles on the major and minor axes using them as diameters and draw any diagonal XX through center O. From the points X, in which the diagonal intersects the large circle, draw lines XE parallel to the minor axis, and from the points H, in which it intersects the small circle, draw lines HE parallel to the major axis. The intersections E are points on the ellipse. Two additional points, S and R, can be found by extending lines XE and HE, giving a total of four points from the one diagonal XX.

II. Draw as many additional diagonals as needed to provide a sufficient number of points for a smooth and symmetrical ellipse, each diagonal accounting for four points on the ellipse. Notice that where the curve is sharpest (near the ends of the ellipse), the points are constructed closer together to better determine the curve.

III. Sketch the ellipse lightly through the points, then heavy in the final ellipse with the aid of the irregular curve.

NOTE It is evident at I, Fig. 5.50, that the ordinate EZ of the ellipse is to the corresponding ordinate XZ of the circle as b is to a, where b represents the semiminor axis and a the semimajor axis. Thus, the area of the ellipse is equal to the area of the circumscribed circle multiplied by $\dfrac{b}{a}$; hence, it is equal to πab.

5.52 To Draw an Ellipse on Conjugate Diameters— Oblique-Circle Method

Fig. 5.51 Let AB and DE be the given conjugate diameters. *Two diameters are conjugate when each is parallel to the tangents at the extremities of the other.* With center at C and radius CA, draw a circle; draw the diameter GF perpendicular to AB, and draw lines joining points D and F and points G and E.

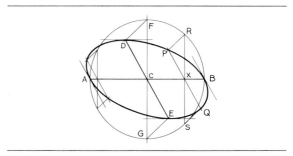

Fig. 5.51 Oblique-Circle Ellipse (§5.52).

Assume that the required ellipse is an oblique projection of the circle just drawn; the points D and E of the ellipse are the oblique projections of the points F and G of the circle, respectively; similarly, the points P and Q are the oblique projections of the points R and S, respectively. The points P and Q are determined by assuming the point X at any point on AB and drawing the lines RS and PQ, and RP and SQ, parallel, respectively, to GF and DE and FD and GE.

Determine at least five points in each quadrant (more for larger ellipses) by assuming additional points on the major axis and proceeding as explained for point X. Sketch the ellipse lightly through the points; then heavy in the final ellipse with the aid of the irregular curve, Fig. 2.66.

5.53   To Draw a Parallelogram Ellipse

Fig. 5.52 (a) and (b)   Given the major and minor axes, or the conjugate diameters AB and CD, draw a rectangle or parallelogram with sides parallel to the axes, respectively. Divide AO and AJ into the same number of equal parts, and draw *light* lines through these points from the ends of the minor axis, as shown. The intersection of like-numbered lines will be points on the ellipse. Locate points in the remaining three quadrants in a similar manner. Sketch the ellipse lightly through the points; then heavy in the final ellipse with the aid of the irregular curve, Fig. 2.66.

5.54   To Find the Axes of an Ellipse, with Conjugate Diameters Given

Fig. 5.53 (a)   Conjugate diameters AB and CD and the ellipse are given. With intersection O of the conjugate diameters (center of ellipse) as center, and any convenient radius, draw a circle to intersect the ellipse in four points. Join these points with straight lines, as shown; the resulting quadrilateral will be a rectangle whose sides are parallel, respectively, to the required major and minor axes. Draw the axes EF and GH parallel to the sides of the rectangle.

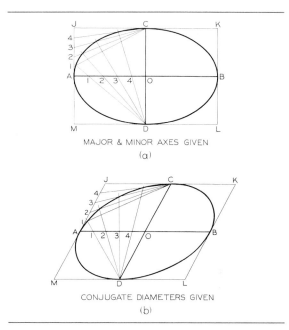

MAJOR & MINOR AXES GIVEN
(a)

CONJUGATE DIAMETERS GIVEN
(b)

Fig. 5.52   Parallelogram Ellipse (§5.53).

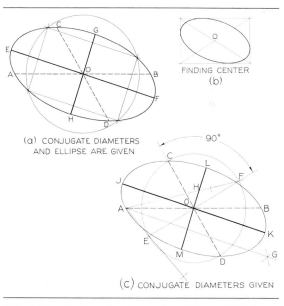

(a) CONJUGATE DIAMETERS AND ELLIPSE ARE GIVEN

FINDING CENTER
(b)

(c) CONJUGATE DIAMETERS GIVEN

Fig. 5.53   Finding the Axes of an Ellipse (§5.54).

Fig. 5.53 (b)   Ellipse only is given. To find the center of the ellipse, draw a circumscribing rectangle or parallelogram about the ellipse; then, draw diagonals to intersect at center O as shown. The axes are then found as shown at (a).

145

Fig. 5.53 (c) Conjugate diameters AB and CD only are given. With O as center and CD as diameter, draw a circle. Through center O and perpendicular to CD, draw line EF. From points E and F, where this perpendicular intersects the circle, draw lines FA and EA to form angle FAE. Draw the bisector AG of this angle. The major axis JK will be parallel to this bisector, and the minor axis LM will be perpendicular to it. The length AH will be one-half the major axis, and HF one-half the minor axis. The resulting major and minor axes are JK and LM, respectively.

5.55   To Draw a Tangent to an Ellipse

CONCENTRIC CIRCLE CONSTRUCTION   Fig. 5.54 (a) To draw a tangent at any point on the ellipse, as E, draw the ordinate at E to intersect the circle at V. Draw a tangent to the circle at V, §5.34, and extend it to intersect the major axis extended at G. The line GE is the required tangent.

To draw a tangent from a point outside the ellipse, as P, draw the ordinate PY and extend it.

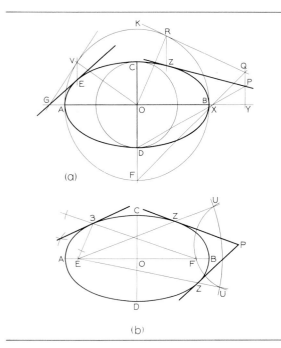

(a)

(b)

Fig. 5.54   Tangents to an Ellipse (§5.55).

Draw DP, intersecting the major axis at X. Draw FX and extend it to intersect the ordinate through P at Q. Then, from similar triangles, QY:PY = OF:OD. Draw tangent to the circle from Q, §5.34, find the point of tangency R, and draw the ordinate at R to intersect the ellipse at Z. The line ZP is the required tangent. As a check on the drawing, the tangents RQ and ZP should intersect at a point on the major axis extended. Two tangents to the ellipse can be drawn from point P.

FOCI CONSTRUCTION   Fig. 5.54 (b) To draw a tangent at any point on the ellipse, such as point 3, draw the focal radii E–3 and F–3, extend one, and bisect the exterior angle, as shown. The bisector is the required tangent.

To draw a tangent from any point outside the ellipse, such as point P, with center at P and radius PF, strike an arc as shown. With center at E and radius AB, strike an arc to intersect the first arc at points U. Draw the lines EU to intersect the ellipse at the points Z. The lines PZ are the required tangents.

5.56   Ellipse Templates

To save time in drawing ellipses, and to insure uniform results, ellipse templates, Fig. 5.55 (a), are often used. These are plastic sheets with elliptical openings in a wide variety of sizes, and they usually come in sets of six or more sheets.

Ellipse guides are usually designated by the ellipse angle, the angle at which a circle is viewed to appear as an ellipse. In Fig. 5.55 (b) the angle between the line of sight and the edge view of the plane of the circle is found to be about 49°; hence, the 50° ellipse template is indicated. Ellipse templates are generally available in ellipse angles at 5° intervals, such as 15°, 20°, and 25°. On this 50° template a variety of sizes of 50° ellipses is provided, and it is only necessary to select the one that fits. If the ellipse angle is not easily determined, you can always look for the ellipse that is approximately as long and as "fat" as the ellipse to be drawn.

A simple construction for finding the ellipse angle when the views are not available is shown

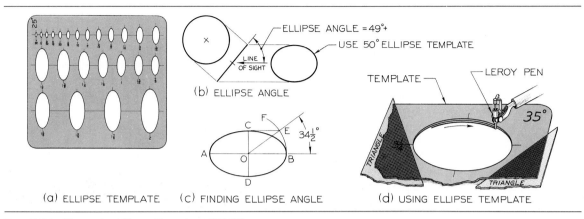

Fig. 5.55 Using the Ellipse Template (§5.56).

at (c). Using center O, strike arc BF; then draw CE parallel to the major axis. Draw diagonal OE, and measure angle EOB with the protractor, §2.18. Use the ellipse template nearest to this angle; in this case a 35° template is selected.

Since it is not feasible to have ellipse openings for every exact size that may be required, it is often necessary to use the template somewhat in the manner of an irregular curve. For example, if the opening is too long and too "fat" for the required ellipse, one end may be drawn and then the template shifted slightly to draw the other end. Similarly, one long side may be drawn and then the template shifted slightly to draw the opposite side. In such cases, leave gaps between the four segments, to be filled in freehand or with the aid of an irregular curve. When the differences between the ellipse openings and the required ellipse are small, it is only neces-

sary to lean the pencil slightly outward or inward from the guiding edge to offset the differences.

For inking the ellipses, the Leroy, Rapidograph, or Wrico pens are recommended. The Leroy pen is shown in Fig. 5.55 (d).

5.57 To Draw an Approximate Ellipse

Fig. 5.56 For many purposes, particularly where a small ellipse is required, the approximate circular-arc method is perfectly satisfactory. Such an ellipse is sure to be symmetrical and may be quickly drawn.

Given axes AB and CD.

I.   Draw line AC. With O as center and OA as radius, strike the arc AE. With C as center and CE as radius, strike the arc EF.

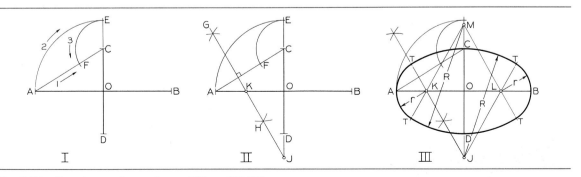

Fig. 5.56 Drawing an Approximate Ellipse (§5.57)

II.   Draw perpendicular bisector GH of the line AF; the points K and J, where it intersects the axes, are centers of the required arcs.

III.   Find centers M and L by setting off OL = OK and OM = OJ. Using centers K, L, M, and J, draw circular arcs as shown. The points of tangency T are at the junctures of the arcs on the lines joining the centers.

5.58   To Draw a Parabola

The curve of intersection between a right circular cone and a plane parallel to one of its elements, Fig. 5.46 (d), is a parabola. *A parabola is generated by a point moving so that its distances from a fixed point, the focus, and from a fixed line, the directrix, remain equal.* For example:

Fig. 5.57 (a)   Given focus F and directrix AB. A parabola may be generated by a pencil guided by a string, as shown. Fasten the string at F and C; its length is GC. The point C is selected at random, its distance from G depending on the desired extent of the curve. Keep the string taut and the pencil against the T-square, as shown.

Fig. 5.57 (b)   Given focus F and directrix AB. Draw a line DE parallel to the directrix and at any distance CZ from it. With center at F and radius CZ, strike arcs to intersect the line DE in the points Q and R, which are points on the parabola. Determine as many additional points as are necessary to draw the parabola accurately, by drawing additional lines parallel to line AB and proceeding in the same manner.

A tangent to the parabola at any point G bisects the angle formed by the focal line FG and the line SG perpendicular to the directrix.

Fig. 5.57 (c)   Given the rise and span of the parabola. Divide AO into any number of equal parts, and divide AD into a number of equal parts amounting to the square of that number. From line AB, each point on the parabola is offset by a number of units equal to the square of the number of units from point O. For example,

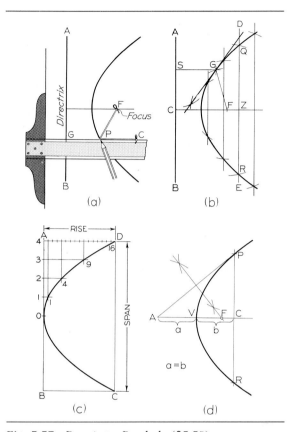

Fig. 5.57   Drawing a Parabola (§5.58).

point 3 projects 9 units (the square of 3). This method is generally used for drawing parabolic arches.

Fig. 5.57 (d)   Given points P, R, and V of a parabola, to find the focus F. Draw tangent at P, making a = b. Draw perpendicular bisector of AP, which intersects the axis at F, the focus of the parabola.

Fig. 5.58 (a) or (b)   Given rectangle or parallelogram ABCD. Divide BC into any even number of equal parts, and divide the sides AB and DC each into half as many parts, and draw lines as shown. The intersections of like-numbered lines are points on the parabola.

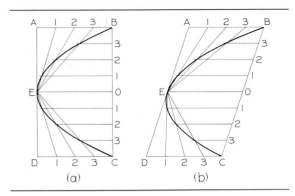

Fig. 5.58   Drawing a Parabola (§5.58).

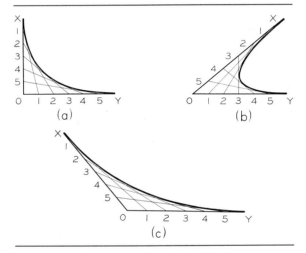

Fig. 5.59   Parabolic Curves (§5.59).

PRACTICAL APPLICATIONS   The parabola is used to reflect surfaces for light and sound, for vertical curves in highways, for forms of arches, and approximately for forms of the curves of cables for suspension bridges. It is also used to show the bending moment at any point on a uniformly loaded beam or girder.

5.59   To Join Two Points by a Parabolic Curve

Fig. 5.59   Let X and Y be the given points. Assume any point O, and draw tangents XO and YO. Divide XO and YO into the same number of

equal parts, number the division points as shown, and connect corresponding points. These lines are tangents of the required parabola and form its envelope. Sketch a light smooth curve, and then heavy in the curve with the aid of the irregular curve, §2.54.

These parabolic curves are more pleasing in appearance than circular arcs and are useful in machine design. If the tangents OX and OY are equal, the axis of the parabola will bisect the angle between them.

5.60   To Draw a Hyperbola

The curve of intersection between a right circular cone and a plane making an angle with the axis smaller than that made by the elements, Fig. 5.46 (e), is a hyperbola. *A hyperbola is generated by a point moving so that the difference of its distances from two fixed points, the foci, is constant and equal to the transverse axis of the hyperbola.*

Fig. 5.60 (a)   Let F and F′ be the foci and AB the transverse axis. The curve may be generated by a pencil guided by a string, as shown. Fasten a string at F′ and C; its length is FC minus AB. The point C is chosen at random; its distance from F depends on the desired extent of the curve.

Fasten the straightedge at F. If it is revolved about F, with the pencil point moving against it and with the string taut, the hyperbola may be drawn as shown.

Fig. 5.60 (b)   To construct the curve geometrically, select any point X on the transverse axis produced. With centers at F and F′ and BX as radius, strike the arcs DE. With the same centers, F and F′, and AX as radius, strike arcs to intersect the arcs first drawn in the points Q, R, S, and T, which are points of the required hyperbola. Find as many additional points as are necessary to draw the curves accurately by selecting other points similar to point X along the transverse axis, and proceeding as described for point X.

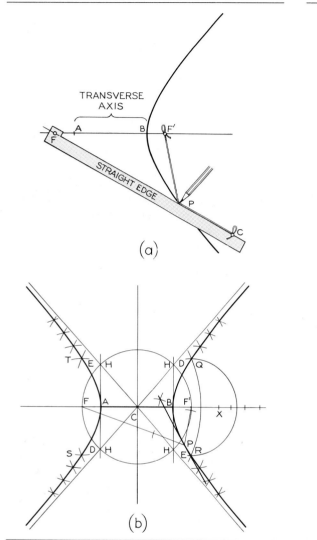

Fig. 5.60   Drawing a Hyperbola (§5.60).

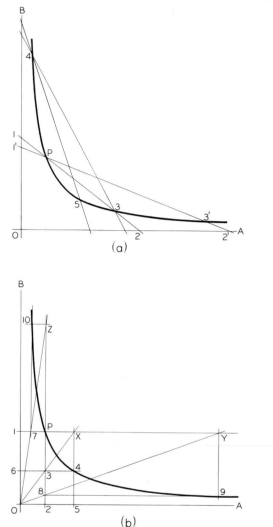

Fig. 5.61   Equilateral Hyperbola (§5.61).

To draw the tangent to a hyperbola at a given point P, bisect the angle between the focal radii FP and F′P. The bisector is the required tangent.

To draw the asymptotes HCH of the hyperbola, draw a circle with the diameter FF′ and erect perpendiculars to the transverse axis at the points A and B to intersect the circle in the points H. The lines HCH are the required asymptotes.

5.61   To Draw an Equilateral Hyperbola

Fig. 5.61   Let the asymptotes OB and OA, at right angles to each other, and the point P on the curve be given.

Fig. 5.61 (a)   In an equilateral hyperbola the asymptotes, at right angles to each other, may be used as the axes to which the curve is re-

150

ferred. If a chord of the hyperbola is extended to intersect the axes, the intercepts between the curve and the axes are equal. For example, a chord through given point P intersects the axes at points 1 and 2, intercepts P–1 and 2–3 are equal, and point 3 is a point on the hyperbola. Likewise, another chord through P provides equal intercepts P–1′ and 3′–2′, and point 3′ is a point on the curve. Not all chords need be drawn through given point P, but as new points are established on the curve, chords may be drawn through them to obtain more points. After enough points are found to insure an accurate curve, the hyperbola is drawn with the aid of the irregular curve, §2.54.

Fig. 5.61 (b) In an equilateral hyperbola, the coordinates are related so that their products remain constant. Through given point P, draw lines 1–P–Y and 2–P–Z parallel, respectively, to the axes. From the origin of coordinates O, draw any diagonal intersecting these two lines at points 3 and X. At these points draw lines parallel to the axes, intersecting at point 4, a point on the curve. Likewise, another diagonal from O intersects the two lines through P at points 8 and Y, and lines through these points parallel to the axes intersect at point 9, another point on the curve. A third diagonal similarly produces point 10 on the curve, and so on. Find as many points as necessary for a smooth curve, and draw the parabola with the aid of the irregular curve, §2.54. It is evident from the similar triangles O–X–5 and O–3–2 that lines P–1 × P–2 = 4–5 × 4–6.

The equilateral hyperbola can be used to represent varying pressure of a gas as the volume varies, since the pressure varies inversely as the volume; that is, pressure × volume is constant.

5.62   To Draw a Spiral of Archimedes

Fig. 5.62 To find points on the curve, draw lines through the pole C, making equal angles with each other, such as 30° angles, and beginning with any one line, set off any distance, such

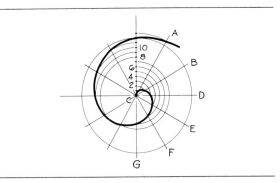

Fig. 5.62   Spiral of Archimedes (§5.62).

as 2 mm or $\frac{1}{16}''$; set off twice that distance on the next line, three times on the third, and so on. Through the points thus determined, draw a smooth curve, using the irregular curve, §2.54.

5.63   To Draw a Helix

Fig. 5.63 *A helix is generated by a point moving around and along the surface of a cylinder or cone with a uniform angular velocity about the axis, and with a uniform linear velocity in the direction of the axis.* A cylindrical helix is generally known simply as a helix. The distance measured parallel to the axis traversed by the point in one revolution is called the lead.

If the cylindrical surface on which a helix is generated is rolled out onto a plane, the helix becomes a straight line as shown in Fig. 5.63 (a), and the portion below the helix becomes a right triangle, the altitude of which is equal to the lead of the helix and the length of the base equal to the circumference of the cylinder. Such a helix can, therefore, be defined as the shortest line that can be drawn on the surface of a cylinder connecting two points not on the same element.

To draw the helix, draw two views of the cylinder on which the helix is generated, (b), and divide the circle of the base into any number of equal parts. On the rectangular view of the cylinder, set off the lead and divide it into the same number of equal parts as the base. Number the

151

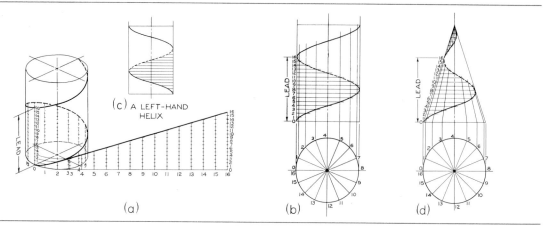

Fig. 5.63   Helix (§5.63).

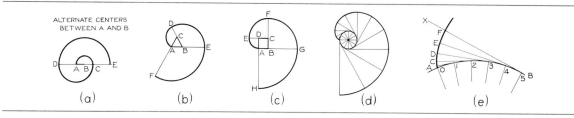

Fig. 5.64   Involutes (§5.64).

divisions as shown, in this case 16. When the generating point has moved one-sixteenth of the distance around the cylinder, it will have risen one-sixteenth of the lead; when it has moved halfway around the cylinder, it will have risen half the lead; and so on. Points on the helix are found by projecting up from point 1 in the circular view to line 1 in the rectangular view, from point 2 in the circular view to line 2 in the rectangular view, and so on.

The helix shown at (b) is a right-hand helix. In a left-hand helix, (c), the visible portions of the curve are inclined in the opposite direction, that is, downward to the right. The helix shown at (b) can be converted into a left-hand helix by interchanging the visible and hidden lines.

The helix finds many applications in industry, as in screw threads, worm gears, conveyors, spiral stairways, and so on. The stripes of a barber pole are helical in form.

The construction for a right-hand conical helix is shown at (d).

5.64   To Draw an Involute

Fig. 5.64   The path of a point on a string, as the string unwinds from a line, a polygon, or a circle, is an involute.

To Draw an Involute of a Line   *Fig. 5.64 (a)* Let AB be the given line. With AB as radius and B as center, draw the semicircle AC. With AC as radius and A as center, draw the semicircle CD. With BD as radius and B as center, draw the semicircle DE. Continue similarly, alternating centers between A and B, until a figure of the required size is completed.

To Draw an Involute of a Triangle   *Fig. 5.64 (b)*   Let ABC be the given triangle. With

152

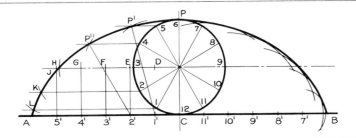

Fig. 5.65 Cycloid (§5.65).

CA as radius and C as center, strike the arc AD. With BD as radius and B as center, strike the arc DE. With AE as radius and A as center, strike the arc EF. Continue similarly until a figure of the required size is completed.

To Draw an Involute of a Square *Fig. 5.64 (c)* Let ABCD be the given square. With DA as radius and D as center, draw the 90° arc AE. Proceed as for the involute of a triangle until a figure of the required size is completed.

To Draw an Involute of a Circle *Fig. 5.64 (d)* A circle may be regarded as a polygon with an infinite number of sides. The involute is constructed by dividing the circumference into a number of equal parts, drawing a tangent at each division point, setting off along each tangent the length of the corresponding circular arc, Fig. 5.45 (c), and drawing the required curve through the points set off on the several tangents.

Fig. 5.64 (e) The involute may be generated by a point on a straight line that is rolled on a fixed circle. Points on the required curve may be determined by setting off equal distances 0–1, 1–2, 2–3, and so on, along the circumference, drawing a tangent at each division point, and proceeding as explained for (d).

The involute of a circle is used in the construction of involute gear teeth. In this system, the involute forms the face and a part of the flank of the teeth of gear wheels; the outlines of the teeth of racks are straight lines.

5.65 To Draw a Cycloid
Fig. 5.65 *A cycloid may be generated by a point P in the circumference of a circle that rolls along a straight line.*

Given the generating circle and the straight line AB tangent to it, make the distances CA and CB each equal to the semicircumference of the circle, Fig. 5.45 (c). Divide these distances and the semicircumference into the same number of equal parts, six, for instance, and number them consecutively as shown. Suppose the circle to roll to the left; when point 1 of the circle reaches point 1' of the line, the center of the circle will be at D, point 7 will be the highest point of the circle, and the generating point 6 will be at the same distance from the line AB as point 5 is when the circle is in its central position. Hence, to find the point P', draw a line through point 5 parallel to AB and intersect it with an arc drawn from the center D with a radius equal to that of the circle. To find point P", draw a line through point 4 parallel to AB, and intersect it with an arc drawn from the center E, with a radius equal to that of the circle. Points J, K, and L are found in a similar manner.

Another method that may be employed is shown in the right half of the figure. With center at 11' and the chord 11–6 as radius, strike an arc. With 10' as center and the chord 10–6 as radius, strike an arc. Continue similarly with centers 9', 8', and 7'. Draw the required cycloid tangent to these arcs.

Either method may be used; however, the second is the shorter one and is preferred. It is evident, from the tangent arcs drawn in the man-

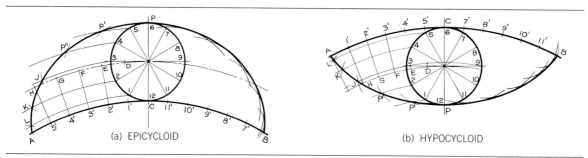

Fig. 5.66   Epicycloid and Hypocycloid (§5.66).

ner just described, that the line joining the generating point and the point of contact for the generating circle is a normal of the cycloid. The lines 1′–P″ and 2′–P′, for instance, are normals; this property makes the cycloid suitable for the outlines of gear teeth.

5.66   To Draw an Epicycloid or a Hypocycloid

Fig. 5.66   If the generating point P is on the circumference of a circle that rolls along the convex side of a larger circle, (a), the curve generated is an epicycloid. If the circle rolls along the concave side of a larger circle, (b), the curve generated is a hypocycloid. These curves are drawn in a manner similar to the cycloid, Fig. 5.65. These curves, like the cycloid, are used to form the outlines of certain gear teeth and are, therefore, of practical importance in machine design.

5.67   Computer Graphics

Through the use of various application programs and routines available in computer graphics, it is possible to accurately establish the various geometric constructions shown in Chapter 5. CAD programs are particularly well suited for repetitive operations such as dividing a line into a number of equal parts, and for generating lines representing mathematical curves such as the hyperbola and parabola. Examples of CAD-produced geometric shapes and surfaces are shown in Fig. 5.67.

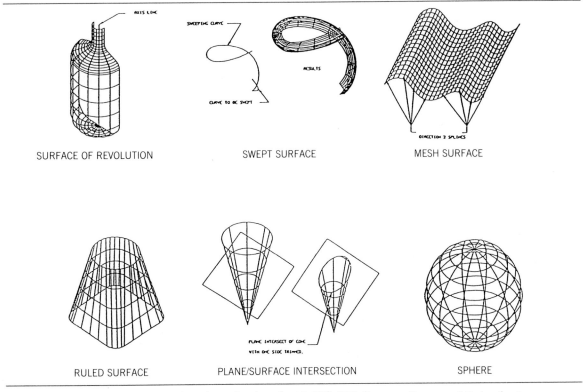

Fig. 5.67   Geometric Shapes and Surfaces Produced with TRI-CAD System. *Courtesy of Cadgrafix, Inc.*

GEOMETRIC CONSTRUCTION PROBLEMS

Geometric constructions should be made very accurately, with a hard pencil (2H to 4H) having a long, sharp, conical point. Draw given and required lines dark and medium in thickness, and draw construction lines *very light*. Do not erase construction lines. Indicate points and lines as described in §5.1.

The instructor will make assignments from the many problems that follow. Use Layout A–2 (see inside of back cover) divided into four parts, as shown in Fig. 5.68, or Layout A4–2 (adjusted). Additional sheets with other problems selected from Figs. 5.69–5.80 and drawn on the same sheet layout, may be assigned by the instructor.

Many problems are dimensioned in the metric system. The instructor may assign the student to convert the remaining problems to metric measure. See inside front cover for decimal and millimeter equivalents.

The student should exercise care in setting up each problem so as to make the best use of the space available, to present the problem to best advantage, and to produce a pleasing appearance. Letter the principal points of all constructions in a manner similar to the various illustrations in this chapter.

Since many of the problems in this chapter are of a general nature, they can also be solved on most computer graphics systems. If a system is available, the instructor may choose to assign specific problems to be completed by this method.

Problems in convenient form for solution may be found in *Technical Drawing Problems*, Series 1, by Giesecke, Mitchell, Spencer, Hill, Dygdon, and Novak; *Technical Drawing Problems*, Series 2, by Spencer, Hill, Dygdon, and Novak; and *Technical Drawing Problems*, Series 3, by Spencer, Hill, Dygdon, and Novak; all designed to accompany this text and published by Macmillan Publishing Company.

Fig. 5.68 Geometric Constructions. Layout A–2 or A4–2 (adjusted). (Probs. 5.1–5.4)

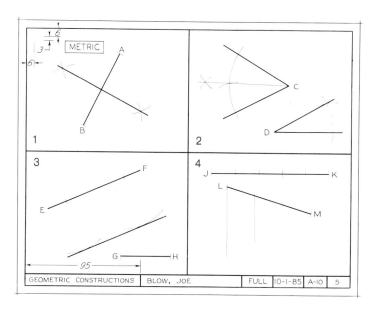

The first four problems are shown in Fig. 5.68.

Prob. 5.1 Draw an inclined line AB 65 mm long and bisect it, Fig. 5.8.

Prob. 5.2 Draw any angle with vertex at C. Bisect it, Fig. 5.10, and transfer one-half in new position at D, Fig. 5.11.

Prob. 5.3 Draw an inclined line EF and assume distance GH = 42 mm. Draw a line parallel to EF and at the distance GH from it, Fig. 5.13 (a).

Prob. 5.4 Draw the line JK 95 mm long and divide it into five equal parts with the dividers, §2.40. Draw a line LM 58 mm long and divide it into three equal parts by the parallel-line method, Fig. 5.15.

Prob. 5.5 Draw a line OP 92 mm long and divide it into three proportional parts to 3, 5, and 9, Fig. 5.17 (a).

Prob. 5.6 Draw a line 87 mm long and divide it into parts proportional to the square of x where x = 1, 2, 3, and 4, Fig. 5.17 (c).

Prob. 5.7 Draw a triangle having sides 76 mm, 85 mm, and 65 mm, Fig. 5.19. Bisect the three interior angles, Fig. 5.10. The bisectors should meet at a point. Draw the inscribed circle with the point as center.

Prob. 5.8 Draw a right triangle having the hypotenuse 65 mm and one leg 40 mm, Figs. 5.3 and 5.20, and draw a circle through the three vertexes, Fig. 5.32.

Prob. 5.9 Draw an inclined line QR 84 mm long. Select a point P on the line 32 mm from Q, and erect a perpendicular, Fig. 5.18 (c). Assume a point S about 45.5 mm from the line, and erect a perpendicular from S to the line, Fig. 5.18 (b).

Prob. 5.10 Draw two lines making an angle of $35\frac{1}{2}°$ with each other, using the tangent method, Fig. 5.21 (a). Check with protractor, §2.18.

Prob. 5.11 Draw two lines making an angle of 33°16′ with each other, using the sine method, Fig. 5.21 (b). Check with protractor, §2.18.

Prob. 5.12 Draw an equilateral triangle, Fig. 5.3 (a), having 63.5 mm sides, Fig. 5.22 (a). Bisect the interior angles, Fig. 5.10. Draw the inscribed circle, using the intersection of the bisectors as center.

Prob. 5.13 Draw inclined line TU 55 mm long, and then draw a square on TU as a given side, Fig. 5.23 (a).

Prob. 5.14 Draw a 54 mm diameter circle (lightly); then inscribe a square in the circle and circumscribe a square on the circle, Fig. 5.23 (c) and (d).

Prob. 5.15 Draw a 65 mm diameter circle (lightly), find the vertexes of a regular inscribed pentagon, Fig. 5.24 (a), and join the vertexes to form a five-pointed star.

Prob. 5.16 Draw a 65 mm diameter circle (lightly), inscribe a hexagon, Fig. 5.25 (b), and circumscribe a hexagon, Fig. 5.26 (d).

Prob. 5.17 Draw a square (lightly) with 63.5 mm sides, Fig. 5.23 (b), and inscribe an octagon, Fig. 5.28 (b).

Prob. 5.18 Draw a triangle similar to that in Fig. 5.29 (a), having sides 50 mm, 38 mm, and 73 mm long, and transfer the triangle to a new location and turned 180° similar to that in Fig. 5.29 (b). Check by the pricked-point method, §5.29.

Prob. 5.19 In center of space, draw a rectangle 88 mm wide and 61 mm high. Show construction for reducing this rectangle to 70 mm wide and then to 58 mm wide, Fig. 5.31 (b).

Prob. 5.20 Draw three points arranged approximately as in Fig. 5.32 (a), and draw a circle through the three points.

Prob. 5.21   Draw a 58 mm diameter circle. Assume a point S on the left side of the circle and draw a tangent at that point, Fig. 5.34 (a). Assume a point T to the right of the circle 50 mm from its center, and draw two tangents to the circle through the point, Fig. 5.34 (b).

Prob. 5.22   Through center of space, draw a horizontal center line; then draw two circles 50 mm diameter and 38 mm diameter, respectively, with centers 54 mm apart. Locate the circles so that the construction will be centered in the space. Draw "open-belt" tangents to the circles, Fig. 5.35 (a).

Prob. 5.23   Same as Prob. 5.21, except draw "crossed-belt" tangents to the circle, Fig. 5.35 (b).

Prob. 5.24   Draw vertical line VW 33 mm from left side of space. Assume point P 44 mm farther to the right and 25 mm down from top of space. Draw a 56 mm diameter circle through P, tangent to VW, Fig. 5.36 (a).

Prob. 5.25   Draw vertical line XY 35 mm from left side of space. Assume point P 44 mm farther to the right and 25 mm down from top of space. Assume point Q on line XY and 50 mm from P. Draw a circle through P and tangent to XY at Q, Fig. 5.36 (b).

Prob. 5.26   Draw a 64 mm diameter circle with center C 16 mm directly to left of center of space. Assume point P at the lower right and 60 mm from C. Draw an arc with 25 mm radius through P and tangent to the circle, Fig. 5.36 (c).

Prob. 5.27   Draw a vertical line and a horizontal line, each 65 mm long, Fig. 5.37 (I). Draw an arc with 38 mm radius, tangent to the lines.

Prob. 5.28   Draw a horizontal line 20 mm up from bottom of space. Select a point on the line 50 mm from the left side of space, and through it draw a line upward to the right at 60° to horizontal. Draw arcs with 35 mm radius within obtuse angle and acute angle, respectively, tangent to the two lines, Fig. 5.38.

Prob. 5.29   Draw two intersecting lines making an angle of 60° with each other similar to Fig. 5.38 (a). Assume a point P on one line at a distance of 45 mm from the intersection. Draw an arc tangent to both lines with one point of tangency at P, Fig. 5.33.

Prob. 5.30   Draw vertical line AB 32 mm from left side of space. Draw an arc of 42 mm radius with center 75 mm to right of line and in lower right portion of space. Draw an arc of 25 mm radius tangent to AB and to the first arc, Fig. 5.39.

Prob. 5.31   With centers 20 mm up from bottom of space, and 86 mm apart, draw arcs of radii 44 mm and 24 mm, respectively. Draw an arc of 32 mm radius tangent to the two arcs, Fig. 5.40.

Prob. 5.32   Draw two circles as in Prob. 5.22. Draw an arc of 70 mm radius tangent to the upper sides of, and enclosing, the circles, Fig. 5.41 (a). Draw an arc of 50 mm radius tangent to the circles but enclosing only the smaller circle, Fig. 5.41 (b).

Prob. 5.33   Draw two parallel inclined lines 45 mm apart. Choose a point on each line and connect them with an ogee curve tangent to the two parallel lines, Fig. 5.43 (a).

Prob. 5.34   Draw an arc of 54 mm radius that subtends an angle of 90°. Find the length of the arc by two methods, Fig. 5.45 (a) and (c). Calculate the length of the arc and compare with the lengths determined graphically. See note at end of §5.45.

Prob. 5.35   Draw a major axis 102 mm long (horizontally) and a minor axis 64 mm long, with their intersection at the center of the space. Draw ellipse by foci method with at least five points in each quadrant, Fig. 5.48.

Prob. 5.36 Draw axes as in Prob. 5.35, and draw ellipse by trammel method, Fig. 5.49.

Prob. 5.37 Draw axes as in Prob. 5.35, and draw ellipse by concentric-circle method, Fig. 5.50.

Prob. 5.38 Draw axes as in Prob. 5.35, and draw ellipse by parallelogram method, Fig. 5.52 (a).

Prob. 5.39 Draw conjugate diameters intersecting at center of space. Draw 88 mm diameter horizontally, and 70 mm diameter at 60° with horizontal. Draw oblique-circle ellipse, Fig. 5.51. Find at least five points in each quadrant.

Prob. 5.40 Draw conjugate diameters as in Prob. 5.39, and draw ellipse by parallelogram method, Fig. 5.52 (b).

Prob. 5.41 Draw axes as in Prob. 5.35, and draw an approximate ellipse, Fig. 5.56.

Prob. 5.42 Draw a parabola with a vertical axis, and the focus 12 mm from the directrix, Fig. 5.57 (b). Find at least nine points on the curve.

Prob. 5.43 Draw a hyperbola with a horizontal transverse axis 25 mm long and the foci 38 mm apart, Fig. 5.60 (b). Draw the asymptotes.

Prob. 5.44 Draw a horizontal line near bottom of space, and a vertical line near left side of space. Assume point P 16 mm to right of vertical line and 38 mm above horizontal line. Draw equilateral hyperbola through P and with reference to the two lines as asymptotes. Use either method of Fig. 5.61.

Prob. 5.45 Using the center of the space as the pole, draw a spiral of Archimedes with the generating point moving in a counterclockwise direction and away from the pole at the rate of 25 mm in each convolution, Fig. 5.62.

Prob. 5.46 Through center of space, draw a horizontal center line, and on it construct a right-hand helix 50 mm diameter, 64 mm long, and with a lead of 25 mm, Fig. 5.63. Draw only a half-circular end view.

Prob. 5.47 Draw the involute of an equilateral triangle with 15 mm sides, Fig. 5.64 (b).

Prob. 5.48 Draw the involute of a 20 mm diameter circle, Fig. 5.64 (d).

Prob. 5.49 Draw a cycloid generated by a 30 mm diameter circle rolling along a horizontal straight line, Fig. 5.65.

Prob. 5.50 Draw an epicycloid generated by a 38 mm diameter circle rolling along a circular arc having a radius of 64 mm, Fig. 5.66 (a).

Prob. 5.51 Draw a hypocycloid generated by a 38 mm diameter circle rolling along a circular arc having a radius of 64 mm, Fig. 5.66 (b).

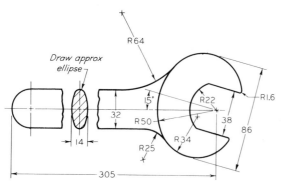

Fig. 5.69   Spanner.*

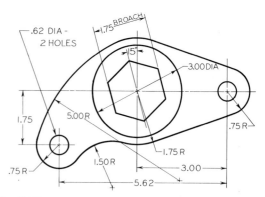

Fig. 5.70   Rocker Arm.*

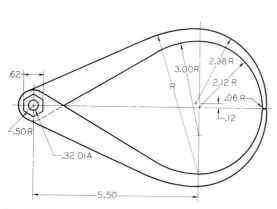

Fig. 5.71   Outside Caliper.*

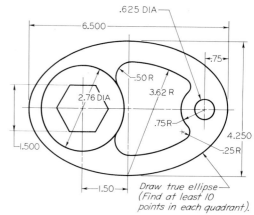

Fig. 5.72   Special Cam.*

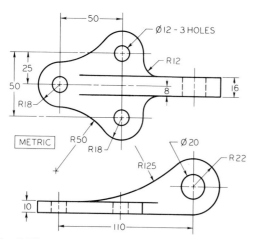

Fig. 5.73   Boiler Stay.*

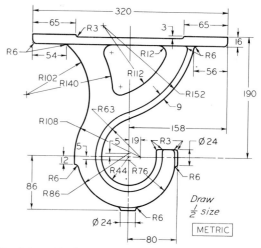

Fig. 5.74   Shaft Hanger Casting.*

*Using Layout A–2 or A4–2 (adjusted), draw assigned problem with instruments. Omit dimensions and notes unless assigned by instructor.

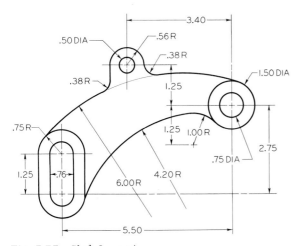

Fig. **5.75**   Shift Lever.*

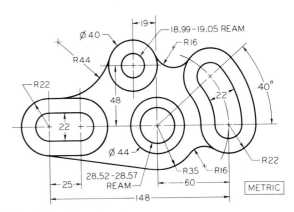

Fig. **5.76**   Gear Arm.*

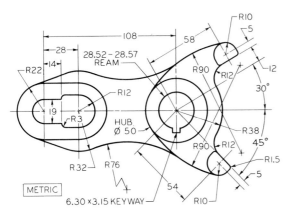

Fig. **5.77**   Form Roll Lever.*

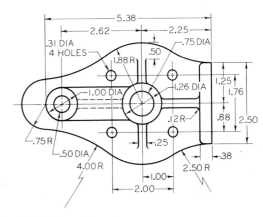

Fig. **5.78**   Press Base.*

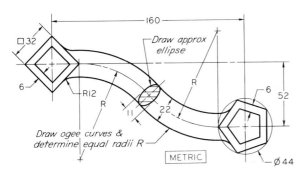

Fig. **5.79**   Special S-Wrench.*

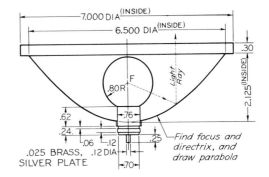

Fig. **5.80**   Auto Headlight Reflector.*

*Using Layout A–2 or A4–2 (adjusted), draw assigned problem with instruments. Omit dimensions and notes unless assigned by instructor.

CHAPTER 6

Sketching and Shape Description

The importance of freehand drawing or sketching in engineering, design, and technical communications cannot be overestimated. To the person who possesses a complete knowledge of drawing as a language, the ability to execute quick, accurate, and clear sketches of ideas and designs constitutes a valuable means of expression. The old Chinese saying that "one picture is worth a thousand words" is not without foundation.

Most original design ideas find their first expression through the medium of a freehand sketch, §16.2. Freehand sketching is a valuable means of amplifying and clarifying, as well as recording, verbal explanations. Executives sketch freehand daily to explain their ideas to subordinates. Engineers often prepare their designs and turn them over to their detailers or designers in this convenient form as shown in the well-executed sketch of details for a steam locomotive, Fig. 6.1.

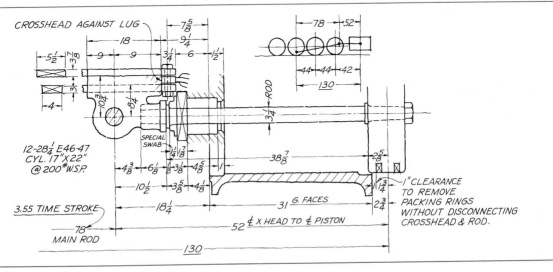

Fig. 6.1 Typical Design Sketch.

163

6.1 Technical Sketching

Freehand sketches are of great assistance to designers in organizing their thoughts and recording their ideas. Sketching is an effective and economical means of formulating various solutions to a given problem so that a choice can be made between them at the outset. Often much time can be lost if the designer starts his or her scaled layout before adequate preliminary study with the aid of sketches. Information concerning changes in design or covering replacement of broken parts or lost drawings is usually conveyed through sketches. Many engineers consider the ability to render serviceable sketches of greater value to them than skill in instrument drawing. The designer, technician, or engineer will find daily use for this valuable means of formulating, expressing, and recording ideas.

The degree of perfection required in a given sketch depends on its use. Sketches hurriedly made to supplement oral description may be rough and incomplete. On the other hand, if a sketch is the medium of conveying important and precise information to engineers, technicians, or skilled workers, it should be executed as carefully as possible under the circumstances.

The term "freehand sketch" is too often understood to mean a crude or sloppy freehand drawing in which no particular effort has been made. On the contrary, a freehand sketch should be made with care and with attention to proportion, clarity, and correct line widths.

6.2 Sketching Materials

One advantage of freehand sketching is that it requires only pencil, paper, and eraser—items that anyone has for ready use.

When sketches are made in the field, where an accurate record is required, a small notebook or sketching pad is frequently used. Often clipboards are employed to hold the paper. Graph paper is helpful to the sketcher, especially to the person who cannot sketch reasonably well without guide lines. Paper with 4, 5, 8, or 10 squares per inch is recommended. Such paper is convenient for sketching to scale since values can be assigned to the squares, and the squares

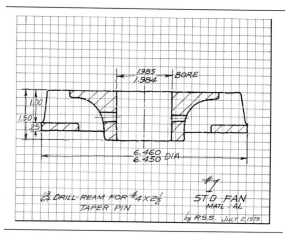

Fig. 6.2 Sketch on Graph Paper.

counted to secure proportional distances, as shown in Fig. 6.2.

Sketching pads of plain tracing paper are available, accompanied by a master cross-section sheet. The drafter places a blank sheet over the master grid and can see the grid through the transparent sheet.

An alternate procedure is to draw, with instruments, a master cross-section sheet, using ink, making the squares either 10 mm or .25″ as desired. Ordinary bond typewriter paper is then placed over the master sheet and the sketch made thereon. Such a sketch is not only more uniform and "true" but shows up better because the cross-section lines are absent.

For isometric sketching, a specially ruled isometric paper is available, Fig. 6.26.

Soft pencils, such as HB or F, should be used for freehand sketching. For carefully made sketches, two soft erasers are recommended, a Pink Pearl and a Mars-Plastic, Fig. 2.16.

6.3 Types of Sketches

Since technical sketches are made of three-dimensional objects, the form of the sketch conforms approximately to one of the four standard types of projection, as shown in Fig. 6.3. In *multiview* projection, (a), the object is described by its necessary views, as discussed in §§6.18–6.24.

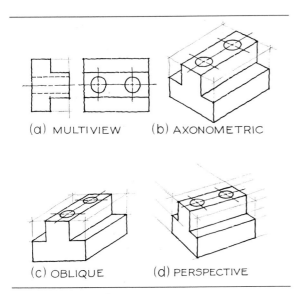

(a) MULTIVIEW   (b) AXONOMETRIC

(c) OBLIQUE   (d) PERSPECTIVE

Fig. 6.3   Types of Projection.

Or the object may be shown pictorially in a single view, by *axonometric, oblique,* or *perspective sketches,* (b), (c), and (d), as discussed in §§6.11–6.17.

6.4   Scale
Sketches usually are *not made to any scale.* Objects should be sketched in their correct proportions as accurately as possible, by eye. However, cross-section paper provides a ready scale (by counting squares) that may be used to assist in sketching to correct proportions. The size of the sketch is purely optional, depending on the complexity of the object and the size of paper available. Small objects are often sketched oversize so as to show the necessary details clearly.

6.5   Technique of Lines
The chief difference between an instrument drawing and a freehand sketch lies in the character or *technique* of the lines. A good freehand line is not expected to be as rigidly straight or exactly uniform as an instrument line. While the effectiveness of an instrument line lies in exacting uniformity, the quality of freehand line lies in its *freedom* and *variety,* Figs. 6.4 and 6.7.

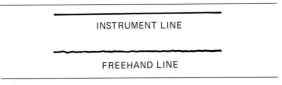

INSTRUMENT LINE

FREEHAND LINE

Fig. 6.4   Comparison of Lines.

Conventional lines, drawn instrumentally, are shown in Fig. 2.14, and the corresponding freehand renderings are shown in Fig. 6.5. The freehand construction line is a very light rough line in which some strokes may overlap. All other lines should be dark and clean-cut. Accent the ends of all dashes, and maintain a sharp contrast between the line thicknesses. In particular, make visible lines **heavy** so the outline will stand out clearly, and make hidden lines, center lines, dimension lines, and extension lines *thin.*

6.6   Sharpening Sketching Pencils
For sketching, use a mechanical pencil with a soft lead, such as HB or F, and sharpen it to a conical point, as shown in Fig. 2.11 (c). Use this

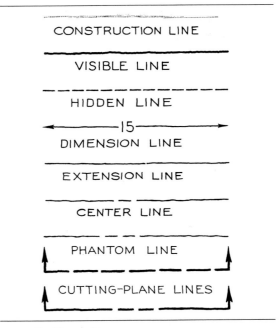

CONSTRUCTION LINE

VISIBLE LINE

HIDDEN LINE

|——————15——————|

DIMENSION LINE

EXTENSION LINE

CENTER LINE

PHANTOM LINE

CUTTING-PLANE LINES

Fig. 6.5   Sketch Lines.

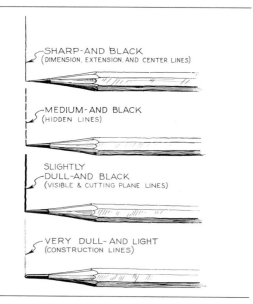

Fig. 6.6   Pencil Points.

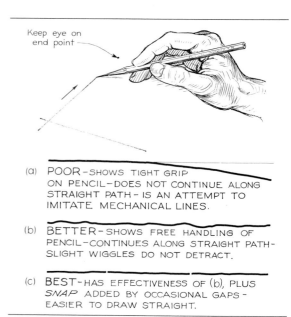

(a)   POOR—SHOWS TIGHT GRIP ON PENCIL—DOES NOT CONTINUE ALONG STRAIGHT PATH—IS AN ATTEMPT TO IMITATE MECHANICAL LINES.

(b)   BETTER—SHOWS FREE HANDLING OF PENCIL—CONTINUES ALONG STRAIGHT PATH—SLIGHT WIGGLES DO NOT DETRACT.

(c)   BEST—HAS EFFECTIVENESS OF (b), PLUS *SNAP* ADDED BY OCCASIONAL GAPS—EASIER TO DRAW STRAIGHT.

Fig. 6.7   Drawing Horizontal Lines.

sharp point for center lines, dimension lines, and extension lines. For visible lines, hidden lines, and cutting-plane lines, round off the point slightly to produce the desired thickness of line, Fig. 6.6. Make all lines dark, with the exception of construction lines, which should be very light.

The use of thin-lead mechanical pencils with suitable diameters and grades of leads minimizes the need for sharpening and point dressing.

6.7   Straight Lines

Since the majority of lines on the average sketch are straight lines, it is necessary to learn to make them well. Hold the pencil naturally, about $1\frac{1}{2}''$ back from the point, and approximately at right angles to the line to be drawn. Draw horizontal lines from left to right with a free and easy wrist-and-arm movement, Fig. 6.7. Draw vertical lines downward with finger and wrist movements, Fig. 6.8.

Inclined lines may be made to conform in direction to horizontal or vertical lines by shifting position with respect to the paper or by turning the paper slightly; hence, they may be drawn with the same general movements, Fig. 6.9.

In sketching long lines, mark the ends of the line with light dots. Then move the pencil back and forth between the dots in long sweeps, keeping the eye always on the dot toward which the pencil is moving, the point of the pencil touching the paper lightly, and each successive stroke correcting the defects of the preceding strokes. When the path of the line has been established sufficiently, apply a little more pressure, replacing the trial series with a distinct line. Then, dim the line with a soft eraser and draw the final line clean-cut and dark, keeping the eye now on the point of the pencil.

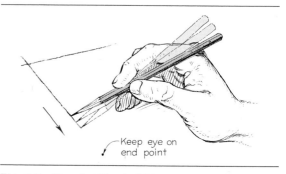

Fig. 6.8   Drawing Vertical Lines.

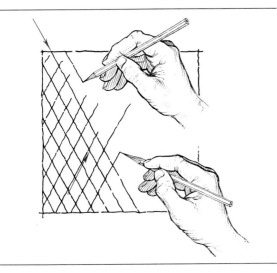

Fig. 6.9   Drawing Inclined Lines.

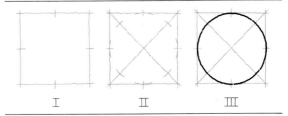

Fig. 6.11   Sketching a Circle.

A common method of finding the midpoint of a line AB, shown at (c), is to hold the pencil in the left hand with the thumb gaging the estimated half-distance. Try this distance on the left and then on the right until the center is located by trial, and mark the center C, as shown. Another method is to mark the total distance AB on the edge of a strip of paper and then to fold the paper to bring points A and B together, thus locating center C at the crease. To find quarter points, the folded strip can be folded once more.

6.8   Circles and Arcs

Small circles and arcs can be easily sketched in one or two strokes, as for the circular portions of letters, without any preliminary blocking in.

One method of sketching a larger circle, Fig. 6.11, is first to sketch lightly the enclosing square, mark the midpoints of the sides, draw light arcs tangent to the sides of the square, and then heavy in the final circle.

An easy method of blocking in horizontal or vertical lines is to hold the hand and pencil rigidly and glide the fingertips along the edge of the pad or board, as shown in Fig. 6.10 (a).

Another method, (b), is to mark the distance on the edge of a card or a strip of paper and transfer this distance at intervals, as shown; then draw the final line through these points. Or the pencil may be held as shown at the lower part of (b), and distance marks made on the paper at intervals by tilting the lead down to the paper. It will be seen that both methods of transferring distances are substitutes for the dividers and will have many uses in sketching.

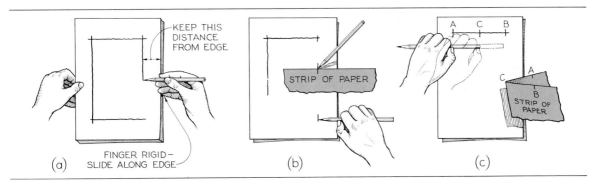

Fig. 6.10   Blocking in Horizontal and Vertical Lines.

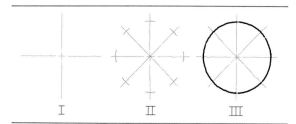

Fig. 6.12 Sketching a Circle.

Another method, Fig. 6.12, is to sketch the two center lines, add light 45° radial lines, sketch light arcs across the lines at the estimated radius distance from the center, and finally sketch the required circle heavily. Dim all construction lines with a soft eraser before heavying in the final circle.

An excellent method, particularly for large circles, Fig. 6.13 (a), is to mark the estimated radius on the edge of a card or scrap of paper, to set off from the center as many points as desired, and to sketch the final heavy circle through these points.

The clever drafter will prefer the method at I and II, in which the hand is used as a compass. Place the tip of the little finger, or the knuckle joint of the little finger, at the center; "feed" the pencil out to the desired radius, hold this position rigidly, and carefully revolve the paper with the other hand, as shown. If you are using a

sketching pad, place the pad on your knee and revolve the entire pad on the knee as a pivot.

At III, two pencils are held rigidly like a compass and the paper is slowly revolved.

Methods of sketching arcs, Fig. 6.14, are adaptations of those used for sketching circles. In general, it is easier to sketch arcs with the hand and pencil on the concave side of the curve. In sketching tangent arcs, always keep in mind the actual geometric constructions, carefully approximating all points of tangency.

6.9 Ellipses

If a circle is viewed obliquely, Fig. 5.50 (b), it appears as an ellipse. With a little practice, you can learn to sketch small ellipses with a free arm movement, Fig. 6.15 (a). Hold the pencil naturally, rest the weight on the upper part of the forearm, and move the pencil rapidly above the paper in the elliptical path desired; then lower the pencil so as to describe several light overlapping ellipses, as shown at I. Dim all lines with a soft eraser and heavy in the final ellipse, II.

Another method, (b), is to sketch lightly the enclosing rectangle, I, mark the midpoints of the sides, and sketch light tangent arcs, as shown. Then, II, complete the ellipse lightly, dim all lines with a soft eraser, and heavy in the final ellipse.

The same general procedure shown at (b) may be used in sketching the ellipse on the given axes, as shown at (c).

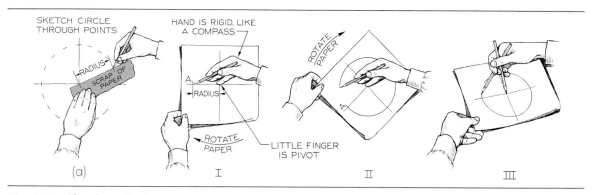

Fig. 6.13 Sketching Circles.

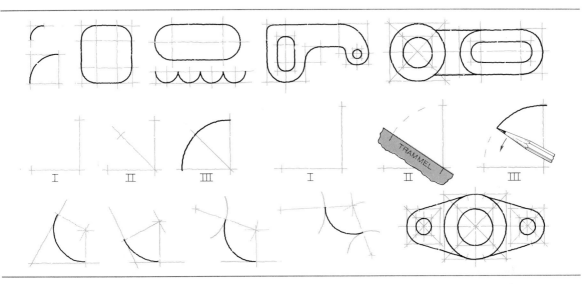

Fig. 6.14 Sketching Arcs.

The trammel method, (d), is excellent for sketching large ellipses. Prepare a "trammel" on the edge of a card or strip of paper, move it to different positions, and mark points on the ellipse at A. The trammel method is explained in §5.50. Sketch the final ellipse through the points, as shown. For sketching isometric ellipses, see §6.13.

6.10 Proportions

The most important rule in freehand sketching is *keep the sketch in proportion*. No matter how brilliant the technique or how well the small details are drawn, if the proportions—especially the large overall proportions—are bad, the sketch will be bad. First, the relative proportions of the height to the width must be carefully es-

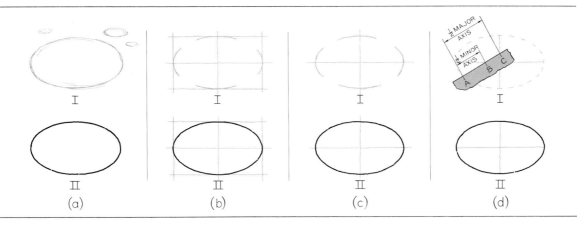

Fig. 6.15 Sketching Ellipses.

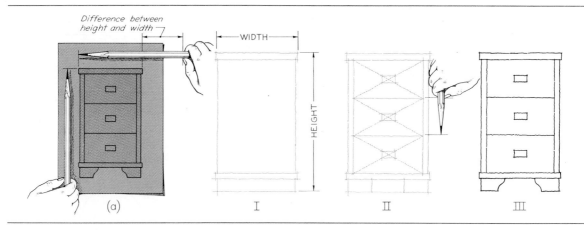

Fig. 6.16   Sketching a Utility Cabinet.

tablished; then as you proceed to the medium-sized areas and the small details, constantly compare each new estimated distance with already established distances.

If you are working from a given picture, such as the utility cabinet in Fig. 6.16 (a), it is first necessary to establish the relative width compared to the height. One way is to use the pencil as a measuring stick, as shown. In this case, the height is about $1\frac{3}{4}$ times the width. Then

I.   Sketch the enclosing rectangle in the correct proportion. In this case, the sketch is to be slightly larger than the given picture.

II.   Divide the available drawer space into three parts with the pencil by trial, as shown. Sketch light diagonals to locate centers of drawers, and block in drawer handles. Sketch all remaining details.

III.   Dim all construction with a soft eraser, and heavy in all final lines.

Another method of estimating distances is illustrated in Fig. 6.17. On the edge of a card or strip of paper, mark an arbitrary unit. Then see how many units wide and how many units high the desk is. If you are working from the actual object, you can use a scale, a piece of paper, or

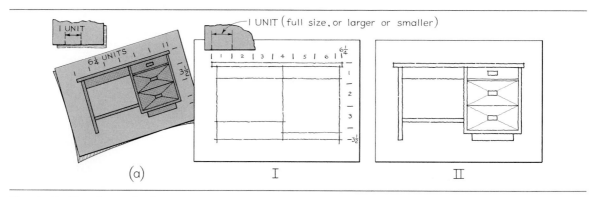

Fig. 6.17   Sketching a Desk.

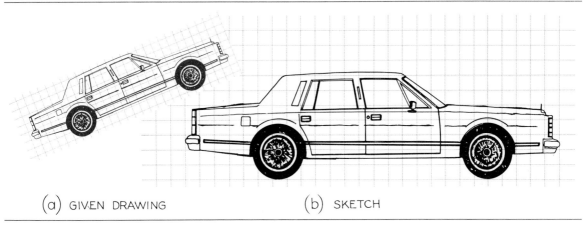

(a) GIVEN DRAWING   (b) SKETCH

Fig. 6.18 Squares Method.

the pencil itself as a unit to determine the proportions.

To sketch an object composed of many curves to the same scale or to a larger or smaller scale, the method of "squares" is recommended, Fig. 6.18. On the given picture, rule accurate grid lines to form squares of any convenient size. It is best to use a scale and some convenient spacing, such as either .50″ or 10 mm. On the new sheet rule a similar grid, making the spacing of the lines proportional to the original, but reduced or enlarged as desired. Make the final sketch by drawing the lines in and across the grid lines as in the original, as near as you can estimate by eye.

In sketching from an actual object, you can easily compare various distances on the object by using the pencil to compare measurements, as shown in Fig. 6.19. While doing this, do not change your position, and always hold your pencil at arm's length. The length sighted can then be compared in similar manner with any other dimension of the object. If the object is small, such as a machine part, you can compare distances in the manner of Fig. 6.16, by actually placing the pencil against the object itself.

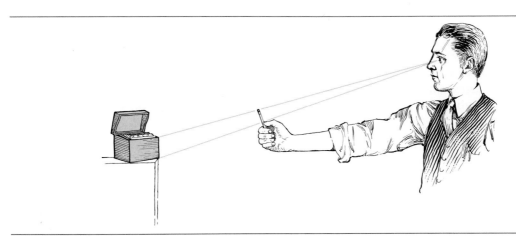

Fig. 6.19 Estimating Dimensions.

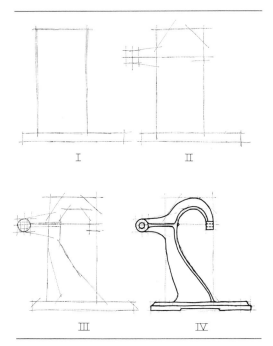

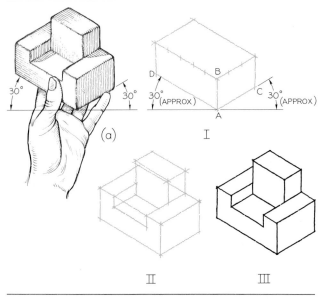

Fig. 6.21   Isometric Sketching.

Fig. 6.20   Blocking in an Irregular Object (Shaft Hanger).

In establishing proportions, the blocking-in method is recommended, especially for irregular shapes. The steps for blocking in and completing the sketch of a Shaft Hanger are shown in Fig. 6.20. As always, first give attention to the main proportions, next to the general sizes and direction of flow of curved shapes, and finally to the snappy lines of the completed sketch.

In making sketches from actual machine parts, it is necessary to use the measuring tools used in the shop, especially those needed to determine dimensions that must be relatively accurate. For a discussion of these methods, see Chapter 12.

6.11   Pictorial Sketching

We will now examine several simple methods of preparing pictorial sketches that will be of great assistance in learning the principles of multiview projection. A detailed and more technical treat-ment of pictorial drawing is given in Chapters 18–20.

6.12   Isometric Sketching

To make an isometric sketch from an actual ob-ject, hold the object in your hand and tilt it to-ward you, as shown in Fig. 6.21 (a). In this po-sition, the front corner will appear vertical, and the two receding bottom edges and those parallel to them, respectively, will appear at about 30° with horizontal, as shown. The steps in sketch-ing are

I.   Sketch the enclosing box lightly, making AB vertical and AC and AD approximately 30° with horizontal. These three lines are the *iso-metric axes*. Make AB, AC, and AD approximately proportional in length to the actual correspond-ing edges on the object. Sketch the remaining lines parallel, respectively, to these three lines.

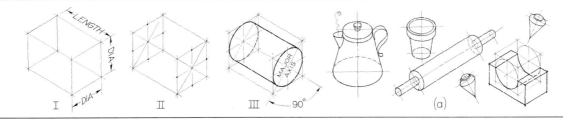

Fig. 6.22 Isometric Ellipses.

II. Block in the recess and the projecting block.

III. Dim all construction lines with a soft eraser, and heavy in all final lines.

NOTE   The angle of the receding lines may be less than 30°, say, 20° or 15°. Although the result will not be an isometric sketch, the sketch may be more pleasing and effective in many cases.

6.13   Isometric Ellipses
As shown in Fig. 5.50 (b), a circle viewed at an angle appears as an ellipse. When objects having cylindrical or conical shapes are placed in the isometric or other oblique positions, the circles will be viewed at an angle and will appear as ellipses, Fig. 6.22.

The most important consideration in sketching isometric ellipses is: *The major axis of the ellipse is always at right angles to the center line of the cylinder, and the minor axis is at right angles to the major axis and coincides with the center line.*

Two views of a block with a large cylindrical hole are shown in Fig. 6.23 (a). The steps in sketching the object are

I.   Sketch the block and the enclosing parallelogram for the ellipse, making the sides of the parallelogram parallel to the edges of the block and equal in length to the diameter of the hole. Draw diagonals to locate the center of the hole, and then draw center lines AB and CD. Points A, B, C, and D will be midpoints of the sides of the parallelogram, and the ellipse will be tangent to the sides at those points. The major axis will be

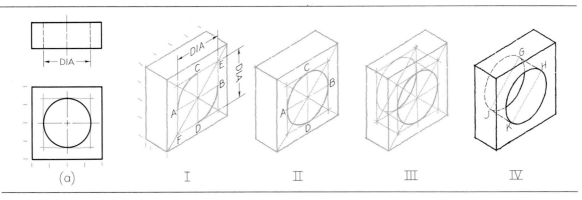

Fig. 6.23 Isometric Ellipses.

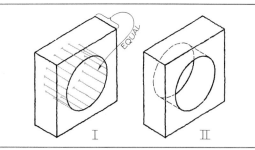

Fig. 6.24   Isometric Ellipses.

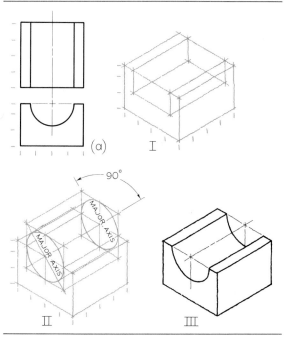

Fig. 6.25   Sketching Semiellipses.

on the diagonal EF, which is at right angles to the center line of the hole, and the minor axis will fall along the short diagonal. Sketch long, flat elliptical sides CA and BD, as shown.

II.   Sketch short small-radius arcs CB and AD to complete the ellipse. Avoid making the ends of the ellipse "squared off" or pointed like a football.

III.   Sketch lightly the parallelogram for the ellipse that lies in the back plane of the object, and sketch the ellipse in the same manner as the front ellipse.

IV.   Draw lines GH and JK tangent to the two ellipses. Dim all construction with a soft eraser, and heavy in all final lines.

Another method for determining the back ellipse is shown in Fig. 6.24.

I.   Select points at random on the front ellipse and sketch "depth lines" equal in length to the depth of the block.

II.   Sketch the ellipse through the ends of the lines, as shown.

Two views of a bearing with a semicylindrical opening are shown in Fig. 6.25 (a). The steps in sketching are

I.   Block in the object, including the rectangular space for the semicylinder.

II.   Block in the box enclosing the complete cylinder. Sketch the entire cylinder lightly.

III.   Dim all construction lines, and heavy in all final lines, showing only the lower half of the cylinder.

6.14   Sketching on Isometric Paper

Two views of a guide block are shown in Fig. 6.26 (a). The steps in sketching illustrate not only the use of isometric paper, but also the sketching of individual planes or faces in order to build up a pictorial visualization from the given views.

I.   Sketch isometric of enclosing box, counting off the isometric grid spaces to equal the corresponding squares on the given views. Sketch surface A, as shown.

II.   Sketch additional surfaces B and C and the small ellipse.

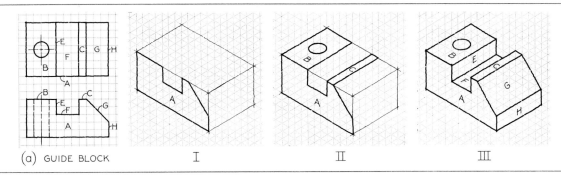

Fig. 6.26   Sketching on Isometric Paper.

III.   Sketch additional surfaces E, F, G, and H to complete the sketch.

6.15   Oblique Sketching

Another simple method for sketching pictorially is oblique sketching, Fig. 6.27. Hold the object in your hand, as shown at (a).

I.   Block in the front face of the bearing, as if you were sketching a front view.

II.   Sketch receding lines parallel to each other and at any convenient angle, say, 30° or 45° with horizontal, approximately. Cut off receding lines so that the depth appears correct. These lines may be full length, but a more natural appear-ance results if they are cut to three-quarters or one-half size, approximately. If they are full length, the sketch is a *cavalier* sketch. If half size, the sketch is a *cabinet* sketch. See §19.4.

III.   Dim all construction lines with a soft eraser and heavy in the final lines.

NOTE   Oblique sketching is a less suitable method for any object having circular shapes in or parallel to more than one plane of the object, because ellipses result when circular shapes are viewed obliquely. Therefore, place the object with most or all of the circular shapes toward you, so that they will appear as true circles and arcs in oblique sketching, as in Fig. 6.27.

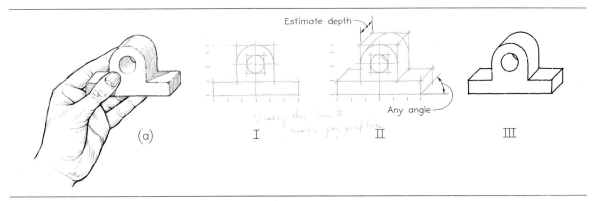

Fig. 6.27   Sketching in Oblique.

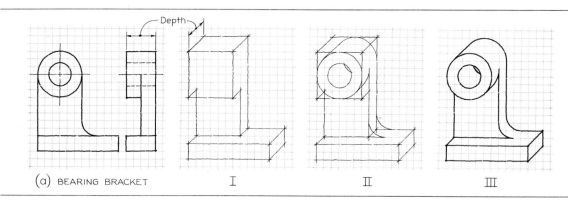

Fig. 6.28   Oblique Sketching on Cross-section Paper.

6.16   Oblique Sketching on Graph Paper

Ordinary graph paper is suitable and convenient for oblique sketching. Two views of a bearing bracket are shown in Fig. 6.28 (a). The dimensions are determined simply by counting the squares.

I.   Sketch lightly the enclosing box construction. Sketch the receding lines at 45° diagonally through the squares. To establish the depth at a reduced scale, sketch the receding lines diagonally through half as many squares as the given number shown at (a).

II.   Sketch all arcs and circles.

III.   Heavy in all final lines.

6.17   Perspective Sketching

The bearing sketched in oblique in Fig. 6.27 can easily be sketched in *one-point perspective* (one vanishing point), as shown in Fig. 6.29.

I.   Sketch the true front face of the object, just as in oblique sketching. Select the vanishing point (VP) for the receding lines. In most cases, it is desirable to place VP above and to the right of the picture, as shown, although it can be placed anywhere in the vicinity of the picture. But if it is placed too close to the center, the lines will converge too sharply, and the picture will be distorted.

II.   Sketch the receding lines toward VP.

III.   Estimate the depth to look well, and sketch in the back portion of the object. Note that the

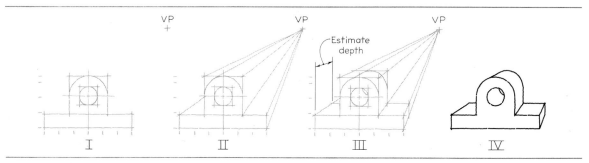

Fig. 6.29   Sketching in One-Point Perspective.

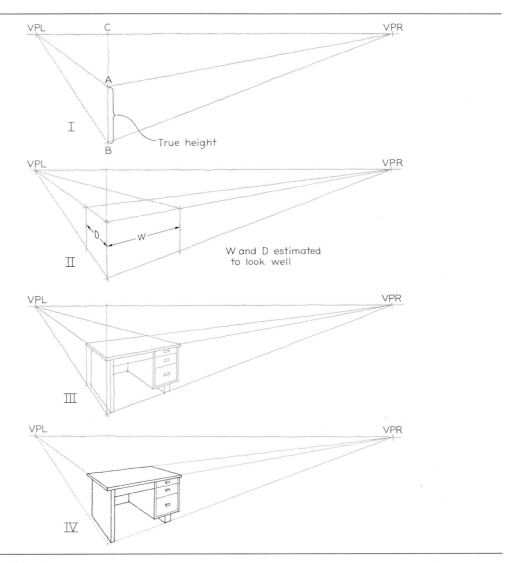

Fig. 6.30 Two-Point Perspective.

back circle and arc will be slightly smaller than the front circle and arc.

IV. Dim all construction lines with a soft eraser, and heavy in all final lines. Note the similarity between the perspective sketch and the oblique sketch in Fig. 6.27.

Two-point perspective (two vanishing points) is the most true to life of all pictorial methods, but it requires some natural sketching

ability or considerable practice for best results. The simple method shown in Fig. 6.30 can be used successfully by the nonartistic student.

I. Sketch front corner of desk in true height, and locate two *vanishing points* (VPL and VPR) on a *horizon* line (eye level). The distance CA may vary—the greater it is, the higher the eye level will be and the more we will be looking down on top of the object. A good rule of thumb

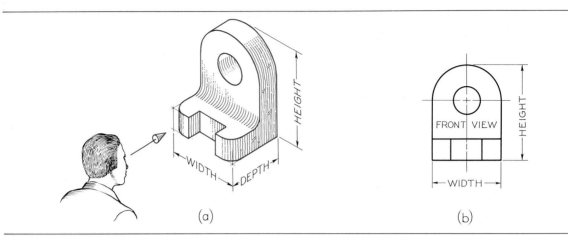

Fig. 6.31   Front View of an Object.

is to make C–VPL one-third to one-fourth of C–VPR.

II.   Estimate depth and width, and sketch enclosing box.

III.   Block in all details. Note that all parallel lines converge toward the same vanishing point.

IV.   Dim the construction lines with a soft eraser as necessary, and heavy in all final lines. Make the outlines thicker and the inside lines thinner, especially where they are close together.

6.18   Views of Objects

A pictorial drawing or a photograph shows an object as it *appears* to the observer, but not as it *is.* Such a picture cannot describe the object fully, no matter from which direction it is viewed, because it does not show the exact shapes and sizes of the several parts.

In industry, a complete and clear description of the shape and size of an object to be made is necessary in order to make certain that the object will be manufactured exactly as intended by the designer. In order to provide this information clearly and accurately, a number of *views,* systematically arranged, are used. This system of

views is called *multiview projection.* Each view provides certain definite information if the view is taken in a direction perpendicular to a principal face or side of the object. For example, as shown in Fig. 6.31 (a), an observer looking perpendicularly toward one face of the object obtains a true view of the shape and size of that side. This view as seen by the observer is shown at (b). (The observer is theoretically at an infinite distance from the object.)

An object has three principal dimensions: *width, height,* and *depth,* as shown at (a). In technical drawing, these fixed terms are used for dimensions taken in these directions, regardless of the shape of the object. The terms "length" and "thickness" are not used because they cannot be applied in all cases. Note at (b) that the front view shows only the height and width of the object and not the depth. In fact, *any one view of a three-dimensional object can show only two dimensions; the third dimension will be found in an adjacent view.*

6.19   Revolving the Object

To obtain additional views, revolve the object as shown in Fig. 6.32. First, hold the object in the front-view position, as shown at (a).

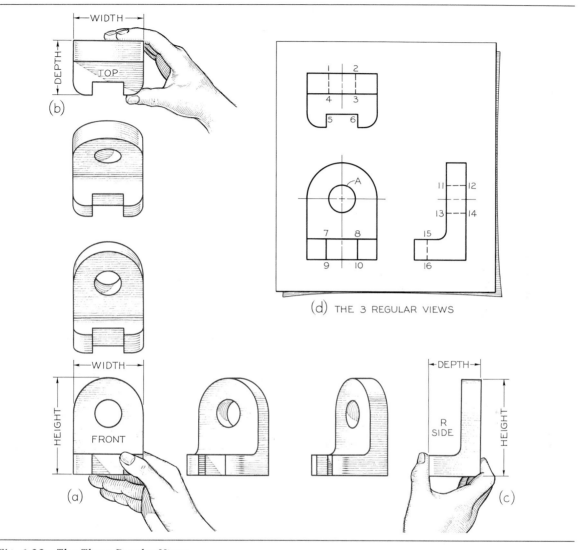

Fig. 6.32 The Three Regular Views.

To get the *top view*, (b), revolve the object so as to bring the *top of the object up and toward you*. To get the *right-side view*, (c), revolve the object so as to bring the *right side to the right and toward you*. To obtain views of any of the other sides, merely turn the object so as to bring those sides toward you.

The top, front, and right-side views, arranged closer together, are shown at (d). These are called the *three regular views* because they are the views most frequently used.

At this stage we can consider spacing between views as purely a matter of appearance. The views should be spaced well apart and yet close enough to appear related to each other. The space between the front and top views may or may not be equal to the space between the front and side views. If dimensions (Chapter 13)

179

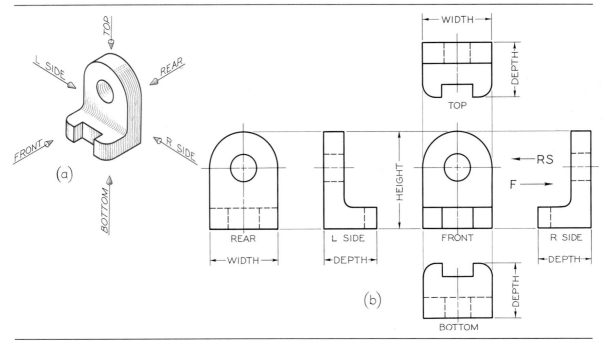

Fig. 6.33   The Six Views.

are to be added to the sketch, sufficient space for them between views will have to be allowed.

An important advantage that a view has over a photograph of an object is that hidden features can be clearly shown by means of *hidden lines,* Fig. 2.14. In Fig. 6.32 (d), surface 7–8–9–10 in the front view appears as a visible line 5–6 in the top view and as a hidden line 15–16 in the side view. Also, hole A, which appears as a circle in the front view, shows as hidden lines 1–4 and 2–3 in the top view, and 11–12 and 13–14 in the side view. For a complete discussion of hidden lines, see §6.25.

Note, too, the use of center lines for the hole in Fig. 6.32 (d). See §6.26.

6.20   The Six Views

Any object can be viewed from six mutually perpendicular directions, as shown in Fig. 6.33 (a). Thus, six views may be drawn if necessary, as shown at (b). Except as explained in §7.8, these

six views are always arranged as shown, which is the American National Standard arrangement of views. The *top, front,* and *bottom views* line up vertically, whereas the *rear, left-side, front,* and *right-side views* line up horizontally. To draw a view out of place is a serious error, generally regarded as one of the worst mistakes one can make in this subject. See Fig. 6.47.

Note that the height is shown in the rear, left-side, front, and right-side views; the width is shown in the rear, top, front, and bottom views; and the depth is shown in the four views that surround the front view, namely, the left-side, top, right-side, and bottom views. In each view, two of the principal dimensions are shown, and the third is not shown. Observe also that in the four views that surround the front view, the front of the object is faced toward the front view.

Adjacent views are reciprocal. If the front view, Fig. 6.33, is imagined to be the object itself, the right-side view is obtained by looking toward the right side of the front view, as shown by the

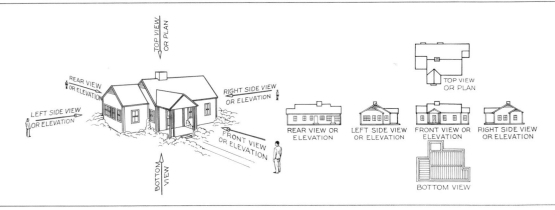

Fig. 6.34   Six Views of a House.

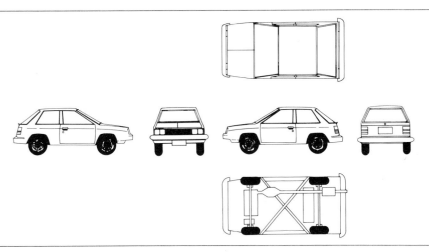

Fig. 6.35   Six Views of a Compact Automobile.

arrow RS. Likewise, if the right-side view is imagined to be the object, the front view is obtained by looking toward the left side of the right-side view, as shown by the arrow F. The same relation exists between any two adjacent views.

Obviously, the six views may be obtained either by shifting the object with respect to the observer, as we have seen, Fig. 6.32, or by shifting the observer with respect to the object, Fig. 6.33. Another illustration of the second method is given in Fig. 6.34, showing six views of a house. The observer can walk around the house and view its front, sides, and rear and can imagine the top view as seen from an airplane and the bottom or "worm's-eye view" as seen from

underneath.* Notice the use of the terms "plan," for the top view, and "elevation," for all views showing the height of the building. These terms are regularly used in architectural drawing and, occasionally, with reference to drawings in other fields.

6.21   Orientation of Front View

Six views of a compact automobile are shown in Fig. 6.35. The view chosen for the front view in this case is the side, not the front of the auto-

*Architects frequently draw the views of a building on separate sheets because of the large sizes of the drawings.

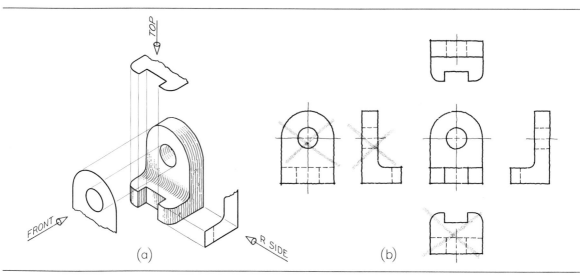

Fig. 6.36   Choice of Views.

mobile. In general, the front view should show the object in its operating position, particularly of familiar objects such as the house shown above and the automobile. A machine part is often drawn in the position it occupies in the assembly. However, in most cases this is not important, and the drafter may assume the object to be in any convenient position. For example, an automobile connecting rod is usually drawn horizontally on the sheet, Fig. 16.38. Also, it is customary to draw screws, bolts, shafts, tubes, and other elongated parts in a horizontal position, not only because they are usually manufactured in this position but also because they can be presented more satisfactorily on paper in this position.

6.22   Choice of Views

A drawing for use in production should contain *only those views* needed for a clear and complete shape description of the object. These minimum required views are referred to as the *necessary views*. In selecting views, the drafter should choose those that show best the essential contours or shapes and should give preference to those with the least number of hidden lines.

As shown in Fig. 6.36 (a), three distinctive features of this object need to be shown on the drawing.

1. Rounded top and hole, seen from the front.
2. Rectangular notch and rounded corners, seen from the top.
3. Right angle with filleted corner, seen from the side.

Another way to choose necessary views is to eliminate unnecessary views. At (b) a "thumbnail sketch" of the six views is shown. Both the front and rear views show the true shapes of the hole and the rounded top, but the front view is preferred because it has no hidden lines. Therefore, the rear view (which is seldom needed) is crossed out.

Both the top and bottom views show the rectangular notch and rounded corners, but the top view is preferred because it has fewer hidden lines.

Both the right-side and left-side views show the right angle with the filleted corner. In fact, in this case the side views are identical, except

182

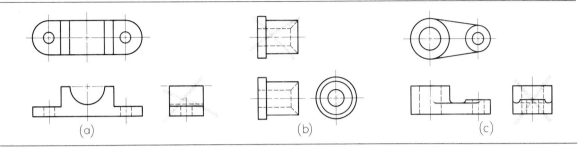

Fig. 6.37   Two Necessary Views.

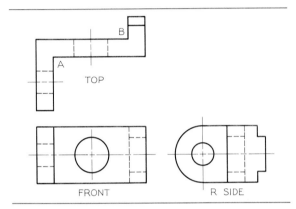

Fig. 6.38   Three Views.

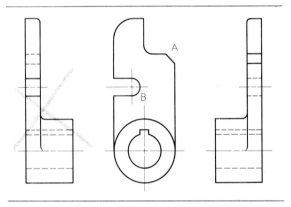

Fig. 6.39   Choice of Right-Side View.

reversed. In such instances, it is customary to choose the right-side view.

The necessary views, then, are the three remaining views: the top, front, and right-side views. These are the three regular views referred to in connection with Fig. 6.32.

More complicated objects may require more than three views or special views such as partial views, §7.9; sectional views, Chapter 9; auxiliary views, Chapter 10.

6.23   Two-View Drawings

Often only two views are needed to describe clearly the shape of an object. In Fig. 6.37 (a), the right-side view shows no significant contours of the object, and is crossed out. At (b) the top and front views are identical, so the top view is eliminated. At (c), no additional information not

already given in the front and top views is shown in the side view, so the side view is unnecessary.

The question often arises: What are the absolute minimum views required? For example, in Fig. 6.38, the top view might be omitted, leaving only the front and right-side views. However, it is more difficult to "read" the two views or visualize the object, because the characteristic "Z" shape of the top view is omitted. In addition, one must assume that corners A and B (top view) are square and not filleted. In this example, all three views are necessary.

If the object requires only two views, and the left-side and right-side views are equally descriptive, the right-side view is customarily chosen, Fig. 6.39. If contour A were omitted, then the presence of slot B would make it necessary to choose the left-side view in preference to the right-side view.

183

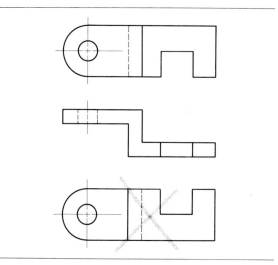

Fig. 6.40   Choice of Top View.

indicating the thickness as 0.25 mm, is sufficient. At (b), the left end is 65 mm square, the next portion is 49.22 mm diameter, the next is 31.75 mm diameter, and the portion with the thread is 20 mm diameter, as indicated in the note. Nearly all shafts, bolts, screws, and similar parts should be represented by single views in this manner.

6.25   Hidden Lines

Correct and incorrect practices in drawing hidden lines are illustrated in Fig. 6.43. In general, a hidden line should join a visible line except when it causes the visible line to extend too far, as shown at (a). In other words, *leave a gap whenever a hidden-line dash forms a continuation of a visible line.* Hidden lines should intersect to form L and T corners, as shown at (b). A hidden line preferably should "jump" a visible line when possible, (c). Parallel hidden lines should be drawn so that the dashes are staggered, in a manner similar to bricklaying, as at (d). When two or three hidden lines meet at a point, the dashes should join, as shown for the bottom of the drilled hole at (e), and for the top of a countersunk hole, (f). The example at (g) is similar to (a) in that hidden lines should not join visible lines when it makes the visible line extend too far. Correct and incorrect methods of drawing hidden arcs are shown at (h).

Poorly drawn hidden lines can easily spoil a drawing. Each dash should be carefully drawn about 5 mm long and spaced only about 1 mm apart, by eye. Accent the beginning and end of

If the object requires only two views, and the top and bottom views are equally descriptive, the top view is customarily chosen, Fig. 6.40.

If only two views are necessary, and the top view and right-side view are equally descriptive, the combination chosen is that which spaces best on the paper, Fig. 6.41.

6.24   One-View Drawings

Frequently, a single view supplemented by a note or lettered symbols is sufficient to describe clearly the shape of a relatively simple object. In Fig. 6.42 (a), one view of the Shim, plus a note

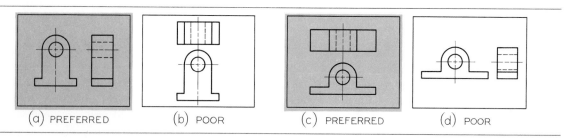

(a) PREFERRED   (b) POOR   (c) PREFERRED   (d) POOR

Fig. 6.41   Choice of Views to Fit Paper.

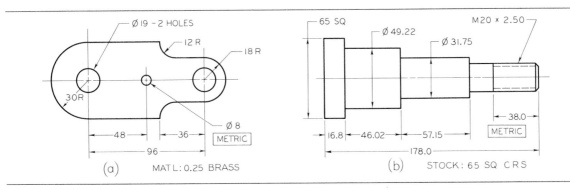

Fig. 6.42 One-View Drawings.

each dash by pressing down on the pencil, whether drawn freehand or mechanically.

In general, views should be chosen that show features with visible lines, so far as possible. After this has been done, hidden lines should be used wherever necessary to make the drawing clear. Where they are not needed for clearness, hidden lines should be omitted, so as not to clutter the drawing any more than necessary and in order to save time. The beginner, however,

would do well to be cautious about leaving out hidden lines until experience shows when they can be safely omitted.

6.26 Center Lines

Center lines (symbol: ℄) are used to indicate axes of symmetrical objects or features, bolt circles, and paths of motion. Typical applications

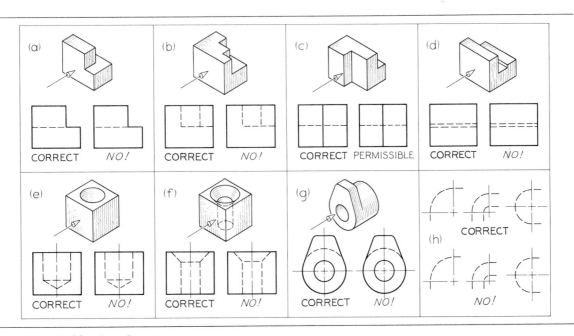

Fig. 6.43 Hidden-Line Practices.

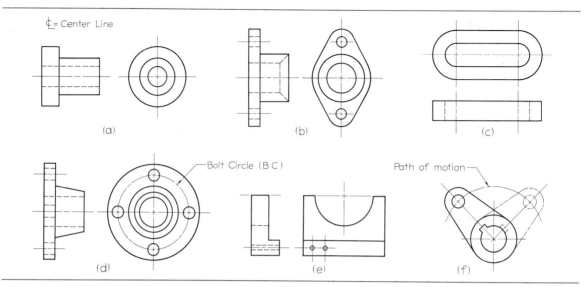

Fig. 6.44   Center-Line Applications.

are shown in Fig. 6.44. As shown at (a), a single center line is drawn in the longitudinal view and crossed center lines in the circular view. The small dashes should cross at the intersections of center lines. Center lines should extend uniformly about 8 mm outside the feature for which they are drawn.

The long dashes of center lines may vary from 20 to 40 mm or more in length, depending on the size of the drawing. The short dashes should be about 5 mm long, with spaces about 2 mm. Center lines should always start and end with long dashes. Short center lines, especially for small holes, as at (e), may be made solid as

shown. Always leave a gap as at (e) when a center line forms a continuation of a visible or hidden line. Center lines should be thin enough to contrast well with the visible and hidden lines but dark enough to reproduce well.

Center lines are useful mainly in dimensioning and should be omitted from unimportant rounded or filleted corners and other shapes that are self-locating.

6.27   Sketching Two Views

The Support Block in Fig. 6.45 (a) requires only two views. The steps in sketching are

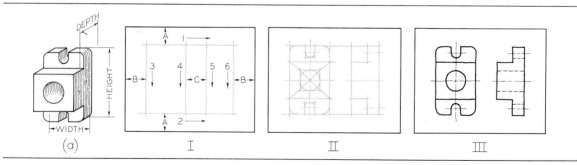

Fig. 6.45   Sketching Two Views of a Support Block.

I. Block in lightly the enclosing rectangles for the two views. Sketch horizontal lines 1 and 2 to establish the height of the object, while making spaces A approximately equal. Sketch vertical lines 3, 4, 5, and 6 to establish the width and depth in correct proportion to the already established height, while making spaces B approximately equal, and space C equal to or slightly less than space B.

II. Block in smaller details, using diagonals to locate the center, as shown. Sketch lightly the circle and arcs.

III. Dim all construction lines with a soft eraser, and heavy in all final lines.

6.28   Sketching Three Views

A Lever Bracket requiring three views is shown in Fig. 6.46 (a). The steps in sketching the three views are as follows.

I. Block in the enclosing rectangles for the three views. Sketch horizontal lines 1, 2, 3, and 4 to establish the height of the front view and the depth of the top view, making spaces A approximately equal and space C equal to or slightly less than space A. Sketch vertical lines 5, 6, 7, and 8 to establish the width of the top and front views, and the depth of the side view. Make sure that this is in correct proportion to the height, while making spaces B approximately equal and space D equal to or slightly less than one space B. Note that spaces C and D are not necessarily equal, but are independent of each other. Similarly, spaces A and B are not necessarily equal. To transfer the depth dimension from the top view to the side view, use the edge of a card or strip of paper, as shown, or transfer the distance by using the pencil as a measuring stick, as shown in Fig. 6.10 (b) and (c). Note that *the depth in the top and side views must always be equal.*

II. Block in all details lightly.

III. Sketch all arcs and circles lightly.

IV. Dim all construction lines with a soft eraser.

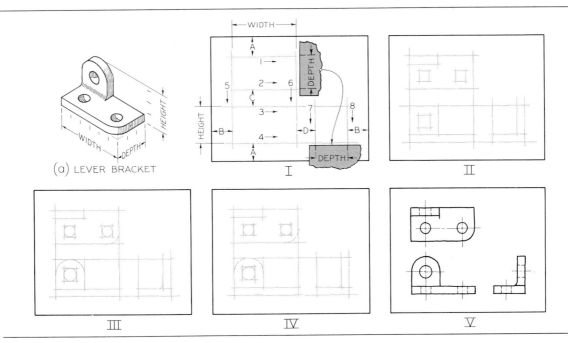

Fig. 6.46   Sketching Three Views of a Lever Bracket.

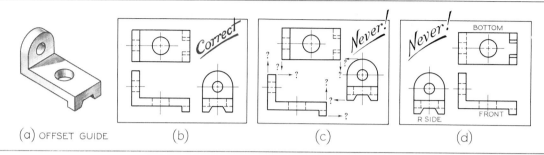

(a) OFFSET GUIDE   (b)   (c)   (d)

Fig. 6.47   Position of Views.

V.   Heavy in all final lines so that the views will stand out clearly.

6.29   Alignment of Views

Errors in arranging the views are so commonly made by students that it is necessary to repeat: The views must be drawn in accordance with the American National Standard arrangement, Fig. 6.33. In Fig. 6.47 (a) an Offset Guide is shown that requires three views. These three views, correctly arranged, are shown at (b). The top view must be directly above the front view, and the right-side view directly to the right of the front view—not out of alignment, as at (c). Also, never draw the views in reversed positions, with the bottom over the front view, or the right-side to the left of the front view, as shown at (d), even though the views do line up with the front view.

6.30   Meaning of Lines

A visible line or a hidden line has three possible meanings, Fig. 6.48: (1) intersection of two surfaces, (2) edge view of a surface, and (3) contour view of a curved surface. Since *no shading is used on a working drawing*, it is necessary to examine all the views to determine the meaning of the lines. For example, the line AB at the top of the front view might be regarded as the edge view of a flat surface if we look at only the front and top views and do not observe the curved surface on top of the object as shown in the right-

side view. Similarly, the vertical line CD in the front view might be regarded as the edge view of a plane surface if we look at only the front and side views. However, the top view shows that the line represents the intersection of an inclined surface.

6.31   Precedence of Lines

Visible lines, hidden lines, and center lines often coincide on a drawing, and it is necessary for the drafter to know which line to show. A visible line always takes precedence over (covers up) a center line or a hidden line, as shown at A and B in Fig. 6.49. A hidden line always takes precedence

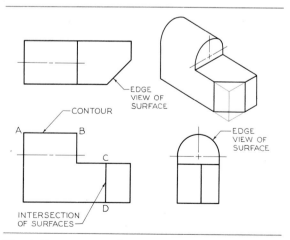

Fig. 6.48   Meaning of Lines.

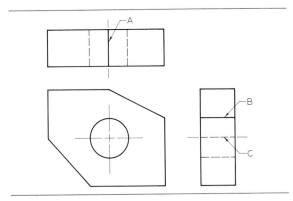

Fig. 6.49   Precedence of Lines.

over a center line, as at **C**. Note that at **A** and **C** the ends of the center line are shown, but are separated from the view by short gaps.

6.32 Computer Graphics

Most preliminary sketches are done on paper with a pencil or pen, but once the user understands how a drawing is made through proper orientation of the object to be depicted and choice of views to be used, CAD programs can be used effectively to easily create pictorials or multiview drawings that can be rapidly modified, Fig. 6.50.

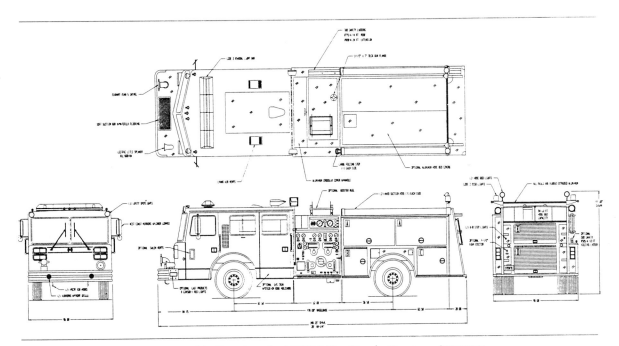

Fig. 6.50   Multiview CAD Assembly Drawing of a MAXIM Fire Truck. *Courtesy of CADKEY.*

SKETCHING PROBLEMS

Figures 6.52 and 6.53 present a variety of objects from which the student is to sketch the necessary views. Using 8.5″ × 11.0″ graph paper, sketch a border and title strip and divide the sheet into two parts as shown in Fig. 6.51. Sketch two assigned problems per sheet, as shown. On the problems in Fig. 6.52, "ticks" are given that indicate .50″ or .25″ spaces. Thus, measurements may be easily spaced off on graph paper having .12″ or .25″ grid spacings.

If desired, the "ticks" on the problems in Fig. 6.52 may be used to indicate 10 mm and 5 mm spaces. Thus, metric measurements may be easily utilized on appropriate metric-grid graph paper.

On the problems in Fig. 6.53 no indications of size are given. The student is to sketch the necessary views of assigned problems to fit the spaces comfortably, as shown in Fig. 6.51. It is suggested that the student prepare a small paper scale, making the divisions equal those on the paper scale in Prob. 1. This scale can be used to determine the approximate sizes. Let each division equal either .50″ or 10 mm on your sketch.

Missing-line and missing-view problems are given in Figs. 6.54 and 6.55, respectively. These are to be sketched, two problems per sheet, in the arrangement shown in Fig. 6.51. If the instructor so assigns, the missing lines or views may be sketched with a colored pencil. The problems given in Figs. 6.54 and 6.55 may be sketched in isometric on isometric paper or in oblique on graph paper.

Since many of the problems in this chapter are of a general nature, they can also be solved on most computer graphics systems. If a system is available, the instructor may choose to assign specific problems to be completed by this method.

Sketching problems in convenient form for solution are available in *Technical Drawing Problems,* Series 1, by Giesecke, Mitchell, Spencer, Hill, Dygdon, and Novak; *Technical Drawing Problems,* Series 2, by Spencer, Hill, Dygdon, and Novak; and *Technical Drawing Problems,* Series 3, by Spencer, Hill, Dygdon and Novak, all designed to accompany this text and published by Macmillan Publishing Company.

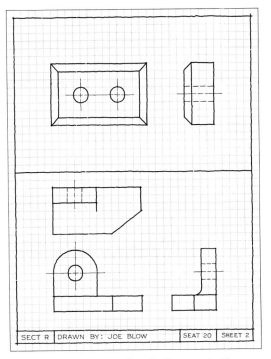

Fig. 6.51 Multiview Sketch (Layout A–1).

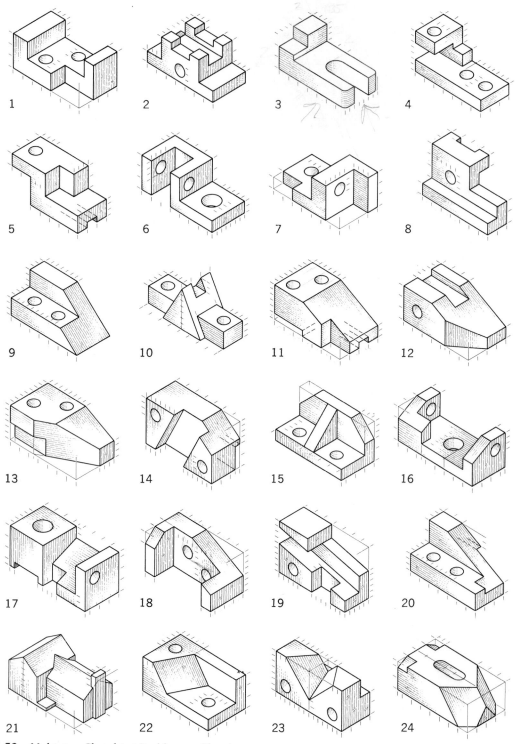

Fig. 6.52 Multiview Sketching Problems.   Sketch necessary views, using Layout A–1 or A4–1 adjusted (free-hand), on graph paper or plain paper, two problems per sheet as in Fig. 6.51. The units shown may be either .50″ and .25″ or 10 mm and 5 mm. See instructions on page 190. All holes are through holes.

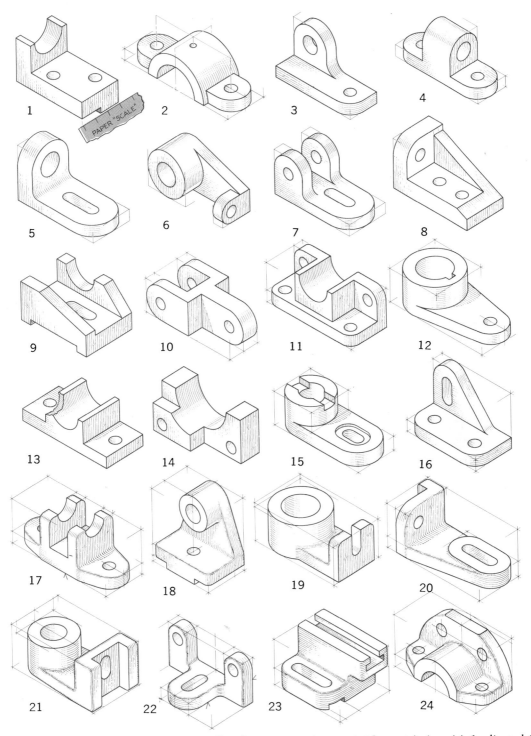

Fig. 6.53 Multiview Sketching Problems. Sketch necessary views, using Layout A–1 or A4–1 adjusted (freehand), on graph paper or plain paper, two problems per sheet as in Fig. 6.51. Prepare paper scale with divisions equal to those in Prob. 1, and apply to problems to obtain approximate sizes. Let each division equal either .50″ or 10 mm on your sketch. See instructions on page 190. For Probs. 17–24, study §§7.34–7.36.

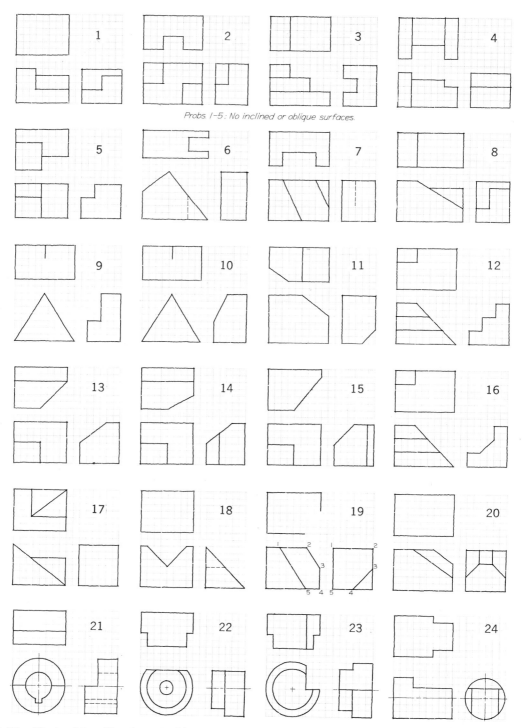

Probs. I–5 : No inclined or oblique surfaces.

Fig. 6.54 Missing-Line Sketching Problems. (1) Sketch given views, using Layout A–1 or A4–1 adjusted (freehand), on graph paper or plain paper, two problems per sheet as in Fig. 6.51. Add missing lines. The squares may be either .25″ or 5 mm. See instructions on page 190. (2) Sketch in isometric on isometric paper or in oblique on cross-section paper.

Fig. 6.55 Third-View Sketching Problems. (1) Using Layout A–1 or A4–1 adjusted (freehand), on graph paper or plain paper, two problems per sheet as in Fig. 6.51, sketch the two given views and add the missing views, as indicated. The squares may be either .25″ or 5 mm. See instructions on page 190. The given views are either front and right-side views or front and top views. Hidden holes with center lines are drilled holes. (2) Sketch in isometric on isometric paper or in oblique on cross-section paper.

195

C H A P T E R **7**

Multiview Projection

A view of a part for a design is known technically as a *projection*. A projection is a view conceived to be drawn or projected onto a plane known as the *plane of projection*. A system of views of an object formed by projectors from the object perpendicular to the desired planes of projection is known as orthographic or multiview projection. See ANSI Y14.3–1975 (R1987). This system of required views provides for the shape description of the object.

7.1 Projection Method

The method of viewing the part to obtain a *multiview projection* is illustrated for a front view in Fig. 7.1 (a). Between the observer and the part, a transparent plane or pane of glass representing a plane of projection is located parallel to the front surfaces of the part. Shown on the plane of projection in outline is how the design appears to the observer. Theoretically, the observer is at an infinite distance from the part or object, so that the *lines of sight* are parallel.

In more precise terms, this view is obtained by drawing perpendiculars, called *projectors,* from all points on the edges or contours of the part or object to the plane of projection, (b). The collective piercing points of these projectors, being infinite in number, form lines on the pane of glass, as shown at (c).

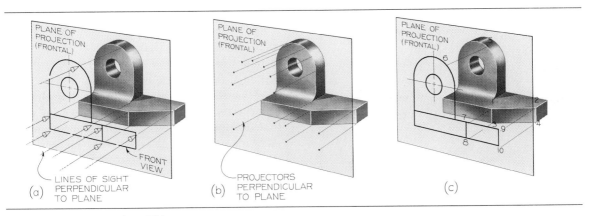

Fig. 7.1 Projection of an Object.

197

Thus, as shown at (c), a projector from point 1 on the object pierces the plane of projection at point 7, which is a view or projection of the point. The same procedure applies to point 2, whose projection is point 9. Since 1 and 2 are endpoints of a straight line on the object, the projections 7 and 9 are joined to give the projection of the line 7–9. Similarly, if the projections of the four corners 1, 2, 3, and 4 are found, the projections 7, 9, 10, and 8 may be joined by straight lines to form the projection of the rectangular surface.

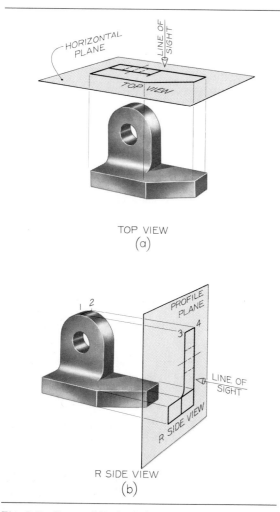

TOP VIEW
(a)

R SIDE VIEW
(b)

Fig. 7.2   Top and Right-Side Views.

The same procedure can be applied to curved lines—for example, the top curved contour of the object. A point, 5, on the curve is projected to the plane at 6. The projection of an infinite number of such points, a few of which are shown at (b), on the plane of projection results in the projection of the curve. If this procedure of projecting points is applied to all edges and contours of the object, a complete view or projection of the object results. This view is necessary in the shape description because it shows the true curvature of the top and the true shape of the hole.

A similar procedure may be used to obtain the top view, Fig. 7.2 (a). This view is necessary in the shape description because it shows the true angle of the inclined surface. In this view, the hole is invisible and its extreme contours are represented by hidden lines, as shown.

The right-side view, (b), is necessary because it shows the right-angled characteristic shape of the object and shows the true shape of the curved intersection. Note how the cylindrical contour on top of the object appears when viewed from the side. The extreme or contour element 1–2 on the object is projected to give the line 3–4 on the view. The hidden hole is also represented by projecting the extreme elements.

The plane of projection on which the front view is projected is called the *frontal plane;* that on which the top view is projected, the *horizontal plane;* and that on which the side view is projected, the *profile plane.*

7.2   The Glass Box

If planes of projection are placed parallel to the principal faces of the object, they form a "glass box," as shown in Fig. 7.3 (a). Notice that the observer is always *on the outside looking in,* so that the object is seen through the planes of projection. Since the glass box has six sides, six views of the object can be obtained.

Note that the object has three principal dimensions: *width, height,* and *depth.* These are fixed terms used for dimensions in these directions, regardless of the shape of the object. See §6.18.

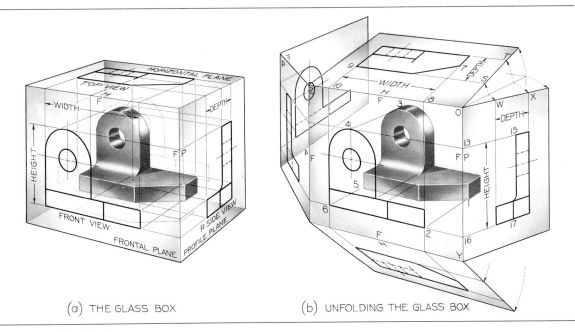

(a) THE GLASS BOX   (b) UNFOLDING THE GLASS BOX

Fig. 7.3   The Glass Box.

Since it is required to show the views of a solid or three-dimensional object on a flat sheet of paper, it is necessary to unfold the planes so that they will all lie in the same plane, Fig. 7.3 (b). All planes except the rear plane are hinged on the frontal plane, the rear plane being hinged to the left-side plane, except as explained in §7.8. Each plane revolves outwardly from the original box position until it lies in the frontal plane, which remains stationary. The hinge lines of the glass box are known as *folding lines*.

The positions of these six planes, after they have been revolved, are shown in Fig. 7.4. Carefully identify each of these planes and corresponding views with its original position in the glass box, and repeat this mental procedure, if necessary, until the revolutions are thoroughly understood.

In Fig. 7.3 (b), observe that lines extend around the glass box from one view to another on the planes of projection. These are the *pro-jections of the projectors* from points on the object to the views. For example, the projector 1–2 is projected on the horizontal plane at 7–8 and on the profile plane at 16–17. When the top plane is folded up, lines 9–10 and 7–8 will become vertical and line up with 10–6 and 8–2, respectively. Thus, 9–10 and 10–6 form a single straight line 9–6, and 7–8 and 8–2 form a single straight line 7–2, as shown in Fig. 7.4. This explains why the top view is the same width as the front view and why it is placed directly above the front view. The same relation exists between the front and bottom views. Therefore, *the front, top, and bottom views all line up vertically and are the same width.*

In Fig. 7.3 (b), when the profile plane is folded out, lines 4–13 and 13–15 become a single straight line 4–15, and lines 2–16 and 16–17 become a single straight line 2–17 as shown in Fig. 7.4. The same relation exists between the front, left-side, and rear views. Therefore, *the rear, left-*

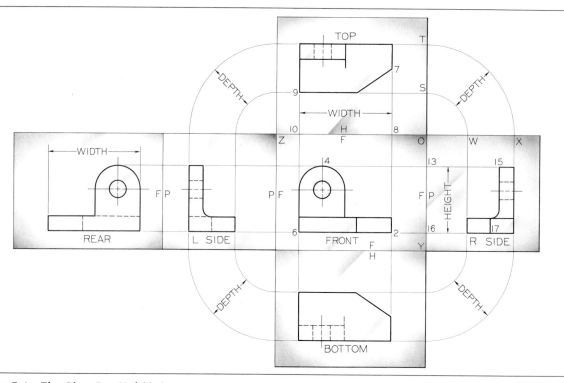

Fig. 7.4   The Glass Box Unfolded.

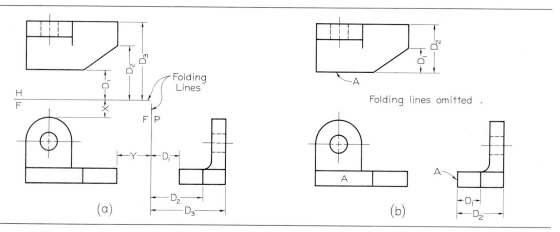

Fig. 7.5   Folding Lines.

side, front, and right-side views all line up horizontally and are the same height.

In Fig. 7.3 (b), note that lines OS and OW and lines ST and WX are respectively equal. These lines of equal length are shown in the unfolded position in Fig. 7.4. Thus, it is seen that the top view must be the same distance from the folding line OZ as the right-side view is from the folding line OY. Similarly, the bottom view and the left-side view are the same distance from their respective folding lines as are the right-side view and the top view. Therefore, *the top, right-side, bottom, and left-side views are all equidistant from the respective folding lines, and are the same depth.* Note that in these four views that surround the front view, the front surfaces of the object are faced inward or toward the front view. Observe also that the left-side and right-side views and the top and bottom views are the reverse of each other in outline shape. Similarly, the rear and front views are the reverse of each other.

7.3 Folding Lines

The three views of the object just discussed are shown in Fig. 7.5 (a), with folding lines between the views. These folding lines correspond to the hinge lines of the glass box, as we have seen. The H/F folding line, between the top and front views, is the intersection of the horizontal and frontal planes. The F/P folding line, between the front and side views, is the intersection of the frontal and profile planes. See Figs. 7.3 and 7.4.

The distances X and Y, from the front view to the respective folding lines, are not necessarily equal, since they depend on the relative distances of the object from the horizontal and profile planes. However, as explained in §7.2, distances D_1, from the top and side views to the respective folding lines, must always be equal. Therefore, the views may be any desired distance apart, and the folding lines may be drawn anywhere between them, as long as distances D_1 are kept equal and the folding lines are at right angles to the projection lines between the views.

It will be seen that distances D_2 and D_3, respectively, are also equal and that the folding lines H/F and F/P are in reality reference lines for making equal *depth* measurements in the top and side views. Thus, any point in the top view is the same distance from H/F as the corresponding point in the side view is from F/P.

While it is necessary to understand the folding lines, particularly because they are useful in solving graphical problems in descriptive geometry, they are as a rule omitted in industrial drafting. The three views, with the folding lines omitted, are shown in Fig. 7.5 (b). Again, the distances between the top and front views and between the side and front views are not necessarily equal. Instead of using the folding lines as reference lines for setting off depth measurements in the top and side views, we may use the front surface A of the object as a reference line. In this way, D_1, D_2, and all other depth measurements are made to correspond in the two views in the same manner as if folding lines were used.

7.4 Two-View Instrumental Drawing

Let it be required to draw, full size with instruments on Layout A–2 (inside back cover), the necessary views of the Operating Arm shown in Fig. 7.6 (a). In this case, as shown by the arrows, only the front and top views are needed.

I. Determine the spacing of the views. The width of the front and top views is approximately 152 mm (6″; 25.4 mm = 1″), and the width of the working space is approximately 266 mm (10½″). As shown at (b), subtract 152 mm from 266 mm and divide the result by 2 to get the value of space A. To set off the spaces, place the scale horizontally along the bottom of the sheet and make short vertical marks.

The depth of the top view is approximately 64 mm (2½″) and the height of the front view is 45 mm (1¾″), while the height of the working space is 194 mm (7⅝″). Assume a space C, say 25 mm (1″), between views that will look well and that will provide sufficient space for dimensions, if any.

As shown at (b), add 64 mm, 25 mm, and 45 mm, subtract the total from 194 mm, and

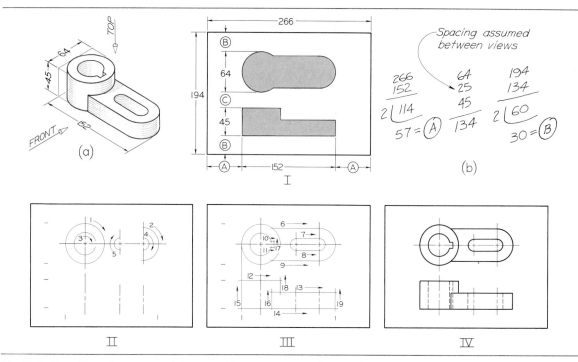

Fig. 7.6 Two-View Instrumental Drawing (dimensions in millimeters).

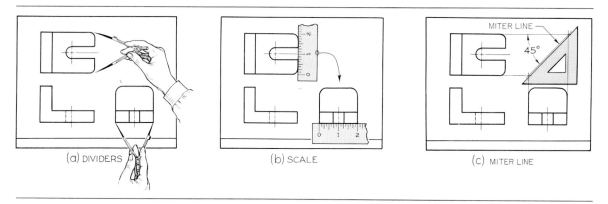

Fig. 7.7 Transferring Depth Dimensions.

divide the result by 2 to get the value of space B. To set off the spaces, place the scale vertically along the left side of the sheet with the full-size scale on the left, and make short marks perpendicular to the scale. See Fig. 2.54 (III).

II. Locate center lines from spacing marks. Construct arcs and circles lightly.

III. Draw horizontal and then vertical construction lines in the order shown. Allow construction lines to cross at corners.

IV. Add hidden lines and heavy in all final lines, clean-cut and dark. The visible lines should be heavy enough to make the views stand out. The hidden lines and center lines should be sharp in contrast to the visible lines, but dark enough to reproduce well. See §2.46 for technique of pencil drawing. Construction lines need not be erased if drawn lightly. If you are working on tracing paper, hold the sheet up to the light to see if the density of your lines is sufficient to reproduce well.

7.5 Transferring Depth Dimensions

Since all depth dimensions in the top and side views must correspond point for point, accurate methods of transferring these distances, such as D_1 and D_2, Fig. 7.5 (b), must be used.

Professional drafters transfer dimensions between the top and side views either by dividers or scale, as shown in Fig. 7.7 (a) and (b). The scale method is especially convenient when the drafting machine, Fig. 2.71, is used because both vertical and horizontal scales are readily available. Beginners might find it convenient to use a 45° miter line to project dimensions between top and side views, as in Fig. 7.7 (c). Note that the right-side view may be moved to the right or left, or the top view may be moved upward or downward, by shifting the 45° line accordingly. It is not necessary to draw continuous lines between the top and side views via the miter line. Instead, make short dashes across the miter line and project from these.

The 45° miter-line method, Fig. 7.7 (c), is also convenient for transferring a large number of points, as when plotting a curve, Fig. 7.35.

7.6 Projecting a Third View

In Fig. 7.8 (top) is a pictorial drawing of a given object, three views of which are required. Each corner of the object is given a number, as shown. At I, the top and front views are shown, with each corner properly numbered in both views. Each number appears twice, once in the top view and once again in the front view.

If a point is *visible* in a given view, the number is placed *outside* the corner, but if the point

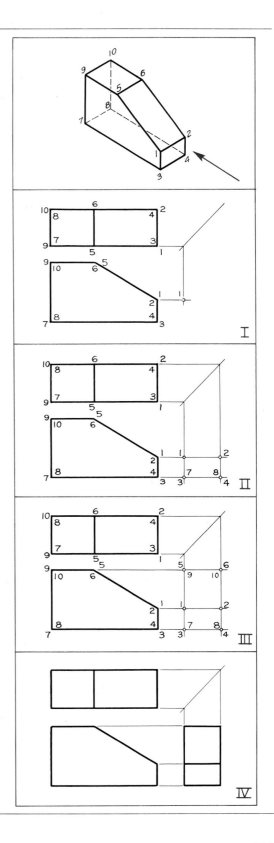

Fig. 7.8 Use of Numbers.

203

is hidden, the numeral is placed *inside* the corner. For example, at I point 1 is visible in both views and is therefore placed outside the corners in both views. However, point 2 is visible in the top view and the number is placed outside, while in the front view it is hidden and is placed inside.

This numbering system, in which points are identified by the same numbers in all views, is useful in projecting known points in two views to unknown positions in a third view. Note that this numbering system assigns the same number to a given point in all views and should not be confused with the numbering system, used in Fig. 7.23 and others, in which a point has different numbers in each view.

Before starting to project the right-side view in Fig. 7.8, try to visualize the view as seen in the direction of the arrow (see pictorial drawing). Then construct the right-side view point by point, using a hard pencil and very light lines.

As shown at I, locate point 1 in the side view by projecting from point 1 in the top view and point 1 in the front view. In space II, project points 2, 3, and 4 in a similar manner to complete the vertical end surface of the object. In space III, project points 5 and 6 to complete the side view of the inclined surface 5–6–2–1. This completes the right-side view, since invisible points 9, 10, 8, and 7 are directly behind visible corners 5, 6, 4, and 3, respectively. Note that in the side view the invisible points are lettered *inside* and the visible points *outside*.

As shown in space IV, the drawing is completed by heavying in the lines in the right-side view.

7.7   Three-View Instrumental Drawing

Let it be required to draw, full size with instruments on Layout A–2, the necessary views of the V-block in Fig. 7.9 (a). In this case, as shown by the arrows, three views are needed.

I.   Determine the spacing of the views. The width of the front view is 108 mm and the depth of the side view is 58 mm, while the width of the working space is 266 mm. Assume a space C between views, say, 32 mm, that will look well and will allow sufficient space for dimensions, if any.

As shown at (b), add 108 mm, 32 mm, and 58 mm, subtract the total from 266 mm, and divide the result by 2 to get the value of space A. To set off these horizontal spacing measurements, place the scale along the bottom of the sheet and make short vertical marks.

The depth of the top view is 58 mm and the height of the front view is 45 mm, while the height of the working space is 194 mm. Assume a space D between views, say, 25 mm. As shown in §7.3, space D need not be the same as space C. As shown at (b), add 58 mm, 25 mm, and 45 mm, subtract the total from 194 mm, and divide the result by 2 to get the value of space B. To set off these vertical spacing measurements, place the scale along the left side of the sheet with the scale used on the left, and make short marks perpendicular to the scale. Allow for dimensions, if any.

II.   Locate the center lines from the spacing marks. Construct lightly the arcs and circles.

III.   Draw horizontal, then vertical, then inclined construction lines, in the order shown. Allow construction lines to cross at the corners. Do not complete one view at a time; construct the views simultaneously.

IV.   Add hidden lines and heavy in all final lines, clean-cut and dark. A convenient method of transferring a hole diameter from the top view to the side view is to use the compass with the same setting used for drawing the hole. The visible lines should be heavy enough to make the views stand out. The hidden lines and center lines should be sharp in contrast to the visible lines, but dark enough to reproduce well. Construction lines need not be erased if they are drawn lightly. If you are working on tracing paper, hold the sheet up to the light to see if the density of your lines is sufficient to reproduce well. See §2.46.

7.8   Alternate Positions of Views

If three views of a wide flat object are drawn, using the conventional arrangement of views, Fig. 7.10 (a), a large wasted space is left on the

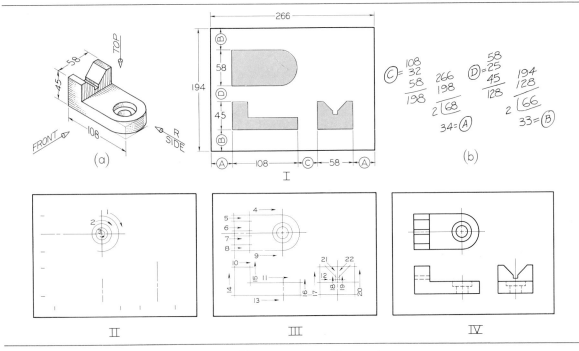

Fig. 7.9 Three-View Instrumental Drawing (dimensions in millimeters).

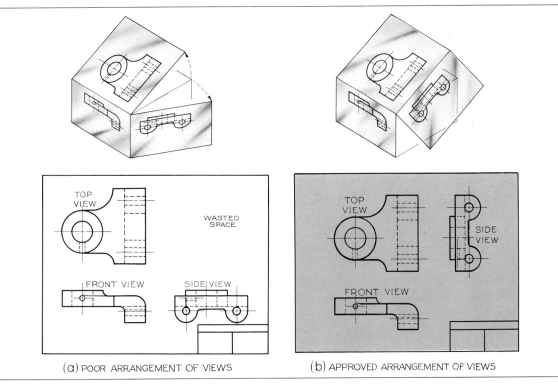

(a) POOR ARRANGEMENT OF VIEWS   (b) APPROVED ARRANGEMENT OF VIEWS

Fig. 7.10 Position of Side View.

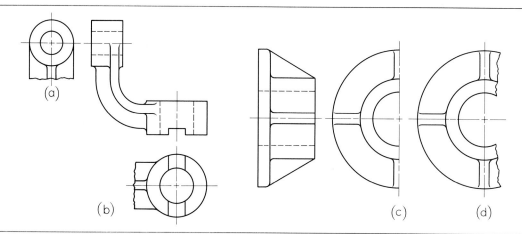

Fig. **7.11**   Partial Views.

paper, as shown. In such cases, the profile plane may be considered hinged to the horizontal plane instead of the frontal plane, as shown at (b). This places the side view beside the top view, which results in better spacing and sometimes makes the use of a reduced scale unnecessary.

It is also permissible in extreme circumstances to place the side view across horizontally from the bottom view. In this case the profile plane is considered hinged to the bottom plane of projection. Similarly, the rear view may be placed directly above the top view or under the bottom view, if necessary As a result, the rear plane is considered hinged to the horizontal or bottom plane, as the case may be, and then rotated into coincidence with the frontal plane.

7.9   Partial Views

Fig. 7.11   A view may not need to be complete but may show only what is necessary in the clear description of the object. Such a view is a partial view. A break line, (a), may be used to limit the partial view; the contour of the part shown may limit the view, (b); or if symmetrical, a half-view may be drawn on one side of the center line, (c), or a partial view, "broken out," may be drawn as at (d). The half shown at (c) and (d) should be the near side, as shown. For half-views in connection with sections, see Fig. 9.31.

Do not place a break line where it will coincide with a visible or hidden line.

Occasionally the distinctive features of an object are on opposite sides, so that in either complete side view there will be a considerable overlapping of shapes, resulting in an unintelligible view. In such cases two side views are often the best solution, Fig. 7.12. Observe that the views are partial views, in both of which certain visible and invisible lines have been omitted for clearness.

7.10   Revolution Conventions

Regular multiview projections are sometimes awkward, confusing, or actually misleading. For example, Fig. 7.13 (a) shows an object that has three triangular ribs, three holes equally spaced in the base, and a keyway. The right-side view at (b) is a regular projection and is not recommended. The lower ribs appear in a foreshortened position, the holes do not appear in their true relation to the rim of the base, and the keyway is projected as a confusion of hidden lines.

The conventional method shown at (c) is preferred, not only because it is simpler to read, but also because it requires less drafting time. Each of the features mentioned has been revolved in the front view to lie along the vertical center line from where it is projected to the correct side view at (c).

At (d) and (e) are shown regular views of a flange with many small holes. The hidden holes at (e) are confusing and take unnecessary time to draw. The preferred representation at (f)

206

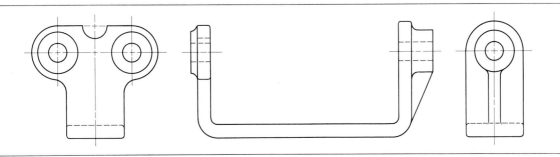

Fig. 7.12   Incomplete Side Views.

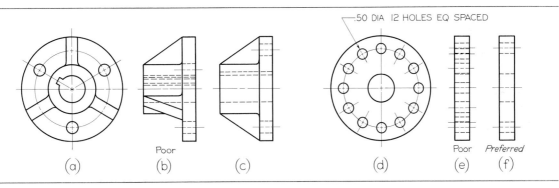

Fig. 7.13   Revolution Conventions.

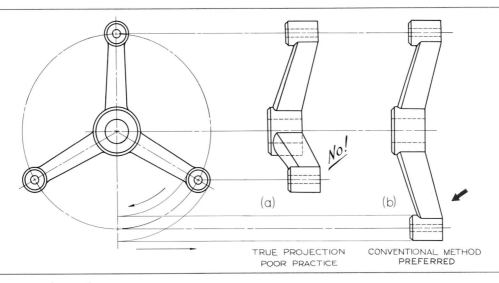

Fig. 7.14   Revolution Conventions.

shows the holes revolved, and the drawing is clear.

Another example is given in Fig. 7.14. As shown at (a), a regular projection results in a confusing foreshortening of an inclined arm. In order to preserve the appearance of symmetry about the common center, the lower arm is revolved to line up vertically in the front view so that it projects true length in the side view at (b).

207

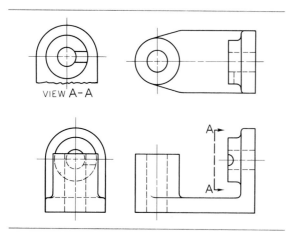

VIEW A-A

A

A

Fig. 7.15   Removed View.

Revolutions of the type discussed here are frequently used in connection with sectioning. Such sectional views are called *aligned sections,* §9.13.

7.11   Removed Views
Fig. 7.15   A removed view is a complete or partial view removed to another place on the sheet so that it no longer is in direct projection with any other view. Such a view may be used to show some feature of the object more clearly, possibly to a larger scale, or to save drawing a complete regular view. A viewing-plane line is used to indicate the part being viewed, the arrows at the corners showing the direction of sight. See §9.5. The removed view should be labeled VIEW A–A or VIEW B–B and so on, the letters referring to those placed at the corners of the viewing-plane line.

7.12   Visualization
As stated in §1.9, the ability to *visualize* or *think in three dimensions* is one of the most important requisites of the successful engineer or scientist. In practice, this means the ability to study the views of an object and to form a mental picture of it—to *visualize* its three-dimensional shape. To the designer it means the ability to *synthesize* or form a mental picture before the object even exists and the ability to express this image in terms of views. The engineer is the master planner in the construction of new machines, structures, or processes. The ability to visualize, and to use the language of drawing as a means of communication or recording of mental images, is indispensable.

Even the experienced engineer or designer cannot look at a multiview drawing and instantly visualize the object represented (except for the simplest shapes) any more than we can grasp the ideas on a book page merely at a glance. It is necessary to *study* the drawing, to read the lines in a logical way, to piece together the little things until a clear idea of the whole emerges. How this is done is described in §§7.13–7.32.

7.13   Visualizing the Views
A method of reading drawings that is essentially the reverse mental process to that of obtaining the views by projection is illustrated in Fig. 7.16. The given views of an Angle Bracket are shown at (a).

I.   The front view shows that the object is L-shaped, the height and width of the object, and the thickness of the members. The meaning of the hidden and center lines is not yet clear, nor do we yet know the depth of the object.

II.   The top view tells us that the horizontal member is rounded on the end and has a round hole. Some kind of slot is indicated at the left end. The depth and width of the object are shown.

III.   The right-side view tells us that the left end of the object has rounded corners at the top and has an open-end slot in a vertical position. The height and depth of the object are shown.

Thus, each view provides certain definite information regarding the shape of the object. All views must be considered in order to visualize the object completely.

7.14   Models
One of the best aids to visualization is an actual model of the object. Such a model need not be made accurately to scale and may be made of

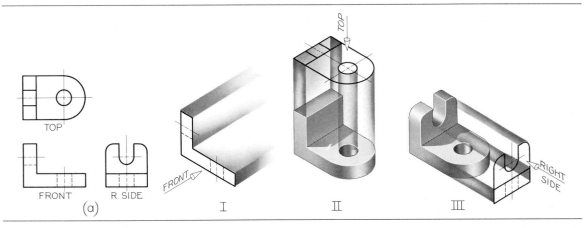

Fig. 7.16 Visualizing from Given Views.

Fig. 7.17 Use of Model to Aid Visualization.

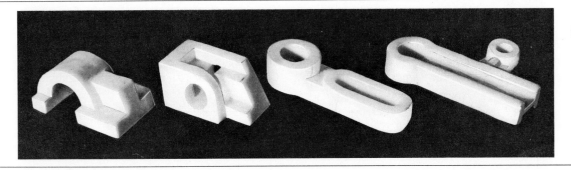

Fig. 7.18 Soap Models.

any convenient material, such as modeling clay, soap, wood, styrofoam, or any material that can be easily carved or cut.

A typical example of the use of soap or clay models is shown in Fig. 7.17, in which three views of an object are given, (a), and the student is to supply a missing line. The model is carved as shown in I, II, and III, and the "missing" line, discovered in the process, is added to the drawing as shown at (b).

Some typical examples of soap models are shown in Fig. 7.18.

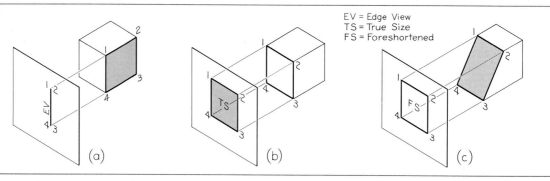

Fig. 7.19 Projections of Surfaces.

7.15 Surfaces, Edges, and Corners

In order to analyze and synthesize multiview projections, it is necessary to consider the component elements that make up most solids. A *surface* (plane) may be bounded by straight lines or curves, or a combination of them. A surface may be *frontal, horizontal,* or *profile,* according to the plane of projection to which it is parallel. See §7.1.

If a plane surface is perpendicular to a plane of projection, it appears as a line, *edge view* (EV), Fig. 7.19 (a). If it is parallel, it appears as a surface, *true size* (TS), (b). If it is situated at an angle, it appears as a surface, *foreshortened* (FS), (c). Thus, *a plane surface always projects as a line or a surface.*

The intersection of two plane surfaces produces an *edge,* or a straight line. Such a line is common to both surfaces and forms a boundary line for each. If an edge is perpendicular to a plane of projection, it appears as a point, Fig. 7.20 (a); otherwise it appears as a line, (b) and (c). If it is parallel to the plane of projection, it shows true length, (b); if not parallel, it appears foreshortened, (c). Thus, a *straight line always projects as a straight line or as a point.* A line may be *frontal, horizontal,* or *profile,* according to the plane of projection to which it is parallel.

A *corner,* or point, is the common intersection of three or more surfaces or edges. A point always appears as a point in every view.

7.16 Adjacent Areas

Consider a given top view, as shown at Fig. 7.21 (a). Lines divide the view into three areas. Each of these must represent a surface *at a different level.* Surface A may be high and surfaces B and C lower, as shown at (b). Or B may be lower than C, as shown at (c). Or B may be highest, with C

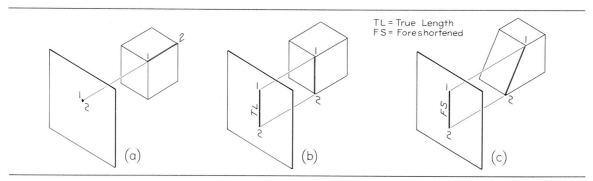

Fig. 7.20 Projections of Lines.

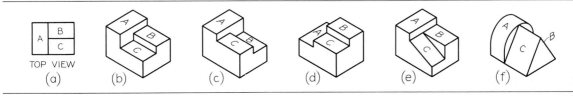

Fig. 7.21 Adjacent Areas.

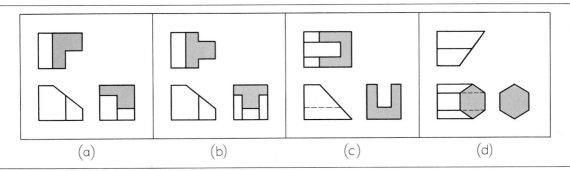

Fig. 7.22 Similar Shapes.

and A each lower, (d). Or one or more surfaces may be inclined, as at (e). Or one or more surfaces may be cylindrical, as at (f), and so on. Hence the rule: *No two adjacent areas can lie in the same plane.*

The same reasoning can apply, of course, to the adjacent areas in any given view. Since an area (surface) on a view can be interpreted in several different ways, it is necessary to observe other views also in order to determine which interpretation is correct.

7.17 Similar Shapes of Surfaces

If a surface is viewed from several different positions, it will in each case be seen to have a certain number of sides and to have a certain characteristic shape. An L-shaped surface, Fig. 7.22 (a), will appear as an L-shaped figure in every view in which it does not appear as a line. A T-shaped surface, (b) a U-shaped surface, (c), or a hexagonal surface, (d), will in each case have the same number of sides and the same characteristic shape in every view in which it appears as a surface.

This repetition of shapes is one of our best means for analyzing views.

7.18 Reading a Drawing

Let it be required to read or visualize the object shown by three views in Fig. 7.23. Since no lines are curved, the object is made up of plane surfaces.

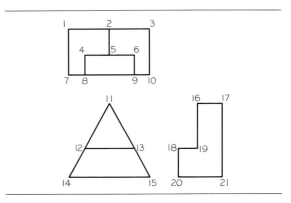

Fig. 7.23 Reading a Drawing.

Surface 2–3–10–9–6–5 in the top view is an L-shaped surface of six sides. It appears in the side view at 16–17–21–20–18–19 and is L-shaped and six-sided. No such shape appears in the front view, but we note that points 2 and 5 line up with 11 in the front view, points 6 and 9 line up with 13, and points 3 and 10 line up with 15. Evidently line 11–15 in the front view is the edge view of the L-shaped surface.

Surface 11–13–12 in the front view is triangular in shape, but no corresponding triangles appear in either the top or the side view. We note that point 12 lines up with 8 and 4 and that point 13 lines up with 6 and 9. However, surface 11–13–12 of the front view cannot be the same as surface 4–6–9–8 in the top view because the former has three sides and the latter has four. Obviously, the triangular surface appears as a line 4–6 in the top view and as a line 16–19 in the side view.

Surface 12–13–15–14 in the front view is trapezoidal in shape. But there are no trapezoids in the top and side views, so the surface evidently appears in the top view as line 7–10 and in the side view as line 18–20.

The remaining surfaces can be identified in the same manner, whence it will be seen that the object is bounded by seven plane surfaces, two of which are rectangular, two triangular, two L-shaped, and one trapezoidal.

Note that the numbering system used in Fig. 7.23 is different from that in Fig. 7.8 in that different numbers are used for all points and there is no significance in a point being inside or outside a corner.

7.19   Normal Surfaces

A normal surface is *a plane surface that is parallel to a plane of projection.* It appears in true size and shape on the plane to which it is parallel, and as a vertical or a horizontal line on adjacent planes of projection.

In Fig. 7.24 four stages in machining a block of steel to produce the final Tool Block in space IV are shown. All surfaces are normal surfaces.

In space I, normal surface A is parallel to the horizontal plane and appears true size in the top view at 2–3–7–6, as line 9–10 in the front view, and as line 17–18 in the side view. Normal surface B is parallel to the profile plane and appears true size in the side view at 17–18–20–19, as line 3–7 in the top view, and as line 10–13 in the front view. Normal surface C, an inverted T-shaped surface, is parallel to the frontal plane and appears true size in the front view at 9–10–13–14–16–15–11–12, as line 5–8 in the top view, and as line 17–21 in the side view.

All other surfaces of the object may be visualized in a similar manner. In the four stages of Fig. 7.24, observe carefully the changes in the views produced by the machining operations, including the introduction of new surfaces, new visible edges and hidden edges, and the dropping out of certain lines as the result of a new cut.

The top view in space I is cut by lines 2–6 and 3–7, which means that there are three surfaces, 1–2–6–5, 2–3–7–6, and 3–4–8–7. In the front view, surface 9–10 is seen to be the highest, and surfaces 11–12 and 13–14 are at the same lower level. In the side view both of these latter surfaces appear as one line 19–20. Surface 11–12 might appear as a hidden line in the side view, but surface 13–14 appears as a visible line 19–20, which covers up the hidden line and takes precedence over it. See §6.31.

7.20   Normal Edges

A normal edge is *a line that is perpendicular to a plane of projection.* It will appear as a point on the plane of projection to which it is perpendicular and as a line in true length on adjacent planes of projection. In space I of Fig. 7.24, edge D is perpendicular to the profile plane of projection and appears as point 17 in the side view. It is parallel to the frontal and horizontal planes of projection, and is shown true length at 9–10 in the front view and 6–7 in the top view. Edges E and F are perpendicular, respectively, to the frontal and horizontal planes of projection, and their views may be similarly analyzed.

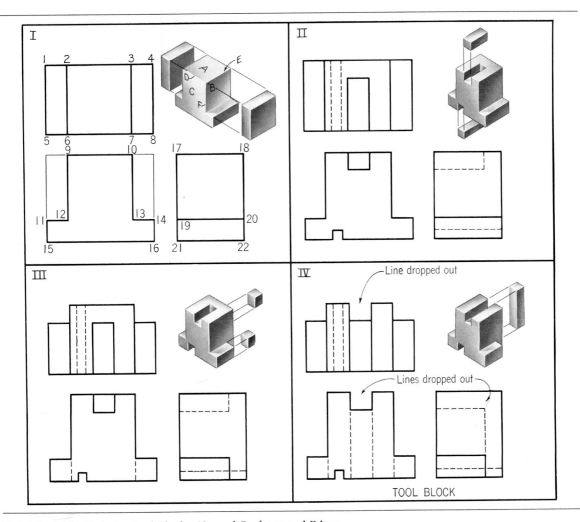

Fig. 7.24   Machining a Tool Block—Normal Surfaces and Edges.

7.21   Inclined Surfaces

An inclined surface is *a plane surface that is perpendicular to one plane of projection but inclined to adjacent planes.* An inclined surface will project as a straight line on the plane to which it is perpendicular, and it will appear foreshortened (FS) on planes to which it is inclined, the degree of foreshortening being proportional to the angle of inclination.

In Fig. 7.25 four stages in machining a Locating Finger are shown, producing several inclined surfaces. In space I, inclined surface A is

perpendicular to the horizontal plane of projection and appears as line 5–3 in the top view. It is shown as a foreshortened surface in the front view at 7–8–11–10 and in the side view at 12–13–16–15. Note that the surface is more foreshortened in the side view than in the front view because the plane makes a greater angle with the profile plane of projection than with the frontal plane of projection.

In space III, edge 23–24 in the front view is the edge view of an inclined surface that appears in the top view as 21–2–3–22 and in the side view

213

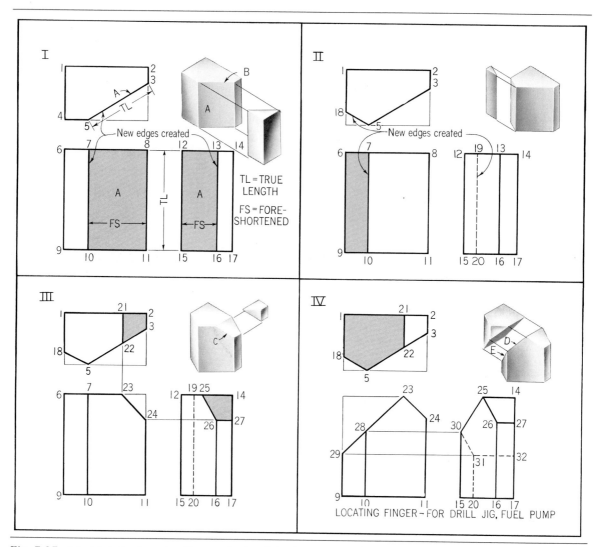

Fig. 7.25   Machining a Locating Finger—Inclined Surfaces.

as 25–14–27–26. Note that 25–14 is equal in length to 21–22 and that the surface has the same number of sides (four) in both views in which it appears as a surface.

In space IV, edge 29–23 in the front view is the edge view of an inclined surface that appears in the top view as visible surface 1–21–22–5–18 and in the side view as invisible surface

25–14–32–31–30. While the surface does not appear true size in any view, it does have the same characteristic shape and the same number of sides (five) in the views in which it appears as a surface.

In order to obtain the true size of an inclined surface, it is necessary to construct an auxiliary view (Chapter 10) or to revolve the surface until

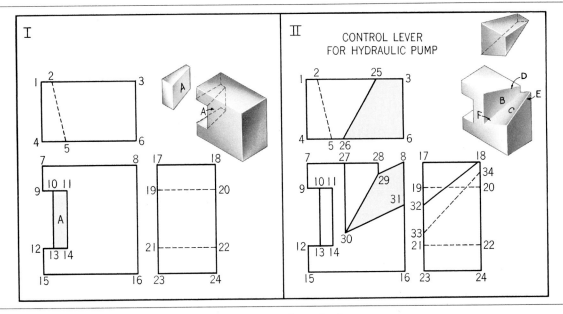

Fig. 7.26   Machining a Control Lever—Inclined and Oblique Surfaces.

it is parallel to a plane of projection (Chapter 11).

7.22   Inclined Edges

An inclined edge is *a line that is parallel to a plane of projection but inclined to adjacent planes.* It will appear true length on the plane to which it is parallel and foreshortened on adjacent planes, the degree of foreshortening being proportional to the angle of inclination. The true-length view of an inclined line is always inclined, while the foreshortened views are either vertical or horizontal lines.

In space I of Fig. 7.25, inclined edge B is parallel to the horizontal plane of projection, and appears true length in the top view at 5–3. It is foreshortened in the front view at 7–8 and in the side view at 12–13. Note that plane A produces two normal edges and two inclined edges.

In spaces III and IV, some of the sloping lines are not inclined lines. In space III, the edge that appears in the top view at 21–22, in the front view at 23–24, and in the side view at 14–27 is an inclined line. However, the edge that appears in the top view at 22–3, in the front view at 23–24, and in the side view at 25–26 is not an inclined line by the definition given here. Actually, it is an oblique line, §7.24.

7.23   Oblique Surfaces

An oblique surface is *a plane that is oblique to all planes of projection.* Since it is not perpendicular to any plane, it cannot appear as a line in any view. Since it is not parallel to any plane, it cannot appear true size in any view. Thus, an oblique surface always appears as a foreshortened surface in all three views.

In space II of Fig. 7.26, oblique surface C appears in the top view at 25–3–6–26 and in the front view at 29–8–31–30. What are its numbers in the side view? Note that any surface appearing as a line in any view cannot be an oblique sur-

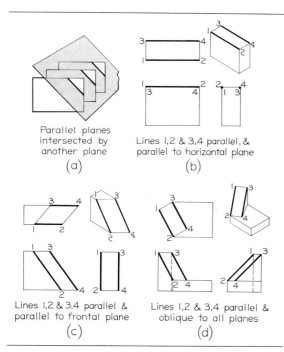

Parallel planes
intersected by
another plane
(a)

Lines 1,2 & 3,4 parallel, &
parallel to horizontal plane
(b)

Lines 1,2 & 3,4 parallel &
parallel to frontal plane
(c)

Lines 1,2 & 3,4 parallel &
oblique to all planes
(d)

Fig. 7.27 Parallel Lines.

face. How many inclined surfaces are there? How many normal surfaces?

To obtain the true size of an oblique surface, it is necessary to construct a secondary auxiliary view, §§10.21 and 10.22, or to revolve the surface until it is parallel to a plane of projection, §11.11.

7.24 Oblique Edges

An oblique edge is *a line that is oblique to all planes of projection.* Since it is not perpendicular to any plane, it cannot appear as a point in any view. Since it is not parallel to any plane, it cannot appear true length in any view. An oblique edge appears foreshortened and in an inclined position in every view.

In space II of Fig. 7.26, oblique edge F appears in the top view at 26–25, in the front view at 30–29, and in the side view at 33–34.

7.25 Parallel Edges

If a series of parallel planes is intersected by another plane, the resulting lines of intersection will be parallel, Fig. 7.27 (a). At (b) the top plane of the object intersects the front and rear planes, producing the parallel edges 1–2 and 3–4. If two lines are parallel in space, their projections in any view are parallel. The example in (b) is a special case in which the two lines appear as points in one view and coincide as a single line in another and should not be regarded as an exception to the rule. Note that even in the pictorial drawings the lines are shown parallel.

Parallel inclined lines are shown in (c), and parallel oblique lines in (d).

In Fig. 7.28 it is required to draw three views of the object after a plane has been passed through the points A, B, and C. As shown at (b), only points that lie in the same plane are joined. In the front view, join points A and C, which are in the same plane, extending the line to P on the vertical front edge of the block extended. In the side view, join P to B, and in the top view, join B to A. Complete the drawing by applying the rule: *Parallel lines in space will be projected as parallel lines in any view.* The remaining lines are thus drawn parallel to lines AP, PB, and BA.

7.26 Angles

If an angle is in a normal plane—that is, parallel to a plane of projection—the angle will be shown true size on the plane of projection to which it is parallel, Fig. 7.29 (a).

If the angle is in an inclined plane, (b) and (c), the angle may be projected either larger or smaller than the true angle, depending on its position. At (b) the 45° angle is shown *oversize* in the front view, and at (c) the 60° angle is shown *undersize* in both views.

A 90° angle will be projected true size, even though it is in an inclined plane, provided one leg of the angle is a normal line, as shown at (d). In this figure, the 60° angle is projected *oversize* and the 30° angle *undersize*. Study these relations, using your own 30° × 60° triangle as a model.

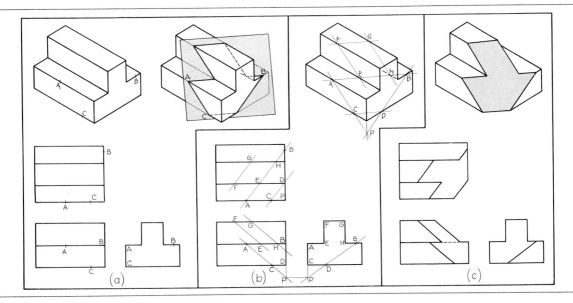

Fig. 7.28   Oblique Surface.

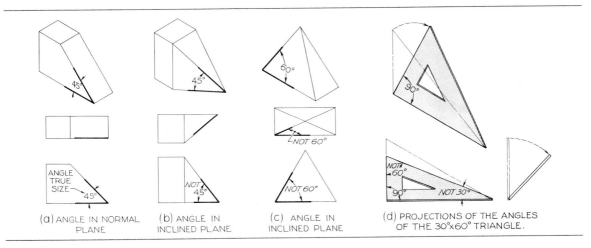

Fig. 7.29   Angles.

7.27   Curved Surfaces

Rounded surfaces are common in engineering practice because they are easily formed on the lathe, the drill press, and other machines using the principle of rotation either of the "work" or of the cutting tool. The most common are the cylinder, cone, and sphere, a few of whose applications are shown in Fig. 7.30. For other geometric solids, see Fig. 5.7.

217

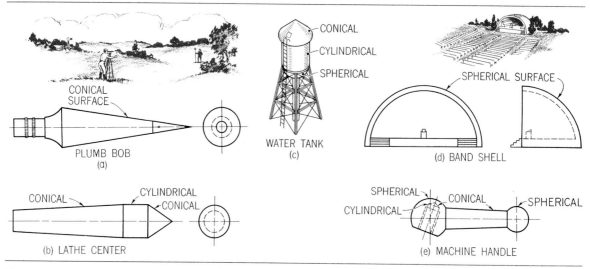

CONICAL

CONICAL
SURFACE

PLUMB BOB
(a)

WATER TANK
(c)

CONICAL
CYLINDRICAL
SPHERICAL

SPHERICAL SURFACE

(d) BAND SHELL

CONICAL
CYLINDRICAL
CONICAL

(b) LATHE CENTER

SPHERICAL
CYLINDRICAL
CONICAL
SPHERICAL

(e) MACHINE HANDLE

Fig. 7.30 Curved Surfaces.

7.28 Cylindrical Surfaces

Three views of a *right-circular cylinder,* the most common type, are shown in Fig. 7.31 (a). The single cylindrical surface is intersected by two plane (normal) surfaces, forming two curved lines of intersection or *circular edges* (the bases of the cylinder). These circular edges are the only actual edges on the cylinder. Fig. 7.31 (b) shows a cylindrical hole in a right square prism.

The cylinder is represented on a drawing by its circular edges and the contour elements. An *element* is a straight line on the cylindrical surface, parallel to the axis, as shown in the pictorial view of the cylinder at (a). In this figure, at both (a) and (b), the circular edges appear in the top views as circles A, in the front views as horizontal lines 5–7 and 8–10, and in the side views as horizontal lines 11–13 and 14–16.

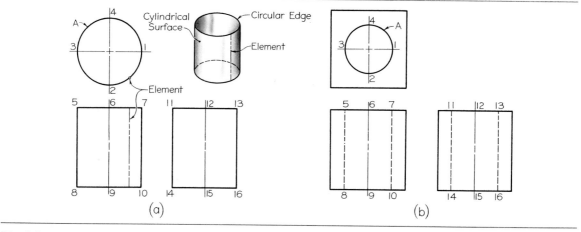

Fig. 7.31 Cylindrical Surfaces.

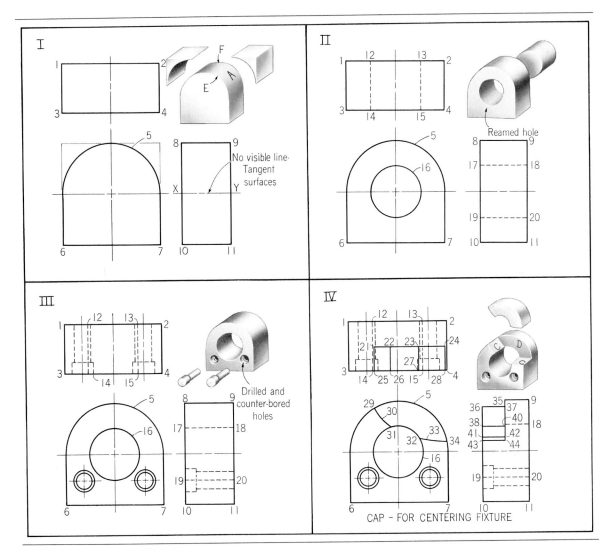

Fig. 7.32 Machining a Cap—Cylindrical Surfaces.

The contour elements 5–8 and 7–10 in the front views appear as points 3 and 1 in the top views. The contour elements 11–14 and 13–16 in the side views appear as points 2 and 4 in the top views.

In Fig. 7.32 four possible stages in machining a Cap are shown, producing several cylindrical surfaces. In space I, the removal of the two upper corners forms cylindrical surface A which ap-

pears in the top view as surface 1–2–4–3, in the front view as arc 5, and in the side view as surface 8–9–Y–X.

In space II, a large reamed hole shows in the front view as circle 16, in the top view as cylindrical surface 12–13–15–14, and in the side view as cylindrical surface 17–18–20–19.

In space III, two drilled and counterbored holes are added, producing four more cylindrical

219

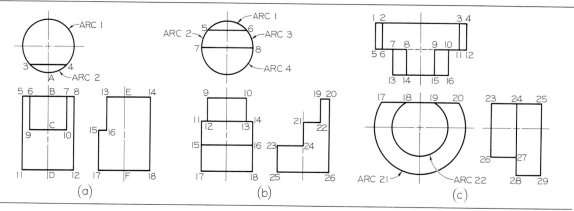

Fig. 7.33   Deformities of Cylinders.

surfaces and two normal surfaces. The two normal surfaces are those at the bottoms of the counterbores.

In space IV, a cylindrical cut is added, producing two cylindrical surfaces that appear edgewise in the front view as arcs 30 and 33, in the top view as surfaces 21–22–26–25 and 23–24–28–27 and in the side view as surfaces 36–37–40–38 and 41–42–44–43.

7.29   Deformities of Cylinders

In shop practice, cylinders are usually machined or formed so as to introduce other surfaces, usually plane surfaces. Fig. 7.33 (a) shows a cut that introduces two normal surfaces. One surface appears as line 3–4 in the top view, as surface 6–7–10–9 in the front view, and as line 13–16 in the side view. The other appears as line 15–16 in the side view, as line 9–10 in the front view, and as surface 3–4, arc 2 in the top view.

All elements touching arc 2, between 3 and 4 in the top view, become shorter as a result of the cut. For example, element A, which shows as a point in the top view, now becomes CD in the front view and 15–17 in the side view. As a result of the cut, the front half of the cylindrical surface has changed from 5–8–12–11 to 5–6–9–10–7–8–12–11 (front view). The back half remains unchanged.

At (b) two cuts introduce four normal surfaces. Note that surface 7–8 (top view) is through the center of the cylinder, producing in the side

view line 21–24 and in the front view surface 11–14–16–15 equal in width to the diameter of the cylinder. Surface 15–16 (front view) is read in the top view as 7–8–ARC 4. Surface 11–14 (front view) is read in the top view as 5–6–ARC 3 –8–7–ARC 2.

At (c) two cylinders on the same axis are shown, intersected by a normal surface parallel to the axis. Surface 17–20 (front view) is 23–25 in the side view, and 2–3–11–9–15–14–8–6 in the top view. A common error is to draw a visible line in the top view between 8 and 9. However, this would produce two surfaces 2–3–11–6 and 8–9–15–14 not in the same plane. In the front view, the larger surface appears as line 17–20 and the smaller as line 18–19. These lines coincide; hence, they are all one surface, and there can be no visible line joining 8 and 9 in the top view.

The vertical surface that appears in the front view at 17–18–ARC 22–19–20–ARC 21 appears as a line in the top view at 5–12, which explains the hidden line 8–9 in the top view.

7.30   Cylinders and Ellipses

If a cylinder is cut by an inclined plane, as in Fig. 7.34 (a), the inclined surface is bounded by an ellipse. The ellipse appears as circle 1 in the top view, as straight line 2–3 in the front view, and as ellipse ADBC in the side view. Note that circle 1 in the top view would remain a circle regardless of the angle of the cut. If the cut is 45° with horizontal, the ellipse will appear as a

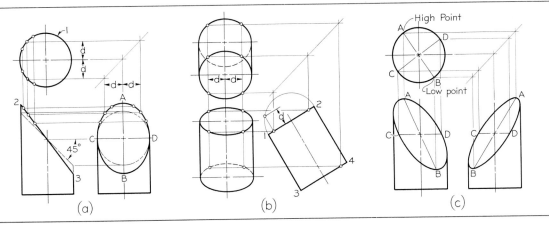

Fig. 7.34 Cylinders and Ellipses.

circle in the side view (see phantom lines) since the major and minor axes in that view would be equal. To find the true size and shape of the ellipse, an auxiliary view will be required, with the line of sight perpendicular to surface 2–3 in the front view, §10.12.

Since the major and minor axes AB and CD are known, the ellipse can be drawn by any of the methods in Figs. 5.48–5.50 and 5.52 (a) (true ellipses) or by the aid of an ellipse template, Fig. 5.55.

If the cylinder is tilted forward, (b), the bases or circular edges 1–2 and 3–4 (side view) become ellipses in the front and top views. Points on the ellipses can be plotted from the semicircular end view of the cylinder, as shown, distances d being equal. Since the major and minor axes for each ellipse are known, the ellipses can be drawn with the aid of an ellipse template, or by any of the true ellipse methods, or by the approximate method.

If the cylinder is cut by an oblique plane, (c), the elliptical surface appears as an ellipse in two views. In the top view, points A and B are selected, diametrically opposite, as the high and low points in the ellipse, and CD is drawn perpendicular to AB. These are the projections of the major and minor axes, respectively, of the actual ellipse in space. In the front and side views, points A and B are assumed at the desired altitudes. Since CD appears true length in the top view, it will appear horizontal in the front and

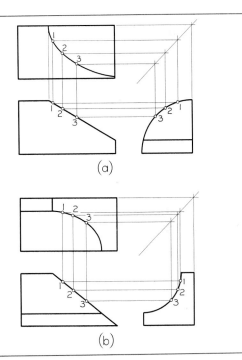

Fig. 7.35 Plotting Elliptical Curves.

side views, as shown. These axes in the front and side views are the conjugate axes of the ellipses. The ellipses may be drawn on these axes by the method of Fig. 5.51 or 5.52 (b) or by trial with the aid of an ellipse template, Fig. 5.55.

In Fig. 7.35, the intersection of a plane and a quarter-round molding is shown at (a), and

221

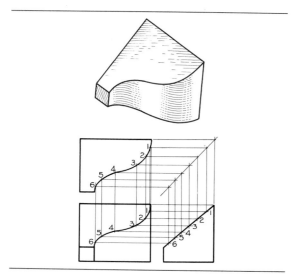

Fig. 7.36 Space Curve.

with a cove molding at (b). In both figures, as-sume points 1, 2, 3, . . ., at random in the side views in which the cylindrical surfaces appear as curved lines, and project the points to the front and top views, as shown. A sufficient number of points should be used to insure smooth curves.

Draw the final curves through the points with the aid of the irregular curve, §2.54.

7.31 Space Curves

The views of a space curve are established by the projections of points along the curve, Fig. 7.36. In this figure any points 1, 2, 3, . . ., are selected along the curve in the top view and then projected to the side view (or the reverse), and points are located in the front view by projecting downward from the top view and across from the side view. The resulting curve in the front view is drawn with the aid of the irregular curve, §2.54.

7.32 Intersections and Tangencies

No line should be drawn where a curved surface is tangent to a plane surface, Fig. 7.37 (a), but when a curved surface *intersects* a plane surface, a definite edge is formed, (b). If curved surfaces are arranged as at (c), no lines appear in the top view, as shown. If the surfaces are arranged as at (d), a vertical surface in the front view pro-duces a line in the top view. Other typical inter-

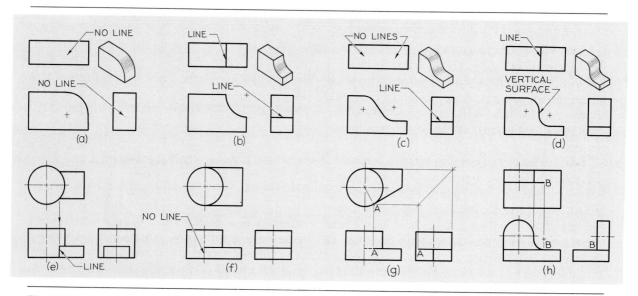

Fig. 7.37 Intersections and Tangencies.

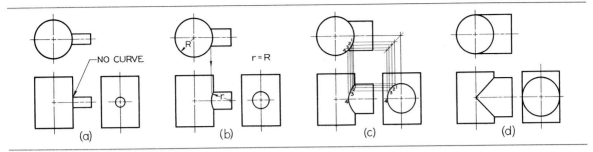

Fig. 7.38 Intersections of Cylinders.

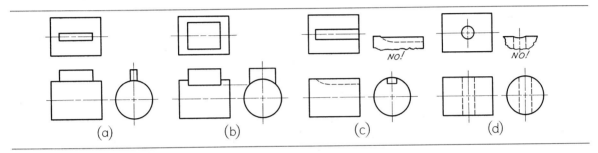

Fig. 7.39 Intersections.

sections and tangencies of surfaces are shown from (e) to (h). To locate the point of tangency A in (g), refer to Fig. 5.34 (b).

The intersection of a small cylinder with a large cylinder is shown in Fig. 7.38 (a). The intersection is so small that it is not plotted, a straight line being used instead. At (b) the intersection is larger, but still not large enough to justify plotting the curve. The curve is approximated by drawing an arc whose radius r is the same as radius R of the large cylinder.

The intersection at (c) is significant enough to justify constructing the true curve. Points are selected at random in the circle in the side or top view, and these are then projected to the other two views to locate points on the curve in the front view, as shown. A sufficient number of points should be used, depending on the size of

the intersection, to insure a smooth and accurate curve. Draw the final curve with the aid of the irregular curve, §2.54.

At (d), the cylinders are the same diameter. The figure of intersection consists of two semi-ellipses that appear as straight lines in the front view.

If the intersecting cylinders are holes, the intersections will be similar to those for the external cylinders in Fig. 7.38. See also Fig. 9.34 (d).

In Fig. 7.39 (a), a narrow prism intersects a cylinder, but the intersection is insignificant and is ignored. At (b) the prism is larger and the intersection is noticeable enough to warrant construction, as shown. At (c) and (d) a keyseat and a small drilled hole, respectively, are shown; in both cases the intersection is not important enough to construct.

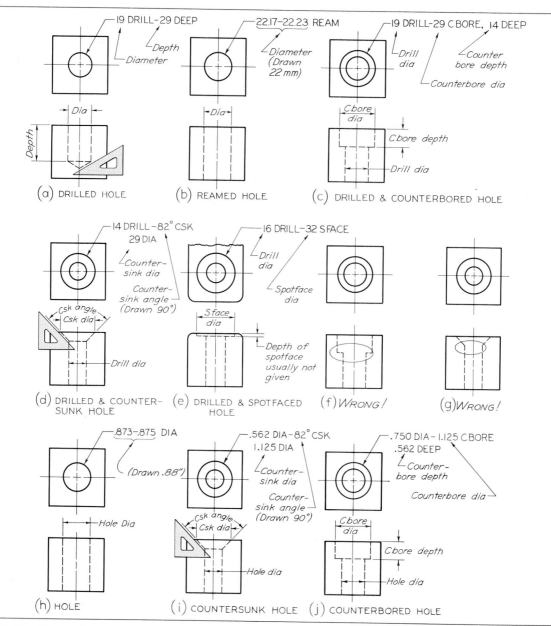

Fig. 7.40   How to Represent Holes.   Dimensions for (a)–(e) in metric. For threaded holes, see §15.24.

7.33   How to Represent Holes

The correct methods of representing most common types of machined holes are shown in Fig. 7.40. Instructions to the machinist are given in the form of notes, and the drafter represents the holes in conformity with these specifications. In general, the notes tell the machine operator what to do and in which order it is to be done. Hole sizes are always specified by diameter—never by radius. For each operation specified,

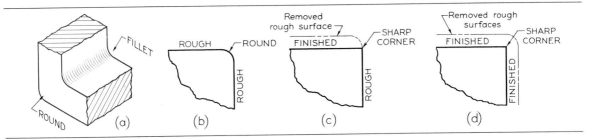

Fig. 7.41   Rough and Finished Surfaces.

the diameter is given first, followed by the method such as drill, ream, and so on, as shown in (a) and (b).

The size of the hole may be specified as a diameter without the specific method such as drill, ream, and so on, since the selection of the method will depend on available production facilities. See (h) to (j).

A drilled hole is a *through* hole if it goes through a member. If the hole has a specified depth, as shown at (a), the hole is called a *blind* hole. The depth includes the cylindrical portion of the hole only. The point of the drill leaves a conical bottom in the hole, drawn approximately with the 30° × 60° triangle, as shown. For drill sizes, see Appendix 16 (Twist Drill Sizes). For abbreviations, see Appendix 4.

A through-drilled or reamed hole is drawn as shown at (b). The note tells how the hole is to be produced—in this case by reaming. Note that tolerances are ignored in actually laying out the diameter of a hole.

At (c) a hole is drilled, and then the upper part is enlarged cylindrically to a specified diameter and depth.

At (d) a hole is drilled, and then the upper part is enlarged conically to a specified angle and diameter. The angle is commonly 82° but is drawn 90° for simplicity.

At (e) a hole is drilled, and then the upper part is enlarged cylindrically to a specified diameter. The depth usually is not specified, but is left to the shop to determine. For average cases, the depth is drawn 1.5 mm ($\frac{1}{16}''$).

For complete information about how holes are made in the shop, see §12.20. For further information on notes, see §13.24.

7.34   Fillets and Rounds

A rounded interior corner is called a fillet, and a rounded exterior corner a round, Fig. 7.41 (a). Sharp corners should be avoided in designing parts to be cast or forged not only because they are difficult to produce but also because, in the case of interior corners, they are a source of weakness and failure. See §12.5 for shop processes involved.

Two intersecting rough surfaces produce a rounded corner, (b). If one of these surfaces is machined, (c), or if both surfaces are machined, (d), the corner becomes sharp. Therefore, on drawings a rounded corner means that both intersecting surfaces are rough, and a sharp corner means that one or both surfaces are machined. On working drawings, fillets and rounds are never shaded. The presence of the curved surfaces is indicated only where they appear as arcs, except as shown in Fig. 7.45.

Fillets and rounds should be drawn with the filleted corners of the triangle, a special fillets and rounds template, or a circle template.

7.35   Runouts

The correct method of representing fillets in connection with plane surfaces tangent to cylinders is shown in Fig. 7.42. These small curves are

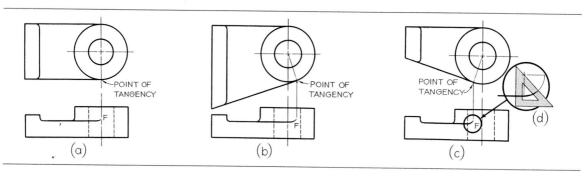

Fig. 7.42 Runouts.

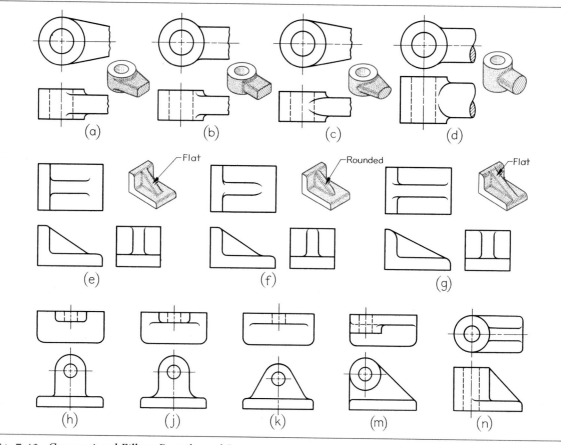

Fig. 7.43 Conventional Fillets, Rounds, and Runouts.

called runouts. Note that the runouts F have a radius equal to that of the fillet and a curvature of about one-eighth of a circle, (d).

Typical filleted intersections are shown in Fig. 7.43. The runouts from (a) to (d) differ be-

cause of the different shapes of the horizontal intersecting members. At (e) and (f) the runouts differ because the top surface of the web at (e) is flat, with only slight rounds along the edge, while the top surface of the web at (f) is consid-

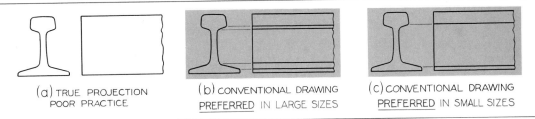

Fig. 7.44 Conventional Repression of a Rail.

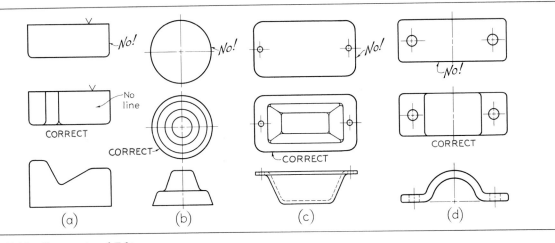

Fig. 7.45 Conventional Edges.

erably rounded. When two different sizes of fillets intersect, as at (g) and (j), the direction of the runout is dictated by the larger fillet, as shown.

7.36 Conventional Edges
Rounded and filleted intersections eliminate sharp edges and sometimes make it difficult to present a clear shape description. In fact, true projection in some cases may be actually misleading, as in Fig. 7.44 (a), in which the side view of the railroad rail is quite blank. A clearer representation results if lines are added for rounded and filleted edges, as shown at (b) and (c). The added lines are projected from the actual intersections of the surfaces as if the fillets and rounds were not present.

Figure 7.45 shows two top views for each given front view. The upper top views are nearly devoid of lines that contribute to the shape descriptions, while the lower top views, in which lines are used to represent the rounded and filleted edges, are quite clear. Note, in the lower top views at (a) and (c), the use of small Ys where rounded or filleted edges meet a rough surface. If such an edge intersects a finished surface, no Y is shown.

7.37 Right-Hand and Left-Hand Parts
In industry many individual parts are located symmetrically so as to function in pairs. These opposite parts are often exactly alike, as for example, the hub caps used on the left and right

227

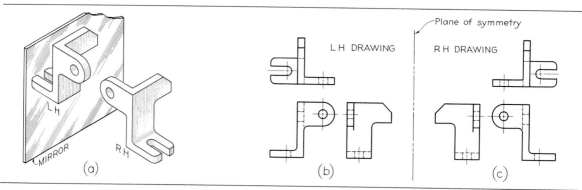

Fig. 7.46 Right-Hand and Left-Hand Parts.

sides of the automobile. In fact, whenever possible, for economy's sake the designer will design identical parts for use on both the right and left. But opposite parts often cannot be exactly alike, such as a pair of gloves or a pair of shoes. Similarly, the right-front fender of an automobile cannot be the same shape as the left-front fender. Therefore, a left-hand part is not simply a right-hand part turned around; the two parts will be opposite and not interchangeable.

A left-hand part is referred to as an LH part, and a right-hand part as an RH part. In Fig. 7.46 (a), the image in the mirror is the "other hand" of the part shown. If the part in front of the mirror is an RH part, the image shows the LH part. No matter how the object is turned, the image will show the LH part. At (b) and (c) are shown LH and RH drawings of the same object, and it will be seen that the drawings are also symmetrical with respect to a reference-plane line between them.

If you hold a drawing faced against a windowpane or a light table so that the lines can be seen through the paper, you can trace the reverse image of the part on the back or on tracing paper, which will be a drawing of the opposite part.

It is customary to draw only one of two opposite parts and to label the one that is drawn with a note, such as LH PART SHOWN, RH OPPOSITE. If the opposite-hand shape is not clear, a separate drawing must be made for it and properly identified.

7.38   First-Angle Projection

If the vertical and horizontal planes of projection are considered indefinite in extent and intersecting at 90° with each other, the four dihedral angles produced are the *first, second, third,* and *fourth* angles, Fig. 7.47 (a). The profile plane intersects these two planes and may extend into all angles. If the object is placed below the horizontal plane and behind the vertical plane as in the glass box, Fig. 7.3, the object is said to be in the third angle. In this case, as we have seen, the observer is always "outside, looking in," so that for all views the lines of sight proceed from the eye *through the planes of projection and to the object.*

If the object is placed above the horizontal plane and in front of the vertical plane, the object is in the first angle. In this case, the observer always looks *through the object and to the planes of projection.* Thus, the right-side view is still obtained by looking toward the right side of the object, the front by looking toward the front, and the top by looking down toward the top; but the views are projected from the object onto a plane in each case. When the planes are unfolded, as at (b), the right-side view falls at the left of the front view, and the top view falls below the front view, as shown. A comparison between first-angle orthographic projection and third-angle orthographic projection is shown in Fig. 7.48. The front, top, and right-side views shown in Fig. 7.47 (b) for first-angle projection are re-

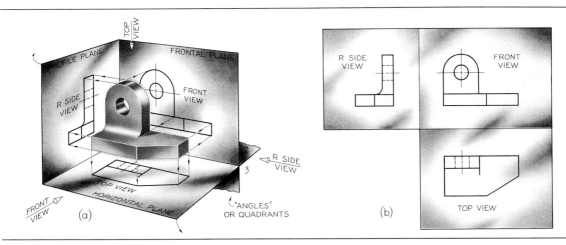

Fig. 7.47   First-Angle Projection.

peated in Fig. 7.48 (a). The front, top, and right-side views for third-angle projection of Fig. 7.4 are repeated at (b). Ultimately, the only difference between third-angle and first-angle projection is in the arrangement of the views. Still, confusion and possibly manufacturing errors may result when the user reading a first-angle drawing thinks it is a third-angle drawing, or vice versa. To avoid misunderstanding, international projection symbols, shown in Fig. 7.48, have been developed to distinguish between first-an-gle and third-angle projections on drawings. On drawings where the possibility of confusion is anticipated, these symbols may appear in or near the title box.

In the United States and Canada, and to some extent in England, third-angle projection is standard, while in most of the rest of the world, first-angle projection is used. First-angle projection was originally used all over the world, including the United States, but in this country it was abandoned around 1890.

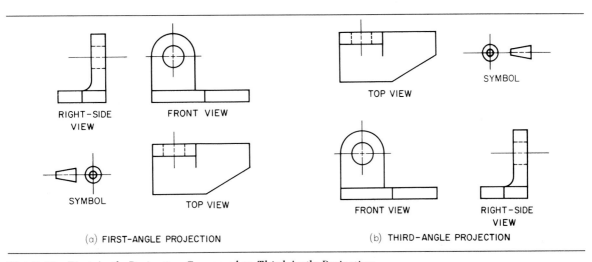

Fig. 7.48   First-Angle Projection Compared to Third-Angle Projection.

229

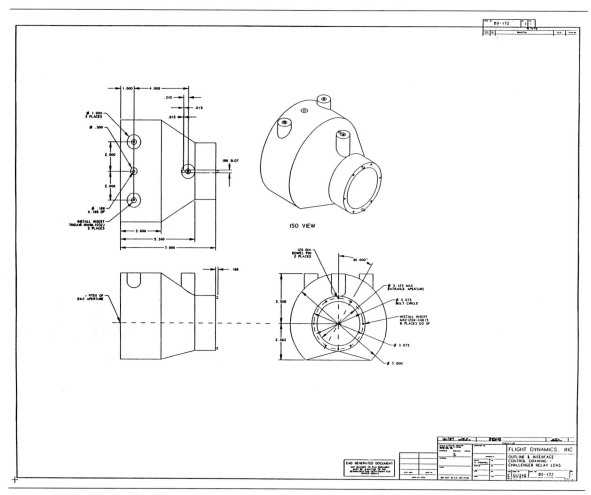

Fig. 7.49 Computer-Generated Multiview and Pictorial Drawing. *Courtesy of Computervision Corporation, a subsidiary of Prime Computer, Inc.*

7.39 Computer Graphics

CAD programs allow the drafter to rapidly create multiview drawings with a minimum amount of effort, Fig. 7.49. Symmetrical features and "mirror image" parts, §7.37, need only be drawn once and can be duplicated as many times as needed and moved around anywhere on the drawing. Features such as fillets, rounds, and runouts, tangencies, and complex curves can be added, changed or removed as desired by the CAD operator.

MULTIVIEW PROJECTION PROBLEMS

The following problems are intended primarily to afford practice in instrumental drawing, but any of them may be sketched freehand on graph paper or plain paper. Sheet layouts, Figs. 7.50 and 7.51, or inside back cover, are suggested, but the instructor may prefer a different sheet size or arrangement.

Dimensions may or may not be required by the instructor. If they are assigned, the student should study §§13.1–13.25. In the given problems, whether in multiview or in pictorial form, it is often not possible to give dimensions in the preferred places or, occasionally, in the standard manner. The student is expected to move dimensions to the preferred locations and otherwise to conform to the dimensioning practices recommended in Chapter 13.

For the problems in Figs. 7.55–7.90, it is suggested that the student make a thumbnail sketch of the necessary views, in each case, and obtain his instructor's approval before starting the mechanical drawing.

For additional problems, see Fig. 10.29. Draw top views instead of auxiliary views.

Problems in convenient form for solution may be found in *Technical Drawing Problems,* Series 1, by Giesecke, Mitchell, Spencer, Hill, Dygdon, and Novak; *Technical Drawing Problems,* Series 2, by Spencer, Hill, Dygdon, and Novak; and *Technical Drawing Problems,* Series 3, by Spencer, Hill, Dygdon, and Novak, all designed to accompany this text and published by Macmillan Publishing Company.

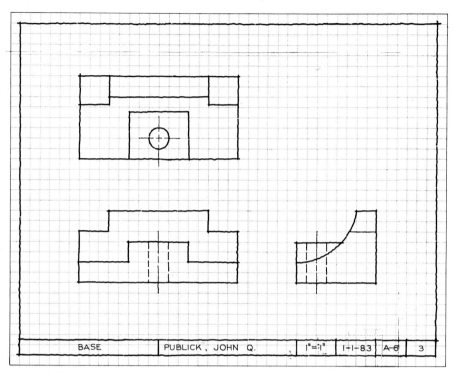

| BASE | PUBLICK , JOHN Q. | 1"=1" | 1-1-83 | A-6 | 3 |

Fig. 7.50   Freehand Sketch   (Layout A–2 or A4–2 adjusted).

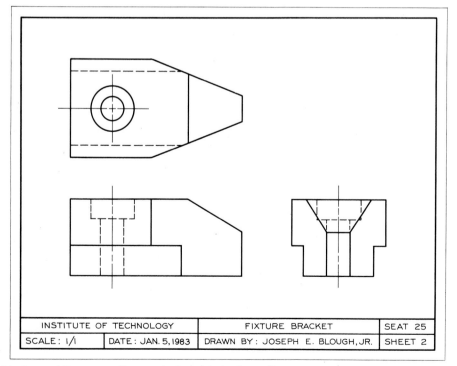

| INSTITUTE OF TECHNOLOGY | | FIXTURE BRACKET | SEAT 25 |
| SCALE: 1/1 | DATE: JAN. 5, 1983 | DRAWN BY: JOSEPH E. BLOUGH, JR. | SHEET 2 |

Fig. 7.51   Mechanical Drawing   (Layout A–3 or A4–3 adjusted).

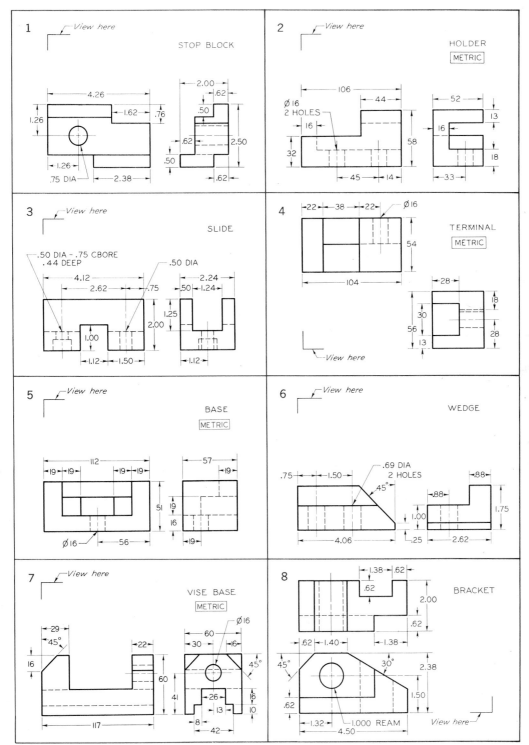

Fig. 7.52 Missing-View Problems. Using Layout A–2 or 3 or Layout A4–2 or 3 (adjusted), sketch or draw with instruments the given views, and add the missing view, as shown in Figs. 7.50 and 7.51. If dimensions are required, study §§13.1–13.25. Use metric or decimal-inch dimensions as assigned by the instructor. Move dimensions to better locations where possible. In Probs. 1–5, all surfaces are normal surfaces.

233

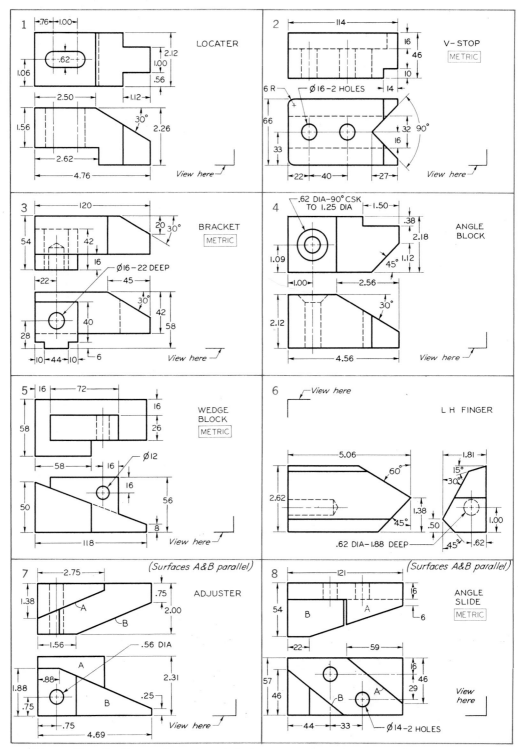

Fig. 7.53 Missing-View Problems. Using Layout A–2 or 3 or Layout A4–2 or 3 (adjusted), sketch or draw with instruments the given views, and add the missing view, as shown in Figs. 7.50 and 7.51. If dimensions are required, study §§13.1–13.25. Use metric or decimal-inch dimensions as assigned by the instructor. Move dimensions to better locations where possible.

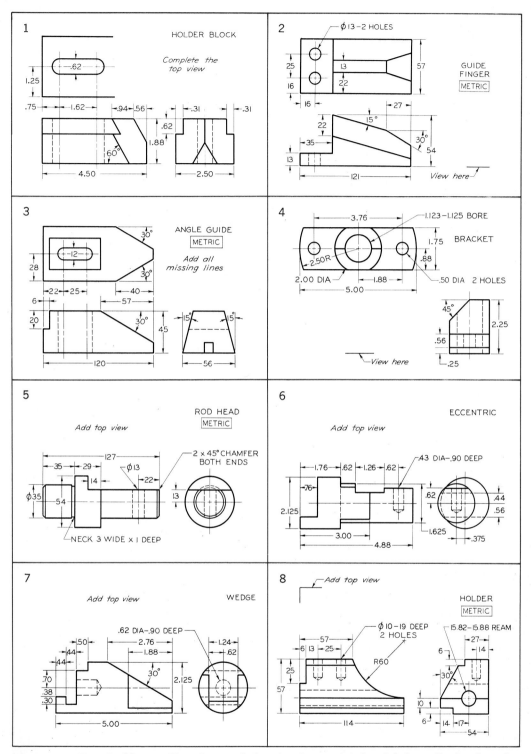

Fig. 7.54 Missing-View Problems. Using Layout A–2 or 3 or Layout A4–2 or 3 (adjusted), sketch or draw with instruments the given views, and add the missing view, as shown in Figs. 7.50 and 7.51. If dimensions are required, study §§13.1–13.25. Use metric or decimal-inch dimensions as assigned by the instructor. Move dimensions to better locations where possible.

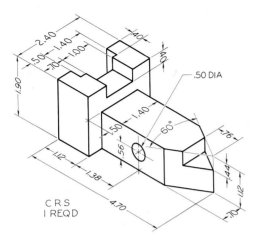

Fig. 7.55   Safety Key   (Layout A–3).*

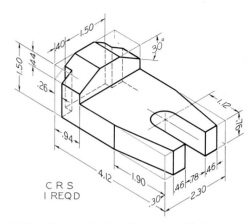

Fig. 7.56   Finger Guide   (Layout A–3).*

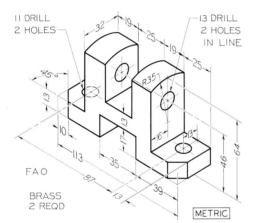

Fig. 7.57   Rod Support   (Layout A–3).*

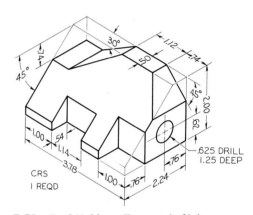

Fig. 7.58   Tool Holder   (Layout A–3).*

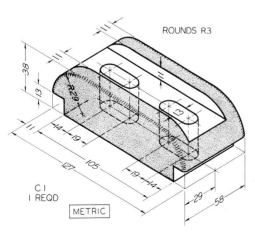

Fig. 7.59   Tailstock Clamp   (Layout A–3).*

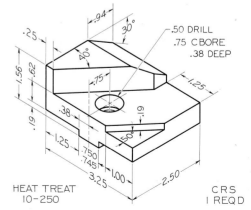

Fig. 7.60   Index Feed   (Layout A–3).*

*Draw or sketch necessary views. Layout A4–3 (adjusted) may be used. If dimensions are required, study §§13.1–13.25. Use metric or decimal-inch dimensions as assigned by the instructor.

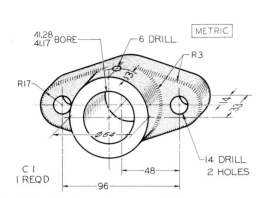

Fig. 7.61 Bearing (Layout A–3).*

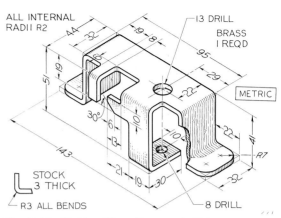

Fig. 7.62 Holder Clip (Layout A–3).*

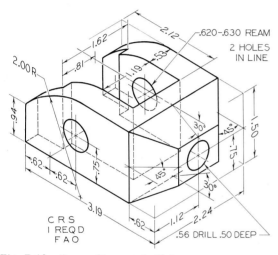

Fig. 7.63 Cam (Layout A–3).*

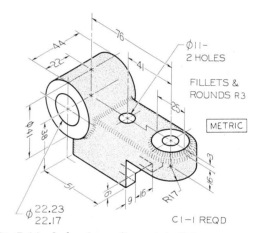

Fig. 7.64 Index Arm (Layout A–3).*

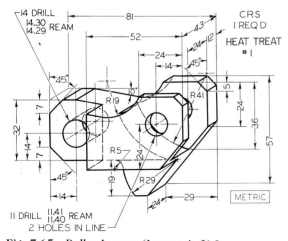

Fig. 7.65 Roller Lever (Layout A–3).*

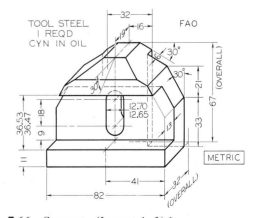

Fig. 7.66 Support (Layout A–3).*

*Draw or sketch necessary views. Layout A4–3 (adjusted) may be used. If dimensions are required, study §§13.1–13.25. Use metric or decimal-inch dimensions as assigned by the instructor.

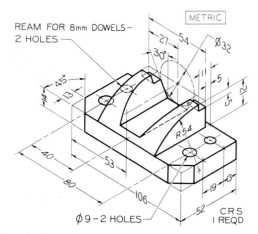

Fig. 7.67   Locating Finger   (Layout A–3).*

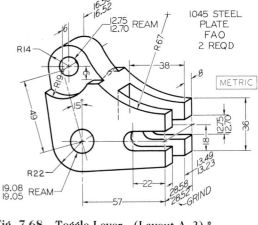

Fig. 7.68   Toggle Lever   (Layout A–3).*

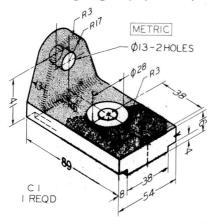

Fig. 7.69   Cut-off Holder   (Layout A–3).*

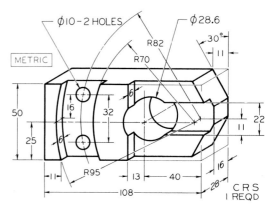

Fig. 7.70   Index Slide   (Layout A–3).*

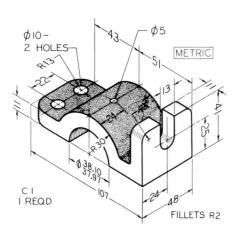

Fig. 7.71   Frame Guide   (Layout A–3).*

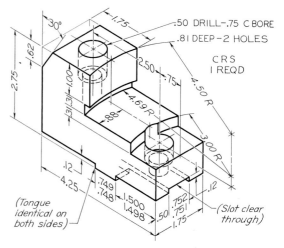

Fig. 7.72   Chuck Jaw   (Layout A–3).*

*Draw or sketch necessary views. Layout A4–3 (adjusted) may be used. If dimensions are required, study §§13.1–13.25. Use metric or decimal-inch dimensions as assigned by the instructor.

238

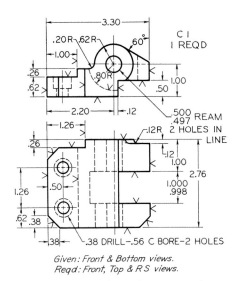

Given: Front & Bottom views.
Reqd: Front, Top & R S views.

Fig. 7.73   Hinge Bracket   (Layout A–3).*

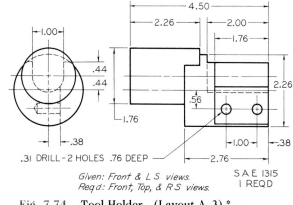

Given: Front & L S views.
Reqd: Front, Top, & R S views.

S A E 1315
1 REQD

Fig. 7.74   Tool Holder   (Layout A–3).*

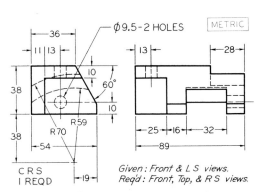

METRIC

Given: Front & L S views.
Reqd: Front, Top, & R S views.

C R S
1 REQD

Fig. 7.75   Shifter Block   (Layout A–3).*

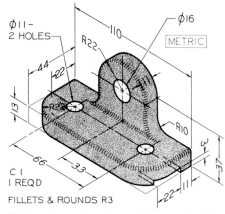

METRIC

C I
1 REQD

FILLETS & ROUNDS R3

Fig. 7.76   Cross-feed Stop   (Layout A–3).*

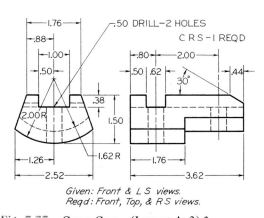

.50 DRILL–2 HOLES
C R S–1 REQD

Given: Front & L S views.
Reqd: Front, Top, & R S views.

Fig. 7.77   Cross Cam   (Layout A–3).*

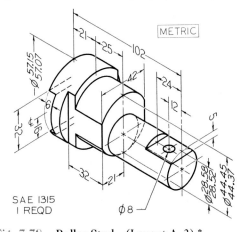

METRIC

S A E 1315
1 REQD

Fig. 7.78   Roller Stud   (Layout A–3).*

*Draw or sketch necessary views. Layout A4–3 (adjusted) may be used. If dimensions are required, study §§13.1–13.25. Use metric or decimal-inch dimensions as assigned by the instructor.

239

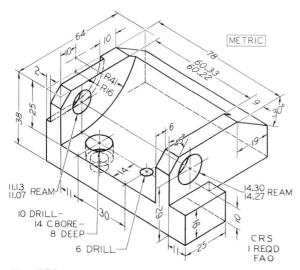

Fig. 7.79 Hinge Block (Layout A–3).*

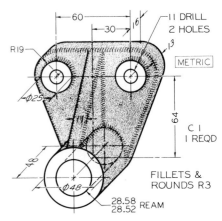

Fig. 7.80 Feed Rod Bearing (Layout A–3).*

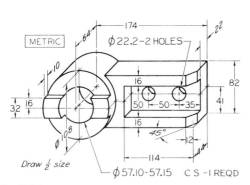

Fig. 7.81 Lever Hub (Layout A–3).*

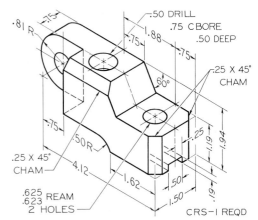

Fig. 7.82 Vibrator Arm (Layout A–3).*

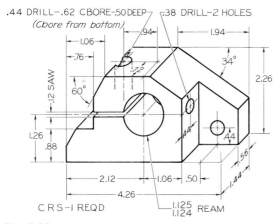

Fig. 7.83 Clutch Lever (Layout A–3).*

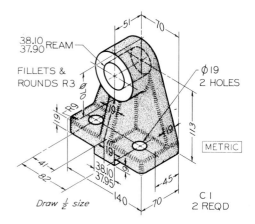

Fig. 7.84 Counter Bearing Bracket (Layout A–3).*

*Draw or sketch necessary views. Layout A4–3 (adjusted) may be used. If dimensions are required, study §§13.1–13.25. Use metric or decimal-inch dimensions as assigned by the instructor.

240

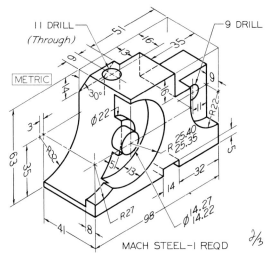

Fig. 7.85 Tool Holder (Layout A–3).*

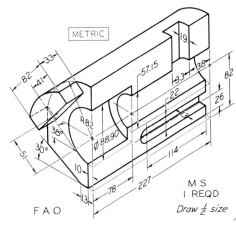

Fig. 7.86 Control Block (Layout A–3).*

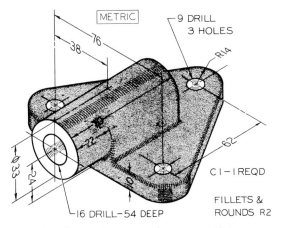

Fig. 7.87 Socket Bearing (Layout A–3).*

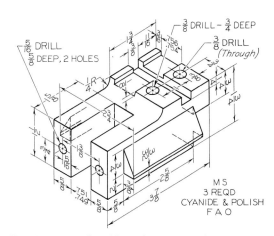

Fig. 7.88 Tool Holder (Layout A–3).*

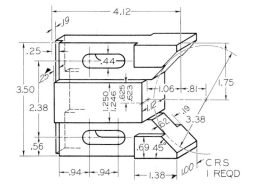

Fig. 7.89 Locating V-Block (Layout A–3).*

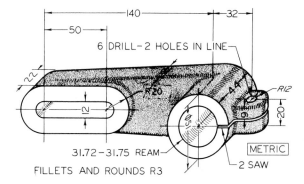

Fig. 7.90 Anchor Bracket (Layout A–3).*

*Draw or sketch necessary views. Layout A4–3 (adjusted) may be used. If dimensions are required, study §§13.1–13.25. Use metric or decimal-inch dimensions as assigned by the instructor.

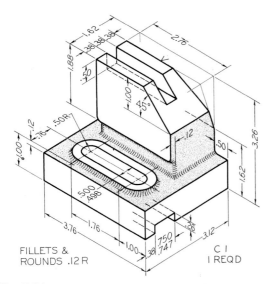

Fig. 7.91 Door Bearing (Layout B–3).*

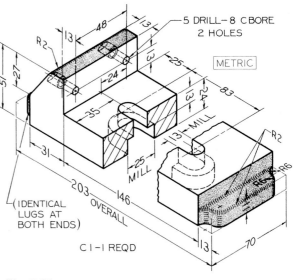

Fig. 7.92 Vise Base (Layout B–3).*

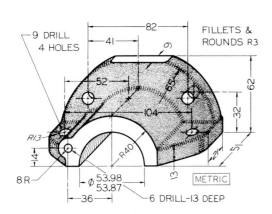

Fig. 7.93 Dust Cap (Layout B–3).*

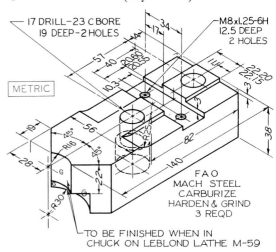

Fig. 7.94 Chuck Jaw (Layout B–3).* For threads, see §§15.9 and 15.10.

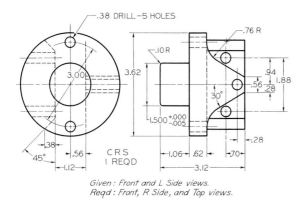

Given : Front and L Side views.
Reqd : Front, R Side, and Top views.

Fig. 7.95 Holder (Layout B–3).*

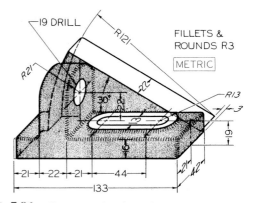

Fig. 7.96 Centering Wedge (Layout B–3).*

*Draw or sketch necessary views. Layout A4–3 (adjusted) may be used. If dimensions are required, study §§13.1–13.25. Use metric or decimal-inch dimensions as assigned by the instructor.

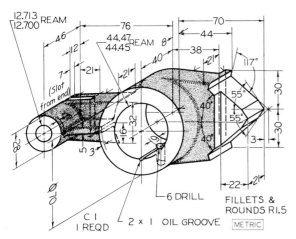

Fig. 7.97 Motor Switch Lever. Draw or sketch necessary views (Layout B–3 or A3–3).*

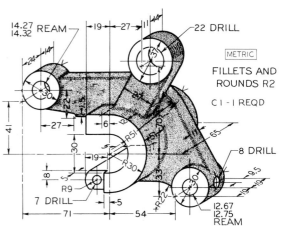

Fig. 7.98 Socket Form Roller—LH. Draw or sketch necessary views (Layout B–4 or A3–4 adjusted).*

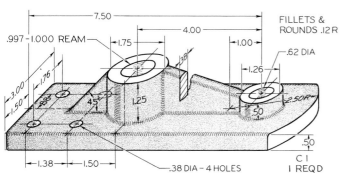

Fig. 7.99 Stop Base. Draw or sketch necessary views (Layout B–3 or A3–3).*

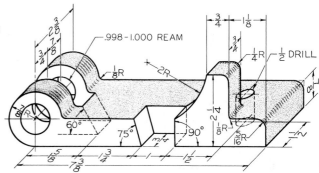

Fig. 7.100 Hinge Base. Draw or sketch necessary views (Layout B–3 or A3–3).*

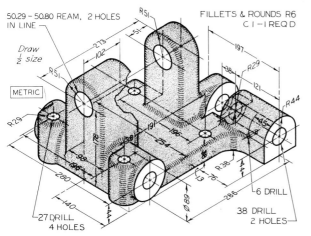

Fig. 7.101 Automatic Stop Base. Draw or sketch necessary views (Layout C–3 or A2–3).*

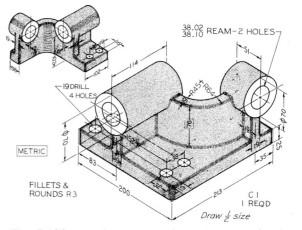

Fig. 7.102 Lead Screw Bracket. Draw or sketch necessary views (Layout C–3 or A2–3).*

*If dimensions are required, study §§13.1–13.25. Use metric or decimal-inch dimensions as assigned by the instructor.

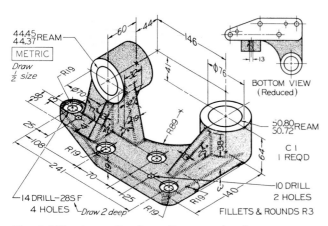

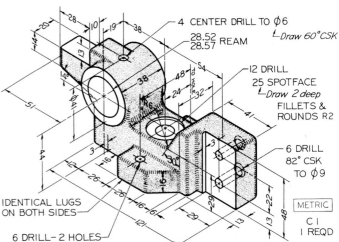

Fig. 7.103 Lever Bracket. Draw or sketch necessary views (Layout C–3 or A2–3).*

Fig. 7.104 Gripper Rod Center. Draw or sketch necessary views (Layout B–3 or A3–3).*

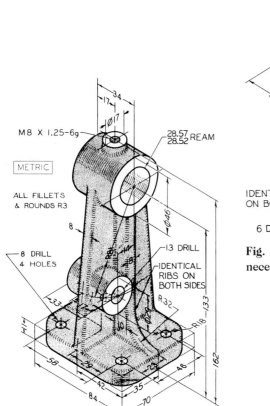

Fig. 7.105 Bearing Bracket. Draw or sketch necessary views (Layout B–3 or A3–3).* For threads, see §§15.9 and 15.10.

*If dimensions are required, study §§13.1–13.25. Use metric or decimal-inch dimensions as assigned by the instructor.

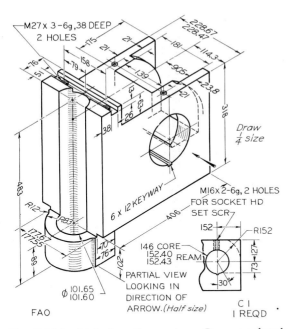

Fig. 7.106 Link Arm Connector. Draw or sketch necessary views (Layout B–3 or A3–3).* For threads, see §§15.9 and 15.10.

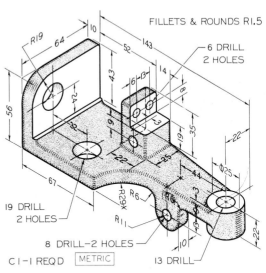

Fig. 7.107 Mounting Bracket. Draw or sketch necessary views (Layout B–3 or A3–3).*

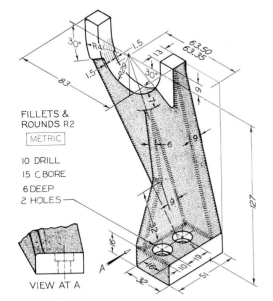

Fig. 7.108 LH Shifter Fork. Draw or sketch necessary views (Layout B–3 or A3–3).*

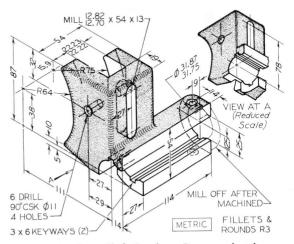

Fig. 7.109 Gear Shift Bracket Draw or sketch necessary views (Layout C–4).*

*If dimensions are required, study §§13.1–13.25. Use metric or decimal-inch dimensions as assigned by the instructor.

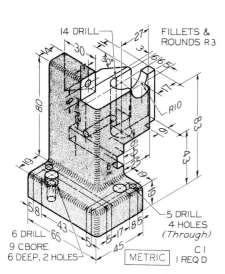

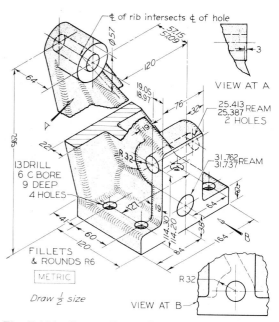

Fig. 7.110   Fixture Base   (Layout C–4).*

Fig. 7.111   Ejector Base   (Layout C–4).*

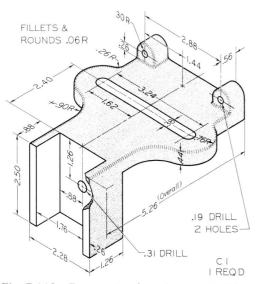

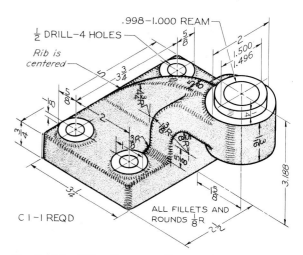

Fig. 7.112   Tension Bracket   (Layout C–4).*

Fig. 7.113   Offset Bearing   (Layout C–4 or A2–4).*

*Draw or sketch necessary views. Layout A2–4 may be used. If dimensions are required, study §§13.1–13.25. Use metric or decimal-inch dimensions as assigned by the instructor.

246

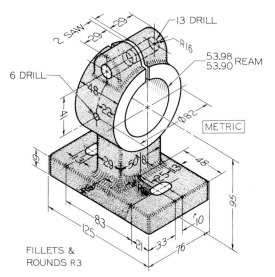

Fig. 7.114 Feed Guide (Layout C–4 or A2–4).*

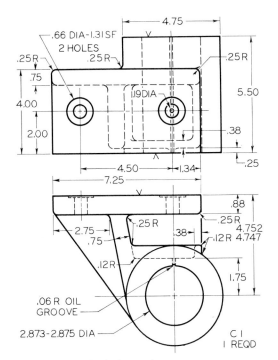

Fig. 7.115 Feed Shaft Bracket.
Given: Front and top views.
Required: Front, Top, and right-side views, half size (Layout B–3 or A3–3).*

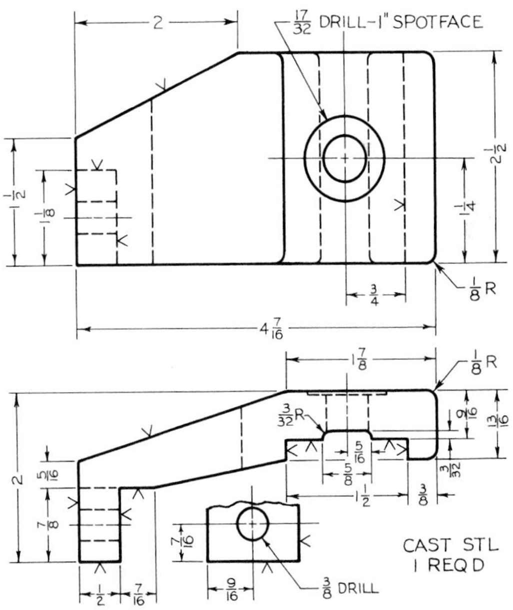

Fig. 7.116 Trip Lever.
Given: Front, Top, and partial side views.
Required: Front, bottom, and left-side views, drawn completely (Layout B–3, or A3–3).*

*Draw or sketch necessary views. Layout A2–4 may be used. If dimensions are required, study §§13.1–13.25. Use metric or decimal-inch dimensions as assigned by the instructor.

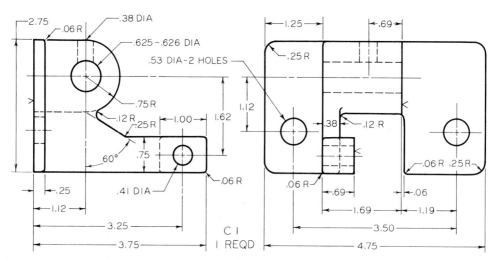

Fig. 7.117 Knurl Bracket Bearing.
Given: Front and left-side views.
Required: Take front as top view on new drawing, and add front and right-side views (Layout B–3 or A3–3).*

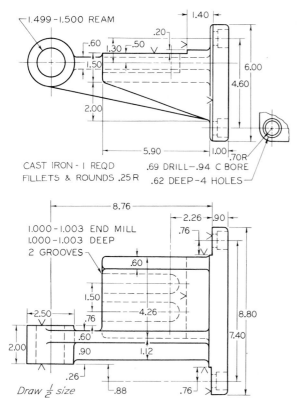

Fig. 7.118 Horizontal Bracket for Broaching Machine.
Given: Front and top views.
Required: Take top as front view in new drawing; then add top and left-side views (Layout C–4 or A2–4).*

*Draw or sketch necessary views. If dimensions are required, study §§13.1–13.25. Use metric or decimal-inch dimensions as assigned by the instructor.

Fig. 7.119 Boom Swing Bearing for a Power Crane.
Given: Front and Bottom views.
Required: Front, top, and left-side views (Layout C–4 or A2–4).*

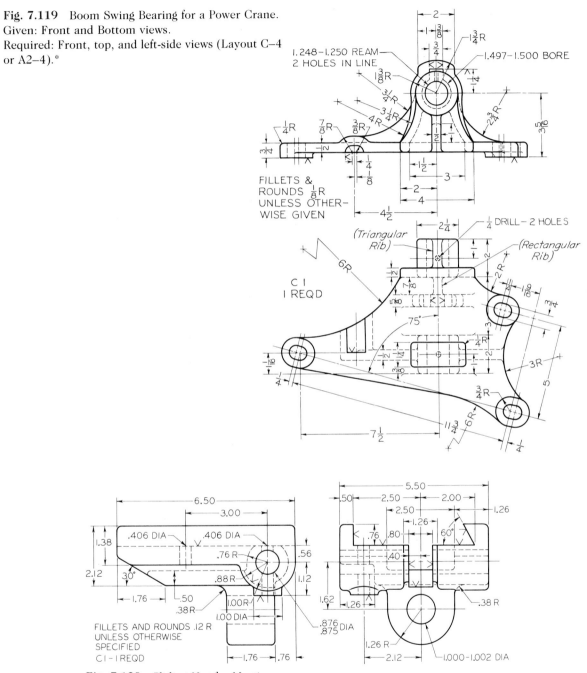

Fig. 7.120 Sliding Nut for Mortiser.
Given: Top and right-side views.
Required: Front, top, and left-side views, full size (Layout C–4 or A2–4).*

*Draw or sketch necessary views. If dimensions are required, study §§13.1–13.25. Use metric or decimal-inch dimensions as assigned by the instructor.

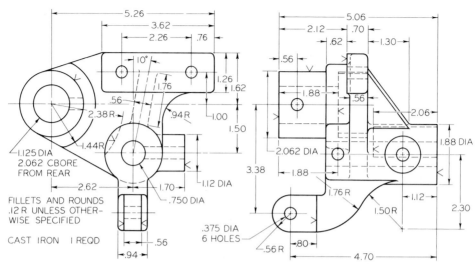

Fig. 7.121   Power Feed Bracket for Universal Grinder.
Given: Front and right-side views.
Required: Front, top, and left-side views, full size (Layout C–4 or A2–4).*

*Draw or sketch necessary views. If dimensions are required, study §§13.1–13.25. Use metric or decimal-inch dimensions as assigned by the instructor.

CHAPTER 8

Using a CAD System

By Byron W. Urbanick*

Computer-aided drafting (CAD) is a practical interactive tool, operable by the senior designer on the engineering team and every member of the staff. It is a useful tool that performs work in the engineering design office or in the manufacturing plant. The software programs for CAD and computer-aided manufacturing (CAM) can interact with the full engineering scheme of concept development to integrate manufacturing and the delivery of a product. The practical applications of such a tool may never be exhausted, for they are limited only by the imagination and creativity of the user.

Interactive CAD is a human-oriented tool. The images and details projected on the computer's monitor are visual evidence to the design engineer or CAD drafter that the all-important CAD database understands the operator's instruction, design, or input request. CAD is a practical and powerful tool because of the almost instant response to the operator's input or request for stored information. It is a high-tech system for creating new designs, developing technical documents, and managing documentation—it is more than a simple electronic drafting machine. The development of networked CAD systems has unleashed the creative imagination and design interaction ability of every individual on the engineering design team.

It is the purpose of this chapter to show that at any moment in the creation of CAD images, the CAD user is building a detailed graphic document using what the computer knows. That is, the CAD software allows you to work sequentially with the programs stored in the central processing unit (CPU). The potential CAD user is introduced to several CAD software "working procedures" and can examine the basic interactive screen formats presented by AutoCAD, CADKEY, and FastCAD. The text describes typical procedures for selecting a drawing command and reviews the menu options available for geometric constructions and detailing on the display screen of the CAD system. These somewhat generic descriptions of CAD functions presented on a personal computer can provide an introduction to the interaction and operations of the CAD user. More specific software instruction is available from individual CAD software developers and of course is more complete and extensive in the instructional format.

*Design Consultant & Vice President, Paneltech LTD., Lombard, IL 60148.

The CAD system on the personal computer as an automation tool is replacing many of the drafting instruments, drafting tables, and drafting files. But, like no other drafting tool before, it raises engineering productivity without replacing the basic functions of the designer, engineer, and drafting technician. CAD developers, in their quest to harness computer technologies, have had a profound impact on the high-tech teams as they resolve problems in research, development, design, production, and operation (the five basic engineering functions).

8.1   CAD Compatibility

How does a CAD system store your current drawing? What is RAM? CAD software stores your drawing in your computer's random access memory (RAM). In Chapter 3 the CAD system is illustrated and the various input and output devices (called hardware) are presented as the tools for the CAD user. The CAD software programs have to be compatible with specific hardware, and the CAD software programmer wants the software "commands and menus" to be compatible in ways that are similar to patterns of manual drafting work. The skills learned "on the board" are related and complementary to those needed by the CAD user. To learn the performance skills needed for creating drawings with CAD tools is time consuming and requires practice and manual dexterity, Fig. 8.1. Both methods of drafting use simple and familiar geometric terminology for structuring the graphic production of technical documents, and both have the same goal—drawings that will meet industry standards.

The basic principles of drafting are common to traditional drafting and computer-aided drafting. The American National Standards Institute (ANSI) has well-established standards for shaping engineering drawings. The knowledge of drafting principles from the alphabet of lines to dimensioning and sectioning procedures continues to be essential in shaping CAD documents. CAD is compatible because it can produce consistent lettering and regulates line work to improve the production of working drawings better

Fig. 8.1   As a human-oriented tool the CAD system is a "hands-on" high-tech feature of Electronic Interaction. *Courtesy of Computervision.*

than any other tool. Drafting continues to be a "hands on" system with CAD software serving as the main form of communication between networked CAD workstations and plotter or printers, Fig. 8.2.

8.2   Software—Invisible

Though invisible to the eye, software is the prerecorded instruction that makes the hardware operate. Today, most CAD software is stored on floppy disks, or transferred from floppy disks ($5\frac{1}{4}''$ or $3\frac{1}{2}''$) to the hard disk within the computer where it becomes quickly accessible. Software tells the computer to direct the flow of data entered by the CAD user either to working storage memory or to the disk for instant recall. Software has the ability to "remember" formulas, for example, the center of gravity for a truncated cone. As shown in Fig. 8.3, software can create a wire-frame model, (a), or a shaded solid model, (b). Software helps the computer find symbols stored for use in creating or drawing and will help you create symbols for storage to be retrieved for specific types of drawings. Software can be programmed to exchange files with other software programs.

Fig. 8.2 This is a high-tech system for developing technical documents. Networked personal computer systems are emerging in the classrooms. *Courtesy of Moraine Valley Community College Technical Center.*

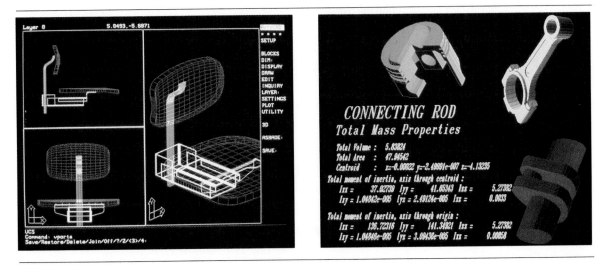

Fig. 8.3 The CAD software increases visualization through wireframe models, (a), and solids modeling, (b), to enhance pictorials. *Courtesy of AutoDesk, Inc., and CADKEY.*

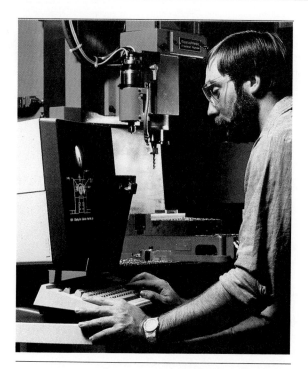

The invisible software helps the computer compile data and arrange it on various drawing layers or levels. The invisible software can remember CAD user passwords for security and controls input of specific data and regulates output of specific data when charged (programmed) with that responsibility. Software can also count, measure, and direct devices to print or plot drawings or to create a bill of materials. CAD software has been designed to serve all the major branches of engineering and manufacturing, Fig. 8.4. The CAD user must understand how the basic software information is installed, stored, transferred, and exchanged.

Fig. 8.4 The CAD software interacts with manufacturing processes and the engineered product has been improved through quality control of computer-aided manufacturing (CAM). *Courtesy of Computervision.*

8.3   CAD Software

All CAD software generates familiar geometric terminology for creating drawings. But even though the geometry is common and the procedures for construction are similar, every CAD software program will vary in operational procedures, Fig. 8.5. The variations may be considered a hierarchy of command structure and may be reviewed with AutoCAD, CADKEY, and FastCAD screens used in this chapter.

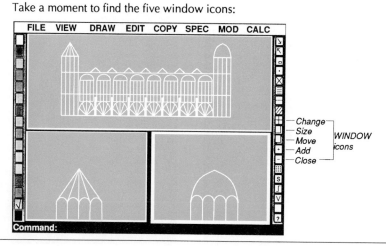

Fig. 8.5   The CAD screen display provides selection procedures that vary with each software program. FastCAD has an icon bar for use of multiple windows. *Courtesy of Evolution Computing.*

FastCAD presents a dialog box:

| FILE | VIEW | DRAW | EDIT | COPY | SPEC | MOD | CALC |

Specify text:

| Height | Angle | Font | Spacing |

Justify text:

| Left | Center | Right |

Command: TSPEC

Fig. 8.6 What do you want to do and how do you want to do it? The text specifications are generated for selection in a "how to" dialog box. *Courtesy of Evolution Computing.*

Three features are found in all CAD software, and the programmer creates these so they work interactively with basic commands and menu options.

1. Commands for geometry generators (basic geometric construction).
2. Functions to control the size of generators (scale of the drawing, paper size, etc.).
3. Modifiers for changing the drawing or editing variations in the drawing (rotate, mirror, delete, group, etc.).

The normal text specification command is structured to two questions for the FastCAD user, Fig. 8.6.

1. What do you want to do?
2. How do you want to do it? For example, in the CADKEY structure, CREATE (the command), LINE (an option), ENDPOINTS (an option) assists the CAD user in creating a line.

The commands and menu options may be selected in two basic ways: the function keys

Fig. 8.7 The digitizer or mouse can select the step-by-step process rather quickly with reasonable practice. *Courtesy of Tektronix.*

(F1–F10 and PageUp and PageDown keys), and the tablet (digitizer) or mouse. With the tablet or mouse, specific areas on the monitor screen can be quickly selected, Fig. 8.7. The Cartesian coordinates may be accessed with the keyboard or the mouse. You may switch between these function keys and input devices at any time to issue commands and selected options. In the CADKEY software example cited earlier, the CAD user would press the F1 function key to obtain the command: F1 to create, then F1 to select the line option, then F1 again to select endpoints. The sequence of selection is called the hierarchy of command structure and provides an ease of operation that is the basis for selecting one software program over another.

8.4   CAD Capability Checklist

The software can offer the following characteristics required for creating technical documents.

1. Draw construction lines at any convenient spacing through any points, at any angles, and create tangent lines to one or more arcs.
2. Draw any type of line such as visible, center, hidden, or section.

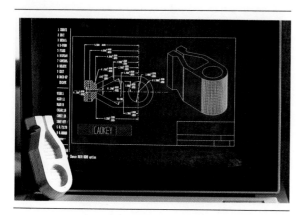

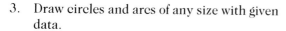

Fig. 8.8   A Casting Profile. As a high-tech tool, CAD enhances the production of design and detailing in all branches of engineering. *Courtesy of CADKEY.*

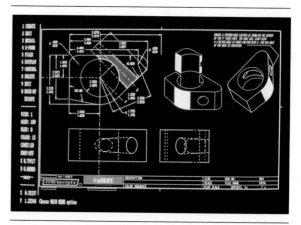

Fig. 8.9   Isometrics–Pictorials–Advanced programs generate 3D graphics. *Courtesy of CADKEY.*

3. Draw circles and arcs of any size with given data.

4. Perform cross-hatching within specified boundaries.

5. Establish a scale or set a new scale within a drawing for various drawings within a document.

6. Calculate or list pertinent data of graphic construction such as actual distances or angles.

7. Create a group of geometric figures for editing or copying.

8. Relocate a drawing to any new position. Correct or change additions in stored document.

9. Edit all or erase (delete) any part of a line, arc, or any geometric form on a drawing. Correct dimensions.

10. Make a mirrored image or create symmetrical forms.

11. Perform associative or datum unit dimensioning, Fig. 8.8.

12. Label drawings with notes and create title blocks and bills of material.

13. Save the entire drawing or any part for use on other documents.

14. Create pictorials from three-view drawings, Fig. 8.9.

15. Retrieve and use stored drawings.

List additional capabilities of your system or think of one more you need. Software programmers are constantly adding new capabilities and options to CAD software.

AutoCAD has extensive capabilities built into over 125 commands, and many commands have more than one complementary option. This CAD software has more than 20 ways to edit a drawing and over 12 dimension options. After reviewing the checklist above, you realize that the list can be expanded when you have over 125 commands in one interactive program.

8.5   Microcomputer Operations

Begin by turning on the computer and all the peripheral tools, including the monitor, digitizer/mouse, and plotter/printer. In order to run your CAD programs on the microcomputer you must have a copy of the disk operating system (DOS) already loaded into it. System software is used to control the operation of the microcomputer, and it will manage data input and output between the microcomputer and other hardware such as disk drives. In order to operate the CAD software programs on the PC workstation, the system must first be under control of the system software.

8.6   Loading System Software

The following steps are used for booting up the operating system. Every CAD user should be well acquainted with the disk operating system (DOS) used on the computer. The ability to interact with major commands of the system software enables the CAD user to become interactive with CAD software changes.

1. For floppy disk systems (no hard disk), open the latch for the diskette drive A, and carefully insert the DOS system disk into the drive slot with the label up or facing you, Fig. 8.10. Close the drive latch. (*Note:* The A drive is usually on the left side or on top.) Diskettes should only be inserted or removed when the drive indicator light is not lit. Always wait until the light goes off, indicating that the drive is not reading or writing data.

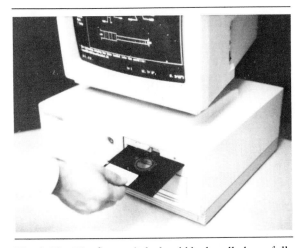

Fig. 8.10 The floppy disk should be handled carefully with the label side up when it is inserted into the computer drive opening. *Courtesy of Michael Eleder Associates.*

2. Turn on the PC (floppy and hard disk systems). If the PC is already turned on, you can boot the DOS operating system by holding down the Ctrl and Alt keys, and then pressing the Del key.

3. After DOS has been loaded into the computer memory, the system monitor will prompt you for the DATE and TIME. Type the correct month, day, and year (for example, 10-30-92) and press RETURN. Then type the correct time (for example, 9:00) and press RETURN. Some systems have a battery clock and will not require you to enter the time and date.

4. DOS has been successfully loaded when the DOS command prompt appears on the monitor as
    A:>   (for non-hard drive systems) or
    C:>   (for hard drive systems).

5. You can type any DOS command or software program name at this time.

8.7   The Floppy Drives

When you are using a dual floppy drive computer, generally, DOS must be loaded each time the computer is turned on. The original DOS diskette should be stored in a safe place, and should not be used for daily operations. You should make a backup copy for daily use. You can make a backup copy using the following steps.

1. Start up DOS. The system is ready when you see the DOS prompt A:> or C:> on the monitor.

2. Type

    FORMAT B: /S   <return> (press RETURN key)

3. The monitor will respond with

    Insert new diskette in drive B
    and strike any key when ready.

    Place a new blank diskette into the B drive, close the latch, and press any key. The disk-

257

ette light will go on, and the blank diskette in drive B will be formatted so it can receive data. When the light goes off, the monitor message will read

Format another (Y/N)?

Type

N

4. The A:> or C:> prompt will be displayed. Now type

COPY A:*.* B: <return>

(Be sure to leave a space before the letter A and before the letter B.)

5. The program files that are copied from the diskette in drive A and transferred to drive B will be displayed on the monitor. After the copy is complete, store the original DOS diskette in a safe place, and use the new copy for daily operations. (Hard disk systems have their working copy on drive C.)

The CAD program needs a place to store the drawings. Blank diskettes without programs may be used. They need to be formatted before drawings can be stored on them. When a previously used diskette is formatted, any information that was on the diskette will be erased. Some new diskettes can be purchased that are preformatted for DOS and do not require formatting before they can be used.

8.8   Access to Drive Commands

When the computer is turned on, the default drive is A for dual floppy systems or C for hard disk systems. When it is necessary to access a diskette drive other than the current drive, type the drive name, colon, and press return. For example, to change from drive A to drive B, type

B: <return>

at the A:> prompt. The prompt on the monitor changes to B:>. To change from the B drive to the C drive, type

C: <return>

at the B:> prompt. The prompt on the monitor changes to C:>. Whatever letter is displayed at the prompt is the current drive you will access for disk operations. Hard disk systems may also have drives D and E depending how the hard disk is partitioned and how large the hard disk is. The prompt, whether it is A:>, B:>, or C:>, is commonly called the DOS prompt.

8.9   Access to Directory Display

CAD drawing files that are stored on diskette can be viewed alphabetically on the monitor using the directory command. The directory for the current drive may be displayed by typing

DIR <return>

at the DOS prompt. The directory that is displayed lists the files that reside on the current drive, Fig. 8.11. To see the directory of another drive, drive B for example, type

DIR B: <return>

Sometimes, the list of files is so long that it scrolls vertically off the screen and cannot be read. You can pause the scrolling by typing a modified command for directory. Type

DIR B:

then

DIR /P <return>

at the DOS prompt and the list will pause. Press any key to view the next group of file names.

When the directory is displayed, the left column is always the file name, which may be up

FastCAD displays a text screen like this:

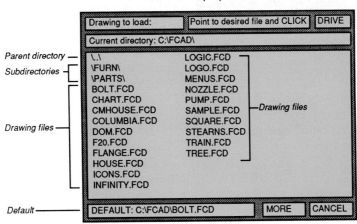

Fig. 8.11   When you access the CAD software directory, you will be able to examine the files that are created to serve the CAD user. *Courtesy of Evolution Computing.*

to eight letters, numbers, or symbols long. The next three letters are called the file extension and describe the type of file. The file name and file extension are separated by a period. Some common file extension abbreviations are

com	command file
exe	executable file
drv	driver file for peripheral device
ovl	overlay file for software program
dwg	drawing file
bak	backup file
mnu	menu file (common to AutoCAD)
doc	word processing file
txt	text file

Files are the management system common to most CAD software programs.

8.10   Deleting CAD Files

The DOS system allows CAD files to be deleted by entering the command DEL or ERASE at the DOS prompt, followed by the complete name of the file to be deleted. For example, if you type

DEL BEARING.BAK   <return>

at the A:> prompt, the file bearing.bak will be erased from the diskette in drive A. If you have limited storage space on a diskette or hard drive, you may need to delete backup files one at a time as shown above. The reason for a backup file is to keep up to date with the latest copy of a drawing and provide a backup in case the current version is damaged or inadvertently erased. If a problem occurs with the current version, you have access to the backup file. To delete all of the backup files in drive B in one step, type

DEL B:*.BAK   <return>

at the DOS prompt. Any file on drive B with a file extension bak will be erased with this command. Then the current drawing is loaded as an updated document in storage.

8.11   Copying CAD Files

In order to copy a drawing file from one disk to another, use the COPY command, followed by the drive and complete file name. For example, to copy the bearing.dwg file from the diskette in drive A to the diskette in drive B, type

COPY A:BEARING.DWG B: <return>

at the DOS prompt.

If you want to change the name of a file, you can use the copy command. For example, to retrieve a backup drawing called bearing.bak, you would rename it by typing

COPY A:BEARING.BAK BRG1.DWG B:
<return>

at the DOS prompt. This would copy the bearing backup file on drive A, change the name to BRG1, change the extension to DWG so the file could be read, and put the copy on drive B. This is typical of AutoCAD retrieval. The new drawing may be accessed now as BRG1.

8.12   CAD Files on the Hard Disk

The student of CAD is generally instructed to save drawings on a personal floppy diskette. The following procedure should be used to run the CAD program (AutoCAD in this example) and load or save drawings on diskette.

1. Type

CD\ACAD   <return>

at the C:> prompt.

2. Type

ACAD   <return>

at the C:> prompt to load the AutoCAD program.

The CAD program is now loaded and ready for use. Drawing names that students use should be preceded with B: or A: (when two drives are available for $5\frac{1}{4}''$ or $3\frac{1}{2}''$ floppy diskettes).

The CAD files stored on the hard disk can be viewed using the directory (DIR) command men-

tioned earlier. Students should keep their personal files on floppy diskettes (drive A or B), and not on the hard disk (drive C). When saving personal files, precede your file name with the drive letter A: or B: when beginning the CAD experience, and save your work on your personal floppy diskette.

8.13   Loading CAD Software

To install or load an additional CAD software program onto a computer's hard disk (you must have adequate storage), it must be compatible or configured with the DOS of the computer workstation. The peripheral input devices, such as the mouse, tablet with puck or cursor, must be specified or selected when the CAD software is configured during the installation of these "input tools," Fig. 8.12.

The "output tools" such as the monitor (CRT), printer, and plotter, are also specified and selected when the devices are configured for the system. CAD software is generally programmed to work with a variety of input and output devices available to the design team. CAD

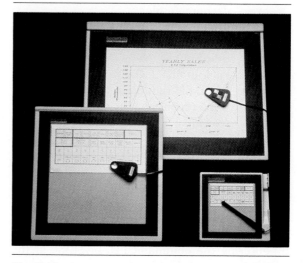

Fig. 8.12   Various input tools and output devices are configured into the CAD program before systems can start up. *Courtesy of Houston Instruments, Division of AMETEK.*

```
          A U T O C A D
Copyright (C) 1982,83,84,85,86,87 Autodesk, Inc.
Release 9.0 (9/17/87) IBM PC
Advanced Drafting Extensions 3
Serial Number:

Main Menu

   0.  Exit AutoCAD
   1.  Begin a NEW drawing +————————START HERE
   2.  Edit an EXISTING drawing
   3.  Plot a drawing
   4.  Printer Plot a drawing

   5.  Configure AutoCAD
   6.  File Utilities
   7.  Compile shape/font description file
   8.  Convert old drawing file

Enter selection: _
```

THE MAIN MENU

Fig. 8.13 AutoCAD screens allow you to start a new drawing or select a drawing from the file. *Courtesy of AutoDesk, Inc.*

software is designed to work with various disk operating systems such as MS-DOS (Microsoft DOS) and UNIX. A typical example of software configuration options would list five or more plotters that would be compatible with a specific CAD software program (called environment) and ten or more different types of monitors.

Versatility and compatibility in CAD tools are needed for varying performance standards that bring high quality to drawing documents. One typical configuration might include a high resolution monitor and a high speed E size (36″ × 48″) plotter.

8.14 Software—Start-up

Three CAD start-up displays are now reviewed for three common software programs. After the CAD software is loaded and all peripheral devices have been installed and configured, the operation of the CAD program must be addressed. The CAD software can be loaded from sophisticated program management systems, or it may simply be loaded with keyed-in commands at the DOS system prompt.

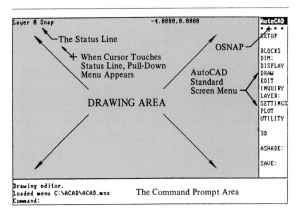

FOUR AREAS OF SCREEN

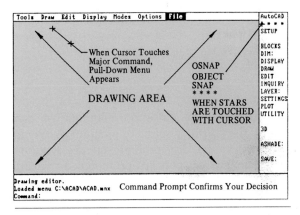

Fig. 8.14 The display screen of AutoCAD has a fixed menu area and a pull-down menu bar. Note the specific area of current info and prompt area. *Courtesy of AutoDesk, Inc.*

AUTOCAD DISPLAY To load and start AutoCAD, type the symbol of the AutoCAD program at the DOS prompt

C:> ACAD

and then press the RETURN key to bring up the AutoCAD options, Fig. 8.13. This may be referred to as "boot-up" of the program, and marks the beginning of the hands-on operation. If you select the appropriate number for a new or existing drawing, you will be viewing the program's display screen, Fig. 8.14. Examine the specific areas of the screen and note the drawing area limits.

261

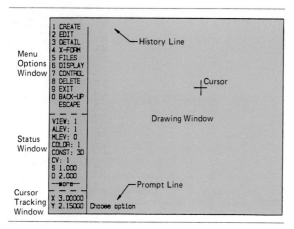

Menu Options Window

```
1 CREATE
2 EDIT
3 DETAIL
4 X-FORM
5 FILES
6 DISPLAY
7 CONTROL
8 DELETE
9 EXIT
0 BACK-UP
  ESCAPE
```

—History Line

┼—Cursor

Drawing Window

Status Window

```
VIEW: 1
ALEV: 1
MLEV: 0
COLOR: 1
CONST: 3D
CV: 1
S 1.000
D 2.000
—more—
```

Cursor Tracking Window

```
X 3.00000
Y 2.15000
```

—Prompt Line

Choose option

Fig. 8.15   The CADKEY display screen functions from left to right with menus located on the left. *Courtesy of CADKEY.*

CADKEY DISPLAY   In a like manner, you can call up the CADKEY software program by typing CADKEY at the C:> prompt and pressing the ENTER key. When you are prompted for a part name, type in the name blank and press ENTER again to start a new drawing, or you can recall an existing drawing from the disk file. When the CADKEY screen display becomes available, you can compare the initial features for the two CAD programs, AutoCAD, Fig. 8.14, and CADKEY, Fig. 8.15.

Note that the menu area for CADKEY is on the left side of the screen and dotted lines separate menu options from the status window (referred to as the intermediate modes). Another dotted line separates the status from the cursor tracking.

FASTCAD DISPLAY   After your FastCAD software program has been installed, the DOS prompt will appear. To load the FastCAD program, type

FCAD   <return>

at the DOS prompt. The software package will generate a display screen as shown in Fig. 8.16. This display screen has a color palette on the left margin of the screen, and the right margin contains icons for intermediate functions available while using the main menu. Now you can access these functions from the icon bar while working any of the drop down menus available from the top of the screen. Note the draw menu for geometric generators and the file menu at the upper left in Fig. 8.16. FastCAD uses an arrow for a cursor point and also has long Cartesian coordinates that serve as a cursor. (AutoCAD and CADKEY use a small cross cursor and also have a long horizontal and vertical line that serve as a Cartesian cursor.)

8.15   CAD Functions: Input

The three CAD displays reviewed have unique features for moving from one command to another and all have specific functions that may

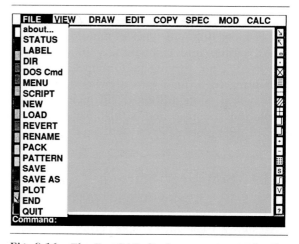

Fig. 8.16   The FastCAD display screen provides the user with a main menu bar across the top of the screen that has pull-down menus and functional icons on the right side. *Courtesy of Evolution Computing.*

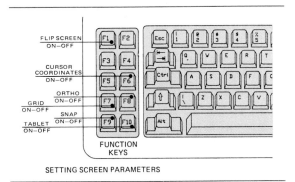

FUNCTION
KEYS

SETTING SCREEN PARAMETERS

Fig. 8.17 Review the function keys used with AutoCAD. They speed up command changes.

be selected with minimal effort. The keyboard of any computer contains the letters of the alphabet, numbers, the function keys (labeled F1–F10), and cursor keys labeled with arrows, Fig. 8.17. They are called cursor keys because they can move the cursor to specified locations on the computer display. The backspace key is useful for changing typed-in commands, altering the text composed for notes on a drawing, undoing characters, or erasing a whole line of text.

The RETURN key is generally a larger key, and is used frequently for entering information that serves as a decision maker. Know your keyboard for effective control of the CAD software design program.

AUTOCAD   Some of the AutoCAD functions are noted in Fig. 8.17 and are effective in setting up the working conditions of the screen work area for construction of geometric features. Note that F6 controls the cursor coordinates and F7 is able to turn the grid on and off. The control key can also manipulate functions when used with other specified keys. The AutoCAD user must know the following key combinations: CTRL-B for snap features that allows the cursor to move to units of measure (e.g., every quarter-inch), CTRL-C (to cancel), CTRL-D (coordinate dis-

play), CTRL-E (toggle crosshairs—the toggle acts as an on-off switch), CTRL-Q ("printer echo," which prints text only), CTRL-S (stop scrolling), and CTRL-O (ortho on or off—controls horizontal and vertical lines). The control key and the letter key have to be pressed simultaneously to effect a response. An example is pressing CTRL and Q at the same time for the text to echo to the printer. These combinations, when pressed, act as toggles, turning features on and off. AutoCAD has a user coordinate system (UCS) for 30 axes input.

CADKEY   This software program uses the keyboard for special descriptive functions.

1. ESC (escape) key will allow you to exit any function and return to the main menu.

2. CTRL (control) key is used to move into the intermediate mode commands. It must be pressed in combination with other specified keys, Fig. 8.18.

3. ALT (alternate) key is also used in conjunction with other keys to initiate intermediate mode commands.

4. BKSP (backspace) key deletes letters and numbers entered at the prompt line.

5. Space bar <space> is used for identifying or designating locations for general entry functions. The space bar can serve as a cursor button for entry.

The function keys F1–F10 can be used to select menus displayed on the screen. The successive changes of menu options are aligned numerically from top to bottom. With F1 on the top, press F1 three times, and successively, CREATE F1, LINE F1, and ENDPOINTS you can locate for lines will appear. This is only one of the three ways to enter commands. F10 returns you to the previous command or cancels the last selection mode.

KEYS WITH SPECIAL FUNCTIONS

There are a number of assigned keys that will increase your productivity within the system. Keep in mind that some keyboards may have different names or symbols assigned to these keys (e.g. Carriage Return may appear as RETURN, ENTER, or <--). Take some time to review the following keys and their functions, and locate them on the keyboard.

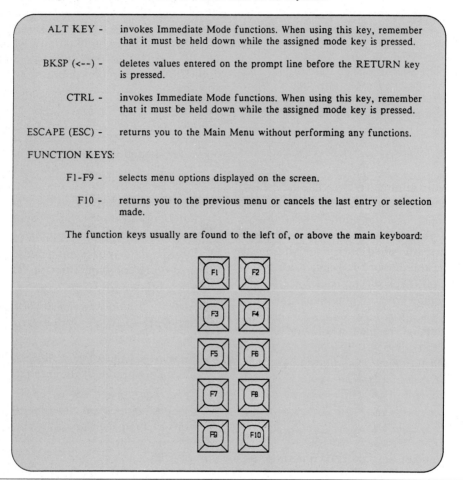

ALT KEY -   invokes Immediate Mode functions. When using this key, remember that it must be held down while the assigned mode key is pressed.

BKSP (<--) -   deletes values entered on the prompt line before the RETURN key is pressed.

CTRL -   invokes Immediate Mode functions. When using this key, remember that it must be held down while the assigned mode key is pressed.

ESCAPE (ESC) -   returns you to the Main Menu without performing any functions.

FUNCTION KEYS:

F1-F9 -   selects menu options displayed on the screen.

F10 -   returns you to the previous menu or cancels the last entry or selection made.

The function keys usually are found to the left of, or above the main keyboard:

Fig. 8.18   Function keys F1–F10 are related to the menus in CADKEY and move quickly to process commands. *Courtesy of CADKEY.*

Review the CADKEY display in Fig. 8.15 and note the optional control and alternate keys in Fig. 8.19 that allow the user to exercise specific functions. With the keyed-in responses, the user quickly modifies the CAD drafting tasks. The views in the intermediate mode are uniquely numbered: View 1 = top, 2 = front, 3 = back, 4 = bottom, 5 = right side, 6 = left side, 7 = isometric, and 8 = axonometric. The status win-dow has three pages (which moves three options), and controls many performance features called attributes, such as layer, line style, color choice, and pen number. The VIEW and WORLD option allows 3D pictorial representation.

FastCAD   The function keys for FastCAD serve as a special kind of macro name. The program will recognize 40 function keys that include the

FUNCTION	IMMEDIATE MODE COMMAND
ARROWS IN/OUT	CTRL-A
AUTO scale	ALT-A
BACK-1 window	ALT-B
change DEPTH	CTRL-D
change VIEW	ALT-V
COLOR bar	ALT-X
cursor SNAPPING on/off	CTRL-X
CURSOR TRACKING display	CTRL-T
DELETE (single)	CTRL-Q
FILE PART (save)	CTRL-F
GRID ON/OFF	CTRL-G
LEVEL masking	ALT-N
LINE LIMITS switch	ALT-L
LINE TYPE menu	ALT-T
LINE WIDTH menu	ALT-Y
PAN	ALT-P
PEN # assignment	ALT-Z
RECALL last (undelete)	CTRL-U
REDRAW	CTRL-R
SCALE with center	ALT-S
set working LEVEL	CTRL-L
2D/3D switch	CTRL-W
type MASKING	ALT-M
VIEW/WORLD coordinates	CTRL-V
WINDOW	ALT-W
WITNESS LINE	CTRL-B
ZOOM-DOUBLE	ALT-D
ZOOM-HALF	ALT-H

Fig. 8.19 The ALT and CTRL keys are used to access functional changes in CADKEY screen display interaction. *Courtesy of CADKEY.*

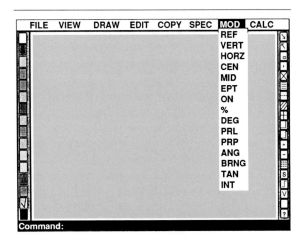

MOD menu includes several *modifiers* that let you refer to exact points on existing entities.

Fig. 8.20 The function keys are used to modify the options in the MOD menu of FastCAD. *Courtesy of Evolution Computing.*

ten standard (F1–F10) and the same keys in combinations with the CTRL (control) key, the SHF (shift) key, and the ALT (alternate) key. You can refer to a function key by its name (F1, CTRL-F1, ALT-F1, etc.) or just press a key or combination of keys. In the standard macro file FCAD.MAC, several function keys have been defined to select modifiers. For example, F1 enters the CEN (center) modifier; the function key F3 will locate the midpoint, and F5 will locate the endpoint, of entities.

Note from the main menu bar across the top of the screen that the pull down option MOD (modify) was selected with a cursor arrow touching and lighting up the MOD portion of the bar,

Fig. 8.20. The control keys serve to support the user as follows: CTRL-R, reenter or get cursor back; CTRL-S, scroll lock; CTRL-C, cancel current command; CTRL-D, pause a macro program for direct input; and CTRL-A, pause macro for input with optional prompt.

FastCAD software has pull down menus, icons, and dialog boxes or the normal type-in keyboard equivalents. The CAD user can select drawing locations with a mouse or digitizer, or type the numeric coordinates, mixing methods at any time. New menus can be customized and scripts and macro libraries developed for repetitive tasks. The four windows are simultaneously active, and the user can draw a line from one window to another and watch the effects in both windows. The software is an object-oriented system. When you draw a circle on the screen, it stores the mathematical definition of a circle rather than just the screen image. When you are ready to change the circle after drawing it, simply touch a point on the circle and the

265

software knows exactly what object to change. The programmer uses floating point numbers to represent entities (any geometric form) and their exact placement. Therefore, the largest entity on your drawing can be at least a million times larger than your smallest. The CAD user can zoom in or out at any selected multiple and place entities with precision.

Dimensioning is truly associative, so when you stretch or shrink a part of your drawing with editing commands, dimensions are automatically updated. No erasing is necessary to reenter the descriptive new dimensions. The CAD user has editing commands such as ERASE, MOVE, SCALE, ROTATE, MIRROR (any angle), COPY, ROTATE COPY, MIRROR COPY, BREAK, TRIM, BEND, STRETCH, and CONNECT. The modifiers refer to specific points on entities such as center, Fig. 8.20, endpoint, percent, degree, tangent, intersection, or perpendicular to, parallel to, or any point at any angle to existing entities.

The capabilities of a specific software are numerous and cannot be addressed in the limits of one chapter. When you examine the commands available on any CAD screen illustrated in this chapter, you have an overview of the work of a programmer who keeps the CAD user provided with drawing techniques for graphic document development. FastCAD has produced 3D generated programs that produce wire-frame solids and modeling.

8.16 Drawing with CAD Software
The geometry that is created, drawn, or generated with CAD programs is generally referred to as entities or elements. The geometric elements (or bits and pieces) are individually constructed figures or groups of elements that consist of points, lines, arcs, circles, rectangles, polygons, splines, cross-hatching, dimensions, and notes. These basic building units are selected from the menu and constructed on specific graphic locations of the monitor screen by the CAD user. Many CAD users can store standard drawing symbols (called blocks) and use overlays on the digitizer for retrieval, Fig. 8.21.

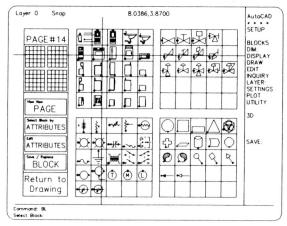

Store your own BLOCKs in the library for future use.

Fig. 8.21 CAD users can develop library symbols for their own software program or use software libraries available from software developers. *Courtesy of Softsource.*

8.17 Typical CAD Construction
The following sequential experiences will introduce the CAD user to the basic interaction necessary for constructing a CAD drawing.

Typical constructions
   Creating line and point.
   Creating arc and circle.
   Creating polygons and splines.
   Creating dimensions.
   Creating text and notes.
   Creating data definition with LIST or STATUS
      commands.
   Creating cross-hatching

The CAD user will learn how to manipulate the features of screen display.

Typical screen display features
   Cursor control.
   Windows.
   Zoom.
   Ortho-Cartesian layout.

Grid–snap.
Command prompts.
Layers–redraw.

The CAD user will learn how to change graphic documents.

Typical modifying (editing) tasks
Delete.
Trim–extend.
Move–scale.
Copy–mirror.
Rotate.

Typical utilities
Print/Plot (creates a hard copy).

The examples that follow are considered typical, although each CAD program will use its own selective terminology and sequences for execution.

8.18 Drawing a Line
The simplest entity to construct is a line. The techniques for construction are similar for all CAD software. Choose the command LINE from the DRAW menu. FastCAD uses a pointer cursor (other CAD software uses a crosshair cursor) to select the pull down DRAW menu.

1. Move the mouse or digitizer stylus until the pointer on the screen is over the word DRAW in the menu bar at the top of the screen. Press the left mouse button, then release the button. A new column of words drops down on the screen, Fig. 8.22. Note that the top word is POINT, and it is highlighted. You have just "pulled down" the DRAW menu. If an error occurs, press the right mouse button and try again. As you pull the mouse toward you, FastCAD highlights each option as you move down the menu.

2. Choose the command LINE from the DRAW menu. The prompt reads "1st point" and displays a crosshairs cursor.

Your screen looks like this:

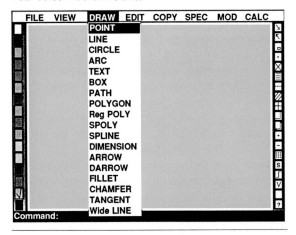

Fig. 8.22 The pull-down menu provides access to 18 DRAW features. *Courtesy of Evolution Computing.*

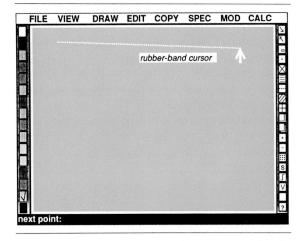

Fig. 8.23 The creation of a line can be sequentially developed with prompts "next point." *Courtesy of Evolution Computing.*

3. Select a point near the upper left corner of the drawing window by moving the crosshairs there, and press the left mouse button.

4. The prompt changes to "Next point," Fig. 8.23. The arrow pointer is now back. Move the arrow to the right. The line stretches to follow the mouse movement. Select a second

endpoint by moving the mouse to the right side of the screen, and press the left button to accept the line. The software draws a line entity, replacing the rubber band cursor. It is easy to draw a series of connected lines. You can move the mouse and repeat the line process as many times as you want. Press the right mouse button to exit the command.

HORIZONTAL–VERTICAL   Lines may be drawn and controlled as horizontal in the x-axis or as vertical lines in the y-axis. In AutoCAD it is the ORTHO option that controls these lines; in CAD-KEY it is the LINE option HRZ/VRT; and in FastCAD the icon bar for ORTHO switch is touched with the cursor.

8.19   Line Types

There are generally six or more line styles on a CAD drawing, and they can be changed by selecting one of the options. The CAD designer can change line styles, colors, and line widths. The designer can also create those lines at various levels for various pen widths in the plotter. Construction lines may be used on a specific level and masked or erased before plotting the drawing. Therefore, the layout makes it easier to assemble the graphic geometry in a technical document.

8.20   Drawing Points

This option is selected from the DRAW menu and is the first option.

1. Select POINT.
2. The software response is "Point at."
3. Move the crosshairs anywhere within the drawing area and press left mouse button.
4. The program draws the point. You can draw as many points (often called pixels) as you need without reselecting the command.
5. Press the right mouse button to stop drawing points.

Some software programs allow you to select various types of points. AutoCAD has an assortment

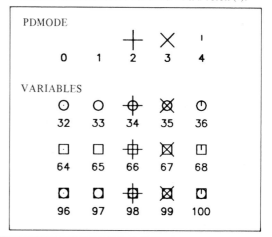

COMMAND EXAMPLES

NOTE: Commands are followed with a colon (:).

Fig. 8.24   The CAD designer has a variety of ways to represent a point with AutoCAD software. *Courtesy of AutoDesk, Inc.*

of points you can select as you create your graphic communication, Fig. 8.24.

8.21   Drawing Circles

Software programs allow you to choose among five methods for entering or creating circles. Three ways are illustrated in the dialog box in Fig. 8.25. This is a typical construction procedure for the CAD user.

1. Select CIRCLE from the DRAW menu and the dialog box appears.
2. Select P|Center & Point from dialog box. The prompt will read "Center point (prior)." FastCAD remembers the x and y coordinates of the last circle drawn, and will offer it as the default (suggested) center for this circle.
3. Press the right mouse button to accept the default instead of a new center. The prompt reads "Point on circle," Fig. 8.26, and then you rubber band the cursor from the center to the new location.
4. Select a point inside the existing circle. The circles can be concentric with the same circle.

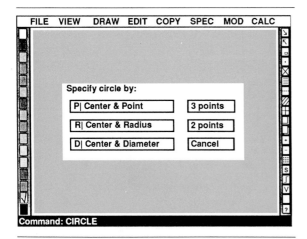

Fig. 8.25   The designer is able to respond to the construction techniques of a circle through the dialog box. *Courtesy of Evolution Computing.*

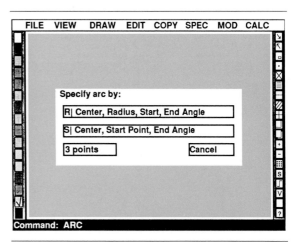

Fig. 8.27   The construction of arcs is easily controlled through CAD software options. *Courtesy of Evolution Computing.*

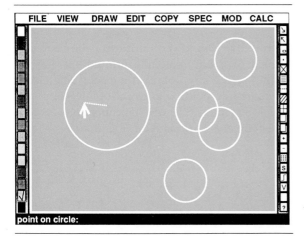

Fig. 8.26   CAD offers visual control to designers as they move the cursor around the display area to create circles. *Courtesy of Evolution Computing.*

5. The prompt may read "center point," but you may exit by pressing the right mouse button twice to cancel the command.

8.22   Drawing Arcs

CAD software allows you to draw arcs by

1. Specifying the center, radius, and the starting and ending angles.

2. Specifying the center, start point, and end angle.

3. Selecting three points on the arc.

THREE POINT ARC   This method is used as a typical example.

1. Select ARC from the DRAW menu. A dialog box is displayed with options, Fig. 8.27.

2. Select 3 points from the lower left corner of the screen.

3. The CAD program displays crosshairs and prompts "1st point."

4. Select the points as noted in Fig. 8.28. The program will mark the 3 points you locate "2nd point" and "3rd point." The prompt will initiate "1st point" again, and you can draw a series of arcs.

5. Press the right mouse button to leave the ARC command.

8.23   Drawing Polygons

A polygon is a closed many-sided geometric figure. The simplest or base polygon is a triangle, and the system shown can create figures with up to 126 sides.

269

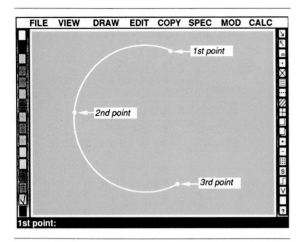

Fig. 8.28   Interactive CAD is prompting decisions for geometric construction. *Courtesy of Evolution Computing.*

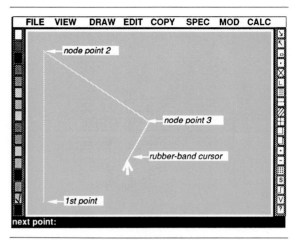

Fig. 8.29   The software allows screen display to move from crosshairs to a pointed cursor that can rubber band (move a line) to form a polygon. *Courtesy of Evolution Computing.*

1. Select POLYGON from the DRAW menu. The prompt reads "1st point."

2. Use the crosshairs to locate a point near the lower left corner of the window. The CAD program replaces the crosshairs with a rubber band cursor, Fig. 8.29.

3. Use the cursor to locate node points 2 and 3. When the cursor is moved back to point 1, the prompt "next point" is displayed. The CAD program closes the polygon automatically when the right mouse button is pressed.

There are other ways to create a polygon, and CAD programs vary in the specific methods that are available. Specifying the center, a specified radius, and the number of sides is another common construction method.

8.24   The LIST Command

CAD software can generate data that provides the designer with information about the geometry created. The LIST command in this program is located in the EDIT menu. Mathematical data

is stored to define the entities of the geometry formed. It is automatically printed out. For example, select any point on a circle. The circle, layer, center point, radius, circumference, and area are printed. The STATUS command also provides current working data, Fig. 8.30.

```
Command: STATUS

49 entities in SAMPLE
Limits are      X:   0.0000    11.0000    (Off)
                Y:   0.0000     8.5000
Drawing uses    X:   3.1250    12.5000    ** Over
                Y:   1.3333     7.4500
Display shows   X:   2.0000     7.3333
                Y:   4.0000     8.0000
Insertion base is  X:  5.0000  Y:  6.0000  Z:  0.0000
Snap resolution is X:  0.2500  Y:  0.2500
Grid spacing is    X:  0.5000  Y:  0.5000

Current layer:     FOUNDATION
Current color:     BYLAYER -- 1 (red)
Current linetype:  CONTINUOUS
Current elevation:    0.0000    thickness:    0.0000
Axis off   Fill on   Grid on   Ortho off   Qtext off   Snap off   Tablet off
Object snap modes:  Endpoint, Midpoint
Free RAM: 13283 bytes        Free disk: 867584 bytes
I/O page space: 124K bytes   Extended I/O page space: 1024K bytes
```

Fig. 8.30   Computers track the graphic information displayed on the screen (STATUS) and can provide data resulting from construction. AutoCAD, CADKEY, and FastCAD are capable of providing calculations to the designer. *Courtesy of AutoDesk, Inc.*

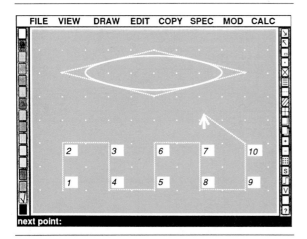

Fig. 8.31   The computer is capable of generating smooth curves from data points provided on the screen. *Courtesy of Evolution Computing.*

FastCAD draws the spline curve:

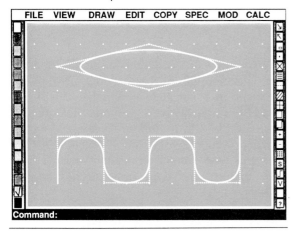

Fig. 8.32   The spline or smooth curve polygon provides a method of drawing irregular curves. *Courtesy of Evolution Computing.*

8.25 Drawing Smooth Curves (Splines)

Drawing smooth and regular curves (splines) is like drawing smooth polygons. You can enter a series of node points, making a frame for the curve. This CAD program, shown in Figs. 8.31 and 8.32, calculates and draws a curve that goes through specified points.

The prompt reads "text: ".

Type **This is more text** and press **RETURN**.

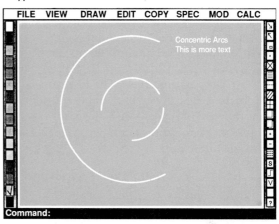

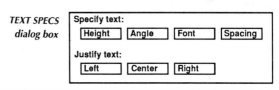

Fig. 8.33   The TEXT menu provides options for placement of notes and has a choice of sizes. *Courtesy of Evolution Computing.*

1. Select SPOLY from the DRAW menu. The screen prompts you to locate "1st point."

2. Use the crosshairs to select the first grid point. Use the rubber band cursor to locate the "next point." Form the diamond shape for the ellipse and the square units to form the spline curve. A smooth spline must have at least 3 points, but no more than 126 node points in this program.

8.26 Creating Text

1. From the DRAW menu select TEXT. The prompt reads "Text." Type the words "Concentric Arcs" and press RETURN to end the line. The prompt reads "Text origin [next]:."

2. Select origin (below "Concentric Arcs"), Fig. 8.33. The CAD program draws the text at the point selected.

271

3. Press the left mouse button to repeat the text command. The prompt reads "Text."

4. Type "This is more text" and press RETURN. The prompt reads "Text origin [next]:"; again, locate a starting point. Press RETURN to accept the location of the text.

SETTING TEXT PROPERTIES   The text selected in the previous figure was horizontal and left justified (aligned). You can use the SPEC (specifications) menu to set these properties for the text you want to create.

1. Select TEXT SPECS from the SPEC menu.

2. Review Specify text: and Justify text:, and select the appropriate features for your design as you are prompted by the dialog box.

Other CAD programs offer various font types and sizes, and provide the user with other interactive alternatives.

8.27   Dimensioning

CAD automatically calculates linear and angular dimensions and can provide the DIM FORMAT from the SPEC menu. You can select the Dimension Format from a dialog box. For example, you may choose inches for this experience and move to the next dialog box for Dimension Properties. This will allow you to set the new properties.

LINEAR DIMENSIONS   You may select Horizontal, Vertical, or Parallel dimensioning. Fig. 8.34 illustrates the typical results from the placement of these dimensions. Horizontal dimensioning measures only the distance along the x-axis between two fixed points. It ignores any vertical distance between the points. Vertical dimensions can only be measured along the y-axis between two fixed points. The parallel dimension can measure actual distances at any given angle between two fixed points.

ANGULAR DIMENSIONS   CAD software can provide information that measures angles (in degrees) and locates dimensions when prompts are

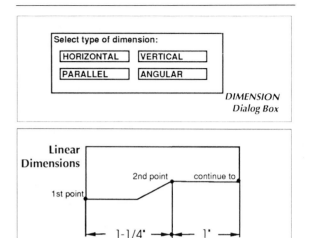

Fig. 8.34   The placement of dimensions is entirely up to CAD operators and they must know how to describe the detailed design. *Courtesy of Evolution Computing.*

entered with Center, Start Point, and End Angle ARC. This is similar to creating arcs. You can edit the location of an angular dimension with the STRETCH option in the EDIT menu. The angular dimension is drawn counterclockwise from the starting to ending point (Fig. 8.35). Other CAD programs offer similar dimension routines from their menus, and each program has specific options.

DRAWING ARROWS   The leader line common to machine operations or notes is created in this FastCAD program with the location of three points, as illustrated in Fig. 8.36.

8.28   Overview

There are many pages written on CAD instruction for creating the geometry of a technical document. This chapter has attempted to deliver several of the basic offerings of some of the leading CAD programs available for the microcomputer workstation. Among other important features of CAD programs is their ability to

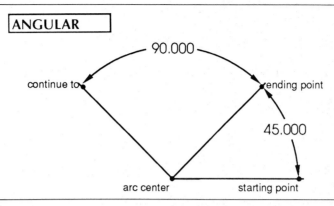

Fig. 8.35 Angular features are easily measured and described with the dimension menus. *Courtesy of Evolution Computing.*

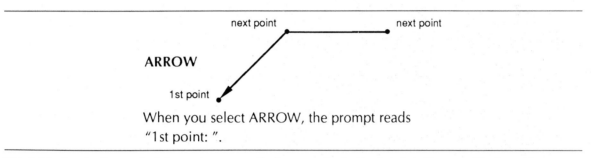

Fig. 8.36 Leader lines take information to specific features of a design drawing. *Courtesy of Evolution Computing.*

maneuver and display features on the screen for easier construction, some of which are

1. Digitizing overlay–tablet menu.
2. Pan.
3. Redraw.
4. Snap–grid–object.
5. Build layers.
6. Move.
7. Scale.
8. Copy.
9. Mirror.
10. Rotate.
11. Mask.
12. Zoom.

There also are several modifying or editing features, for example,

1. Fillets.
2. Trim and extend (sketch).
3. Break.
4. Group.
5. Create segments.
6. Recall.
7. Chamfer.
8. Stretch.

The illustrations for some of these display screen features and modifying or editing capabilities are shown in Figs. 8.37–8.45.

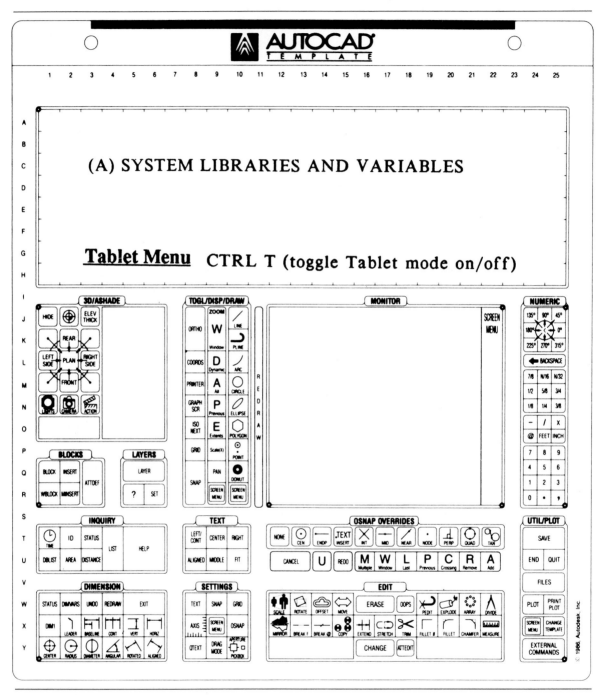

Fig. 8.37 The tablet menu for use with AutoCAD's standard tablet template uses four menu areas. It operates in conjunction with the screen menu to provide easy access to all AutoCAD facilities. *Courtesy of AutoDesk, Inc.*

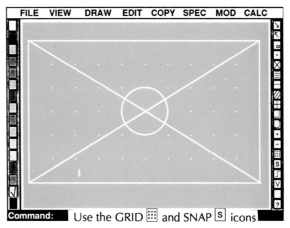

Fig. 8.38 The grid features are used for layout and snap grid for control of input. *Courtesy of Evolution Computing.*

Draw a box, two lines, and a center & point circle, like the illustration. (Make the box 10 units wide and 6 units high. Put the circle's center where the lines cross, and the "point on circle" one unit to the right.)

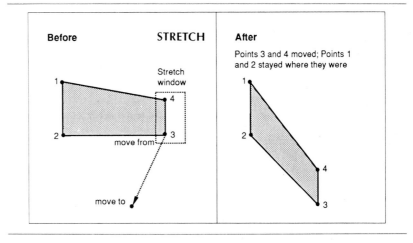

Fig. 8.39 STRETCH reshapes most entities by moving some of their end or node points in relation to the others. (STRETCH simply moves points, circles, and text entities without reshaping them.) *Courtesy of Evolution Computing.*

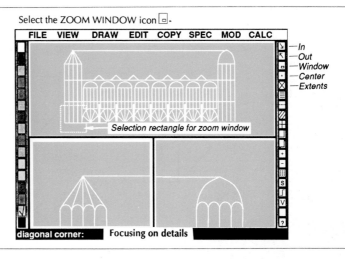

Fig. 8.40 The zoom option is capable of moving in and out and facilitates a close-up view of small details. Windows can be created and are edited interactively as shown. *Courtesy of Evolution Computing.*

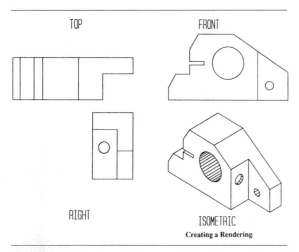

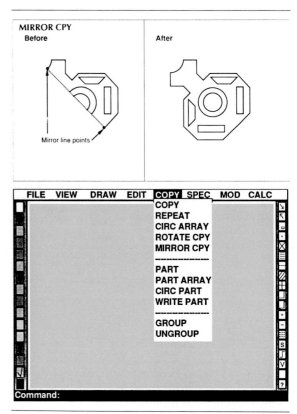

Fig. 8.41   The three-dimensional drawing is achieved with the three-axes development in CAD software. *Courtesy of CADKEY.*

The CAD user is responsible for preparing engineering documents that are an integral part of the total manufacturing process, Fig. 8.46. The ability to interact with all forms of technical information increases the significant role of the drafting technician.

Fig. 8.42   (a) The mirror imaging of a design is simply developed for accurate layout of symmetrical designs. (b) Commands in the COPY menu add new entities to your drawing by duplicating existing entities. *Courtesy of Evolution Computing.*

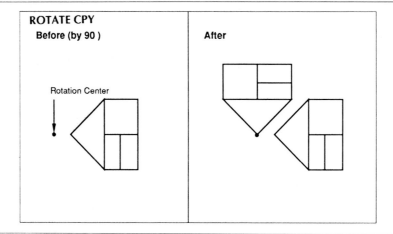

Fig. 8.43   The copy and rotate modifiers are capable of providing easy repetitive images.

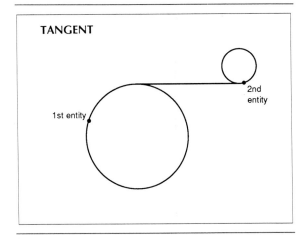

Fig. 8.44 When you select TANGENT, FastCAD prompts you for "1st entity:" and "2nd entity:". Use the crosshairs and left button to select two circles or arcs. FastCAD draws the tangent line (on the side of the circle or arc that you selected). *Courtesy of Evolution Computing.*

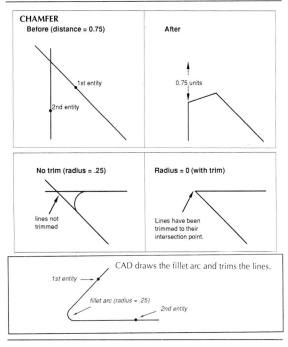

Fig. 8.45 The modifiers of software can create chamfers and fillets. *Courtesy of Evolution Computing.*

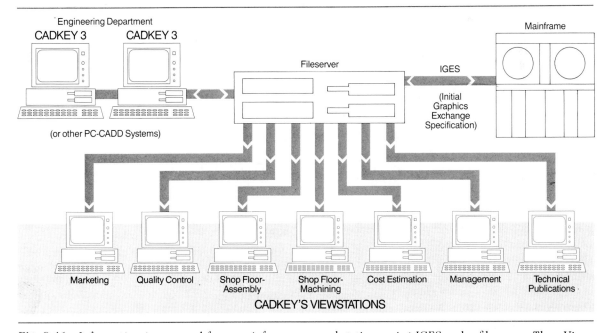

Fig. 8.46 Information is accessed from mainframes or workstations using IGES and a fileserver. Then Viewstations in key departments allow for design input. This is an overview of the total Engineering Team at work. *Courtesy of CADKEY.*

CAD PROBLEMS

The problems in this chapter will allow the future CAD system user to become aware of the way CAD software solves some typical geometric construction that is similar from one software program to another. However, these problems are included for the student who is prepared to solve some problems of change. All the problems illustrated have been prepared on a CAD system. Prepare the required CAD drawing problems as shown with your CAD system and produce a hard copy with a printer or plotter for approval.

Prob. 8.1 Prepare a list of the software commands illustrated in this chapter. Each CAD software program is different. How many commands have been created for the AutoCAD user?

Prob. 8.2 Prepare a list of modifying (or editing) illustrations shown in this chapter. How many editing features are listed on the AutoCAD template illustrated in this chapter?

Prob. 8.3 The unknown distance KA in Fig. 8.47 has been determined and the angle measured with the use of AutoCAD. You are to recreate this problem with your CAD system, changing the 90° angle at H to 75°, and then determine the angles at K and A and length of line KA.

Prob. 8.4 Prepare a revised version of the CAD drawing, Fig. 8.48, by increasing the radius 0.40 to 0.4375 and changing the slot dimension 1.60 to 1.70.

Prob. 8.5 Prepare a detailed CAD drawing of the Safety Key, Fig. 8.49, with the following changes. Correct the right-side view and add the missing dimension 0.40. Examine the placement of dimensions and relocate where necessary. Change 1.12 to 1.25 and add the difference to dimension 4.70.

Prob. 8.6 Revise Fig. 8.50 using the "F" notation for the frontal projections instead of the V as shown. Change the 45° angle to 40°, and determine how much the dimension 2.121 changes.

Prob. 8.7 Using a CAD system, determine the true length of lines AD and CD, Fig. 8.51, when the horizontal projection of point A is relocated to a new coordinate reading of (0, 3.125) and the horizontal projection of point D is relocated to a Cartesian coordinate of (1.75, 1.625). Revise the drawing using the "F" notation for the frontal projections instead of the V notation as shown. What is the new slope of line CD?

Prob. 8.8 Create a revised CAD drawing from Fib. 8.52, stretching the 5.5000 inch dimension to 5.75 and adjusting dimensions as necessary.

Prob. 8.9 Develop a CAD working drawing of the 2″ thick Cam Lock, Fig. 8.53, and create necessary views to clarify the detailing (pictorial, sections, or auxiliary views).

Prob. 8.10 Using a CAD system, create the profile of the Swivel Joint, Fig. 8.54. Examine the proportions of the drawing and develop alternative dimensions if a three-view drawing is assigned. A pictorial is optional.

Prob. 8.11 Using a CAD system, produce a hard copy drawing of the Plastics Molder Plate shown in Fig. 2.77. Omit all dimensions on the final drawing.

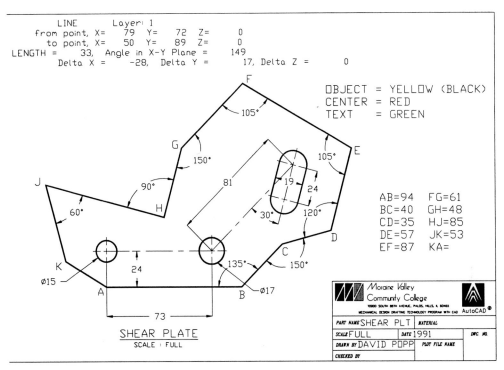

Fig. 8.47 Shear Plate (Prob. 8.3). *Courtesy of Moraine Valley Community College, David Popp.*

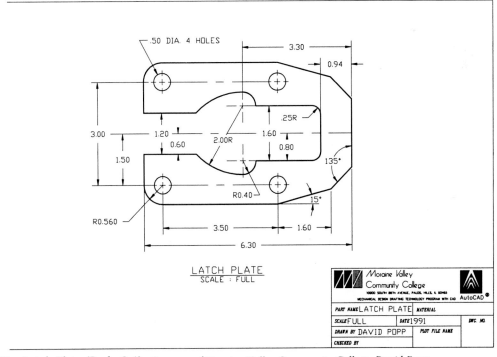

Fig. 8.48 Latch Plate (Prob. 8.4). *Courtesy of Moraine Valley Community College, David Popp.*

279

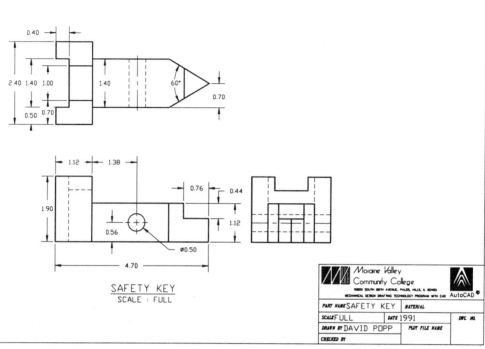

Fig. 8.49   Safety Key (Prob. 8.5). *Courtesy of Moraine Valley Community College, David Popp.*

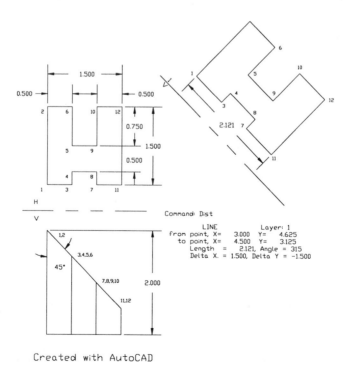

Fig. 8.50   H-Block (Prob. 8.6).

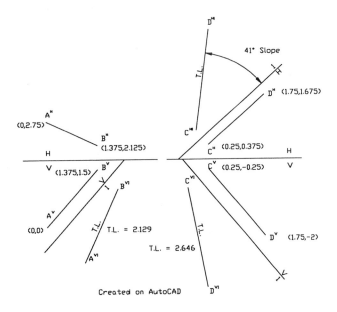

Fig. 8.51 (Prob. 8.7).

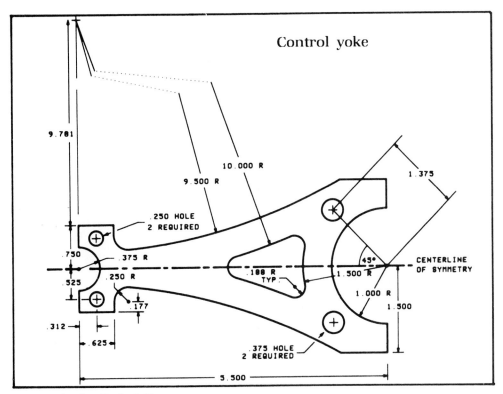

Fig. 8.52 Control Yoke (Prob. 8.8).

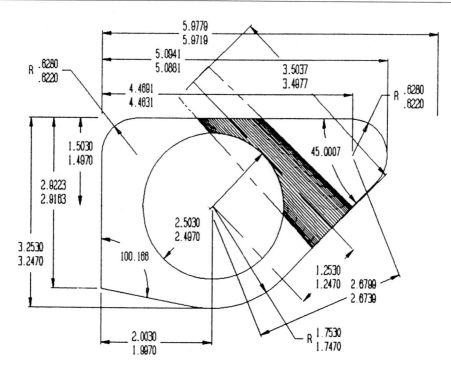

R .6280
.6220

5.9779
5.9719

5.0941
5.0881

3.5037
3.4977

R .6280
.6220

4.4691
4.4631

45.0007

1.5030
1.4970

2.8223
2.8163

2.5030
2.4970

3.2530
3.2470

100.168

1.2530
1.2470

2.6799
2.6739

1.7530
R 1.7470

2.0030
1.9970

CREATE A TAPERED HOLE LOCATED AS SHOWN ON THE CENTER
OF THE 2" THICK PART. THE HOLE WILL START WITH
A 1.00 DIA HOLE AT FRONT AND END WITH A .750 DIA HOLE
AT THE BACK AS INDICATED.

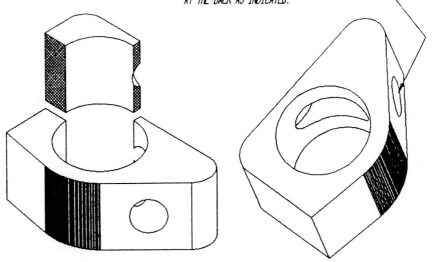

Courtesy of PFB Concepts

Fig. 8.53   Cam Lock (Prob. 8.9).

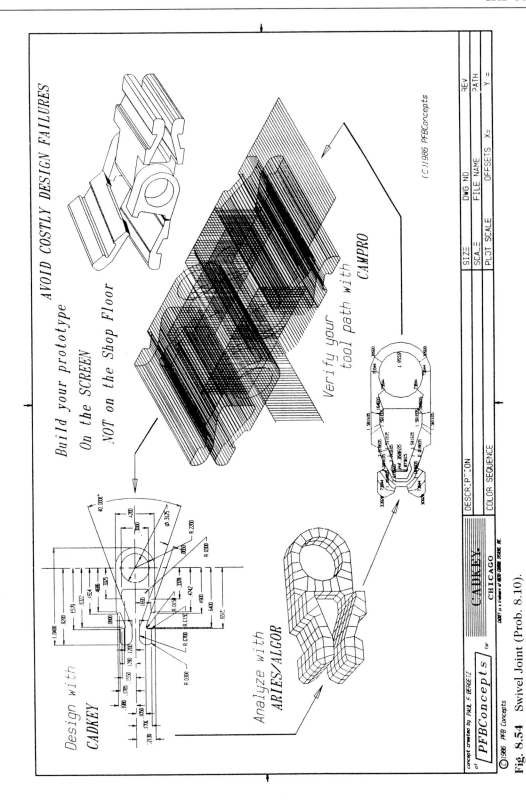

Fig. 8.54 Swivel Joint (Prob. 8.10).

Sectional Views

The basic method of representing parts for designs by views, or projections, has been explained in previous chapters. By means of a limited number of carefully selected views, the external features of the most complicated designs can be fully described.

However, we are frequently confronted with the necessity of showing more or less complicated interiors of parts that cannot be shown clearly by means of hidden lines. We accomplish this by slicing through the part much as one would cut through an apple or a melon. A cutaway view of the part is then drawn; it is called a *sectional view,* a *cross section,* or simply a *section.* See ANSI Y14.2M–1979 (R1987) and Y14.3–1975 (R1987) for complete standards for multiview and sectional-view drawings.

9.1   Sectioning

To produce a sectional view, a *cutting plane,* §9.5, is assumed to be passed through the part for the design, as shown in Fig. 9.1 (a). Then, at (b) the cutting plane is removed and the two halves drawn apart, exposing the interior construction. In this case, the direction of sight is toward the left half, as shown, and for purposes of the section the right half is mentally discarded. The sectional view will be in the position of a right-side view.

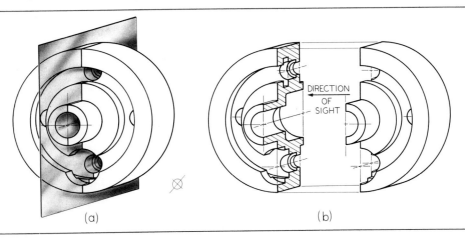

DIRECTION
OF
SIGHT

(a)   (b)

Fig. 9.1   A Section.

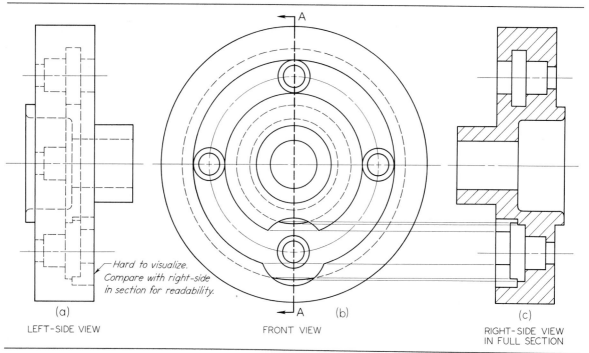

Fig. 9.2 Full Section.

9.2 Full Sections

The sectional view obtained by passing the cutting plane fully through the object is called a *full section,* Fig. 9.2 (c). A comparison of this sectional view with the left-side view, (a), emphasizes the advantage in clearness of the former. The left-side view would naturally be omitted. In the front view, the cutting plane appears as a line, called a *cutting-plane line,* §9.5. The arrows at the ends of the cutting-plane line indicate the direction of sight for the sectional view.

Note that in order to obtain the sectional view, the right half is only *imagined* to be removed and not actually shown removed anywhere except in the sectional view itself. In the sectional view, the section-lined areas are those portions that have been in actual contact with the cutting plane. Those areas are *crosshatched* with thin parallel section lines spaced carefully by eye. In addition, the visible parts behind the cutting plane are shown but not crosshatched.

As a rule, the location of the cutting plane is obvious from the section itself, and, therefore, the cutting-plane line is omitted. It is shown in Fig. 9.2 for illustration only. Cutting-plane lines should, of course, be used wherever necessary for clearness, as in Figs. 9.21, 9.22, 9.24, and 9.25.

9.3 Lines in Sectioning

A correct front view and sectional view are shown in Fig. 9.3 (a) and (b). In general, *all visible edges and contours behind the cutting plane should be shown;* otherwise a section will appear to be made up of disconnected and unrelated parts, as shown in (c). Occasionally, however, visible lines behind the cutting plane are not necessary for clearness and should be omitted.

Sections are used primarily to replace hidden-line representation; hence, as a rule, *hidden*

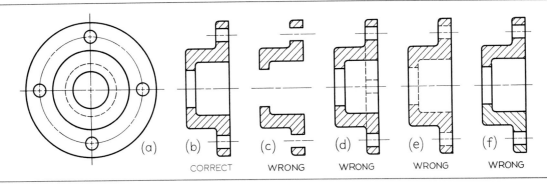

Fig. 9.3   Lines in Sectioning.

lines should be omitted in sectional views. As shown in Fig. 9.3 (d), the hidden lines do not clarify the drawing; they tend to confuse, and they take unnecessary time to draw. Sometimes hidden lines are necessary for clearness and should be used in such cases, especially if their use will make it possible to omit a view, Fig. 9.4.

A section-lined area is always completely bounded by a visible outline—never by a hidden line as in Fig. 9.3 (e), since in every case the cut surfaces and their boundary lines will be visible. Also, a visible line can never cut across a section-lined area.

In a sectional view of a part, alone or in assembly, the section lines in all sectioned areas must be parallel, not as shown in Fig. 9.3 (f). The use of section lining in opposite directions is an indication of different parts, as when two or more parts are adjacent in an assembly drawing, Fig. 16.42.

9.4   Section Lining

Symbolic section-lining symbols, Fig. 9.5, have been used to indicate the material to be used in producing the object. These symbols represented the general types only, such as cast-iron, brass, and steel. Now, however, there are so many different materials, and each general type has so many subtypes, that a general name or symbol is not enough. For example, there are hundreds of different kinds of steel alone. Since detailed specifications of material must be lettered in the form of a note or in the title strip, the general-purpose (cast-iron) section lining is used for all materials on detail drawings (single parts).

Symbolic section lining may be used in assembly drawings in cases where it is desirable to distinguish the different materials; otherwise, the general-purpose symbol is used for all parts. For assembly sections, see §16.22.

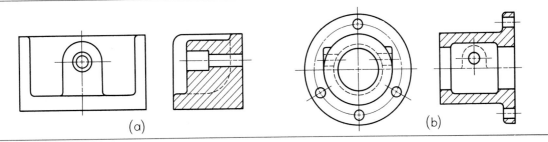

Fig. 9.4   Hidden Lines in Sections.

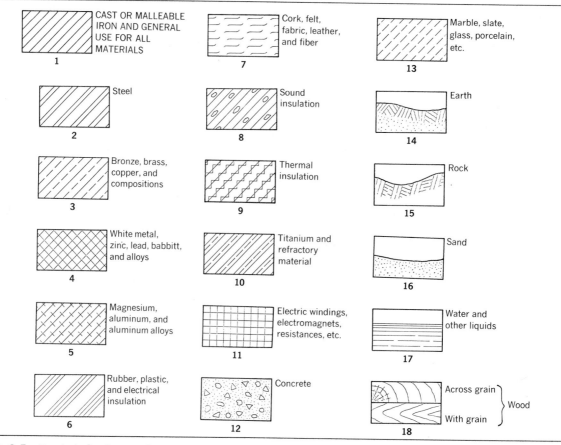

Fig. 9.5 Symbols for Section Lining.

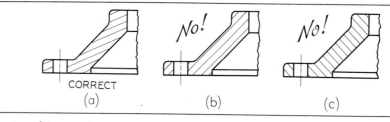

Fig. 9.6 Section-Lining Technique.

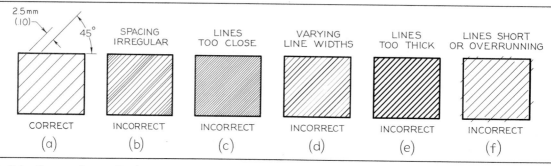

Fig. 9.7 Direction of Section Lines.

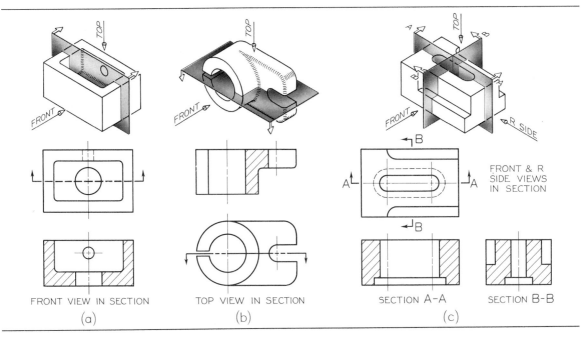

Fig. 9.8   Cutting Planes and Sections.

The correct method of drawing section lines is shown in Fig. 9.6 (a). Draw the section lines with a sharp medium-grade pencil (H or 2H) with a conical point as shown in Fig. 2.11 (c). Always draw the lines at 45° with horizontal as shown, unless there is some advantage in using a different angle. Space the section lines as uniformly as possible by eye from about approximately 1.5 mm ($\frac{1}{16}''$) to 3 mm ($\frac{1}{8}''$) or more apart, depending on the size of the drawing or of the sectioned area. *For average drawings, space the lines about 2.5 mm ($\frac{3}{32}''$) or more apart.* As a rule, space the lines as generously as possible and yet close enough to distinguish clearly the sectioned areas.

After the first few lines have been drawn, look back repeatedly at the original spacing to avoid gradually increasing or decreasing the intervals, Fig. 9.6 (b). Beginners almost invariably draw section lines too close together, (c). This is very tedious because with small spacing the least inaccuracy in spacing is conspicuous.

Section lines should be uniformly thin, never varying in thickness, as in (d). There should be a marked contrast in thickness of the visible outlines and the section lines. Section lines should not be too thick, as in (e). Also avoid running section lines beyond the visible outlines or stopping the lines too short, as in (f).

If section lines drawn at 45° with horizontal would be parallel or perpendicular (or nearly so) to a prominent visible outline, the angle should be changed to 30°, 60°, or some other angle, Fig. 9.7.

Dimensions should be kept off sectioned areas, but when this is unavoidable the section lines should be omitted where the dimension figure is placed. See Fig. 13.13.

Section lines may be drawn adjacent to the boundaries of the sectioned areas (outline sectioning) providing that clarity is not sacrificed. See Fig. 16.6.

9.5   The Cutting Plane

The cutting plane is indicated in a view adjacent to the sectional view, Fig. 9.8. In this view, the cutting plane appears edgewise, as a line called

289

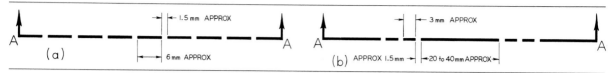

Fig. 9.9   Cutting-Plane Lines (Full Size).

the *cutting-plane line.* Alternate styles of cutting-plane lines are shown in Fig. 9.9. See also Fig. 2.14. The first form, Fig. 9.9 (a), composed of equal dashes each about 6 mm ($\frac{1}{4}''$) or more long plus the arrowheads, is the standard in the automotive industry. This form without the dashes between the ends is especially desirable on complicated drawings. The form shown in (b), composed of alternate long dashes and pairs of short dashes plus the arrowheads, has been in general use for a long time. Both lines are drawn the same thickness as visible lines. Arrowheads indicate the direction in which the cutaway object is viewed.

Capital letters are used at the ends of the cutting-plane line when necessary to identify the cutting-plane line with the indicated section. This most often occurs in the case of multiple sections, Fig. 9.25, or removed sections, Fig. 9.21.

As shown in Fig. 9.8, sectional views occupy normal projected positions in the standard arrangement of views. At (a) the cutting plane is a frontal plane, §7.15, and appears as a line in the top view. The front half of the object (lower half in the top view) is imagined removed. The arrows at the ends of the cutting-plane line point in the direction of sight for a front view, that is, away from the front view or section. Note that the arrows do not point in the direction of withdrawal of the removed portion. The resulting full section may be referred to as the "front view in section," since it occupies the front view position.

In Fig. 9.8 (b), the cutting plane is a horizontal plane, §7.15, and appears as a line in the front view. The upper half of the object is imagined

removed. The arrows point toward the lower half in the same direction of sight as for a top view, and the resulting full section is a "top view in section."

In Fig. 9.8 (c), two cutting planes are shown, one a frontal plane and the other a profile plane, §7.15, both of which appear edgewise in the top view. Each section is completely independent of the other and drawn as if the other were not present. For section A–A, the front half of the object is imagined removed. The back half is then viewed in the direction of the arrows for a front view, and the resulting section is a "front view in section." For section B–B, the right half of the object is imagined removed. The left half then is viewed in the direction of the arrows for a right-side view, and the resulting section is a "right-side view in section." The cutting-plane lines are preferably drawn through an exterior view, in this case the top view, as shown, instead of a sectional view.

The cutting-plane lines in Fig. 9.8 are shown for purposes of illustration only. They are generally omitted in cases such as these, in which the location of the cutting plane is obvious. When a cutting-plane line coincides with a center line, the cutting-plane line takes precedence.

Correct and incorrect relations between cutting-plane lines and corresponding sectional views are shown in Fig. 9.10.

9.6   Visualizing a Section

Two views of an object to be sectioned, having a drilled and counterbored hole, are shown in Fig. 9.11 (a). The cutting plane is assumed along the

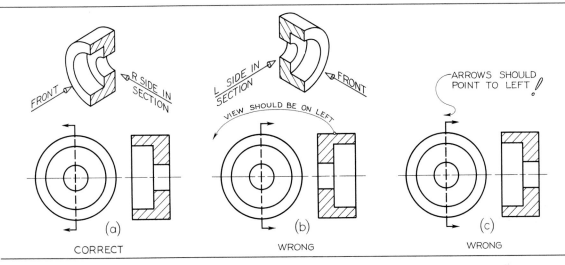

Fig. 9.10   Cutting Planes and Sections.

horizontal center line in the top view, and the front half of the object (lower half of the top view) is imagined removed. A pictorial drawing of the remaining back half is shown at (b). The two cut surfaces produced by the cutting plane are 1–2–5–6–10–9 and 3–4–12–11–7–8. However, the corresponding section at (c) is incomplete because certain visible lines are missing.

If the section is viewed in the direction of sight, as shown at (b), arcs A, B, C, and D will be visible. As shown at (d), these arcs will appear as straight lines 2–3, 6–7, 5–8, and 10–11. These lines may also be accounted for in other ways. The top and bottom surfaces of the object appear in the section as lines 1–4 and 9–12. The bottom surface of the counterbore appears in the section as line 5–8. Also, the semicylindrical surfaces for the back half of the counterbore and of the drilled hole will appear as rectangles in the section at 2–3–8–5 and 6–7–11–10.

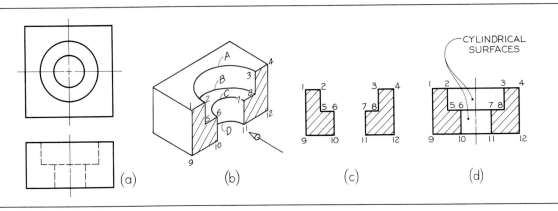

Fig. 9.11   Visualizing a Section.

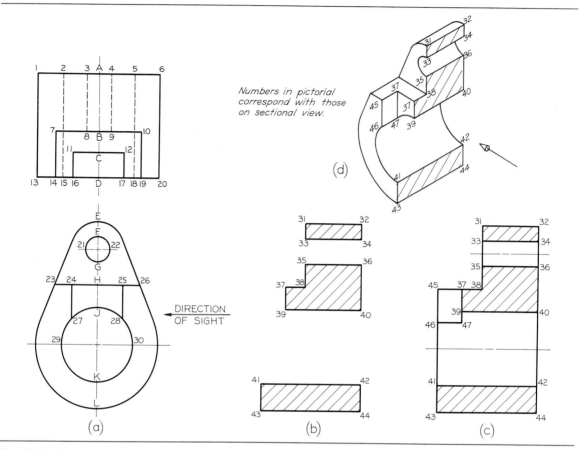

Numbers in pictorial correspond with those on sectional view.

DIRECTION OF SIGHT

(a)   (b)   (c)   (d)

Fig. 9.12 Drawing a Full Section.

The front and top views of a Collar are shown in Fig. 9.12 (a), and a right-side view in full section is required. The cutting plane is understood to pass along the center lines AD and EL. If the cutting plane were drawn, the arrows would point to the left in conformity with the direction of sight (see arrow) for the right-side view. The right side of the object is imagined removed, and the left half will be viewed in the direction of the arrow, as shown pictorially at (d). The cut surfaces will appear edgewise in the top and front views along AD and EL; and since the direction of sight for the section is at right angles to them,

they will appear in true size and shape in the sectional view. Each sectioned area will be completely enclosed by a boundary of visible lines. The sectional view will show, in addition to the cut surfaces, all visible parts behind the cutting plane. No hidden lines will be shown.

Whenever a surface of the object (plane or cylindrical) appears as a line and is intersected by a cutting plane that also appears as a line, a new edge (line of intersection) is created that will appear as a *point* in that view. Thus, in the front view, the cutting plane creates new edges appearing as points at E, F, G, H, J, K, and L. In

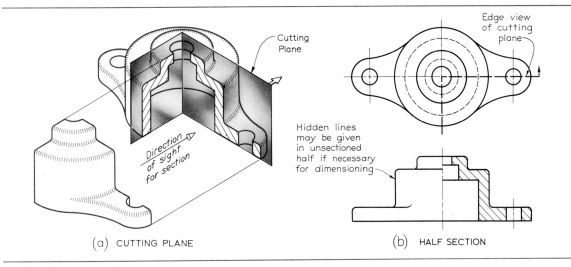

(a) CUTTING PLANE   (b) HALF SECTION

Fig. 9.13   Half Section.

the sectional view, (b), these are horizontal lines 31–32, 33–34, 35–36, 37–38, 39–40, 41–42, and 43–44.

Whenever a surface of the object appears as a surface (that is, not as a line) and is cut by a cutting plane that appears as a line, a new edge is created that will appear as a line in the view, coinciding with the cutting-plane line, and as a line in the section.

In the top view, D is the *point view* of a vertical line KL in the front view and 41–43 in the section at (b). Point C is the point view of a vertical line HJ in the front view and 37–39 in the section. Point B is the point view of two vertical lines EF and GH in the front view, and 31–33 and 35–38 in the section. Point A is the point view of three vertical lines EF, GJ, and KL in the front view, and 32–34, 36–40, and 42–44 in the section. This completes the boundaries of three sectioned areas 31–32–34–33, 35–36–40–39–37–38, and 41–42–44–43. It is only necessary now to add the visible lines beyond the cutting plane.

The semicylindrical left half F–21–G of the small hole (front view) will be visible as a rectangle in the sections at 33–34–36–35, as shown

at (c). The two semicircular arcs will appear as straight lines in the section at 33–35 and 34–36.

Surface 24–27, appearing as a line in the front view, appears as line 11–16 in the top view and as surface 45–37–47–46, true size, in the section at (c).

Cylindrical surface J–29–K, appearing as an arc in the front view, appears in the top view as 2–A–C–11–16–15, and in the section as 46–47–39–40–42–41. Thus, arc 27–29–K (front view) appears in the section, (c), as straight lines 46–41; and arc J–29–K appears as straight line 40–42.

All cut surfaces here are part of the same object; hence, the section lines must all run in the same direction, as shown.

9.7   Half Sections

If the cutting plane passes halfway through the object, the result is a half section, Fig. 9.13. A half section has the advantage of exposing the interior of one half of the object and retaining the exterior of the other half. Its usefulness is, therefore, largely limited to symmetrical objects.

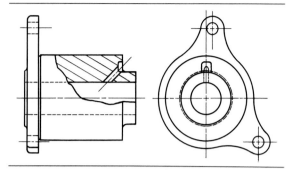

Fig. 9.14 Broken-Out Section.

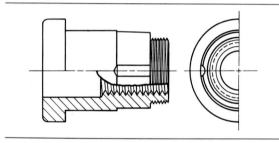

Fig. 9.15 Break Around Keyway.

It is not widely used in detail drawings (single parts) because of this limitation of symmetry and also because of difficulties in dimensioning internal shapes that are shown in part only in the sectioned half, Fig. 9.13 (b).

In general, hidden lines should be omitted from both halves of a half section. However, they may be used in the unsectioned half if necessary for dimensioning.

The greatest usefulness of the half section is in assembly drawing, Fig. 16.42, in which it is often necessary to show both internal and external construction on the same view, but without the necessity of dimensioning.

As shown in Fig. 9.13 (b), a center line is used to separate the halves of the half section. The American National Standards Institute recommends a center line for the division line between the sectioned half and the unsectioned half of a half-sectional view, although in some cases the

same overlap of the exterior portion, as in a broken-out section, is preferred. See Fig. 9.33 (b). Either form is acceptable.

9.8 Broken-Out Sections
It often happens that only a partial section of a view is needed to expose interior shapes. Such a section, limited by a break line, Fig. 2.14, is called a broken-out section. In Fig. 9.14, a full or half section is not necessary, a small broken-out section being sufficient to explain the construction. In Fig. 9.15, a half section would have caused the removal of half the keyway. The keyway is preserved by breaking out around it. Note that in this case the section is limited partly by a break line and partly by a center line.

9.9 Revolved Sections
The shape of the cross section of a bar, arm, spoke, or other elongated object may be shown in the longitudinal view by means of a revolved section, Fig. 9.16. Such sections are made by assuming a plane perpendicular to the center line or axis of the bar or other object, as shown in Fig. 9.17 (a), then revolving the plane through 90° about a center line at right angles to the axis, as at (b) and (c).

The visible lines adjacent to a revolved section may be broken out if desired, as shown in Figs. 9.16 (k) and 9.18.

The superimposition of the revolved section requires the removal of all original lines covered by it, Fig. 9.19. The true shape of a revolved section should be retained after the revolution of the cutting plane, regardless of the direction of the lines in the view, Fig. 9.20.

9.10 Removed Sections
A removed section is one that is not in direct projection from the view containing the cutting plane—that is, it is not positioned in agreement with the standard arrangement of views. This displacement from the normal projection position should be made without turning the section from its normal orientation.

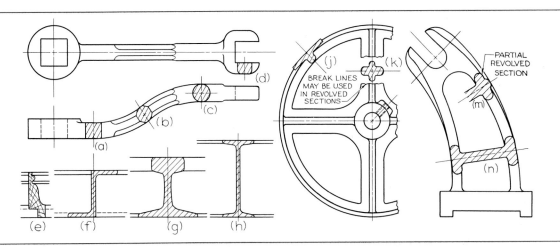

Fig. 9.16   Revolved Sections.

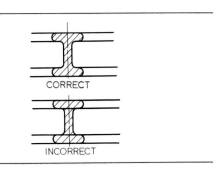

Fig. 9.17   Use of the Cutting Plane in Revolved Sections.

Fig. 9.18   Conventional Breaks Used with Revolved Sections.

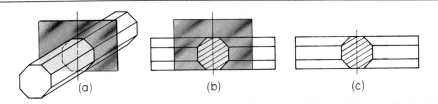

Fig. 9.19   A Common Error in Drawing Revolved Sections.

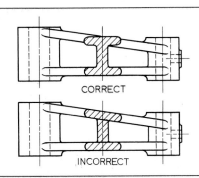

Fig. 9.20   A Common Error in Drawing Revolved Sections.

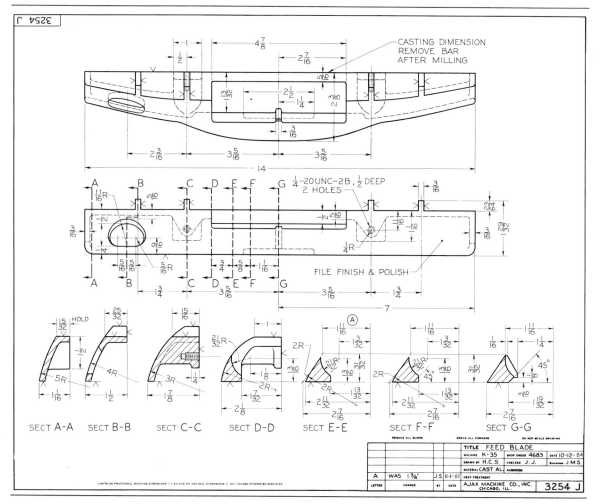

Fig. 9.21 Removed Sections.

Removed sections, Fig. 9.21, should be labeled, such as SECTION A–A and SECTION B–B, corresponding to the letters at the ends of the cutting-plane line. They should be arranged in alphabetical order from left to right on the sheet. Section letters should be used in alphabetical order, but letters I, O, and Q should not be used because they are easily confused with the numeral 1 or the zero.

A removed section is often a partial section. Such a removed section, Fig. 9.22, is frequently drawn to an enlarged scale, as shown. This is often desirable in order to show clear delineation of some small detail and to provide sufficient space for dimensioning. In such a case the enlarged scale should be indicated beneath the section title.

A removed section should be placed so that it no longer lines up in projection with any other view. It should be separated clearly from the standard arrangement of views. See Fig. 14.9.

Whenever possible removed sections should be on the same sheet with the regular views. If a section must be placed on a different sheet, cross-references should be given on the related sheets. A note should be given below the section title, such as

SECTION B–B ON SHEET 4, ZONE A3

A similar note should be placed on the sheet on which the cutting-plane line is shown, with a leader pointing to the cutting-plane line and referring to the sheet on which the section will be found.

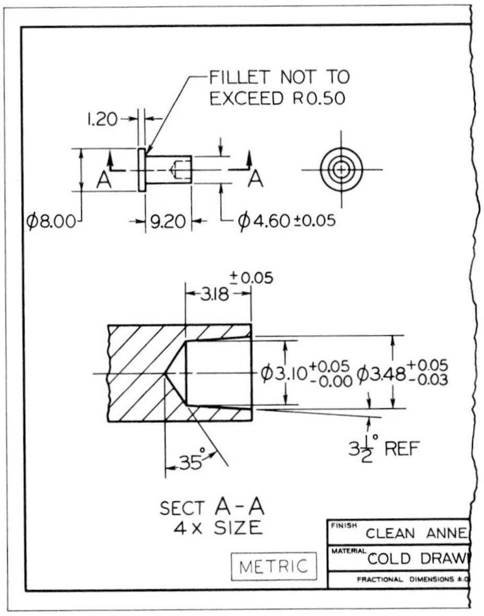

Sometimes it is convenient to place removed sections on center lines extended from the section cuts, Fig. 9.23.

9.11   Offset Sections

In sectioning through irregular objects, it is often desirable to show several features that do not lie in a straight line, by "offsetting" or bending the cutting plane. Such a section is called an offset section. In Fig. 9.24 (a) the cutting plane is offset

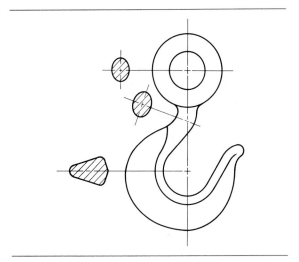

Fig. 9.22   Removed Section.

Fig. 9.23   Removed Sections.

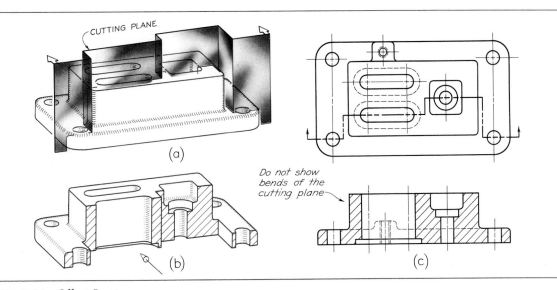

Fig. 9.24   Offset Section.

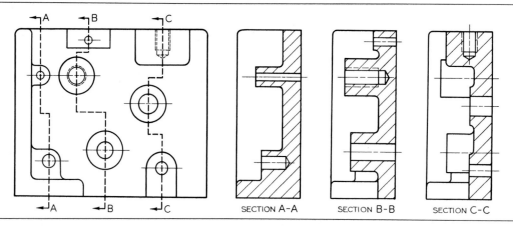

Fig. 9.25   Three Offset Sections.

in several places in order to include the hole at the left end, one of the parallel slots, the rectangular recess, and one of the holes at the right end. The front portion of the object is then imagined to be removed, (b). The path of the cutting plane is shown by the cutting-plane line in the top view at (c), and the resulting offset section is shown in the front view. The offsets or bends in the cutting plane are all 90° and are *never shown in the sectional view.*

Figure 9.24 also illustrates how hidden lines in a section eliminate the need for an additional view. In this case, an extra view would be needed to show the small boss on the back if hidden lines were not shown.

An example of multiple offset sections is shown in Fig. 9.25. Notice that the visible background shapes without hidden lines appear in each sectional view.

9.12   Ribs in Section

To avoid a false impression of thickness and solidity, ribs, webs, gear teeth, and other similar flat features are not sectioned even though the cutting plane passes along the center plane of the feature. For example, in Fig. 9.26, the cutting plane A–A passes flatwise through the vertical web, or rib, and the web is not section-lined, (a). *Such thin features should not be section-lined,*

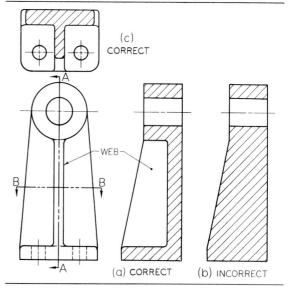

Fig. 9.26   Webs in Section.

even though the cutting plane passes through them. The incorrect section is shown at (b). Note the false impression of thickness or solidity resulting from section lining the rib.

If the cutting plane passes *crosswise* through a rib or any thin member, as does the plane B–B in Fig. 9.26, the member should be section-lined in the usual manner, as shown in the top view at (c).

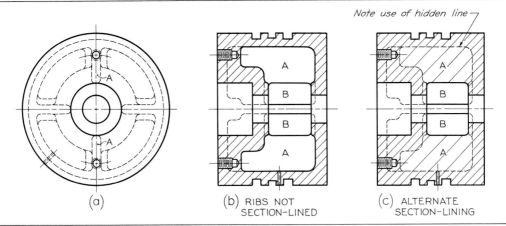

Fig. 9.27 Alternate Section Lining.

In some cases, if a rib is not section-lined when the cutting plane passes through it flatwise, it is difficult to tell whether the rib is actually present, as, for example, ribs A in Fig. 9.27 (a) and (b). It is difficult to distinguish spaces B as open spaces and spaces A as ribs. In such cases, double-spaced *section lining* of the ribs should be used, (c). This consists simply in continuing alternate section lines through the ribbed areas, as shown.

9.13 Aligned Sections

In order to include in a section certain angled elements, the cutting plane may be bent so as to pass through those features. The plane and feature are then imagined to be revolved into the original plane. For example, in Fig. 9.28, the cutting plane was bent to pass through the angled arm and then revolved to a vertical position (aligned), from where it was projected across to the sectional view.

In Fig. 9.29 the cutting plane is bent so as to include one of the drilled and counterbored holes in the sectional view. The correct section view at (b) gives a clearer and more complete description than does the section at (c), which was taken along the vertical center line of the front view—that is, without any bend in the cutting plane.

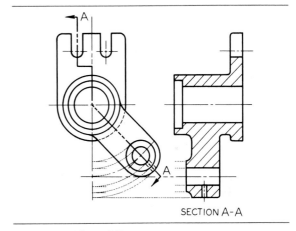

SECTION A-A

Fig. 9.28 Aligned Section.

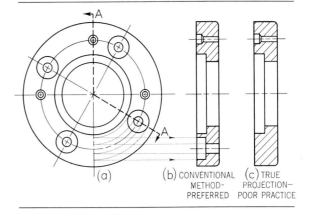

Fig. 9.29 Aligned Section.

299

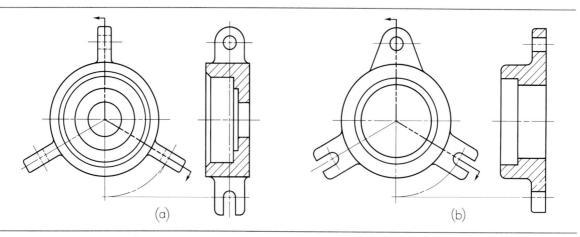

Fig. 9.30   Aligned Sections.

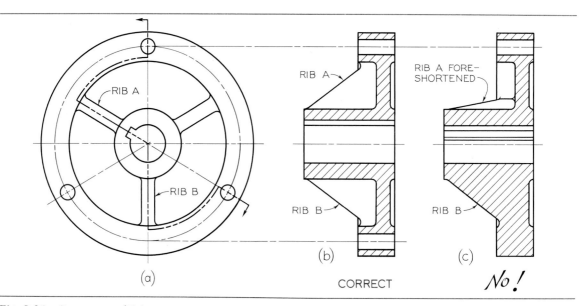

Fig. 9.31   Symmetry of Ribs.

In such cases, the angle of revolution should always be less than 90°.

The student is cautioned *not to revolve* features when clearness is not gained. In some cases the revolving features will result in a loss of clarity. Examples in which revolution should not be used are Fig. 9.40, Probs. 17 and 18.

In Fig. 9.30 (a) is an example in which the projecting lugs were not sectioned on the same basis that ribs are not sectioned. At (b) the projecting lugs are located so that the cutting

plane would pass through them crosswise; hence, they are sectioned.

Another example involving rib sectioning and also aligned sectioning is shown in Fig. 9.31. In the circular view, the cutting plane is offset in circular-arc bends to include the upper hole and upper rib, the keyway and center hole, the lower rib, and one of the lower holes. These features are imagined to be revolved until they line up vertically and are then projected from that position to obtain the section at (b). Note that the

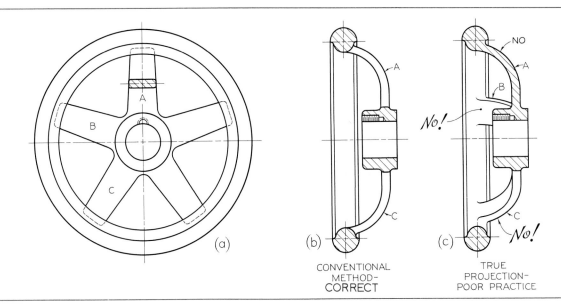

Fig. 9.32   Spokes in Section.

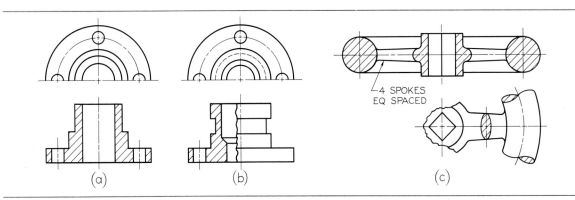

Fig. 9.33   Partial Views.

ribs are not sectioned. If a regular full section of the object were drawn, without the use of conventions discussed here, the resulting section, (c), would be both incomplete and confusing and, in addition, would take more time to draw.

In sectioning a pulley or any spoked wheel, it is standard practice to revolve the spokes if necessary (if there is an odd number) and not to section-line the spokes, Fig. 9.32 (b). If the spoke is sectioned, as shown at (c), the section gives a false impression of continuous metal. If the lower spoke is not revolved, it will be foreshortened in the sectional view in which it

presents an "amputated" and wholly misleading appearance.

Figure 9.32 also illustrates correct practice in omitting visible lines in a sectional view. Notice that spoke **B** is omitted at (b). If it were included, (c), the spoke would be foreshortened, difficult and time consuming to draw, and confusing to the reader of the drawing.

9.14   Partial Views

If space is limited on the paper or if it is necessary to save drafting time, *partial views* may be used in connection with sectioning, Fig. 9.33.

301

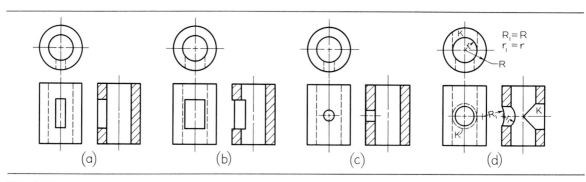

Fig. 9.34   Intersections.

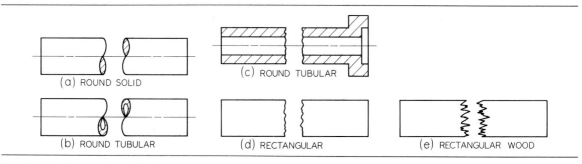

Fig. 9.35   Conventional Breaks.

Half views are shown at (a) and (b) in connection with a full section and a half section, respectively. Note that in each case the back half of the object in the circular view is shown, in conformity with the idea of removing the front portion of the object in order to expose the back portion for viewing in section. See also §7.9.

Another method of drawing a partial view is to break out much of the circular view, retaining only those features that are needed for minimum representation, Fig. 9.33 (c).

9.15   Intersections in Sectioning

Where an intersection is small or unimportant in a section, it is standard practice to disregard the true projection of the figure of intersection, as shown in Fig. 9.34 (a) and (c). Larger figures of intersection may be projected, as shown at (b), or approximated by circular arcs, as shown for the smaller hole at (d). Note that the larger hole K is the same diameter as the vertical hole.

In such cases the curves of intersection (ellipses) appear as straight lines, as shown. See also Figs. 7.38 and 7.39.

9.16   Conventional Breaks

In order to shorten a view of an elongated object, conventional breaks are recommended, as shown in Fig. 9.35. For example, the two views of a garden rake are shown in Fig. 9.36 (a), drawn to a small scale to get it on the paper. At (b) the handle was "broken," a long central portion removed, and the rake then drawn to a larger scale, producing a much clearer delineation.

Parts thus broken must have the same section throughout, or if tapered they must have a uniform taper. Note at (b) the full-length dimension is given, just as if the entire rake were shown.

The breaks used on cylindrical shafts or tubes are often referred to as "S-breaks" and in

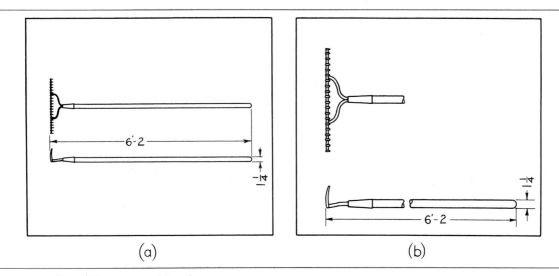

Fig. 9.36   Use of Conventional Breaks.

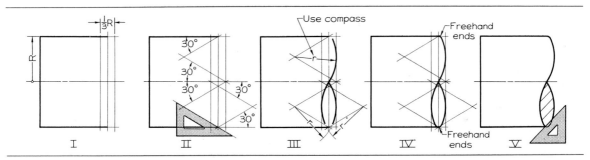

Fig. 9.37   Steps in Drawing S-Break for Solid Shaft.

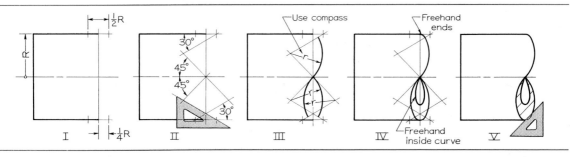

Fig. 9.38   Steps in Drawing S-Breaks for Tubing.

the industrial drafting room are usually drawn entirely freehand or partly freehand and partly with the irregular curve or the compass. By these methods, the result is often very crude, especially when attempted by beginners. Simple methods of construction for use by the student or the industrial drafter are shown in Figs. 9.37 and 9.38 and will always produce a professional result. Excellent S-breaks are also obtained with an S-break template.

Breaks for rectangular metal and wood sections are always drawn freehand, as shown in

303

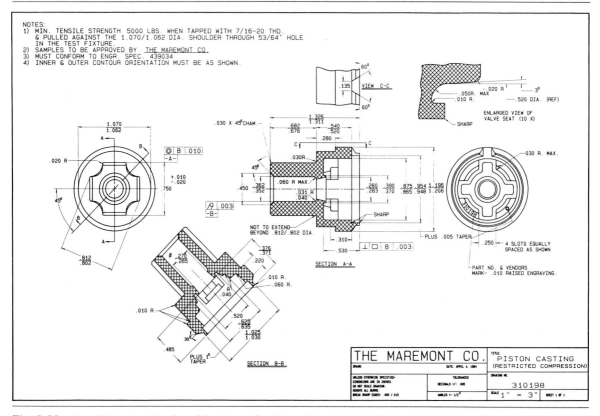

Fig. 9.39   Detail Drawing Produced by Using the VersaCAD Advanced System. *Courtesy of VersaCAD.*

Fig. 9.35. See also Fig. 9.18, which illustrates the use of breaks in connection with the revolved sections.

9.17 Computer Graphics

Sectioning of even the most complex objects can be readily accomplished with CAD, providing its user understands the theory of sectioning and how it is to be applied. Using computer graphics, one can quickly and efficiently explore a variety of possible solutions to a sectioning problem and choose the best representation without making tedious changes to a paper drawing. With CAD, it is relatively simple to examine various cross-sections of an object, depict removed sections, and add appropriate section lining, Fig. 9.39.

SECTIONING PROBLEMS

Any of the following problems, Figs. 9.40–9.70, may be drawn freehand or with instruments, as assigned by the instructor. However, the problems in Fig. 9.40 are especially suitable for sketching on 8.5″ × 11.0″ graph paper with appropriate grid squares. Two problems can be drawn on one sheet, using Layout A–1 similar to Fig. 6.51, with borders drawn freehand. If desired, the problems may be sketched on plain drawing paper. Before making any sketches, the student should study carefully §§6.1–6.10.

The problems in Figs. 9.41–9.60 are intended to be drawn with instruments, but may be drawn freehand, if desired. If metric or decimal dimensions are required, the student should first study §§13.1–13.25. If an ink tracing is required, the student is referred to §§2.48–2.52.

Sectioning problems in convenient form for solution are available in *Technical Drawing Problems*, Series 1, by Giesecke, Mitchell, Spencer, Hill, Dygdon, and Novak; *Technical Drawing Problems*, Series 2, by Spencer, Hill, Dygdon, and Novak; and *Technical Drawing Problems*, Series 3, by Spencer, Hill, Dygdon, and Novak, all designed to accompany this text and published by Macmillan Publishing Company.

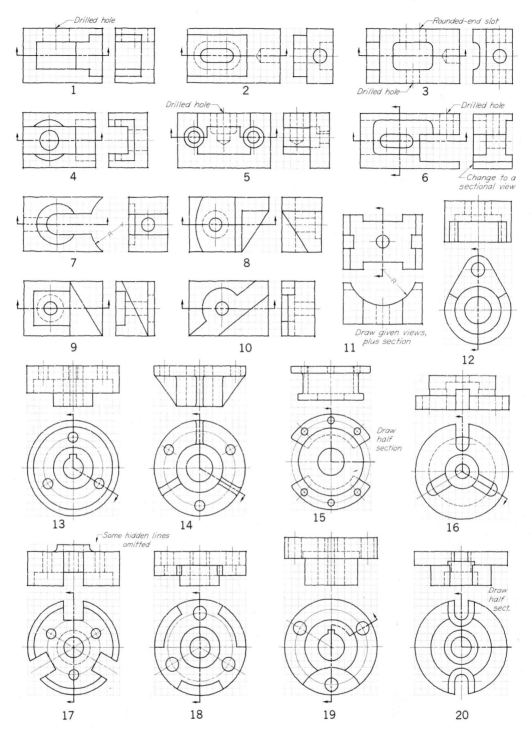

Fig. 9.40 Freehand Sectioning Problems. Using Layout A–1 or A4–1 adjusted (freehand) on graph paper or plain paper, two problems per sheet as in Fig. 6.51, sketch views with sections as indicated. Each grid square = 6 mm ($\frac{1}{4}''$). In Probs. 1–10, top and right-side views are given. Sketch front sectional views and then move right-side views to line up horizontally with front sectional views. Omit cutting planes except in Probs. 5 and 6.

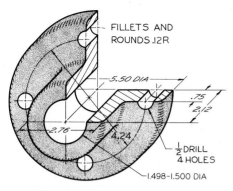

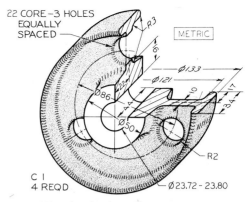

Fig. 9.41 Bearing. Draw necessary views, with full section. (Layout A–3).*

Fig. 9.42 Truck Wheel. Draw necessary views, with half section (Layout A–3).*

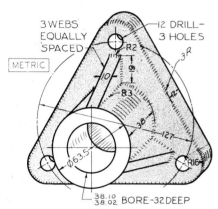

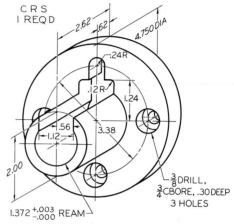

Fig. 9.43 Column Support. Draw necessary views, with full section (Layout A–3).*

Fig. 9.44 Centering Bushing. Draw necessary views, with full section (Layout A–3).*

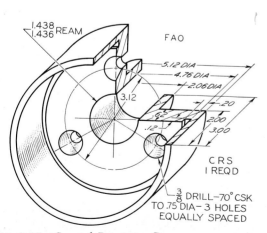

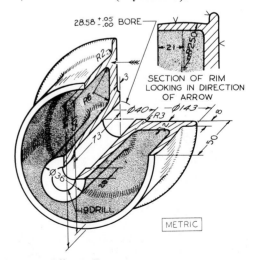

Fig. 9.45 Special Bearing. Draw necessary views, with full section (Layout A–3).*

Fig. 9.46 Idler Pulley. Draw necessary views, with full section (Layout A–3).*

*Layout A4–3 (adjusted) may be used. If dimensions are required, study §§13.1–13.25. Use metric or decimal-inch dimensions as assigned by the instructor.

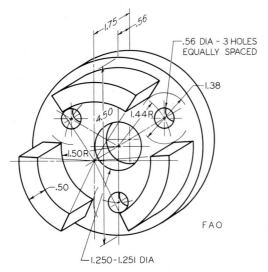

Fig. 9.47 Cup Washer. Draw necessary views, with full section (Layout A–3 or A4–3 adjusted).*

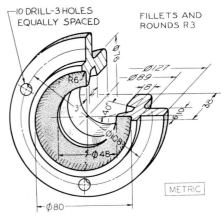

Fig. 9.48 Fixed Bearing Cup. Draw necessary views, with full section (Layout A–3 or A4–3 adjusted).*

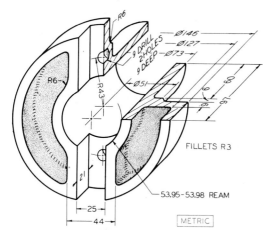

Fig. 9.49 Stock Guide. Draw necessary views, with half section (Layout B–4 or A3–4 adjusted).*

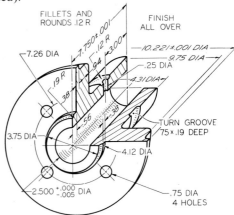

Fig. 9.50 Bearing. Draw necessary views with half section. Scale: half size (Layout B–4 or A3–4 adjusted).*

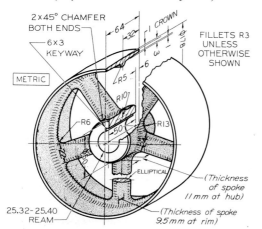

Fig. 9.51 Pulley. Draw necessary views, with full section, and revolved section of spoke (Layout B–4 or A3–4 adjusted).*

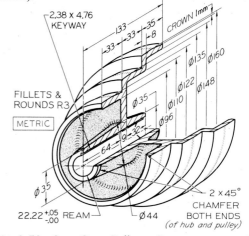

Fig. 9.52 Step-Cone Pulley. Draw necessary views, with full section (Layout B–4 or A3–4 adjusted).*

*If dimensions are required, study §§13.1–13.25. Use metric or decimal-inch dimensions as assigned by the instructor.

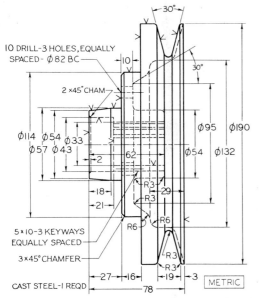

10 DRILL-3 HOLES, EQUALLY
SPACED - ⌀82 BC

2 × 45° CHAM

⌀114 ⌀54 ⌀33
⌀57 ⌀43

5 × 10-3 KEYWAYS
EQUALLY SPACED

3 × 45° CHAMFER

CAST STEEL-1 REQD

Fig. 9.53 Sheave. Draw two views, including half section (Layout B–4).*

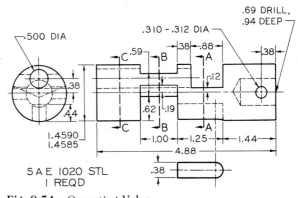

S A E 1020 STL
1 REQD

.500 DIA

.310 - .312 DIA

.69 DRILL,
.94 DEEP

Fig. 9.54 Operating Valve.
Given: Front, left-side, and partial bottom views.
Required: Front, right-side, and full bottom views, plus indicated removed sections (Layout B–4).*

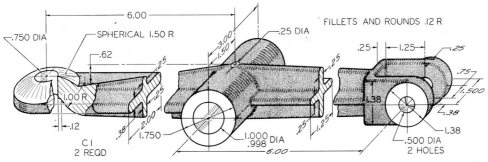

.750 DIA

SPHERICAL 1.50 R

FILLETS AND ROUNDS .12 R

.25 DIA

1.00 R

C 1
2 REQD

1.000
.998 DIA

.500 DIA
2 HOLES

Fig. 9.55 Rocker Arm. Draw necessary views, with revolved sections (Layout B–4).*

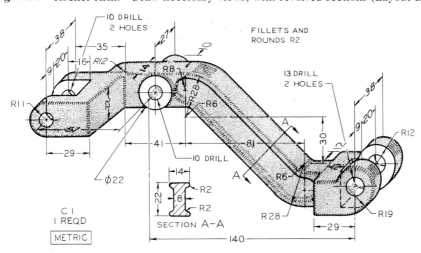

10 DRILL
2 HOLES

FILLETS AND
ROUNDS R2

13 DRILL
2 HOLES

R11

10 DRILL

⌀22

C 1
1 REQD

METRIC

SECTION A–A

Fig. 9.56 Dash Pot Lifter. Draw necessary views, using revolved section instead of removed section (Layout B–4).*

*Layout A3–4 (adjusted) may be used. If dimensions are required, study §§13.1–13.25. Use metric or decimal-inch dimensions as assigned by the instructor.

309

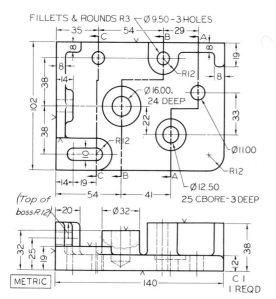

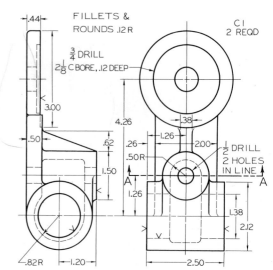

Fig. 9.57   Adjuster Base.
Given: Front and top views.
Required: Front and top views and sections A–A, B–B, and C–C. Show all visible lines (Layout B–4).*

Fig. 9.58   Mobile Housing.
Given: Front and left-side views.
Required: Front view, right-side view in full section, and removed section A–A (Layout B–4).*

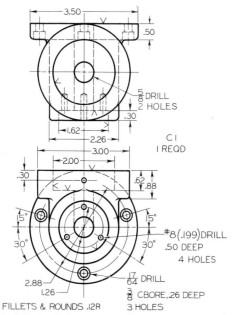

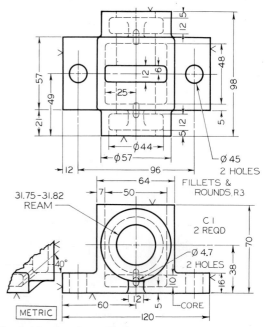

Fig. 9.59   Hydraulic Fitting.
Given: Front and top views.
Required: Front and top views and right-side view in full section (Layout B–4).*

Fig. 9.60   Auxiliary Shaft Bearing.
Given: Front and top views.
Required: Front and top views and right-side view in full section (Layout B–4).*

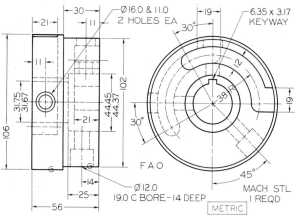

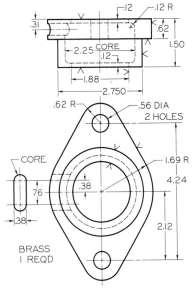

Fig. 9.61 Traverse Spider.
Given: Front and left-side views.
Required: Front and right-side views and top view in
full section (Layout B–4 or A3–4 adjusted).*

Fig. 9.62 Gland.
Given: Front, top, and partial left-side views.
Required: Front view and right-side view in full sec-
tion (Layout A–3 or A4–3 adjusted).*

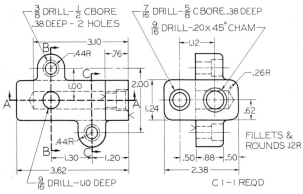

Fig. 9.63 Bracket.
Given: Front and right-side views.
Required: Take front as new top; then add right-side
view, front view in full section A–A, and sections B–B
and C–C (Layout B–4 or A3–4 adjusted).*

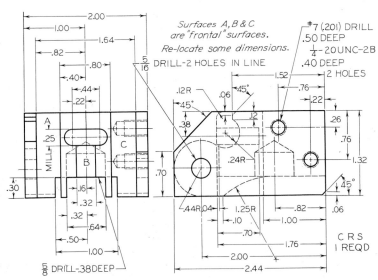

Fig. 9.64 Cocking Block.
Given: Front and right-side views.
Required: take front as new top view;
then add new front view, and right-
side view in full section. Draw double
size on Layout C–4 or A2–4.*

*Layout A3–4 (adjusted) may be used. If dimensions are required, study §§13.1–13.25. Use metric or decimal-
inch dimensions as assigned by the instructor.

311

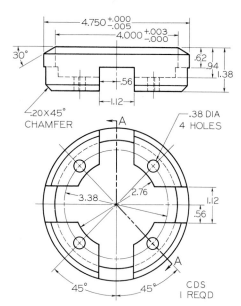

Fig. 9.65 Packing Ring.
Given: Front and top views.
Required: Front view and section A–A (Layout A–3 or A4–3 adjusted).*

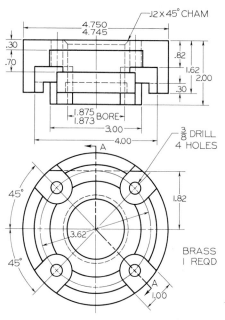

Fig. 9.67 Oil Retainer.
Given: Front and top views.
Required: Front view and section A–A (Layout B–4 or A3–4 adjusted).*

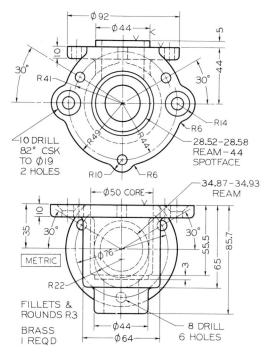

Fig. 9.66 Strainer Body.
Given: Front and bottom views.
Required: Front and top views and right-side view in full section (Layout C–4 or A2–4).*

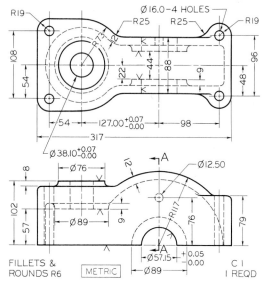

Fig. 9.68 Gear Box.
Given: Front and top views.
Required: Front in full section, bottom view, and right-side section **A–A**. Draw half size on Layout B–4 or A3–4 (adjusted).*

*If dimensions are required, study §§13.1–13.25. Use metric or decimal-inch dimensions as assigned by the instructor.

312

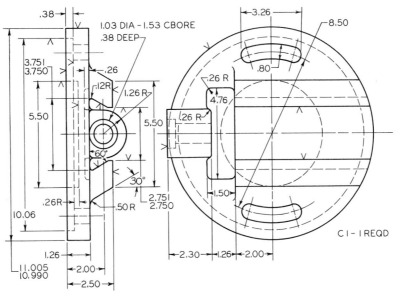

Fig. 9.69 Slotted Disk for Threading Machine.
Given: Front and left-side views.
Required: Front and right-side views and top full-section view. Draw half size on Layout B–4 or A3–4 (adjusted).*

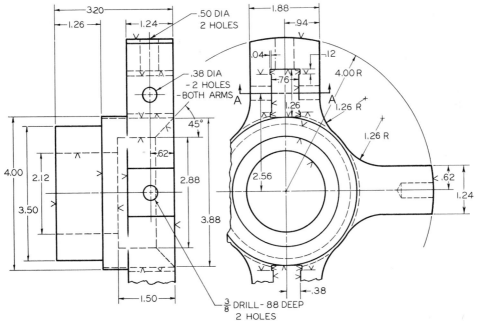

Fig. 9.70 Web for Lathe Clutch.
Given: Partial front and left-side views.
Required: Full front view, right-side view in full section, and removed section A–A (Layout C–4 or A2–4).*

*If dimensions are required, study §§13.1–13.25. Use metric or decimal-inch dimensions as assigned by the instructor.

313

Auxiliary Views

Many objects are of such shape that their principal faces cannot always be assumed parallel to the regular planes of projection. For example, in Fig. 10.1 (a), the base of the design for the bearing is shown in its true size and shape, but the rounded upper portion is situated at an angle with the planes of projection and does not appear in its true size and shape in any of the three regular views.

In order to show the true circular shapes, it is necessary to assume a direction of sight perpendicular to the planes of those curves, as shown at (b). The resulting view is known as an *auxiliary view*. This view, together with the top view, completely describes the object. The front and right-side views are not necessary.

10.1 Definitions

A view obtained by a projection on any plane other than the horizontal, frontal, and profile projection planes is called an *auxiliary view*. A *primary* auxiliary view is projected on a plane that is perpendicular to one of the principal planes of projection and is inclined to the other two. A *secondary* auxiliary view is projected

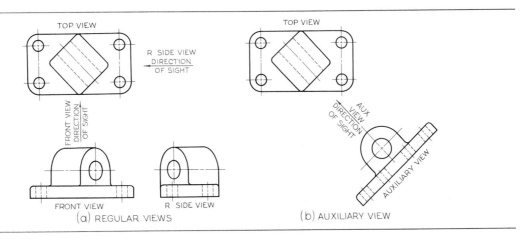

Fig. 10.1   Regular Views and Auxiliary Views.

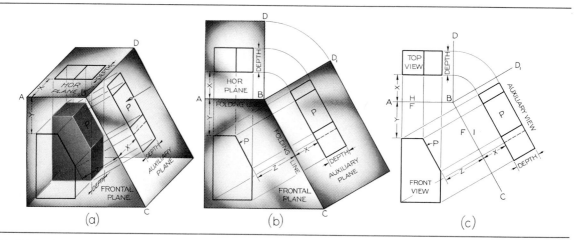

Fig. 10.2   An Auxiliary View.

from a primary auxiliary view and on a plane inclined to all three principal projection planes. See §10.19.

10.2   The Auxiliary Plane

In Fig. 10.2 (a), the object shown has an inclined surface that does not appear in its true size and shape in any regular view. The auxiliary plane is assumed parallel to the inclined surface P, that is, perpendicular to the line of sight, which is at right angles to that surface. The auxiliary plane is then perpendicular to the frontal plane of projection and hinged to it.

When the horizontal and auxiliary planes are unfolded to appear in the plane of the front view, as shown at (b), the *folding lines* represent the hinge lines joining the planes. The drawing is simplified, as shown at (c), by retaining the folding lines, H/F and F/1, and omitting the planes. As will be shown later, the folding lines may themselves be omitted in the actual drawing. The inclined surface P is shown in its true size and shape in the auxiliary view, the long dimension of the surface being projected directly from the front view and the *depth* from the top view.

It should be observed that the positions of the folding lines depend on the relative positions of

the planes of the glass box at (a). If the horizontal plane is moved upward, the distance Y is increased. If the frontal plane is brought forward, the distances X are increased but remain *equal*. If the auxiliary plane is moved to the right, the distance Z is increased. Note that both the top and auxiliary views show the *depth* of the object.

10.3   To Draw an Auxiliary View— Folding-Line Method

As shown in Fig. 10.2 (c), the folding lines are the hinge lines of the glass box. Distances X must be equal, since they both represent the distance of the front surface of the object from the frontal plane of projection.

Although distances X must remain equal, distances Y and Z, from the front view to the respective folding lines, may or may not be equal.

The steps in drawing an auxiliary view with the aid of the folding lines, shown in Fig. 10.3, are described as follows.

I.   The front and top views are given. It is required to draw an auxiliary view showing the true size and shape of inclined surface P. Draw the folding line H/F between the views at right

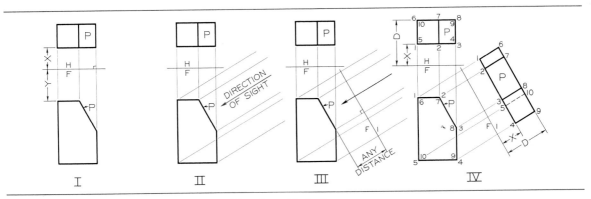

Fig. 10.3   To Draw an Auxiliary View—Folding-Line Method.

angles to the projection lines. Distances X and Y may or may not be equal, as desired.

NOTE   In the following steps, manipulate the triangle (either triangle) as shown in Fig. 10.4 to draw lines parallel or perpendicular to the inclined face.

II.   Assume arrow, indicating direction of sight, perpendicular to surface P. Draw light projection lines from the front view parallel to the arrow, or perpendicular to surface P.

III.   Draw folding line F/1 for the auxiliary view at right angles to the projection lines and at any convenient distance from the front view.

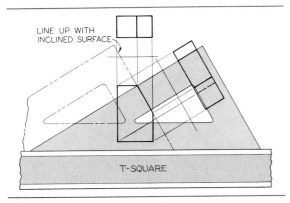

Fig. 10.4   Drawing Parallel or Perpendicular Lines.

IV.   Draw the auxiliary view, using the numbering system explained in §7.6. Locate all points the same distances from folding line F/1 as they are from folding line H/F in the top view. For example, points 1 to 5 are distance X from the folding lines in both the top and auxiliary views, and points 6 to 10 are distance D from the corresponding folding lines. Since the object is viewed in the direction of the arrow, it will be seen that edge 5–10 will be hidden in the auxiliary view.

10.4   Reference Planes

In the auxiliary views shown in Figs. 10.2 (c) and 10.3, the folding lines represent the edge views of the frontal plane of projection. In effect, the frontal plane is used as a *reference plane,* or *datum plane,* for transferring distances (*depth measurements*) from the top view to the auxiliary view.

Instead of using one of the planes of projection as a reference plane, it is often more convenient to assume a reference plane inside the glass box parallel to the plane of projection and touching or cutting through the object. For example, Fig. 10.5 (a), a reference plane is assumed to coincide with the front surface of the object. This plane appears edgewise in the top and auxiliary views, and the two reference lines are then used in the same manner as folding lines. Dimensions D, to the reference lines, are

317

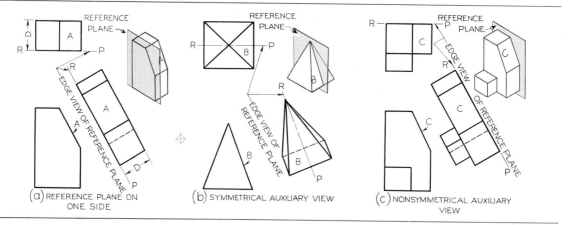

Fig. 10.5 Position of the Reference Plane.

equal. The advantage of the reference-plane method is that fewer measurements are required, since some points of the object lie in the reference plane.

The reference plane may coincide with the front surface of the object as at (a), or it may cut through the object as at (b) if the object is symmetrical, or the reference plane may coincide with the back surface of the object as at (c), or through any intermediate point of the object.

The reference plane should be assumed in the position most convenient for transferring distances with respect to it. Remember the following.

1. Reference lines, like folding lines, are always at right angles to the projection lines between the views.

2. A reference plane appears as a line in two alternate views, never in adjacent views.

3. Measurements are always made at right angles to the reference lines or parallel to the projection lines.

4. In the auxiliary view, all points are at the same distances from the reference line as the corresponding points are from the reference line in the *second previous view,* or alternate view.

10.5 To Draw an Auxiliary View—Reference-Plane Method

The object shown in Fig. 10.6 (a) is numbered as explained in §7.6. To draw the auxiliary view, proceed as follows.

I. Draw two views of the object, and assume an arrow indicating the direction of sight for the auxiliary view of surface A.

II. Draw projection lines parallel to the arrow.

III. Assume a reference plane coinciding with back surface of object as shown at (a). Draw reference lines in the top and auxiliary views at right angles to the projection lines: *these are the edge views of the reference plane.*

IV. Draw auxiliary view of surface A. It will be true size and shape because the direction of sight was taken perpendicular to that surface. Transfer depth measurements from the top view to the auxiliary view with dividers or scale. Each point in the auxiliary view will be on its projection line from the front view and the same distance from the reference line as it is in the top view to the corresponding reference line.

V. Complete the auxiliary view by adding other visible edges and surfaces of the object. Each numbered point in the auxiliary view lies on its projection line from the front view and is

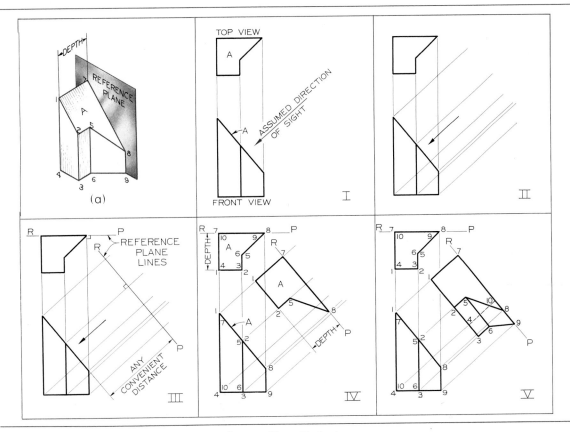

Fig. 10.6   To Draw an Auxiliary View—Reference-Plane Method.

the same distance from the reference line as it is in the top view. Note that two surfaces of the object appear as lines in the auxiliary view.

10.6 Classification of Auxiliary Views

Auxiliary views are classified and named according to the principal dimensions of the object shown in the auxiliary view. For example, the auxiliary view in Fig. 10.6 is a *depth auxiliary view* because it shows the principal dimension of the object, *depth.* Any auxiliary view projected from the front view, also known as front adjacent view, will show the depth of the object and is a depth auxiliary view.

Similarly, any auxiliary view projected from the top view, also known as top adjacent view, is a height auxiliary view, and any auxiliary view projected from the side view (either side), also known as side adjacent view, is a width auxiliary view. For examples of height auxiliary views, see Figs. 10.1 (b) and 10.13 (b). Depth auxiliary views are illustrated in Figs. 10.27 and 10.33.

10.7 Depth Auxiliary Views

An infinite number of auxiliary planes can be assumed perpendicular to, and hinged to, the frontal plane (F) of projection. Five such planes are shown in Fig. 10.7 (a), the horizontal plane being included to show that it is similar to the

319

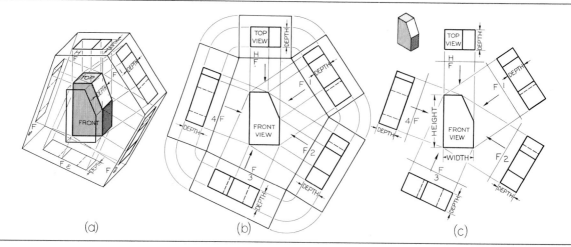

Fig. 10.7   Depth Auxiliary Views.

others. In all these views the principal dimension, *depth,* is shown; hence all the auxiliary views are depth auxiliary views.

The unfolded auxiliary planes are shown in (b), which also shows how the depth dimension may be projected from the top view to all auxiliary views. The arrows indicate the directions of sight for the several views, and the projection lines are respectively parallel to these arrows. The arrows may be assumed but need not be actually drawn, since the projection lines determine the direction of sight. The folding lines are perpendicular to the arrows and the corresponding projection lines. Since the auxiliary planes can be assumed at any distance from the object, it follows that the folding lines may be any distance from the front view.

The complete drawing, with the outlines of the planes of projection omitted, is shown at (c). This shows the drawing as it would appear on paper, in which use is made of reference planes as described in §10.4, all depth dimensions being measured perpendicular to the reference line in each view.

Note that the front view shows the *height* and the *width* of the object, *but not the depth.* The

depth is shown in all views that are projected from the front view; hence, this rule: *The principal dimension shown in an auxiliary view is that one which is not shown in the adjacent view from which the auxiliary view was projected.*

10.8   Height Auxiliary Views

An infinite number of auxiliary planes can be assumed perpendicular to, and hinged to, the horizontal plane (H) of projection, several of which are shown in Fig. 10.8 (a). The front view and all the auxiliary views show the principal dimension, *height.* Hence, all the auxiliary views are height auxiliary views.

The unfolded projection planes are shown at (b), and the complete drawing, with the outlines of the planes of projection omitted, is shown at (c). All reference lines are perpendicular to the corresponding projection lines, and all height dimensions are measured parallel to the projection lines, or perpendicular to the reference lines, in each view. Note that in the view projected from, which is the top view, the only dimension *not shown* is height.

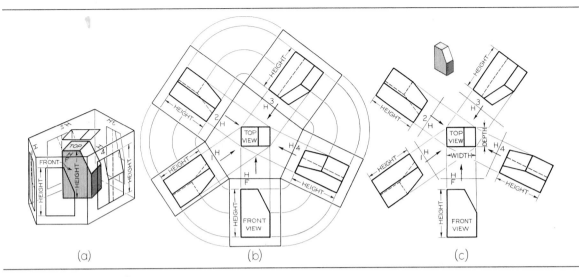

Fig. 10.8   Height Auxiliary Views.

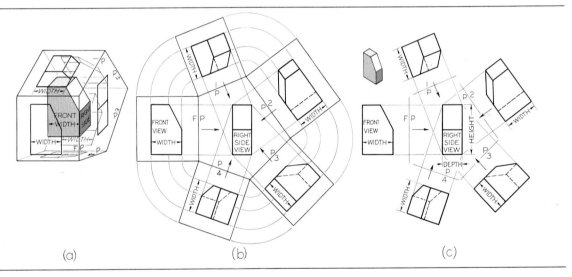

Fig. 10.9   Width Auxiliary Views.

10.9   Width Auxiliary Views

An infinite number of auxiliary planes can be assumed perpendicular to, and hinged to, the profile plane (P) of projection, several of which are shown in Fig. 10.9 (a). The front view and all the auxiliary views show the principal dimension, *width*. Hence, all the auxiliary views are width auxiliary views.

The unfolded planes are shown at (b), and the complete drawing, with the outlines of the planes of projection omitted, is shown at (c). All reference lines are perpendicular to the corresponding projection lines, and all width dimensions are measured parallel to the projection lines, or perpendicular to the reference lines, in each view. Note that in the right-side view, from

321

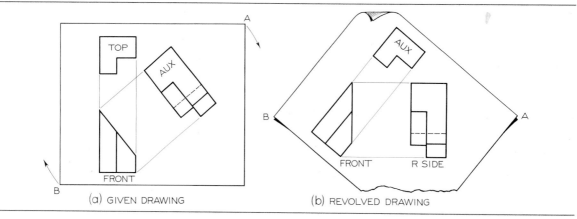

Fig. 10.10   Revolving a Drawing.

which the auxiliary views are projected, the only dimension *not shown* is width.

10.10   Revolving a Drawing

In Fig. 10.10 (a) is a drawing showing top, front, and auxiliary views. At (b) the drawing is shown revolved, as indicated by the arrows, until the auxiliary view and the front view line up horizontally. Although the views remain exactly the same, the names of the views are changed if drawn in this position. The auxiliary view now becomes a right-side view, and the top view becomes an auxiliary view. Some students find it easier to visualize and draw an auxiliary view when revolved to the position of a regular view in this manner. In any case, it should be understood that an auxiliary view basically is like any other view.

10.11   Dihedral Angles

The angle between two planes is a dihedral angle. One of the principal uses of auxiliary views is to show dihedral angles in true size, mainly for dimensioning purposes. In Fig. 10.11 (a) a block is shown with a V-groove situated so that the true dihedral angle between inclined surfaces A and B is shown in the front view.

Assume a line in a plane. For example, draw a straight line on a sheet of paper; then hold the paper so as to view the line as a point. You will observe that when the line appears as a point, the plane containing the line appears as a line. Hence, this rule: *To get the edge view of a plane, find the point view of any line in that plane.*

In Fig. 10.11 (a), line 1–2 is the line of intersection of planes A and B. Now, line 1–2 lies in both planes at the same time; therefore, a point view of this line will show both planes as lines, and the angle between them is the dihedral angle between the planes. Hence, this rule: *To get the true angle between two planes, find the point view of the line of intersection of the planes.*

At (b), the line of intersection 1–2 does not appear as a point in the front view; hence, planes A and B do not appear as lines, and the true dihedral angle is not shown. Assuming that the actual angle is the same as at (a), does the angle show larger or smaller than at (a)? The drawing at (b) is unsatisfactory. The true angle does not appear because the direction of sight (see arrow) is not parallel to the line of intersection 1–2.

At (c) the direction of sight arrow is taken parallel to the line 1–2, producing an auxiliary view in which line 1–2 appears as a point, planes A and B appear as lines, and the true dihedral angle is shown. *To draw a view showing a true*

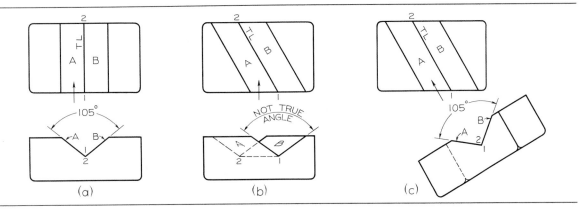

Fig. 10.11   Dihedral Angles.

dihedral angle, assume the direction of sight parallel to the line of intersection between the planes of the angle.

10.12   Plotted Curves

As shown in §7.30, if a cylinder is cut by an inclined plane, the inclined surface is elliptical in shape. In Fig. 7.34 (a), such a surface is produced, but the ellipse does not show true size and shape because the plane of the ellipse is not seen at right angles in any view.

In Fig. 10.12 (a), the line of sight is taken perpendicular to the edge view of the inclined surface, and the resulting ellipse is shown in true size and shape in the auxiliary view. The major axis is found by direct projection from the front view, and the minor axis is equal to the diameter of the cylinder. The left end of the cylinder (a circle) will appear as an ellipse in the auxiliary view, the major axis of which is equal to the diameter of the cylinder.

Since this is a symmetrical object, the reference plane is assumed to be located through

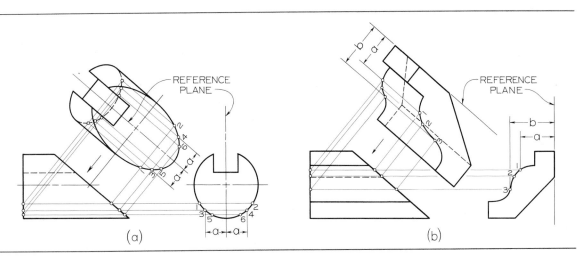

Fig. 10.12   Plotted Curves.

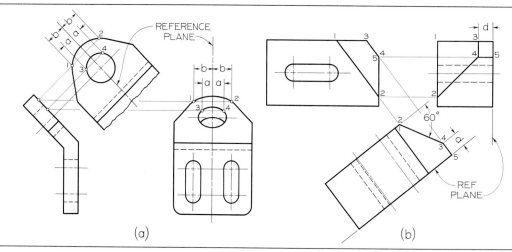

Fig. 10.13   Reverse Construction.

the center, as shown. To plot points on the el- lipses, select points on the circle of the side view, and project them across to the inclined surface or to the left-end surface, and then upward to the auxiliary view. In this manner, two points can be projected each time, as shown for points 1–2, 3–4, and 5–6. Distances a are equal and are transferred from the side view to the auxiliary view with the aid of dividers. A sufficient number of points must be projected to establish the curves accurately. Use the irregular curve as de- scribed in §2.54.

Since the major and minor axes are known, any of the true ellipse methods of Figs. 5.48–5.50 and 5.52 (a) may be used. If an approximate el- lipse is adequate for the job in hand, the method of Fig. 5.56 can be used. But the quickest and easiest method is to use an ellipse template, as explained in §5.56.

In Fig. 10.12 (b), the auxiliary view shows the true size and shape of the inclined cut through a piece of molding. The method of plotting points is similar to that explained for the ellipse in Fig. 10.12 (a).

10.13   Reverse Construction

In order to complete the regular views, it is often necessary to construct an auxiliary view first. For example, in Fig. 10.13 (a) the upper portion

of the right-side view cannot be constructed until the auxiliary view is drawn and points are estab- lished on the curves and then projected back to the front view as shown.

At (b), the 60° angle and the location of line 1–2 in the front view are given. In order to locate line 3–4 in the front view, the lines 2–4, 3–4, and 4–5 in the side view, it is necessary first to con- struct the 60° angle in the auxiliary view and project back to the front and side views, as shown.

10.14   Partial Auxiliary Views

The use of an auxiliary view often makes it pos- sible to omit one or more regular views and thus to simplify the shape description, as shown in Fig. 10.1 (b). In Fig. 10.14 three complete aux- iliary-view drawings are shown. Such drawings take a great deal of time to draw, particularly when ellipses are involved, as is so often the case, and the completeness of detail may add nothing to clearness or may even detract from it because of the clutter of lines. However, in these cases some portion of every view is needed—no view can be completely eliminated, as was done in Fig. 10.1 (b).

As described in §7.9, *partial views* are often sufficient, and the resulting drawings are consid- erably simplified and easier to read. Similarly, as

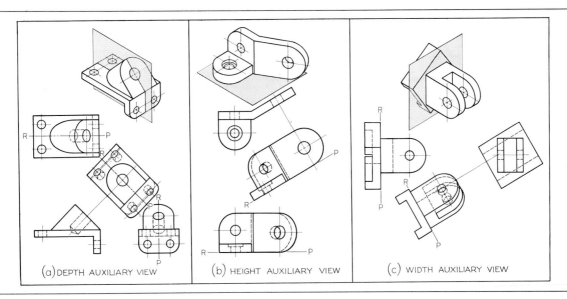

Fig. 10.14 Primary Auxiliary Views.

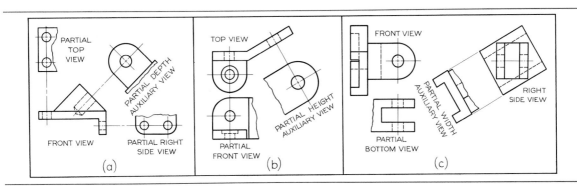

Fig. 10.15 Partial Views.

shown in Fig. 10.15, partial regular views and partial auxiliary views are used with the same result. Usually a break line is used to indicate the imaginary break in the views. *Do not draw a break line coinciding with a visible line or a hidden line.*

In order to clarify the relationship of views, the auxiliary views should be connected to the views from which they are projected, either with a center line or with one or two projection lines.

This is particularly important with regard to partial views that often are small and easily appear to be "lost" and not related to any view.

10.15 Half Auxiliary Views

If an auxiliary view is symmetrical, and if it is necessary to save space on the drawing or to save time in drafting, only half of the auxiliary view may be drawn, as shown in Fig. 10.16. In this

325

case, half of a regular view is also shown, since the bottom flange is also symmetrical. See §§7.9 and 9.14. Note that in each case the *near half* is shown.

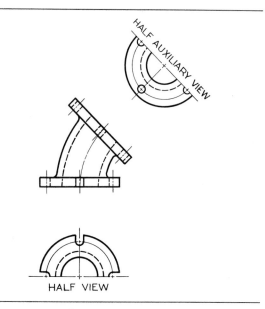

Fig. 10.16   Half Views.

10.16   Hidden Lines in Auxiliary Views

In practice, hidden lines should be omitted in auxiliary views, as in ordinary views, §6.25, unless they are needed for clearness. The beginner, however, should show all hidden lines, especially if the auxiliary view of the entire object is shown. Later, in advanced work, it will become clearer as to when hidden lines can be omitted.

10.17   Auxiliary Sections

An auxiliary section is simply an auxiliary view in section. In Fig. 10.17 (a), note the cutting-plane line and the terminating arrows that indicate the direction of sight for the auxiliary section. Observe that the section lines are drawn at approximately 45° with visible outlines. In an auxiliary section drawing, the entire portion of the object behind the cutting plane may be shown, as at (a), or the cut surface alone, as at (b) and (c).

An auxiliary section through a cone is shown in Fig. 10.18. This is one of the conic sections, §5.47, in this case a parabola. The parabola may be drawn by other methods, Figs. 5.57 and 5.58,

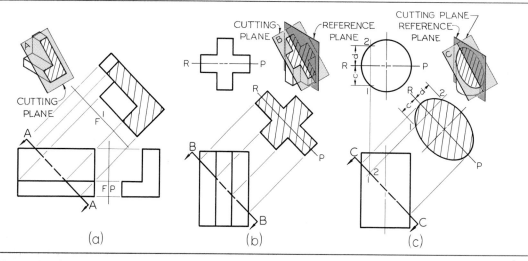

Fig. 10.17   Auxiliary Sections.

but the method shown here is by projection. In Fig. 10.18, elements of the cone are drawn in the front and top views. These intersect the cutting plane at points 1, 2, 3, and so on. These points are established in the top view by projecting upward to the top views of the corresponding elements. In the auxiliary section, all points on the parabola are the same distance from the reference plane RP as they are in the top view.

A typical example of an auxiliary section in machine drawing is shown in Fig. 10.19. Here, there is not sufficient space for a revolved section, §9.9, although a removed section, §9.10, could have been used instead of an auxiliary section.

10.18   True Length of Line—Auxiliary-View Method

A line will show in true length when projected to a projection plane parallel to the line.

In Fig. 10.20, let it be required to find the true length of the hip rafter 1–2 by means of a depth auxiliary view.

I.   Assume an arrow perpendicular to 1–2 (front view) indicating the direction of sight, and place the H/F folding line as shown.

II.   Draw the F/1 folding line perpendicular to the arrow and at any convenient distance from 1–2 (front view), and project the points 1 and 3 toward it.

III.   Set off the points 1 and 2 in the auxiliary view at the same distance from the folding line as they are in the top view. The triangle 1–2–3 in the auxiliary view shows the true size and shape of the roof section 1–2–3, and the distance 1–2 in the auxiliary view is the true length of the hip rafter 1–2.

To find the true length of a line by revolution, see §11.10.

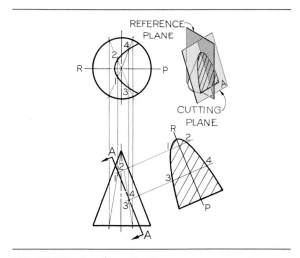

Fig. 10.18   Auxiliary Section.

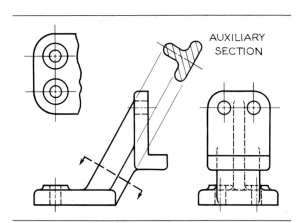

Fig. 10.19   Auxiliary Section.

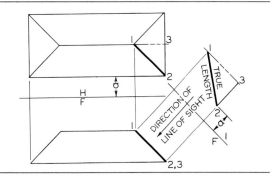

Fig. 10.20   True Length of a Line by Means of an Auxiliary View.

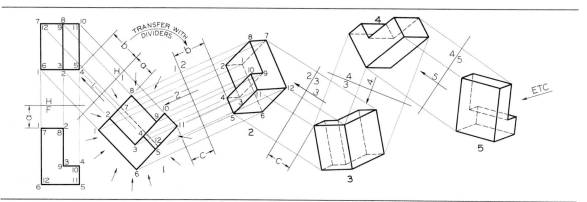

Fig. 10.21   Successive Auxiliary Views.

10.19   Successive Auxiliary Views

Up to this point we have dealt with *primary* auxiliary views, that is, single auxiliary views projected from one of the regular views. In Fig. 10.21, auxiliary view 1 is a primary auxiliary view, projected from the top view.

From primary auxiliary view 1, a *secondary* auxiliary view 2 can be drawn; then from it a third auxiliary view 3, and so on. An infinite number of such successive auxiliary views may be drawn, a process that may be likened to the "chain reaction" in a nuclear explosion.

However, secondary auxiliary view 2 is not the only one that can be projected from primary auxiliary view 1 and thus start an independent "chain reaction." As shown by the arrows around view 1, an infinite number of secondary auxiliary views, with different lines of sight, may be projected. *Any auxiliary view projected from a primary auxiliary view is a secondary auxiliary view.* Furthermore, any succeeding auxiliary view may be used to project an infinite number of "chains" of views from it.

In this example, folding lines are more convenient than reference-plane lines. In auxiliary view 1, all numbered points of the object are the same distance from folding line H/1 as they are in the front view from folding line H/F. These distances, such as distance a, are transferred from the front view to the auxiliary view with the aid of dividers.

To draw the secondary auxiliary view 2, drop the front view from consideration, and center attention on the sequence of three views: the top view, view 1, and view 2. Draw arrow 2 toward view 1 in the direction desired for view 2, and draw light projection lines parallel to the arrow. Draw folding line 1/2 perpendicular to the projection lines and at any convenient distance from view 1. Locate all numbered points in view 2 from folding line 1/2 at the same distances they are in the top view from folding line H/1, using the dividers to transfer distances. For example, transfer distance b to locate points 4 and 5. Connect points with straight lines, and determine visibility. The corner nearest the observer (11) for view 2 will be visible, and the one farthest away (1) will be hidden, as shown.

To draw views 3, 4, and so on, repeat this procedure, remembering that each time we will be concerned only with a sequence of three views. In drawing any auxiliary view, the paper may be revolved so as to make the last two views line up as regular views.

10.20   Uses of Auxiliary Views

Generally, auxiliary views are used to show the true shape or true angle of features that appear distorted in the regular views. Basically, auxiliary views have the following four uses.

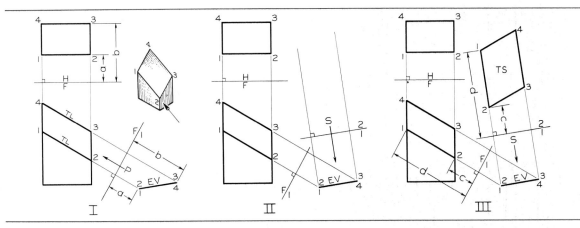

Fig. 10.22   True Size of Oblique Surface—Folding-Line Method.

1.   True length of line (TL), §10.18.
2.   Point view of line, §10.11.
3.   Edge view of plane (EV), §10.21.
4.   True size of plane (TS), §10.21.

10.21   True Size of an Oblique Surface—Folding-Line Method

A typical requirement of a secondary auxiliary view is to show the true size and shape of an oblique surface, such as surface 1–2–3–4 in Fig. 10.22. In this case folding lines are used, but the same results can be obtained with reference lines. Proceed as follows.

I.   Draw the primary auxiliary view showing surface 1–2–3–4 as a line. As explained in §10.11, the edge view (EV) of a plane is found by getting the point view of a line in that plane. To get the point view of a line, the line of sight must be assumed parallel to the line. Therefore, draw arrow P parallel to lines 1–2 and 3–4, which are true length (TL) in the front view, and draw projection lines parallel to the arrow. Draw folding line H/F between the top and front views and F/1 between the front and auxiliary views, perpendicular to the respective projection lines. All

points in the auxiliary view will be the same distance from the folding line F/1 as they are in the top view from folding line H/F. Lines 1–2 and 3–4 will appear as points in the auxiliary view, and plane 1–2–3–4 will therefore appear *edgewise*, that is, as a line.

II.   Draw arrow S perpendicular to the edge view of plane 1–2–3–4 in the primary auxiliary view, and draw projection lines parallel to the arrow. Draw folding line 1/2 perpendicular to these projection lines and at a convenient distance from the primary auxiliary view.

III.   Draw the secondary auxiliary view. Locate each point (transfer with dividers) the same distance from the folding line 1/2 as it is in the front view to the folding line F/1, as for example, dimensions c and d. The true size (TS) of the surface 1–2–3–4 will be shown in the secondary auxiliary view, since the direction of sight, arrow S, was taken perpendicular to it.

10.22   True Size of an Oblique Surface—Reference-Plane Method

In Fig. 10.23 (a), it is required to draw an auxiliary view in which triangular surface 1–2–3 will appear in true size and shape. In order for the true size of the surface to appear in the second-

329

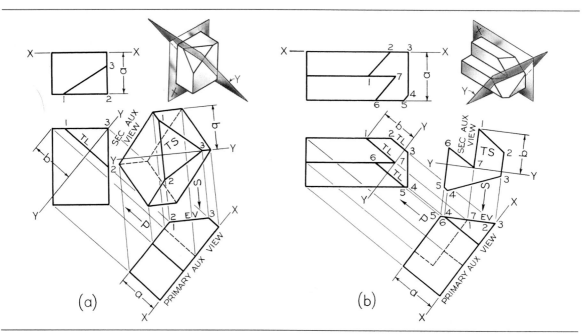

Fig. 10.23   True Size of an Oblique Surface—Reference-Plane Method.

ary auxiliary view, arrow S must be assumed perpendicular to the edge view of that surface; it is therefore necessary to have the edge view of surface 1–2–3 in the primary auxiliary view first. In order to do this, the direction of sight, arrow P, must be parallel to a line in surface 1–2–3 that appears true length (TL) in the front view. Hence, arrow P is drawn parallel to line 1–2 of the front view, line 1–2 will appear as a point in the primary auxiliary view, and surface 1–2–3 must therefore appear edgewise in that view.

In this case it is convenient to use reference lines and to assume the reference plane X (for drawing the primary auxiliary view) coinciding with the back surface of the object, as shown. For the primary auxiliary view, all depth measurements, as a in the figure, are transferred with dividers from the top view with respect to the reference line X–X.

For the secondary auxiliary view, reference plane Y is assumed cutting through the object for convenience in transferring measurements. All measurements perpendicular to Y–Y in the secondary auxiliary view are the same as be-

tween the reference plane and the corresponding points in the front view. Note that corresponding measurements must be *inside* (toward the central view in the sequence of three views) or *outside* (away from the central view). For example, dimension b is on the side of Y–Y *away* from the primary auxiliary view in both places.

In Fig. 10.23 (b) it is required to find the true size and shape of surface 1–2–3–4–5–6–7 and not to draw the complete secondary auxiliary view. The method is similar to that just described.

10.23   Secondary Auxiliary View, Oblique Direction of Sight Given

In Fig. 10.24 two views of a block are given, with two views of an arrow indicating the direction in which it is desired to look at the object to obtain a view. Proceed as follows.

I.   *Draw primary auxiliary view of both the object and the assumed arrow,* which will show the true length *of the arrow.* In order to do this, assume a horizontal reference plane X–X in the

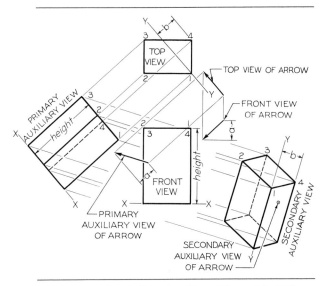

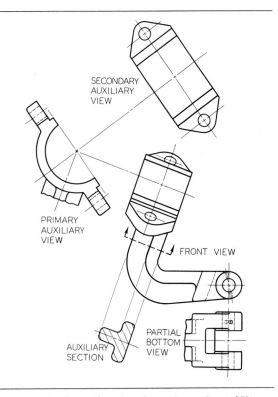

Fig. 10.24 Secondary Auxiliary View with Oblique Direction of Sight Given.

Fig. 10.25 Secondary Auxiliary View—Partial Views.

front and auxiliary views, as shown. Then assume a direction of sight perpendicular to the given arrow. In the front view, the butt end of the arrow is a distance **a** higher than the arrow point, and this distance is transferred to the primary auxiliary view as shown. All *height* measurements in the auxiliary view correspond to those in the front view.

II. *Draw secondary auxiliary view,* which will show the arrow as a point. This can be done because the arrow shows in true length in the primary auxiliary view, and projection lines for the secondary auxiliary view are drawn parallel to it. Draw reference line Y–Y, for the secondary auxiliary view perpendicular to these projection lines. In the top view, draw Y–Y perpendicular to the projection lines to the primary auxiliary view. All measurements, such as **b**, with respect to Y–Y correspond in the secondary auxiliary view and the top view.

It will be observed that the secondary auxiliary views of Figs. 10.23 (a) and 10.24 have considerable pictorial value. These are trimetric projections, §18.30. However, the direction of

sight could be assumed, in the manner of Fig. 10.24, to produce either isometric or dimetric projections. If the direction of sight is assumed parallel to the diagonal of a cube, the resulting view is an *isometric projection,* §18.3.

A typical application of a secondary auxiliary view in machine drawing is shown in Fig. 10.25. All views are partial views, except the front view. The partial secondary auxiliary view illustrates a case in which break lines are not needed. Note the use of an auxiliary section to show the true shape of the arm.

10.24 Ellipses

As shown in §7.30, if a circle is viewed obliquely, the result is an ellipse. This often occurs in successive auxiliary views, because of the variety of

331

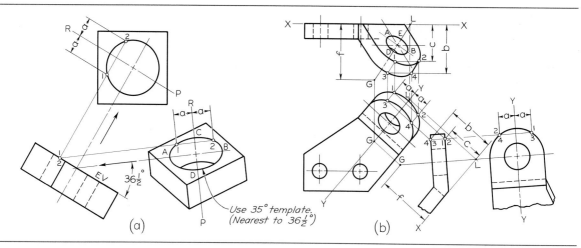

Fig. 10.26 Ellipses.

directions of sight. In Fig. 10.26 (a) the hole appears as a true circle in the top view. The circles appear as straight lines in the primary auxiliary view and as ellipses in the secondary auxiliary view. In the latter, the major axis **AB** of the ellipse is parallel to the projection lines and equal in length to the true diameter of the circle as shown in the top view. The minor axis **CD** is perpendicular to the major axis, and its foreshortened length is projected from the primary auxiliary view.

The ellipse can be completed by projecting points, such as 1 and 2, symmetrically located about the reference plane RP coinciding with CD with distances *a* equal in the top and secondary auxiliary views as shown, and finally, after a sufficient number of points have been plotted, by applying the irregular curve, §2.54

Since the major and minor axes are easily found, any of the true-ellipse methods of Figs. 5.48–5.50 and 5.52 (a) may be used, or an approximate ellipse, Fig. 5.56, may be found sufficiently accurate for a particular drawing. Or the ellipses may be easily and rapidly drawn with the aid of an ellipse template, §5.56. The "angle" of ellipse to use is the one that most closely matches the angle between the direction of sight arrow and the plane (EV) containing the circle,

as seen in this case in the primary auxiliary view. Here the angle is $36\frac{1}{2}°$, so a 35° ellipse is selected.

At (b) successive auxiliary views are shown in which the true circular shapes appear in the secondary auxiliary view, and the elliptical projections in the front and top views. It is necessary to construct the circular shapes in the secondary auxiliary view, then to project plotted points back to the primary auxiliary view, the front view, and finally to the top view, as shown in the figure for points 1, 2, 3, and 4. The final curves are then drawn with the aid of the irregular curve.

If the major and minor axes are found, any of the true-ellipse methods may be used; or better still, an ellipse template, §5.56, may be employed. The major and minor axes are easily established in the front view, but in the top view, they are more difficult to find. The major axis **AB** is at right angles to the center line GL of the hole, and equal in length to the true diameter of the hole. The minor axis **ED** is at right angles to the major axis. Its length is found by plotting several points in the vicinity of one end of the minor axis, or by using descriptive geometry to find the angle between the line of sight and the inclined surface, and by this angle selecting the ellipse guide required.

10.25 Computer Graphics

If the user has a thorough working knowledge of the principles of projection and can visualize which views of an object are required to best depict it in a drawing, computer graphics can provide a powerful tool allowing the operator to quickly manipulate the views of a drawing to show any desired viewing orientation.

CAD programs make it easy to create secondary auxiliary views, partial views, and auxiliary sections, Fig. 10.25.

AUXILIARY VIEW PROBLEMS

The problems in Figs. 10.27–10.58 are to be drawn with instruments or freehand. If partial auxiliary views are not assigned, the auxiliary views are to be complete views of the entire object, including all necessary hidden lines.

If is often difficult to space properly the views of an auxiliary view drawing. In some cases it may be necessary to make a trial blocking out on a preliminary sheet before starting the actual drawing. Allowances for dimensions must be made if metric or decimal dimensions are to be included. In such case, the student should study §§13.1–13.25.

Problems in convenient form for solution may be found in *Technical Drawing Problems*, Series 1, by Giesecke, Mitchell, Spencer, Hill, Dygdon, and Novak; *Technical Drawing Problems*, Series 2, by Spencer, Hill, Dygdon, and Novak; and *Technical Drawing Problems*, Series 3, by Spencer, Hill, Dygdon, and Novak; all designed to accompany this text and published by Macmillan Publishing Company.

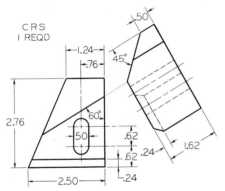

Fig. 10.27   RH Finger.
Given: Front and auxiliary views.
Required: Complete front, auxiliary, left-side, and top views (Layout A–3 or A4–3 adjusted).

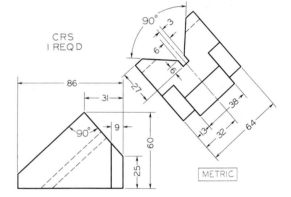

Fig. 10.28   V-Block.
Given: Front and auxiliary views.
Required: Complete front, top, and auxiliary views (Layout A–3 or A4–3 adjusted).

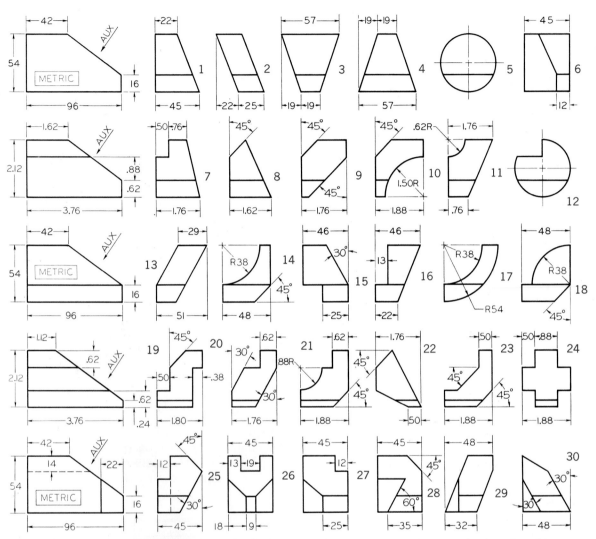

Fig. 10.29 Auxiliary View Problems. Make freehand sketch or instrument drawing of selected problem as assigned by instructor. Draw given front and right-side views, and add incomplete auxiliary view, including all hidden lines (Layout A–3 or A4–3 adjusted). If assigned, design your own right-side view consistent with given front view, and then add complete auxiliary view. Problems 1–6, 13–18, and 25–30 are given in metric dimensions. Problems 7–12 and 19–24 are given in decimal-inch dimensions.

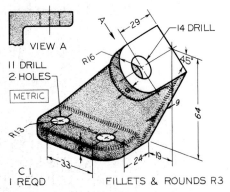

VIEW A

11 DRILL
2 HOLES

METRIC

R13

C 1
1 REQD

FILLETS & ROUNDS R3

29 14 DRILL 45° R16 64 9 33 24 19

Fig. 10.30. Anchor Bracket. Draw necessary views or partial views (Layout A–3 or A4–3 adjusted).*

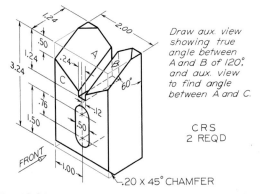

Draw aux. view showing true angle between A and B of 120°, and aux. view to find angle between A and C.

C R S
2 REQD

FRONT

1.24 2.00 .50 1.24 3.24 .24 60° .76 12 1.50 50 1.00 .20 X 45° CHAMFER

Fig. 10.31 Centering Block. Draw complete front, top, and right-side views, plus indicated auxiliary views (Layout B–3 or A3–3).*

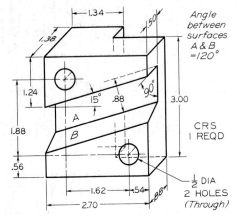

Angle between surfaces A & B =120°

1.34 .50 1.38 1.24 15° .88 90° 3.00 1.88 A B .56 1.62 .54 .88 2.70

C R S
1 REQD

½ DIA
2 HOLES
(Through)

Fig. 10.32 Clamp Slide. Draw necessary views completely (Layout B–3 or A3–3).*

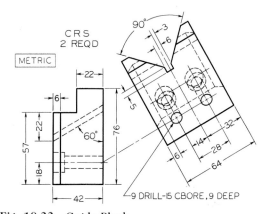

C R S
2 REQD

METRIC

90° 3 6 22 6 5 22 57 60° 76 18 14 28 32 6 64 42 9 DRILL-15 CBORE, 9 DEEP

Fig. 10.33 Guide Block.
Given: Right-side and auxiliary views.
Required: Right-side, auxiliary, plus front and top views—all complete (Layout B–3 or A3–3).*

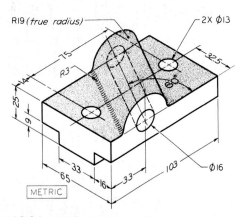

R19 (true radius) 2X Ø13 75 32.5 R3 80° 25 9 33 103 65 16 33 Ø16

METRIC

Fig. 10.34 Angle Bearing. Draw necessary views, including a complete auxiliary view (Layout A–3 or A4–3 adjusted).*

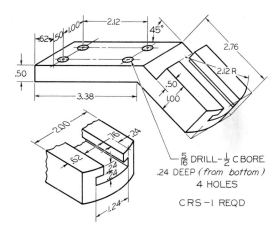

.62 .50 1.00 2.12 45° 2.76 .50 3.38 .50 1.00 2.12 R 2.00 .76 .24 .62 .24 .54 1.24 $\frac{5}{16}$ DRILL-½ C BORE .24 DEEP (from bottom) 4 HOLES

C R S – 1 REQD

Fig. 10.35 Guide Bracket. Draw necessary views or partial views (Layout B–3 or A3–3).*

*If dimensions are required, study §§13.1–13.25. Use metric or decimal-inch dimensions as assigned by the instructor.

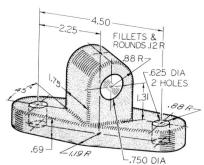

Fig. 10.36 Rod Guide. Draw necessary views, including complete auxiliary view showing true shape of upper rounded portion (Layout B–4 or A3–4 adjusted).*

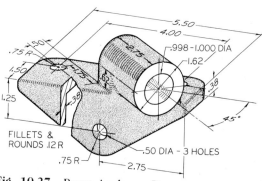

Fig. 10.37 Brace Anchor. Draw necessary views, including partial auxiliary view showing true shape of cylindrical portion (Layout B–4 or A3–4 adjusted).*

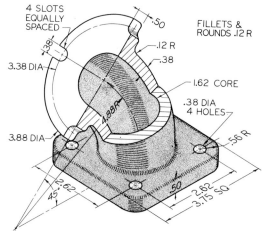

Fig. 10.38 45° Elbow. Draw necessary views, including a broken section and two half views of flanges (Layout B–4 or A3–4 adjusted).*

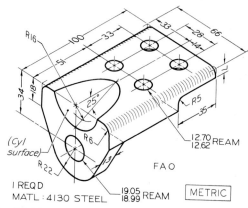

Fig. 10.39 Angle Guide. Draw necessary views, including a partial auxiliary view of cylindrical recess (Layout B–4 or A3–4 adjusted).*

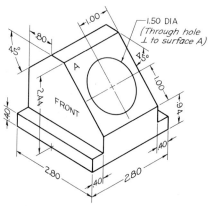

Fig. 10.40 Holder Block. Draw front and right-side views (2.80″ apart) and complete auxiliary view of entire object showing true shape of surface A and all hidden lines (Layout A–3 or A4–3 adjusted).*

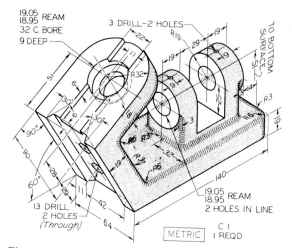

Fig. 10.41 Control Bracket. Draw necessary views, including partial auxiliary views and regular views (Layout C–4 or A2–4).*

*If dimensions are required, study §§13.1–13.25. Use metric or decimal-inch dimensions as assigned by the instructor.

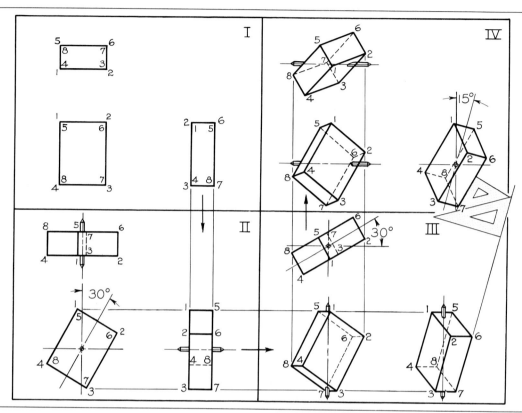

Fig. 11.5 Successive Revolutions of a Prism.

limited to three or four stages, offers excellent practice in multiview projection. While it is possible to make several revolutions of a simple object without the aid of a system of numbers, it is absolutely necessary in successive revolutions to assign a number or a letter to every corner of the object. See §7.6.

The numbering or lettering must be consistent in the various views of the several stages of revolution. Figure 11.5 shows four sets of multiview drawings numbered I, II, III, and IV, respectively. These represent the same object in different positions with reference to the planes of projection.

In space I, the object is represented in its normal position, with its faces parallel to the planes of projection. In space II, the object is

represented after it has been revolved clockwise through an angle of 30° about an axis perpendicular to the frontal plane. The drawing in space II is placed under space I so that the side view, whose width remains unchanged, can be projected from space I to space II as shown.

During the revolution, all points of the object describe circular arcs parallel to the frontal plane of projection and remain at the same distance from that plane. The side view, therefore, may be projected from the side view of space I and the front view of space II. The top view may be projected in the usual manner from the front and side views of space II.

In space III, the object is taken as represented in space II and is revolved counterclockwise through an angle of 30° about an axis per-

346

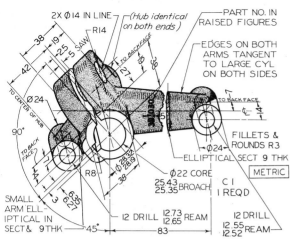

Fig. 10.46 Brake Control Lever. Draw necessary views and partial views (Layout B–4 or A3–4 adjusted).*

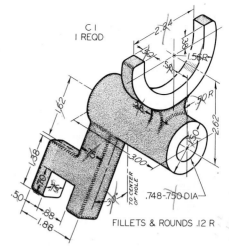

Fig. 10.47 Shifter Fork. Draw necessary views, including partial auxiliary view showing true shape of inclined arm (Layout B–4 or A3–4 adjusted).*

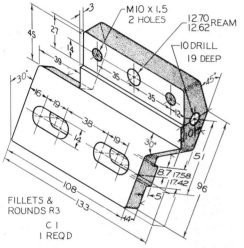

Fig. 10.48 Cam Bracket. Draw necessary views or partial views as needed. For threads, see §§15.9 and 15.10 (Layout B–4 or A3–4 adjusted).*

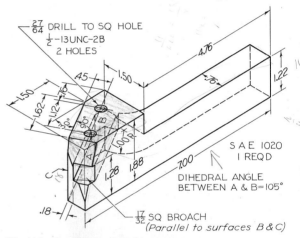

Fig. 10.49 RH Tool Holder. Draw necessary views, including partial auxiliary views showing 105° angle and square hole true size. For threads, see §§15.9 and 15.10 (Layout B–4 or A3–4 adjusted).*

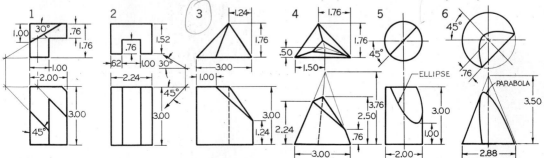

Fig. 10.50 Draw secondary auxiliary views, complete, which (except Prob. 2) will show the true sizes of the inclined surfaces. In Prob. 2, draw secondary auxiliary view as seen in direction of arrow (Layout B–3 or A3–3).*

*If dimensions are required, study §§13.1–13.25. Use metric or decimal-inch dimensions as assigned by the instructor.

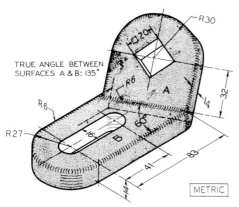

Fig. 10.51 Control Bracket. Draw necessary views including primary and secondary auxiliary views so that the latter shows true shape of oblique surface A (Layout B–4 or A3–4 adjusted).*

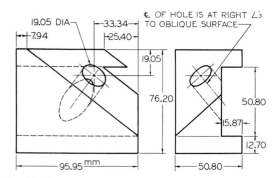

Fig. 10.52 Holder Block. Draw given views and primary and secondary auxiliary views so that the latter shows true shape of oblique surface (Layout B–4 or A3–4 adjusted).*

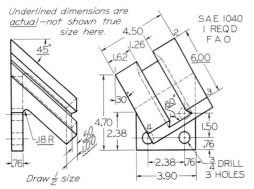

Fig. 10.53 Dovetail Slide. Draw complete given views and auxiliary views, including view showing true size of surface 1–2–3–4 (Layout B–4 or A3–4 adjusted).*

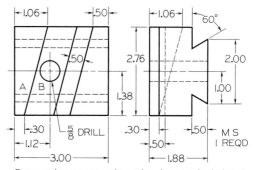

Draw primary aux. view showing angle between planes A and B; then secondary auxiliary view showing true size of surface A.

Fig. 10.54 Dovetail Guide. Draw given views plus complete auxiliary views as indicated (Layout B–4 or A3–4 adjusted).*

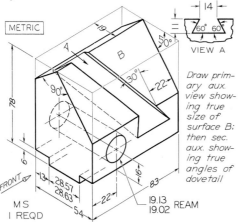

Fig. 10.55 Adjustable Stop. Draw complete front and auxiliary views plus partial right-side view. Show all hidden lines (Layout C–4 or A2–4).*

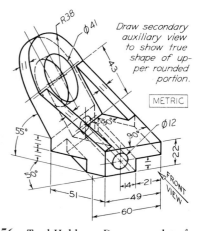

Fig. 10.56 Tool Holder. Draw complete front view, and primary and secondary auxiliary views as indicated (Layout B–4 or A3–4 adjusted).*

*If dimensions are required, study §§13.1–13.25. Use metric or decimal-inch dimensions as assigned by the instructor.

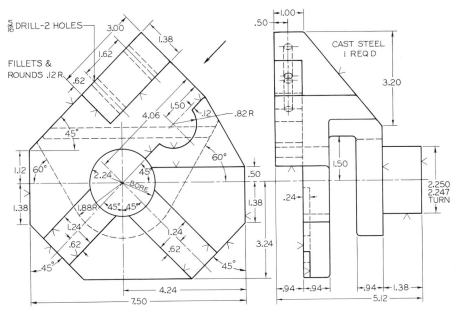

Fig. 10.57 Box Tool Holder for Turret Lathe.
Given: Front and right-side views.
Required: Front and left-side views, and complete auxiliary view as indicated by arrow (Layout C–4).*

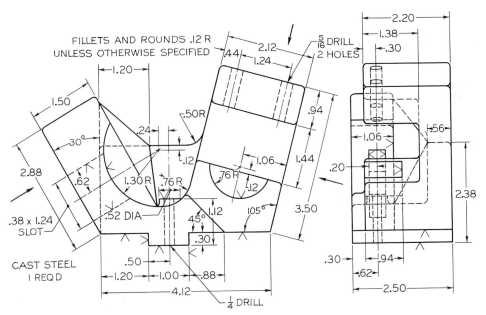

Fig. 10.58 Pointing Tool Holder for Automatic Screw Machine.
Given: Front and right-side views.
Required: Front view and three partial auxiliary views (Layout C–4).*

*Layout A2–4 may be used. If dimensions are required, study §§13.1–13.25. Use metric or decimal-inch dimensions as assigned by the instructor.

341

Revolutions

To obtain an auxiliary view, the observer changes position with respect to the object, as shown by the arrow in Fig. 11.1 (a). The auxiliary view shows the true size and shape of surface A. Exactly the same view of the object can also be obtained by moving the object with respect to the observer, as shown at (b). Here the object is revolved until surface A appears in its true size and shape in the right-side view.

11.1 Axis of Revolution

In Fig. 11.1 (b) the axis of revolution is assumed perpendicular to the frontal plane of projection. Note that the view in which the axis of revolution appears as a point (in this case the front view) revolves but does not change shape and that in the views in which the axis is shown as a line in true length the *dimensions of the object parallel to the axis do not change*.

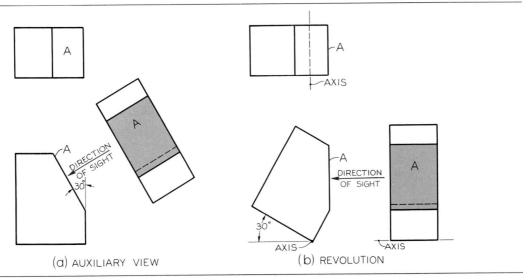

Fig. 11.1 Auxiliary View and Revolution Compared.

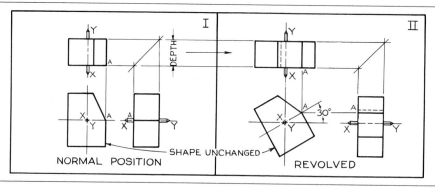

Fig. 11.2   Primary Revolution About an Axis Perpendicular to Frontal Plane.

To make a revolution drawing, the view on the plane of projection that is perpendicular to the axis of revolution is drawn first, since it is the only view that remains unchanged in size and shape. This view is drawn revolved either *clockwise* or *counterclockwise* about a point that is the end view, or point view, of the axis of revolution. This point may be assumed at any convenient point on or outside the view. The other views are then projected from this view.

The axis of revolution is usually considered perpendicular to one of the three principal planes of projection. Thus, an object may be revolved about an axis perpendicular to the horizontal, frontal, or profile planes of projection, and the views drawn in the new positions. Such a process is called a *primary revolution*. If this drawing is then used as a basis for another revolution, the operation is called *successive revolutions*. Obviously, this process may be continued indefinitely, which reminds us of the "chain reaction" in successive auxiliary views, Fig. 10.21.

the frontal plane of projection, and during the revolution all points of the object describe circular arcs parallel to that plane. The axis may pierce the object at any point or may be exterior to it. In space II, the front view is drawn *revolved* (but not changed in shape) through the angle desired (30° in this case), and the top and side views are obtained by projecting from the front view. The *depth* of the top view and the side view is found by projecting from the top view of the first unrevolved position (space I) *because the depth, since it is parallel to the axis, remains unchanged.* If the front view of the revolved position is drawn directly without first drawing the normal unrevolved position, the depth of the object, as shown in the revolved top and side views, may be drawn to known dimensions. No difficulty should be encountered by the student who understands how to obtain projections of points and lines, §7.6.

Note the similarity between the top and side views in space II of Fig. 11.2 and some of the auxiliary views of Fig. 10.7 (e).

11.2 Revolution About Axis Perpendicular to Frontal Plane

A primary revolution is illustrated in Fig. 11.2. An imaginary axis XY is assumed, about which the object is to revolve to the desired position. In this case the axis is selected perpendicular to

11.3 Revolution About Axis Perpendicular to Horizontal Plane

A revolution about an axis perpendicular to the horizontal plane of projection is shown in Fig. 11.3. An imaginary axis XY is assumed perpendicular to the top plane of projection, the top

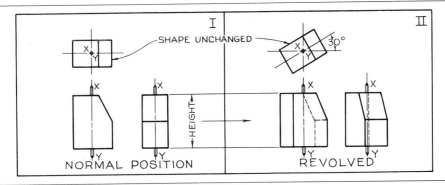

Fig. 11.3   Primary Revolution About an Axis Perpendicular to Horizontal Plane.

view is drawn revolved (but not changed in shape) to the desired position (30° in this case), and the other views are obtained by projecting from this view. During the revolution, all points of the object describe circular arcs parallel to the horizontal plane. The *heights* of all points in the front and side views in the revolved position remain unchanged, since they are measured parallel to the axis, and may be drawn by projecting from the initial front and side views of space I.

Note the similarity between the front and side views in space II of Fig. 11.3 and some of the auxiliary views of Fig. 10.8 (c).

11.4   Revolution About Axis Perpendicular to Profile Plane

A revolution about an axis XY perpendicular to the profile plane of projection is illustrated in Fig. 11.4. During the revolution, all points of the object describe circular arcs parallel to the profile plane of projection. The widths of the top and front views in the revolved position remain unchanged, since they are measured parallel to the axis, and may be obtained by projection from the top and front views of space I, or may be set off by direct measurement.

Note the similarity between the top and front views in space II of Fig. 11.4 and some of the auxiliary views of Fig. 10.9 (c).

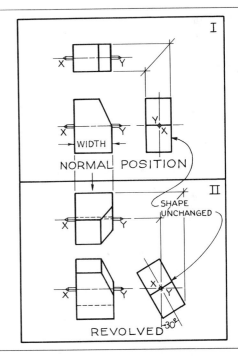

Fig. 11.4   Primary Revolution About an Axis Perpendicular to Profile Plane.

11.5   Successive Revolutions

It is possible to draw an object in an infinite number of revolved positions by making successive revolutions. Such a procedure, Fig. 11.5,

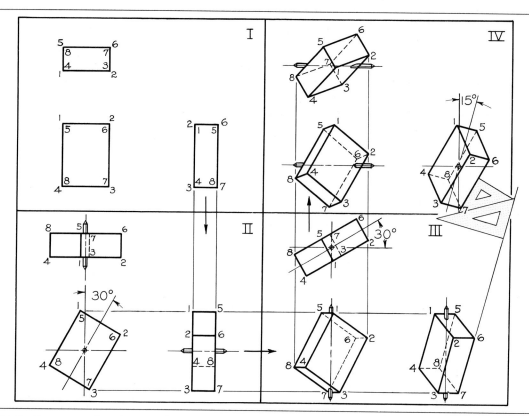

Fig. 11.5   Successive Revolutions of a Prism.

limited to three or four stages, offers excellent practice in multiview projection. While it is possible to make several revolutions of a simple object without the aid of a system of numbers, it is absolutely necessary in successive revolutions to assign a number or a letter to every corner of the object. See §7.6.

The numbering or lettering must be consistent in the various views of the several stages of revolution. Figure 11.5 shows four sets of multiview drawings numbered I, II, III, and IV, respectively. These represent the same object in different positions with reference to the planes of projection.

In space I, the object is represented in its normal position, with its faces parallel to the planes of projection. In space II, the object is

represented after it has been revolved clockwise through an angle of 30° about an axis perpendicular to the frontal plane. The drawing in space II is placed under space I so that the side view, whose width remains unchanged, can be projected from space I to space II as shown.

During the revolution, all points of the object describe circular arcs parallel to the frontal plane of projection and remain at the same distance from that plane. The side view, therefore, may be projected from the side view of space I and the front view of space II. The top view may be projected in the usual manner from the front and side views of space II.

In space III, the object is taken as represented in space II and is revolved counterclockwise through an angle of 30° about an axis perpendicular to the horizontal plane of projection. During the revolution, all points describe *horizontal* circular arcs and remain at the same distance from the horizontal plane of projection. The top view is copied from space II but is revolved through 30°. The front and side views are obtained by projecting from the front and side views of space II and from the top view of space III.

In space IV, the object is taken as represented in space III and is revolved clockwise through 15° about an axis perpendicular to the profile plane of projection. During the revolution, all points of the object describe circular arcs parallel to the profile plane of projection and remain at the same distance from that plane. The side view is copied, §5.28, from the side view of space III but revolved through 15°. The front and top views are projected from the side view of space IV and from the top and front views of space III.

Another convenient method of copying a view in a new revolved position is to use tracing paper as described in §5.29. Either a tracing can be made and transferred by rubbing, or the prick points may be made and transferred, as shown.

In spaces III and IV of Fig. 11.5, each view is an axonometric projection, §18.2. An isometric projection can be obtained by revolution, as shown in Fig. 18.3, and a dimetric projection, §18.28, can be constructed in a similar manner. If neither an isometric nor a dimetric projection is specifically sought, the successive revolution will produce a trimetric projection, §18.30, as shown in Fig. 11.5.

11.6   Revolution of a Point— Normal Axis

Examples of the revolution of a point about a straight-line axis are often found in design problems that involve pulleys, gears, cranks, linkages, and so on. For example, in Fig. 11.6 (a), as the disk is revolved, point 3 moves in a circular path lying in a plane perpendicular to the axis 1–2. This relationship is represented in the two views at (b). Note in this instance that the axis

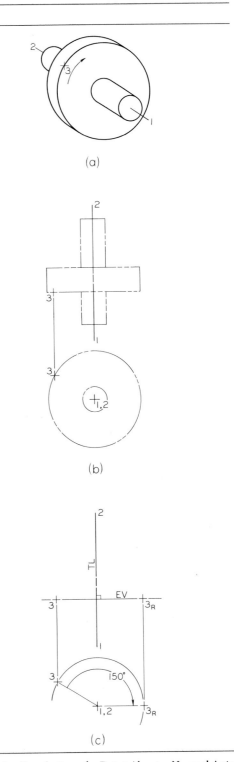

Fig. 11.6   Revolution of a Point About a Normal Axis.

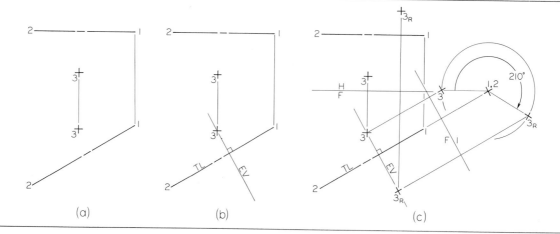

Fig. 11.7 Revolution of a Point About an Inclined Axis.

is normal or perpendicular to the frontal plane of projection, resulting in a front view that shows a point view of the axis and a true-size view of the circular path of revolution for point 3. The top view shows the path of revolution in edge view and perpendicular to the true-length view of the axis. Similar two-view relationships would occur if the axis were perpendicular or normal to either the horizontal or profile planes of projection.

The clockwise revolution through 150° for point 3 is illustrated at (c).

11.7 Revolution of a Point—Inclined Axis

In Fig. 11.7 (a) the axis of revolution for point 3 is positioned parallel to the frontal plane and inclined to the horizontal and profile projection planes. Since the axis 1–2 is true length in the front view, the edge view of the path of revolution can be located as at (b). In order to establish the circular path of revolution for point 3, an auxiliary view showing the axis in point view is required as at (c). The required revolution of point 3 (in this case, 210°) is now performed in this circular view. The revolved position of the point is projected to the given front and top views, as shown.

Note the similarity of the relationships of the front view and auxiliary view and the constructions shown in Fig. 11.6 (c).

11.8 Revolution of a Point—Oblique Axis

In Fig. 11.8 (a), the axis of revolution for point 3 is oblique to all principal planes of projection and, therefore, is shown neither in true length nor as a point view in the top, front, or profile views. To establish the necessary true length and point view of the axis 1–2 in adjacent views, two successive auxiliary views are required, as shown at (b). The required revolved position of point 3 can now be located and then projected back to complete the given front and top views.

11.9 Revolution of a Line

The procedure for the revolution of a line about an axis is very similar to that required for the revolution of a point, §11.6. All points on a line must revolve through the same angle, or the revolved line becomes altered.

In Fig. 11.9 (a), the line 1–2 is to be revolved through 150° about the inclined axis 3–4.

Since the axis 3–4 is given in true length in the top view, an auxiliary view is required to

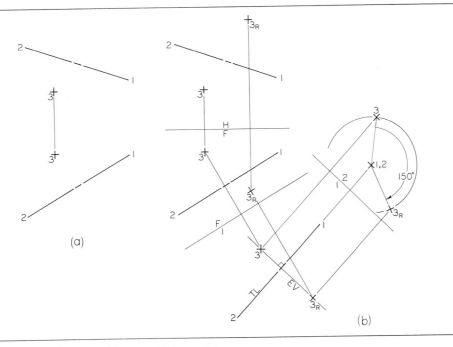

Fig. 11.8   Revolution of a Point About an Oblique Axis.

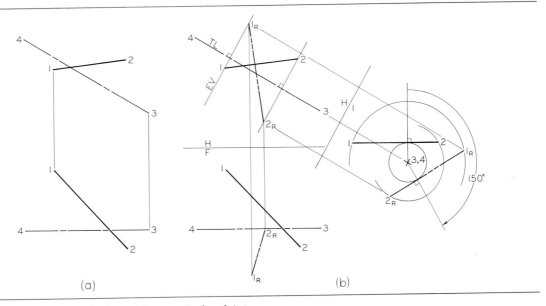

Fig. 11.9   Revolution of a Line About an Inclined Axis.

provide a point view of the axis, as shown in Fig. 11.9 (b). The necessary revolution can then be made about point view 3–4. In order to insure that all points on the line rotate through the same number of degrees, note that a construction circle tangent to line 1–2 is drawn and that a perpendicular through the tangency point becomes the reference for measuring the angle of rotation. The circular arc paths for points 1 and 2 locate the points 1_R and 2_R, as the revolved position of the line is drawn perpendicular to the radial line subtending the 150° arc of revolution, and tangent to the smaller circle. The alternate-position line is used to distinguish the revolved-position line from the original given line.

11.10   True Length of a Line—Revolution Method

If a line is parallel to one of the planes of projection, its projection on that plane is equal in length to the line, Fig. 7.20. In Fig. 11.10 (a), the element AB of the cone is oblique to the planes of projection; hence its projections are foreshortened. If AB is revolved about the axis of the cone until it coincides with either of the contour elements, for example AB_R, it will be shown in its true length in the front view for it will then be parallel to the frontal plane of projection.

Likewise, at (b), the edge of the pyramid CD is shown in its true length CD_R when it has been revolved about the axis of the pyramid until it is parallel to the frontal plane of projection. At (c), the line EF is shown in its true length at EF_R when it has been revolved about a vertical axis until it is parallel to the frontal plane of projection.

The true length of a line may also be found by constructing a right triangle or a true-length diagram, as shown at (d), whose base is equal to the top view of the line and whose altitude is the difference in elevation of the ends. The hypotenuse of the triangle is equal to the true length of the line.

In these cases the lines are revolved until parallel to a plane of projection. The true length of a line may also be found by leaving the line stationary but shifting the position of the observer—that is, the method of auxiliary views, §10.18.

11.11   True Size of a Plane Surface—Revolution Method

If a surface is parallel to one of the planes of projection, its projection on that plane is true size, Fig. 7.19. In Fig. 11.11 (a), the inclined surface 1–2–3–4 is foreshortened in the top and side views and appears as a line in the front view. Line 2–3 is taken as the axis of revolution, and the surface is revolved clockwise in the front view to the position 4_R–3 and projected to the side view at 4_R–1_R–2–3, which is the true size of the surface. In this case the surface was revolved until parallel to the profile plane of projection.

At (b), triangular surface 1–2–3 is revolved until parallel to the horizontal plane of projection so that the surface appears true size in the top view, as shown.

At (c), the true size of the oblique surface 1–2–3–4–5 cannot be found by a simple primary revolution. The true size can be found by two successive revolutions or by a combination of an auxiliary view and a primary revolution. The latter is shown at (c). First, draw an auxiliary view that will show the edge view (EV) of the plane. See Fig. 10.22. Second, revolve the edge view of the surface until it is parallel to the folding line F/1, as shown. All points in the front view, except those in the axis of revolution line 4–5, will describe circular arcs parallel to the reference plane F/1. These arcs will appear in the front view as lines parallel to the folding line, such as 2–2_R and 3–3_R. The true size of the surface is found by connecting the points with straight lines.

11.12   Revolution of Circles

As shown in Fig. 5.50 (a) to (c), a circle, when viewed obliquely, appears as an ellipse. In that case the coin is revolved by the fingers. The geometric construction of this revolution is shown in Fig. 11.12 (a). In the front view the circle appears as ACBD, and in the side view as line CD,

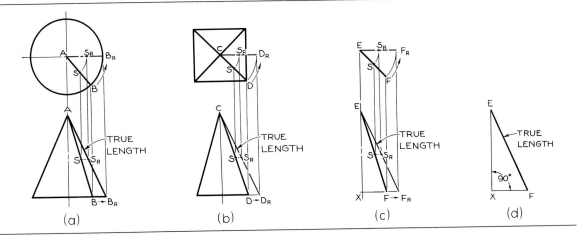

Fig. 11.10   True Length of a Line—Revolution Method.

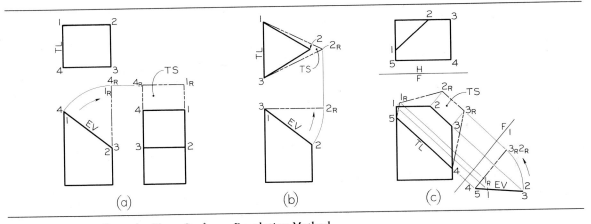

Fig. 11.11   True Size of a Plane Surface—Revolution Method.

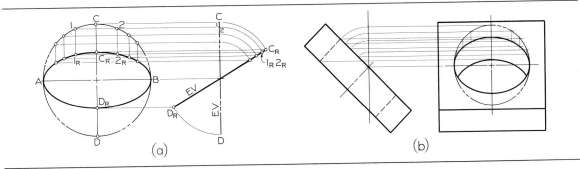

Fig. 11.12   Revolution of a Circle.

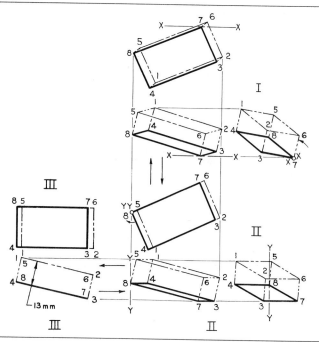

Fig. 11.13   Counterrevolution of a Prism.

which is really the edge view of the plane containing the circle. In the side view, CD (the side view of the circle) is revolved through any desired acute angle to C_RD_R.

To find points on the ellipse, draw a series of horizontal lines across the circle in the front view. Each line will cut the circle at two points, as 1 and 2. Project these points across to the vertical line representing the unrevolved circle; then, revolve each point and project horizontally to the front view to establish points on the ellipse. Plot as many points as necessary to secure a smooth curve.

An application of this construction to the representation of a revolved object with a large hole is shown at (b).

11.13   Counterrevolution

The reverse procedure to revolution is *counterrevolution*. For example, if the three views of space II in Fig. 11.2 are given, the object can be drawn in the unrevolved position of space I by

counterrevolution. The front view is simply counterrevolved back to its normal upright position, and the top and side views are drawn as shown. Similarly, in Fig. 11.5, the object may be counterrevolved from its position of space IV to its unrevolved position of space I by simply reversing the process.

In practice, it sometimes becomes necessary to draw the views of an object located on or parallel to a given oblique surface. In such an oblique position, it is very difficult to draw the views of the object because of the foreshortening of lines. The work is greatly simplified by counterrevolving the oblique surface to a simple position, completing the drawing, and then revolving to the original given position.

An example is shown in Fig. 11.13. Assume that the oblique surface 8–4–3–7 (three views in space I) is given and that it is required to draw the three views of a prism 13 mm high, having the given oblique surface as its base. Revolve the surface about any horizontal axis XX, perpendicular to the side view, until the edges 8–4 and 3–7 are horizontal, as shown in space II. Then re-volve the surface about any vertical axis YY, which appears as a point in the top view, until the edges 8–7 and 4–3 are parallel to the frontal plane, as shown in space III. In this position the given surface is perpendicular to the frontal plane, and the front and top views of the re-quired prism can be drawn, as shown by phan-tom lines in the figure, because the edges 4–1 and 3–2, for example, are parallel to the frontal plane and, therefore, are shown in their true lengths, 13 mm. When the two views in space III have been drawn, counterrevolve the object from III to II and then from II to I to find the required views of the given object in space I.

11.14   Computer Graphics

Computer graphics programs provide the user with simple and fast ways of revolving objects about any desired axis. Successive revolutions can be easily accomplished using CAD. Com-puter-generated revolutions also enable a drafter to readily depict circular features, which appear elliptical when viewed obliquely, Fig. 11.14.

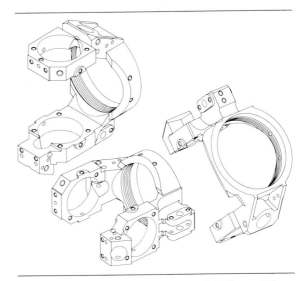

Fig. 11.14   Pictorial Drawings With Different View-ing Angles Created by Using Computervision De-signer System   The system provides complete flexi-bility to manipulate the original illustration to display any desired viewing orientation. *Courtesy of Computer-vision Corporation, a subsidiary of Prime Computer, Inc.*

REVOLUTION PROBLEMS

In Figs. 11.15–11.19 are problems covering primary revolutions, successive revolutions, and counterrevolutions.

Since many of the problems in this chapter are of a general nature, they can also be solved on most computer graphics systems. If a system is available, the instructor may choose to assign specific problems to be completed by this method.

Additional problems, in convenient form for solution, are available in *Technical Drawing Problems,* Series 1, by Giesecke, Mitchell, Spencer, Hill, Dygdon, and Novak; *Technical Drawing Problems,* Series 2, by Spencer, Hill, Dygdon, and Novak; and *Technical Drawing Problems,* Series 3, by Spencer, Hill, Dygdon, and Novak; designed to accompany this text and published by Macmillan Publishing Company.

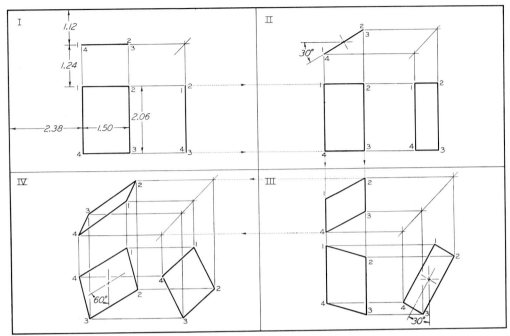

Fig. 11.15 Using size B or A3 sheet, divide working area into four equal parts, as shown. Draw given views of rectangle, and then the primary revolution in space II, followed by successive revolutions in spaces III and IV. Number points as shown. Omit dimensions. Use Form 3 title box.

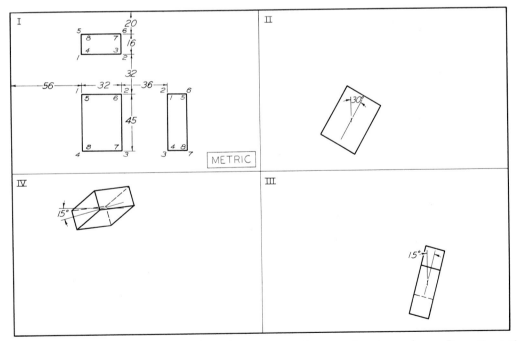

Fig. 11.16 Using size B or A3 sheet, divide working area into four equal parts, as shown. Draw given views of prism as shown in space I; then draw three views of the revolved prism in each succeeding space, as indicated. Number all corners. Omit dimensions. Use Form 3 title box.

355

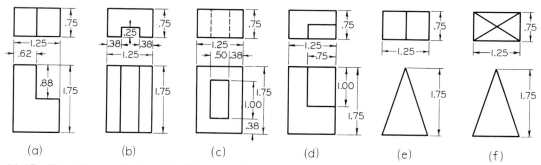

(a)   (b)   (c)   (d)   (e)   (f)

Fig. 11.17 Using Layout B–4 or A3–4 (adjusted) sheet, divide into four equal parts as in Fig. 11.15. In the upper two spaces, draw a simple revolution as in Fig. 11.2, and in the lower two spaces, draw a simple revolution as in Fig. 11.3, but for each problem use a block assigned from Fig. 11.17.

Alternative Assignment: Using Layout B–4 or A3–4 (adjusted) sheet, divide into four equal parts as in Fig. 11.15. In the two left-hand spaces, draw a simple revolution as in Fig. 11.4, but use an object assigned from Fig. 11.17. In the two right-hand spaces, draw another simple revolution as in Fig. 11.4, but use a different object taken from Fig. 11.17 and revolve through 45° instead of 30°.

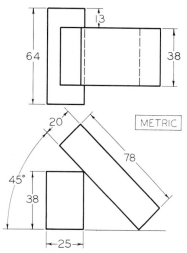

Fig. 11.18 Using Layout A–2 or A–3 or Layout A4–2 or A4–3 (adjusted), draw three views of the blocks but revolved 30° clockwise about an axis perpendicular to the top plane of projection. Do not change the relative positions of the blocks.

Fig. 11.19 *opposite* Use Layout A–1 or A4–1 (adjusted), and divide the working area into four equal areas for four problems per sheet to be assigned by the instructor. Data for the layout of each problem are given by a coordinate system in metric dimensions. For example, in Prob. 1, point 1 is located by the scale coordinates (28 mm, 38 mm, 76 mm). The first coordinate locates the front view of the point from the left edge of the problem area. The second one locates the front view of the point from the bottom edge of the problem area. The third one locates either the top view of the point from the bottom edge of the problem area or the side view of the point from the left edge of the problem area. Inspection of the given problem layout will determine which application to use.

1. Revolve clockwise point 1(28, 38, 76) through 210° about the axis 2(51, 58, 94)–3(51, 8, 94).
2. Revolve point 3(41, 38, 53) about the axis 1(28, 64, 74)–2(28, 8, 74) until point 3 is at the farthest distance behind the axis.
3. Revolve point 3(20, 8, 84) about the axis 1(10, 18, 122)–2(56, 18, 76) through 210° and to the rear of line 1–2.
4. Revolve point 3(5, 53, 53) about the axis 1(10, 13, 71)–2(23, 66, 71) to its extreme position to the left in the front view.

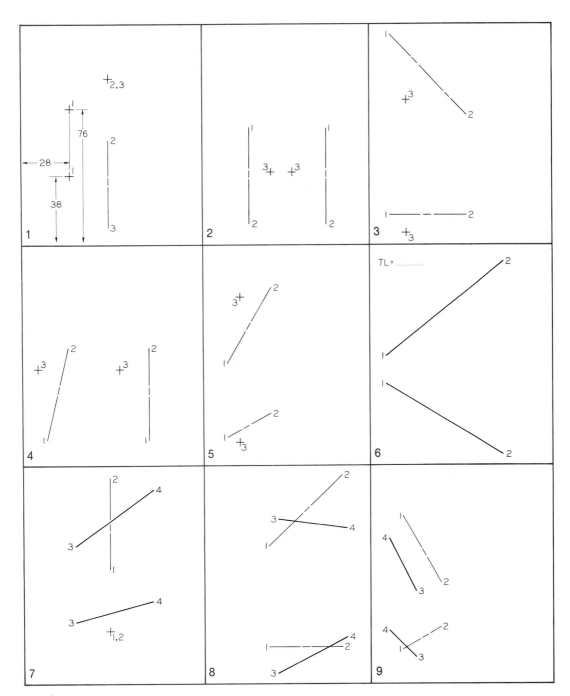

Fig. 11.19

5. Revolve point 3(15, 8, 99) about the axis 1(8, 10, 61)–2(33, 25, 104) through 180°.
6. By revolution find the true length of line 1(8, 48, 64)–2(79, 8, 119). Scale: 1:100.
7. Revolve line 3(30, 38, 81)–4(76, 51, 114) about axis 1(51, 33, 69)–2(51, 33, 122) until line 3–4 is shown true length and below the axis 1–2. Scale: 1:20.
8. Revolve line 3(53, 8, 97)–4(94, 28, 91) about the axis 1(48, 23, 81)–2(91, 23, 122) until line 3–4 is in true length and above the axis.
9. Revolve line 3(28, 15, 99)–4(13, 30, 84) about the axis 1(20, 20, 97)–2(43, 33, 58) until line 3–4 is level and above the axis.

357

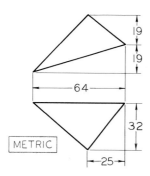

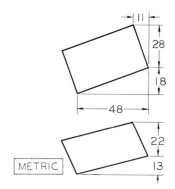

Fig. 11.20 Using Layout B–3 or A3–3, draw three views of a right prism 38 mm high that has as its lower base the triangle shown above. See §11.13.

Fig. 11.21 Using Layout B–3 or A3–3, draw three views of a right pyramid 51 mm high, having as its lower base the parallelogram shown above. See §11.13.

CHAPTER 12

Manufacturing Processes

BY J. GEORGE H. THOMPSON*
AND JOHN GILBERT MCGUIRE†
REVISED BY STEPHEN A. SMITH‡

The test of the usefulness of any working drawing is whether the object described can be satisfactorily produced without further information than that furnished on the drawing. The drawing must give information as to shape, size, material, and finish and, where necessary, indicate the manufacturing processes required to produce the desired object.

The purpose of this chapter is to provide beginning engineers with some information about certain fundamental terms and processes and to assist them in using this information on their drawings. The drafter and engineer in any organization must have a thorough knowledge of manufacturing processes and methods before they can properly indicate on drawings the machining operations, heat-treatment, finish, and the accuracy desired on each part.

12.1 Production Processes

A manufacturing department starts with what might be called *raw stock* and modifies this until it agrees with the detail drawing. The shape of raw stock usually has to be altered.

Changing the shape and size of the material of which a part is being made requires one or more of the following processes: (1) removing part of the original material, (2) adding more material, and (3) redistributing original material. Cutting, such as turning on a lathe, punching holes by means of a power press, or cutting with a laser system, removes material. Welding, braz-ing, soldering, metal spraying, and electrochem-ical plating add material. Forging, pressing, drawing, extruding, spinning, and plastics processing redistribute material.

12.2 Manufacturing Methods and the Drawing

Before preparing a drawing for the production of a part, the drafter should consider what manu-facturing processes are to be used. These processes will determine the representation of the detailed features of the part, the choice of

*Late Professor Emeritus of Mechanical Engineering, Texas A&M University.
†Assistant Dean of Engineering Emeritus, Texas A&M University.
‡Manager, Product Design and Development, Packaging Corporation of America, A Tenneco Company.

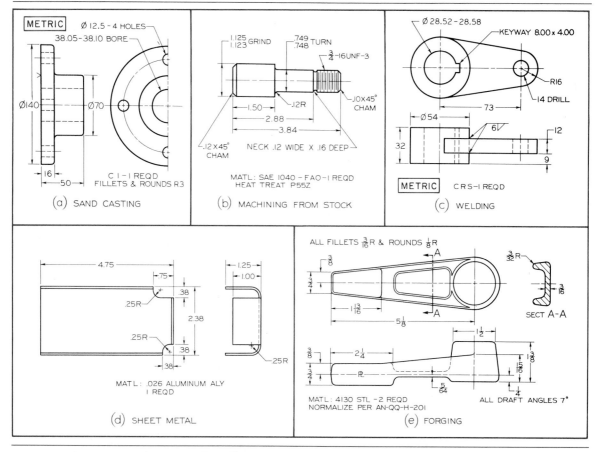

Fig. 12.1   Comparison of Drawings for Different Manufacturing Processes.

dimensions, and the machining or processing accuracy. Principal types of metal forming are (1) casting, (2) machining from standard stock, (3) welding, (4) forming from sheet stock, and (5) forging. A knowledge of these processes, along with a thorough understanding of the intended use of the part, will help determine some basic manufacturing processes. Drawings that reflect these manufacturing methods are shown in Fig. 12.1.

In sand casting, Fig. 12.1 (a), all cast surfaces remain rough textured, with all corners filleted or rounded. Sharp corners indicate at least one of the surfaces is finished, §7.34, and finish marks are shown on the edge view of the finished surface. See §§13.16 and 13.17.

In drawings of parts machined from standard stock, Fig. 12.1 (b), most surfaces are represented as machined. In some cases, as on shafting, the surface existing on the raw stock is often accurate enough without further finishing. Corners are usually sharp, but fillets and rounds are machined when necessary. For example, an interior corner may be machined with a radius to provide greater strength.

On welding drawings, (c), the several pieces are cut to size, brought together, and then welded. Welding symbols, Appendix 32, are used to indicate the welds required. Generally, there are no fillets and rounds other than those generated during the welding process itself. Certain surfaces may be machined after welding or, in some cases, before welding. Notice that lines are shown where the separate pieces are joined.

On sheet-metal drawings, (d), the thickness of material is uniform and is usually given in the material specification note rather than by a dimension on the drawing. Bend radii and bend reliefs at corners are specified according to standard practice. For dimensions, either the decimal-inch or metric dimensioning systems may be used. Allowances of extra material for joints may be required when the flat blank size is being determined.

For forged parts, §12.27, separate drawings are usually made for the diemaker and for the machinist. Thus, a forging drawing, (e), provides only the information to produce the forging, and the dimensions given are those needed by the diemaker. All corners are rounded and filleted and are so shown on the drawing. The draft is drawn to scale, and is usually specified by degrees in a note.

12.3   Sand Casting

Although a number of different casting processes are used, sand molds are the most common. Sand molds are made by ramming sand around a pattern, and then carefully removing it, leaving a cavity that exactly matches the pattern to receive the molten metal, as shown in Fig. 12.2 (a). The sand is contained in a two-part box called a *flask,* (b), the upper part of which is called the *cope* and the lower the *drag.* For more complex work, one or more intermediate boxes, called *cheeks,* may be introduced between the cope and the drag. The pattern must be of such a shape that it will "pull away" from both the cope and the drag. The plane of separation of the two halves is the *parting line* on the pattern.

On each side of the parting line the pattern must be tapered slightly to permit the withdrawal of the pattern from the sand, unless a

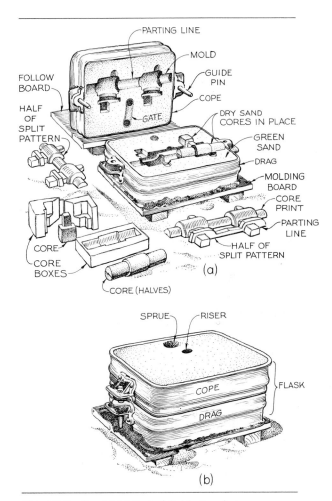

Fig. 12.2   Sand Molding of Table Bracket.

segmented pattern is used. This taper is known as *draft.* Although draft is usually not shown, and dimensions are not given for it on the working drawing, the design must be such that draft can be properly built into the pattern by the patternmaker. When large quantities of castings are to be made, a matchplate pattern is often used. In this process, the pattern is split along the parting line, mounted, and aligned on both sides of an aluminum plate that fits between the cope and the drag. Withdrawal of the matchplate pattern is facilitated by the guide pins on the cope, and the integrity of the mold is protected despite rapid withdrawal of the pattern.

A *sprue stick,* or round peg, is placed in a position during the ramming process and then removed to leave a vertical hole, called the *sprue,* through which the metal may be poured.

The passageway for the metal from the sprue to the mold is called the *gate,* Fig. 12.2.

On some molds another hole, known as the *riser,* is provided to allow gases to escape and to provide a reservoir of metal that feeds back to the casting as it cools and shrinks.

Since shrinkage occurs when metal cools, patterns are made slightly oversize. The patternmaker accomplishes this by increasing the pattern size by predetermined amounts that depend on the kind of metal being used in the casting. Allowance for shrinkage is not shown on the working drawing but is taken care of entirely by the pattern shop. The patternmaker must refer to the drawing to identify the kind of metal to be used in the casting.

A *core print* is a projection added to a pattern for the purpose of forming a cavity in a mold into which a corresponding portion of a *core* will rest, as shown in Fig. 12.2 thus forming an anchor to hold the core in place. Cores are made of sand and are used to provide certain hollow portions of the casting. The most common use of a core is to extend through a casting to form a *cored hole.* When it is necessary to form the sand into shapes that would ordinarily not permit the necessary adhesion and strength, or in instances where the shape of the casting would interfere with the removal of the pattern, a *dry sand core* is used. Dry sand cores, Fig. 12.2, are made by ramming a prepared mixture of sand and a binding substance into a *core box;* the core is then removed and baked in a *core oven* to make it sufficiently rigid. A *green sand core* (not baked) is used when it is practical to make the core along with the mold, as for example the central hole shown in Fig. 12.3.

Because of shrinkage and draft, small holes are better drilled in the casting, and large holes are better cored (cast-in) and then bored.

12.4   The Patternmaker and the Drawing

The patternmaker receives the working drawing showing the object in its completed state, including all dimensions and finish marks. Usually the same drawing is used by the patternmaker and the machinist; hence, it should contain all dimensions and notes needed by both, as shown in Fig. 12.3. Some companies follow the practice of dimensioning a copy of the drawing in colored pencil with the patternmaker's information. Pattern dimensions, §13.27, typically need to be accurate only to within $\frac{1}{32}''$ or $\frac{1}{16}''$. More critical tolerances are required for machining, however.

Finish marks, §13.17, are as important to the patternmaker as to the machinist because additional material must be provided on each surface that will eventually be machined. For small and medium-sized castings, 1.5 mm ($\frac{1}{16}''$ approximate) to 3 mm ($\frac{1}{8}''$ approximate) is usually sufficient; larger allowances are made if there is a probability of distortion or warping. On the Flange, Fig. 12.3, it is necessary for the patternmaker to provide material for finish on all surfaces, since the note indicates finish all over (FAO).

Sometimes patternmakers work directly from the drawing; in other cases, they may find it desirable to make their own pattern layout. Except for simple objects, it is common practice for the patternmaker to scribe full-sized (to shrink rule) pattern layout drawings on white pine boards or on a piece of steel or aluminum plate. Wood or metal is used instead of paper because it is more durable and because paper stretches and shrinks excessively. The pattern is then checked against this pattern layout drawing on which draft, shrinkage, core print dimensions, and other such information are clearly shown.

12.5   Fillets and Rounds

Fillets (inside rounded corners) and rounds (outside rounded corners) must be included in the pattern, in order to provide for maximum strength as well as a pleasing appearance in the

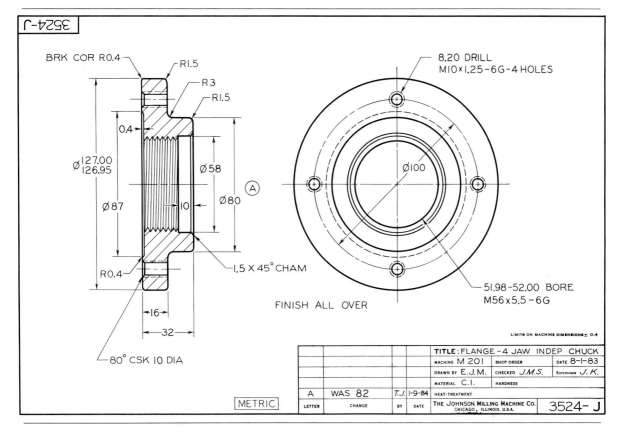

Fig. 12.3 A Detail Working Drawing.

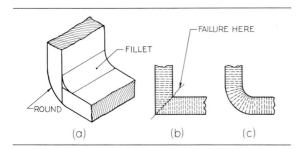

Fig. 12.4 Fillets and Rounds.

finished casting, Fig. 12.4. Crystals of cooling metal tend to arrange themselves perpendicular to the exterior surfaces as indicated at (b) and (c). If the corners of a casting are rounded as shown at (c), a much stronger casting results. It has been demonstrated that the load that a part can safely carry may sometimes be doubled simply by increasing the radius of a fillet. A fillet should therefore have as large a radius as possible.

Rounds are constructed by rounding off the corners of the pattern by sanding, planing, or turning, while fillets are constructed of preformed leather, wax, or wood.

For some cylindrical patterns it is possible to form on the lathe the proper radius for the fillets and rounds as an integral part of the wood pattern.

All fillets and rounds should be shown on the drawing, either freehand or drawn to scale with the use of a compass or with a circle template. For a general discussion of fillets and rounds from the standpoint of representation on the drawing, see §§7.34–7.36. For dimensioning of fillets and rounds, see §13.16.

12.6   Construction of the Pattern

After having examined the drawing and determined such things as material and finish, the patternmaker constructs the pattern of a durable and dimensionally stable wood (usually basswood or mahogany) or of acetate (a plastic), using dimensions given on the working drawing or pattern layout drawing and making proper allowances for metal shrinkage, machining, and draft.

If a quantity of molds were to be made from the pattern, the pattern stock would probably be glued up from several pieces of wood with the grain arranged to minimize warpage. For a large number of molds, a *master pattern* would be constructed of wood or acetate and a duplicate working pattern cast in aluminum or epoxy for actual use in the foundry to protect the master pattern from excessive wear. In this case the patternmaker would employ "double" shrinkage, one for aluminum or epoxy and one for the material to be cast.

12.7   The Foundry

After the pattern has been completed, it is sent to the foundry for casting. Sand is rammed around the pattern, and then the pattern is carefully removed, leaving a cavity for the molten metal.

After the cope has been returned to position on the drag, molten metal is poured through the sprue (hole leading to cavity) into the mold. After sufficient cooling time, the casting is removed from the broken mold.

Although sand molds are the least expensive to make, sometimes plaster of Paris molds are used when it is desired to produce castings with smoother surfaces. Stainless steel, for example, can be successfully cast in plaster of Paris molds.

The *investment casting process* (also called the "lost wax process") produces castings of great dimensional accuracy and detail, and cast sections 0.38 mm (.015″) in thickness may be produced. In this process, the wax pattern is melted after the mold has been formed. Thus, none of the details of the mold are injured by the removal of the pattern, and for that reason shapes may be molded that would be impossible if a pattern had to be drawn out of the mold.

A casting process of importance is *centrifugal casting*. Molten metal is poured into a mold that is already rotating and that may continue to rotate until the metal has solidified. Centrifugal force produces a less porous casting than that produced in a sand mold. This process is extensively used in the manufacture of cast iron pipe, and for steel gears and disks.

Die casting is the process of forcing molten metal into metallic dies under pressure. These castings are accurate in size and shape and possess surfaces superior in appearance and accuracy to those produced by other casting processes. Die castings, as a rule, require little or no machining. Although die castings generally cannot be heat-treated or put into service at high temperatures, this process is the fastest of the casting processes and consequently the least expensive for mass-produced items, such as aluminum automobile engine blocks, carburetors, door handles, and home appliances.

12.8   Powder Metallurgy

Powder metallurgy is an old process (dating back to 300 B.C.), but today it is an important technique of producing metallic parts from powders of a single metal, of two or more metals, or of a combination of metals and nonmetals. It consists of successive steps of mixing powders, compressing them at high pressures (30 or 40 tons per square inch) into a preliminary shape, and heating them at high temperatures, but below the melting point of the principal metal. This process bonds the individual particles and produces a part with smooth surfaces that are quite accurate in dimension.

Some of the advantages of this process include the elimination of scrap, and of machining, suitability to mass production, better control of composition and structure of a part, and the abil-

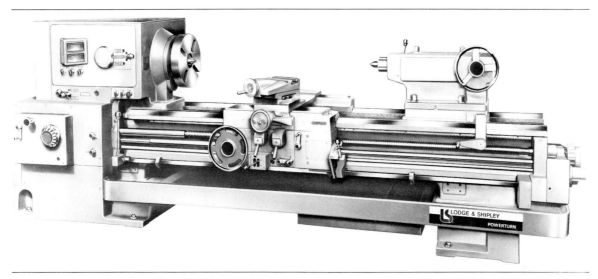

Fig. 12.5   Engine Lathe. *Courtesy of Lodge & Shipley.*

ity to produce parts made from a mixture of metals and nonmetals. Some of the disadvantages include high cost of dies, lower physical properties, higher cost of raw materials, and the limitation on design of a part. Powders when pressed do not flow around corners, nor do they transmit pressures as a liquid; therefore, a part designed for casting will normally require redesigning in order to be produced by powder metallurgy.

12.9   The Machinist and the Drawing

The casting is sent from the foundry to the machine shop for machining. The drawing is used in the machine shop to obtain information for the machining operations. In some cases, a special drawing is made for machining, and, in rare instances, a drawing is produced for each machining operation. Also, in some production facilities, a special set of instructions indicates routing among machines and individual operations for each particular machine. As a rule, however, working drawings used in the preceding shops are also used by the machinist to bring the product to completion.

It is important to note that machined surfaces are usually held to much greater accuracy than are cast surfaces. Dimensions should reflect the desired accuracy for machining.

12.10   Machine Tools

Some of the more common machine tools are the engine lathe, drill press, milling machine, shaper, planer, grinding machine, and boring mill. Brief descriptions of these machines follow. See also §12.29 for examples of how these machine tools can be blended into automated processing.

12.11   Engine Lathe

The engine lathe, Fig. 12.5, is one of the most versatile machines used in the machine shop, and on it are performed such operations as turning, boring, reaming, facing, threading, and knurling.

The workpiece is held in the lathe in a chuck (essentially a rotating vise). The cutting tool is fastened in the tool holder of the lathe and fed mechanically into the work as required.

12.12   Drill Press

The drill press, Fig. 12.6, is one of the most frequently used machine tools. Some of the operations that may be performed on this machine are drilling, reaming, boring, tapping, spot facing, counterboring, and countersinking.

The *radial* drill press, shown in Fig. 12.7, with its adjustable head and spindle, is very versatile and is especially suitable for large work.

A *multiple-spindle* drill press supports a number of spindles driven from the same shaft and is used in mass production. A multiple-spindle drilling head, Fig. 12.8, mounts on a conventional drill press to enable the simultaneous drilling of up to 21 holes.

12.13   Milling Machine

On the milling machine, Fig. 12.9, cutting is accomplished by feeding the work into a rotating cutter. Milling machines are economical metal removers because of their high efficiencies. Their versatility allows their use for machining slabs, cutting contours such as keyways, or even gear manufacture, as shown in Fig. 12.10. Large-scale production of gears, however, is more often done by special machine tools, such as gear hobbers. Large plane milling machines, the basic components of which are similar to the planer shown in Fig. 12.14, are also important production machines. In such plane mills, the single-point cutting tools used on planers are replaced with rotating milling cutters, Fig. 12.11. In addi-

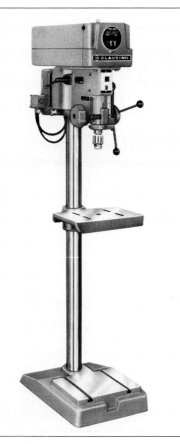

Fig. **12.6**   Drill Press. *Courtesy of Clausing Machine Tools.*

Fig. **12.7**   Radial Drill Press. *Courtesy of Clausing Machine Tools.*

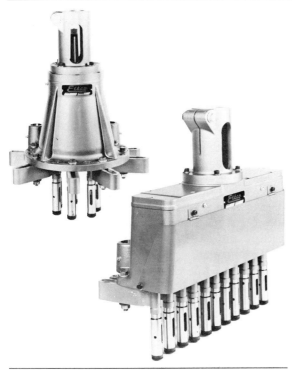

Fig. 12.8 Multiple-Spindle Drilling Heads. *Courtesy of Ettco Tool & Machine Co., Inc.*

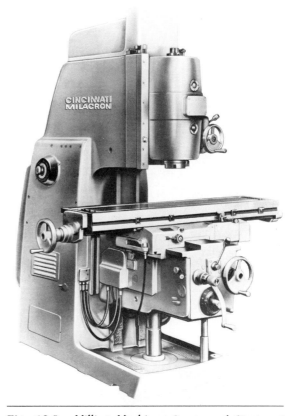

Fig. 12.9 Milling Machine. *Courtesy of Cincinnati Milacron.*

Fig. 12.10 Cutting Teeth on Gear in Milling Machine.

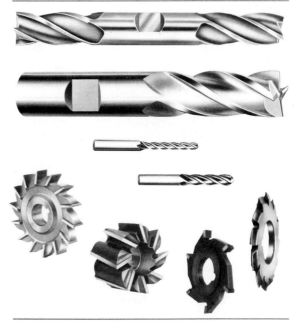

Fig. 12.11 Typical Milling Cutters. *Courtesy of Sharpaloy Division, Precision Industries, Inc.*

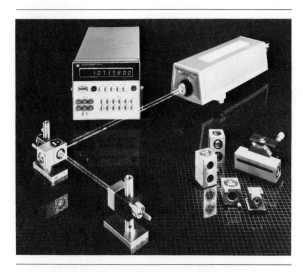

Fig. 12.12   Laser Machine Tool Calibration System. *Courtesy of Hewlett-Packard Company.*

Fig. 12.13   Machining Plane Surface on Shaper.

tion, it is practical to drill, ream, and bore on the milling machine rather than on a drill press.

The accuracy of many types of machine tools is dependent on the degree of accuracy to which the machine is set up and maintained. A laser-based measurement system, Fig. 12.12, is used

in the calibration of the positioning accuracy of numerically controlled machine tools such as milling machines.

12.14   Shaper

On the shaper, Fig. 12.13, work is held in a vise while a single-pointed nonrotating cutting tool mounted in a reciprocating head is forced to move in a straight line past the stationary work. Between succeeding strokes of the tool, the vise that holds the work is fed mechanically into the path of the tool for the next cut alongside the one just completed. Shapers are often used to cut external and internal keyways, gear racks, dovetails, and T-slots.

12.15   Planer

On the planer, Fig. 12.14, work is fixed to a table that is moved mechanically so that the cutting takes place between stationary tools and moving work. This machine, with its reciprocating bed, and a tool head that is adjustable both horizontally and vertically, is used principally for machining large plane surfaces, or surfaces on a large number of pieces, as shown. In extremely large planers, the table is stationary and the cutting tool is carried along a track past the work.

12.16   Boring Mill

The vertical boring mill, Fig. 12.15, is used for facing, turning, and boring heavy work weighing up to 20 tons. The vertical boring mill has a large rotating table and a nonrotating cutting tool that moves mechanically into the work.

The *horizontal boring machine* and the *jig borer* are similar to the milling machine in that the cutting action is between a rotating tool and nonrotating work. The horizontal boring machine is suitable for accurate boring, reaming, facing, counterboring, and milling of pieces larger than could be handled on the typical milling machine. The jig borer is a precision machine that somewhat resembles a drill press in its basic features of a rotating vertical spindle

Fig. 12.14   Planer.

supporting a cutting tool and a stationary horizontal table for holding the work.

The precision jig borer, however, is equipped with a table that can be locked in position while a hole is being cut but may be moved between cutting operations so as to locate one hole with respect to another. The work is not moved with respect to the table; instead, the table may be accurately positioned in two perpendicular directions by means of two accurate lead screws or by micrometer measuring bars.

Since conditions for boring vary, it is difficult to give a figure for accuracy of a bored hole. On S.I.P. (Swiss) jig-boring machines, holes are said to be bored within 0.002 mm (.00008″) of true size and are said to be located to within 0.003 mm (.0001″) accuracy. Boring machines with a diamond-tipped cutting tool produce holes to the following limits: 0.0003 mm (.00001″) for out-of-round, 0.003 mm (.0001″) for straightness, and 0.0003 mm (.00001″) for size. Thus, the boring operation at its best is capable of very accurate results. The tables in Appendixes 5–14, however, are more representative of current practice.

Fig. 12.15   Vertical Boring Mill.

Fig. 12.16   Vertical and Horizontal Turning Center. *Courtesy of Ingersoll Milling Machine Co.*

12.17   Vertical and Horizontal Turning Center

Vertical and horizontal turning centers, Fig. 12.16, eliminate the need for individual shaping, planing, and boring machines as previously described. These computer-driven machining centers not only replace planer mills, but they also do away with the need for moving a machined part from a milling machine to a drilling machine to a boring machine, and so on.

These machines are used primarily in manufacture of heavy parts such as large valves, pump bodies, machine tool beds, and diesel engine frames. The worktable sizes range from 49″ × 78″ to 88″ × 318″ depending on machine models. When the two independently controlled tables are joined together to perform as a single unit, the workpiece length can range from 185″ to 657″. The automatic tool changer on this machine stores 48 different cutters weighing up to 120 pounds each.

12.18   Grinding Machine

A grinding machine is used for removing a relatively small amount of material to bring the work to a very fine and accurate finish. In grinding, the work is fed mechanically against the rapidly rotating grinding wheel, and the depth of cut may be varied from 0.03 mm (.001″) to 0.0064 mm (.00025″).

Surface grinders, Fig. 12.17, reciprocate the work on a table that is simultaneously fed transverse to the grinding wheel. The unit shown lets the operator dial into the electronic memory the precise amount of metal to be removed during

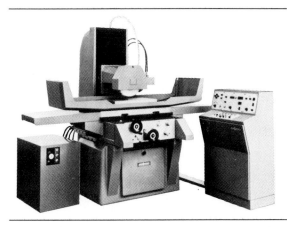

Fig. 12.17 Electronic Sequence-Controlled Surface Grinder. *Courtesy of Clausing/Jakobsen.*

each of the roughing and finishing cuts and the amount of metal to be removed with each pass and even allows the operator to dial in the amount of wheel dressing to be done each cycle. The electronic memory automatically compensates for the loss of abrasive on the grinding wheel as the work proceeds.

12.19 Broaching

Broaching is similar to a single-stroke filing operation. As the broach is forced through the work, each succeeding tooth bites deeper and

deeper into the metal, thus enlarging and forming the keyway as the broach passes through the hole.

A typical broach, along with the corresponding drawing calling for its use, is shown in Fig. 12.18.

Figure 12.18 illustrates *internal broaching.* Square, cylindrical, hexagonal, and other shaped ho'es are produced by this process. An initial hole, produced by drilling, punching, or some other means, is necessary in order to permit the broach to enter the work. Thus, internal broaching only enlarges and varies the shape of a hole that has already been produced by some other operation.

In addition to internal broaching, *surface broaching* is used to produce flat surfaces on such parts as engine blocks.

Broaching is distinctly a mass-production process in which a large expenditure of money is usually involved both for the broaches themselves and for the special machines developed for their use in mass production. Work produced by broaching is of a high order of precision. With this process it is common to produce holes that are within 0.013 mm (.0005″) of the desired size.

12.20 Holes

Rough holes are produced in metal by coring, piercing (punching), and flame cutting. Drilling, Fig. 12.19 (a), though superior, does not produce

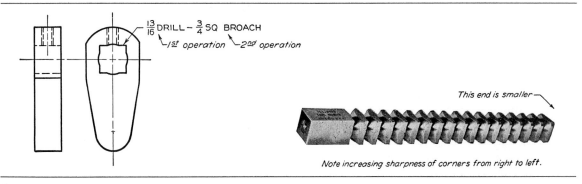

$\frac{13}{16}$ DRILL — $\frac{3}{4}$ SQ BROACH
1st operation   2nd operation

This end is smaller

Note increasing sharpness of corners from right to left.

Fig. 12.18 Broaching.

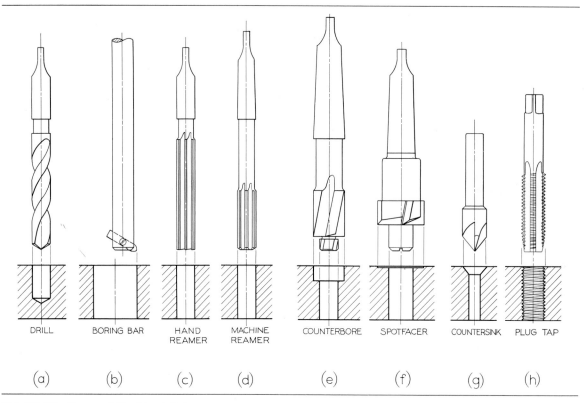

Fig. 12.19   Types of Machined Holes.

a hole of extreme accuracy in roundness, straightness, or size.

Drills frequently cut holes slightly larger than their nominal size. A twist drill (Appendix 16) is somewhat flexible, which makes it tend to follow a path of least resistance. For work that demands greater accuracy, drilling is followed by boring, Fig. 12.19 (b), or by reaming, (c) or (d). When a drilled hole is to be finished by boring, it is drilled slightly undersize, and the boring tool, which is supported by a relatively rigid bar, generates a hole that is round and straight. Reaming is also used for enlarging and improving the surface quality of a drilled or bored hole. Reamers are a finishing tool, and best results are achieved by limiting the material removed to .004″ to .012″ on the diameter.

Good practice is to drill, bore, and then ream to produce an accurate and finely finished hole. Standard reamers are available in 0.4 mm ($\frac{1}{64}''$) increments of diameter.

Counterboring, Fig. 12.19 (e), is the cutting of an enlarged cylindrical portion of a previously produced hole, usually to receive the head of a fillister-head or socket-head screw, Figs. 15.33 and 15.34.

Spotfacing is similar to counterboring, but is quite shallow, usually about 1.5 mm ($\frac{1}{16}''$) deep or just deep enough to clean a rough surface, Fig. 12.19 (f), or to finish the top of a boss to form a bearing surface. Although the depth of a spotface is commonly drawn 1.5 mm ($\frac{1}{16}''$) deep, the actual depth is usually left up to the machinist. It is also good practice to include a note, "spotface

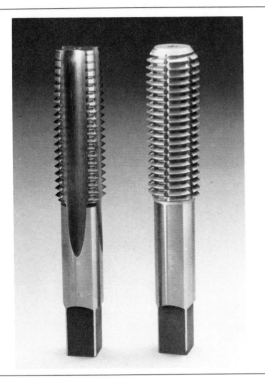

Fig. 12.20   Comparison of Form Tap and Regular Tap. *Courtesy of Cleveland Twist Drill.*

to clean." A spotface provides an accurate bearing surface for the underside of a bolt or screw head.

Countersinking, Fig. 12.19 (g), is the process of cutting a conical taper at one end of a hole, usually to receive the head of a flat-head screw, Figs. 15.32 and 15.33, or to provide a seat for a lathe center, as shown on part No. 9 of Figs. 16.88–16.90. Actually, this operation is an example of drilling and countersinking, which is accomplished in one operation with a tool called a *combined drill and countersink,* made especially for the purpose.

Tapping is the threading of small holes by the use of one or more styles of taps, Fig. 12.19 (h). Before tapping may be accomplished, a hole must be drilled, as shown in Fig. 12.19.

Another means of obtaining a threaded hole is with the form tap, Fig. 12.20. The advantage of the form tap is that metal is not removed from the predrilled hole but is compressed, thereby work-hardening the thread flank for stronger threads that have a greater resistance to stripping.

The location of holes is a subject that is as important as the production of the holes themselves. Details are beyond the scope of this chapter, but it should be mentioned that the layout of the part to be machined is usually a necessary operation before cutting begins.

For the production of holes, the layout is customarily accomplished by first treating the surface with layout bluing or other suitable substances so that scratches will be clearly visible. Sharp-pointed instruments (scribers) are used to scratch center lines on the work. The surface plate or layout table is often used for such layouts. Frequently the work is clamped to a toolmaker's angle block and the center lines scratched on with a vernier height gage.

Of the several methods commonly used to locate work relative to the cutting tool, the use of toolmaker's buttons is probably the most common. Buttons are small hardened-steel rings that may be temporarily attached to the work so that the centers of the rings are precisely where the centers of the holes are to be. The work is then set up with the center of one of the buttons in the center of rotation of, say, the faceplate of an engine lathe. The button is then unscrewed and a hole is drilled, bored, and reamed in proper location. By repeating this process, the required holes may be produced in their proper locations.

12.21   Measuring Devices Used in Manufacturing

Although the machinist uses various measuring devices depending on the kind of dimensions (fractional, decimal, or metric) shown on the drawing, it is evident that to dimension correctly, the engineering designer must have at least a working knowledge of the common measuring tools. The machinists' steel rule, or scale, is a commonly used measuring tool in the shop, Fig. 12.21 (a). The smallest division on one

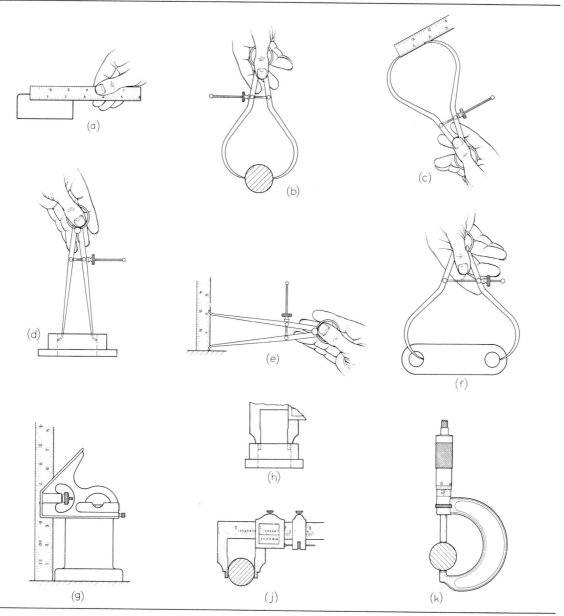

Fig. 12.21   Measuring Devices Used by the Machinist.

scale of this rule is $\frac{1}{64}''$, and such a scale is used for common fractional dimensions. Also, many machinists' rules have a decimal scale with the smallest division of .01″, which is used for dimensions given on the drawing by the decimal system, §13.10. For checking the nominal size of outside diameters, the outside spring caliper and steel scale are used as shown at (b) and (c). Likewise, the inside spring caliper is used for checking nominal dimensions, as shown at (d) and (e). Another use for the outside caliper, (f), is to check the nominal distance between holes

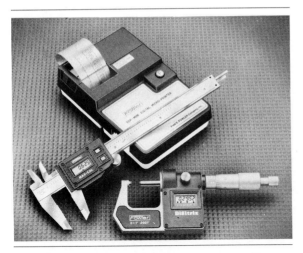

Fig. 12.22   Computerized Measurement System. *Courtesy of Fred V. Fowler Co., Inc.*

(center-to-center). The combination square may be used for checking height, as shown at (g), and for a variety of other measurements. Measuring devices are also available that have metric scales.

For dimensions that require more precise measurements, the vernier caliper, (h) and (j), or the micrometer caliper, (k), may be used. It is common practice to check measurements to 0.025 mm (.001″) with these instruments, and in some instances they are used to measure directly to 0.0025 mm (.0001″).

Computerized measuring devices have broadened the range of accuracy previously attainable. Figure 12.22 illustrates an ultraprecision electronic digital readout micrometer and caliper that contain integral microprocessors. In addition to the hand-held printer/recorder providing a hard-copy output of measurements, the printer also calculates and lists statistical mean, minimum, and maximum values as well as standard deviation.

Most measuring devices in manufacturing are adjustable so they can each be employed to measure any size within their range of designed usage. There is also a need for measuring devices

designed to be used for only one particular dimension. These are called *fixed gages* because their setting is fixed and cannot be changed.

A common type of fixed gage consists of two carefully finished rounds. One might think of each of these rounds as being 25.4 mm (1.00″) in diameter and 38 mm (1.500″) long. Let one of these diameters be slightly larger than the other. One can see that, for a certain range of hole sizes, the smaller round will enter the hole but the larger will not. If the larger round diameter is made slightly greater than the largest acceptable hole diameter and if the diameter of the smaller round is made slightly less than the smallest acceptable hole diameter, then the large round *will never go* into any acceptable hole but the small round *will go* into any acceptable hole. A fixed gage consisting of two such rounds is called a "go–no go" gage. There are, of course, many kinds of "go–no go" gages.

The subject of gages and gaging is a specialized field and involves so many technical considerations that many large companies employ highly trained workers to attend to nothing but this one feature of their operations.

12.22   Chipless Machining

Several manufacturing processes do not employ the cutting action described in most of the previous sections.

Chemical milling removes material through chemical reactions at the surface of the workpiece. An etchant, acidic or basic, is agitated over an immersed workpiece that is masked where no metal removal is desired. For a material with high homogeneity, the tolerances obtainable in such an operation are ±0.08 mm (±.003″).

Electrodischarge machining (EDM) utilizes high-energy electric sparks that build up high charge densities on the surface of the work material. Thermal stresses, exceeding the strength of the work material, cause minute particles to break away from the work. Although this process is slow, it allows intricate shapes to be cut in hard materials, such as tungsten carbides.

Fig. 12.23   Laser Processing Center. *Courtesy of Coherent, Inc.*

Laser machine tools are helping designers and manufacturing engineers increase productivity over a wide range of machine applications. Laser systems reduce costs and improve the quality of many manufactured products that require cutting, engraving, scribing, drilling, welding, perforating, or heat treating.

Metal cutting is the single largest application for most lasers. Laser processing offers several advantages over typical cutting processes, including greater accuracy and flexibility, reduced costs, and higher material throughput. Lasers produce an extremely narrow kerf (slit or notch), providing unmatched precision for thermal cutting of small holes, narrow slots, and closely spaced patterns. Kerf widths of 0.10 mm to 0.25 mm (.004″ to .010″) for metals up to 9.5 mm (.375″) thick are typical.

A typical laser system, Fig. 12.23, includes the laser, work-handling station, computer numerical control, and a beam delivery system.

The computerized hole-making system shown in Fig. 12.24 combines a precision punching machine to produce conventional round and shaped holes using punch and die tooling with a laser beam. The laser is used to cut holes that

are impractical or impossible for conventional tooling. Linear cutting rates range from 200 IPM (inches per minute) for .125″ acrylic and 160 IPM for .039″ low-carbon steel to 20 IPM for .250″ low-carbon steel. Another advantage of this combination of a punching and laser system is that the laser can cut materials that cannot be successfully punched, such as acrylics, printed circuit board materials, plywoods and hardboards, titanium, and rubber.

12.23   Welding

Welding is a process of joining metals by fusion. Although welding by lasers is becoming popular in sophisticated plants, arc welding, gas welding, resistance welding, and atomic hydrogen welding of stock plates, tubing, and angles are more commonly used. These procedures and the use of standardized symbols and techniques for their representation are described in Chapter 27.

Welded structures are built up in most cases from stock forms, particularly plate, tubing, and angles. Often both heat-treating and machine shop operations must be performed on welded machine parts. Since welding frequently distorts

Figure 12.24 Numerically Controlled Hole Punching and Laser Cutting System. *Courtesy of Strippit/Di-Acro Houdaille.*

a part or structure enough to alter permanently the dimensions to which it has been cut, it is usual practice not to weld accurate work after finish machining has been done unless the volume of the welded material is very small in comparison to the volume of the part.

12.24   Stock Forms

Many standardized structural shapes are available in stock sizes for the fabrication of parts or structures. Among these are bars of various shapes, flat stock, rolled structural shapes, and extrusions, Fig. 12.25. The tube shown at (e) is often round, square, or rectangular in shape and can be formed by seamless extrusion or by rolling flat stock and welding the seam.

The manufacturing processes of rolling, drawing, and extruding often add a great deal of toughness and strength to a metal, and these stock forms so processed are very useful in the manufacture of small machined parts, such as screws, and for the fabrication of welded and riveted structures.

12.25   Jigs and Fixtures

A general-purpose machine tool may have its effectiveness on a specific job increased by means of jigs or fixtures. A *jig* is a device that holds the

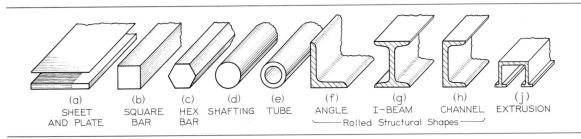

(a) SHEET AND PLATE	(b) SQUARE BAR	(c) HEX BAR	(d) SHAFTING	(e) TUBE	(f) ANGLE	(g) I–BEAM	(h) CHANNEL	(j) EXTRUSION

Rolled Structural Shapes

Fig. 12.25 Common Stock Forms.

Fig. 12.26   Use of Drilling Jig.

example of this is a fixture for holding a connecting rod in a milling machine when the ends are being faced by straddle milling.

Assembly tooling consists of devices to hold work and guide tools as parts are being assembled. For example, the center section of the wing of a large airplane was assembled in a special jig in which parts were held in place, drilled when together, and riveted. The assembly jig was built so that the component parts could fit together only when correctly located and so that no measurement was ever made and no blueprint ever referred to by workers assembling the wing.

Large quantities of precision products may be produced and assembled by less skilled labor than will otherwise be required if a relatively small group of highly skilled workers first produce the required manufacturing tooling and assembly tooling.

work and guides the tool; it is usually not rigidly fixed to a machine. A drilling jig, Fig. 12.26, is a common device by means of which holes on many duplicate parts may be drilled exactly alike. A *fixture* is rigidly attached to the machine, becoming in reality an extension of it, and holds the work in position for the cutting tools without acting as a guide for them. Drawings of one of the important fixtures used in the production of a connecting rod, Figs. 16.36–16.38, are shown in Figs. 16.107–16.109. This fixture was built at considerable expense for the single purpose of holding the connecting rod in the exact position required for the efficient and speedy execution of a single operation.

Jigs and fixtures are usually designed in a tooling department by tool designers. Usually they are built by highly skilled machinists using especially accurate equipment. Such tooling devices are commonly held to tolerances one-tenth of those applied to the parts to be produced on these jigs and fixtures.

Jigs and fixtures may be grouped into two general classes: manufacturing tooling and assembly tooling. *Manufacturing tooling* consists of devices used in producing individual parts. An

12.26   Hot and Cold Working of Metals

Recall that in each of the casting operations, previously discussed, §§12.3 and 12.7, the casting assumes its shape by filling a mold as a liquid and does not appreciably change its shape as it solidifies. In many manufacturing processes, however, the shape of the part is changed after the metal has solidified. If these processes involve slowly pressing the part (squeezing) or rapidly and repeatedly striking the part (hammering), then the term *mechanical working* is used to describe the operation.

The practical effects of mechanical working depend a great deal on the temperature at which the metal is squeezed or hammered. The terms *hot working* and *cold working* are used widely in engineering today, and the student of metallurgy and machine design will spend a lot of time on the hot working and cold working of metals.

12.27   Forging

Forging is the process of shaping metal to a desired form by means of pressing or hammering. Generally, forging is hot forging, Fig. 12.27, in which the metal is heated to a required temperature before forging. Some softer metals can be forged without heating, and this is cold forging.

When metal is worked under pressure, as in forging, the material is compressed and greatly strengthened. Complicated shapes are extremely expensive because of costly dies. Forgings are made of steel, copper, brass, bronze, aluminum, magnesium, lightweight titanium, and other alloys.

Closed-die forgings, or drop forgings, are formed between dies that are machined to the desired shapes. Open-die forgings, or hand forgings, are formed between flat dies and manipulated by hand during forging to obtain the required shape.

Usually the first step in the design of a forging is to locate the *parting plane*, which separates the upper and lower dies. As a rule, this surface follows the center line of the piece, as shown in Fig. 12.28 (a), and should pass through the largest portion of the object.

Draft, or taper, must be provided on all forgings. Although draft is not usually shown on drawings for sand casting, it is always shown on forging drawings. A typical forging drawing is shown in Fig. 16.36. For outside surfaces, the standard draft angle is 7° for all materials. For inside surfaces in aluminum and magnesium, the draft is also 7°, but in steel the angle is usually 10°.

Small fillets and rounds in forgings reduce die life and decrease the quality of the forging. Minimum rounds or corner radii are shown in inches

Fig. 12.27   Drop Forging. *Courtesy of Jervis B. Webb Co.*

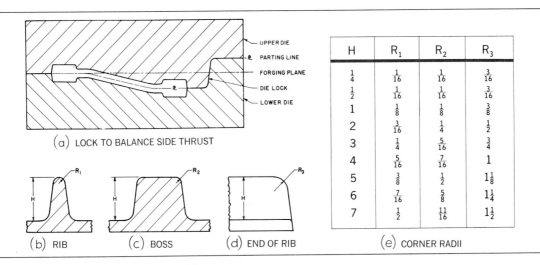

(a) LOCK TO BALANCE SIDE THRUST

UPPER DIE
PARTING LINE
FORGING PLANE
DIE LOCK
LOWER DIE

(b) RIB   (c) BOSS   (d) END OF RIB

(e) CORNER RADII

H	R_1	R_2	R_3
$\frac{1}{4}$	$\frac{1}{16}$	$\frac{1}{16}$	$\frac{3}{16}$
$\frac{1}{2}$	$\frac{1}{16}$	$\frac{1}{16}$	$\frac{3}{16}$
1	$\frac{1}{8}$	$\frac{1}{8}$	$\frac{3}{8}$
2	$\frac{3}{16}$	$\frac{1}{4}$	$\frac{1}{2}$
3	$\frac{1}{4}$	$\frac{5}{16}$	$\frac{3}{4}$
4	$\frac{5}{16}$	$\frac{7}{16}$	1
5	$\frac{3}{8}$	$\frac{1}{2}$	$1\frac{1}{8}$
6	$\frac{7}{16}$	$\frac{5}{8}$	$1\frac{1}{4}$
7	$\frac{1}{2}$	$\frac{11}{16}$	$1\frac{1}{2}$

Fig. 12.28   Forging Design [ANSI Y14.9].

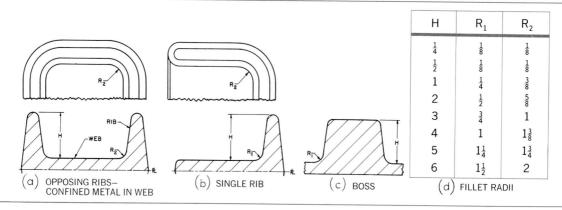

H	R_1	R_2
$\frac{1}{4}$	$\frac{1}{8}$	$\frac{1}{8}$
$\frac{1}{2}$	$\frac{1}{8}$	$\frac{1}{8}$
1	$\frac{1}{4}$	$\frac{3}{8}$
2	$\frac{1}{2}$	$\frac{5}{8}$
3	$\frac{3}{4}$	1
4	1	$1\frac{3}{8}$
5	$1\frac{1}{4}$	$1\frac{3}{4}$
6	$1\frac{1}{2}$	2

(a) OPPOSING RIBS—CONFINED METAL IN WEB   (b) SINGLE RIB   (c) BOSS   (d) FILLET RADII

Fig. 12.29   Minimum Radii for Forgings [ANSI Y14.9].

in Fig. 12.28 (b) to (e) and minimum fillet radii in inches in Fig. 12.29. Whenever possible, larger radii than those shown should be used.

The minimum thickness of a web depends on the area at the parting line. Minimum web-thickness values (in inches) are shown in the following table.

Area, in.^{2}	Thickness, in.
4	$\frac{3}{32}$
20	$\frac{1}{8}$
80	$\frac{3}{16}$
200	qb
400	$\frac{5}{16}$
600	$\frac{3}{4}$
800	$\frac{7}{16}$
1000	$\frac{1}{2}$

Ribs will increase in thickness as the height increases. Minimum values conform to radius R_1 in Fig. 12.28 (b).

The working drawing, as it comes to the shop, may show the completed or machined part, in which case the necessary forging and machining allowances are made by the diemaker. However, since dies are so expensive and the forging itself is so important, it is common practice to make separate forging and machining drawings, as shown in Figs. 16.36 and 16.37. It is standard practice to show visible lines for rounded corners as if the corners were sharp. See §7.36. Note that the parting plane is represented by a center line.

12.28   Heat-Treating

Heat-treating is when heat is used to alter the properties of metal. Different procedures affect metal in different ways. Annealing and normalizing involve heating to a critical temperature range, and then slowly cooling to soften the metal, thereby releasing internal stresses developed in previous manufacturing processes.

Hardening requires heating to above the critical temperature followed by rapid cooling—quenching in oil, water, brine, or in some instances in air. *Tempering* reduces internal stresses caused by hardening and also improves the toughness and ductility. *Surface hardening* is a way of hardening the surface of a steel part while leaving the inside of the piece soft. Surface hardening is accomplished by *carburizing*, followed by heat-treatment; by *cyaniding*, followed by heat-treatment; by *nitriding*; by *induction hardening*; and by *flame hardening*.

Lasers are widely used for transformation hardening of selected surface areas of metal parts. Laser hardening applies less heat to the part than do traditional methods, and thermal distortion is greatly reduced.

Fig. 12.30   Industrial Robots. *Courtesy of Cincinnati Milacron.*

12.29   Automation

Automation is the term applied to systems of automatic machines and processes. These machines and processes are essentially the same as previously discussed, with the addition of mechanisms to control the sequence of operation, movement of tool, flow, and so forth. Little, if any, operator interaction is required once the equipment has been "set up."

Most contain built-in computers to control all their functions, such as the surface grinder in Fig. 12.17. Some machines are used in combination with an industrial robot, Fig. 12.30, where the robot takes the place of an operator for the load and unload functions for the machine. Robots are also used extensively for spray painting and welding applications as well as for around-the-clock performance in environments where a human operator could not exist.

When a machine's cycle is relatively long and the robot is underutilized, two or more machines can be grouped together in such a fashion that the robot can service each. The flexible manufacturing system in Fig. 12.31 consists of a

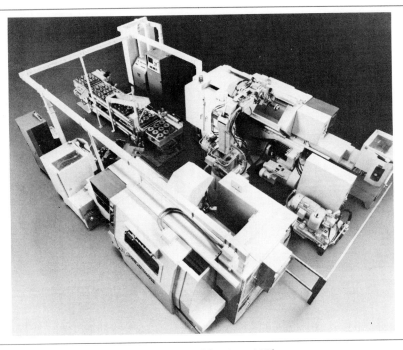

Fig. 12.31   Flexible Manufacturing System. *Courtesy of Cincinnati Milacron.*

Fig. 12.32   Automated Manufacturing System. *Courtesy of Cargill Detroit.*

Fig. 12.33   Automated Material Conveying. *Courtesy of Jervis B. Webb Co.*

unique computer numerically controlled (CNC) turning system equipped with fully automated features, designed for reliable batch manufacturing of quality parts while virtually unattended. The two turning centers are rear loaded by a heavy-duty industrial robot. The system also features automatic tool changing from an 84-tool storage magazine as well as automatic postprocess gaging and a take-away parts conveyor.

The manufacturing system in Fig. 12.32 is based on a different concept. The parts being produced are the front- and rear-end housings for automotive air-conditioning compressors. A conveyor system circulates the housing parts throughout the 20-station system, where various facing, boring, drilling, chamfering, probing, reaming, gaging, wirebrushing, washing, blow drying, and unloading and reject-diverting operations are performed. The disadvantage of such a system is that it is usually custom-made for the specific application and, therefore, is not considered flexible enough for most smaller manufacturing operations.

Large-scale manufacturing operations such as automotive plants must be automated to remain competitive. Figure 12.33 shows Pontiac automobile bodies being carried on a conveyor into a one-of-a-kind milling and drilling machine.

12.30   Plastics Processing

The growth of the plastics industry has expanded in recent decades to the point that the domestic demand for plastics reached approximately 58 billion pounds in 1989. Plastic resins used in motor vehicle production alone accounted for about 1.9 billion pounds. As an example of this tremendous increase, the 3000-pound Chevrolet Corvette contains about 580 pounds of various plastic parts.

The two main families of plastics are known as *thermosetting* and *thermoplastic*. The thermosetting plastics will take a set and will not soften when reheated, whereas thermoplastics will soften whenever heat is applied. Some of the common thermoplastic materials are ABS (acrylonitrile-butadiene-styrene), acetals, cellulosics, nylons, polycarbonates, PET (polyethyl-

ene terephthalate), polypropylene, polyethylene, polystyrene, sulfones, and vinyls. Common thermosetting plastics include epoxies, melamines, phenolics, polyesters, polyurethanes, silicones, and ureas. Thermoplastic polyesters are also manufactured.

Typical plastic processing operations include extrusion, blow molding, compression molding, transfer molding, injection molding, and thermoforming.

12.31   Extrusion Molding

An extrusion-molding machine transfers solid plastic particles, additives, colorants, and regrinds from a feed hopper into a heating chamber. The extruder screw conveys the material forward through a heated barrel toward a die. The contour of the die will determine the cross-sectional shape of the extruded member. Typical uses of the extrusion process are in compounding and pelletizing of bulk plastics, pipe and profile extrusion, blown film and sheet manufacturing, and blow molding.

12.32   Blow Molding

Blow-molding operations are classified as *extrusion, injection,* and *stretch* blow molding. In an extrusion blow-molding machine, Fig. 12.34, an

Fig. 12.34   Extrusion Blow-Molding Machine. *Courtesy of Hoover Universal.*

Fig. 12.35   Blow-Molded Aerodynamic Bumper System. *Courtesy of Navistar International Transportation Corp.*

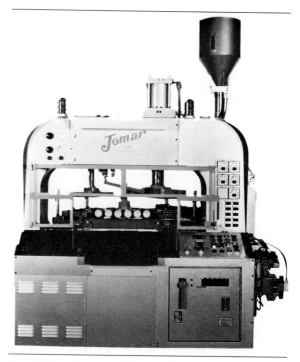

Fig. 12.36   Injection Blow-Molding Machine. *Courtesy of Jomar.*

open-ended parison (hollow tube) is extruded, and the mold is closed on the parison, thus closing the bottom of the tube. When compressed air is introduced through the blow pin or blow needles, the parison expands to take the shape of the mold. The part is cooled by being in contact with the cooler mold surface and then is removed from the mold. Automotive coolant and windshield washer fluid reservoirs, seat backs, ductwork, stadium seats, wheels, toy parts, infant seats, and juice and handled bottles are common extrusion blow-molded shapes.

With larger, more sophisticated machines, and with higher performance engineering resins, new shapes and applications are possible. Blow-molded bumpers and other aerodynamic external truck components are shown in Fig. 12.35. The design of the bumper system, with its seven different parts that are up to 7 feet long and weigh a total of approximately 75 pounds, improves the fuel economy by reducing air turbulence and drag.

In injection blow molding, Fig. 12.36, the parison is formed by injection molding a test-tube-shaped preform on a core pin and is then transferred to the blow-molding station. An advantage of the injection blow-molding process is that there is no flash to trim, and any thread detail on the neck is of injection-molding quality.

In stretch blow molding, also called orientation blow molding, a parison is formed by either an extrusion- or injection-molding operation and is subsequently stretched and blown in both an axial and radial direction. Advantages of this process are greater strength, better clarity, improved impact strength, and increased stiffness.

12.33   Compression Molding

In compression molding, a charge of plastic material, usually thermosetting, is placed into a mold, the mold is closed, and a plunger compresses the material so that heat and pressure modifies the plastic to take the shape of the mold

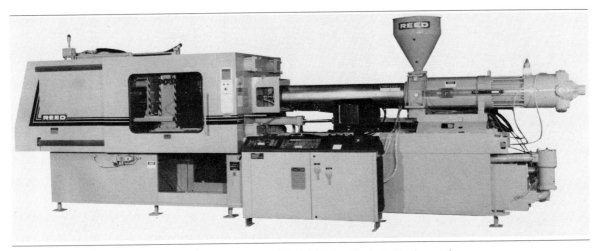

Fig. 12.37   Injection-Molding Machine. *Courtesy of Reed Division, Package Machinery Co.*

cavity. The pressure is maintained during the curing cycle, after which the plunger is removed and the part is ejected. Preformed, preheated disks are often used for better control of the volume of material, thereby reducing variations in part thickness or part density.

12.34   Transfer Molding

In the transfer-molding process, an amount of thermosetting material is placed in a chamber (usually referred to as a pot) and then is forced out of the pot and into the closed cavities where polymerization takes place. The plastics formulations used are generally of a softer plastic than those that are used in compression molding. This enables the design of the part to include ribs and thin sections that are not formable in compression molding.

12.35   Injection Molding

Injection molding, Fig. 12.37, is used primarily to manufacture thermoplastic products. Granular or powdered materials are fed from the machine hopper into the barrel where the material is preheated to its melting temperature range, transported through the barrel by the screw, and

injected through a nozzle into the mold where the plastic solidifies into the shape of the mold cavities. The press is opened after the plastic has sufficiently cooled, and the parts are then ejected. Owing to the relatively high costs of tooling, the process is typically and only economically suited for mass production.

The injection-molding process is also becoming popular in the molding of thermoset plastics. The main difference is in the hardening process for each material. Thermosetting materials are heated to their liquid state and are then rapidly flowed into the mold cavities where the mold is held at elevated temperatures, during which time an irreversible chemical curing takes place. The hot, cured, hardened parts are then removed from the mold with no need for cooling.

12.36   Thermoforming

Thermoforming is a process in which one or both sides of a thermoplastic sheet are heated so that it can be formed into various products. Figure 12.38 represents a thermoformer that feeds roll stock up to .050″ through the heater zone and into a form area. Once the heated sheet is clamped in place over the mold (some machines have a form area of up to 50″ wide by 50″ long),

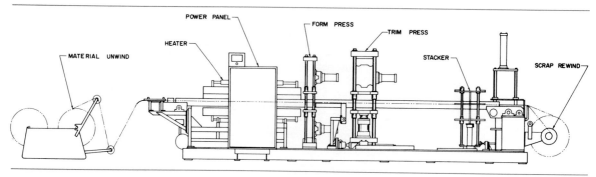

Fig. 12.38   Thermoforming Machine. *Courtesy of Sencorp Systems, Inc.*

pressure is exerted on the outside of the sheet, and a vacuum is drawn to evacuate the air that is trapped between the sheet and the mold. The sheet is held in place on the mold until the plastic is cooled sufficiently to maintain the part shape, and compressed air is then introduced through the vacuum holes to lift the parts off the mold. The sheet is then indexed to the die cutting and stack stations.

Figure 12.39 shows a machine that is geared toward high-volume production runs. The parts being molded are high-impact foam polystyrene egg cartons and are being formed at about 200 cartons per minute, cut on a matched metal punch press, and then stacked automatically for packing and shipping.

12.37   Do's and Don'ts of Practical Design

Figures 12.40 and 12.41 contain a number of examples in which knowledge of manufacturing processes and limitations is essential for good design.

Fig. 12.39   Thermoforming Machine. *Courtesy of Brown Machine Division, John Brown Plastics Machinery, Inc.*

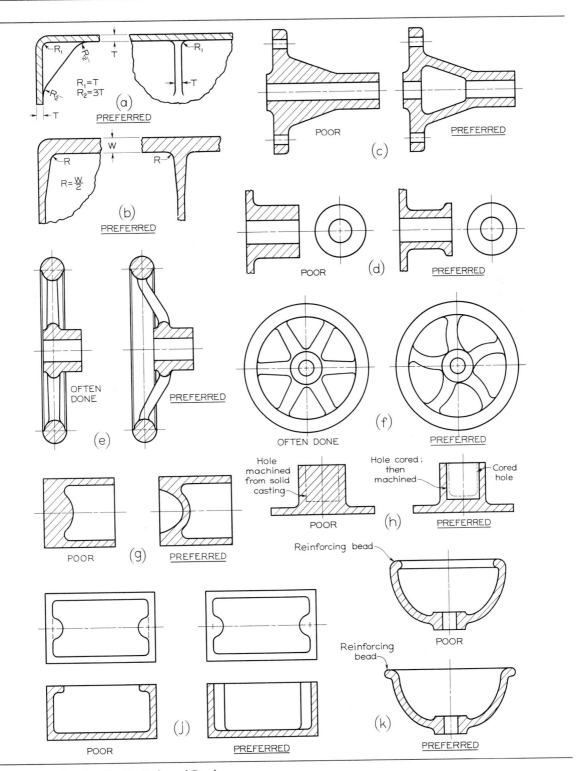

Fig. 12.40 Casting Design Do's and Don'ts.

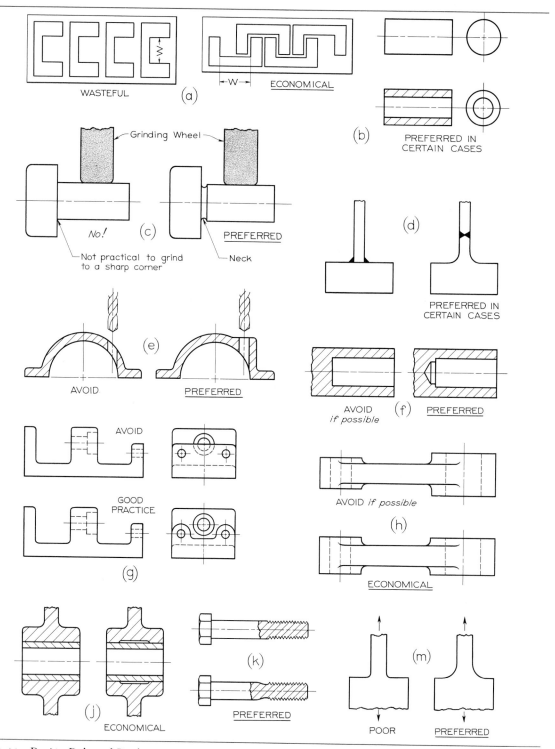

Fig. 12.41 Design Do's and Don'ts.

Many difficulties in producing good castings result from abrupt changes in section or thickness. In Fig. 12.40 (a) rib thicknesses are uniform so that the metal will flow easily to all parts. Fillet radii are equal to the rib thickness—a good general rule to follow. When it is necessary to join a thin member to a thicker member, the thin member should be thickened as it approaches the intersection, as shown at (b).

At (c), (g), and (h) coring is used to produce walls of more uniform sections. At (d) an abrupt change in sections is avoided by making thinner walls and leaving a collar, as shown.

At (e) and (f) are shown examples in which the preferred design will tend to allow the castings to cool without introducing internal stresses. The less desirable design is more likely to crack as it cools, since there is no "give" in the design. Curved spokes are preferable to straight spokes, and an odd number of spokes is better than an even number because direct stresses along opposite spokes are avoided.

The design of a part may cause unnecessary trouble and expense for the pattern shop and foundry without any gain in the usefulness of the design. For example, in the "poor" designs at (j) and (k), one-piece patterns would not withdraw from the sand, and two-piece patterns would be necessary. In the "preferred" examples, the design is just as useful and is conducive to economical work in the pattern shop and foundry.

As shown in Fig. 12.41 (a), a narrower piece of stock sheet metal can be used for certain designs that can be linked or overlapped. In this case, the stampings may be overlapped if dimension W is increased slightly, as shown. By such an arrangement, great savings in scrap metal can often be effected.

The maximum hardness that can be obtained in the heat treatment of steel depends on the carbon content of the steel. To get this hardness, it is necessary to cool rapidly (quench) after heating to the temperature required. In practice it is often impossible to quench uniformly because of the design. In the design at (b), the piece is solid and will harden well on the outside, but will remain soft and relatively weak on the inside. However, as shown in the preferred example, a hollow piece can be quenched from both the outside and inside. Thus, it is possible for a hardened hollow shaft to be stronger than a hardened solid shaft of the same diameter.

As shown at (c), the addition of a rounded groove, called a *neck,* around a shaft next to a shoulder will eliminate a practical difficulty in precision grinding. It is not only more expensive to grind a sharp internal corner, but such sharp corners often lead to cracking and failure.

The design at the right at (d) eliminates a costly "reinforced" weld, which would be required by the design on the left. The strong virgin metal with a generous radius is present at the point at which the stress is likely to be most severe. It is possible to make the design on the left as strong as the design on the right, but it is more expensive and requires expert skill and special equipment.

It is difficult to drill into a slanting surface, as shown at the left at (e). The drilling will be greatly facilitated if a boss is provided, as shown at the right.

At (f), the design at the left requires accurately boring or reaming a blind hole all the way to a flat bottom, which is difficult and expensive. It is better to drill deeper than the hole is to be finished, as shown at the right, in order to provide room for tool clearance and for chips.

In the upper example at (g), the drill and counterbore cannot be used for the hole in the center piece because of the raised portion at the right end. In the approved example, the end is redesigned to provide access for the drill and counterbore.

In the design above at (h), the ends are not the same height. As a result, each flat surface must be machined separately. In the design below, the ends are the same height, the surfaces are in line horizontally, and only two machining operations are necessary. It is always good design to simplify and limit the machining as much as possible.

The design at the left at (j) requires that the housing be bored for the entire length in order to receive a pressed bushing. If the cored recess is made as shown, machining time can be decreased. This assumes that average loads would be applied in use.

At (k), the lower bolt is encircled by a rounded groove no deeper than the root of the thread. This makes a gentle transition from the small diameter at the root of the threads and the large diameter of the body of the bolt, producing less stress concentration and a stronger bolt. In general, sharp internal corners should be avoided as points of stress concentration and possible failure.

At (m) a $\frac{1''}{4}$ steel plate is being pulled, as shown by the arrows. Increasing the radius of the inside corners increases the strength of the plate by distributing the load over a greater area.

CHAPTER 13

Dimensioning

We have all heard of "rule of thumb." Actually, at one time an inch was defined as the width of a thumb, and a foot was simply the length of a man's foot. In old England, an inch used to be "three barley corns, round and dry." In the time of Noah and the Ark, the *cubit* was the length of a man's forearm, or about 18".

In 1791, France adopted the *meter** (1 meter = 39.37"; 1" = 25.4 mm), from which the decimalized metric system evolved. In the meantime England was setting up a more accurate determination of the *yard,* which was legally defined in 1824 by act of Parliament. A foot was $\frac{1}{3}$ yard, and an inch was $\frac{1}{36}$ yard. From these specifications, graduated rulers, scales, and many types of measuring devices have been developed to achieve even more accuracy of measurement and inspection.

Until this century, common fractions were considered adequate for dimensions. Then, as designs became more complicated and it became necessary to have interchangeable parts in order to support mass production, more accurate specifications were required, and it became necessary to turn to the decimal-inch system or the SI system. See §§13.9 and 13.10.

13.1 Metric Units

The current rapid growth of worldwide science and commerce has fostered an international system of units (SI), based on the meter and suitable for measurements in physical science and engineering. The seven basic units of measure are the meter (length), kilogram (mass), second (time), ampere (electric current), kelvin (thermodynamic temperature), mole (amount of substance), and candela (luminous intensity).

The SI system is gradually coming into use in the United States, especially by the many multinational companies in the chemical, electronic, and mechanical industries. A tremendous effort is now under way to convert all standards of the American National Standards Institute (ANSI) to the SI units in conformity with the International Standards Organization (ISO) standards.

See inside the front cover of this book for a table converting fractional inches to decimal inches or millimeters. For the International System of Units and their United States equivalents, see Appendix 31.

---

*In the SI system the meter is now defined as a length equal to the distance traveled by light in a vacuum during a time interval of $\dfrac{1}{299\ 792\ 458}$ second.

13.2   Size Description

In addition to a complete *shape description* of an object, as discussed in previous chapters, a drawing of the design must also give a complete *size description;* that is, it must be *dimensioned.* See ANSI Y14.5M–1982 (R1988).

The need for *interchangeability* of parts is the basis for the development of modern methods of size description. Drawings today must be dimensioned so that production personnel in widely separated places can make mating parts that will fit properly when brought together for final assembly or when used as repair or replacement parts by the customer, §13.26.

The increasing need for precision manufacturing and the necessity to control sizes for interchangeability has shifted responsibility for size control to the designing engineer and the drafter. The production worker no longer exercises judgment in engineering matters, but only in the proper execution of instructions given on the drawings. Therefore, engineers and designers should be familiar with materials and methods of construction and with production requirements. The engineering student or designer should seize every opportunity to become familiar with the fundamental manufacturing processes, especially *patternmaking, foundry, forging,* and *machine shop practice.*

The drawing should show the object in its completed condition and should contain all necessary information to bring it to that final state. Therefore, in dimensioning a drawing, the designer and the drafter should keep in mind the finished piece, the production processes required, and above all the function of the part in the total assembly. Whenever possible—that is, when there is no conflict with functional dimensioning, §13.4—dimensions should be given that are convenient for the individual worker or the production engineer. These dimensions should be given so that it will not be necessary to scale or assume any dimensions. Do not give dimensions to points or surfaces that are not accessible to the worker.

Dimensions should not be duplicated or superfluous, §13.30. Only those dimensions should be given that are needed to produce and inspect the part exactly as intended by the designer. The student often makes the mistake of giving the dimensions that are used *to make the drawing.* These are not necessarily the dimensions required. There is much more to the theory of dimensioning, as we will see.

13.3   Scale of Drawing

Drawings should be made to scale, and the scale should be indicated in the title block even though the worker is never expected to scale the drawing or print for a needed dimension. See §2.31 for indications of scales.

A heavy straight line should be drawn under any dimension that is not to scale, Fig. 13.15, or the abbreviation *NTS* (not to scale) should be indicated. This procedure may be necessary when a change is made in the drawing that is not important enough to justify making an entirely new drawing.

13.4   Learning to Dimension

Dimensions are given in the form of linear distances, angles, or notes regardless of the dimensioning units being used. The ability to dimension properly in millimeters, decimal inch, or fractional inch requires the following.

1. The student must learn the *technique of dimensioning:* the character of the lines, the spacing of dimensions, the making of arrowheads, and so forth. A typical dimensioned drawing is shown in Fig. 13.1. Note the strong contrast between the visible lines of the object and the thin lines used for the dimensions.

2. The student must learn the rules of *placement of dimensions* on the drawing. These practices assure a logical and practical arrangement with maximum legibility.

3. The student should learn the *choice of dimensions.* Formerly, manufacturing processes were considered the governing factor in dimensioning. Now function is considered

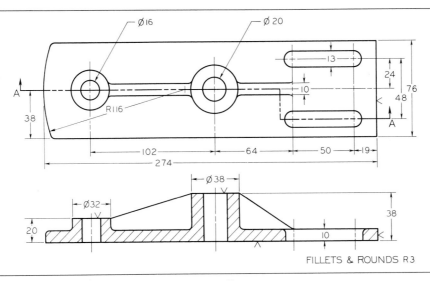

Fig. 13.1 Dimensioning Technique.   Dimensions in millimeters.

first and the manufacturing processes second. The proper procedure is to dimension tentatively for function and then review the dimensioning to see if any improvements from the standpoint of production can be made without adversely affecting the functional dimensioning. A "geometric breakdown," §13.20, will assist the beginner in selecting dimensions. In most cases dimensions thus determined will be functional, but this method should be accompanied by a logical analysis of the functional requirements.

13.5 Lines Used in Dimensioning

A dimension line, Fig. 13.2 (a), is a thin, dark, solid line terminated by arrowheads, which indicates the direction and extent of a dimension. In machine drawing, the dimension line is broken, usually near the middle, to provide an open space for the dimension figure. In structural and architectural drawing, it is customary to place the dimension figure above an unbroken dimension line.

As shown in Fig. 13.2 (b), the dimension line nearest the object outline should be spaced at least 10 mm ($\frac{3''}{8}$) away. All other parallel dimen-

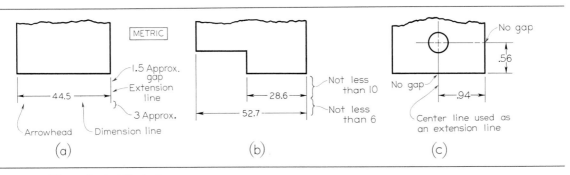

Fig. 13.2 Dimensioning Technique.

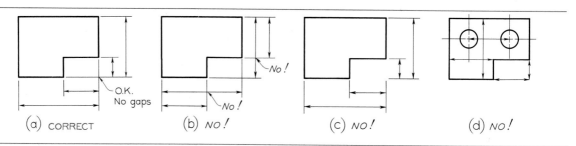

Fig. 13.3   Dimension and Extension Lines.

sion lines should be at least 6 mm ($\frac{1}{4}''$) apart, and more if space is available. *The spacing of dimension lines should be uniform throughout the drawing.*

An *extension line,* (a), is a thin, dark, solid line that "extends" from a point on the drawing to which a dimension refers. The dimension line meets the extension lines at right angles except in special cases, as in Fig. 13.6 (a). A gap of about 1.5 mm ($\frac{1}{16}''$) should be left where the extension line would join the object outline. The extension line should extend about 3 mm ($\frac{1}{8}''$) beyond the outermost arrowhead, (a) and (b).

The foregoing dimensions for lettering height, spacing, and so on should be increased approximately 50 percent for drawings that are to be microfilmed and blown back to one-half size for the working print. Otherwise the lettering and dimensioning often are not legible.

A *center line* is a thin, dark line composed of alternate long and short dashes and is used to represent axes of symmetrical parts and to denote centers. As shown in Fig. 13.2 (c), center lines are commonly used as extension lines in locating holes and other features. When so used, the center line crosses over other lines of the drawing without gaps. A center line should always end in a long dash.

13.6   Placement of Dimension and Extension Lines

A correct example of the placement of dimension lines and extension lines is shown in Fig. 13.3 (a). The shorter dimensions are nearest to the object outline. Dimension lines should not cross

extension lines as at (b), which results from placing the shorter dimensions outside. Note that it is perfectly satisfactory to cross extension lines, as shown at (a). They should never be shortened, as at (c). A dimension line should never coincide with, or form a continuation of, any line of the drawing, as shown at (d). Avoid crossing dimension lines wherever possible.

Dimensions should be lined up and grouped together as much as possible, as shown in Fig. 13.4 (a), and not as at (b).

In many cases, extension lines and center lines must cross visible lines of the object, Fig. 13.5 (a). When this occurs, gaps should not be left in the lines, as at (b).

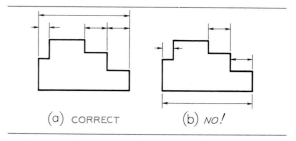

Fig. 13.4   Grouped Dimensions.

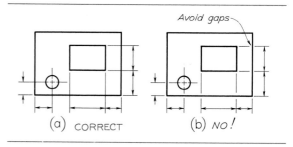

Fig. 13.5   Crossing Lines.

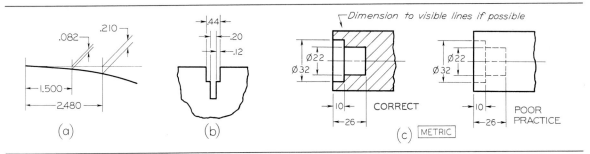

Fig. 13.6   Placement of Dimensions.

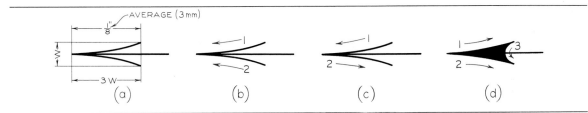

Fig. 13.7   Arrowheads.

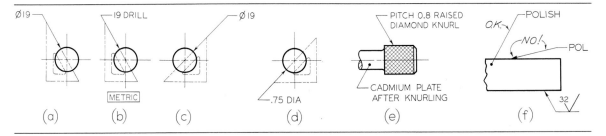

Fig. 13.8   Leaders.

Dimension lines are normally drawn at right angles to extension lines, but an exception may be made in the interest of clearness, as shown in Fig. 13.6 (a). In crowded conditions, gaps in extension lines near arrowheads may be left, in order to clarify the dimensions, as shown at (b). In general, avoid dimensioning to hidden lines, as shown at (c).

13.7   Arrowheads

Arrowheads, Fig. 13.7, indicate the extent of dimensions. They should be uniform in size and style throughout the drawing and not varied according to the size of the drawing or the length of dimensions. Arrowheads should be drawn freehand, and the length and width should be in a ratio of 3:1. The length of the arrowhead should be equal to the height of the dimension whole numbers. For average use, make arrowheads about 3 mm ($\frac{1}{8}''$) long and very narrow, (a). Use strokes toward the point or away from the point desired, (b) to (d). The method at (b) is easier when the strokes are drawn toward the drafter. For best appearance, fill in the arrowhead as at (d).

13.8   Leaders

A leader, Fig. 13.8, is a thin solid line leading from a note or dimension, and terminated by an arrowhead or a dot touching the part to which

395

attention is directed. Arrowheads should always terminate on a line such as the edge of a hole; dots should be within the outline of the object. A leader should generally be an inclined straight line, if possible, except for the short horizontal shoulder (6 mm or $\frac{1}{4}''$, approx.) extending from midheight of the lettering *at the beginning or end of a note.*

A leader to a circle should be radial; that is, if extended, it would pass through the center. A drawing presents a more pleasing appearance if leaders near each other are drawn parallel. Leaders should cross as few lines as possible and should never cross each other. They should not be drawn parallel to nearby lines of the drawing, allowed to pass through a corner of the view, drawn unnecessarily long, or drawn horizontally or vertically on the sheet. A leader should be drawn at a large angle and terminate with the appropriate arrowhead, or with a dot, as shown in Fig. 13.8 (f).

13.9   Fractional, Decimal, and Metric Dimensions

In the early days of machine manufacturing in this country, the worker would scale the undimensioned design drawing to obtain any needed dimensions, and it was the worker's responsibility to see to it that the parts fitted together properly. Workers were skilled, and it should not be thought that very accurate and excellent fits were not obtained. Hand-built machines were often beautiful examples of precision craftsmanship.

The system of units and common fractions is still used in architectural and structural work where close accuracy is relatively unimportant and where the steel tape or framing square is used to set off measurements. Architectural and structural drawings are therefore often dimensioned in this manner. Also, certain commercial commodities, such as pipe and lumber, are identified by standard nominal designations that are close approximations of actual dimensions.

As industry has progressed, there has been greater and greater demand for more accurate

specifications of the important functional dimensions—more accurate than the $\frac{1}{64}''$ permitted by the engineers', architects', and machinists' scale. Since it was cumbersome to use still smaller fractions, such as $\frac{1}{128}$ or $\frac{1}{256}$, it became the practice to give decimal dimensions, such as 4.2340 and 3.815, for the dimensions requiring accuracy. However, some dimensions, such as standard nominal sizes of materials, punched holes, drilled holes, threads, keyways, and other features produced by tools that are so designated are still expressed in whole numbers and common fractions.

Thus, drawings may be dimensioned entirely with whole numbers and common fractions, or entirely with decimals, or with a combination of the two. However, more recent practice adopted the decimal-inch system, and current practice also utilizes the metric system as recommended by ANSI. Millimeters and inches in the decimal form can be added, subtracted, multiplied, and divided more easily than can fractions.

Also the decimal system is compatible with the calibrations of machine tool controls, the requirements for numerically controlled machine tools, and for computer-programmed digital plotting. For examples of computer-made drawings, see Figs. 8.47–8.54.

For inch–millimeter equivalents of decimal and common fractions, see inside the front cover of this book. For additional metric equivalents, see Appendix 31. For rounding off decimals, see §13.10.

13.10   Decimal Systems

A decimal system, based on the decimal inch or the millimeter as a linear unit of measure, has many advantages and is compatible with most measuring devices and machine tools. Metric measurement is based on the meter as a linear unit of measure, but the millimeter is used on most engineering drawings. To facilitate the changeover to metric dimensions during the transition, some drawings are dual-dimensioned in millimeters and decimal inches. See §13.11.

Total decimal dimensioning is *preferred* by the American National Standards Institute.

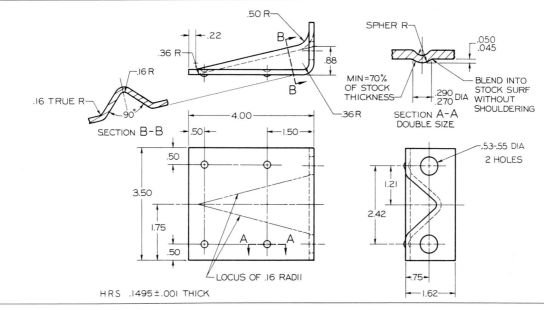

Fig. 13.9   Complete Decimal Dimensioning.

Complete decimal dimensioning employs decimals for all dimensions and designations except where certain commercial commodities, such as pipe and lumber, are identified by standardized nominal designations. *Combination dimensioning* employs decimals for all dimensions except the designations of nominal sizes of parts or features such as bolts, screw threads, keyseats, or other standardized fractional designations. [ANSI Y14.5M–1982 (R1988)].

In these systems, two-place inch or one-place millimeter decimals are used when a common fraction has been regarded as sufficiently accurate.

In the combination dimensioning system, common fractions may continue to be used to indicate nominal sizes of materials, drilled holes, punched holes, threads, keyways, and other standard features.

One-place millimeter decimals are used when tolerance limits of ±0.1 mm or more can be permitted. Two (or more)-place millimeter decimals

are used for tolerance limits less than ±0.1 mm. Fractions are considered to have the same tolerance as two-place decimal-inch dimensions when determining the number of places to retain in the conversion to millimeters. Keep in mind that 0.1 mm is approximately equal to .004 inch.

Two-place inch decimals are used when tolerance limits of ±.01″, or more, can be permitted. Three or more decimal places are used for tolerance limits less than ±.01″. In two-place decimals, the second place preferably should be an even digit (for example, .02, .04, and .06 are preferred to .01, .03, or .05) so that when the dimension is divided by 2, as is necessary in determining the radius from a diameter, the result will be a decimal of two places. However, odd two-place decimals are used when required for design purposes, such as in dimensioning points on a smooth curve or when strength or clearance is a factor.

A typical example of the use of the complete decimal-inch system is shown in Fig. 13.9. The use of the preferred decimal-millimeter system is shown in Fig. 13.10.

397

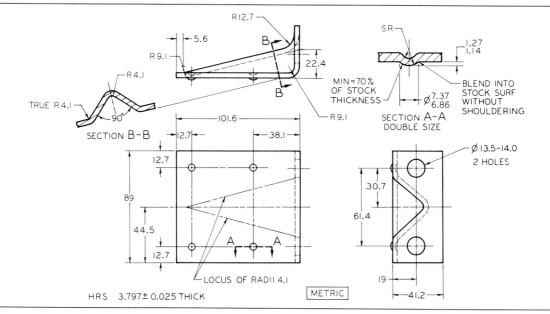

Fig. 13.10   Complete Metric Dimensioning.

When a decimal value is to be rounded off to fewer places than the calculated number, regardless of the unit of measurement involved, the method prescribed is as follows.

The last figure to be retained should not be changed when the figure beyond the last figure to be retained is less than 5.

EXAMPLE   3.46325, if rounded off to three places, should be 3.463.

The last figure to be retained should be increased by 1 when the figure beyond the last figure to be retained is greater than 5 (or two figures are greater than 50).

EXAMPLE   8.37652, if rounded off to three places, should be 8.377.

The last figure to be retained should be unchanged if it is even, or increased by 1 if odd, when followed by exactly 5.

EXAMPLE   4.365 becomes 4.36 when rounded off to two places. Also 4.355 becomes 4.36 when cut off to two places.

The use of the metric system means not only a changeover of measuring equipment but also a changeover in thinking on the part of drafters and designers. They must stop thinking in terms of inches and common fractions and think in terms of millimeters and other SI units. Dimensioning practices remain essentially the same; only the units are changed. Compare Figs. 13.9 and 13.10.

Shop scales and drafting scales for use in the decimal systems are available in a variety of forms. The drafting scale is known as the *decimal scale,* and is discussed in §2.29. See inside the front cover for a two-, three-, and four-place decimal equivalent table.

Once the metric system is installed, the advantages in computation, in checking, and in simplified dimensioning techniques are considerable.

13.11   Dimension Figures

The importance of good lettering of dimension figures cannot be overstated. The shop produces according to the directions on the drawing, and to save time and prevent costly mistakes, all let-

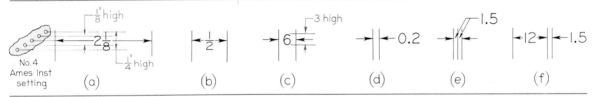

Fig. 13.11   Dimension Figures.   Metric dimensions (c)–(f).

tering should be perfectly legible. A complete discussion of numerals is given in §§4.18–4.20. The standard height for whole numbers is $\frac{1}{8}''$ (3 mm), and for fractions double that, or $\frac{1}{4}''$ (6 mm). Beginners should use guide lines, as shown in Figs. 4.27 and 4.28. The numerator and denominator of a fraction should be clearly separated from the fraction bar, and the fraction bar should always be horizontal, as in Fig. 4.29 (c). An exception to this may be made in crowded places, such as parts lists, but never in dimensioning.

Legibility should never be sacrificed by crowding dimension figures into limited spaces. For every such case there is a practical and effective method, as shown in Fig. 13.11. At (a) and (b) there is enough space for the dimension line, the numeral, and the arrowheads. At (c) there is only enough room for the figure, and the arrowheads are placed outside. At (d) both the arrowheads and the figure are placed outside. Other methods are shown at (e) and (f).

If necessary, a removed partial view may be drawn to an enlarged scale to provide the space needed for clear dimensioning, Fig. 9.22.

Methods of lettering and displaying decimal dimension figures are shown in Fig. 13.12. All numerals are 3 mm ($\frac{1}{8}''$) high whether on one or two lines. The space between lines of numerals is 1.5 mm ($\frac{1}{16}''$) or 0.8 mm ($\frac{1}{32}''$) on each side of a dimension line. To draw guide lines with the Ames Lettering Guide, Fig. 4.24, use the No. 4 setting or appropriate metric setting and the center column of holes.

Make all decimal points bold, allowing ample space. Where the metric dimension is a whole number, neither a decimal point nor a zero is given, Fig. 13.11 (c) and (f). Where the metric dimension is less than 1 millimeter, a zero precedes the decimal point, as at (d). Where the dimension exceeds a whole number by a fraction of 1 millimeter, the last digit to the right of the decimal point is not followed by a zero, (e) and (f), except when expressing tolerances.

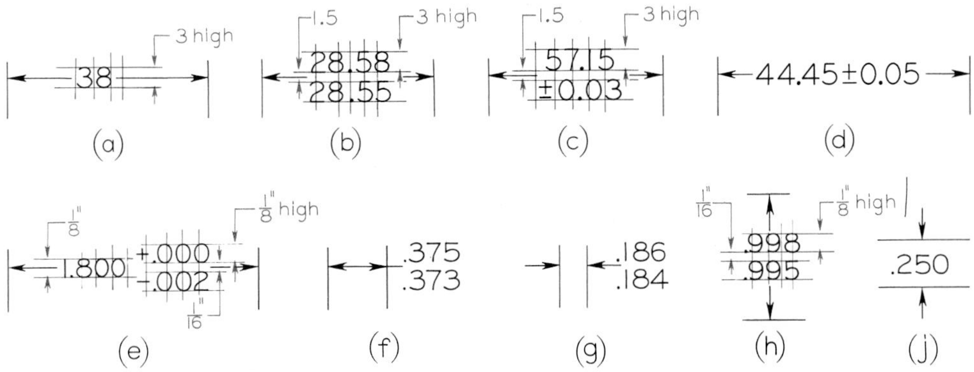

Fig. 13.12   Decimal Dimension Figures.   Metric dimensions (a)–(d).

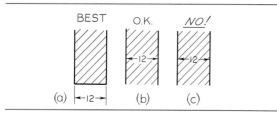

Fig. 13.13   Dimensions and Section Lines.   Metric.

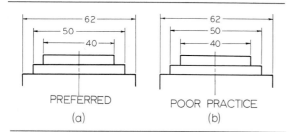

Fig. 13.14   Staggered Numerals.   Metric.

Where the decimal-inch dimension is used on drawings, a zero is not used before the decimal point of values less than 1 inch, Fig. 13.12 (f)–(j). The decimal-inch dimension is expressed to the same number of decimal places as its tolerance. Thus, zeros are added to the right of the decimal point as necessary, as at (e).

Never letter a dimension figure over any line on the drawing, but break the line if necessary. Place dimension figures outside a sectioned area if possible, Fig. 13.13 (a). When a dimension must be placed on a sectioned area, leave an opening in the section lining for the dimension figure, (b).

In a group of parallel dimension lines, the numerals should be staggered, as in Fig. 13.14 (a), and not stacked up one above the other, as at (b).

Dual dimensioning, though not recommended, is used to show metric and decimal-inch dimensions on the same drawing. Two methods of displaying the dual dimensions are as follows.

POSITION METHOD   The millimeter value is placed above the inch dimension and is sepa-

rated by a dimension line or an added line for some dimensions when the unidirectional system of dimensioning is used. An alternative arrangement in a single line places the millimeter dimension to the left of the inch dimension, separated by a slash line (virgule). Each drawing should illustrate the dimension identification as $\frac{\text{MILLIMETER}}{\text{INCH}}$ or MILLIMETER/INCH. (Placement of the inch dimension above or to the left of the millimeter is also acceptable.)

EXAMPLES

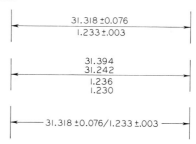

BRACKET METHOD   The millimeter dimension is enclosed in square brackets, [ ]. The location of this dimension is optional but should be uniform on any drawing, that is, above or below or to the left or right of the inch dimension. Each drawing should include a note to identify the dimension values as DIMENSIONS IN [ ] ARE MILLIMETERS.

EXAMPLES

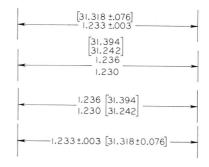

When converting a decimal-inch dimension to millimeters, multiply the inch dimension by 25.4 and round off to one less digit to the right of the decimal point than for the inch value (see §13.10). When converting a millimeter dimen-

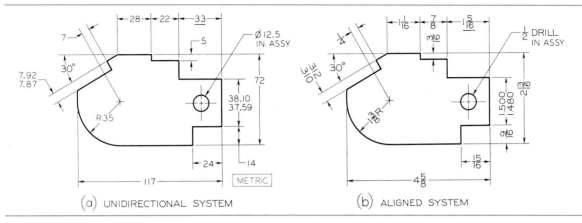

(a) UNIDIRECTIONAL SYSTEM   (b) ALIGNED SYSTEM

Fig. 13.15   Directions of Dimension Figures.

sion to inches, divide the millimeter dimension by 25.4 and round off to one more digit to the right of the decimal point than for the millimeter value.

13.12   Direction of Dimension Figures

Two systems of reading direction for dimension figures are available. In the preferred *unidirectional system,* approved by ANSI, Fig. 13.15 (a), all dimension figures and notes are lettered horizontally on the sheet and are read from the bottom of the drawing. The unidirectional system has been extensively adopted in the aircraft, automotive, and other industries because it is easier to use and read, especially on large drawings. In the *aligned system,* Fig. 13.15 (b), all dimension figures are aligned with the dimension lines

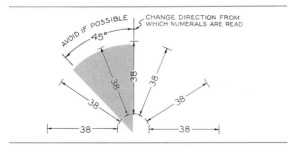

Fig. 13.16   Directions of Dimensions.

so that they may be read from the right side of the sheet. Dimension lines in this system should not run in the directions included in the shaded area of Fig. 13.16, if avoidable.

In both systems, dimensions and notes shown with leaders are aligned with the bottom of the drawing. Notes without leaders should also be aligned with the bottom of the drawing.

13.13   Millimeters and Inches

Millimeters are indicated by the lowercase letters mm placed one space to the right of the numeral; thus, 12.5 mm. *Meters* are indicated by the lowercase letter m placed similarly; thus, 50.6 m. *Inches* are indicated by the symbol " placed slightly above and to the right of the numeral; thus, $2\frac{1}{2}''$. *Feet* are indicated by the symbol ' similarly placed; thus, $3'-0$, $5'-6$, $10'-0\frac{1}{4}$. It is customary in such expressions to omit the inch marks.

It is standard practice to omit mm designations and inch marks on a drawing except when there is a possibility of misunderstanding. For example, 1 VALVE should be 1″ VALVE, and 1 DRILL should be 1″ DRILL or 1 mm DRILL. Where some inch dimensions are shown on a millimeter-dimensioned drawing, the abbreviation IN. follows the inch values.

In some industries all dimensions, regardless of size, are given in inches; in others dimensions

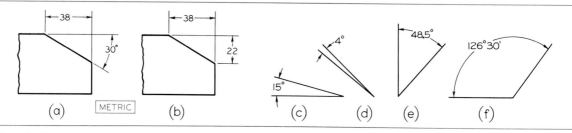

Fig. 13.17   Angles.

13.14   Dimensioning Angles

Angles are dimensioned, preferably, by means of an angle in degrees and a linear dimension, Fig. 13.17 (a), or by means of coordinate dimensions of the two legs of a right triangle, (b). The coordinate method is more suitable for work requiring a high degree of accuracy. Variations of angle (in degrees) are hard to control because the amount of variation increases with the distance from the vertex of the angle. Methods of indicating various angles are shown from (c) to (f). Tolerances of angles are discussed in §14.18.

When degrees alone are indicated, the symbol ° is used. When minutes alone are given, the number should be preceded by 0°.

EXAMPLE   0° 23′.

In all cases, whether in the unidirectional system or in the aligned system, the dimension figures for angles are lettered on horizontal guide lines. For a general discussion of angles, see §5.2.

In civil engineering drawings, *slope* represents the angle with the horizontal, whereas *batter* is the angle referred to the vertical. Both are expressed by making one member of the ratio equal to 1, as shown in Fig. 13.18. *Grade*, as of a highway, is similar to slope but is expressed in percentage of rise per 100′ of run. Thus a 20′ rise in a 100′ run is a grade of 20 percent.

In structural drawings, angular measurements are made by giving the ratio of "run" to

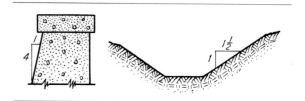

Fig. 13.18   Angles in Civil Engineering Projects.

"rise," with the larger size being 12″. These right triangles are referred to as *bevels*.

13.15   Dimensioning Arcs

A circular arc is dimensioned in the view in which its true shape is shown by giving the numerical denoting its radius, preceded by the abbreviation R, as shown in Fig. 13.19. The centers may be indicated by small crosses to clarify the drawing but not for small or unimportant radii. Crosses should not be shown for undimensioned arcs. As shown at (a) and (b), when there is room enough, both the numeral and the arrowhead are placed inside the arc. At (c) the arrowhead is left inside, but the numeral had to be moved outside. At (d) both the arrowhead and the numeral had to be moved outside. At (e) is shown an alternate method to (c) or (d) to be used when section lines or other lines are in the way. Note that in the unidirectional system, all of these numerals are lettered horizontally on the sheet.

For a long radius, as shown at (f), when the center falls outside the available space, the dimension line is drawn toward the actual center; but a false center may be indicated and dimension line "jogged" to it, as shown.

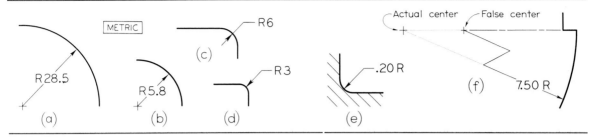

Fig. 13.19   Dimensioning Arcs

13.16   Fillets and Rounds

Individual fillets and rounds are dimensioned as any arc, as shown in Fig. 13.19 (c), (d), and (e). If there are only a few and they are obviously the same size, as in Fig. 13.43 (5), one typical radius is sufficient. However, fillets and rounds are often quite numerous on a drawing, and most of them are likely to be some standard size, as R3 and R6 when dimensioning in metric or .125R and .250R when using the decimal-inch system.

In such cases it is customary to give a note in the lower portion of the drawing to cover all uniform fillets and rounds, thus,

> FILLETS R6 AND ROUNDS R3 UNLESS
> OTHERWISE SPECIFIED

or

> ALL CASTING RADII R6 UNLESS NOTED

or simply

> ALL FILLETS AND ROUNDS R6

For a discussion of fillets and rounds in the shop, see §12.5.

13.17   Finish Marks

A *finish mark* is used to indicate that a surface is to be machined, or finished, as on a rough casting or forging. To the patternmaker or die-maker, a finish mark means that allowance of extra metal in the rough workpiece must be provided for the machining. See §12.4. On drawings of parts to be machined from rolled stock, finish marks are generally unnecessary, for it is obvious that the surfaces are finished. Similarly, it is not necessary to show finish marks when an operation is specified in a note that indicates machining, such as drilling, reaming, boring, countersinking, counterboring, and broaching, or when the dimension implies a finished surface, such as ∅6.22–6.35 (metric) or 2.45–2.50 DIA (decimal-inch).

Three styles of finish marks, the general ∨ symbol, the new basic √ symbol, and the traditional ƒ symbol, are used to indicate an ordinary smooth machined surface. The ∨ symbol, Fig. 13.20 (a), is like a capital V, made about 3 mm ($\frac{1}{8}''$) high in conformity with the height of dimensioning lettering. The extended √ symbol, preferred by ANSI, Fig. 13.20 (b), is like a

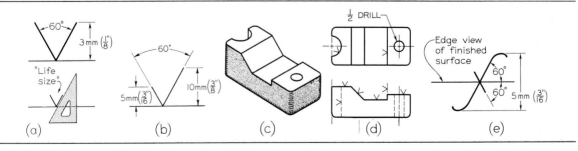

Fig. 13.20   Finish Marks.

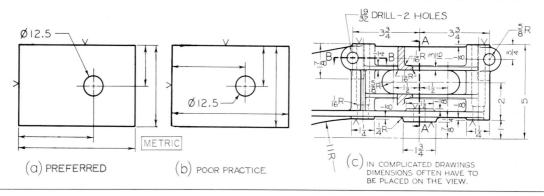

Fig. 13.21   Dimensions On or Off the Views.

larger capital with the right leg extended. The short leg is made about 5 mm ($\frac{3}{16}''$) high and the height of the long leg is about 10 mm ($\frac{3}{8}''$). The basic symbol may be altered for more elaborate surface texture specifications; see §14.22.

For best results all finished marks should be drawn with the aid of a template or the 30° × 60° triangle. The point of the ∨ symbol should be directed inward toward the body of metal in a manner similar to that of a tool bit. The √ symbol is not shown upside down; see Figs. 13.35 and 14.45.

The preferred form and placement for the ⨍ symbol are described and given in Fig. 13.20 (e). The ⨍ symbol is in limited use and found mainly on drawings made in accordance with earlier drafting standards.

At (c) is shown a simple casting having several finished surfaces, and at (d) are shown two views of the same casting, showing how the finish marks are indicated on a drawing. *The finish mark is shown only on the edge view of a finished surface and is repeated in any other view in which the surface appears as a line, even if the line is a hidden line.*

The several kinds of finishes are detailed in machine shop practice manuals. The following terms are commonly used: *finish all over, rough finish, file finish, sand blast, pickle, scrape, lap, hone, grind, polish, burnish, buff, chip, spot-* *face, countersink, counterbore, core, drill, ream, bore, tap, broach, knurl,* and so on. When it is necessary to control the surface texture of finished surfaces beyond that of an ordinary machine finish, the new basic √ symbol is used as a base for the more elaborate surface quality symbols as discussed in §14.22.

If a part is to be finished all over, finish marks should be omitted, and a general note, such as FINISH ALL OVER or FAO, should be lettered on the lower portion of the sheet.

13.18   Dimensions On or Off Views

Dimensions should not be placed on a view unless doing so promotes the clearness of the drawing. The ideal form is shown in Fig. 13.21 (a), in which all dimensions are placed outside the view. Compare this with the evidently poor practice shown at (b). This is not to say that a dimension should never be placed on a view, for in many cases, particularly in complicated drawings, this is necessary, as shown at (c). Certain radii and other dimensions are given on the views, but in each case investigation will reveal a good reason for placing the dimension on the view. *Place dimensions outside of views, except where directness of application and clarity are gained by placing them on the views where they will be closer to the features dimensioned.* When

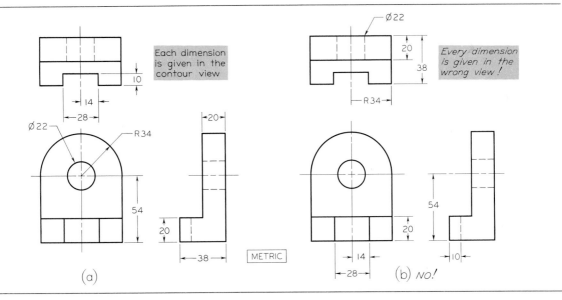

Fig. 13.22   Contour Dimensioning.

a dimension must be placed in a sectioned area or on the view, leave an opening in the sectioned area or a break in the lines for the dimension figures, Figs. 13.13 (b) and 13.21 (c).

13.19 Contour Dimensioning

Views are drawn to describe the shapes of the various features of the object, and dimensions are given to define exact sizes and locations of those shapes. It follows that *dimensions should be given where the shapes are shown,* that is, in the views where the contours are delineated, as shown in Fig. 13.22 (a). Incorrect placement of the dimensions is shown at (b).

If individual dimensions are attached directly to the contours that show the shapes being dimensioned, this will automatically prevent the attachment of dimensions to hidden lines, as shown for the depth 10 of the slot at (b). It will also prevent the attachment of dimensions to a visible line, the meaning of which is not clear in a particular view, such as dimension 20 for the height of the base at (b).

Although the placement of notes for holes follows the contour rule wherever possible, as shown at (a), the diameter of an external cylindrical shape is preferably given in the rectangular view where it can be readily found near the dimension for the length of the cylinder, as shown in Figs. 13.23 (b), 13.26, and 13.27.

13.20 Geometric Breakdown

Engineering structures are composed largely of simple geometric shapes, such as the prism, cylinder, pyramid, cone, and sphere, as shown in Fig. 13.23 (a). They may be exterior (positive) or interior (negative) forms. For example, a steel shaft is a positive cylinder, and a round hole is a negative cylinder.

These shapes result directly from the design necessity to keep forms as simple as possible and from the requirements of the fundamental manufacturing operations. Forms having plane surfaces are produced by planing, shaping, milling, and so forth, while forms having cylindrical, conical, or spherical surfaces are produced by turn-

405

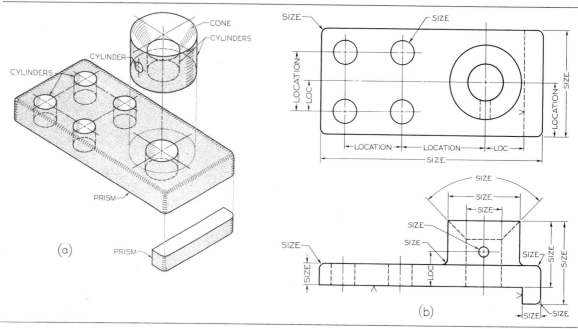

Fig. 13.23   Geometric Breakdown.

ing, drilling, reaming, boring, countersinking, and other rotary operations. See Chapter 12, "Manufacturing Processes."

The dimensioning of engineering structures involves two basic steps.

1. Give the dimensions showing the *sizes* of the simple geometric shapes, called *size dimensions*.

2. Give the dimensions *locating* these elements with respect to each other, called *location dimensions*.

The process of geometric analysis is very helpful in dimensioning any object, but must be modified when there is a conflict either with the function of the part in the assembly or with the manufacturing requirements in the shop.

Figure 13.23 (b) is a multiview drawing of the object shown in isometric at (a). Here it will be seen that each geometric shape is dimensioned with size dimensions and that these shapes are then located with respect to each other with lo-

cation dimensions. Note that a *location dimension locates a three-dimensional geometric element* and not just a surface; otherwise, all dimensions would have to be classified as location dimensions.

13.21   Size Dimensions—Prisms

The right rectangular prism, Fig. 5.7, is probably the most common geometric shape. Front and top views are dimensioned as shown in Fig. 13.24 (a) or (b). The height and width are given in the front view and the depth in the top view. The vertical dimensions can be placed on the left or right provided both of them are placed in line. The horizontal dimension applies to both the front and top views and should be placed between them, as shown, and not above the top or below the front view.

Front and side views should be dimensioned as at (c) or (d). The horizontal dimensions can be placed above or below the views, provided both are placed in line. The dimension between

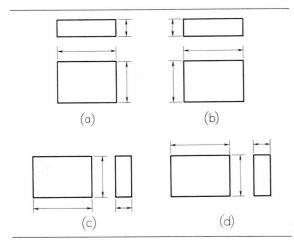

Fig. 13.24   Dimensioning Rectangular Prisms.

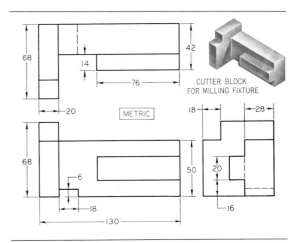

Fig. 13.25   Dimensioning a Machine Part That Is Composed of Prismatic Shapes.

views applies to both views and should not be placed elsewhere without a special reason.

An application of size dimensions to a machine part composed entirely of rectangular prisms is shown in Fig. 13.25.

13.22   Size Dimensions—Cylinders

The right circular cylinder is the next most common geometric shape and is commonly seen as a shaft or a hole. The general method of dimensioning a cylinder is to give both its diameter and its length in the rectangular view, Fig. 13.26. If the cylinder is drawn in a vertical position, the length or altitude of the cylinder may be given at the right as at (a), or on the left as at (b). If the cylinder is drawn in a horizontal position, the length may be given above the rectangular view as at (c) or below as at (d). An application showing the dimensioning of cylindrical shapes is shown in Fig. 13.27. The use of a diagonal diameter in the circular view, in addition to the method shown in Fig. 13.26, is not recommended except in special cases when clearness is gained thereby. The use of several diagonal diameters on the same center is definitely to be discouraged, since the result is usually confusing.

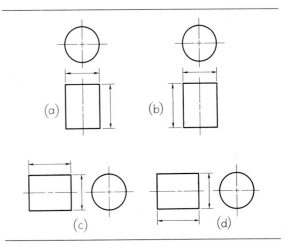

Fig. 13.26   Dimensioning Cylinders.

The radius of a cylinder should never be given, since measuring tools, such as the micrometer caliper, are designed to check diameters.

Small cylindrical holes, such as drilled, reamed, or bored holes, are usually dimensioned by means of notes specifying the diameter and the depth, with or without manufacturing operations, Figs. 13.27 and 13.32.

407

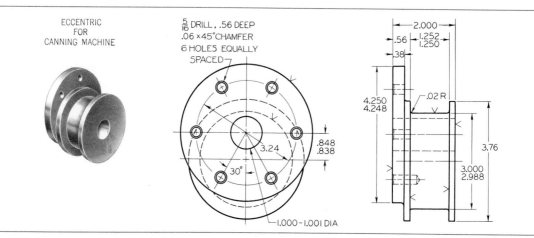

Fig. 13.27   Dimensioning a Machine Part That Is Composed of Cylindrical Shapes.

The diameter symbol ∅ should be given before all diametral dimensions on metric drawings rather than the abbreviation DIA following the dimension that is used on decimal-inch drawings, Fig. 13.28 (a) [ANSI Y14.5M–1982 (R1988)]. In some cases, the symbol on metric drawings or the abbreviation on decimal-inch drawings may be used to eliminate the circular view, as in (b).

13.23   Symbols and Size Dimensions—Miscellaneous Shapes

Traditional terms and abbreviations used to describe various shapes and manufacturing processes, in addition to size specifications, are employed throughout this text. A variety of dimensioning symbols, introduced by ANSI [Y14.5M–1982 (R1988)] to replace traditional terms or abbreviations, are given with construc-

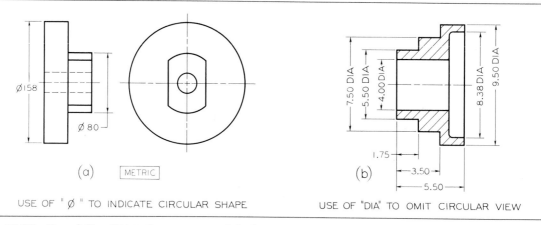

Fig. 13.28   Use of ∅ or DIA in Dimensioning Cylinders.

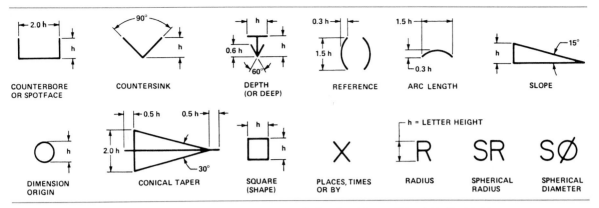

COUNTERBORE OR SPOTFACE — COUNTERSINK — DEPTH (OR DEEP) — REFERENCE — ARC LENGTH — SLOPE

DIMENSION ORIGIN — CONICAL TAPER — SQUARE (SHAPE) — PLACES, TIMES OR BY — RADIUS — SPHERICAL RADIUS — SPHERICAL DIAMETER

h = LETTER HEIGHT

Fig. 13.29 Form and Proportion of Dimensioning Symbols [ANSI Y14.5M–1982 (R1988)].

tion details in Fig. 13.29. Traditional terms and abbreviations are suitable for use where the symbols are not desired. Typical applications of some of these symbols are given in Fig. 13.30.

A triangular prism is dimensioned, Fig. 13.31 (a) by giving the height, width, and displacement of the top edge in the front view and the depth in the top view.

A rectangular pyramid is dimensioned, (b), by giving the heights in the front view, and the

dimensions of the base and the centering of the vertex in the top view. If the base is square, (c), it is necessary to give the dimensions for only one side of the base, provided it is labeled SQ as shown or preceded by the square symbol ☐. See Fig. 13.29 for form and proportions of dimensioning symbols.

A cone is dimensioned, (d), by giving its altitude and diameter of the base in the triangular view. A frustum of a cone may be dimensioned, (e), by giving the vertical angle and the diameter

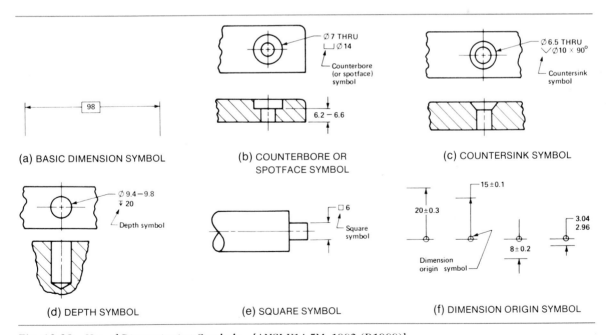

(a) BASIC DIMENSION SYMBOL

(b) COUNTERBORE OR SPOTFACE SYMBOL

(c) COUNTERSINK SYMBOL

(d) DEPTH SYMBOL

(e) SQUARE SYMBOL

(f) DIMENSION ORIGIN SYMBOL

Fig. 13.30 Use of Dimensioning Symbols [ANSI Y14.5M–1982 (R1988)].

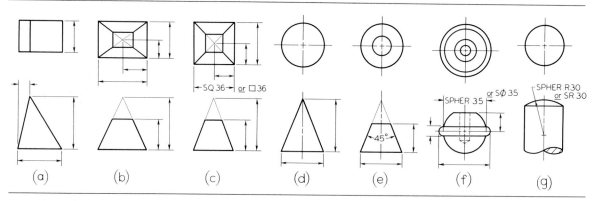

Fig. 13.31   Dimensioning Various Shapes.

of one of the bases. Another method is to give the length and the diameters of both ends in the front view. Still another is to give the diameter at one end and the amount of taper per foot in a note, §13.33.

In Fig. 13.31 (f) is shown a two-view drawing of a plastic knob. The main body is spherical and is dimensioned by giving its diameter preceded by the abbreviation and symbol for spherical diameter S⌀ or followed by the abbreviation SPHER. A bead around the knob is in the shape of a torus, Fig. 5.7, and it is dimensioned by giving the thickness of the ring and the outside diameter, as shown. At (g) a spherical end is dimensioned by a radius preceded by the abbreviation SR or followed by SPHER R.

Internal shapes corresponding to the external shapes in Fig. 13.31 would be dimensioned in a similar manner.

13.24   Size Dimensioning of Holes

Holes that are to be drilled, bored, reamed, punched, cored, and so on are usually specified by standard notes, as shown in Figs. 7.40, 13.32 (a), and 13.44. The order of items in a note corresponds to the order of procedure in the shop in producing the hole. Two or more holes are dimensioned by a single note, the leader pointing to one of the holes, as shown at the top of Fig. 13.32.

As illustrated in Figs. 7.40 and 13.32, the leader of a note should, as a rule, point to the circular view of the hole. It should point to the rectangular view only when clearness is promoted thereby. When the circular view of the hole has two or more concentric circles, as for counterbored, countersunk, or tapped holes, the arrowhead should touch the outer circle, Fig. 13.44 (b), (c), and (e)–(j).

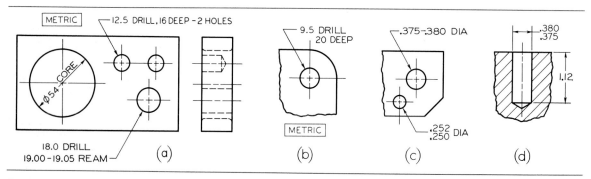

Fig. 13.32   Dimensioning Holes.

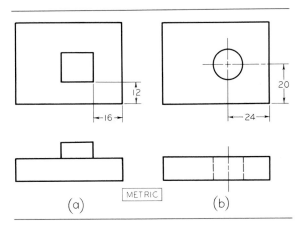

Fig. 13.33 Location Dimensions.

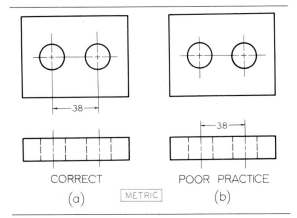

Fig. 13.34 Locating Holes.

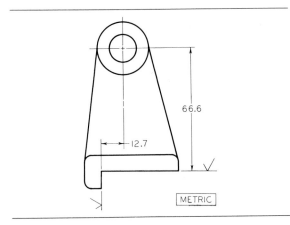

Fig. 13.35 Dimensions to Finished Surfaces.

Notes should always be lettered horizontally on the drawing paper, and guide lines should always be used.

The use of decimal fractions to designate metric or inch drill sizes has gained wide acceptance,* Fig. 13.32 (b). For numbered or letter-size drills, Appendix 16, it is recommended that the decimal size be given in this manner, or given in parentheses; thus, #28 (.1405) DRILL, or "P" (.3230) DRILL. Metric drills are all decimal size and are not designated by number or letter.

On drawings of parts to be produced in large quantity for interchangeable assembly, dimensions and notes may be given without specification of the manufacturing process to be used. Only the dimensions of the holes are given, without reference to whether the holes are to be drilled, reamed, or punched, as in Fig. 13.32 (c) and (d). It should be realized that even though manufacturing operations are omitted from a note, the tolerances indicated would tend to dictate the manufacturing processes required.

13.25 Location Dimensions

After the geometric shapes composing a structure have been dimensioned for *size,* as discussed, *location dimensions* must be given to show the relative positions of these geometric shapes, as shown in Fig. 13.23. Fig. 13.33 (a) shows that rectangular shapes, whether in the form of solids or of recesses, are located with reference to their faces. At (b), cylindrical or conical holes or bosses, or other symmetrical shapes, are located with reference to their center lines.

As shown in Fig. 13.34, location dimensions for holes are preferably given in the circular view of the holes.

Location dimensions should lead to finished surfaces wherever possible, Fig. 13.35, because

*Although drills are still listed fractionally in manufacturers' catalogs, many companies have supplemented drill and wire sizes with a decimal value. In many cases the number, letter, or common fraction has been replaced by the decimal-inch size. Metric drills are usually listed separately with a decimal-millimeter value.

411

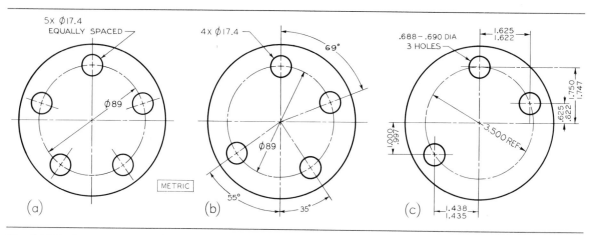

Fig. 13.36　Locating Holes About a Center.

rough castings and forgings vary in size, and unfinished surfaces cannot be relied on for accurate measurements. Of course, the *starting dimension,* used in locating the first machined surface on a rough casting or forging, must necessarily lead from a rough surface, or from a center or a center line of the rough piece. See Figs. 16.102 and 16.103.

In general, location dimensions should be built from a finished surface as a datum plane, or from an important center or center line.

When several cylindrical surfaces have the same center line, as in Fig. 13.28 (b), it is not necessary to locate them with respect to each other.

Holes equally spaced about a common center may be dimensioned, Fig. 13.36 (a), by giving the diameter (diagonally) of the *circle of centers,* or *bolt circle,* and specifying "equally spaced" in the note. Repetitive features or dimensions may be specified by the use of an X preceded with a numeral to indicate the number of times or places the feature is required. Allow a space between the letter X and the dimension as at (a) and (b).

Holes unequally spaced, (b), are located by means of the bolt circle diameter plus angular

measurements with reference to *only one* of the center lines, as shown.

Where greater accuracy is required, coordinate dimensions should be given, as at (c). In this case, the diameter of the bolt circle is marked REF, or enclosed in parentheses, ( ), Fig. 13.29, to indicate that it is to be used only as a *reference dimension.* Reference dimensions are given for information only. They are not intended to be measured and do not govern the manufacturing operations. They represent calculated dimensions and are often useful in showing the intended design sizes. See Fig. 13.36 (c).

When several nonprecision holes are located on a common arc, they are dimensioned, Fig. 13.37 (a), by giving the radius and the angular measurements from a *base line,* as shown. In this case, the base line is the horizontal center line.

At (b) the three holes are on a common center line. One dimension locates one small hole from the center; the other gives the distance between the small holes. Note the omission of a dimension at X. This method is used when (as is usually the case) the distance between the small holes is the important consideration. If the relation between the center hole and each of the

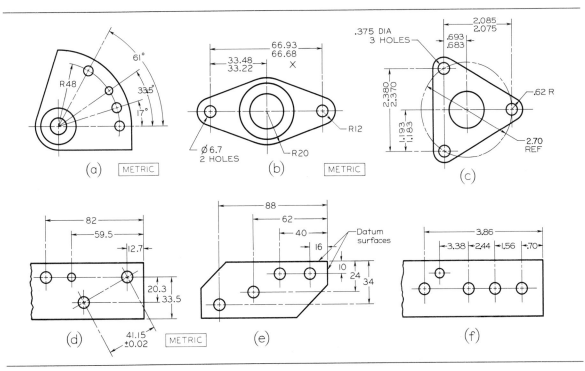

Fig. 13.37 Locating Holes.

small holes is more important, then include the distance at X, and mark the overall dimension REF.

At (c) is another example of coordinate dimensioning. The three small holes are on a bolt circle whose diameter is marked REF for reference purposes only. From the main center, the small holes are located in two mutually perpendicular directions.

Another example of locating holes by means of linear measurements is shown at (d). In this case, one such measurement is made at an angle to the coordinate dimensions because of the direct functional relationship of the two holes.

At (e) the holes are located from two *baselines* or *datums*. When all holes are located from a common datum, the sequence of measuring and machining operations is controlled, overall tolerance accumulations are avoided, and proper functioning of the finished part is assured as intended by the designer. The datum surfaces selected must be more accurate than any measurement made from them, must be accessible during manufacture, and must be arranged so as to facilitate tool and fixture design. Thus, it may be necessary to specify accuracy of the datum surfaces in terms of straightness, roundness, flatness, and so forth. See §14.16.

At (f) is shown a method of giving, in a single line, all the dimensions from a common datum. Each dimension except the first has a single arrowhead and is accumulative in value. The final and longest dimension is separate and complete.

These methods of locating holes are equally applicable to locating pins or other symmetrical features.

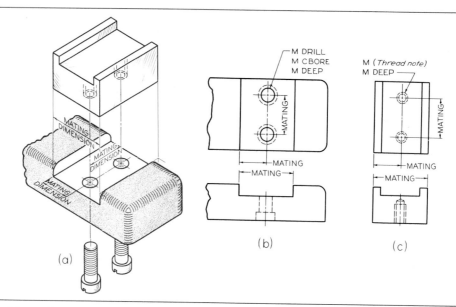

Fig. 13.38   Mating Dimensions.

13.26   Mating Dimensions

In dimensioning a single part, its relation to mating parts must be taken into consideration. For example, in Fig. 13.38 (a), a guide block fits into a slot in a base. Those dimensions common to both parts are *mating dimensions,* as indicated.

These mating dimensions should be given on the multiview drawings in the corresponding locations, as shown at (b) and (c). Other dimensions are not mating dimensions since they do not control the accurate fitting together of two parts. The actual *values* of two corresponding mating dimensions may not be exactly the same. For example, the width of the slot at (b) may be dimensioned $\frac{1}{32}''$ (0.8 mm) or several thousandths of an inch larger than the width of the block at (c), but these are mating dimensions figured from a single basic width. It will be seen that the mating dimensions shown might have been arrived at from a geometric breakdown, §13.20. However, the mating dimensions need to be identified so that they can be specified in the corresponding locations on the two parts and so that they can be given with the degree of accu-

racy commensurate with the proper fitting of the parts.

In Fig. 13.39 (a) the dimension A should appear on both the drawings of the bracket and of the frame and, therefore, is a necessary mating dimension. At (b), which shows a redesign of the bracket into two parts, dimension A is not used on either part, as it is not necessary to control closely the distance between the cap screws. But dimensions F are now essential mating dimensions and should appear correspondingly on the drawings of both parts. The remaining dimensions E, D, B, and C, at (a), are not considered to be mating dimensions, since they do not directly affect the mating of the parts.

13.27   Machine, Pattern, and Forging Dimensions

In Fig. 13.38 (a), the base is machined from a rough casting; the patternmaker needs certain dimensions to make the pattern, and the machinist needs certain dimensions for the machining. In some cases one dimension will be used

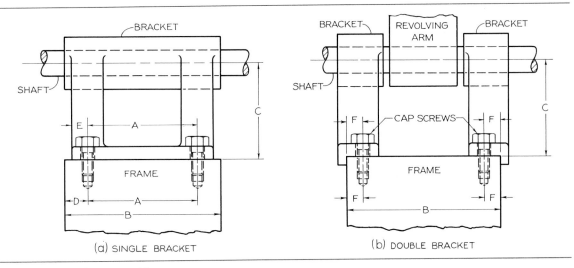

Fig. **13.39**   Bracket Assembly.

by both. Again, in most cases, these dimensions will be the same as those resulting from a geometric breakdown, §13.20, but it is important to identify them in order to assign values to them.

The same part is shown in Fig. 13.40, with the machine dimensions and pattern dimensions identified by the letters M and P. The patternmaker is interested only in the dimensions required to make the pattern, and the machinist, in general, is concerned only with the dimensions needed to machine the part. Frequently, a dimension that is convenient for the machinist is not convenient for the patternmaker, or vice versa. Since the patternmaker uses the drawing only once, while making the pattern, and the machinist refers to it continuously, the dimensions should be given primarily for the convenience of the machinist.

If the part is large and complicated, two separate drawings are sometimes made, one showing the pattern dimensions and the other the machine dimensions. The usual practice, however, is to prepare one drawing for both the patternmaker and the machinist. See §12.4.

For forgings, it is common practice to make separate forging drawings and machining drawings. A forging drawing of a connecting rod, showing only the dimensions needed in the forge shop, is shown in Fig. 16.36. A machining drawing of the same part, but containing only the dimensions needed in the machine shop, is shown in Fig. 16.37.

Unless a decimal system is used, §13.10, the pattern dimensions are nominal, usually to the nearest $\frac{1}{16}''$, and given in whole numbers and common fractions. If a machine dimension is

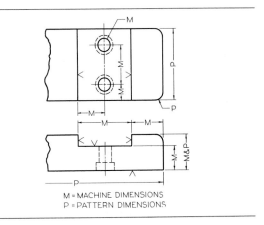

M = MACHINE DIMENSIONS
P = PATTERN DIMENSIONS

Fig. **13.40**   Machine and Pattern Dimensions.

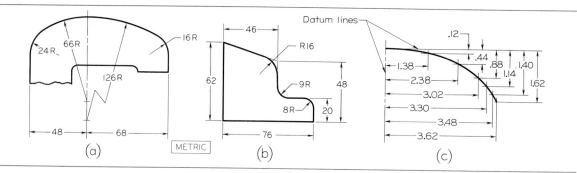

Fig. 13.41   Dimensioning Curves.

given in whole numbers and common fractions, the machinist is usually allowed a tolerance (permissible in variation in size) of $\pm\frac{1}{64}''$. Some companies specify a tolerance of $\pm.010''$ on all common fractions. If greater accuracy is required, the dimensions are given in decimal form. Metric dimensions are given to one or more places, and decimal-inch dimensions are given to three or more places, §§13.10 and 14.1. Remember that 0.1 mm is approximately .004 inch.

13.28   Dimensioning of Curves

Curved shapes may be dimensioned by giving a group of radii, as shown in Fig. 13.41 (a). Note that in dimensioning the R126 arc whose center is inaccessible, the center may be moved inward along a center line, and a jog made in the dimension line. See also Fig. 13.19 (f). Another method is to dimension the outline envelope of a curved shape so that the various radii are self-locating from "floating centers," as at (b). Either a circular or a noncircular curve may be dimensioned by means of coordinate dimensions referred to datums, as shown at (c). See also Fig. 13.6 (a).

13.29   Dimensioning of Rounded-End Shapes

The method used for dimensioning rounded-end shapes depends on the degree of accuracy required, Fig. 13.42. When precision is not nec-

essary, the methods used are those that are convenient for manufacturing, as at (a), (b), and (c).

At (a) the link, to be cast or to be cut from sheet metal or plate, is dimensioned as it would be laid out for manufacture, by giving the center-to-center distance and the radii of the ends. Note that only one such radius dimension is necessary, but that the number of places may be included with the size dimension.

At (b) the pad on a casting, with a milled slot, is dimensioned from center to center for the convenience of both the patternmaker and the machinist in layout. An additional reason for the center-to-center distance is that it gives the total travel of the milling cutter, which can be easily controlled by the machinist. The width dimension indicates the diameter of the milling cutter; hence, it is incorrect to give the radius of a machined slot. On the other hand, a cored slot (see §12.3) should be dimensioned by radius in conformity with the patternmaker's layout procedure.

At (c) the semicircular pad is laid out in a similar manner to the pad at (b), except that angular dimensions are used. Angular tolerances, §14.18, can be used if necessary.

When accuracy is required, the methods shown at (d) to (g) are recommended. Overall lengths of rounded-end shapes are given in each case, and radii are indicated, but without specific values. In the example at (f), the center-to-center distance is required for accurate location of the holes.

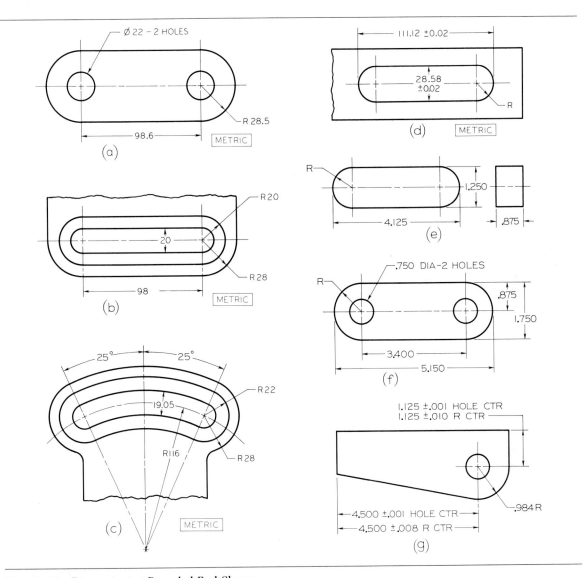

Fig. 13.42 Dimensioning Rounded-End Shapes.

At (g) the hole location is more critical than the location of the radius; hence, the two are located independently, as shown.

13.30 Superfluous Dimensions

Fig. 13.43 All necessary dimensions must be shown, but the designer should avoid giving unnecessary or superfluous dimensions. Dimen-

sions should not be repeated on the same view or on different views, nor should the same information be given in two different ways.

Figure 13.43 (2) illustrates a type of superfluous dimensioning that should generally be avoided, especially in machine drawing where accuracy is important. The production personnel should not be allowed a choice between two dimensions. *Avoid "chain" dimensioning,* in

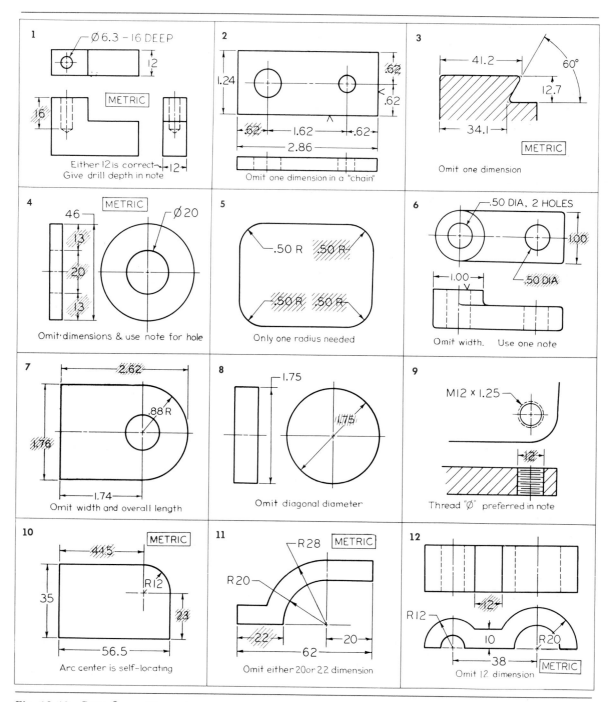

Fig. 13.43 Superfluous Dimensions.

which a complete series of detail dimensions is given, together with an overall dimension. In such cases, one dimension of the chain should be omitted, as shown, so that the machinist is obliged to work from one surface only. This is particularly important in tolerance dimensioning, §14.1, where an accumulation of tolerances can cause serious difficulties. See also §14.9.

Some inexperienced detailers have the habit of omitting both dimensions, such as those at the right at (2), on the theory that the holes are symmetrically located and will be understood to be centered. One of the two location dimensions should be given.

As shown at (5), when one dimension clearly applies to several identical features, it need not be repeated, but the number of places should be indicated. Dimensions for fillets and rounds and other noncritical features need not be repeated nor number of places specified, as at (5).

For example, the radii of the rounded ends in Fig. 13.42 (a)–(f) need not be repeated, and in Fig. 13.1 both ribs are obviously the same thickness so it is unnecessary to repeat the 10 mm dimension.

13.31   Notes

Fig. 13.44   It is usually necessary to supplement the direct dimensions with notes. They should be brief and should be carefully worded so as to be capable of only one interpretation. *Notes should always be lettered horizontally on the sheet, with guide lines, and arranged in a systematic manner.* They should not be lettered in crowded places. Avoid placing notes between views, if possible. They should not be lettered closely enough to each other to confuse the reader or close enough to another view or detail to suggest application to the wrong view. Leaders should be as short as possible and cross as few lines as possible. They should never run through a corner of a view or through any specific points or intersections.

Notes are classified as *general notes* when they apply to an entire drawing and as *local notes* when they apply to specific items.

GENERAL NOTES   General notes should be lettered in the lower right-hand corner of the drawing, above or to the left of the title block, or in a central position below the view to which they apply.

EXAMPLES

FINISH ALL OVER (FAO)

BREAK SHARP EDGES TO R0.8

G33106 ALLOY STEEL–BRINELL 340–380

ALL DRAFT ANGLES 3° UNLESS OTHERWISE SPECIFIED

DIMENSIONS APPLY AFTER PLATING

In machine drawings, the title strip or title block will carry many general notes, including material, general tolerances, heat treatment, and pattern information. See Fig. 16.25.

LOCAL NOTES   Local notes apply to specific operations only and are connected by a leader to the point at which such operations are performed.

EXAMPLES

6.30 DRILL–4 HOLES

45° x 1.6 CHAMFER

96 DP DIAMOND KNURL, RAISED

The leader should be attached at the front of the first word of a note, or just after the last word, and not at any intermediate place.

For information on notes applied to holes, see §13.24.

Certain commonly used abbreviations may be used freely in notes, such as THD, DIA, MAX. The less common abbreviations should be avoided as much as possible. All abbreviations should conform to ANSI Y1.1–1972 (R1984). See Appendix 4 for American National Standard abbreviations. See Figs. 13.29 and 13.30 for form and use of alternative dimensioning symbols.

In general, leaders and notes should not be placed on the drawing until the dimensioning is substantially completed. If notes are lettered

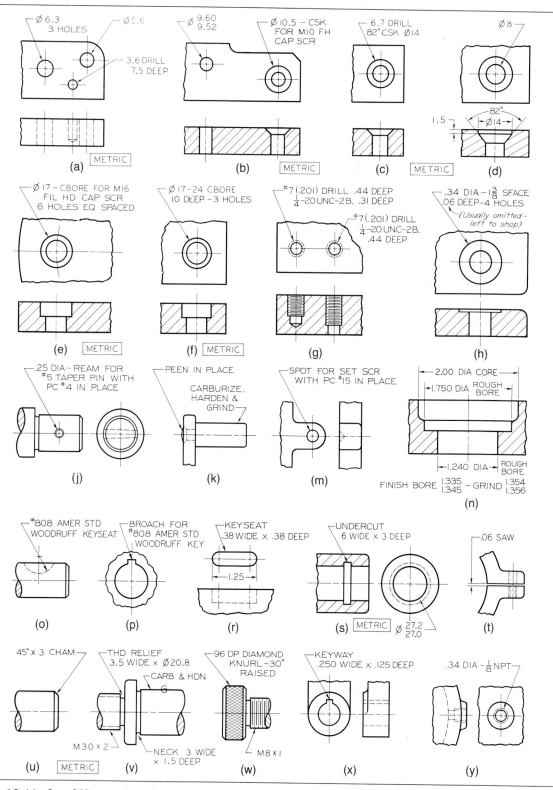

Fig. 13.44 Local Notes. See also Figs. 7.40 and 13.32.

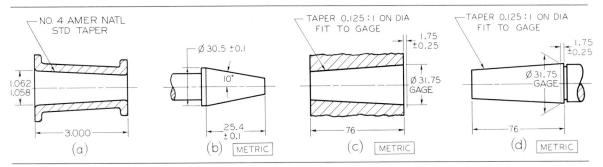

Fig. 13.45 Dimensioning Tapers.

first, almost invariably they will be in the way of necessary dimensions and will have to be moved.

13.32 Dimensioning of Threads

Local notes are used to specify dimensions of threads. For tapped holes the notes should, if possible, be attached to the circular views of the holes, as shown in Fig. 13.44 (g). For external threads, the notes are usually placed in the longitudinal views where the threads are more easily recognized, as at (v) and (w). For a detailed discussion of thread notes, see §15.21.

13.33 Dimensioning of Tapers

A *taper* is a conical surface on a shaft or in a hole. The usual method of dimensioning a taper is to give the amount of taper in a note such as TAPER 0.167 ON DIA (often TO GAGE added), and then give the diameter at one end, plus the length, or give the diameter at both ends and omit the length. *Taper on diameter means the difference in diameter per unit of length.*

Standard machine tapers are used on machine spindles, shanks of tools, pins, for example, and are described in "Machine Tapers," ANSI B5.10–1981 (R1987). Such standard tapers are dimensioned on a drawing by giving the diameter, usually at the large end, the length, and a note, such as NO. 4 AMERICAN NATIONAL STANDARD TAPER. See Fig. 13.45 (a).

For not-too-critical requirements, a taper may be dimensioned by giving the diameter at the large end, the length, and the included angle, all with proper tolerances, (b). Or the diameters of both ends, plus the length, may be given with necessary tolerances.

For close-fitting tapers, the amount of *taper per unit on diameter* is indicated as shown at (c) and (d). A gage line is selected and located by a comparatively generous tolerance, while other dimensions are given appropriate tolerances as required.

13.34 Dimensioning of Chamfers

A *chamfer* is a beveled or sloping edge, and it is dimensioned by giving the length of the offset and the angle, as in Fig. 13.46 (a). A 45° chamfer also may be dimensioned in a manner similar to that shown at (a), but usually it is dimensioned by note without or with the word CHAMFER as at (b).

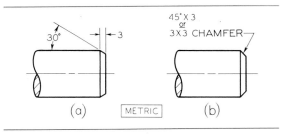

Fig. 13.46 Dimensioning Chamfers.

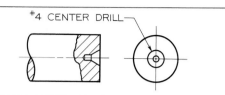

Fig. 13.47 Shaft Center.

13.35 Shaft Centers

Shaft centers are required on shafts, spindles, and other conical or cylindrical parts for turning, grinding, and other operations. Such a center may be dimensioned as shown in Fig. 13.47. Normally the centers are produced by a combined drill and countersink.

13.36 Dimensioning Keyways

Methods of dimensioning keyways for Woodruff keys and stock keys are shown in Fig. 13.48. Note in both cases, the use of a dimension to center the keyway in the shaft or collar. The preferred method of dimensioning the depth of a keyway is to give the dimension from the bottom of the keyway to the opposite side of the shaft or hole, as shown. The method of computing such a dimension is shown at (d). Values for A may be found in machinists' handbooks.

For general information about keys and keyways, see §15.34.

13.37 Dimensioning of Knurls

A *knurl* is a roughened surface to provide a better handgrip or to be used for a press fit between two parts. For handgripping purposes, it is necessary only to give the pitch of the knurl, the type of knurling, and the length of the knurled area, Fig. 13.49 (a) and (b). To dimension a knurl for a press fit, the toleranced diameter before knurling should be given, (c). A note should be added giving the pitch and type of knurl and the minimum diameter after knurling. See ANSI B94.6–1984.

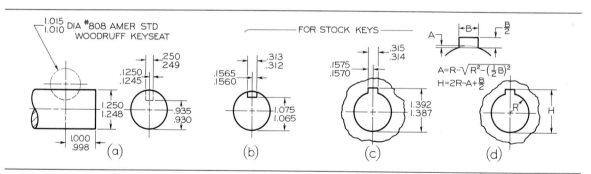

Fig. 13.48 Dimensioning Keyways.

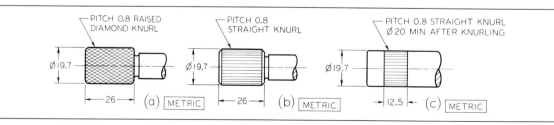

Fig. 13.49 Dimensioning Knurls.

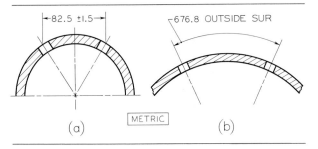

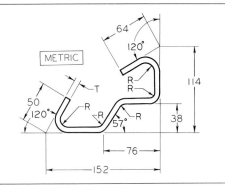

Fig. 13.50   Dimensioning Along Curved Surfaces.

Fig. 13.51   Profile Dimensioning.

13.38   Dimensioning Along Curved Surfaces

When angular measurements are unsatisfactory, chordal dimensions, Fig. 13.50 (a), or linear dimensions on the curved surfaces, as shown at (b), may be given.

13.39   Sheet-Metal Bends

In sheet-metal dimensioning, allowance must be made for bends. The intersection of the plane surfaces adjacent to a bend is called the *mold line,* and this line, rather than the center of the arc, is used to determine dimensions, Fig. 13.51. The following procedure for calculating bends is typical. If the two inner plane surfaces of an angle are extended, their line of intersection is called the IML or *inside mold line,* Fig. 13.52 (a) to (c). Similarly, if the two outer plane surfaces are extended, they produce the OML or *outside mold line.* The *center line of bend* ($\mathcal{C}$B) refers primarily to the machine on which the bend is made and is at the center of the bend radius.

The length, or *stretchout,* of the pattern equals the sum of the flat sides of the angle plus the distance around the bend measured along the *neutral axis.* The distance around the bend is called the *bend allowance.* When metal bends, it compresses on the inside and stretches on the outside. At a certain zone in between, the metal is neither compressed nor stretched. This is called the neutral axis. See Fig. 13.52 (d). The neutral axis is usually assumed to be 0.44 of the thickness from the inside surface of the metal.

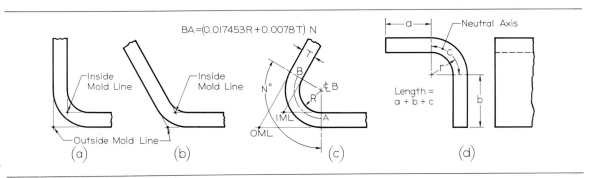

Fig. 13.52   Bends.

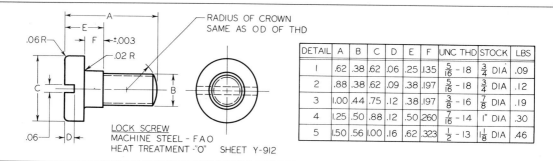

Fig. 13.53 Tabular Dimensioning.

DETAIL	A	B	C	D	E	F	UNC THD	STOCK	LBS
I	.62	.38	.62	.06	.25	.135	$\frac{5}{16}$ – 18	$\frac{3}{4}$ DIA	.09
2	.88	.38	.62	.09	.38	.197	$\frac{5}{16}$ – 18	$\frac{3}{4}$ DIA	.12
3	1.00	.44	.75	.12	.38	.197	$\frac{3}{8}$ – 16	$\frac{7}{8}$ DIA	.19
4	1.25	.50	.88	.12	.50	.260	$\frac{7}{16}$ – 14	1" DIA	.30
5	1.50	.56	1.00	.16	.62	.323	$\frac{1}{2}$ – 13	$1\frac{1}{8}$ DIA	.46

The developed length of material, or bend allowance (BA), to make the bend is computed from the empirical formula

$$BA = (0.017453R + 0.0078T)N$$

where R = radius of bend, T = metal thickness, and N = number of degrees of bend. See Fig. 13.52 (c).

13.40 Tabular Dimensions

A series of objects having like features but varying in dimensions may be represented by one drawing, Fig. 13.53. Letters are substituted for dimension figures on the drawing, and the varying dimensions are given in tabular form. The dimensions of many standard parts are given in this manner in the various catalogs and handbooks.

13.41 Standards

Dimensions should be given, wherever possible, to make use of readily available materials, tools, parts, and gages. The dimensions for many commonly used machine elements, such as bolts, screws, nails, keys, tapers, wire, pipes, sheet metal, chains, belts, ropes, pins, and rolled metal shapes, have been standardized, and the drafter must obtain these sizes from company standards manuals, from published handbooks, from American National Standards, or from manufacturers' catalogs. Tables of some of the more common items are given in the Appendix of this text.

Such standard parts are not delineated on detail drawings unless they are to be altered for use, but are drawn conventionally on assembly drawings and are listed in parts lists, §16.14. Common fractions are often used to indicate the nominal sizes of standard parts or tools. If the complete decimal-inch system is used, all such sizes ordinarily are expressed by decimals; for example, .250 DRILL instead of $\frac{1}{4}$ DRILL. If the all-metric system of dimensioning is used, then the *preferred* metric drill of the approximate same size (.2480″) will be indicated as a **6.30 DRILL**.

13.42 Coordinate Dimensioning

In general, the basic coordinate dimensioning practices are compatible with the data requirements for tape or computer-controlled automatic production machines. However, to design for automated production, the designer and/or drafter should first consult the manufacturing machine manuals before making the drawings for production. Certain considerations should be noted.

1. A set of three mutually perpendicular datum or reference planes is usually required for coordinate dimensioning. These planes either must be obvious or clearly identified. See Fig. 13.54.

2. The designer selects as origins for dimensions those surfaces or other features most important to the functioning of the part. Enough of these features are selected to position the part in relation to the set of mutually perpendicular planes. All related di-

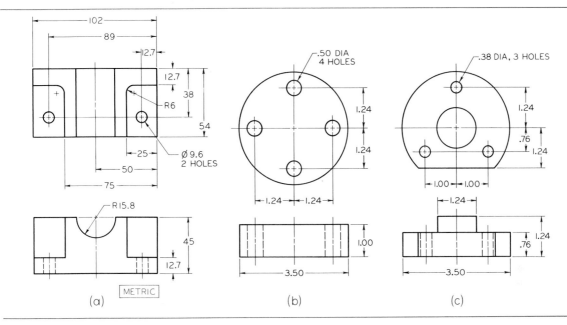

Fig. 13.54   Coordinate Dimensioning.

mensions on the part are then made from
these planes.

3. All dimensions should be in decimals.

4. Angles should be given, where possible, in
degrees and decimal parts of degrees.

5. Standard tools, such as drills, reamers, and
taps, should be specifed wherever possible.

6. All tolerances should be determined by the
design requirements of the part, not by the
capability of the manufacturing machine.

13.43   Do's and Don'ts
of Dimensioning

The following checklist summarizes briefly most
of the situations in which a beginning designer
is likely to make a mistake in dimensioning. The
student should check the drawing by this list before submitting it to the instructor.

1. Each dimension should be given clearly, so
that it can be interpreted in only one way.

2. Dimensions should not be duplicated or the
same information be given in two different
ways—dual dimensioning excluded—and
no dimensions should be given except
those needed to produce or inspect the
part.

3. Dimensions should be given between
points or surfaces that have a functional
relation to each other or that control the
location of mating parts.

4. Dimensions should be given to finished surfaces or important center lines in preference to rough surfaces wherever possible.

5. Dimensions should be so given that it will
not be necessary for the machinist to calculate, scale, or assume any dimension.

6. Dimensions should be attached to the view
where the shape is best shown (contour
rule).

7. Dimensions should be placed in the views
where the features dimensioned are shown
true shape.

425

8. Dimensioning to hidden lines should be avoided wherever possible.

9. Dimensions should not be placed on a view unless clearness is promoted and long extension lines are avoided.

10. Dimensions applying to two adjacent views should be placed between views, unless clearness is promoted by placing some of them outside.

11. The longer dimensions should be placed outside all intermediate dimensions, so that dimension lines will not cross extension lines.

12. In machine drawing, all unit marks should be omitted, except when necessary for clearness; for example, 1″ VALVE or 1 mm DRILL.

13. Production personnel should not be expected to assume that a feature is centered (as a hole on a plate), but a location dimension should be given from one side. However, if a hole is to be centered on a symmetrical rough casting, mark the center line and omit the locating dimension from the center line.

14. A dimension should be attached to only one view, not to extension lines connecting two views.

15. Detail dimensions should "line up" in chain fashion.

16. A complete chain of detail dimensions should be avoided; it is better to omit one; otherwise REF (reference) should be added to one detail dimension or the overall dimension or the dimension should be enclosed within parentheses, ( ).

17. A dimension line should never be drawn through a dimension figure. A figure should never be lettered over any line of the drawing. The line can be broken if necessary.

18. Dimension lines should be spaced uniformly throughout the drawing. They

should be at least 10 mm ($\frac{3}{8}$″) from the object outline and 6 mm ($\frac{1}{4}$″) apart.

19. No line of the drawing should be used as a dimension line or coincide with a dimension line.

20. A dimension line should never be joined end to end (chain fashion) with any line of the drawing.

21. Dimension lines should not cross, if avoidable.

22. Dimension lines and extension lines should not cross, if avoidable. (Extension lines may cross each other.)

23. When extension lines cross extension lines or visible lines, no break in either line should be made.

24. A center line may be extended and used as an extension line, in which case it is still drawn like a center line.

25. Center lines should generally not extend from view to view.

26. Leaders for notes should be straight, not curved, and pointing to the center of circular views of holes wherever possible.

27. Leaders should slope at 45°, or 30°, or 60° with horizontal but may be made at any convenient angle except vertical or horizontal.

28. Leaders should extend from the beginning or from the end of a note, the horizontal "shoulder" extending from midheight of the lettering.

29. Dimension figures should be approximately centered between the arrowheads, except that in a "stack" of dimensions, the figures should be "staggered."

30. Dimension figures should be about 3 mm ($\frac{1}{8}$″) high for whole numbers and 6 mm ($\frac{1}{4}$″) high for fractions.

31. Dimension figures should never be crowded or in any way made difficult to read.

32. Dimension figures should not be lettered over lines or sectioned areas unless necessary, in which case a clear space should be reserved for the dimension figures.

33. Dimension figures for angles should generally be lettered horizontally.

34. Fraction bars should never be inclined except in confined areas, such as in tables.

35. The numerator and denominator of a fraction should never touch the fraction bar.

36. Notes should always be lettered horizontally on the sheet.

37. Notes should be brief and clear, and the wording should be standard in form, Fig. 13.44.

38. Finish marks should be placed on the edge views of all finished surfaces, including hidden edges and the contour and circular views of cylindrical surfaces.

39. Finish marks should be omitted on holes or other features where a note specifies a machining operation.

40. Finish marks should be omitted on parts made from rolled stock.

41. If a part is finished all over, all finish marks should be omitted, and the general note FINISH ALL OVER or FAO should be used.

42. A cylinder is dimensioned by giving both its diameter and length in the rectangular view, except when notes are used for holes. A diagonal diameter in the circular view may be used in cases where clearness is gained thereby.

43. Holes to be bored, drilled, reamed, and so on are size-dimensioned by notes in which the leaders preferably point toward the center of the circular views of the holes. Indications of manufacturing processes may be omitted from notes.

44. Drill sizes are preferably expressed in decimals. For drills designated by number or letter, the decimal size must also be given.

45. In general, a circle is dimensioned by its diameter, an arc by its radius.

46. Diagonal diameters should be avoided, except for very large holes and for circles of centers. They may be used on positive cylinders when clearness is gained thereby.

47. A metric diameter dimension value should always be preceded by the symbol $\varnothing$, and the diameter dimension value in inches should be followed by DIA.

48. A metric radius dimension should always be preceded by the letter R and the radius value in inches should be followed by the letter R. The radial dimension line should have only one arrowhead, and it should pass through or point through the arc center and touch the arc.

49. Cylinders should be located by their center lines.

50. Cylinders should be located in the circular views, if possible.

51. Cylinders should be located by coordinate dimensions in preference to angular dimensions where accuracy is important.

52. When there are several rough, noncritical features obviously the same size (fillets, rounds, ribs, etc.), it is necessary to give only typical (abbreviation TYP) dimensions, or to use a note.

53. When a dimension is not to scale, it should be underscored with a heavy straight line or marked NTS or NOT TO SCALE.

54. Mating dimensions should be given correspondingly on drawings of mating parts.

55. Pattern dimensions should be given in two-place decimals or in common whole numbers and fractions to the nearest $\frac{1}{16}''$.

56. Decimal dimensions should be used for all machining dimensions.

57. Cumulative tolerances should be avoided, especially in limit dimensioning, described in §14.9.

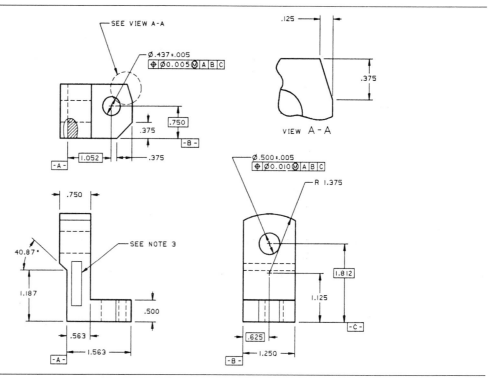

Fig. 13.55   Dimensioned Detail Drawing Produced by the Computervision CADDS 4X Production Drafting System. *Courtesy of Computervision Corporation, a subsidiary of Prime Computer, Inc.*

13.44   Computer Graphics

A drafter with thorough knowledge of the principles and practices of dimensioning will find that modern CAD software provides an easy means of applying the desired dimensional data to a drawing, Fig. 13.55. Linear dimensions in either the metric or inch systems can be shown in either a horizontal, vertical, or parallel dimensioning format. Angles, in degrees, can be calculated and displayed. Other dimensioning requirements such as arrows, leaders, extension lines, etc. are standard features of CAD dimensioning programs.

DIMENSIONING PROBLEMS

It is expected that most of the student's practice in dimensioning will be in connection with working drawings assigned from other chapters. However, a limited number of special dimensioning problems are available here in Figs. 13.56 and 13.57. The problems are designed for Layout A–3 (8.5″ × 11.0″) and are to be drawn with instruments and dimensioned to a full-size scale. Layout A4–3 (297 mm × 420 mm) may be used with appropriate adjustments in the title strip layout.

Since many of the problems in this and other chapters are of a general nature, they can also be solved on most computer graphics systems. If a system is available, the instructor may choose to assign specific problems to be completed by this method.

Problems in convenient form for solution may be found in *Technical Drawing Problems*, Series 1, by Giesecke, Mitchell, Spencer, Hill, Dygdon, and Novak; *Technical Drawing Problems*, Series 2, by Spencer, Hill, Dygdon, and Novak; and *Technical Drawing Problems*, Series 3, by Spencer, Hill, Dygdon, and Novak; all designed to accompany this text and published by Macmillan Publishing Company.

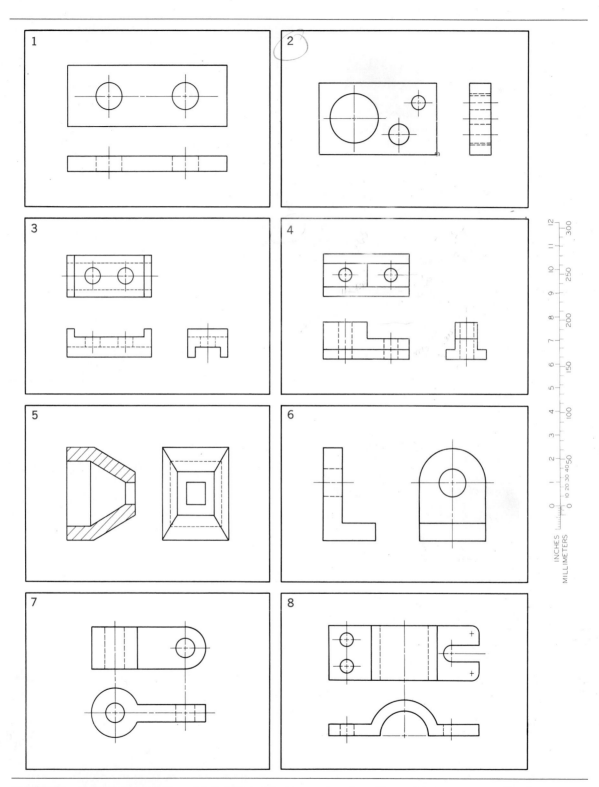

Fig. 13.56 Using Layout A–3 or A4–3 (adjusted), draw assigned problem with instruments. To obtain sizes, place bow dividers on the views on this page and transfer to scale at the side to obtain values. Dimension drawing completely in one-place millimeters or two-place inches as assigned, full size. See inside front cover for decimal-inch and millimeter equivalents.

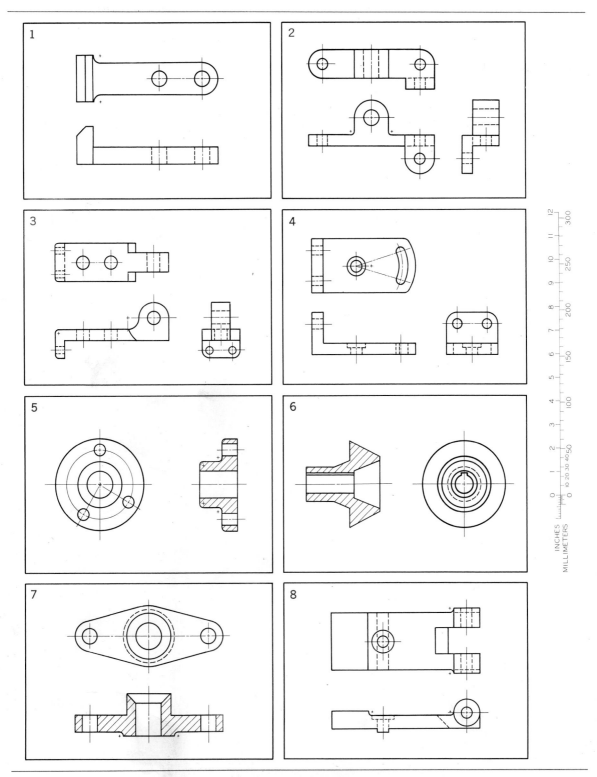

Fig. 13.57 Using Layout A–3 or A4–3 (adjusted), draw assigned problem with instruments. To obtain sizes, place bow dividers on the views on this page and transfer to scale at the side to obtain values. Dimension drawing completely in one-place millimeters or two-place inches as assigned, full size. See inside front cover for decimal-inch and millimeter equivalents.

CHAPTER 14

Tolerancing

Interchangeable manufacturing, by means of which parts can be made in widely separated localities and then be brought together for assembly, where the parts will all fit together properly, is an essential element of mass production. Without interchangeable manufacturing, modern industry could not exist, and without effective size control by the engineer, interchangeable manufacturing could not be achieved.

For example, an automobile manufacturer not only subcontracts the manufacture of many parts of a design to other companies but also must provide parts for replacement. All parts in each category must be near enough alike so that any one of them will fit properly in any assembly. Unfortunately, *it is impossible to make anything to exact size*. Parts can be made to very close dimensions, even to a few millionths of an inch or thousandths of a millimeter (e.g., gage blocks), but such accuracy is extremely expensive.

Fortunately, *exact* sizes are not needed. The need is for varying degrees of accuracy according to functional requirements. A manufacturer of children's tricycles would soon go out of business if the parts were made with jet-engine accuracy, as no one would be willing to pay the price. So what is wanted is a means of specifying dimensions with whatever degree of accuracy may be required. The answer to the problem is the specification of a *tolerance* on each dimension.

14.1 Tolerance Dimensioning

Tolerance is the total amount a specific dimension is permitted to vary, which is the difference between the maximum and the minimum limits [ANSI Y14.5M–1982 (R1988)]. For example, a dimension given as 1.625 ± .002 means that it may be (on the manufactured part) 1.627″ or 1.623″, or anywhere between these *limit dimensions*. The tolerance, or total amount of variation "tolerated," is .004″. Thus, it becomes the function of the detailer or designer to specify the allowable error that may be tolerated for a given dimension and still permit the satisfactory functioning of the part. Since greater accuracy costs more money, the detailer or designer will not specify the closest tolerance, but instead will specify as generous a tolerance as possible.

In order to control the dimensions of quantities of the two parts so that any two mating parts will be interchangeable, it is necessary to

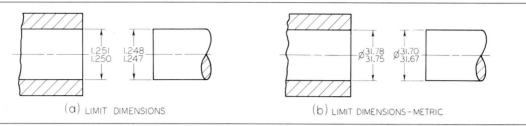

Fig. 14.1 Fits Between Mating Parts.

assign tolerances to the dimensions of the parts, as shown in Fig. 14.1 (a). The diameter of the hole may be machined not less than 1.250″ and not more than 1.251″, these two figures representing the *limits* and the difference between them, .001″, being the *tolerance*. Likewise, the shaft must be produced between the limits of 1.248″ and 1.247″, the tolerance on the shaft being the difference between these, or .001″. The metric versions for these limit dimensions for the hole and shaft are shown at (b). The difference in the dimensions for either the hole or shaft is 0.03 mm, the total *tolerance*.

A pictorial illustration of the dimensions in Fig. 14.1 (a) is shown in Fig. 14.2 (a). The maximum shaft is shown solid, and the minimum shaft is shown in phantom. The difference in diameters, .001″, is the tolerance on the shaft. Similarly, the tolerance on the hole is the difference between the two limits shown, or .001″. The loosest fit, or maximum clearance, occurs when the smallest shaft is in the largest hole, as shown at (b). The tightest fit, or minimum clearance,

occurs when the largest shaft is in the smallest hole, as shown at (c). The difference between these, .002″, is the *allowance*. The average clearance is .003″, which is the same difference as allowed in the example of Fig. 14.1 (a); thus, any shaft will fit any hole interchangeably.

When expressed in metric dimensions, the limits for the hole are 31.75 mm and 31.78 mm, the difference between them, 0.03 mm, being the tolerance. Similarly, the limits for the shaft are 31.70 mm and 31.67 mm, the tolerance on the shaft being the difference between them, or 0.03 mm.

When parts are required to fit properly in assembly but not to be interchangeable, the size of one part need not be toleranced, but indicated to be made to fit at assembly, Fig. 14.3.

14.2   Definitions of Terms

At this point, it is well to fix in mind the definitions of certain terms [ANSI Y14.5M–1982 (R1988)].

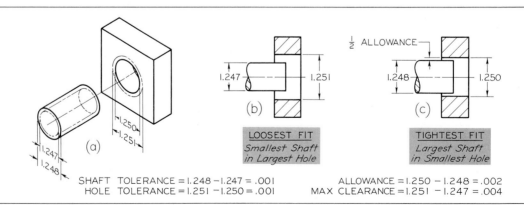

Fig. 14.2 Limit Dimensions.

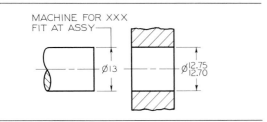

Fig. 14.3   Noninterchangeable Fit.

Nominal size   The designation that is used for the purpose of general identification is usually expressed in common fractions. In Fig. 14.1, the nominal size of both hole and shaft, which is 1¼″, would be 1.25″ or 31.75 mm in a decimal system of dimensioning.

Basic size or dimension   The theoretical size from which limits of size are derived by the application of allowances and tolerances. It is the size from which limits are determined for the size, shape, or location of a feature. In Fig. 14.1 (a), the basic size is the decimal equivalent of the nominal size 1¼″, or 1.250″ or 31.75 mm at (b).

Actual size   The measured size of the finished part.

Tolerance   The total amount by which a given dimension may vary, or the difference between the limits. In Fig. 14.2 (a) the tolerance on either the shaft or hole is the difference between the limits, or .001″.

Limits   The maximum and minimum sizes indicated by a toleranced dimension. In Fig. 14.2 (a) the limits for the hole are 1.250″ and

1.251″, and for the shaft they are 1.248″ and 1.247″.

Allowance   The minimum clearance space (or maximum interference) intended between the maximum material condition (MMC) of mating parts. In Fig. 14.2 (c) the allowance is the difference between the smallest hole, 1.250″, and the largest shaft, 1.248″, or .002″. Allowance, then, represents the tightest permissible fit and is simply the smallest hole minus the largest shaft. For clearance fits, this difference will be positive, while for interference fits it will be negative.

14.3   Fits Between Mating Parts

"Fit is the general term used to signify the range of tightness or looseness that may result from the application of a specific combination of allowances and tolerances in mating parts." [ANSI Y14.5M–1982 (R1988)]. There are four general types of fits between parts.

Clearance fit   In which an internal member fits in an external member (as a shaft in a hole), and always leaves a space or clearance between the parts. In Fig. 14.2 (c) the largest shaft is 1.248″ and the smallest hole is 1.250″, which permits a minimum air space of .002″ between the parts. This space is the allowance, and in a clearance fit it is always positive.

Interference fit   In which the internal member is larger than the external member such that there is always an actual interference of metal. In Fig. 14.4 (a) the smallest shaft is 1.2513″, and

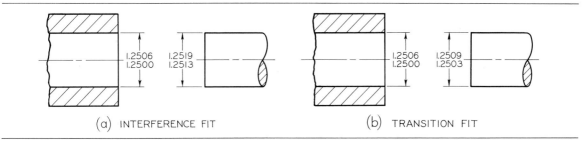

Fig. 14.4   Fits Between Parts.

435

the largest hole is 1.2506″, so that there is an actual interference of metal amounting to at least .0007″. Under maximum material conditions the interference would be .0019″. This interference is the allowance, and in an interference fit it is always negative.

Transition fit   In which the fit might result in either a clearance or interference condition. In Fig. 14.4 (b) the smallest shaft, 1.2503″, will fit in the largest hole, 1.2506″, with .0003″ to spare. But the largest shaft, 1.2509″, will have to be forced into the smallest hole, 1.2500″, with an interference of metal (negative allowance) of .0009″.

Line fit   In which limits of size are so specified that a clearance or surface contact may result when mating parts are assembled.

14.4   Selective Assembly

If allowances and tolerances are properly given, mating parts can be completely interchangeable. But for close fits, it is necessary to specify very small allowances and tolerances, and the cost may be very high. In order to avoid this expense, either manual or computer-controlled *selective assembly* is often used. In selective assembly, all parts are inspected and classified into several grades according to actual sizes, so that "small" shafts can be matched with "small" holes, "medium" shafts with "medium" holes, and so on. In this way, very satisfactory fits may be obtained at much less expense than by machining all mating parts to very accurate dimensions. Since a transition fit may or may not represent an interference of metal, interchangeable assembly generally is not as satisfactory as selective assembly.

14.5   Basic Hole System

Standard reamers, broaches, and other standard tools are often used to produce holes, and standard plug gages are used to check the actual sizes. On the other hand, shafting can easily be

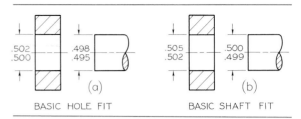

Fig. 14.5   Basic Hole and Basic Shaft Systems.

machined to any size desired. Therefore, toleranced dimensions are commonly figured on the so-called *basic hole system*. In this system, the *minimum hole is taken as the basic size*, an allowance is assigned, and tolerances are applied on both sides of, and away from, this allowance.

In Fig. 14.5 (a) the minimum size of the hole, .500″, is taken as the basic size. An allowance of .002″, is decided on and subtracted from the basic hole size, giving the maximum shaft, .498″. Tolerances of .002″ and .003″, respectively, are applied to the hole and shaft to obtain the maximum hole of .502″ and the minimum shaft of .495″. Thus, the minimum clearance between the parts becomes .500″ − .498″ = .002″ (smallest hole minus largest shaft), and the maximum clearance is .502″ − .495″ = .007″ (largest hole minus smallest shaft).

In the case of an interference fit, the maximum shaft size would be found by *adding the desired allowance* (maximum interference) to the basic hole size. In Fig. 14.4 (a), the basic size is 1.2500″. The maximum interference decided on was .0019″, which added to the basic size gives 1.2519″, the largest shaft size.

The basic hole size can be changed to the basic shaft size by subtracting the allowance for a clearance fit, or adding it for an interference fit. The result is the largest shaft size, which is the new basic size.

14.6   Basic Shaft System

In some branches of industry, such as textile machinery manufacturing, in which use is made of a great deal of cold-finished shafting, the *basic shaft system* is often used. This system should be used only when there is a reason for it. For example, it is advantageous when several parts having different fits, but one nominal size, are required on a single shaft. In this system, *the maximum shaft is taken as the basic size,* an allowance for each mating part is assigned, and tolerances are applied on both sides of, and away from, this allowance.

In Fig. 14.5 (b) the maximum size of the shaft, .500″, is taken as the basic size. An allowance of .002″ is decided on and added to the basic shaft size, giving the minimum hole, .502″. Tolerances of .003″ and .001″, respectively, are applied to the hole and shaft to obtain the maximum hole, .505″, and the minimum shaft, .499″. Thus, the minimum clearance between the parts is .502″ − .500″ = .002″ (smallest hole minus largest shaft), and the maximum clearance is .505″ − .499″ = .006″ (largest hole minus smallest shaft).

In the case of an interference fit, the minimum hole size would be found by *subtracting the desired allowance from the basic shaft size.*

The basic shaft size may be changed to the basic hole size by adding the allowance for a clearance fit or by subtracting it for an interference fit. The result is the smallest hole size, which is the new basic size.

14.7   Specification of Tolerances

A tolerance of a decimal dimension must be given in decimal form to the same number of places. See Fig. 14.8.

General tolerances on decimal dimensions in which tolerances are not given may also be covered in a printed note, such as

DECIMAL DIMENSIONS TO BE HELD TO ± .001.

Thus if a dimension 3.250 is given, the worker machines between the limits 3.249 and 3.251. See Fig. 14.9.

Tolerances for metric dimensions may be covered in a note, such as the commonly used

METRIC DIMENSIONS TO BE HELD TO ± 0.08.

Thus, when the given dimension of 3.250″ is converted to millimeters, the worker machines between the limits of 82.63 mm and 82.47 mm.

Every dimension on a drawing should have a tolerance, either direct or by general tolerance note, except that commercial material is often assumed to have the tolerances set by commercial standards.

It is customary to indicate an overall general tolerance for all common fraction dimensions by means of a printed note in or just above the title block.

EXAMPLES

ALL FRACTIONAL DIMENSIONS $\pm\frac{1}{64}$″
UNLESS OTHERWISE SPECIFIED.

HOLD FRACTIONAL DIMENSIONS TO $\pm\frac{1}{64}$″
UNLESS OTHERWISE NOTED.

See Fig. 14.9. General angular tolerances also may be given as

ANGULAR TOLERANCE ± 1°.

Several methods of expressing tolerances in dimensions are approved by ANSI [Y14.5M–1982 (R1988)] as follows.

1. *Limit Dimensioning.* In this preferred method, the maximum and minimum limits of size and location are specified, as shown in Fig. 14.6. The high limit (maximum value) is placed above the low limit (minimum value). See Fig. 14.6 (a). In single-line note form, the low limit precedes the high limit separated by a dash, Fig. 14.6 (b).

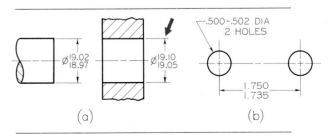

Fig. 14.6   Method of Giving Limits.

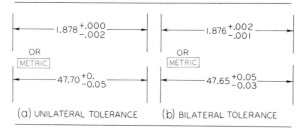

Fig. 14.7   Tolerance Expression.

2. *Plus and Minus Dimensioning.* In this method the basic size is followed by a plus and minus expression of tolerance resulting in either a unilateral or bilateral tolerance as in Fig. 14.7. If two unequal tolerance numbers are given, one plus and one minus, the plus is placed above the minus. One of the numbers may be zero, if desired. If a single tolerance value is given, it is preceded by the plus-or-minus symbol ($\pm$), Fig. 14.8. This method should be used when the plus and minus values are equal.

The *unilateral system* of tolerances allows variations in only one direction from the basic size. This method is advantageous when a critical size is approached as material is removed during manufacture, as in the case of close-fitting holes and shafts. In Fig. 14.7 (a) the basic size is 1.878″ (47.70 mm). The tolerance .002″ (0.05 mm) is all in one direction—toward the smaller size. If this is a shaft diameter, the basic size 1.878″ (47.70 mm) is the size nearest the critical size be-

cause it is nearest to the tolerance zone; hence, the tolerance is taken *away* from the critical size. A unilateral tolerance is always all plus or all minus; that is, either the plus or the minus value must be zero. However, the zeros should be given as shown at (a).

The *bilateral system* of tolerances allows variations in both directions from the basic size. Bilateral tolerances are usually given with location dimensions or with any dimensions that can be allowed to vary in either direction. In Fig. 14.7 (b) the basic size is 1.876″ (47.65 mm), and the actual size may be larger by .002″ (0.05 mm) or smaller by .001″ (0.03 mm). If it is desired to specify an equal variation in both directions, the combined plus or minus symbol ($\pm$) is used with a single value, as shown in Fig. 14.8.

A typical example of limit dimensioning is given in Fig. 14.9.

3. *Single-Limit Dimensioning.* It is not always necessary to specify both limits. MIN or MAX is often placed after a number to indicate minimum or maximum dimensions desired where other elements of design determine the other unspecified limit. For example, a thread length may be dimensioned thus: |←——1.500——→| MIN FULL THD or a radius dimensioned: .05 R MAX—⟍ Other applications include depths of holes, chamfers, and so on.

4. *Angular tolerances* are usually bilateral and in terms of degrees, minutes, and seconds.

EXAMPLES $25° \pm 1°$, $25° 0′ \pm 0° 15′$, or $25° \pm 0.25°$. See also §14.18.

14.8   American National Standard Limits and Fits

The American National Standards Institute has issued the ANSI B4.1–1967 (R1987), "Preferred Limits and Fits for Cylindrical Parts," defining terms and recommending preferred standard sizes, allowances, tolerances, and fits in terms of the decimal inch. This standard gives

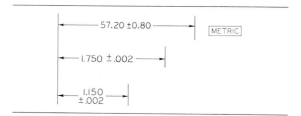

Fig. 14.8   Bilateral Tolerances.

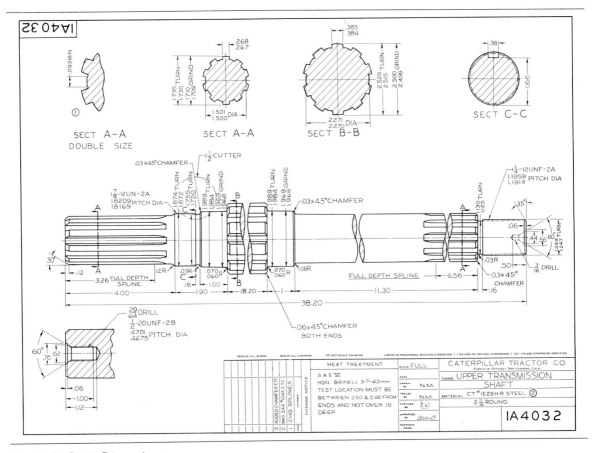

Fig. 14.9 Limit Dimensions.

a series of standard types and classes of fits on a unilateral hole basis such that the fit produced by mating parts in any one class will produce approximately similar performance throughout the range of sizes. These tables prescribe the fit for any given size, or type of fit; they also prescribe the standard limits for the mating parts which will produce the fit.

The tables are designed for the basic hole system, §14.5. See Appendixes 5–9. For coverage of the metric system of tolerances and fits, see §§14.11–14.13, and Appendixes 11–14.

Letter symbols to identify the five types of fits are

RC   Running or Sliding Clearance Fits
LC   Locational Clearance Fits
LT   Transition Clearance or Interference Fits
LN   Locational Interference Fits
FN   Force or Shrink Fits

These letter symbols, plus a number indicating the class of fit within each type, are used to indicate a complete fit. Thus, FN 4 means a Class 4 Force Fit. The fits are described [ANSI B4.1–1967 (R1987)] as follows.

RUNNING AND SLIDING FITS

Running and sliding fits, for which description of classes of fits and limits of clearance are given [Appendix 5], are intended to provide a similar running performance, with suitable lubrication allowance, throughout the range of sizes. The clearances for the first two classes, used chiefly as slide fits, increase more slowly with diameter than the other classes, so that accurate location is maintained even at the expense of free relative motion.

LOCATIONAL FITS

Locational fits [Appendixes 6–8] are fits intended to determine only the location of the mating parts; they may provide rigid or accurate location, as with interference fits, or provide some freedom of location, as with clearance fits. Accordingly they are divided into three groups: clearance fits, transition fits, and interference fits.

FORCE FITS

Force or shrink fits [Appendix 9] constitute a special type of interference fit, normally characterized by maintenance of constant bore pressures throughout the range of sizes. The interference therefore varies almost directly with diameter, and the difference between its minimum and maximum value is small, to maintain the resulting pressures within reasonable limits.

In the tables for each class of fit, the range of nominal sizes of shafts or holes is given in inches. To simplify the tables and reduce the space required to present them, the other values are given in thousandths of an inch. Minimum and maximum limits of clearance are given, the top number being the least clearance, or the allowance, and the lower number the maximum clearance, or the greatest looseness of fit. Then, under the heading "Standard Limits" are given the limits for the hole and for the shaft that are

to be applied algebraically to the basic size to obtain the limits of size for the parts, using the basic hole system.

For example, take a 2.0000″ basic diameter with a Class RC 1 fit. This fit is given in Appendix 5. In the column headed "Nominal Size Range, Inches," find 1.97–3.15, which embraces the 2.0000″ basic size. Reading to the right we find under "Limits of Clearance" the values 0.4 and 1.2, representing the maximum and minimum clearance between the parts *in thousandths of an inch*. To get these values in inches, simply multiply by one thousandth; thus, $\frac{4}{10} \times \frac{1}{1000} = .0004″$. To convert 0.4 thousandths to inches, simply move the decimal point three places to the left; thus: .0004″. Therefore, for this 2.0000″ diameter, with a Class RC 1 fit, the minimum clearance, or allowance, is .0004″, and the maximum clearance, representing the greatest looseness, is .0012″.

Reading farther to the right, we find under "Standard Limits" the value +0.5, which converted to inches is .0005″. Add this to the basic size thus: 2.0000″ + .0005″ = 2.0005″, the upper limit of the hole. Since the other value given for the hole is zero, the lower limit of the hole is the basic size of the hole, or 2.0000″. The hole would then be dimensioned as

$$\begin{array}{ccc} 2.0005 & & \\ 2.0000 & \text{or} & 2.0000 \begin{array}{l} +.0005 \\ -.0000 \end{array} \end{array}$$

The limits for the shaft are read as −.0004″ and −.0007″. To get the limits of the shaft, subtract these values from the basic size; thus,

$$2.0000″ - .0004″ = 1.9996″ \text{ (upper limit)}$$
$$2.0000″ - .0007″ = 1.9993″ \text{ (lower limit)}$$

The shaft would then be dimensioned in inches as follows.

$$\begin{array}{ccc} 1.9996 & & \\ 1.9993 & \text{or} & 1.9996 \begin{array}{l} +.0000 \\ -.0003 \end{array} \end{array}$$

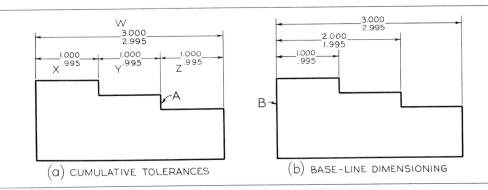

Fig. 14.10 Cumulative Tolerances.

14.9 Accumulation of Tolerances

In tolerance dimensioning, it is very important to consider the effect of one tolerance on another. When the location of a surface in a given direction is affected by more than one tolerance figure, the tolerances are *cumulative*. For example, in Fig. 14.10 (a), if dimension Z is omitted, surface A will be controlled by both dimensions X and Y, and there can be a total variation of .010″ instead of the variation of .005″ permitted by dimension Y, which is the dimension directly applied to surface A. Furthermore, if the part is made to all the minimum tolerances of X, Y, and Z, the total variation in the length of the part will be .015″, and the part can be as short as 2.985″. However, the tolerance on the overall dimension W is only .005″, permitting the part to be only as short as 2.995″. The part is superfluously dimensioned.

In some cases, for functional reasons, it may be desired to hold all three small dimensions X, Y, and Z closely without regard to the overall length. In such a case the overall dimension is just a *reference dimension* and should be marked REF. In other cases it may be desired to hold two small dimensions X and Y and the overall closely without regard to dimension Z. In that case, dimension Z should be omitted, or marked REF.

As a rule, it is best to dimension each surface so that it is affected by only one dimension. This can be done by referring all dimensions to a sin-

gle datum surface, such as B, as shown at (b). See also Fig. 13.37 (d)–(f).

14.10 Tolerances and Machining Processes

As has been repeatedly stated in this chapter, tolerances should be as coarse as possible and still permit satisfactory use of the part. If this is done, great savings can be effected as a result of the use of less expensive tools, lower labor and inspection costs, and reduced scrapping of material.

Figure 14.11 shows a chart of tolerance grades obtainable in relation to the accuracy of machining processes that may be used as a guide by the designer. Metric values may be ascertained by multiplying the given decimal-inch values by 25.4 and rounding off the product to one less place to the right of the decimal point than given for the decimal-inch value. See §13.10. For detailed information on manufacturing processes and measuring devices see Chapter 12.

14.11 Metric System of Tolerances and Fits

The preceding material on limits and fits between mating parts is suitable, without need of conversion, for the decimal-inch system of measurement. A system of preferred metric limits and fits by the International Organization for

Range of Sizes		Tolerances								
From	To & Incl.									
.000	.599	.00015	.0002	.0003	.0005	.0008	.0012	.002	.003	.005
.600	.999	.00015	.00025	.0004	.0006	.001	.0015	.0025	.004	.006
1.000	1.499	.0002	.0003	.0005	.0008	.0012	.002	.003	.005	.008
1.500	2.799	.00025	.0004	.0006	.001	.0015	.0025	.004	.006	.010
2.800	4.499	.0003	.0005	.0008	.0012	.002	.003	.005	.008	.012
4.500	7.799	.0004	.0006	.001	.0015	.0025	.004	.006	.010	.015
7.800	13.599	.0005	.0008	.0012	.002	.003	.005	.008	.012	.020
13.600	20.999	.0006	.001	.0015	.0025	.004	.006	.010	.015	.025

Lapping & Honing

Grinding, Diamond
  Turning & Boring

Broaching

Reaming

Turning, Boring,
  Slotting, Planing &
  Shaping

Milling

Drilling

Fig. 14.11   Tolerances Related to Machining Processes.

Standardization (ISO) is in the ANSI B4.2 standard. The system is specified for holes, cylinders, and shafts, but it is also adaptable to fits between parallel surfaces of such features as keys and slots. The following terms for metric fits, although somewhat similar to those for decimal-inch fits, are illustrated in Fig. 14.12.

Basic size   The size from which limits or deviations are assigned. Basic sizes, usually diameters, should be selected from a table of preferred sizes. See Fig. 14.17.

Deviation   The difference between the basic size and the hole or shaft size. (This is equivalent to the tolerance in the decimal-inch system.)

Upper deviation   The difference between the basic size and the permitted maximum size of

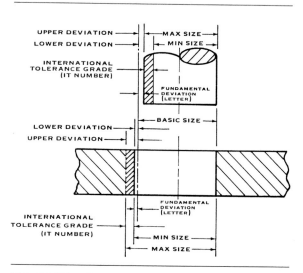

Fig. 14.12   Terms Related to Metric Limits and Fits [ANSI B4.2–1978 (R1984)].

the part. (This compares with the maximum tolerance in the decimal-inch system.)

Lower deviation   The difference between the basic size and the minimum permitted size of the part. (This compares with the minimum tolerance in the decimal-inch system.)

Fundamental deviation   The deviation closest to the basic size. (This compares with the minimum allowance in the decimal-inch system.)

Tolerance   The difference between the permitted minimum and maximum size of a part.

International tolerance grade (IT)   A set of tolerances that varies according to the basic size and provides a uniform level of accuracy within the grade. For example, in the dimension 50H8 for a close-running fit, the IT grade is indicated by the numeral 8. (The letter H indicates the tolerance is on the hole for the 50 mm dimension.) In all, there are 18 IT grades—IT01, IT0, and IT1 through IT16. See Figs. 14.13 and 14.14 for IT grades related to machining processes and the practical use of the IT grades. See also Appendix 10.

Tolerance zone   The tolerance and its position in relation to basic size. It is established by a combination of the fundamental deviation indicated by a letter and the IT grade number. In the dimension 50H8, for the close running fit, the H8 specifies the tolerance zone. See Fig. 14.14.

Hole-basis system of preferred fits   A system based on the basic diameter as the minimum size. For the generally preferred hole-basis system, the fundamental deviation is specified by the uppercase letter H. See Fig. 14.15 (a).

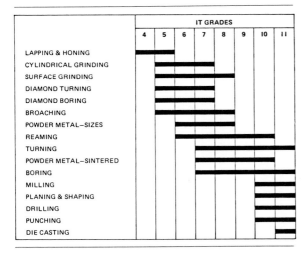

Fig. 14.13   International Tolerance Grades Related to Machining Processes   [ANSI B4.2–1978 (R1984)].

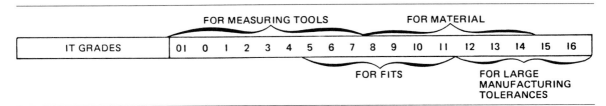

Fig. 14.14   Practical Use of International Tolerance Grades.

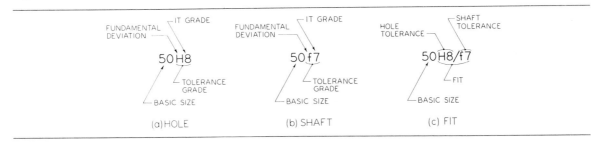

Fig. 14.15   Application of Definitions and Symbols to Holes and Shafts   [ANSI B4.2–1978 (R1984)].

Shaft-basis system of preferred fits A system based on the basic diameter as the maximum size of the shaft. The fundamental deviation is given by the lowercase letter f. See Fig. 14.15 (b).

Interference fit A fit that results in an interference fit between two mating parts under *all* tolerance conditions.

Transition fit A fit that results in either a clearance or an interference condition between two assembled parts.

Tolerance symbols Symbols used to specify the tolerances and fits for mating parts. See Fig. 14.15 (c). For the hole-basis system, the 50 indicates the diameter in millimeters; the fundamental deviation for the hole is indicated by the capital letter H, and for the shaft it is indicated by the lowercase letter f. The numbers following the letters indicate this IT grade. Note that the symbols for the hole and shaft are separated by the slash (slanting line). Tolerance symbols may be given in several acceptable forms as in Fig. 14.16 for a 50 mm diameter hole. The values in parentheses are for reference only and may be omitted. The upper and lower limit values may be found in Appendix 11.

14.12 Preferred Sizes
The preferred basic sizes for computing tolerances are given in Table 14.1. Basic diameters should be selected from the first choice column since these are readily available stock sizes for round, square, and hexagonal products.

Table 14.1 *Preferred Sizes* [ANSI B4.2–1978 (R1984)]

Basic Size, mm		Basic Size, mm		Basic Size, mm	
First Choice	Second Choice	First Choice	Second Choice	First Choice	Second Choice
1		10		100	
	1.1		11		110
1.2		12		120	
	1.4		14		140
1.6		16		160	
	1.8		18		180
2		20		200	
	2.2		22		220
2.5		25		250	
	2.8		28		280
3		30		300	
	3.5		35		350
4		40		400	
	4.5		45		450
5		50		500	
	5.5		55		550
6		60		600	
	7		70		700
8		80		800	
	9		90		900
				1000	

14.13 Preferred Fits
The symbols for either the hole-basis or shaft-basis preferred fits (clearance, transition, and interference) are given in Table 14.2. Fits should be selected from this table for mating parts where possible.

The values corresponding to the fits are found in Appendixes 11–14. Although second and third choice basic size diameters are possible, they must be calculated from tables not included in this text. For the generally preferred hole-basis system, note that the ISO symbols range from H11/c11 (loose running) to H7/u6 (force fit). For the shaft-basis system, the preferred symbols range from C11/h11 (loose fit) to U7/h6 (force fit).

50 H8   50H8$\left(\dfrac{50.039}{50.000}\right)$ $\dfrac{50.039}{50.000}$(50H8)

(a) PREFERRED   (b)   (c)

Fig. 14.16 Acceptable Methods of Giving Tolerance Symbols   [ANSI Y14.5M–1982 (R1988)].

Table 14.2 *Preferred Fits* [ANSI B4.2–1978 (R1984)]

ISO Symbol		Description
Hole Basis	**Shaft[a] Basis**	
H11/c11	C11/h11	***Loose running*** fit for wide commercial tolerances or allowances on external members.
H9/d9	D9/h9	***Free running*** fit not for use where accuracy is essential, but good for large temperature variations, high running speeds, or heavy journal pressures.
H8/f7	F8/h7	***Close running*** fit for running on accurate machines and for accurate location at moderate speeds and journal pressures.
H7/g6	G7/h6	***Sliding*** fit not intended to run freely, but to move and turn freely and locate accurately.
H7/h6	H7/h6	***Locational clearance*** fit provides snug fit for locating stationary parts; but can be freely assembled and disassembled.
H7/k6	K7/h6	***Locational transition*** fit for accurate location, a compromise between clearance and interference.
H7/n6	N7/h6	***Locational transition*** fit for more accurate location where greater interference is permissible.
H7/p6	P7/h6	***Locational interference*** fit for parts requiring rigidity and alignment with prime accuracy of location but without special bore pressure requirements.
H7/s6	S7/h6	***Medium drive*** fit for ordinary steel parts or shrink fits on light sections, the tightest fit usable with cast iron.
H7/u6	U7/h6	***Force*** fit suitable for parts which can be highly stressed or for shrink fits where the heavy pressing forces required are impractical.

Left margin labels: Clearance Fits, Transition Fits, Interference Fits

Right margin labels: More clearance ↑, More interference ↓

[a]The transition and interference shaft basis fits shown do not convert to exactly the same hole basis fit conditions for basic sizes in range from Q through 3 mm. Interference fit P7/h6 converts to a transition fit H7/p6 in the above size range.

Assume that it is desired to use the symbols to specify the dimensions for a free-running fit (hole basis) for a proposed diameter of 48 mm. Since 48 mm is not listed as a preferred size in Table 14.1, the design is altered to use the acceptable 50 mm diameter. From the preferred fits descriptions in Table 14.2, the free-running fit (hole-basis) is H9/d9. To determine the upper and lower deviation limits of the hole as given in the preferred hole-basis table, Appendix 11, follow across from the basic size of 50 to H9 under "Free running." The limits for the hole are 50.000 and 50.062 mm. Then, the upper and lower limits of deviation for the shaft are found in the d9 column under "Free running." They are 49.920 and 49.858 mm, respectively. Limits for other fits are established in a similar manner.

The limits for the shaft basis dimensioning are determined similarly from the preferred shaft basis table in Appendix 13. See Figs. 14.16

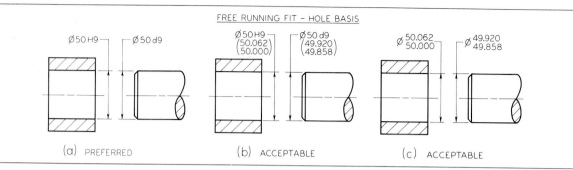

Fig. 14.17   Methods of Specifying Tolerances with Symbols for Mating Parts.

and 14.17 for acceptable methods of specifying tolerances by symbols on drawings. A single note for the mating parts (free-running fit, hole basis) would be ∅50 H9/d9, Fig. 14.17.

14.14   Geometric Tolerancing

Geometric tolerances state the maximum allowable variations of a form or its position from the perfect geometry implied on the drawing. The term "geometric" refers to various forms, such as a plane, a cylinder, a cone, a square, a hexagon. Theoretically these are perfect forms, but, because it is impossible to produce perfect forms, it may be necessary to specify the amount of variation permitted. These tolerances specify either the diameter or the width of a tolerance zone within which a surface or the axis of a cylinder or a hole must be if the part is to meet the required accuracy for proper function and fit. When tolerances of form are not given on a drawing, it is customary to assume that, regardless of form variations, the part will fit and function satisfactorily.

Tolerances of form and position or location control such characteristics as straightness, flatness, parallelism, perpendicularity (squareness), concentricity, roundness, angular displacement, and so on.

Methods of indicating geometric tolerances by means of *geometric characteristic symbols*, as recommended by ANSI, rather than by traditional notes, are discussed and illustrated subsequently. See the latest Dimensioning and Tol-

erancing Standard [Y14.5M–1982 (R1988)] for more complete coverage.

14.15   Symbols for Tolerances of Position and Form

Since traditional narrative notes for specifying tolerances of *position* (location) and *form* (shape) may be confusing or not clear, may require much of the space available on the drawing, and often may not be understood internationally, most multinational companies have adopted symbols for such specifications [ANSI Y14.5M–1982 (R1988)]. These ANSI symbols provide an accurate and concise means of specifying geometric characteristics and tolerances in a minimum of space, Table 14.3. The symbols may be supplemented by notes if the precise geometric requirements cannot be conveyed by the symbols. For construction details of the geometric tolerancing symbols, see Appendix 37.

Combinations of the various symbols and their meanings are given in Fig. 14.18. Application of the symbols to a drawing are illustrated in Fig. 14.43.

The geometric characteristic symbols plus the supplementary symbols are further explained and illustrated with material adapted from ANSI Y14.5M–1982 (R1988), as follows.

BASIC DIMENSION SYMBOL   The basic dimension is identified by the enclosing frame symbol, Fig. 14.18 (a). The basic dimension (size) is the value used to describe the theoretically exact size,

Table 14.3 *Geometric Characteristics and Modifying Symbols* [ANSI Y14.5M–1982 (R1988)]

Geometric characteristic symbols

	Type of Tolerance	Characteristic	Symbol
For individual features	Form	Straightness	—
		Flatness	▱
		Circularity (roundness)	○
		Cylindricity	/○/
For individual or related features	Profile	Profile of a line	⌒
		Profile of a surface	⌓
For related features	Orientation	Angularity	∠
		Perpendicularity	⊥
		Parallelism	//
	Location	Position	⊕
		Concentricity	◎
	Runout	Circular runout	↗
		Total runout	↗↗ ᵃ

ᵃArrowhead(s) may be filled in.

Modifying symbols

Term	Symbol
At maximum material condition	Ⓜ
Regardless of feature size	Ⓢ
At least material condition	Ⓛ
Projected tolerance zone	Ⓟ
Diameter	⌀
Spherical diameter	S⌀
Radius	R
Spherical radius	SR
Reference	( )
Arc length	⌒

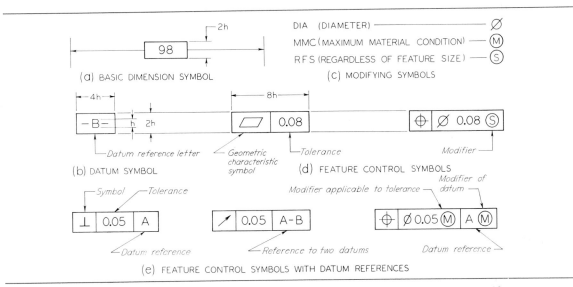

Fig. 14.18 Use of Symbols for Tolerance of Position and Form   [ANSI Y14.5M–1982 (R1988)].

shape, or location of a feature. It is the basis from which permissible variations are established by tolerances on other dimensions in notes, or in feature control frames.

DATUM IDENTIFYING SYMBOL   The datum identifying symbol consists of a frame containing a reference letter preceded and followed by a dash, Fig. 14.18 (b). A point, line, plane, cylinder, or other geometric form assumed to be exact for purposes of computation may serve as a datum from which the location or geometric relationship of features of a part may be established.

SUPPLEMENTARY SYMBOLS   The symbols for MMC (maximum material condition, i.e., minimum hole diameter, maximum shaft diameter) and RFS (regardless of feature size—the tolerance applies to any size of the feature within its size tolerance and or the actual size of a datum feature) are illustrated in Fig. 14.18 (c). The abbreviations MMC and RFS are also used in notes. See also Table 14.3.

The symbol for diameter is used instead of the abbreviation DIA to indicate a diameter, and it precedes the specified tolerance in a feature control symbol, Fig. 14.18 (d). This symbol for diameter instead of the abbreviation DIA may be used on a drawing, and it should precede the dimension. For narrative notes, the abbreviation DIA is preferred.

COMBINED SYMBOLS   Individual symbols, datum reference letters, needed tolerances, and so on may be combined in a single frame, Fig. 14.18 (e).

A position of form tolerance is given by a feature control symbol made up of a frame about the appropriate geometric characteristic symbol plus the allowable tolerance. A vertical line separates the symbol and the tolerance, Fig. 14.18 (d). Where needed, the tolerance should be preceded by the symbol for diameter and followed by the symbol for MMC or RFS.

A tolerance of position or form related to a datum is so indicated in the feature control symbol by placing the datum reference letter following either the geometric characteristic symbol or the tolerance. Vertical lines separate the entries, and, where applicable, the datum reference letter entry includes the symbol for MMC or RFS. See Fig. 14.18.

14.16   Positional Tolerances

In §13.25 are shown a number of examples of the traditional methods of locating holes, that is, by means of rectangular coordinates or angular dimensions. Each dimension has a tolerance, either given directly or indicated on the completed drawing by a general note.

For example, Fig. 14.19 (a) shows a hole located from two surfaces at right angles to each

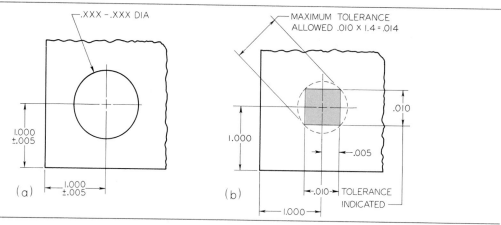

Fig. 14.19   Tolerance Zones.

other. As shown at (b), the center may lie anywhere within a square tolerance zone, the sides of which are equal to the tolerances. Thus, the total variations along either diagonal of the square by the coordinate method of dimensioning will be 1.4 times greater than the indicated tolerance. Hence, a .014 diameter tolerance zone would increase the square tolerance zone area 57 percent without exceeding the tolerance permitted along the diagonal of the square tolerance zone.

Features located by toleranced angular and radial dimensions will have a wedge-shaped tolerance zone. See Fig. 14.28.

If four holes are dimensioned with rectangular coordinates as in Fig. 14.20 (a), acceptable patterns for the square tolerance zones for the holes are shown at (b) and (c). The locational tolerances are actually greater than indicated by the dimensions.

Feature control symbols are related to the feature by one of several methods illustrated in Fig. 14.43. The following methods are preferred.

1. Adding the symbol to a note or dimension pertaining to the feature.

2. Running a leader from the symbol to the feature.

3. Attaching the side, end, or corner of the symbol frame to an extension line from the feature.

4. Attaching a side or end of the symbol frame to the dimension line pertaining to the feature.

In Fig. 14.20 (a), hole A is selected as a datum, and the other three are located from it. The square tolerance zone for hole A results from the tolerances on the two rectangular coordinate dimensions locating hole A. The sizes of the tolerance zones for the other three holes result from the tolerances between the holes, while their locations will vary according to the actual location of the datum hole A. Two of the many possible zone patterns are shown at (b) and (c).

Thus, with the dimensions shown at (a), it is difficult to say whether the resulting parts will actually fit the mating parts satisfactorily even though they conform to the tolerances shown on the drawing.

These disadvantages are overcome by giving exact theoretical locations by untoleranced dimensions and then specifying by a note how far actual positions may be displaced from these locations. This is called *true-position dimensioning*. It will be seen that the tolerance zone for each hole will be a circle, the size of the circle depending on the amount of variation permitted from "true position."

A true-position dimension denotes the theoretically exact position of a feature. The location of each feature such as a hole, slot, stud, and so

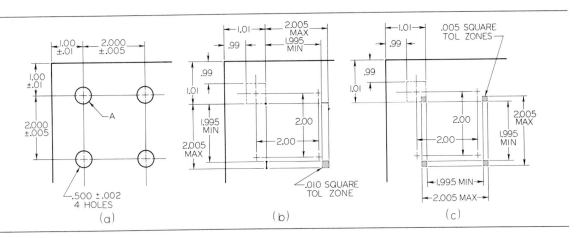

Fig. 14.20   Tolerance Zones.

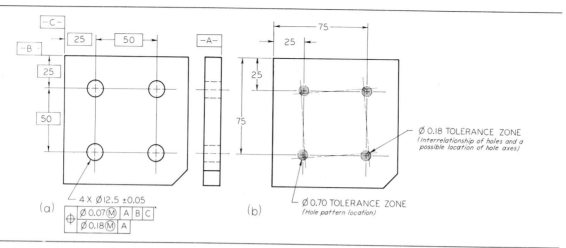

Fig. 14.21 True-Position Dimensioning [ANSI Y14.5M–1982 (R1988)].

on, is given by untoleranced basic dimensions identified by the enclosing frame or symbol. To prevent misunderstandings, true position should be established with respect to a datum.

In simple arrangements, the choice of a datum may be obvious and not require identification.

Positional tolerancing is identified by a characteristic symbol directed to a feature, which establishes a circular tolerance zone, Fig. 14.21.

Actually, the "circular tolerance zone" is a cylindrical tolerance zone (the diameter of which is equal to the positional tolerance and its length is equal to the length of the feature unless otherwise specified), and its axis must be within this cylinder, Fig. 14.22.

The center line of the hole may coincide with the center line of the cylindrical tolerance zone, (a), or it may be parallel to it but displaced so as to remain within the tolerance cylinder, (b), or it may be inclined while remaining within the tolerance cylinder, (c). In this last case we see that the positional tolerance also defines the limits of squareness variation.

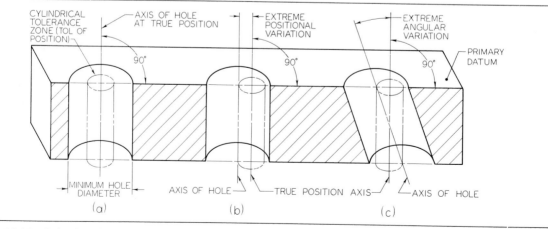

Fig. 14.22 Cylindrical Tolerance Zone [ANSI Y14.5M–1982 (R1988)].

In terms of the cylindrical surface of the hole, the positional tolerance specification indicates that all elements on the hole surface must be on or outside a cylinder whose diameter is equal to the minimum diameter (MMC; §14.17) or the maximum diameter of the hole minus the positional tolerance (diameter, or twice the radius), with the center line of the cylinder located at true position, Fig. 14.23.

The use of basic untoleranced dimensions to locate features at true position avoids one of the chief difficulties in tolerancing—the accumulation of tolerances, §14.9, even in a chain of dimensions. Fig. 14.24.

While features, such as holes and bosses, may vary in any direction from the true-position axis, other features, such as slots, may vary on either side of a true-position plane, Fig. 14.25.

Since the exact locations of the true positions are given by untoleranced dimensions, it is important to prevent the application of general tolerances to these. A note should be added to the drawing such as

GENERAL TOLERANCES DO NOT APPLY TO BASIC TRUE–POSITION DIMENSIONS.

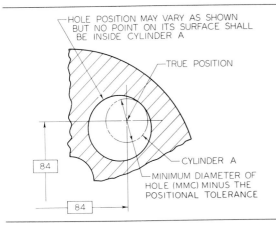

Fig. 14.23   True Position Interpretation [ANSI Y14.5M–1982 (R1988)].

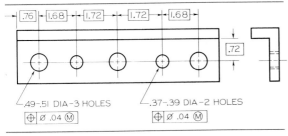

Fig. 14.24   No Tolerance Accumulation.

14.17   Maximum Material Condition

Maximum material condition, usually abbreviated to MMC, means that a feature of a finished product contains the maximum amount of material permitted by the toleranced size dimensions shown for that feature. Thus, we have MMC when holes, slots, or other internal features are at minimum size, or when shafts, pads, bosses, and other external features are at their maximum size. We have MMC for both mating parts when the largest shaft is in the smallest hole and there is the least clearance between the parts.

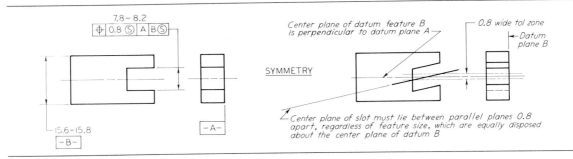

Fig. 14.25   Positional Tolerancing for Symmetry   [ANSI Y14.5M–1982 (R1988)].

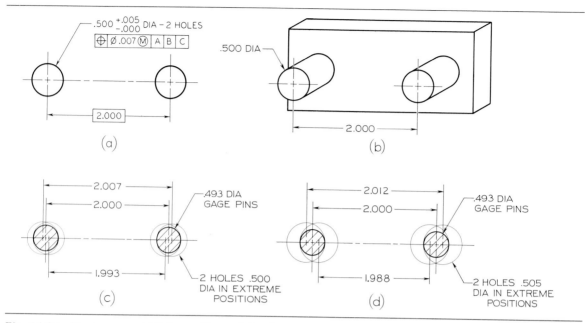

Fig. 14.26   Maximum and Minimum Material Conditions—Two-Hole Pattern   [ANSI Y14.5M–1982 (R1988)].

In assigning positional tolerance to a hole, it is necessary to consider the size limits of the hole. If the hole is at MMC (smallest size), the positional tolerance is not affected, but if the hole is larger, the available positional tolerance is greater. In Fig. 14.26 (a) two half-inch holes are shown. If they are exactly .500″ in diameter (MMC, or smallest size) and are exactly 2.000″ apart, they should receive a gage, (b), made of two round pins .500″ in diameter fixed in a plate 2.000″ apart. However, the center-to-center distance between the holes may vary from 1.993″ to 2.007″.

If the .500″ diameter holes are at their extreme positions, (c), the pins in the gage would have to be .007″ smaller, or .493″ diameter, to enter the holes. Thus, if the .500″ diameter holes are located at the maximum distance apart, the .493″ diameter gage pins would contact the inner sides of the holes; and if the holes are located at the minimum distance apart, the .493″ diameter pins would contact the outer surfaces of the holes, as shown. If gagemakers' tolerances are not considered, the gage pins would have to be

.493″ diameter and exactly 2.000″ apart if the holes are .500″ diameter, or MMC.

If the holes are .505″ diameter—that is, at maximum size, as at (d)—they will be accepted by the same .493″ diameter gage pins at 2.000″ apart if the inner sides of the holes contact the inner sides of the gage pins and the outer sides of the holes contact the outer sides of the gage pins, as shown. Thus the holes may be 2.012″ apart, which is beyond the tolerance permitted for the center-to-center distance between the holes. Similarly, the holes may be as close together as 1.988″ from center to center, which again is outside the specified positional tolerance.

Thus, when the holes are not at MMC—that is, when they are at maximum size—a greater positional tolerance becomes available. Since all features may vary in size, it is necessary to make clear on the drawing at what basic dimension the true position applies. In all but a few exceptional cases, the additional positional tolerance available, when holes are larger than minimum size, is acceptable and desirable. Parts thus accepted

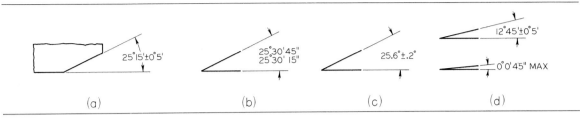

Fig. 14.27 Tolerances of Angles.

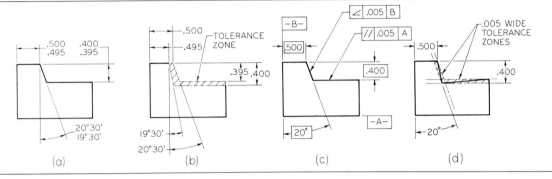

Fig. 14.28 Angular Tolerance Zones [ANSI Y14.5M–1982 (R1988)].

can be freely assembled whether or not the holes or other features are within the specified positional tolerance. This practice has been recognized and used in manufacturing for years, as is evident from the use of fixed-pin gages, which have been commonly used to inspect parts and control the least favorable condition of assembly. Thus it has become common practice for both manufacturing and inspection to assume that positional tolerance applies to MMC and that greater positional tolerance becomes permissible when the part is not at MMC.

To avoid possible misinterpretation as to whether maximum material condition (MMC) or regardless of feature size (RFS) applies, it should be clearly stated on the drawing by the addition of MMC or RFS symbols to each applicable tolerance or by suitable coverage in a document referenced on the drawing.

When MMC or RFS is not specified on the drawing with respect to an individual tolerance, datum reference, or both, the following rules will apply.

1. True-position tolerances and related datum references apply at MMC.

2. Angularity, parallelism, perpendicularity, concentricity, and symmetry tolerances, including related datum references, apply at RFS. No element of the actual feature will extend beyond the envelope of the perfect form at MMC.

14.18 Tolerances of Angles

Bilateral tolerances have traditionally been given on angles as illustrated in Fig. 14.27. Consequently, the wedge-shaped tolerance zone increases as the distance from the vertex of the angle increases. Thus, the tolerance had to be figured after considering the total displacement at the point farthest from the vertex of the angle before a tolerance could be specified that would not exceed the allowable displacement. The use of angular tolerances may be avoided by using gages. Taper turning is often handled by machining to fit a gage or by fitting to the mating part.

If an angular surface is located by a linear and an angular dimension, Fig. 14.28 (a), the surface must lie within a tolerance zone as shown at (b). The angular zone will be wider as the distance from the vertex increases. In order

453

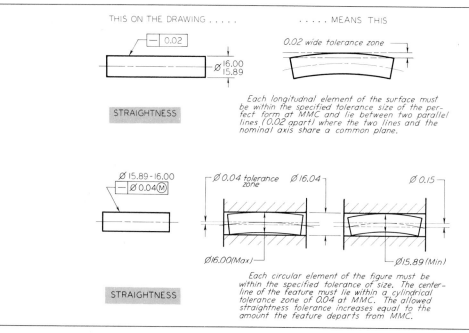

THIS ON THE DRAWING . . . . MEANS THIS

0.02 wide tolerance zone

STRAIGHTNESS

Each longitudnal element of the surface must be within the specified tolerance size of the perfect form at MMC and lie between two parallel lines (0.02 apart) where the two lines and the nominal axis share a common plane.

STRAIGHTNESS

Ø 0.04 tolerance zone   Ø 16.04   Ø 0.15

Ø16.00 (Max)   Ø15.89 (Min)

Each circular element of the figure must be within the specified tolerance of size. The center-line of the feature must lie within a cylindrical tolerance zone of 0.04 at MMC. The allowed straightness tolerance increases equal to the amount the feature departs from MMC.

Fig. 14.29   Specifying Straightness [ANSI Y14.5M–1982 (R1988)].

to avoid the accumulation of tolerances, that is, to decrease the tolerance zone, the *basic angle* tolerancing method of (c) is recommended [ANSI Y14.5M–1982 (R1988)]. The angle is indicated as basic with the proper symbol and no angular tolerance is specified. The tolerance zone is now defined by two parallel planes, resulting in improved angular control, (d).

14.19   Form Tolerances for Single Features

Straightness, flatness, roundness, cylindricity, and, in some instances, profile are form tolerances applicable to single features.

STRAIGHTNESS TOLERANCE   A straightness tolerance specifies a tolerance zone within which an axis or all points of the considered element must lie, Fig. 14.29. Straightness is a condition where an element of a surface or an axis is a straight line.

FLATNESS TOLERANCE   A flatness tolerance specifies a tolerance zone defined by two parallel planes within which the surface must lie, Fig. 14.30. Flatness is the condition of a surface having all elements in one plane.

ROUNDNESS (CIRCULARITY) TOLERANCE   A roundness tolerance specifies a tolerance zone bounded by two concentric circles within which each circular element of the surface must lie, Fig. 14.31. Roundness is a condition of a surface of revolution where, for a cone or cylinder, all points of the surface intersected by any plane perpendicular to a common axis are equidistant from that axis. For a sphere, all points of the surface intersected by any plane passing through a common center are equidistant from that center.

CYLINDRICITY TOLERANCE   A cylindricity tolerance specifies a tolerance zone bounded by two concentric cylinders within which the surface

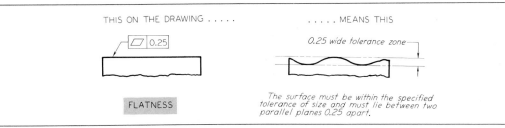

THIS ON THE DRAWING . . . . MEANS THIS

FLATNESS

0.25 wide tolerance zone

The surface must be within the specified
tolerance of size and must lie between two
parallel planes 0.25 apart.

Fig. 14.30   Specifying Flatness [ANSI Y14.5M–1982 (R1988)].

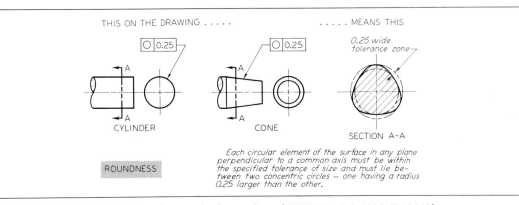

THIS ON THE DRAWING . . . . MEANS THIS

CYLINDER   CONE   SECTION A-A

0.25 wide
tolerance zone

ROUNDNESS

Each circular element of the surface in any plane
perpendicular to a common axis must be within
the specified tolerance of size and must lie between two concentric circles — one having a radius
0.25 larger than the other.

Fig. 14.31   Specifying Roundness for a Cylinder or Cone [ANSI Y14.5M–1982 (R1988)].

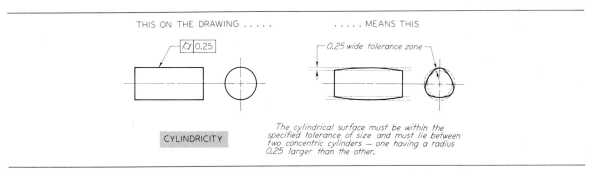

THIS ON THE DRAWING . . . . MEANS THIS

0.25 wide tolerance zone

CYLINDRICITY

The cylindrical surface must be within the
specified tolerance of size and must lie between
two concentric cylinders — one having a radius
0.25 larger than the other.

Fig. 14.32   Specifying Cylindricity [ANSI Y14.5M–1982 (R1988)].

must lie, Fig. 14.32. This tolerance applies to
both circular and longitudinal elements of the
entire surface. Cylindricity is a condition of a
surface of revolution in which all points of the
surface are equidistant from a common axis.
When no tolerance of form is given, many pos-

sible shapes may exist within a tolerance zone,
as illustrated in Fig. 14.33.

PROFILE TOLERANCE   A profile tolerance specifies a uniform boundary or zone along the true
profile within which all elements of the surface

455

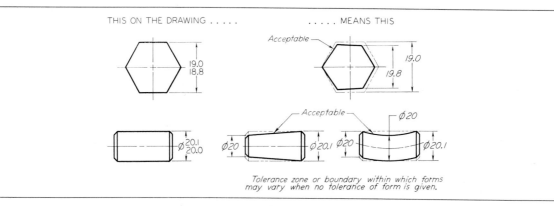

Fig. 14.33 Acceptable Variations of Form—No Specified Tolerance of Form.

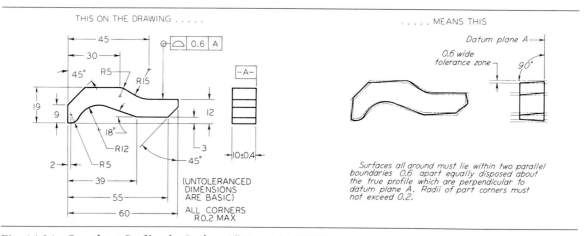

Fig. 14.34 Specifying Profile of a Surface All Around [ANSI Y14.5M–1982 (R1988)].

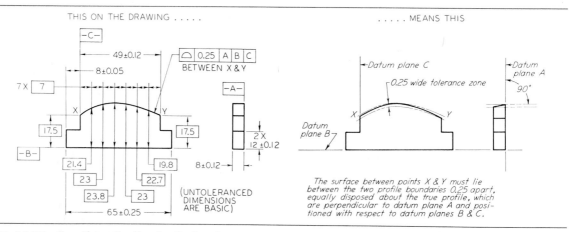

Fig. 14.35 Specifying Profile of a Surface Between Points [ANSI Y14.5M–1982 (R1988)].

must lie, Figs. 14.34 and 14.35. A profile is the outline of an object in a given plane (two-dimensional) figure. Profiles are formed by projecting a three-dimensional figure onto a plane or by taking cross sections through the figure with the resulting profile composed of such elements as straight lines, arcs, or other curved lines.

14.20 Form Tolerances for Related Features

Angularity, parallelism, perpendicularity, and, in some instances, profile are form tolerances applicable to related features. These tolerances control the attitude of features to one another [ANSI Y14.5M–1982 (R1988)].

ANGULARITY TOLERANCE   An angularity tolerance specifies a tolerance zone defined by two parallel planes at the specified basic angle (other than 90°) from a datum plane or axis within which the surface or the axis of the feature must lie, Fig. 14.36.

PARALLELISM TOLERANCE   A parallelism tolerance specifies a tolerance zone defined by two parallel planes or lines parallel to a datum plane or axis within which the surface or axis of the feature must lie, or the parallelism tolerance may specify a cylindrical tolerance zone parallel to a datum axis within which the axis of the feature must lie, Figs. 14.37–14.39.

PERPENDICULARITY TOLERANCE   Perpendicularity is a condition of a surface, median plane, or axis at 90° to a datum plane or axis. A perpendicularity tolerance specifies one of the following.

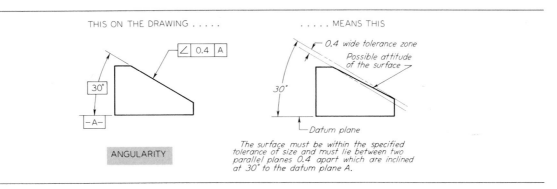

Fig. 14.36   Specifying Angularity for a Plane Surface   [ANSI Y14.5M–1982 (R1988)].

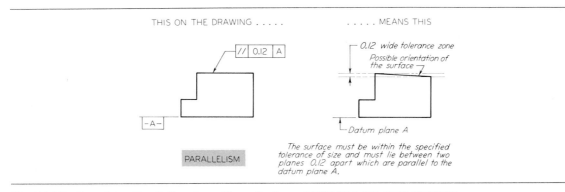

Fig. 14.37   Specifying Parallelism for a Plane Surface   [ANSI Y14.5M–1982 (R1988)].

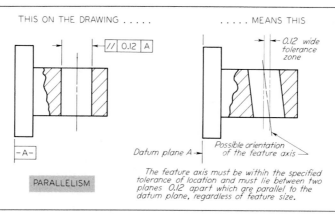

THIS ON THE DRAWING . . . . MEANS THIS

PARALLELISM

*The feature axis must be within the specified
tolerance of location and must lie between two
planes 0.12 apart which are parallel to the
datum plane, regardless of feature size.*

Fig. 14.38   Specifying Parallelism for an Axis Feature RFS   [ANSI Y14.5M–1982 (R1988)].

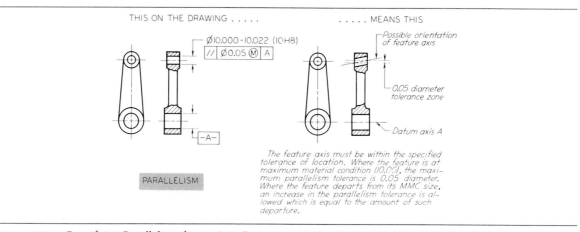

THIS ON THE DRAWING . . . . MEANS THIS

PARALLELISM

*The feature axis must be within the specified
tolerance of location. Where the feature is at
maximum material condition (10.00), the maximum parallelism tolerance is 0.05 diameter.
Where the feature departs from its MMC size,
an increase in the parallelism tolerance is allowed which is equal to the amount of such
departure.*

Fig. 14.39   Specifying Parallelism for an Axis Feature at MMC   [ANSI Y14.5M–1982 (R1988)].

1. A tolerance zone defined by two parallel planes perpendicular to a datum plane, datum axis, or axis within which the surface of the feature must lie, Fig. 14.40.

2. A cylindrical tolerance zone perpendicular to a datum plane within which the axis of the feature must lie, Fig. 14.41.

CONCENTRICITY TOLERANCE   Concentricity is the condition where the axes of all cross-sectional elements of a feature's surface of revolution are common to the axis of a datum fea-

ture. A concentricity tolerance specifies a cylindrical tolerance zone whose axis coincides with a datum axis and within which all cross-sectional axes of the feature being controlled must lie, Fig. 14.42.

14.21   Application of Geometric Tolerancing

The use of various feature control symbols in lieu of notes for position and form tolerance dimen-

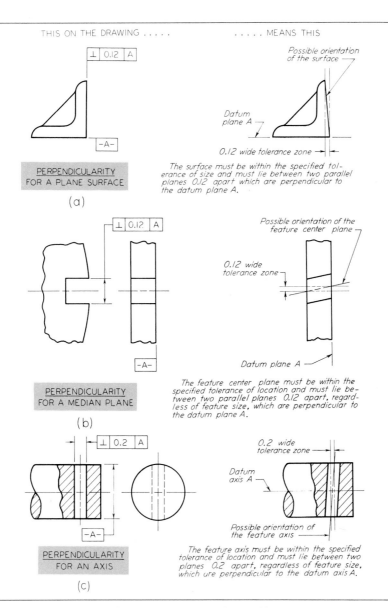

Possible orientation
of the surface

⊥ | 0.12 | A

Datum
plane A

-A-

0.12 wide tolerance zone

PERPENDICULARITY
FOR A PLANE SURFACE

The surface must be within the specified tolerance of size and must lie between two parallel planes 0.12 apart which are perpendicular to the datum plane A.

(a)

⊥ | 0.12 | A

Possible orientation of the
feature center plane

0.12 wide
tolerance zone

Datum plane A

PERPENDICULARITY
FOR A MEDIAN PLANE

The feature center plane must be within the specified tolerance of location and must lie between two parallel planes 0.12 apart, regardless of feature size, which are perpendicular to the datum plane A.

(b)

⊥ | 0.2 | A

0.2 wide
tolerance zone

Datum
axis A

-A-

Possible orientation of
the feature axis

PERPENDICULARITY
FOR AN AXIS

The feature axis must be within the specified tolerance of location and must lie between two planes 0.2 apart, regardless of feature size, which are perpendicular to the datum axis A.

(c)

Fig. 14.40   Specifying Perpendicularity [ANSI Y14.5M–1982 (R1988)].

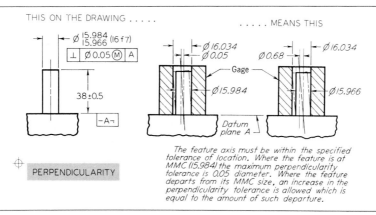

Ø $\frac{15.984}{15.966}$ (16 f 7)

⊥ | Ø 0.05 Ⓜ | A

38 ±0.5

-A-

Ø16.034
Ø0.05

Gage

Ø15.984

Datum
plane A

Ø0.68   Ø16.034

Ø15.966

PERPENDICULARITY

The feature axis must be within the specified tolerance of location. Where the feature is at MMC (15.984) the maximum perpendicularity tolerance is 0.05 diameter. Where the feature departs from its MMC size, an increase in the perpendicularity tolerance is allowed which is equal to the amount of such departure.

Fig. 14.41   Specifying Perpendicularity for an Axis, Pin, or Boss [ANSI Y14.5M–1982 (R1988)].

459

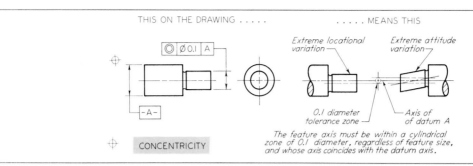

Fig. 14.42   Specifying Concentricity [ANSI Y14.5M–1982 (R1988)].

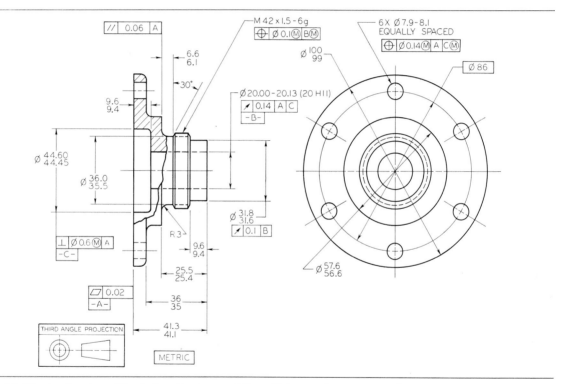

Fig. 14.43   Application of Symbols to Position and Form Tolerance Dimensions   [ANSI Y14.5M–1982 (R1988)].

sions as abstracted from ANSI Y14.5M–1982 (R1988) is illustrated in Fig. 14.43. For a more detailed treatment of geometric tolerancing, consult the latest ANSI Y14.5 Dimensioning and Tolerancing standard.

14.22   Surface Roughness, Waviness, and Lay

The modern demands of the automobile, the airplane, and other machines that can stand heavier loads and higher speeds with less friction and

	Symbol	Meaning
(a)		Basic Surface Texture Symbol. Surface may be produced by any method except when the bar or circle, (b) or (d), is specified.
(b)		Material Removal By Machining Is Required. The horizontal bar indicates that material removal by machining is required to produce the surface and that material must be provided for that purpose.
(c)	3.5	Material Removal Allowance. The number indicates the amount of stock to be removed by machining in millimeters (or inches). Tolerances may be added to the basic value shown or in a general note.
(d)		Material Removal Prohibited. The circle in the vee indicates that the surface must be produced by processes such as casting, forging, hot finishing, cold finishing, die casting, powder metallurgy or injection molding without subsequent removal of material.
(e)		Surface Texture Symbol. To be used when any surface characteristics are specified above the horizontal line or to the right of the symbol. Surface may be produced by any method except when the bar or circle, (b) or (d), is specified.
(f)	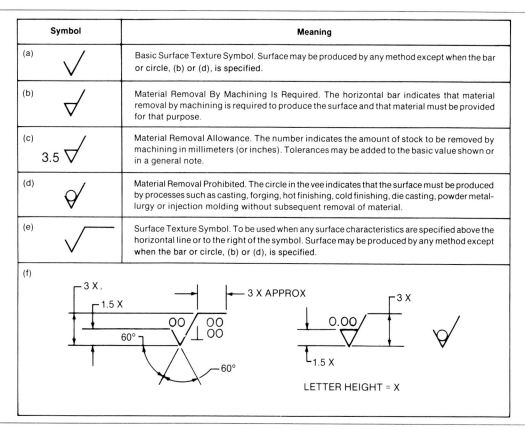	

Fig. 14.44   Surface Texture Symbols and Construction   [ANSI Y14.36–1978 (R1987)].

wear have increased the need for accurate control of surface quality by the designer regardless of the size of the feature. Simple finish marks are not adequate to specify surface finish on such parts.

Surface finish is intimately related to the functioning of a surface, and proper specification of finish of such surfaces as bearings and seals is necessary. Surface quality specifications should be used only where needed, since the cost of producing a finished surface becomes greater as the quality of the surface called for is increased. Generally, the ideal surface finish is the roughest one that will do the job satisfactorily.

The system of surface texture symbols recommended by ANSI [Y14.36–1978 (R1987)] for use on drawings, regardless of the system of measurement used, is now broadly accepted by American industry. These symbols are used to define *surface texture, roughness,* and *lay.* See Fig. 14.44 for meaning and construction of these symbols. The basic surface texture symbol in Fig. 14.45 (a) indicates a finished or machined surface by any method just as does the general V symbol, Fig. 13.20 (a). Modifications to the basic surface texture symbol, (b) through (d), define restrictions on material removal for the finished surface. Where surface texture values other than roughness average (R_a) are specified, the symbol must be drawn with the horizontal extension as shown in (e). Construction details for the symbols are given in (f).

Applications of the surface texture symbols are given in Fig. 14.45 (a). Note that the symbols

461

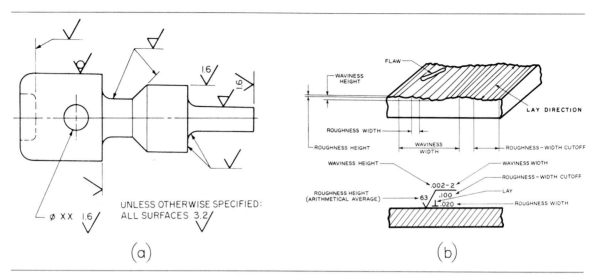

Fig. 14.45   Application of Surface Texture Symbols and Surface Characteristics   [ANSI Y14.36–1978 (R1987)].

Table 14.4   *Preferred Series Roughness Average Values (R_a)* [ANSI Y14.36–1978 (R1987)]

Recommended values are in color.

Micro-meters[a] (μm)	Micro-inches (μin.)	Micro-meters[a] (μm)	Micro-inches (μin.)
0.012	0.5	1.25	50
0.025	1	1.60	63
0.050	2	2.0	80
0.075	3	2.5	100
0.10	4	3.2	125
0.125	5	4.0	180
0.15	6	5.0	200
0.20	8	6.3	250
0.25	10	8.0	320
0.32	13	10.0	400
0.40	16	12.5	500
0.50	20	15	600
0.63	25	20	800
0.80	32	25	1000
1.00	40		

[a]Micrometers are the same as thousandths of a millimeter (1 μm = 0.001 mm).

Table 14.5   *Standard Roughness Sampling Length (Cutoff) Values* [ANSI Y14.36–1978 (R1987)]

Millimeters (mm)	Inches (in.)	Millimeters (mm)	Inches (in.)
0.08	.003	2.5	.1
0.25	.010	8.0	.3
0.80	.030	25.0	1.0

Table 14.6   *Preferred Series Maximum Waviness Height Values* [ANSI Y14.36–1978 (R1987)]

Millimeters (mm)	Inches (in.)	Millimeters (mm)	Inches (in.)
0.0005	.00002	0.025	.001
0.0008	.00003	0.05	.002
0.0012	.00005	0.08	.003
0.0020	.00008	0.12	.005
0.0025	.0001	0.20	.008
0.005	.0002	0.25	.010
0.008	.0003	0.38	.015
0.012	.0005	0.50	.020
0.020	.0008	0.80	.030

LAY SYMBOLS

SYM	DESIGNATION	EXAMPLE	SYM	DESIGNATION	EXAMPLE
═	Lay parallel to the line representing the surface to which the symbol is applied.	 DIRECTION OF TOOL MARKS	X	Lay angular in both directions to line representing the surface to which symbol is applied.	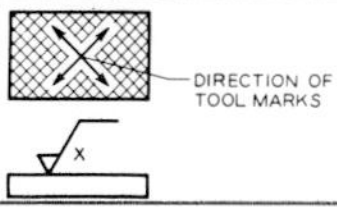 DIRECTION OF TOOL MARKS
⊥	Lay perpendicular to the line representing the surface to which the symbol is applied.	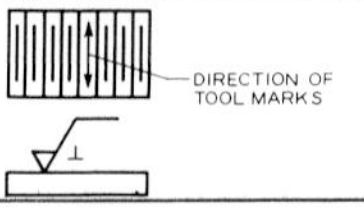 DIRECTION OF TOOL MARKS	M	Lay multidirectional	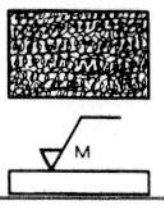
C	Lay approximately circular relative to the center of the surface to which the symbol is applied.	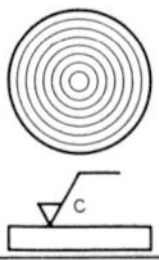	R	Lay approximately radial relative to the center of the surface to which the symbol is applied.	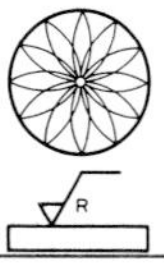

Fig. 14.46 Lay Symbols [ANSI Y14.36–1978 (R1987)].

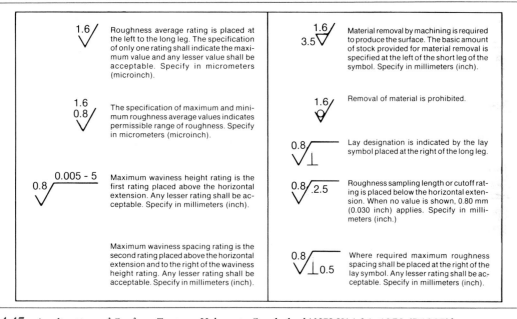

Fig. 14.47 Application of Surface Texture Values to Symbol [ANSI Y14.36–1978 (R1987)].

read from the bottom and/or the right side of the drawing and that they are not drawn at any angle or upside down.

Measurements for roughness and waviness, unless otherwise specified, apply in the direction that gives the maximum reading, usually across the lay. See Fig. 14.45 (b). The recommended roughness height values are given in Table 14.4.

When it is necessary to indicate the roughness-width cutoff values, the standard values to be used are listed in Table 14.5. If no value is specified, the 0.80 value is assumed.

When maximum waviness height values are required, the recommended values to be used are as given in Table 14.6.

When it is desired to indicate lay, the lay symbols in Fig. 14.46 are added to the surface texture symbol as per the examples given. Selected applications of the surface texture values to the symbol are given and explained in Fig. 14.47.

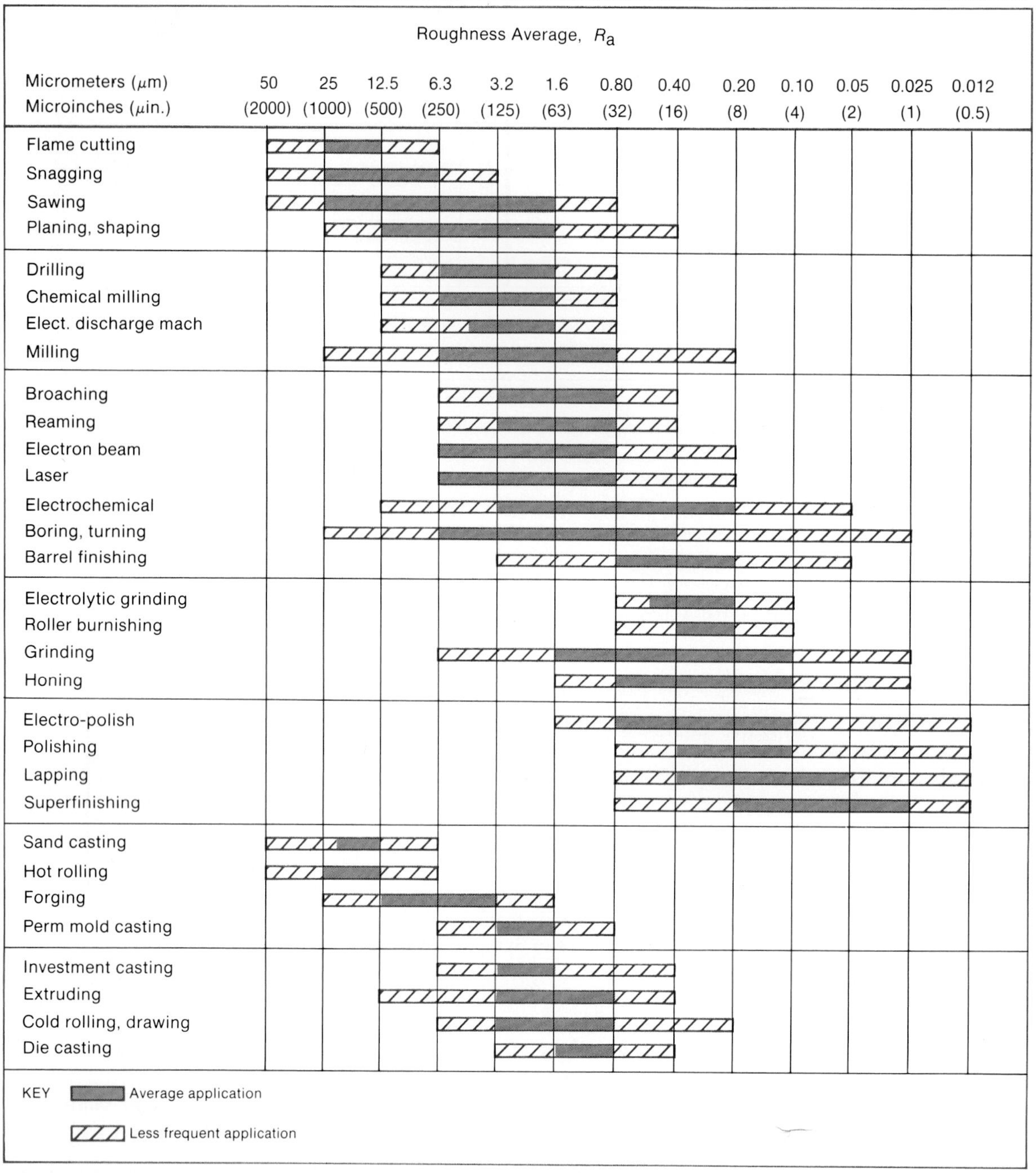

Roughness Average, R_a												
Micrometers (μm) 50	25	12.5	6.3	3.2	1.6	0.80	0.40	0.20	0.10	0.05	0.025	0.012
Microinches (μin.) (2000)	(1000)	(500)	(250)	(125)	(63)	(32)	(16)	(8)	(4)	(2)	(1)	(0.5)

Flame cutting
Snagging
Sawing
Planing, shaping

Drilling
Chemical milling
Elect. discharge mach
Milling

Broaching
Reaming
Electron beam
Laser
Electrochemical
Boring, turning
Barrel finishing

Electrolytic grinding
Roller burnishing
Grinding
Honing

Electro-polish
Polishing
Lapping
Superfinishing

Sand casting
Hot rolling
Forging
Perm mold casting

Investment casting
Extruding
Cold rolling, drawing
Die casting

KEY ▬ Average application

▨ Less frequent application

Fig. 14.48 Surface Roughness Produced by Common Production Methods [ANSI/ASME B46.1–1985]. The ranges shown are typical of the processes listed. Higher or lower values may be obtained under special conditions.

A typical range of surface roughness values that may be obtained from various production methods is shown in Fig. 14.48. Preferred roughness-height values are shown at the top of the chart.

allow the user to add surface finish symbols by inputting a few keystrokes, Fig. 14.49. Geometric tolerancing can be applied to a drawing with the CAD system supplying standard ANSI symbols, Fig. 14.50, as well as user-defined custom symbols.

14.23 Computer Graphics

Computer graphics programs are available that

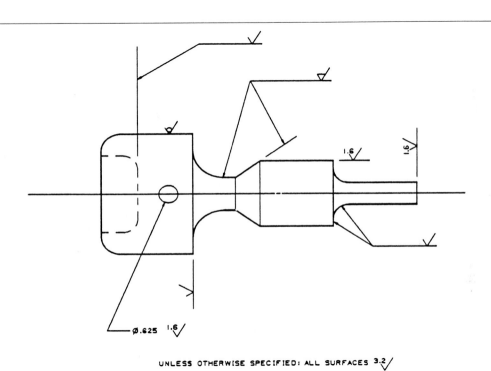

Fig. 14.49   Application of Surface Finish Symbols with Computervision CADDS 4X Production Drafting System. *Courtesy of Computervision Corporation, a subsidiary of Prime Computer, Inc.*

GENERAL TOLERANCE SYMBOLS

Ø — Diameter

XX.XX — Basic

Ⓜ — Maximum Material Condition (MMC)

Ⓛ — Least Material Condition (LMC)

Ⓢ — Regardless of Feature Size

SYMBOL NOT DEFINED — Full Indicator Movement (FIM)

Ⓟ — Projected Tolerance Zone

⊕▸ — Dimension Origin

Ø — All–Around

R — Radius

( ) — Reference Dimension (REF)

SØ — Spherical Diameter (SD)

SYMBOL NOT DEFINED — Spherical Radius (SR)

⌒ — Arc Length

--- — Chain Line

▷ — Conical Taper

◁ — Slope

⊔ — Counterbore/Spotface

V — Countersink

↧ — Depth/Deep (DP)

XX.XX — Dimension Not To Scale

X — Times/Places

FORM AND ORIENTATION TOLERANCE SYMBOLS

— — Straightness

▱ — Flatness

○ — Circularity

⌭ — Cylindricity

⊥ — Perpendicularity

∠ — Angularity

// — Parallelism

⌓ — Surface Profile

⌒ — Line Profile

⌰ — Total Runout

↗ — Circular Runout

SYMBOL NOT DEFINED — Unit Control

TOLERANCE SYMBOLS OF LOCATION

◎ — Concentricity

⊕ — Position

⹀ — Symmetry

Fig. 14.50   Geometric Dimensioning and Tolerancing Symbols Available in Computervision's CADDS 4X Production Drafting System. *Courtesy of Computervision Corporation, a subsidiary of Prime Computer, Inc.*

CHAPTER 15

Threads, Fasteners, and Springs*

The concept of the screw thread seems to have occurred first to Archimedes, the third-century B.C. mathematician, who wrote briefly on spirals and invented or designed several simple devices applying the screw principle. By the first century B.C. the screw was a familiar element, but was crudely cut from wood or filed by hand on a metal shaft. Nothing more was heard of the screw thread in Europe until the fifteenth century.

Leonardo da Vinci understood the screw principle, and he has left sketches showing how to cut screw threads by machine. In the sixteenth century, screws appeared in German watches, and screws were used to fasten suits of armor. In 1569 the screw-cutting lathe was invented by the Frenchman Besson, but the method did not take hold for another century and a half; nuts and bolts continued to be made largely by hand. In the eighteenth century, screw manufacturing got started in England during the Industrial Revolution.

15.1 Standardized Screw Threads

In early times, there was no such thing as standardization. Nuts made by one manufacturer would not fit the bolts of another. In 1841 Sir Joseph Whitworth started crusading for a standard screw thread, and soon the Whitworth thread was accepted throughout England.

In 1864, the United States adopted a thread proposed by William Sellers of Philadelphia, but the Sellers' nuts would not screw on a Whitworth bolt, or vice versa. In 1935 the American Standard thread, with the same 60° V form of the old Sellers' thread, was adopted in the United States. Still there was no standardization among countries. In peacetime it was a nuisance; in

World War I it was a serious inconvenience; and in World War II the obstacle was so great that the Allies decided to do something about it. Talks began among the Americans, British, and Canadians, and in 1948 an agreement was reached on the unification of American and British screw threads. The new thread was called the *Unified screw thread,* and it represents a compromise between the American Standard and Whitworth systems, allowing complete interchangeability of threads in the three countries.

In 1946, an International Organization for Standardization (ISO) committee was formed to establish a single international system of metric screw threads. Consequently, through the co-

---

*For a listing of American National Standards Institute (ANSI) standards for threads, fasteners, and springs, see Appendix 1.

operative efforts of the Industrial Fasteners Institute (IFI), several committees of the American National Standards Institute, and the ISO representatives, a metric fastener standard (IFI-500–1975) was prepared.

Today screw threads are vital to our industrial life. They are designed for hundreds of different purposes, the three basic applications being (1) to *hold parts* together, (2) to *adjust parts* with reference to each other, and (3) to *transmit power*.

15.2 Definitions of Terms
The following definitions apply to screw threads in general, Fig. 15.1.

Screw thread A ridge of uniform section in the form of a helix, §5.63, on the external or internal surface of a cylinder.

External thread A thread on the outside of a member, as on a shaft.

Internal thread A thread on the inside of a member, as in a hole.

Major diameter The largest diameter of a screw thread (applies to both internal and external threads).

Minor diameter The smallest diameter of a screw thread (applies to both internal and external threads).

Pitch The distance from a point on a screw thread to a corresponding point on the next thread measured parallel to the axis. The pitch P is equal to 1 divided by the number of threads per inch.

Pitch diameter The diameter of an imaginary cylinder passing through the threads so as to make equal the widths of the threads and the widths of the spaces cut by the cylinder.

Lead The distance a screw thread advances axially in one turn.

Angle of thread The angle included between the sides of the thread measured in a plane through the axis of the screw.

Crest The top surface joining the two sides of a thread.

Root The bottom surface joining the sides of two adjacent threads.

Side The surface of the thread that connects the crest with the root.

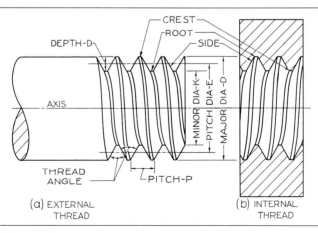

Fig. 15.1   Screw-Thread Nomenclature.

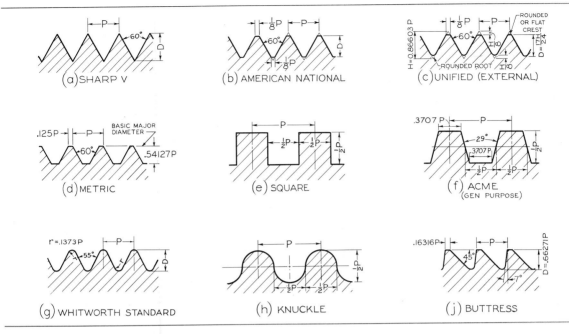

Fig. 15.2 Screw-Thread Forms.

Axis of screw The longitudinal center line through the screw.

Depth of thread The distance between the crest and the root of the thread measured normal to the axis.

Form of thread The cross section of thread cut by a plane containing the axis.

Series of thread Standard number of threads per inch for various diameters.

15.3 Screw-Thread Forms

Fig. 15.2 Various forms of threads are used to hold parts together, to adjust parts with reference to each other, or to transmit power. The 60° *Sharp-V thread* was originally called the United States Standard thread, or the Sellers' thread. For purposes of certain adjustments, the Sharp-V thread is useful with the increased friction resulting from the full thread face. It is also used on brass pipe work.

The *American National thread* with flattened roots and crests is a stronger thread. This form replaced the Sharp-V thread for general use.

The *Unified thread* is the standard thread agreed upon by the United States, Canada, and Great Britain in 1948, and is gradually replacing the American National form. The crest of the external thread may be flat or rounded, and the root is rounded; otherwise, the thread form is essentially the same as the American National.

The *metric thread* is the standard screw thread agreed upon for international screw thread fasteners. The crest and root are flat, but the external thread is often rounded if formed by a rolling process. The form is similar to the American National and the Unified threads but with less depth of thread.

The *square thread* is theoretically the ideal thread for power transmission, since its face is nearly at right angles to the axis, but due to the difficulty of cutting it with dies and because of other inherent disadvantages, such as the fact that split nuts will not readily disengage, the square thread has been displaced to a large ex-

469

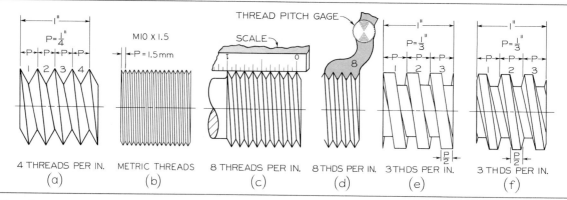

Fig. 15.3   Pitch of Threads.

tent by the Acme thread. The square thread is not standardized.

The *Acme thread* is a modification of the square thread and has largely replaced it. It is stronger than the square thread, is easier to cut, and has the advantage of easy disengagement from a split nut, as on the lead screw of a lathe.

The *standard worm thread* (not shown) is similar to the Acme thread, but is deeper. It is used on shafts to carry power to worm wheels.

The *Whitworth thread* has been the British standard and is being replaced by the Unified thread. Its uses correspond to those of the American National thread.

The *knuckle thread* is usually rolled from sheet metal but is sometimes cast, and is used in modified forms in electric bulbs and sockets, bottle tops, and the like.

The *buttress thread* is designed to transmit power in one direction only and is used in large guns, in jacks, and in other mechanisms of similar high-strength requirements.

15.4   Thread Pitch

The *pitch* of any thread form is the distance parallel to the axis between corresponding points on adjacent threads, Figs. 15.1 (a) and 15.2. For metric threads, this distance is specified in millimeters.

The pitch for a metric thread that is included with the major diameter in the thread designation determines the size of the thread, as shown in Fig. 15.3 (b). For example, M10 × 1.5. See also §15.21, or Appendix 15, for more information.

For threads dimensioned in inches, the pitch is equal to 1 divided by the number of threads per inch. The thread tables give the number of threads per inch for each standard diameter. Thus, a Unified coarse thread, Appendix 15, of 1″ diameter, has eight threads per inch, and the pitch P equals $\frac{1}{8}$″.

As shown in Fig. 15.3 (a), if a thread has only four threads per inch, the pitch and the threads themselves are quite large. If there are, say, 16 threads per inch, the pitch is only $\frac{1}{16}$″, and the threads are relatively small, similar to the threads shown at (b).

The pitch or the number of threads per inch can easily be measured with a scale, (c), or with a *thread-pitch gage,* (d).

15.5   Right-Hand and Left-Hand Threads

A right-hand thread is one that advances into a nut when turned clockwise, and a left-hand thread is one that advances into a nut when turned counterclockwise, Fig. 15.4. A thread is

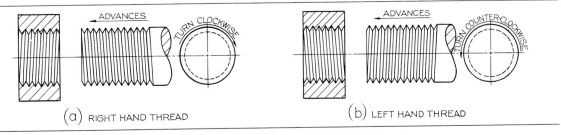

Fig. 15.4   Right-Hand and Left-Hand Threads.

always considered to be right-hand (RH) unless otherwise specified. A left-hand thread is always labeled LH on a drawing. See Fig. 15.19 (a).

15.6   Single and Multiple Threads

A *single* thread, as the name implies, is composed of one ridge, and the lead is therefore equal to the pitch. *Multiple* threads are composed of two or more ridges running side by side. As shown in Fig. 15.5 (a)–(c), the *slope line* is the hypotenuse of a right triangle whose short side equals $\frac{1}{2}$P for single threads, P for double threads, $1\frac{1}{2}$P for triple threads, and so on. This applies to all forms of threads. In *double* threads, the lead is twice the pitch; in *triple* threads the lead is three times the pitch, and so on. On a drawing of a single or triple thread, a root is opposite a crest; in the case of a double or quadruple thread, a root is drawn opposite a root.

Therefore, in one turn, a double thread advances twice as far as a single thread, and a triple thread advances three times as far.

RH double square and RH triple Acme threads are shown at (d) and (e), respectively.

Multiple threads are used wherever quick motion, but not great power, is desired, as on fountain pens, toothpaste caps, valve stems, and the like. The threads on a valve stem are frequently multiple threads, to impart quick action in opening and closing the valve. Multiple threads on a shaft can be recognized and counted by observing the number of thread endings on the end of the screw.

15.7   Thread Symbols

There are three methods of representation for showing screw threads on drawings—the *schematic,* the *simplified,* and the *detailed.* For clar-

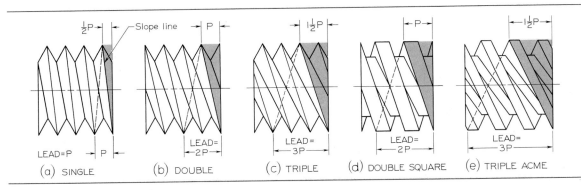

Fig. 15.5   Multiple Threads.

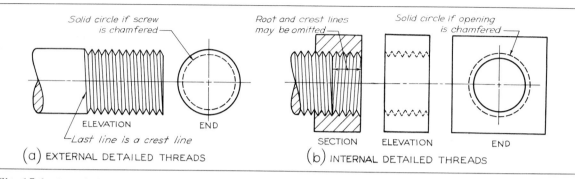

Fig. 15.6   Detailed Metric, American National, and Unified Threads.

ity of representation and where good judgment dictates, schematic, simplified, and detailed thread symbols may be combined on a single drawing.

Two sets of thread symbols, the schematic and the more common simplified, are used to represent threads of small diameter, under approximately 1″ or 25 mm diameter on the drawing. The symbols are the same for all forms of threads, such as metric, Unified, square, and Acme.

The detailed representation is a close approximation of the exact appearance of a screw thread. The true projection of the helical curves of a screw thread, Fig. 15.1, presents a pleasing appearance, but this does not compensate for the laborious task of plotting the helices, §5.63. Consequently, the true projection is rarely used in practice.

When the diameter of the thread on the drawing is over approximately 1″ or 25 mm, a pleasing drawing may be made by the *detailed representation* method, in which the true profiles of

the threads (any form of thread) are drawn; but the helical curves are replaced by straight lines, as shown in Fig. 15.6.* Whether the crests or roots are flat or rounded, they are represented by single lines and not double lines as in Fig. 15.1; consequently, American National and Unified threads are drawn in exactly the same way.

15.8   External Thread Symbols

Simplified external thread symbols are shown in Fig. 15.7 (a) and (b). The threaded portions are indicated by hidden lines parallel to the axis at the approximate depth of the thread, whether in section or in elevation. Use the schematic depth of thread as given in the table in Fig. 15.9 (a), to draw these lines, (d).

When the schematic form is shown in section, (c), it is necessary to show the Vs; other-

---

*A thread 42 mm or 1⅝″ diameter, if drawn half size, would be less than 25 mm or 1″ diameter on the drawing and hence would be too small for this method of representation.

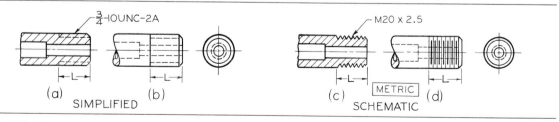

Fig. 15.7   External Thread Symbols.

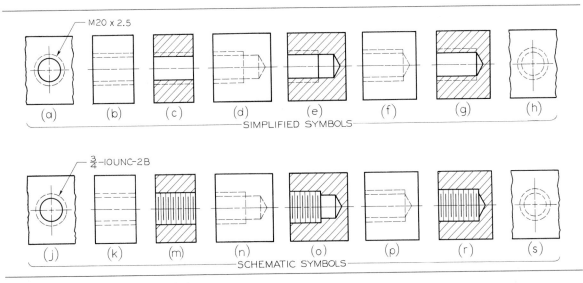

Fig. 15.8   Internal Thread Symbols.

wise no threads would be evident. However, it is not necessary to show the Vs to scale or according to the actual slope of the crest lines. To draw the Vs, use the schematic thread depth, Fig. 15.9 (a), and let the pitch be determined by the 60° Vs.

Schematic threads in elevation, Fig. 15.7 (d), are indicated by alternate long and short lines at right angles to the center line, the root lines being preferably thicker than the crest lines. Although theoretically the crest lines would be spaced according to actual pitch, the lines would often be very crowded and tedious to draw, thus defeating the purpose of the symbol, to save drafting time. In practice, the experienced drafter spaces the crest lines carefully by eye, and then adds the heavy root lines spaced by eye halfway between the crest lines. In general, the spacing should be proportionate for all diameters. For convenience in drawing, proportions for the schematic symbol are given in Fig. 15.9.

15.9   Internal Thread Symbols

Internal thread symbols are shown in Fig. 15.8. Note that the only differences between the schematic and simplified internal thread symbols oc-

cur in the sectional views. The representation of the schematic thread in section, Fig. 15.8 (m), (o), and (r), is exactly the same as the external symbol in Fig. 15.7 (d). Hidden threads, by either method, are represented by pairs of hidden lines. The hidden dashes should be staggered, as shown.

In the case of blind tapped holes, the drill depth normally is drawn at least three schematic pitches beyond the thread length, as shown in Fig. 15.8 (d), (e), (n), and (o). The symbols at (f) and (p) are used to represent the use of a bottoming tap, when the length of thread is the same as the depth of drill. See also §15.24.

15.10   To Draw Thread Symbols

Fig. 15.9 (a) shows a table of values of depth and pitch to use in drawing thread symbols. These values are selected to produce a well-proportioned symbol and to be convenient to set off with the scale. The experienced drafter will carefully space the lines by eye, but the student should use the scale. Note that the values of D and P are for the diameter *on the drawing*. Thus, a $1\frac{1}{2}''$ diameter thread at half-scale would be $\frac{3}{4}''$ diameter on the drawing, and values of D and P

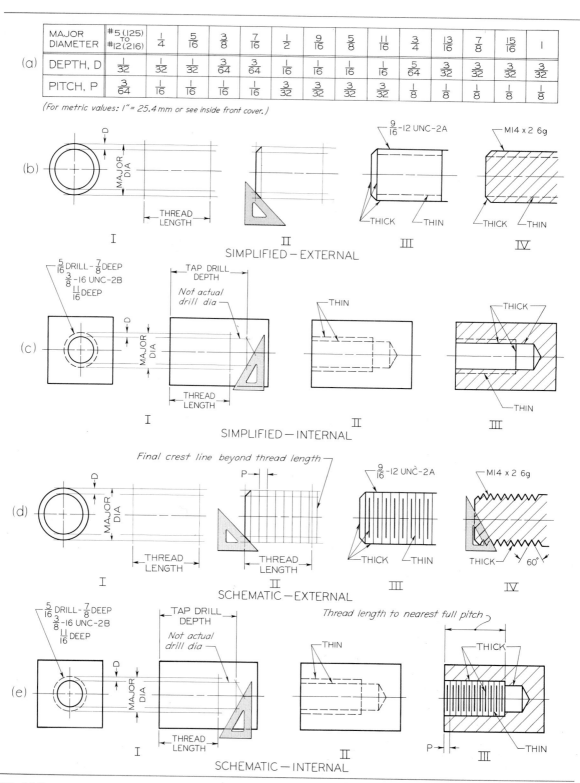

MAJOR DIAMETER	#5 (125) TO #12 (216)	$\frac{1}{4}$	$\frac{5}{16}$	$\frac{3}{8}$	$\frac{7}{16}$	$\frac{1}{2}$	$\frac{9}{16}$	$\frac{5}{8}$	$\frac{11}{16}$	$\frac{3}{4}$	$\frac{13}{16}$	$\frac{7}{8}$	$\frac{15}{16}$	1
DEPTH, D	$\frac{1}{32}$	$\frac{1}{32}$	$\frac{1}{32}$	$\frac{3}{64}$	$\frac{3}{64}$	$\frac{1}{16}$	$\frac{1}{16}$	$\frac{1}{16}$	$\frac{1}{16}$	$\frac{5}{64}$	$\frac{3}{32}$	$\frac{3}{32}$	$\frac{3}{32}$	$\frac{3}{32}$
PITCH, P	$\frac{3}{64}$	$\frac{1}{16}$	$\frac{1}{16}$	$\frac{1}{16}$	$\frac{1}{16}$	$\frac{3}{32}$	$\frac{3}{32}$	$\frac{3}{32}$	$\frac{3}{32}$	$\frac{1}{8}$	$\frac{1}{8}$	$\frac{1}{8}$	$\frac{1}{8}$	$\frac{1}{8}$

(For metric values: 1" = 25.4 mm or see inside front cover.)

Fig. 15.9 To Draw Thread Symbols—Simplified and Schematic.

474

for a $\frac{3}{4}''$ major diameter would be used. Nominal diameters for metric threads are treated in a similar manner.

SIMPLIFIED SYMBOLS The steps for drawing the simplified symbols for an external thread in elevation and in section are shown in Fig. 15.9 (b). The thread depth from the table at (a) is used for establishing the pairs of hidden lines that represent the threads in elevation and in section. No pitch measurement is needed. The completed symbols are shown at III and IV.

The steps for drawing the simplified symbol for an internal thread in section are shown in Fig. 15.9 (c). The simplified representation for the internal thread in elevation is identical to that used for schematic representation, as the threads are indicated by pairs of hidden lines as shown at II. The simplified symbol for an internal thread in section is shown at III. The major diameter of the thread is represented by hidden lines across the sectioned area.

SCHEMATIC SYMBOLS The steps for drawing the schematic symbols for an external thread in elevation and in section are shown in Fig. 15.9 (d). Note that when the pitches P are set off in II, the final crest line for a full pitch may fall beyond the actual thread length as shown. The completed schematic symbol for an external thread in elevation is shown at III. The completed schematic symbol for an external thread in section is shown at IV. The schematic thread depth is used for drawing the Vs, and the pitch is established by the 60° Vs.

The steps for drawing the schematic symbols for an internal thread in elevation and in section are shown at (e). Here again the symbol thread length may be slightly longer than the actual given thread length. If the tap drill depth is known or given, the drill is drawn to that depth, as shown. If the thread note omits this information, as is often done in practice, the drafter merely draws the hole three thread pitches (schematic) beyond the thread length. The tap drill diameter is represented approximately, as shown, and not to actual size. The completed

schematic symbol for an internal thread in elevation is shown at II. Pairs of hidden lines represent the threads, and the hidden-line dashes are staggered. The completed schematic symbol for an internal thread in section is shown at III. The schematic internal thread in section is represented in the same manner as for the schematic external thread.

15.11 Detailed Representation— Metric, Unified, and American National Threads

The detailed representation for metric, unified, and American National threads is the same, since the flats, if any, are disregarded. The steps in drawing these threads are shown in Fig. 15.10.

I.   Draw center line and lay out length and major diameter.

II.   Find the number of threads per inch in Appendix 15 for American National and Unified threads. This number depends on the major diameter of the thread, whether the thread is internal or external. Find P (pitch) by dividing 1 by the number of threads per inch, §15.4. The pitch for metric threads is given directly in the thread designation. For example the M14 × 2 thread has a pitch of 2 mm. See Appendix 15. Establish the slope of the thread by offsetting the slope line $\frac{1}{2}$P for single threads, P for double threads, $1\frac{1}{2}$P for triple threads, and so on.* For right-hand external threads, the slope line slopes upward to the left; for left-hand external threads, the slope line slopes upward to the right. If the number of threads per inch conforms to the scale, the pitch can be set off directly. For example, eight threads per inch can easily be set off with the architects' scale, and ten threads per inch with the engineers' scale. Otherwise, use the bow dividers or use the parallel-line method shown in Fig. 15.10 (II).

---

*These offsets are the same in terms of P for any form of thread.

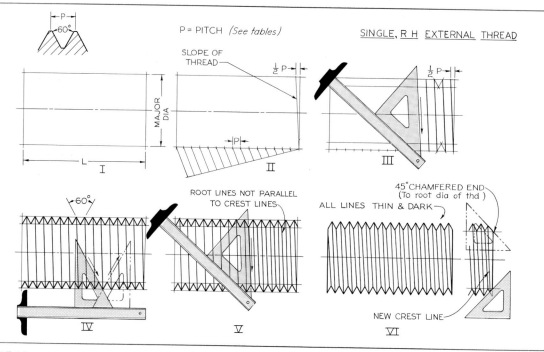

Fig. 15.10   Detailed Representation—External Metric, Unified, and American National Threads.

III.   From the pitch points, draw crest lines parallel to slope line. These should be dark thin lines. Slide triangle on T-square (or another triangle) to make lines parallel. Draw two Vs to establish depth of thread, and draw guide lines for root of thread, as shown.

IV.   Draw 60° Vs finished weight. These Vs should stand vertically; that is, they should not "lean" with the thread.

V.   Draw root lines dark at once. Root lines will *not* be parallel to crest lines. Slide triangle on straightedge to make root lines parallel.

VI.   When the end is chamfered (usually 45° with end of shaft, sometimes 30°), the chamfer extends to the thread depth. The chamfer creates a new crest line, which is then drawn between the two new crest points. It is not parallel to the other crest lines. In the final drawing, all thread lines should be approximately the same weight—thin, but dark.

The corresponding internal detailed threads, in section, are drawn as shown in Fig. 15.11. Notice that for LH threads the lines slope upward to the left, as shown at (a), while for RH threads the lines slope upward to the right, as at (b). Make all final thread lines medium-thin but dark.

15.12   Detailed Representation of Square Threads

The steps in drawing the detailed representation of an external square thread when the major diameter is over 1″ or 25 mm (approx.) on the drawing are shown in Fig. 15.12.

I.   Draw center line, and lay out length and major diameter of thread. Determine P by dividing 1 by the number of threads per inch. See Appendix 22. For a single RH thread, the lines slope upward to the left, and the slope line is

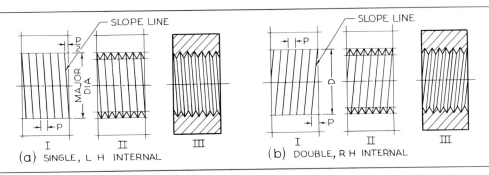

Fig. 15.11 Detailed Representation—Internal Metric, Unified, and American National Threads.

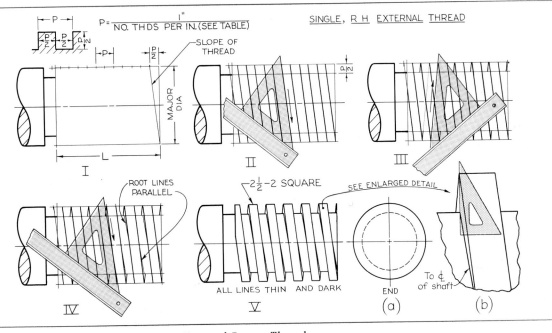

Fig. 15.12 Detailed Representation—External Square Threads.

offset *as for all single threads of any form.* On the upper line, set off spaces equal to $\frac{P}{2}$, as shown, using a scale if possible; otherwise, use the bow dividers or the parallel-line method to space the points.

II.   From the $\frac{P}{2}$ points on the upper line, draw fairly thin lines. Draw guide lines for root of thread, making the depth $\frac{P}{2}$ as shown.

III.   Draw parallel visible back edges of threads.

IV.   Draw parallel visible root lines. Note enlarged detail at (b).

V.   Accent the lines. All lines should be thin and dark.

Note the end view of the shaft at (a). The root circle is hidden; no attempt is made to show the true projection. If the end is chamfered, a solid circle would be drawn instead of the hidden circle.

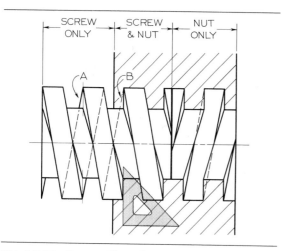

Fig. 15.13 Square Threads in Assembly.

posite direction from those on the front side of the screw.

Steps in drawing a single internal square thread in section are shown in Fig. 15.14. Note in step II that a crest is drawn opposite a root. This is the case for both single and triple threads. For double or quadruple threads, a crest is opposite a crest. Thus, the construction in steps I and II is the same for any multiple of thread. The differences are developed in step III where the threads and spaces are distinguished and outlined.

The same internal thread is shown in elevation (external view) in Fig. 15.14 (a). The profiles of the threads are drawn in their normal position, but with hidden lines, and the sloping lines are omitted for simplicity. The end view of the same internal thread is shown at (b). Note that the hidden and solid circles are opposite those for the end view of the shaft. See Fig. 15.12 (a).

An assembly drawing, showing an external square thread partly screwed into the nut, is shown in Fig. 15.13. The detail of the square thread at **A** is the same as shown in Fig. 15.12. But when the external and internal threads are assembled, the thread in the nut overlaps and covers up half of the V, as shown at **B**.

The internal thread construction is the same as in Fig. 15.14. Note that the thread lines representing the back half of the internal threads (since the thread is in section) slope in the op-

15.13   Detailed Representation of Acme Thread

The steps in drawing the detailed representation of Acme threads when the major diameter is over 1″ or 25 mm (approx.) on the drawing are shown in Fig. 15.15.

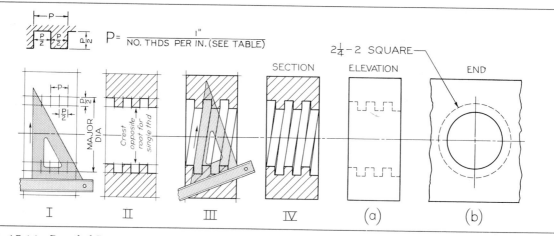

Fig. 15.14 Detailed Representation—Internal Square Threads.

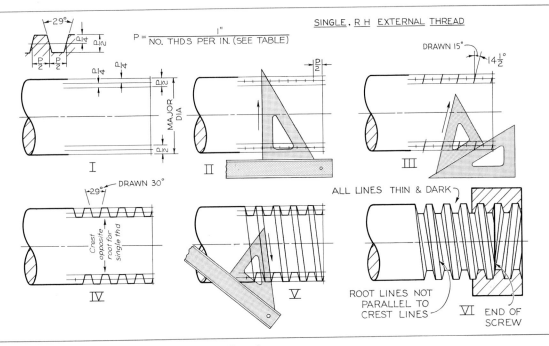

Fig. 15.15   Detailed Representation—Acme Threads.

I.   Draw center line, and lay out length and major diameter of thread. Determine P by dividing 1 by the number of threads per inch. See Appendix 22. Draw construction lines for the root diameter, making the thread depth $\frac{P}{2}$. Draw construction lines halfway between crest and root guide lines.

II.   On the intermediate construction lines, lay off $\frac{P}{2}$ spaces, as shown. Setting off spaces directly with a scale is possible (for example, if $\frac{P}{2} = \frac{1}{10}''$, use the engineers' scale); otherwise, use bow dividers or parallel-line method.

III.   Through alternate points, draw construction lines for sides of threads (draw 15° instead of $14\frac{1}{2}°$).

IV.   Draw construction lines for other sides of threads. Note that for single and triple threads, a crest is opposite a root, while for double and quadruple threads, a crest is opposite a crest. Heavy in tops and bottoms of threads.

V.   Draw parallel crest lines, final weight at once.

VI.   Draw parallel root lines, final weight at once, and heavy in the thread profiles. All lines should be thin and dark. Note that the internal threads in the back of the nut slope in the opposite direction to the external threads on the front side of the screw.

End views of Acme threaded shafts and holes are drawn exactly like those for the square thread, Figs. 15.12 and 15.14.

15.14   Use of Phantom Lines
In representing objects having a series of identical features, phantom lines, Fig. 15.16, may be used to save time. Threaded shafts and springs thus represented may be shortened without the

Fig. 15.16   Use of Phantom Lines.

use of conventional breaks, but must be correctly dimensioned. The use of phantom lines is limited almost entirely to detailed drawings.

15.15   Threads in Section

Detailed representations of large threads in section are shown in Figs. 15.6 and 15.10–15.15. As indicated by the note in Fig. 15.6 (b), the root lines and crest lines may be omitted in internal sectional views, if desired.

External thread symbols are shown in section in Fig. 15.7. Note that in the schematic symbol, the Vs must be drawn. Internal thread symbols in section are shown in Fig. 15.8.

Threads in an assembly drawing are shown in Fig. 15.17. It is customary not to section a stud or a nut, or any solid part, unless necessary to show some internal shapes. See §15.22. Note that when external and internal threads are sec-

tioned in assembly, the Vs are required to show the threaded connection.

15.16   American National Thread

The old American National thread was adopted in 1935. The *form*, or profile, Fig. 15.2 (b), is the same as the old Sellers' profile, or U.S. Standard, and is known as the *National form*. The methods of representation are the same as for the Unified and metric threads. As noted earlier, American National threads are being replaced by the Unified and metric threads. However, they may still be encountered on earlier drawings. Five *series* of threads were embraced in the old ANSI standards.

1. *Coarse thread*—A general-purpose thread for holding purposes. Designated NC (National Coarse).

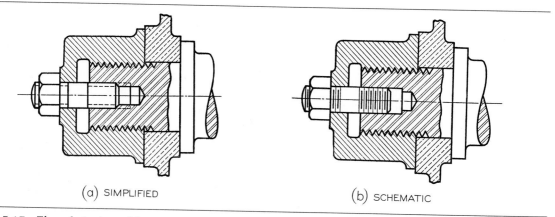

(a) SIMPLIFIED   (b) SCHEMATIC

Fig. 15.17   Threads in Assembly.

2. *Fine thread*—A greater number of threads per inch; used extensively in automotive and aircraft construction. Designated NF (National Fine).

3. *8-pitch thread*—All diameters have 8 threads per inch. Used on bolts for high-pressure pipe flanges, cylinder-head studs, and similar fasteners. Designated 8N (National form, 8 threads per inch).

4. *12-pitch thread*—All diameters have 12 threads per inch; used in boiler work and for thin nuts on shafts and sleeves in machine construction. Designated 12N (National form, 12 threads per inch).

5. *16-pitch thread*—All diameters have 16 threads per inch; used where necessary to have a fine thread regardless of diameter, as on adjusting collars and bearing retaining nuts. Designated 16N (National form, 16 threads per inch).

15.17   Unified Extra Fine Threads

The Unified extra fine thread series has many more threads per inch for given diameters than any series of the American National or Unified.

The form of thread is the same as the American National. These small threads are used in thin metal where the length of thread engagement is small, in cases where close adjustment is required, and where vibration is great. It is designated UNEF (extra fine).

15.18   American National Thread Fits

For general use, three classes of screw-thread fits between mating threads (as between bolt and nut) have been established by ANSI.

These fits are produced by the application of tolerances listed in the standard and are described as follows.

Class 1 fit—Recommended only for screw-thread work where clearance between mating parts is essential for rapid assembly and where shake or play is not objectionable.

Class 2 fit—Represents a high quality of commercial thread product and is recommended for the great bulk of interchangeable screw-thread work.

Class 3 fit—Represents an exceptionally high quality of commercially threaded product and is recommended only in cases where the high cost of precision tools and continual checking are warranted.

The class of fit desired on a thread is indicated in the thread note, as shown in §15.21.

15.19   Metric and Unified Threads

The preferred metric thread for commercial purposes conforms to the International Organization for Standardization publication [ISO 68] basic profile M for metric threads. This M profile design is comparable to the Unified inch profile, but they are not interchangeable. For commercial purposes, two series of metric threads are preferred—coarse (general purpose) and fine—thus drastically reducing the number of previously used thread series. See Appendix 15.

The Unified thread constitutes the present American National Standard. Some of the earlier American National threads are still included in the new standard. See Appendix 15. The standard lists eleven different series of numbers of threads per inch for the various standard diameters, together with selected combinations of special diameters and pitches.

The eleven series are the *coarse thread series* (UNC or NC) recommended for general use corresponding to the old National coarse thread; the *fine thread series* (UNF or NF), for general use in automotive and aircraft work and in applications where a finer thread is required; the *extra fine series* (UNEF or NEF), which is the same as the SAE extra fine series, used particularly in aircraft and aeronautical equipment and generally for threads in thin walls; and the eight series of 4, 6, 8, 12, 16, 20, 28, and 32 threads with constant pitch. The 8UN or 8N, 12UN or 12N, and 16UN or 16N series are recommended for the uses corresponding to the old 8-, 12-, and

16-pitch American National threads. In addition, there are three special thread series—UNS, NS, and UN—that involve special combinations of diameter, pitch, and length of engagement.

15.20   Thread Fits, Metric and Unified

For some specialized metric thread applications, the tolerances and deviations are specified by tolerance grade, tolerance position, class, and length of engagement.* Two classes of metric thread fits are generally recognized. The first class of fits is for general-purpose applications and has a tolerance class of 6H for internal threads and a class of 6g for external threads. The second class of fits is used where closer fits are necessary and has a tolerance class of 6H for internal threads and a class of 5g6g for external threads. Metric thread tolerance classes of 6H/6g are generally assumed if not otherwise designated and are used in applications comparable to the 2A/2B inch classes of fits.

The single-tolerance designation of 6H refers to both the tolerance grade and position for the pitch diameter and the minor diameter for an internal thread. The single-tolerance designation of 6g refers to both the tolerance grade and po-

sition for the pitch diameter and the major diameter of the external thread. A double designation of 5g6g indicates separate tolerance grades for the pitch diameter and for the major diameter of the external thread.

The standard for Unified screw threads specifies tolerances and allowances defining the several classes of fit (degree of looseness or tightness) between mating threads. In the symbols for fit, the letter A refers to external threads and B to internal threads. There are three classes of fit each for external threads (1A, 2A, 3A) and internal threads (1B, 2B, 3B). Classes 1A and 1B have generous tolerances, facilitating rapid assembly and disassembly. Classes 2A and 2B are used in the normal production of screws, bolts, and nuts, as well as a variety of general applications. Classes 3A and 3B provide for applications needing highly accurate and close-fitting threads.

15.21   Thread Notes

Thread notes for metric, Unified, and American National screw threads are shown in Figs. 15.18 and 15.19. These same notes or symbols are

*ISO Standards Handbook No. 18, 1984.

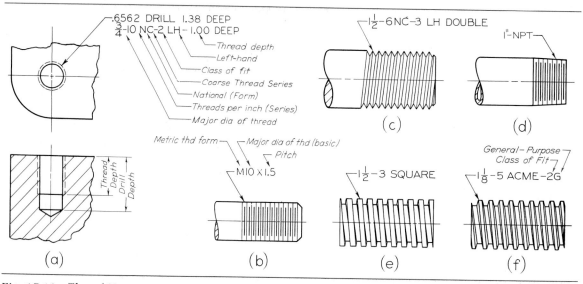

Fig. 15.18   Thread Notes.

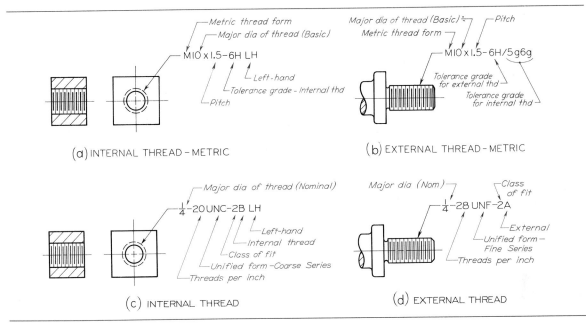

Fig. 15.19 Metric and Unified Thread Notes.

used in correspondence, on shop and storeroom records, and in specifications for parts, taps, dies, tools, and gages.

Metric screw threads are designated basically by the letter M for metric profile followed by the nominal size (basic major diameter) and the pitch, both in millimeters and separated by the symbol ×. For example, the basic thread note M10 × 1.5 is adequate for most commercial purposes, Fig. 15.18 (b). If the generally understood tolerances need to be specified, the tolerance class of 6H for the internal thread or the tolerance class of 6g for the external thread is added to the basic note, Fig. 15.19 (a). Where closer mating threads are desired, a tolerance of 6H for the internal thread and tolerance classes of 5g6g for the external thread are added to the basic note. When the thread note refers to mating parts, a single basic note is adequate with the addition of the internal and external thread tolerance classes separated by the slash. The basic note for mating threads now becomes M10 × 1.5–6H/6g for the general-purpose thread or M10 × 1.5–6H/5g6g for the close-fitting thread, Fig. 15.19 (b). Should the thread be left-hand, LH is

added to the thread note. (Absence of LH indicates an RH thread.) If it is necessary to indicate the length of the thread engagement, the letter S (short), N (normal), or L (long) is added to the thread note. For example, the single note M10 × 1.5–6H/6g–N–LH combines the specifications for internal and external mating left-hand metric threads of 10 mm diameter and 1.5 mm pitch with general-purpose tolerances and normal length of engagement.

A thread note for a blind tapped hole is shown in Fig. 15.18 (a). In a complete note, the tap drill and depth should be given, though in practice they are often omitted and left to the shop. For tap drill sizes, see Appendix 15. If the LH symbol is omitted, the thread is understood to be RH. If the thread is a multiple thread, the word DOUBLE, TRIPLE, or QUADRUPLE should precede the thread depth; otherwise, the thread is understood to be single. Thread notes for holes are preferably attached to the circular views of the holes, as shown.

Thread notes for external threads are preferably given in the longitudinal view of the threaded shaft, as shown in Fig. 15.18 (b)–(f).

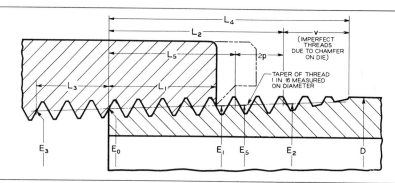

Fig. 15.20   American National Standard Taper Pipe Thread   [ANSI B2.1–1968].

Examples of 8-, 12-, and 16-pitch threads, not shown in the figure, are 2–8N–2, 2–12N–2, and 2–16N–2. A sample special thread designation is $1\frac{1}{2}$–7N–LH.

General-purpose Acme threads are indicated by the letter G, and centralizing Acme threads by the letter C. Typical thread notes are $1\frac{1}{4}$–4 ACME–2G or $1\frac{3}{4}$–6 ACME–4C.

Thread notes for Unified threads are shown in Fig. 15.19 (c) and (d). Unified threads are distinguished from American National threads by the insertion of the letter U before the series letters, and by the letters A or B (for external or internal, respectively) after the numeral designating the class of fit. If the letters LH are omitted, the thread is understood to be RH. Some typical thread notes are

$$\tfrac{1}{4}\text{–20 UNC–2A TRIPLE}$$

$$\tfrac{9}{16}\text{–18 UNF–2B}$$

$$1\tfrac{3}{4}\text{–16 UN–2A}$$

15.22   American National Standard Pipe Threads

The American National Standard for pipe threads, originally known as the Briggs standard, was formulated by Robert Briggs in 1882. Two general types of pipe threads have been approved as American National Standard: *taper pipe threads* and *straight pipe threads*.

The profile of the taper pipe thread is illustrated in Fig. 15.20. The taper of the standard tapered pipe thread is 1 in 16 or .75″ per foot measured on the diameter and along the axis. The angle between the sides of the thread is 60°. The depth of the sharp V is .8660p, and the basic maximum depth of the truncated thread is .800p, where p = pitch. The basic pitch diameters, E_0 and E_1, and the basic length of the effective external taper thread, L_2, are determined by the formulas

$$E_0 = D - (.050D + 1.1)\frac{1}{n}$$
$$E_1 = E + .0625L_1$$
$$L_2 = (.80D + 6.8)\frac{1}{n}$$

where D = basic O.D. of pipe, E_0 = pitch diameter of thread at end of pipe, E_1 = pitch diameter of thread at large end of internal thread, L_1 = normal engagement by hand, and n = number of threads per inch.

The ANSI also recommended two modified taper pipe threads for (1) dryseal pressure-tight joints (.88″ per foot taper) and (2) rail fitting joints. The former is used to provide a metal-to-metal joint, eliminating the need for a sealer, and is used in refrigeration, marine, automotive, aircraft, and ordnance work. The latter is used to provide a rigid mechanical thread joint as required in rail fitting joints.

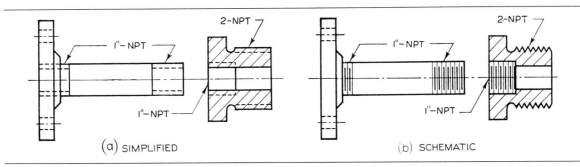

Fig. 15.21   Conventional Representation of Pipe Threads.

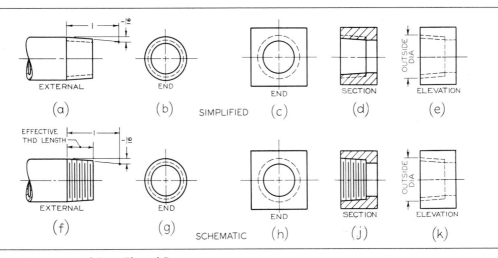

Fig. 15.22   Conventional Pipe Thread Representation.

While taper pipe threads are recommended for general use, there are certain types of joints where straight pipe threads are used to advantage. The number of threads per inch, the angle, and the depth of thread are the same as on the taper pipe thread, but the threads are cut parallel to the axis. Straight pipe threads are used for pressure-tight joints for pipe couplings, fuel and oil line fittings, drain plugs, free-fitting mechanical joints for fixtures, loose-fitting mechanical joints for locknuts, and loose-fitting mechanical joints for hose couplings.

Pipe threads are represented by detailed or symbolic methods in a manner similar to the representation of Unified and American National threads. The symbolic representation (sche-

matic or simplified) is recommended for general use regardless of diameter, Fig. 15.21. The detailed method is recommended only when the threads are large and when it is desired to show the profile of the thread as, for example, in a sectional view of an assembly.

As shown in Fig. 15.21, it is not necessary to draw the taper on the threads unless there is some reason to emphasize it, since the thread note indicates whether the thread is straight or tapered. If it is desired to show the taper, it should be exaggerated, as shown in Fig. 15.22, where the taper is drawn $\frac{1}{16}''$ per $1''$ on *radius* (or $6.75''$ per $1'$ on *diameter*) instead of the actual taper of $\frac{1}{16}''$ on *diameter*. American National Standard taper pipe threads are indicated by a

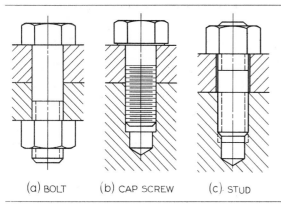

Fig. **15.23**   Bolt, Cap Screw, and Stud.

note giving the nominal diameter followed by the letters NPT (National pipe taper), as shown in Fig. 15.21. When straight pipe threads are specified, the letters NPS (National pipe straight) are used. In practice, the tap drill size is normally not given in the thread note.

15.23   Bolts, Studs, and Screws

The term *bolt* is generally used to denote a "through bolt" that has a head on one end and is passed through clearance holes in two or more aligned parts and is threaded on the other end to receive a nut to tighten and hold the parts together, Fig. 15.23 (a). See also §§15.25 and 15.26.

A hexagon head *cap screw*, (b), is similar to a bolt, except that it generally has greater length of thread, for when it is used without a nut, one

of the members being held together is threaded to act as a nut. It is screwed on with a wrench. Cap screws are not screwed into thin materials if strength is desired. See §15.29.

A *stud*, (c), is a steel rod threaded on both ends. It is screwed into place with a pipe wrench or, preferably, with a stud driver. As a rule, a stud is passed through a clearance hole in one member and screwed into another member, and a nut is used on the free end, as shown.

A *machine screw*, Fig. 15.33, is similar to the slotted-head cap screws but, in general, is smaller. It may be used with or without a nut.

A *set screw*, Fig. 15.34, is a screw with or without a head that is screwed through one member and whose special point is forced against another member to prevent relative motion between the two parts.

It is not customary to section bolts, nuts, screws, and similar parts when drawn in assembly, as shown in Figs. 15.23 and 15.32, because they do not themselves require sectioning for clearness. See §16.22.

15.24   Tapped Holes

The bottom of a drilled hole is conical in shape, as formed by the point of the twist drill, Fig. 15.24 (a) and (b). When an ordinary drill is used in connection with tapping, it is referred to as a *tap drill*. On drawings, an angle of 30° is used to approximate the actual 31°.

The thread length is the length of full or perfect threads. The tap drill depth is the depth of the cylindrical portion of the hole and does not include the cone point, (b). The portion A of the

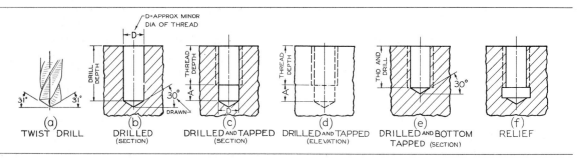

Fig. **15.24**   Drilled and Tapped Holes.

drill depth shown beyond the threads at (c) and (d) includes the several imperfect threads produced by the chamfered end of the tap. This distance A varies according to drill size and whether a plug tap, Fig. 12.19 (h), or a bottoming tap is used to finish the hole. For drawing purposes, when the tap drill depth is not specified, the distance A may be drawn equal to three schematic thread pitches, Fig. 15.9.

A tapped hole finished with a bottoming tap is drawn as shown in Fig. 15.24 (e). Blind bottoming holes should be avoided wherever possible. A better procedure is to cut a relief with its diameter slightly greater than the major diameter of the thread, Fig. 15.24 (f).

One of the chief causes of tap breakage is insufficient tap drill depth, in which the tap is forced against a bed of chips in the bottom of the hole. Therefore, the drafter should never draw a blind hole when a through hole of not much greater length can be used. When a blind hole is necessary, however, the tap drill depth should be generous. Tap drill sizes for Unified, American National, and metric threads are given in Appendix 15. It is good practice to give the tap drill size in the thread note, §15.21.

The thread length in a tapped hole depends on the major diameter and the material being tapped. In Fig. 15.25, the minimum engagement length X, when both parts are steel, is equal to the diameter D of the thread. When a steel screw is screwed into cast iron, brass, or bronze, X = $1\frac{1}{2}$D; when screwed into aluminum, zinc, or plastic, X = 2D.

Since the tapped thread length contains only full threads, it is necessary to make this length only one or two pitches beyond the end of the engaging screw. In simplified or schematic representation, the threads are omitted in the bottoms of tapped holes so as to show the ends of the screws clearly, Fig. 15.25.

When a bolt or a screw is passed through a clearance hole in one member, the hole may be drilled 0.8 mm ($\frac{1}{32}''$) larger than the screw up to $\frac{3}{8}''$ or 10 mm diameter and 1.5 mm ($\frac{1}{16}''$) larger for larger diameters. For more precise work, the clearance hole may be only 0.4 mm ($\frac{1}{64}''$) larger than the screw up to $\frac{3}{8}''$ or 10 mm diameter and

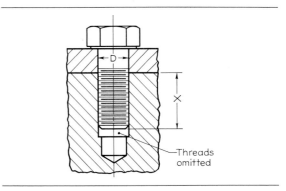

Fig. 15.25   Tapped Holes.

0.8 mm ($\frac{1}{32}''$) larger for larger diameters. Closer fits may be specified for special conditions.

The clearance spaces on each side of a screw or bolt need not be shown on a drawing unless it is necessary to show that there is no thread engagement, in which case the clearance spaces are drawn about 1.2 mm ($\frac{3}{64}''$) wide for clarity.

15.25   Standard Bolts and Nuts

American National Standard bolts and nuts,* metric and inch series, are produced in the hexagon form, and the square form is only produced in the inch series, Fig. 15.26. Square heads and

*The ANSI standards cover several bolts and nuts. For complete details, see the standards.

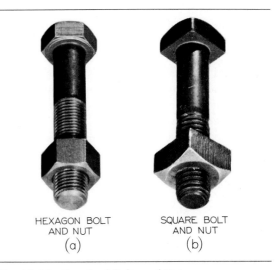

HEXAGON BOLT
AND NUT
(a)

SQUARE BOLT
AND NUT
(b)

Fig. 15.26   Standard Bolts and Nuts.

nuts are chamfered at 30°, and hexagon heads and nuts are chamfered at 15–30°. Both are drawn at 30° for simplicity.

BOLT TYPES   Bolts are grouped according to use: *regular* bolts for general service, and *heavy* bolts for heavier service or easier wrenching. Square bolts come only in the regular type; hexagon bolts, screws, and nuts and square nuts are standard in both types.

FINISH   Square bolts and nuts, hexagon bolts, and hexagon flat nuts are *unfinished*. Unfinished bolts and nuts are not machined on any surface except for the threads. Traditionally, hexagon bolts and hexagon nuts have been available as unfinished, semifinished, or finished. According to the latest standards, hexagon cap screws and finished hexagon bolts have been consolidated into a single product—*hex cap screws*—thus eliminating the regular semifinished hexagon bolt classification. Heavy semifinished hexagon bolts and heavy finished hexagon bolts also have been combined into a single product called *heavy hex screws*. Hexagon cap screws, heavy hexagon screws, and all hexagon nuts, except hexagon flat nuts, are considered *finished* to some degree and are characterized by a "washer face" machined or otherwise formed on the bearing surface. The washer face is $\frac{1}{64}''$ thick (drawn $\frac{1}{32}''$), and its diameter is equal to $1\frac{1}{2}$ times the body diameter D for the inch series.

For nuts the bearing surface also may be a circular surface produced by chamfering. Hexagon screws and hexagon nuts have closer tolerances and may have a more finished appearance, but are not completely machined. There is no difference in the drawing for the degree of finish on finished screws and nuts.

Metric bolts, cap screws, and nuts are produced in the hexagon form, Fig. 15.26 (a). The hexagon heads and nuts are chamfered at 15–30°. Both are drawn at 30° for simplicity.

BOLT TYPES   Metric hexagon bolts are grouped according to use: *regular* and *heavy* bolts and nuts for general service and *high-strength* bolts and nuts for structural bolting.

FINISH   Hexagon cap screws, hexagon bolts, and hexagon nuts (including metric) are considered finished to some degree and are usually characterized by a "washer face" machined or otherwise formed on the bearing surface. The washer face is approximately 0.5 mm thick (drawn 1 mm) and its diameter is equal to $1\frac{1}{2}$ times the body diameter D.

For hexagon nuts the circular bearing surface may be produced by chamfering. Hexagon head bolts (usually the larger sizes) are available without the washer-face bearing surface.

PROPORTIONS   Sizes based on diameter D of the bolt body (including metric), Fig. 15.27, which are either exact formula proportions or close approximations for drawing purposes, are as follows.

Regular hexagon and square bolts and nuts:

$$W = 1\tfrac{1}{2}D \qquad H = \tfrac{2}{3}D \qquad T = \tfrac{7}{8}D$$

where W = width across flats, H = head height, and T = nut height.

Heavy hexagon bolts and nuts and square nuts:

$$W = 1\tfrac{1}{2}D + \tfrac{1}{8}'' \text{ (or } + \text{ 3 mm)}$$
$$H = \tfrac{2}{3}D \qquad T = D$$

The washer face is always included in the head or nut height for finished hexagon screw heads and nuts.

THREADS   Square and hex bolts, hex cap screws, and finished nuts in the inch series are usually Class 2 and may have coarse, fine, or 8-pitch threads. Unfinished nuts have coarse threads, Class 2B. For diameter and pitch specifications for metric threads, see Appendix 18.

THREAD LENGTHS
For bolts or screws up to 6" (150 mm) in length:

Thread length = $2D + \tfrac{1}{4}''$ (or + 6 mm)

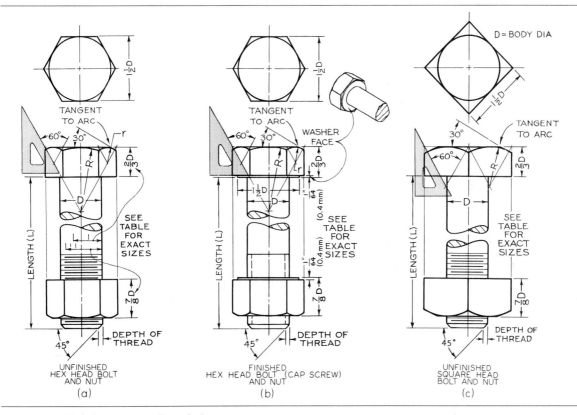

Fig. **15.27**   Bolt Proportions (Regular).

For bolts or screws over 6" in length,

$$\text{Thread length} = 2D + \tfrac{1}{2}" \ (\text{or} \ + \ 12 \ \text{mm})$$

Fasteners too short for these formulas are threaded as close to the head as practicable. For drawing purposes, this may be taken as approximately three pitches. The threaded end may be rounded or chamfered, but is usually drawn with a 45° chamfer from the thread depth, Fig. 15.27.

BOLT LENGTHS   Lengths of bolts have not been standardized because of the endless variety required by industry. Short bolts are typically available in standard length increments of $\frac{1}{4}"$ (6 mm), while long bolts come in increments of $\frac{1}{2}$ to $1"$ (12 to 25 mm). For dimensions of standard bolts and nuts, see Appendix 18.

15.26   To Draw Standard Bolts

In practice, standard bolts and nuts are not shown on detail drawings unless they are to be altered, but they appear so frequently on assembly drawings that a suitable but rapid method of drawing them must be used. Time-saving templates are available, or they may be drawn from exact dimensions taken from tables (see Appendix 18) if accuracy is important, as in figuring clearances, but in the great majority of cases the conventional representation, in which proportions based on the body diameter are used, will be sufficient, and a considerable amount of time may be saved. Three typical bolts illustrating the use of these proportions for the regular bolts are shown in Fig. 15.27.

Although the curves produced by the chamfer on the bolt heads and nuts are hyperbolas, in

489

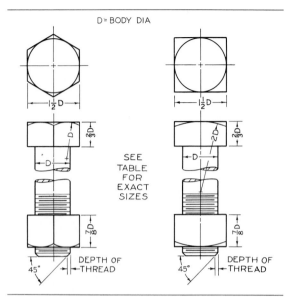

Fig. 15.28   Bolts "Across Flats."

practice these curves are always represented approximately by means of circular arcs, as shown in Fig. 15.27.

Generally, bolt heads and nuts should be drawn "across corners" in all views, regardless of projection. This conventional violation of projection is used to prevent confusion between the square and hexagon heads and nuts and to show actual clearances. Only when there is a special reason should bolt heads and nuts be drawn across flats. In such cases, the conventional proportions shown in Fig. 15.28 are used.

Steps in drawing hexagon bolts, cap screws, and nuts are illustrated in Fig. 15.29, and those for square bolts and nuts in Fig. 15.30. Before drawing a bolt, the diameter of the bolt, the length (from the underside of the bearing surface to the tip), the style of head (square or hexagon), and the type (regular or heavy), as well as the finish, must be known.

If only the longitudinal view of a bolt is needed, it is necessary to draw only the lower half of the top views in Figs. 15.29 and 15.30

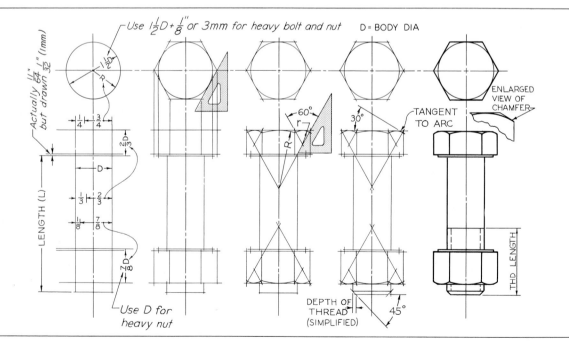

Fig. 15.29   Steps in Drawing a Finished Hexagon Head Bolt (Cap Screw) and Hexagon Nut.

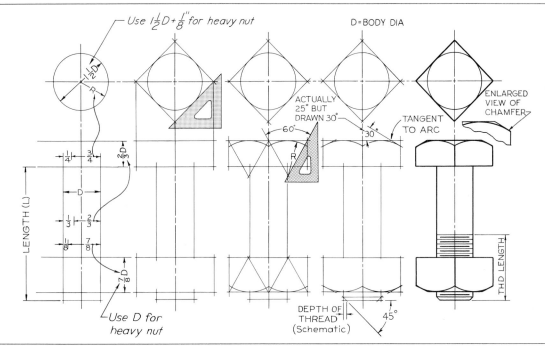

Fig. 15.30   Steps in Drawing Square-Head Bolt and Square Nut.

with light construction lines in order to project the corners of the hexagon or square to the front view. These construction lines may then be erased if desired.

The head and nut heights can be spaced off with the dividers on the shaft diameter and then transferred as shown in both figures, or the scale may be used as in Fig. 5.15. The heights should not be determined by arithmetic.

The $\frac{1}{64}''$ (0.4 mm) washer face has a diameter equal to the distance across flats of the bolt head or nut. It appears only on the metric and finished hexagon screws or nuts, the washer face thickness being drawn at $\frac{1}{32}''$ (1 mm) for clearness. The $\frac{1}{32}''$ (1 mm) is included in the head or nut height.

Threads should be drawn in simplified or schematic form for body diameters of $1''$ (25 mm) or less on the drawing, Fig. 15.9 (b) or (d), and by detailed representation for larger diameters, §§15.7 and 15.8. The threaded end of the screw should be chamfered at 45° from the schematic thread depth, Fig. 15.9 (a).

On drawings of small bolts or nuts under approximately $\frac{1}{2}''$ diameter (12 mm), where the chamfer is hardly noticeable, the chamfer on the head or nut may be omitted in the longitudinal view.

Many styles of templates are available for saving time in drawing bolt heads and nuts. One of these, the Draftsquare, is illustrated in Fig. 2.69 (b).

15.27   Specifications for Bolts and Nuts

In specifying bolts in parts lists, in correspondence, or elsewhere, the following information must be covered in order.

1. Nominal size of bolt body

2. Thread specification or thread note (see §15.21)

3. Length of bolt

4. Finish of bolt

5. Style of head

6. Name

EXAMPLE (Complete)

$$\frac{3}{4}\text{–10 UNC–2A} \times 2\frac{1}{2} \text{ HEXAGON CAP SCREW}$$

EXAMPLE (Abbreviated)

$$\frac{3}{4} \times 2\frac{1}{2} \text{ HEX CAP SCR}$$

EXAMPLE (Metric)

$$\text{M8} \times \text{1.25–40, HEX CAP SCR}$$

Nuts may be specified as follows.

EXAMPLE (Complete)

$$\frac{5}{8}\text{–11 UNC–2B SQUARE NUT}$$

EXAMPLE (Abbreviated) $\frac{5}{8}$ SQ NUT

EXAMPLE (Metric)   M8 × 1.25 HEX NUT

For either bolts or nuts, the words REGULAR or GENERAL PURPOSE are assumed if omitted from the specification. If the heavy series is intended, the word HEAVY should appear as the first word in the name of the fastener. Likewise, HIGH STRENGTH STRUCTURAL should be indicated for such metric fasteners. However, the number of the specific ISO standard is often included in the metric specifications; for example, HEXAGON NUT ISO 4032 M12 × 1.75. Similarly, finish need not be mentioned if the fastener or nut is correctly named. See §15.25.

15.28 Locknuts and Locking Devices

Many types of special nuts and devices to prevent nuts from unscrewing are available, some of the most common of which are illustrated in Fig. 15.31. The American National Standard *jam nuts,* (a) and (b), are the same as the hexagon

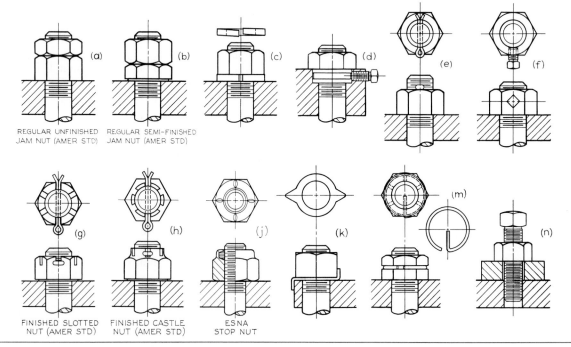

Fig. 15.31 Locknuts and Locking Devices.

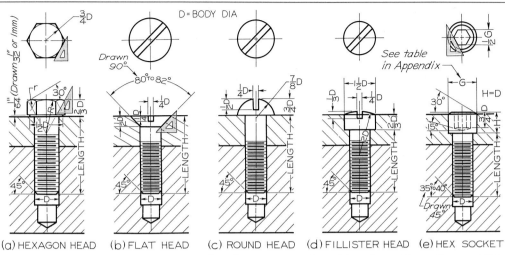

(a) HEXAGON HEAD   (b) FLAT HEAD   (c) ROUND HEAD   (d) FILLISTER HEAD   (e) HEX SOCKET

Hexagon Head Screws   Coarse, Fine, or 8-Thread Series, 2A. Thread length = $2D + \frac{1}{4}''$ up to 6″ long and $2D + \frac{1}{2}''$ if over 6″ long. For screws too short for formula, threads extend to within $2\frac{1}{2}$ threads of the head for diameters up to 1″. Screw lengths not standardized. For suggested lengths for metric Hexagon Head Screws, see Appendix 15.

Slotted Head Screws   Coarse, Fine, or 8-Thread Series, 2A. Thread length = $2D + \frac{1}{4}''$. Screw lengths not standardized. For screws too short for formula, threads extend to within $2\frac{1}{2}$ threads of the head.

Hexagon Socket Screws   Coarse or Fine Threads, 3A. Coarse thread length = $2D + \frac{1}{2}''$ where this would be over $\frac{1}{2}L$; otherwise thread length = $\frac{1}{2}L$. Fine thread length = $1\frac{1}{2}D + \frac{1}{2}''$ where this would be over $\frac{3}{8}L$; otherwise thread length = $\frac{3}{8}L$. Increments in screw lengths = $\frac{1}{8}''$ for screws $\frac{1}{4}''$ to 1″ long, $\frac{1}{4}''$ for screws 1″ to 3″ long, and $\frac{1}{2}''$ for screws $3\frac{1}{2}''$ to 6″ long.

Fig. 15.32   Standard Cap Screws.   See Appendixes 18 and 19.

or hexagon flat nuts, except that they are thinner. The application at (b), where the larger nut is on top and is screwed on more tightly, is recommended. They are the same distance across flats as the corresponding hexagon nuts ($1\frac{1}{2}D$ or $1\frac{1}{2}D + \frac{1}{8}''$). They are slightly over $\frac{1}{2}D$ in thickness, but are drawn $\frac{1}{2}D$ for simplicity. They are available with or without the washer face in the regular and heavy types. The tops of all are flat and chamfered at 30°, and the finished forms have either a washer face or a chamfered bearing surface.

The lock washer, shown at (c), and the cotter pin, (e), (g), and (h), are very common. See Appendixes 27 and 30. The set screw, (f), is often made to press against a plug of softer material, such as brass, which in turn presses against the threads without deforming them.

For use with cotter pins (see Appendix 30), a hex slotted nut, (g), and a hex castle nut, (h),

as well as a hex thick slotted nut and a heavy hex thick slotted nut are recommended.

Similar metric locknuts and locking devices are available. See fastener catalogs for details.

15.29   Standard Cap Screws

Five types of American National Standard cap screws are shown in Fig. 15.32. The first four of these have standard heads, while the socket head cap screws, (e), have several different shapes of round heads and sockets. Cap screws are regularly produced in finished form and are used on machine tools and other machines, for which accuracy and appearance are important. The ranges of sizes and exact dimensions are given in Appendixes 18 and 19. The hexagon head cap screw and hex socket head cap screw in several forms are available in metric. See Appendix 18.

Cap screws ordinarily pass through a clearance hole in one member, as explained in §15.24, and screw into another. The clearance hole need not be shown on the drawing when the presence of the unthreaded clearance hole is obvious.

Cap screws are inferior to studs if frequent removal is necessary; hence, they are used on machines requiring few adjustments. The slotted or socket-type heads are best under crowded conditions.

The actual standard dimensions may be used in drawing the cap screws whenever exact sizes are necessary, but this is seldom the case. In Fig. 15.32 the dimensions are given in terms of body diameter D, and they closely conform to the actual dimensions. The resulting drawings are almost exact reproductions and are easy to draw. The hexagonal head cap screw is drawn in the manner shown in Fig. 15.29. The points are drawn chamfered at 45° from the schematic thread depth.

For correct representation of tapped holes, see §15.24. For information on drilled, countersunk, or counterbored holes, see §§7.33 and 12.20.

In an assembly section, it is customary not to section screws, bolts, shafts, or other solid parts whose center lines lie in the cutting plane. Such parts in themselves do not require sectioning and are, therefore, shown "in the round," Fig. 15.32 and §16.22.

Note that screwdriver slots are drawn at 45° in the circular views of the heads, without regard to true projection, and that threads in the bottom of the tapped holes are omitted so that the ends of the screws may be clearly seen. A typical cap screw note is as follows:

EXAMPLE (Complete)

$$\frac{3}{8}\text{–16 UNC–2A} \times 2\frac{1}{2} \text{ HEXAGON HEAD CAP SCREW}$$

EXAMPLE (Abbreviated)

$$\frac{3}{8} \times 2\frac{1}{2} \text{ HEX HD CAP SCR}$$

EXAMPLE (Metric)

$$\text{M20} \times 2.5 \times 80 \text{ HEX HD CAP SCR}$$

15.30   Standard Machine Screws

Machine screws are similar to cap screws but are in general smaller (.060″ to .750″ dia). There are eight ANSI-approved forms of heads shown in Appendix 20. The hexagonal head may be slotted if desired. All others are available in either slotted or recessed-head forms. Standard machine screws are regularly produced with a naturally bright finish, not heat-treated, and are regularly supplied with plain-sheared ends, not chamfered. For similar metric machine screw forms and specifications, see Appendix 20.

Machine screws are particularly adapted to screwing into thin materials, and all the smaller-numbered screws are threaded nearly to the head. They are used extensively in firearms, jigs, fixtures, and dies. Machine screw nuts are used mainly on the round head, pan head, and flat head types, and are usually hexagonal in form.

Exact dimensions of machine screws are given in Appendix 20, but they are seldom needed for drawing purposes. The four most common types of machine screws are shown in Fig. 15.33, where proportions based on diameter D conform closely to the actual dimensions and produce almost exact drawings. Clearance holes and counterbores should be made slightly larger than the screws, as explained in §15.24.

Note that the threads in the bottom of the tapped holes are omitted so that the ends of the screws will be clearly seen. Observe also that it is conventional practice to draw the screwdriver slots at 45° in the circular view without regard to true projection.

A typical machine screw note is

EXAMPLE (Complete)

$$\text{No. 10 (.1900)–32 NF–3} \times \frac{5}{8} \text{ FILLISTER HEAD}$$
$$\text{MACHINE SCREW}$$

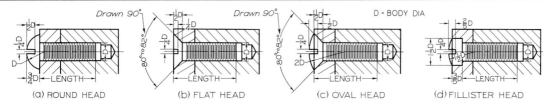

Threads   National Coarse or Fine, Class 2 fit. On screws 2″ long or less, threads extend to within 2 threads of head; on longer screws thread length = $1\frac{3}{4}$″. Screw lengths not standardized.

Fig. 15.33   Standard Machine Screws.   See Appendix 20.

EXAMPLE (Abbreviated)

No. 10 (.1900) $\times \frac{5}{8}$ FILL HD MACH SCR

EXAMPLE (Metric)

M8 $\times$ 1.25 $\times$ 30 SLOTTED PAN HEAD
MACHINE SCREW

15.31   Standard Set Screws

The function of set screws, Fig. 15.34 (a), is to prevent relative motion, usually rotary, between two parts, such as the movement of the hub of a pulley on a shaft. A set screw is screwed into one part so that its point bears firmly against another part. If the point of the set screw is cupped, (e), or if a flat is milled on the shaft, (a), the screw will hold much more firmly. Obviously, set screws are not efficient when the load is heavy or is suddenly applied. Usually they are manufactured of steel and case hardened.

The American National Standard square head set screw and slotted headless set screw are shown in Fig. 15.34 (a) and (b). Two American National Standard socket set screws are illustrated at (c) and (d). American National Standard set screw points are shown at (e)–(k). The headless set screws have come into greater

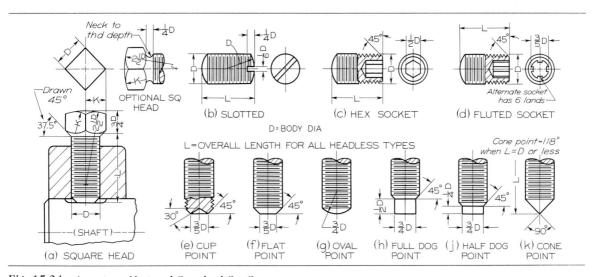

Fig. 15.34   American National Standard Set Screws.

use because the projecting head of headed set screws has caused many industrial casualties; this has resulted in legislation prohibiting their use in many states.

Most of the dimensions in Fig. 15.34 are American National Standard formula dimensions, and the resulting drawings are almost exact representations.

Metric hexagon socket headless set screws with the full range of points are available and are represented in the same manner as given in Fig. 15.34. Nominal diameters of metric hex socket set screws are 1.6, 2, 2.5, 3, 4, 5, 6, 8, 10, 12, 16, 20, and 24 mm.

Square head set screws have coarse, fine, or 8-pitch threads, Class 2A, but are usually furnished with coarse threads, since the square head set screw is generally used on the rougher grades of work. Slotted headless and socket set screws have coarse or fine threads, Class 3A.

Nominal diameters of set screws range from number 0 up through 2″; set screw lengths are standardized in increments of $\frac{1}{32}″$ to 1″ depending on the overall length of the set screw.

Metric set screw length increments range from 0.5 to 4 mm, again depending on overall screw length.

Set screws are specified as follows.

EXAMPLE (Complete)

$$\frac{3}{8}\text{–16 UNC–2a} \times \frac{3}{4} \text{ SQUARE HEAD}$$
$$\text{FLAT POINT SET SCREW}$$

EXAMPLES (Abbreviated)

$$\frac{3}{8} \times 1\frac{1}{4} \text{ SQ HD FL PT SS}$$

$$\frac{7}{16} \times \frac{3}{4} \text{ HEX SOC CUP PT SS}$$

$$\frac{1}{4}\text{–20 UNC 2A} \times \frac{5}{8} \text{ SLOT. HDLS CONE PT SS}$$

EXAMPLE (Metric)

M10 × 1.5 × 12 HEX SOCKET HEAD SET SCREW

15.32   American National Standard Wood Screws

Wood screws with three types of heads—flat, round, and oval—have been standardized, Fig. 15.35. The dimensions shown closely approximate the actual dimensions and are more than sufficiently accurate for use on drawings.

The Phillips style recessed head is also available on several types of fasteners as well as wood screws. Three styles of cross recesses have been standardized by the ANSI. Many examples may be seen on the automobile. A special screwdriver is used, as shown in Fig. 15.36 (q), and results in rapid assembly without damage to the head.

15.33   Miscellaneous Fasteners

Many other types of fasteners have been devised for specialized uses. Some of the more common types are shown in Fig. 15.36. A number of these are American National Standard round head

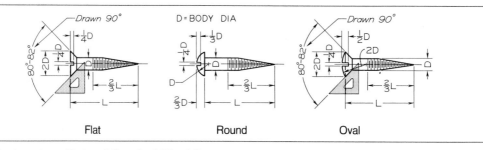

Flat   Round   Oval

Fig. 15.35   American National Standard Wood Screws.

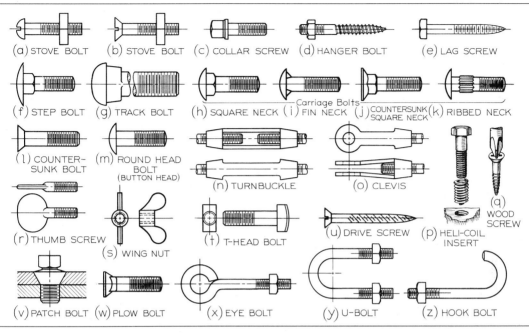

Fig. 15.36 Miscellaneous Bolts and Screws.

bolts including carriage, button head, step, and countersunk bolts.

Aero-thread inserts, or heli-coil inserts, as shown at (p), are shaped like a spring except that the cross section of the wire conforms to threads on the screw and in the hole. These are made of phosphor bronze or stainless steel, and they provide a hard, smooth protective lining for tapped threads in soft metals and in plastics.

15.34 Keys

Keys are used to prevent relative movement between shafts and wheels, couplings, cranks, and similar machine parts attached to or supported by shafts, Fig. 15.37.

For heavy duty, rectangular keys (flat or square) are suitable, and sometimes two rectangular keys are necessary for one connection. For even stronger connections, interlocking *splines*

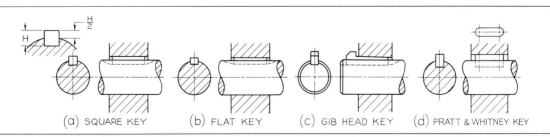

Fig. 15.37 Square and Flat Keys.

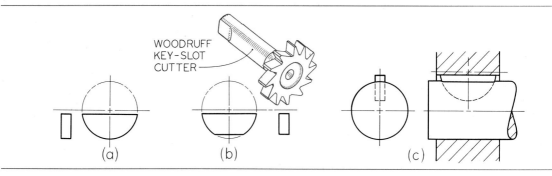

Fig. 15.38 Woodruff Keys and Key-Slot Cutter.

may be machined on the shaft and in the hole. See Fig. 14.9.

A *square key* is shown in Fig. 15.37 (a) and a *flat key* in (b). The widths of keys generally used are about one-fourth the shaft diameter. In either case, one-half the key is sunk into the shaft. The depth of the keyway or the keyseat is measured on the side—not the center, (a). Square and flat keys may have the top surface tapered $\frac{1}{8}''$ per foot, in which case they become square taper or flat taper keys.

A rectangular key that prevents rotary motion but permits relative longitudinal motion is a *feather key,* and is usually provided with *gib heads,* or otherwise fastened so it cannot slip out of the keyway. A *gib head key* is shown at (c). It is exactly the same as the square taper or flat taper key, except that a gib head, which provides for easy removal, is added. Square and flat keys are made from cold-finished stock and are not machined. For dimensions, see Appendix 21.

The *Pratt & Whitney key,* (d), is rectangular in shape, with semicylindrical ends. Two-thirds of the height of the P & W key is sunk into the shaft keyseat. See Appendix 25.

The *Woodruff key* is semicircular in shape, Fig. 15.38. The key fits into a semicircular key slot cut with a Woodruff cutter, as shown, and the top of the key fits into a plain rectangular keyway. Sizes of keys for given shaft diameters are not standardized, but for average conditions it will be found satisfactory to select a key whose

diameter is approximately equal to the shaft diameter. For dimensions, see Appendix 23.

A *keyseat* is in a shaft; a *keyway* is in the hub or surrounding part.

Typical specifications for keys are

$$\frac{1}{4} \times 1\frac{1}{2} \text{ SQ KEY}$$

No. 204 WOODRUFF KEY

$$\frac{1}{4} \times \frac{1}{16} \times 1\frac{1}{2} \text{ FLAT KEY}$$

No. 10 P & W KEY

Notes for nominal specifications of keyways and keyseats are shown in Fig. 13.44 (o), (p), (r), and (x). For production work, keyways and keyseats should be dimensioned as shown in Fig. 13.48. See manufacturers' catalogs for specifications for metric counterparts.

15.35   Machine Pins

Machine pins include taper pins, straight pins, dowel pins, clevis pins, and cotter pins. For light work, the taper pin is effective for fastening hubs or collars to shafts, as shown in Fig. 15.39, in which the hole through the collar and shaft is drilled and reamed when the parts are assembled. For slightly heavier duty, the taper pin may be used parallel to the shaft as for square keys. See Appendix 29.

Dowel pins are cylindrical or conical in shape and are used for a variety of purposes, chief of

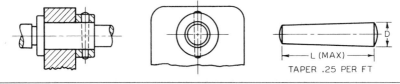

Fig. 15.39   Taper Pin.

which is to keep two parts in a fixed position or to preserve alignment. The dowel pin is most commonly used and is recommended where accurate alignment is essential. Dowel pins are usually made of steel and are hardened and ground in a centerless grinder.

The clevis pin is used in a clevis and is held in place by a cotter pin. For the latter, see Appendix 30.

15.36   Rivets

Rivets are regarded as permanent fastenings as distinguished from removable fastenings, such as bolts and screws. They are generally used to hold sheet metal or rolled steel shapes together and are made of wrought iron, carbon steel, copper, or occasionally other metals.

To fasten two pieces of metal together, holes are punched, drilled, or punched and then reamed, slightly larger in diameter than the shank of the rivet. Rivet diameters in practice are made from $d = 1.2\sqrt{t}$ to $d = 1.4\sqrt{t}$, where d is the rivet diameter and t is the metal thickness. The larger size is used for steel and single-riveted joints, and the smaller may be used for multiple-riveted joints. In structural work it is common practice to make the hole 1.6 mm ($\frac{1}{16}''$) larger than the rivet.

When the red-hot rivet is inserted, a "dolly bar," having a depression the shape of the driven head, is held against the head. A riveting machine is then used to drive the rivet and to form the head on the plain end. This action causes the rivet to swell and fill the hole tightly.

Large rivets or heavy hex structural bolts are often used in structural work of bridges and buildings and in ship and boiler construction, and they are shown in their exact formula proportions in Fig. 15.40. The button head and countersunk head types, (a) and (e), are the rivets most commonly used in structural work. The button head and cone head are commonly used in tank and boiler construction.

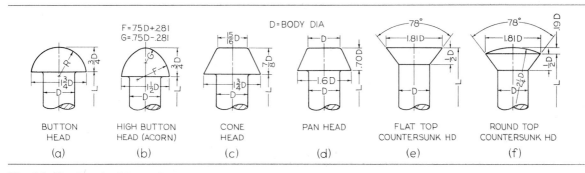

Fig. 15.40   Standard Large Rivets.

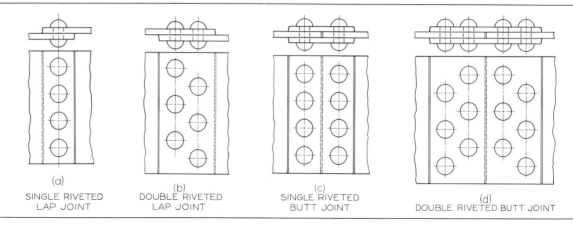

Fig. 15.41 Common Riveted Joints.

Typical riveted joints are illustrated in Fig. 15.41. Notice that the longitudinal view of each rivet shows the shank of the rivet with both heads made with circular arcs, and the circular view of each rivet is represented by only the outside circle of the head. In structural drafting, where there may be many such circles to draw, the drop spring bow, Fig. 2.53, is a convenient instrument.

Since many engineering structures are too large to be built in the shop, they are built in the largest units possible and then are transported to the desired location. Trusses are common examples of this. The rivets driven in the shop are called *shop rivets,* and those driven on the job are called *field rivets.* However, heavy steel bolts are commonly used on the job for structural work. Solid black circles are used to represent field rivets, and other standard symbols are used to show other features, as shown in Fig. 15.42.

For light work, small rivets are used. American National Standard small solid rivets are illustrated with dimensions showing their standard proportions in Fig. 15.43 [ANSI

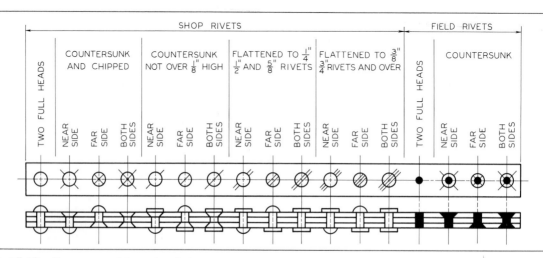

Fig. 15.42 Conventional Rivet Symbols.

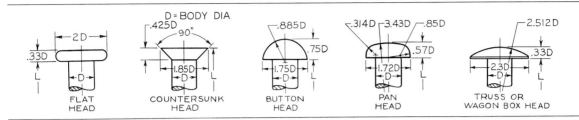

Fig. 15.43   American National Standard Small Solid Rivet Proportions.

B18.1.1–1972 (R1983)]. Included in the same standard are tinners', coppers', and belt rivets. Metric rivets are also available. Dimensions for large rivets can be found in ANSI B18.21.1–1972 (R1983). See manufacturers' catalogs for additional details.

15.37   Springs

A *spring* is a mechanical device designed to store energy when deflected and to return the equivalent amount of energy when released [ANSI Y14.13M–1981]. Springs are classified as *helical springs*, Fig. 15.44, or *flat springs*, Fig. 15.48.

Helical springs may be cylindrical or conical, but are usually the former.

There are three types of helical springs: *compression springs*, which offer resistance to a compressive force, Fig. 15.44 (a) to (e), *extension springs*, which offer resistance to a pulling force, Fig. 15.46, and *torsion springs*, which offer resistance to a torque load or twisting force, Fig. 15.47.

On working drawings, true projections of helical springs are never drawn because of the labor involved. As in the drawing of screw threads, the detailed and schematic methods, employing straight lines in place of helical curves, are used as shown in Fig. 15.44.

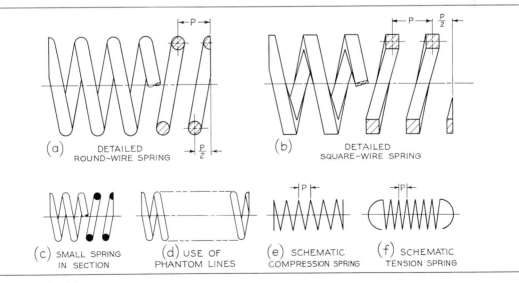

Fig. 15.44   Helical Springs.

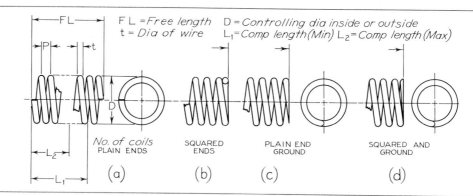

Fig. 15.45   Compression Springs.

The elevation view of the square-wire spring is similar to the square thread with the core of the shaft removed, Fig. 15.12. Standard section lining is used if the areas in section are large, as shown in Fig. 15.44 (a) and (b). If these areas are small, the sectioned areas may be made solid black, (c). In cases where a complete picture of the spring is not necessary, phantom lines may be used to save time in drawing the coils, (d). If the drawing of the spring is too small to be represented by the outlines of the wire, it may be drawn by the schematic method, in which single lines are used, (e) and (f).

Compression springs have *plain ends,* Fig. 15.45 (a), or *squared (closed) ends,* (b). The ends may be *ground* as at (c), or both *squared and ground* as at (d). Required dimensions are indicated in the figure. When required, RH or LH is specified.

Many companies use standard printed spring drawings with a printed form to be filled in by the drafter, providing the necessary information, plus load at a specified deflected length, the load rate, finish, type of service, and other data.

An extension spring may have any one of many types of ends, and it is therefore necessary to draw the spring or at least the ends and a few adjacent coils, Fig. 15.46. Note the use of phantom lines to show the continuity of coils. Printed forms are used when a given form of spring is

produced with differences in verbal specification only.

A typical torsion spring drawing is shown in Fig. 15.47. Here also printed forms are used when there is sufficient uniformity in product to permit a common representation.

A typical flat spring drawing is shown in Fig. 15.48. Other types of flat springs are *power springs* (or flat coil springs), *Belleville springs* (like spring washers), and *leaf springs* (commonly used in automobiles).

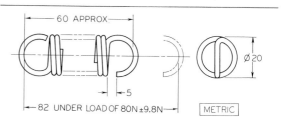

MATERIAL: 2.00 OIL TEMPERED SPRING STEEL WIRE
           14.5 COILS RIGHT HAND
           MACHINE LOOP AND HOOK IN LINE
           SPRING MUST EXTEND TO 110 WITHOUT SET
FINISH: BLACK JAPAN

Fig. 15.46   Extension Spring Drawing.

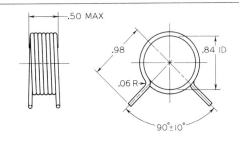

MATERIAL : .059 MUSIC WIRE
               6.75 COILS RIGHT HAND NO INITIAL TENSION
TORQUE : 2.50 INCH LB AT 155° DEFLECTION SPRING MUST
              DEFLECT 180° WITHOUT PERMANENT SET AND
              MUST OPERATE FREELY ON .75 DIAMETER SHAFT
FINISH : CADMIUM OR ZINC PLATE

Fig. 15.47   Torsion Spring Drawing.

15.38   To Draw Helical Springs

The construction for a schematic elevation view of a compression spring having six total coils is shown in Fig. 15.49 (a). Since the ends are closed, or squared, two of the six coils are "dead" coils, leaving only four full pitches to be set off along the top of the spring, as shown.

If there are $6\frac{1}{2}$ total coils, as at (b), the $\frac{P}{2}$ spacings will be on opposite sides of the spring. The construction of an extension spring with six active coils and loop ends is shown at (c).

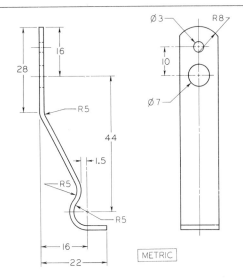

MATERIAL : 1.20 X 14.0 SPRING STEEL
HEAT TREAT : 44-48 C ROCKWELL
FINISH : BLACK OXIDE AND OIL

Fig. 15.48   Flat Spring.

In Fig. 15.50 are shown the steps in drawing a sectional view and an elevation view of a compression spring by detailed representation. The given spring is shown pictorially at (a). At (b) a cutting plane has passed through the center line of the spring, and the front portion of the

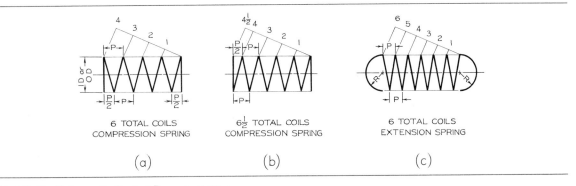

     6 TOTAL COILS $6\frac{1}{2}$ TOTAL COILS   6 TOTAL COILS
COMPRESSION SPRING   COMPRESSION SPRING   EXTENSION SPRING

       (a)   (b)   (c)

Fig. 15.49   Schematic Spring Representation.

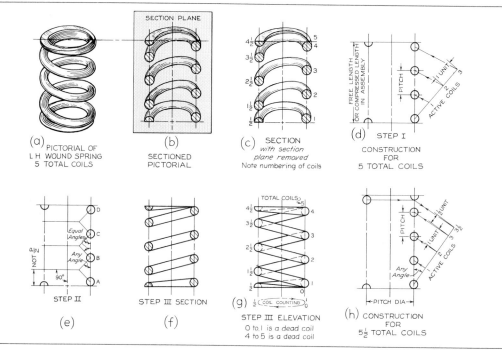

Fig. 15.50   Steps in Detailed Representation of Spring.

spring has been removed. At (c) the cutting plane has been removed. Steps in constructing the spring through several stages to obtain the sectional view are shown at (d)–(f). The corresponding elevation view is shown at (g).

If there is a fractional number of coils, such as $5\frac{1}{2}$ coils at (h), note that the half-rounds of sectional wire are placed on opposite sides of the spring.

15.39   Computer Graphics

Standard representations of threaded fasteners and springs, in both detailed and schematic

forms, are available in CAD program symbol libraries. Use of computer graphics frees the drafter from the need to draw time-consuming repetitive features by hand, Fig. 15.51, and also makes it easy to modify a drawing if required. It is still necessary, however, for the CAD user to have a thorough familiarity with the various types of fasteners and springs available, and their relative usefulness in a particular application.

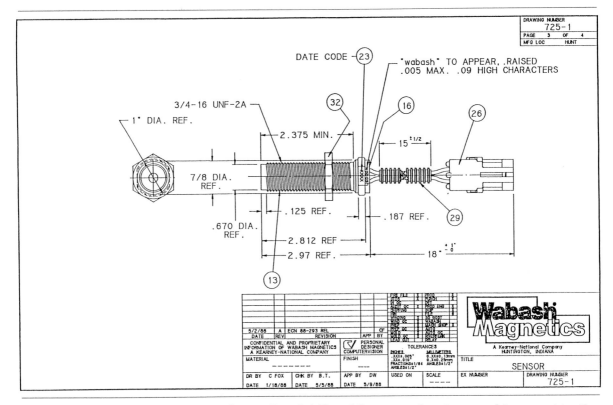

Fig. 15.51 CAD-Generated Drawing Showing Detailed Thread Representation. *Courtesy of Computervision Corporation, a Subsidiary of Prime Computer, Inc.*

505

THREAD AND FASTENER PROBLEMS

It is expected that the student will make use of the information in this chapter and in various manufacturers' catalogs in connection with the working drawings at the end of the next chapter, where many different kinds of threads and fasteners are required. However, several problems are included here for specific assignment in this area, Figs. 15.52–15.55. All are to be drawn on tracing paper or detail paper, size B or A3 sheet (see inside back cover).

Problems in convenient form for solution may be found in *Technical Drawing Problems,* Series 1, by Giesecke, Mitchell, Spencer, Hill, Dygdon, and Novak; *Technical Drawing Problems,* Series 2, by Spencer, Hill, Dygdon, and Novak; and *Technical Drawing Problems,* Series 3, by Spencer, Hill, Dygdon, and Novak; all designed to accompany this text and published by Macmillan Publishing Company.

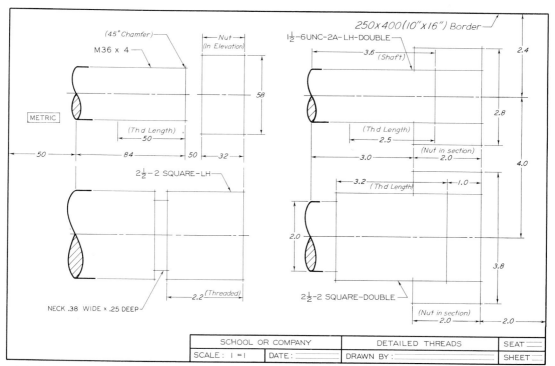

Fig. 15.52 Draw specified detailed threads arranged as shown, Layout B–3 or A3–3. Omit all dimensions and notes given in inclined letters. Letter only the thread notes and the title strip.

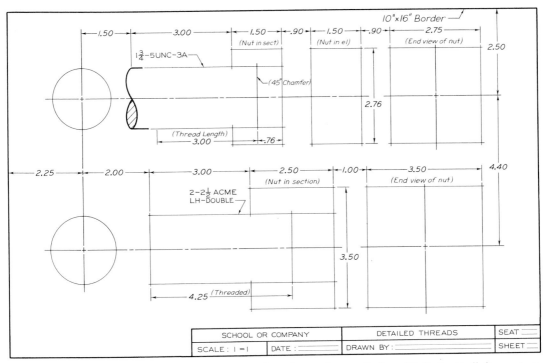

Fig. 15.53 Draw specified detailed threads, arranged as shown, Layout B–3 or A3–3. Omit all dimensions and notes given in inclined letters. Letter only the thread notes and the title strip.

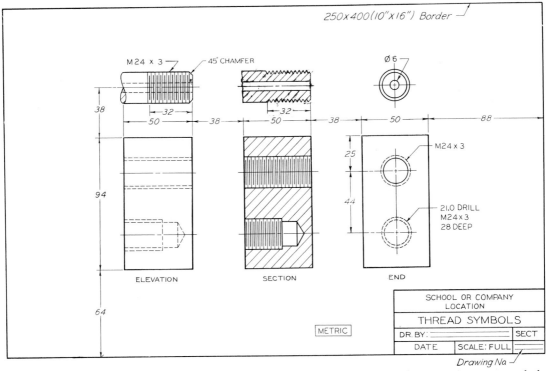

Fig. 15.54 Draw specified thread symbols, arranged as shown. Draw simplified or schematic symbols, as assigned by instructor, Layout B–5 or A3–5. Omit all dimensions and notes given in inclined letters. Letter only the drill and thread notes, the titles of the views, and the title strip.

507

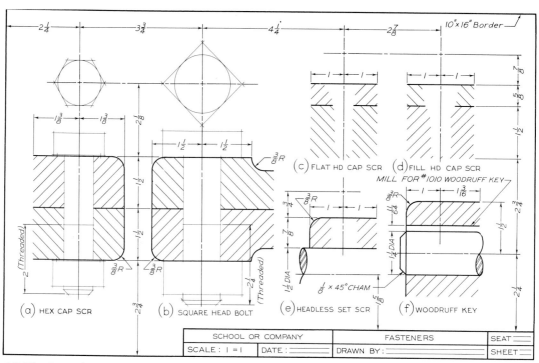

Fig. 15.55 Draw fasteners, arranged as shown, Layout B–3 or A3–3. At (a) draw $\frac{7}{8}$–9 UNC–2A $\times$ 4 Hex Cap Screw. At (b) draw $1\frac{1}{8}$–7 UNC–2A $\times$ $4\frac{1}{4}$ Sq Hd Bolt. At (c) draw $\frac{3}{8}$–16 UNC–2A $\times$ $1\frac{1}{2}$ Flat Hd Cap Screw. At (d) draw $\frac{7}{16}$–14 UNC–2A $\times$ 1 Fill Hd Cap Screw. At (e) draw $\frac{1}{2}$ $\times$ 1 Headless Slotted Set Screw. At (f) draw front view of No. 1010 Woodruff Key. Draw simplified or schematic thread symbols as assigned. Letter titles under each figure as shown.

CHAPTER 16

Design and Working Drawings

The many products, systems, and services that enrich our standard of living are largely the result of the design activities of engineers. It is principally this design activity that distinguishes engineering from science and research; the engineer is a designer, a creator, or a "builder."

The design process is an exciting and challenging effort, and the engineer-designer relies heavily on graphics as a means to create, record, analyze, and communicate to others design concepts or ideas. The ability to communicate verbally, symbolically, and graphically is essential.

16.1 Design Sources

There are two general types of design: *empirical design,* sometimes referred to as conceptual design, and *scientific design.* In scientific design, use is made of the principles of physics, mathematics, chemistry, mechanics, and other sciences in the new or revised design of devices, structures, or systems intended to function under specific conditions. In empirical design, much use is made of the information in handbooks, which in turn has been learned by experience. Nearly all technical design is a combination of scientific and empirical design. Therefore, a competent designer has both adequate engineering and scientific knowledge and access to the many handbooks related to the field.

You may not yet have acquired the necessary educational background and experience to undertake a sophisticated design. And you probably do not have access to the variety of experts on materials, production methods, business eco-

nomics, legal considerations, or other vital areas that probably would be available in a large company. Nevertheless, through conscientious application of your design ability and the sources of information that are available, you can go far toward the creation or improvement of some device, system, or service that *works* and is uniquely your own. The working models of the Postal Scale and the Educational Toys in Fig. 16.1 are examples of student design efforts.

16.2 Design Concepts

New ideas or design concepts must exist initially in the mind of the designer. In order to capture, preserve, and develop these ideas, the designer makes liberal use of freehand sketches of views and pictorials, Chapter 6. These sketches are revised or redrawn as the concept is developed. All sketches should be preserved for reference and dated as a record of the development of the design.

(a) POSTAL SCALE   (b) EDUCATIONAL TOYS

Fig. 16.1   Working Models of Student Designs.

At some point in the development of the idea, you will probably find it to your advantage to pool your ideas with those of others and begin working in a team effort; such a team may include others familiar with problems of materials, production, marketing, and so on. In industry, the project becomes a team effort long before the product is produced and marketed. Obviously, the design process is not a haphazard operation of an inventor working in a garage or basement, although it might well begin in that manner. Industry could not long survive if its products were determined in a haphazard manner. Hence, nearly all successful companies support a well-organized design effort, and the vitality of the company depends to a large extent on the *planned* output of its designers.

Since it is important for you to be able to work effectively with others in a group or team, you must be able to express yourself clearly and concisely. Do not underestimate the importance of your communication skills, your ability to express your ideas verbally (written and spoken), symbolically (equations, formulas, etc.), and *graphically*. See page 542 for a suggested format of a student design report.

The graphical skills include the ability to present information and ideas clearly and effectively in the form of sketches, drawings, graphs, and so on. This textbook is dedicated to helping you develop your communication skills in graphics.

16.3   The Design Process

Design is the ability to combine ideas, scientific principles, resources, and often existing products into a solution of a problem. This ability to solve problems in design is the result of an organized and orderly approach to the problem known as the *design process*.

The design process leading to manufacturing, assembly, marketing, service, and the many activities necessary for a successful product is composed of several easily recognized phases. Although many industrial groups may identify them in their own particular way, a convenient procedure for the design of a new or improved product is in five stages as follows:

1.   Identification of problem.
2.   Concepts and ideas.

510

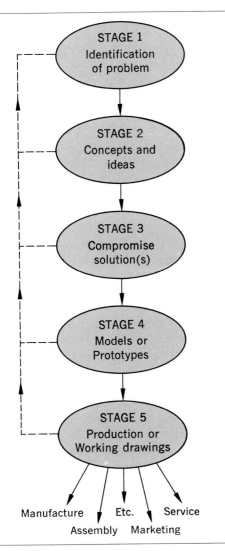

Fig. 16.2   Stages of the Design Process.

3. Compromise solutions.
4. Models and/or prototypes.
5. Production and/or working drawings.

Ideally, the design moves through the stages as shown in Fig. 16.2, but if a particular stage proves unsatisfactory, it may be necessary to return to a previous stage and repeat the procedure

Fig. 16.3   Pull-Tab Can Opener. *John Schultz— PAR/NYC.*

as indicated by the dashed-line paths. This repetitive procedure is often referred to as *looping*.

16.4   Stage 1—Identification of Problem

The design activity begins with the recognition and determination of a need or want for a product, service, or system and the economic feasibility of fulfilling this need. Engineering design problems may range from the need for a simple and inexpensive container opener such as the pull tab commonly used on beverage cans, Fig. 16.3, to the more complex problems associated with the needs of air and ground travel, space exploration, environmental control, and so forth. Although the product may be very simple, such as the pull tab on a beverage can, the production tools and dies require considerable engineering and design effort. The airport auto-

511

Fig. 16.4   Airport Transit System. *Courtesy of Westinghouse Electric Corp.*

mated transit system design, Fig. 16.4, meets the need of moving people efficiently between the terminal areas. The system is capable of moving 3300 people every 10 minutes.

The Lunar Roving Vehicle, Fig. 16.5, is a solution to a need in the space program to explore larger areas of the lunar surface. This vehicle is the end result of a great deal of design work associated with the support systems and the related hardware.

At the problem identification stage, either the designer recognizes that there does exist a need requiring a design solution or, perhaps more often, a directive is received to that effect from management. No attempt is made at this time to set goals or criteria for the solution.

Information concerning the identified problem becomes the basis for a problem proposal, which may be a paragraph or a multipage report presented for formal consideration. A proposal is a plan for action that will be followed to solve the problem. The proposal, if approved, becomes an agreement to follow the plan. In the classroom, the agreement is made between you and your instructor on the identification of the problem and your proposed plan of action.

Following approval of the proposal, further aspects of the problem are explored. Available information related to the problem is collected, and parameters or guidelines for time, cost, function, and so on are defined within which you will work. For example: What is the design expected to do? What is the estimated cost limit? What is the market potential? What can it be sold for? When will the prototype be ready for testing? When must production drawings be ready? When will production begin? When will the product be available on the market?

The parameters of a design problem, including the time schedule, are established at this stage. Nearly all designs represent a compromise, and the amount of time budgeted to a project is no exception.

Fig. 16.5   Lunar Roving Vehicle. *Courtesy of The Boeing Co.*

16.5   Stage 2—Concepts and Ideas

At this stage, many ideas are collected, "wild" or otherwise, for possible solutions to the problem. The ideas are broad and unrestricted to permit the possibility of new and unique solutions. This compilation of ideas may be from individuals, or they may come from group or team "brainstorming" sessions wherein a suggested idea often generates many more ideas from the group. As the ideas are elicited, they are written down and/or recorded in graphic form (multiview or pictorial sketches) for future consideration and refinement.

The larger the collection of ideas, the greater are the chances of finding one or more ideas suitable for further refinement. All sources of ideas such as technical literature, reports, design and

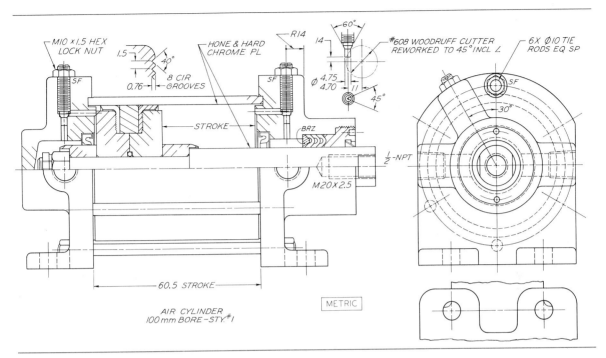

Fig. 16.6   Design Layout.

trade journals, patents, and existing products are explored. The Greenfield Village Museum in Dearborn, Michigan, the Museum of Science and Industry in Chicago, trade exhibitions, large hardware and supply stores, mail order catalogs, and so on are all excellent sources for ideas. Another source is the user of an existing product who often has suggestions for improvement. The potential user may be helpful with specific reactions to the proposed solution.

No attempt is made to evaluate ideas at this stage. All notes and sketches are signed, dated, and retained for possible patent proof.

16.6   Stage 3—Compromise Solutions

Various features of the many conceptual ideas generated in the preceding stages are selected after careful consideration and combined into one or more promising compromise solutions. At this point the best of the solutions is evaluated in detail, and attempts are made to simplify it

and thereby make manufacture and performance more efficient.

The sketches of the design are often followed by a study of suitable materials and of motion problems that may be involved. What source of power is to be used—manual, electric motor, or what? What type of motion is needed? Is it necessary to translate rotary motion to linear motion or vice versa? Many of these problems are solved graphically by means of a schematic drawing in which various parts are shown in skeleton form. A pulley is represented by a circle, meshing gears by tangent pitch circles, an arm by a single line, paths of motion by center lines, and so on. At this time, too, certain basic calculations, such as those related to velocity and acceleration, may be made, if required.

These preliminary studies are followed by a *design layout,* made with instruments, or a *layout sketch* from which a drafter makes an accurate to-scale instrumental drawing so that actual sizes and proportions can be clearly visualized, Fig. 16.6. At this time all parts are carefully de-

513

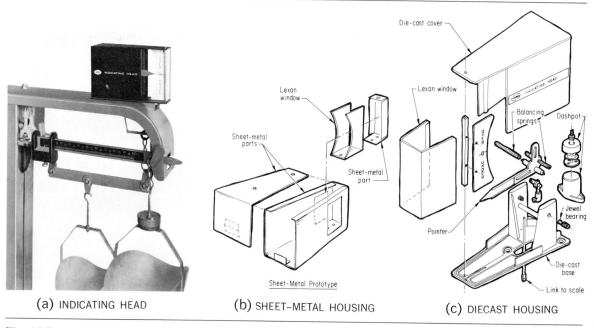

(a) INDICATING HEAD   (b) SHEET–METAL HOUSING   (c) DIECAST HOUSING

Fig. 16.7   Improved Design of Indicating Head. *Courtesy of Ohaus Scale Corp. and* Machine Design.

signed for strength and function. Costs are constantly kept in mind, for no matter how well the device performs, it must be built to sell for a profit or the time and development costs will have been a loss.

During the layout process, great reliance is placed on what has gone on before. Experience provides a sense of proportion and size that permits the noncritical or more standard features to be designed by eye or with the aid of empirical data. Stress analysis and detailed computation may be necessary in connection with high speeds, heavy loads, or special requirements or conditions.

As shown in Fig. 16.6, the layout is an assembly showing how parts fit together and the basic proportions of the various parts. Auxiliary views or sections are used, if necessary. Section lining may be used sparingly to save time. All lines should be sharp and the drawing made as accurately as possible, since most dimensions are omitted except for a few key ones that will be used in the detail or working drawings for production. Any notes or other information related to the detail drawing should be given on the layout.

Special attention is given to clearances of moving parts, ease of assembly, and serviceability. Standard parts are used wherever possible, for it is less costly to use stock items. Most companies maintain some form of an *engineering standards manual,* which contains much of the empirical data and detailed information that is regarded as "company standard." Materials and costs are carefully considered. Although functional considerations must come first, manufacturing problems must be kept constantly in mind.

A great many design problems are concerned with the improvement of an existing product or with the redesign of a device from a different approach in which many of the details will be similar to others previously used. For example, in Fig. 16.7, the Indicating Head is attached to a portable beam scale to add damping, sensitivity, and improved visibility to the weight readout. The original design of the housing was made of three sheet-metal parts with a plastic window, (b). The new two-piece design of a die-cast housing and larger plastic window, (c), provides more resistance to abuse and a drop in the unit cost of the housing after the first 2400 units, to less

than one-third of the cost of the original sheet-metal design. Very often a change in material or a slight change in the shape of some part may be made without any loss of effectiveness and yet may save hundreds or thousands of dollars. The *ideal design* is the one that will do the job required at the lowest possible cost.

The Totalizer Wheel, Fig. 16.8, represents a cost-reducing redesign of an assembly in an adding machine. The redesigned wheel replaces an assembly of the 23 parts shown and continues to act as an indexing gear, integral bearing, integral spring, position stop, and print wheel.

An example of design from a different approach is the Electric Wheel used in heavy-duty, four-wheel drive, earth-moving equipment, Fig. 16.9. This self-contained wheel design eliminates the usual restrictive drive train components, such as the drive shaft, universal joints,

Fig. 16.8   Redesigned Totalizer Wheel for Adding Machine. *Courtesy of Addmaster Corp. and E. I. du Pont de Nemours & Co., Inc.*

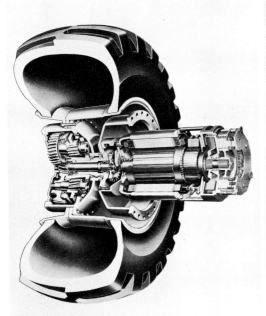

Fig. 16.9   Electric Wheel. Model L-700, LeTro-Loader, manufactured by Marathon LeTourneau Co., Equipment Division, Longview, Texas. *Courtesy of Marathon LeTourneau Co.*

(a) MODEL

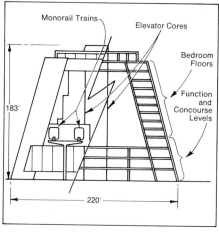

(b) ELEVATION

(c) MONORAIL TRAIN

Fig. 16.10   Recreation Park. *(a, b)* © *Walt Disney Productions; (c) courtesy of* Engineering New-Record.

differential gears, and transmission, and makes possible nonslip traction at each wheel. The motor in the wheel is powered by a heavy-duty diesel electric generating system aboard the equipment.

The designs for a large vacation-area complex include unique and unusual approaches or systems to the problem of housing and transportation. The 14-story A-frame hotel, as shown by the model, Fig. 16.10 (a), is serviced by water craft, surface vehicles, and special monorail trains that pass through the structure as indicated in (b). A pictorial of the train is shown in (c).

In a revised electronic organ design, the room-filling pipes and bellows have been replaced by a digital musical computer composed of aerospace microelectronics that requires about 1 cubic foot, Fig. 16.11. The computer contains some 48,000 transistors and enables the organ to be played in a virtually unlimited number of voices.

An improved design of a freight-handling system for transporting subcompact automobiles includes the unique design of a special railroad car called Vert-A-Pack, Fig. 16.12. The sides of the five compartments on each side, which hold three cars each, are hinged at the bottom for use

516

Fig. 16.11   Electronic Organ. *Courtesy of Allen Organ Co.*

Fig. 16.12   Railway Auto Transport System. *Courtesy of American Iron and Steel Institute.*

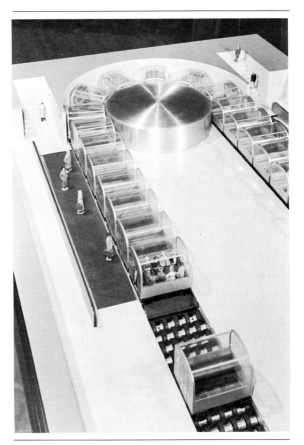

Fig. 16.13   Carveyor. *Courtesy of Goodyear Tire and Rubber Co.*

as ramps for efficient drive-on loading and drive-away unloading. The autos lock into place as the ramps are raised, and when the compartments are closed the autos are protected from vandalism and the weather.

16.7   Stage 4—Models and Prototypes

A model to scale is often constructed to study, analyze, and refine a design, Figs. 16.1, 16.11, and 16.17. The model of the Carveyor, Fig. 16.13, shows how it works and how people may be moved in such areas as airports, shopping centers, and college campuses. It is a loop system and is designed to carry up to 22,000 people per hour.

To instruct the model shop craftsperson in the construction of the prototype or model, dimensioned sketches and/or rudimentary working drawings are required. A full-size working model made to final specifications, except possibly for materials, is known as a *prototype*. The prototype is tested, modified where necessary, and the results noted in the revision of the sketches and working drawings.

If the prototype proves to be unsatisfactory, it may be necessary to return to a previous stage in the design process and repeat the procedures. It must be remembered that time and expense ceilings always limit the duration of this "looping." Eventually, a decision must be reached for the production model.

16.8   Stage 5—Production or Working Drawings

To produce or manufacture a product, a final set of production or working drawings is made, checked, and approved.

In industry the approved production design layouts are turned over to the engineering department for the production drawings. The drafter, or detailers, "pick off" the details from the layouts with the aid of the scale or dividers. The necessary views, §6.22, are drawn for each part to be made, and complete dimensions and notes, Chapter 13, are added so that the drawings will describe these parts completely. These working drawings of the individual parts are also known as *detail drawings*, §16.10.

Unaltered standard parts, §13.41, do not require a detail drawing but are shown conventionally on the assembly drawing and listed with specifications in the parts list, §16.14.

A detail drawing of one of the parts from the design layout of Fig. 16.4 is shown in Fig. 16.14. For details concerning working drawings, see §§16.10–16.18.

After the parts have been detailed, an *assembly drawing* is made, showing how all the parts go together in the complete product. The assembly may be made by tracing the various details in place directly from the detail drawings, or the assembly may be traced from the original design layout, but if either is done, the value of the assembly for checking purposes, §16.24, will be largely lost. The various types of assemblies are discussed in §§16.19–16.24.

Finally, in order to protect the manufacturer, a *patent drawing,* which is often a form of assembly, is prepared and filed with the U.S. Patent Office. Patent drawings are line shaded, often lettered in script, and otherwise follow rules of the Patent Office, §16.25.

16.9   Design of a New Product

An example of the design and development of a new product is that of the Cordless Electric Eraser shown in Fig. 16.15.

STAGE 1—IDENTIFICATION OF THE PROBLEM   In order to determine the feasibility of the *first* cordless eraser, opinions and ideas were solicited from many sources, including engineers, drafters, drafting teachers, drafting supply house managers, drafting supply store owners, and others. Price ranges and estimated sales were also carefully explored. This extensive survey indicated that there was a need and a potential market for a cordless eraser, provided it was convenient to use, durable, lightweight, versatile, maintenance free, safe, and competitively priced.

STAGE 2—CONCEPTS AND IDEAS   Various cord erasers on the market were examined, tested, and analyzed for possible improvements. Several cordless devices on the market were also studied. Power and speed requirements for efficient erasing of pencil and/or ink lines were determined. Several methods of holding the eraser were reviewed. Various eraser refills for electric erasers were tested. With this collection of information as a background, several questions now needed to be considered. Should a direct drive and a large motor or a reduction drive and a small motor be used? What power source

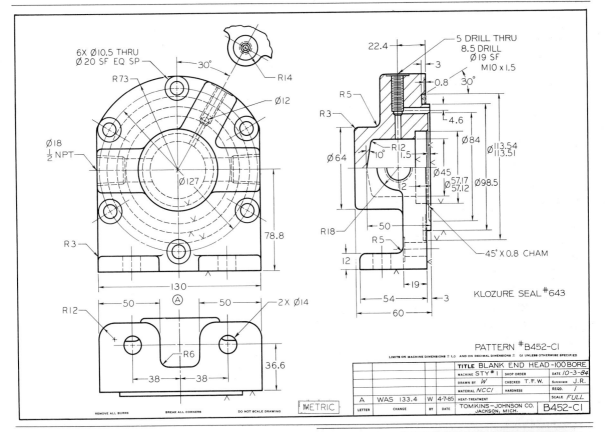

Fig. 16.14 A Detail Drawing.

should be used—replaceable or rechargeable batteries? What recharging arrangements are necessary for rechargeable battery power? What about safety precautions with 110 volt power for the charger? What materials are suitable? What bearings are available? What is a suitable way to chuck the eraser? Should long or short eraser plugs or only short ones be used? What standard parts are available for such items as the motor, bearings, batteries, recharging unit, and power cord?

How many ways can these components be arranged for the solution?

STAGE 3—COMPROMISE SOLUTION The cordless version of the eraser with a charging unit in a separate stand or console was selected as the preferred goal. The power train of a small battery-driven motor with a pinion gear meshed with a larger spur gear on the main shaft pro-

Fig. 16.15 Cordless Eraser with Recharging Console. *Courtesy of Pierce Business Products, Inc.*

519

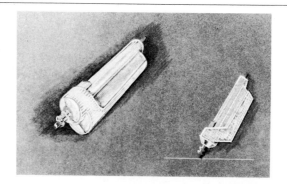
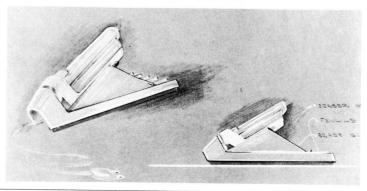

Fig. 16.16   Preliminary Design Pictorials of Cordless Eraser. *Courtesy of Pierce Business Products, Inc.*

vided adequate power and speed for the eraser chuck. The batteries provided power long enough for normal usage. Recharging would occur while the eraser was at rest on the charging console. Careful selection of components and materials could lead to a durable and lightweight unit.

The few simple components of the system could be arranged in several ways. Thus, some flexibility in the final design was possible. Pictorials of several concepts, Fig. 16.16, were made and evaluated for balance, handling qualities, and appearance.

STAGE 4—PROTOTYPE   A Prototype or working model, Fig. 16.17 (a), was built, tested, refined, and restyled into the final production design, (b). The section of the production model, Fig. 16.18, shows the selected arrangements of components;

note that all the components are held in place in one-half of the molded shell without additional parts or fasteners. The Main Shaft assembly is shown in Fig. 16.19. To achieve the goals of lightweight durability and a competitive selling price, the engineers selected a molding of reinforced nylon resin rather than a metal tubing. A manufacturing cost saving of over 50 percent for the main shaft alone was made possible by the elimination of several secondary operations. For example, in Fig. 16.19 note the following features.

1. No thread cutting required—threads are molded.

2. No groove cutting—molded hubs locate the bearings and also eliminate the need for retaining rings.

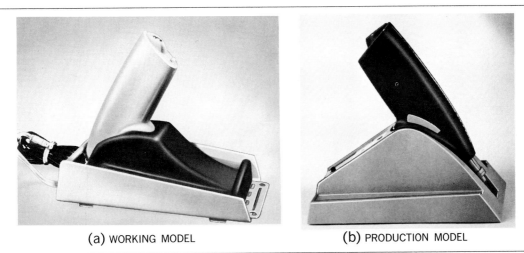

(a) WORKING MODEL   (b) PRODUCTION MODEL

Fig. 16.17   Models of Cordless Eraser. *Courtesy of Pierce Business Products, Inc.*

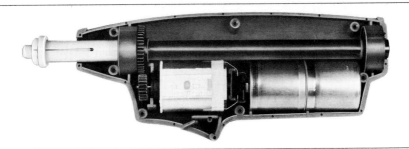

Fig. 16.18   Section of Cordless Eraser Production Model. *Courtesy of Pierce Business Products, Inc.*

Fig. 16.19   Main Shaft Assembly. *Courtesy of Pierce Business Products, Inc.*

3. No slot cutting for chuck removal slot—slot is molded.

4. No external grinding of shaft is necessary—precision molding gives a diameter suitable for installation of the high-speed bearings.

5. No keyseat required—molded-in key eliminates need for a key.

6. No keyway required on gear—molded-in keyway. No key or retaining ring installations facilitate more simplified assembly.

521

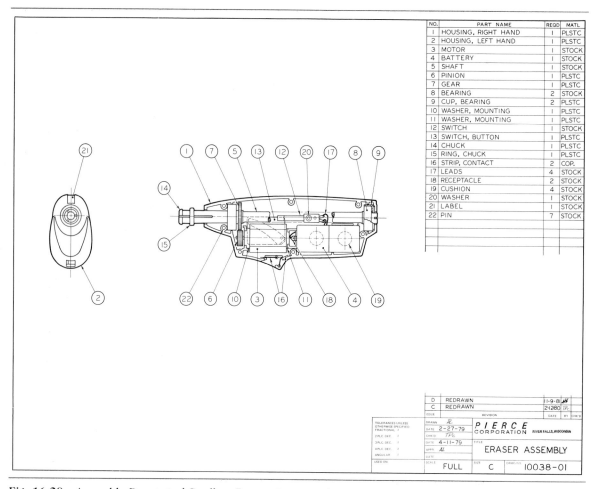

NO.	PART NAME	REQD	MATL
1	HOUSING, RIGHT HAND	1	PLSTC
2	HOUSING, LEFT HAND	1	PLSTC
3	MOTOR	1	STOCK
4	BATTERY	1	STOCK
5	SHAFT	1	STOCK
6	PINION	1	PLSTC
7	GEAR	1	PLSTC
8	BEARING	2	STOCK
9	CUP, BEARING	2	PLSTC
10	WASHER, MOUNTING	1	PLSTC
11	WASHER, MOUNTING	1	PLSTC
12	SWITCH	1	STOCK
13	SWITCH, BUTTON	1	PLSTC
14	CHUCK	1	PLSTC
15	RING, CHUCK	1	PLSTC
16	STRIP, CONTACT	2	COP.
17	LEADS	4	STOCK
18	RECEPTACLE	2	STOCK
19	CUSHION	4	STOCK
20	WASHER	1	STOCK
21	LABEL	1	STOCK
22	PIN	7	STOCK

D	REDRAWN		11-9-81	
C	REDRAWN		24280	
ISSUE		REVISION	DATE	BY	CHK'D

TOLERANCES UNLESS OTHERWISE SPECIFIED

FRACTIONAL
2 PLC DEC.
3 PLC DEC.
4 PLC DEC.
ANGULAR
USED ON

DRAWN 2-27-79
CHK'D TPc
APPR 4-11-79 AL

PIERCE CORPORATION RIVER FALLS, WISCONSIN

TITLE **ERASER ASSEMBLY**

SCALE **FULL** SIZE **C** DRWG NO. **10038-01**

Fig. 16.20 Assembly Drawing of Cordless Eraser. *Courtesy of Pierce Business Products, Inc.*

The foregoing considerations are typical of the attention given all components in the design.

STAGE 5—PRODUCTION DRAWINGS Complete sets of detail and assembly drawings were made for the eraser and the charging console. The assembly drawing for the eraser is shown in Fig. 16.20, and the assembly drawing of the charger base is shown in Fig. 16.21. A detail of the shaft is given in Fig. 16.22. Standard parts were specified on separate sheets or in a parts list. (Space limitations do not permit including more of the drawings necessary for the product.)

16.10 Working Drawings

Working drawings, which normally include assembly and details, are the specifications for the manufacture of a design. Therefore, they must be neatly made and carefully checked. The working drawings of the individual parts are also referred to as *detail drawings*, §§16.11–16.18.

16.11 Number of Details per Sheet

Two general methods are followed in industry regarding the grouping of details on sheets. If the machine or structure is small or composed of few

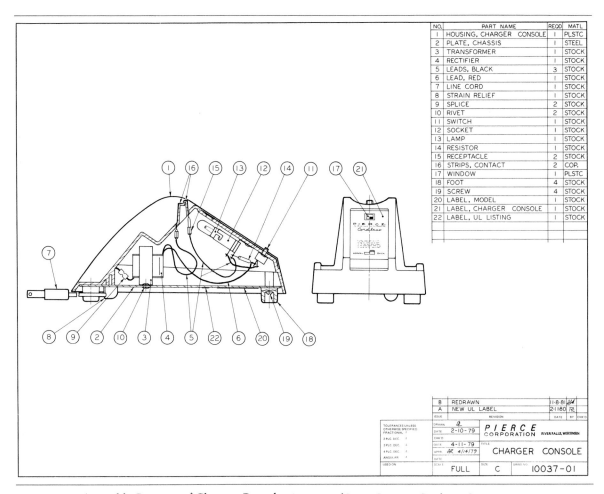

NO.	PART NAME	REQD	MATL
1	HOUSING, CHARGER CONSOLE	1	PLSTC
2	PLATE, CHASSIS	1	STEEL
3	TRANSFORMER	1	STOCK
4	RECTIFIER	1	STOCK
5	LEADS, BLACK	3	STOCK
6	LEAD, RED	1	STOCK
7	LINE CORD	1	STOCK
8	STRAIN RELIEF	1	STOCK
9	SPLICE	2	STOCK
10	RIVET	2	STOCK
11	SWITCH	1	STOCK
12	SOCKET	1	STOCK
13	LAMP	1	STOCK
14	RESISTOR	1	STOCK
15	RECEPTACLE	2	STOCK
16	STRIPS, CONTACT	2	COP.
17	WINDOW	1	PLSTC
18	FOOT	4	STOCK
19	SCREW	4	STOCK
20	LABEL, MODEL	1	STOCK
21	LABEL, CHARGER CONSOLE	1	STOCK
22	LABEL, UL LISTING	1	STOCK

| B | REDRAWN | 11-8-81 |
| A | NEW UL LABEL | 2-1180 |

PIERCE CORPORATION RIVER FALLS, WISCONSIN

CHARGER CONSOLE

SCALE FULL   SIZE C   DWG NO. 10037-01

Fig. 16.21   Assembly Drawing of Charger Console. *Courtesy of Pierce Business Products, Inc.*

parts, all the details may be shown on one large sheet, Fig. 16.23.

When larger or more complicated mechanisms are represented, the details may be drawn on several large sheets, several details to the sheet, and the assembly is drawn on a separate sheet. Most companies have now adopted the practice of drawing only one detail per sheet, however simple or small, Fig. 16.22. The basic 8.5″ × 11.0″ or 210 mm × 297 mm sheet is most commonly used for details, multiples of these sizes being used for larger details or the assembly. For standard sheet sizes, see §2.62.

When several details are drawn on one sheet, careful consideration must be given to spacing. The drafter should determine the necessary views for each detail and *block in all views lightly before beginning to draw any view,* as shown in Fig. 16.23. Ample space should be allowed for dimensions and notes. A simple method to space the views is to cut out rectangular scraps of paper roughly equal to the sizes of the views and to move these around on the sheet until a suitable spacing is determined. The corner locations are then marked on the sheet, and the scraps of paper are discarded.

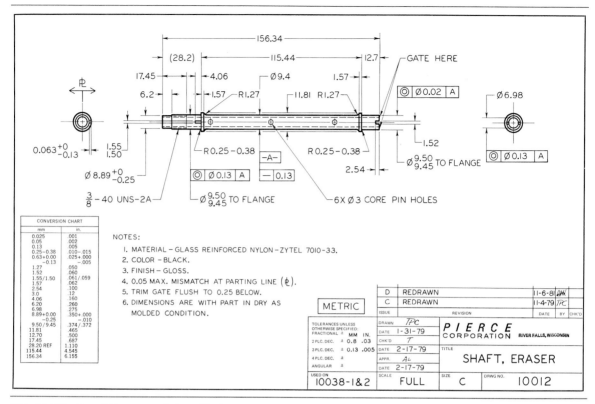

Fig. 16.22 Detail Drawing of Main Shaft. *Courtesy of Pierce Business Products, Inc.*

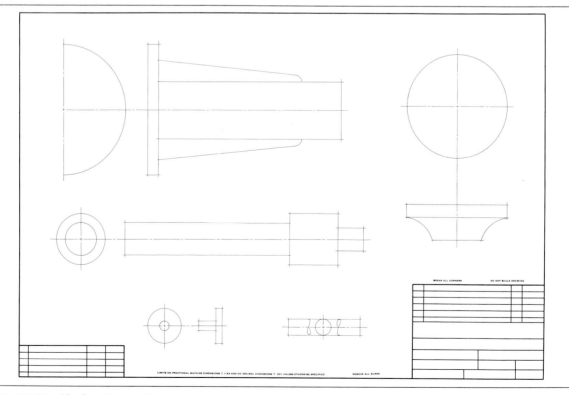

Fig. 16.23 Blocking In the Views.

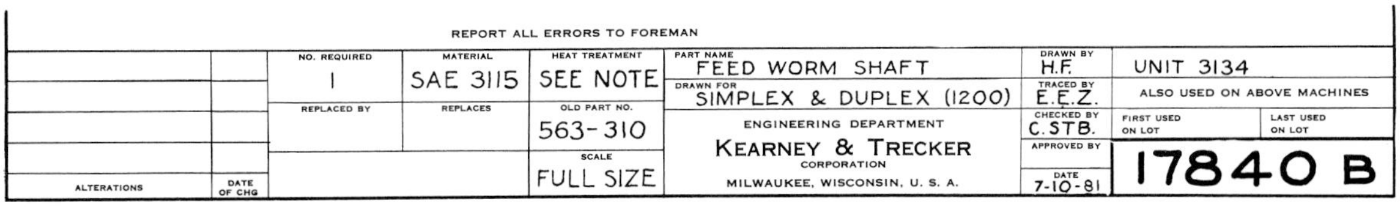

REPORT ALL ERRORS TO FOREMAN							
	NO. REQUIRED	MATERIAL	HEAT TREATMENT	PART NAME FEED WORM SHAFT	DRAWN BY H.F.	UNIT 3134	
	1	SAE 3115	SEE NOTE	DRAWN FOR SIMPLEX & DUPLEX (1200)	TRACED BY E.E.Z.	ALSO USED ON ABOVE MACHINES	
	REPLACED BY	REPLACES	OLD PART NO. 563-310	ENGINEERING DEPARTMENT **KEARNEY & TRECKER** CORPORATION MILWAUKEE, WISCONSIN, U. S. A.	CHECKED BY C.STB.	FIRST USED ON LOT	LAST USED ON LOT
ALTERATIONS	DATE OF CHG		SCALE FULL SIZE		APPROVED BY 7-10-81	17840 B	

Fig. 16.24   Title Strip.

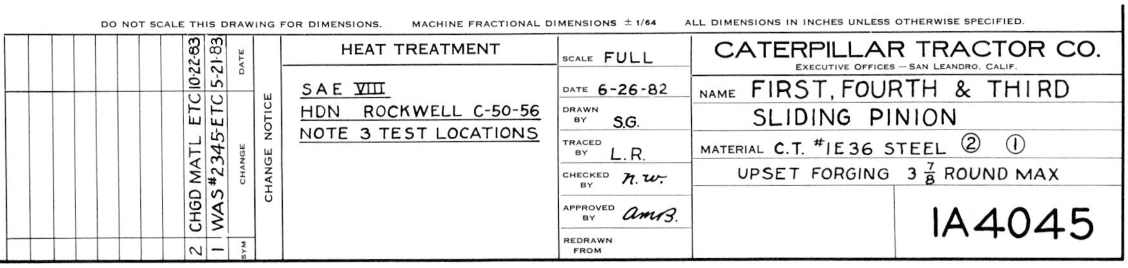

DO NOT SCALE THIS DRAWING FOR DIMENSIONS.   MACHINE FRACTIONAL DIMENSIONS ± 1/64   ALL DIMENSIONS IN INCHES UNLESS OTHERWISE SPECIFIED.

	HEAT TREATMENT	SCALE FULL	**CATERPILLAR TRACTOR CO.** EXECUTIVE OFFICES—SAN LEANDRO, CALIF.
	SAE VIII HDN ROCKWELL C-50-56 NOTE 3 TEST LOCATIONS	DATE 6-26-82	NAME FIRST, FOURTH & THIRD SLIDING PINION
		DRAWN BY S.G.	
		TRACED BY L.R.	MATERIAL C.T. #IE36 STEEL ② ①
		CHECKED BY n.w.	UPSET FORGING $3\frac{7}{8}$ ROUND MAX
		APPROVED BY amB.	
		REDRAWN FROM	IA4045

Fig. 16.25   Title Strip.

The same scale should be used for all details on a single sheet, if possible. When this is not possible, the scales for the dissimilar details should be clearly noted under each.

16.12   Title and Record Strips

The function of the title and record strip is to show, in an organized manner, all necessary information not given directly on the drawing with its dimensions and notes. Obviously, the type of title used depends on the filing system in use, the processes of manufacture, and the requirements of the product. The following information should generally be given in the title form:

1. Name of the object represented.
2. Name and address of the manufacturer.
3. Name and address of the purchasing company, if any.
4. Signature of the drafter who made the drawing and the date of completion.

5. Signature of the checker and the date of completion.
6. Signature of the chief drafter, chief engineer, or other official, and the date of approval.
7. Scale of the drawing.
8. Number of the drawing.

Other information may be given, such as material, quantity, heat treatment, finish, hardness, pattern number, estimated weight, superseding and superseded drawing numbers, symbol of machine, and many other items, depending on the plant organization and the peculiarities of the product. Some typical commercial titles are shown in Figs. 16.24, 16.25, and 16.26. See inside back cover for traditional title forms and ANSI-approved sheet sizes.

The title form is usually placed along the bottom of the sheet, Fig. 16.24, or in the lower right-hand corner of the sheet, Fig. 16.26, because drawings are often filed in flat, horizontal drawers, and the title must be easily found. However,

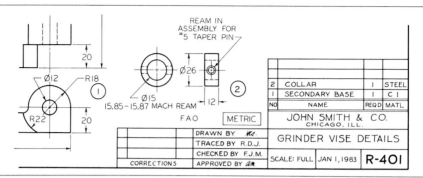

Fig. 16.26   Identification of Details with Parts List.

Table 16.1 *Recommended[a] Minimum Letter Heights*

Use	Minimum Letter Heights		Drawing Size
	Freehand	Instrumental	
Drawing number in title block	.312″ ($\frac{5}{16}$) 7 mm	.290″ 7 mm	Larger than 17″ × 22″
	.250″ ($\frac{1}{4}$) 7 mm	.240″ 7 mm	Up to and including 17″ × 22″
Drawing title	.250″ ($\frac{1}{4}$) 7 mm	.240″ 7 mm	
Section and tabulation letters	.250″ ($\frac{1}{4}$) 7 mm	.240″ 7 mm	All
Zone letters and numerals in borders	.188″ ($\frac{3}{16}$) 5 mm	.175″ 5 mm	
Dimensions, tolerances, limits, notes, subtitles for special views, tables, revisions, and zone letters for the body of the drawing	.125″ ($\frac{1}{8}$) 3.5 mm	.120″ 3.5 mm	Up to and including 17″ × 22″
	.156″ ($\frac{5}{32}$) 5 mm	.140″ 5 mm	Larger than 17″ × 22″

[a]ANSI Y14.2M–1979 (R1987).

many filing systems are in use, and the location of the title form is governed by the system employed.

Lettering should be single-stroke vertical or inclined Gothic capitals, Figs. 4.25 and 4.26. The items in the title form should be lettered in accordance with their relative importance. The drawing number should receive the greatest emphasis, closely followed by the name of the object and the name of the company. The date, scale, and drafter's and checker's names are important, but they do not deserve prominence. Greater importance of items is indicated by heavier lettering, larger lettering, wider spacing of letters, or by a combination of these methods. See Table 16.1 for recommended letter heights.

Many companies have adopted their own title forms or those preferred by ANSI and have them printed on standard-size sheets, so that the drafters need merely fill in the blank spaces.

Drawings constitute important and valuable information regarding the products of a manufacturer. Hence, carefully designed, well-kept, systematic files are generally maintained for the filing of drawings.

16.13   Drawing Numbers

Every drawing should be numbered. Some companies use serial numbers, such as 60412, or a number with a prefix or suffix letter to indicate the sheet size, as A60412 or 60412-A. The size A sheet would probably be the standard 8.5″ × 11.0″ or 9.0″ × 12.0″, and the B size a multiple thereof. Many different numbering schemes are in use in which various parts of the drawing number indicate different things, such as model number of the machine and the general nature or use of the part. In general, it is best to use a simple numbering system and not to load the number with too many indications.

The drawing number should be lettered 7 mm (.250″) high in the lower-right and upper-left corners of the sheet, Fig. 16.33.

16.14   Parts Lists

A bill of material, or *parts list,* consists of an itemized list of the several parts of a structure shown on a detail drawing or an assembly drawing [ANSI Y14.34M–1982 (R1988)]. This list is often given on a separate sheet, but is frequently lettered directly on the drawing, Fig. 16.32. The title strip alone is sufficient on detail drawings of only one part, Fig. 16.22, but a parts list is necessary on detail drawings of several parts, Fig. 16.26.

Parts lists on machine drawings contain the part numbers or symbols, a descriptive title of

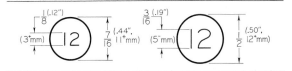

Fig. 16.27   Identification Numbers.

each part, the number required, the material specified, and frequently other information, such as pattern numbers, stock sizes of materials, and weights of parts.

Parts are listed in general order of size or importance. The main castings or forgings are listed first, parts cut from cold-rolled stock second, and standard parts such as fasteners, bushings, and roller bearings third. If the parts list rests on top of the title box or strip, the order of items should be from the bottom upward, Figs. 16.26 and 16.32, so that new items can be added later, if necessary. If the parts list is placed in the upper-right corner, the items should read downward.

Each detail on the drawing may be identified with the parts list by the use of a small circle containing the part number, placed adjacent to the detail, as in Fig. 16.26. One of the sizes in Fig. 16.27 will be found suitable, depending on the size of the drawing.

Standard parts, §13.41, whether purchased or company produced, are not drawn but are included in the parts list. Bolts, screws, bearings, pins, keys, and so on are identified by the part number from the assembly drawing and are specified by name and size or number.

16.15   Zoning

To facilitate locating an item on a large or complex drawing, regular ruled intervals are labeled along the margins, often in the right and lower margins only. The intervals on the horizontal margin are labeled from right to left with numerals, and the intervals on the vertical margin are labeled from bottom to top with letters. See Fig. 16.38.

16.16   Checking

The importance of accuracy in technical drawing cannot be overestimated. In commercial offices, errors sometimes cause tremendous unnecessary expenditures. *The drafter's signature on a drawing identifies who is responsible for the accuracy of the work.*

In small offices, checking is usually done by the designer or by one of the drafters. In large offices, experienced engineers are employed who devote a major part of their time to checking drawings.

The pencil drawing, upon completion, is carefully checked and signed by the drafter who made it. The drawing is then checked by the designer for function, economy, practicability, and so on. Corrections, if any, are then made by the original drafter.

The final checker should be able to discover all remaining errors, and, to be effective, the work must be done in a systematic way. The checker should study the drawing with particular attention to the following points.

1. Soundness of design, with reference to function, strength, materials, economy, manufacturability, serviceability, ease of assembly and repair, lubrication, and so on.

2. Choice of views, partial views, auxiliary views, sections, line work, lettering, and so on.

3. Dimensions, with special reference to repetition, ambiguity, legibility, omissions, errors, and finish marks. Special attention should be given to tolerances.

4. Standard parts. In the interest of economy, as many parts as possible should be standard.

5. Notes, with special reference to clear wording and legibility.

6. Clearances. Moving parts should be checked in all possible positions to assure freedom of movement.

7. Title form information.

16.17   Drawing Revisions

Changes on drawings are necessitated by changes in design, changes in tools, desires of customers, or errors in design or in production. In order that the sources of all changes of information on drawings may be understood, verified, and accessible, an accurate record of all changes should be made on the drawings. The record should show the character of the change, by whom, when, and why made.

The changes are made by erasures directly on the original drawing or by means of erasure fluid on a reproduction print. Additions are simply drawn in on the original. The removal of information by crossing out is not recommended. If a dimension is not noticeably affected by a change, it may be underlined with a heavy line as shown in Fig. 13.15 to indicate that it is not to scale. In any case, prints of each issue or microfilms are kept on file to show how the drawing appeared before the revision. New prints are issued to supersede old ones each time a change is made.

If considerable change on a drawing is necessary, a new drawing may be made and the old one then stamped OBSOLETE and placed in the "obsolete" file. In the title block of the old drawing, the words "SUPERSEDED BY . . ." or "REPLACED BY . . ." are entered followed by the number of the new drawing. On the new drawing, under "SUPERSEDES . . ." or "REPLACES . . .," the number of the old drawing is entered.

Various methods are used to reference the area on a drawing where the change is made, with the entry in the revision block. The most common is to place numbers or letters in small circles near the places where the changes were made and to use the same numbers or letters in the revision block, Fig. 16.28. On zoned draw-

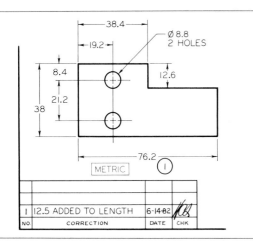

Fig. 16.28   Revisions.

1. Use word description in place of drawing wherever practicable.

2. Never draw an unnecessary view. Often a view can be eliminated by using abbreviations or symbols such as HEX, SQ, DIA, ⌀, ☐, and ℄.

3. Draw partial views instead of full views wherever possible. Draw half views of symmetrical parts.

4. Avoid elaborate, pictorial, or repetitive detail as much as possible. Use phantom lines to avoid drawing repeated features, §15.14.

5. List rather than draw, when possible, standard parts such as bolts, nuts, keys, and pins.

6. Omit unnecessary hidden lines. See §6.25.

7. Use outline section lining in large sectioned areas wherever it can be done without loss of clarity.

8. Omit unnecessary duplication of notes and lettering.

9. Use symbolic representation wherever possible, such as piping symbols and thread symbols.

10. Draw freehand, or mechanically plus freehand, wherever practicable.

11. Avoid hand lettering as much as possible. For example, parts lists should be typed on a separate sheet.

12. Use laborsaving devices wherever feasible, such as templates and plastic overlays.

13. Use electronic devices or computer graphics systems wherever feasible for design, drawing, and repetitive work.

ings, §16.15, the zone of the correction would be shown in the revision block. In addition, the change should be described briefly, and the date and the initials of the person making the change should be given.

16.18   Simplified Drafting

Drafting time is a considerable element of the total cost of a product. Consequently, industry attempts to reduce drawing costs by simplifying its drafting practices, but without loss of clarity to the user.

The American National Standard Drafting Manual, published by the American National Standards Institute, incorporates the best and most representative practices in this country, and the authors are in full accord with them. These standards advocate simplification in many ways, for example, partial views, half views, thread symbols, piping symbols, and single-line spring drawings. Any line or lettering on a drawing that is not needed for clarity should be omitted.

A summary of practices to simplify drafting is as follows.

Some industries have attempted to simplify their drafting practices even more. Until these practices are accepted generally by industry and in time find their way into the ANSI standards, the students should follow the ANSI standards as exemplified throughout this book. Fundamentals should come first—shortcuts perhaps later.

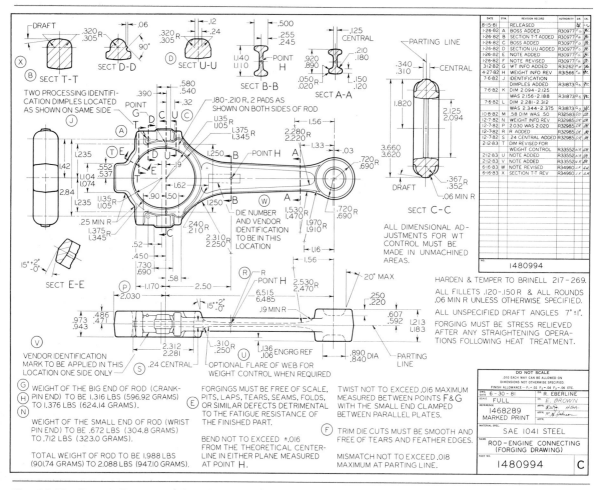

Fig. 16.29   Forging Drawing of Connecting Rod. *Courtesy of Cadillac Motor Car Division.*

16.19   Assembly Drawings

An assembly drawing shows the assembled machine or structure, with all detail parts in their functional positions. Assembly drawings vary in character according to use, as follows: (1) design assemblies, or layouts, discussed in §16.6, (2) general assemblies, (3) working drawing assemblies, (4) outline or installation assemblies, and (5) check assemblies.

16.20   General Assemblies

A set of working drawings includes the *detail drawings* of the individual parts and the *assembly drawing* of the assembled unit. The detail drawings of an automobile connecting rod are shown in Figs. 16.29 and 16.30, and the corresponding assembly drawing is shown in Fig. 16.31. Such an assembly, showing only one unit of a larger machine, is often referred to as a *subassembly*.

An example of a complete general assembly appears in Fig. 16.32, which shows the assembly of a hand grinder. Another example of a subassembly is shown in Fig. 16.33.

1. VIEWS   In selecting the views for an assembly drawing, the purpose of the drawing must be kept in mind: to show how the parts fit together

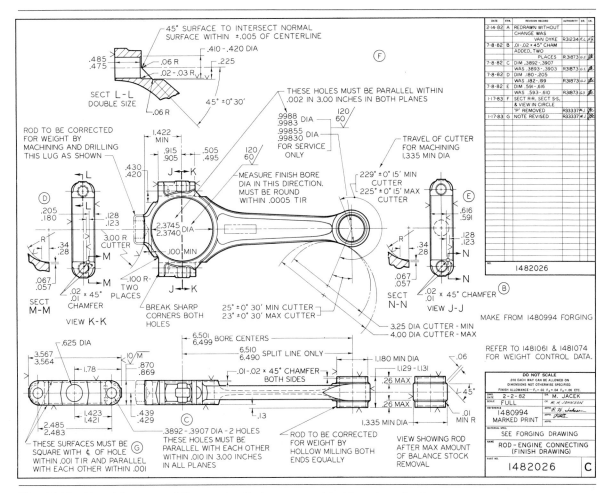

Fig. 16.30 Detail Drawing of Connecting Rod. *Courtesy of Cadillac Motor Car Division.*

in the assembly and to suggest the function of the entire unit, not to describe the shapes of the individual parts. The assembly worker receives the actual finished parts. If more information is needed about a part that cannot be obtained from the part itself, the detail drawing must be checked. Thus, the assembly drawing purports to show *relationships* of parts, *not shapes*. The view or views selected should be the minimum views or partial views that will show how the parts fit together. In Fig. 16.31, only one view is needed, while in Fig. 16.32 only two views are necessary.

2. SECTIONS   Since assemblies often have parts fitting into or overlapping other parts, hidden-line delineation is usually out of the question. Hence, in assemblies, sectioning can be used to great advantage. For example, in Fig. 16.32, try to imagine the right-side view drawn in elevation with interior parts represented by hidden lines. The result would be completely unintelligible.

Any kind of section may be used as needed. A broken-out section is shown in Fig. 16.32, a half section in Fig. 16.33, and several removed sections are shown in Fig. 16.29. For general information on assembly sectioning, see §16.21.

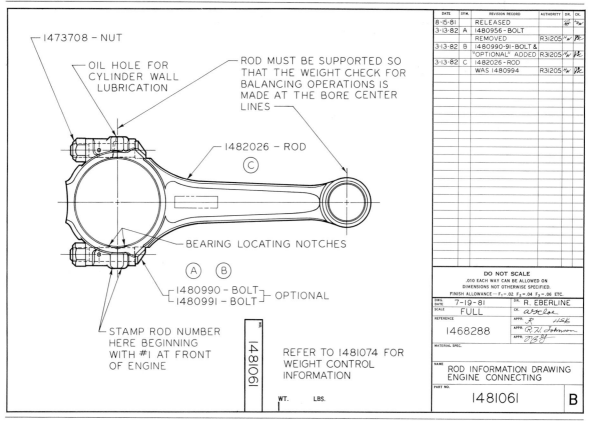

DATE	SYM.	REVISION RECORD	AUTHORITY	DR.	CK.
8-15-81		RELEASED			
3-13-82	A	1480956-BOLT REMOVED	R31205		
3-13-82	B	1480990-91-BOLT & "OPTIONAL" ADDED	R31205		
3-13-82	C	1482026-ROD WAS 1480994	R31205		

1473708 – NUT

OIL HOLE FOR CYLINDER WALL LUBRICATION

ROD MUST BE SUPPORTED SO THAT THE WEIGHT CHECK FOR BALANCING OPERATIONS IS MADE AT THE BORE CENTER LINES

1482026 – ROD

BEARING LOCATING NOTCHES

1480990 – BOLT
1480991 – BOLT — OPTIONAL

STAMP ROD NUMBER HERE BEGINNING WITH #1 AT FRONT OF ENGINE

REFER TO 1481074 FOR WEIGHT CONTROL INFORMATION

1481061

WT.   LBS.

DO NOT SCALE
.010 EACH WAY CAN BE ALLOWED ON DIMENSIONS NOT OTHERWISE SPECIFIED.
FINISH ALLOWANCE—F_1 = .02 F_2 = .04 F_3 = .06 ETC.

DWG. DATE 7-19-81   DR. R. EBERLINE
SCALE FULL   CK.
REFERENCE   APPR. R   HSB
1468288   APPR. R. H. Johnson
MATERIAL SPEC.   APPR.

NAME ROD INFORMATION DRAWING ENGINE CONNECTING

PART NO. 1481061   B

Fig. 16.31 Assembly Drawing of Connecting Rod. *Courtesy of Cadillac Motor Car Division.*

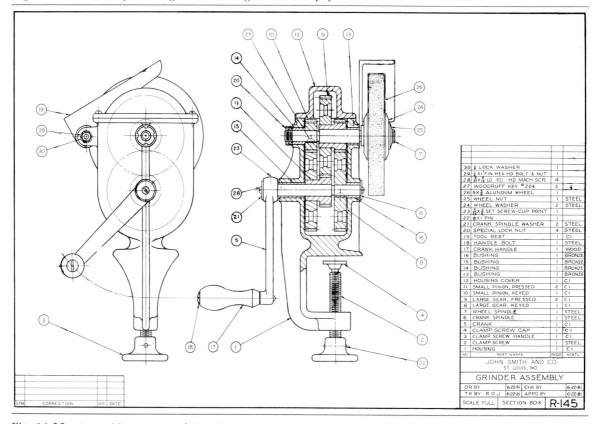

NO.	PART NAME	REQD	MATL
30	¼ LOCK WASHER	1	
29	¼ X 1" FIN HEX HD BOLT & NUT	1	
28	10/32 X ⅝ LG RD HD MACH SCR	4	
27	WOODRUFF KEY #204	2	
26	5X¾ ALUNDUM WHEEL	1	
25	WHEEL NUT	1	STEEL
24	WHEEL WASHER	2	STEEL
23	5/16X⅜ SET SCREW-CUP POINT	1	
22	3/16X1 PIN	1	
21	CRANK SPINDLE WASHER	2	STEEL
20	SPECIAL LOCK NUT	4	STEEL
19	TOOL REST	1	C I
18	HANDLE BOLT	1	STEEL
17	CRANK HANDLE	1	WOOD
16	BUSHING	1	BRONZE
15	BUSHING	1	BRONZE
14	BUSHING	1	BRONZE
13	BUSHING	1	BRONZE
12	HOUSING COVER	1	C I
11	SMALL PINION, PRESSED	2	C I
10	SMALL PINION, KEYED	1	C I
9	LARGE GEAR, PRESSED	2	C I
8	LARGE GEAR, KEYED	1	C I
7	WHEEL SPINDLE	1	STEEL
6	CRANK SPINDLE	1	STEEL
5	CRANK	1	C I
4	CLAMP SCREW CAP	1	C I
3	CLAMP SCREW HANDLE	1	C I
2	CLAMP SCREW	1	STEEL
1	HOUSING	1	C I

JOHN SMITH AND CO.
ST. LOUIS, MO.

GRINDER ASSEMBLY

DR BY   6-22-81   CHK BY   6-22-81
TR BY R.D.J.   6-22-81   APPD BY   6-22-81
SCALE FULL   SECTION 80-X   R-145

Fig. 16.32 Assembly Drawing of Grinder.

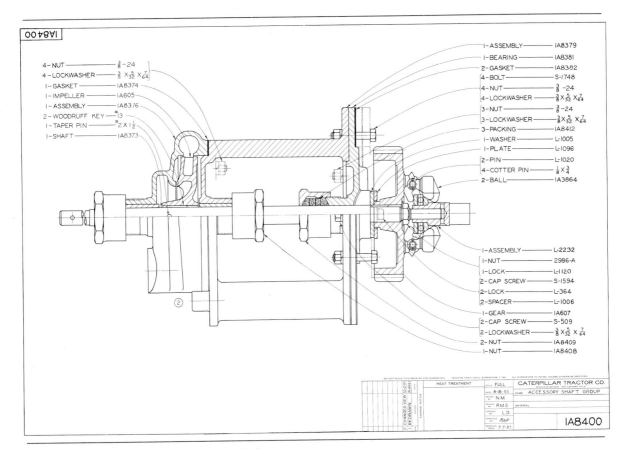

Fig. 16.33   Subassembly of Accessory Shaft Group.

For methods of drawing threads in sections, see §15.15.

3. HIDDEN LINES   As a result of the extensive use of sectioning in assemblies, hidden lines are often not needed. However, they should be used wherever necessary for clearness.

4. DIMENSIONS   As a rule, dimensions are not given on assembly drawings, since they are given completely on the detail drawings. If dimensions are given, they are limited to some function of the object as a whole, such as the maximum height of a jack, or the maximum opening between the jaws of a vise. Or when machining is required in the assembly operation, the necessary dimensions and notes may be given on the assembly drawing.

5. IDENTIFICATION   The methods of identification of parts in an assembly are similar to those used in detail drawings where several details are shown on one sheet, as in Fig. 16.26. Circles containing the part numbers are placed adjacent to the parts, with leaders terminated by arrowheads touching the parts as in Fig. 16.32. The circles shown in Fig. 16.27 for detail drawings are, with the addition of radial leaders, satisfactory for assembly drawings. Note, in Fig. 16.32, that these circles are placed in orderly horizontal or vertical rows and not scattered over the sheet. Leaders are never allowed to cross, and adjacent leaders are parallel or nearly so.

The parts list includes the part numbers or symbols, a descriptive title of each part, the number required per machine or unit, the material specified, and frequently other informa-

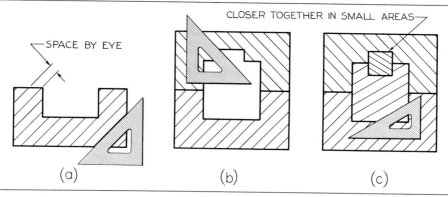

Fig. 16.34   Section Lining (Full Size).

tion, such as pattern numbers, stock sizes, weights, and so on. Frequently the parts list is lettered or typed on a separate sheet.

Another method of identification is to letter the part names, numbers required, and part numbers, at the end of leaders as shown in Fig. 16.33. More commonly, however, only the part numbers are given, together with ANSI-approved straight-line leaders.

6. DRAWING REVISIONS   Methods of recording changes are the same as those for detail drawings, Fig. 16.29, for example. See §16.17.

16.21   Assembly Sectioning

In assembly sections it is necessary not only to show the cut surfaces but also to distinguish between adjacent parts. This is done by drawing the section lines in opposing directions, as shown in Fig. 16.34. The first large area, (a), is section-lined at 45°. The next large area, (b), is section-lined at 45° in the opposite direction. Additional areas are then section-lined at other angles, such as 30° or 60° with horizontal, as shown at (c). If necessary, "odd" angles may be used. Note at (c) that in small areas it is necessary to space the section lines closer together. The section lines in adjacent areas should not meet at the visible lines separating the areas.

For general use, the cast-iron general-purpose section lining is recommended for assem-

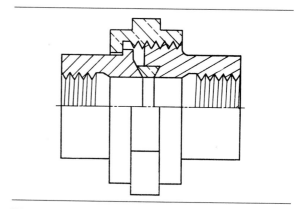

Fig. 16.35   Symbolic Section Lining.

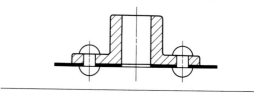

Fig. 16.36   Sectioning Thin Parts.

blies. Wherever it is desired to give a general indication of the materials used, symbolic section lining may be used, as in Fig. 16.35.

In sectioning relatively thin parts in assembly, such as gaskets and sheet-metal parts, section lining is ineffective, and such parts should be shown in solid black, Fig. 16.36.

Often solid objects, or parts that themselves do not require sectioning, lie in the path of the cutting plane. It is customary and standard practice to show such parts unsectioned, or "in the round." These include bolts, nuts, shafts, keys, screws, pins, ball or roller bearings, gear teeth, spokes, and ribs among others. Many are shown in Fig. 16.37, and similar examples are shown in Figs. 16.32 and 16.33.

16.22 Working Drawing Assembly

A working drawing assembly, Fig. 16.38, is a combined detail and assembly drawing. Such drawings are often used in place of separate detail and assembly drawings when the assembly is simple enough for all its parts to be shown

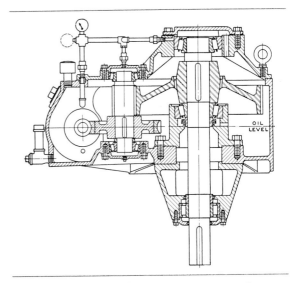

Fig. 16.37   Assembly Section. *Courtesy of Hewitt-Robins, Inc.*

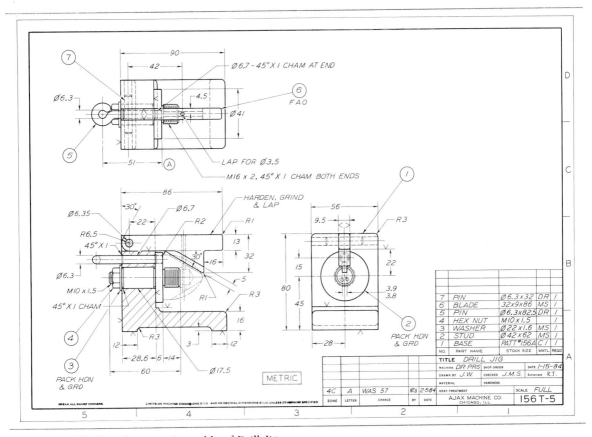

Fig. 16.38   Working Drawing Assembly of Drill Jig.

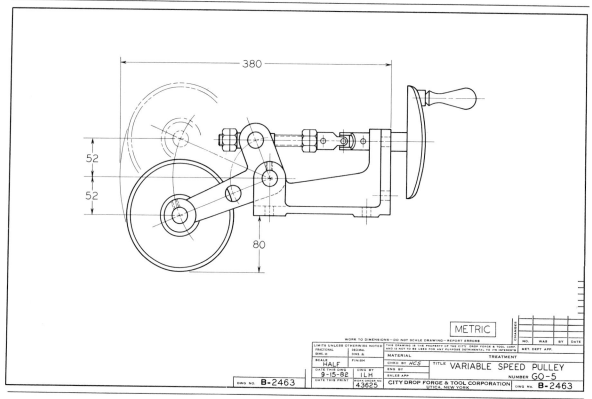

Fig. 16.39   Installation Assembly.

clearly in the single drawing. In some cases, all but one or two parts can be drawn and dimensioned clearly in the assembly drawing, in which event these parts are detailed separately on the same sheet. This type of drawing is common in valve drawings, locomotive subassemblies, aircraft subassemblies, and drawings of jigs and fixtures.

16.23   Installation Assemblies

An assembly made specifically to show how to install or erect a machine or structure is an *installation assembly*. This type of drawing is also often called an *outline assembly*, because it shows only the outlines and the relationships of exterior surfaces. A typical installation assembly is shown in Fig. 16.39. In aircraft drafting, an installation drawing (assembly) gives complete information for placing details or subassemblies in their final positions in the airplane.

16.24   Check Assemblies

After all detail drawings of a unit have been made, it may be necessary to make a *check assembly*, especially if a number of changes were made in the details. Such an assembly is drawn accurately to scale in order to check graphically the correctness of the details and their relationship in assembly. After the check assembly has served its purpose, it may be converted into a general assembly drawing.

16.25   Patent Drawings

The patent application for a machine or device must include drawings to illustrate and explain

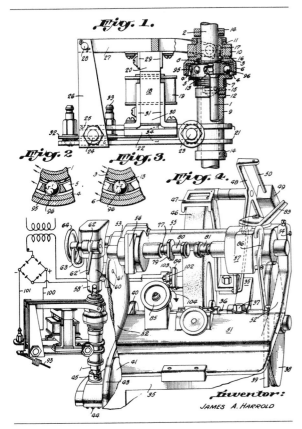

Fig. 16.40 A Well-Executed Patent Drawing.

the invention. It is essential that all patent drawings be mechanically correct and constitute complete illustrations of every feature of the invention claimed. The strict requirements of the U.S. Patent Office in this respect serve to facilitate the examination of applications and the interpretation of patents issued thereon. A typical patent drawing is shown in Fig. 16.40.

The drawings for patent applications are pictorial and explanatory in nature; hence, they are not detailed as are working drawings for production purposes. Center lines, dimensions, notes, and so forth are omitted. Views, features, and parts, for example, are identified by numbers that refer to the descriptions and explanations given in the specification section of the patent application.

Patent drawings are made with India ink on heavy, smooth, white paper, exactly $10.0'' \times 15.0''$ with $1.0''$ borders on all sides. A space of not less than $1.25''$ from the shorter border, which is the top of the drawing, is left blank for the heading of title, name, number, and other data to be added by the Patent Office.

All lines must be solid black and suitable for reproduction at a smaller size. Line shading is used whenever it improves readability.

The drawings must contain as many figures as necessary to show the invention clearly. There is no restriction on the number of sheets. The figures may be plan, elevation, section, pictorial, and detail views of portions or elements, and they may be drawn to an enlarged scale if necessary. The required signatures must be placed in the lower right-hand corner of the drawing, either inside or outside the border line.

Because of the strict requirements of the Patent Office, applicants are advised to employ competent drafters to make their drawings. To aid drafters in the preparation of drawings for submission in patent applications, the *Guide for Patent Draftsmen* has been prepared by and can be obtained from the Superintendent of Documents, U.S. Government Printing Office, Washington, D.C. 20402.

16.26 Computer Graphics

A computer is an electronic machine that is capable of performing specific tasks at incredibly high speeds. The use of the computer today in various business and industrial operations is very well known. Engineers and scientists have used computers for many years to perform the mathematical calculations required in their work. In recent years, new and innovative technological developments in computer applications have expanded the versatility of the computer into a still more useful engineering and design tool.

Computer graphics is a general term used to define any procedure that uses computers to generate, process, and display graphic images.

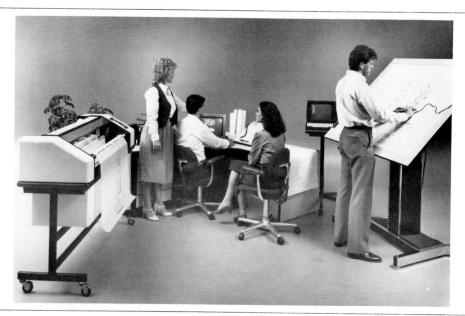

Fig. 16.41   Bausch & Lomb Producer Electronic/CAD System. *Courtesy of Bausch & Lomb.*

Fig. 16.42   Prime CAD Workstation. *Courtesy of Prime Computer, Inc.*

Fig. 16.43 Plotter. *Courtesy of Versatec, Inc.*

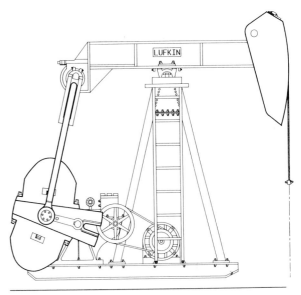

Fig. 16.44 Assembly Drawing Produced with the VersaCAD Advanced System. *Courtesy of VersaCAD.*

Computer-aided design or *computer-aided drafting (CAD)* and *computer-aided design and drafting (CADD)*, or other comparable terms, are used synonymously and refer to a specific process that uses a computer system to assist in the creation, modification, and display of a drawing or design.

All CAD systems have similar hardware components that include input devices, a central processing unit, data storage devices, and output devices. A typical CAD system is shown in Fig. 16.41. All CAD systems must also have software, which are the programs and instructions that permit the computer system to operate.

The designer or drafter can create a drawing by first entering the appropriate data into the system at a CAD workstation, Fig. 16.42, using any one of several different input devices. The immediate result of the input is a graphic display on the screen of a cathode-ray tube (CRT). The

designer can then view and analyze the drawing, make changes, delete sections, and "think out" the design on the tube much the same way as would be done on a drawing board. The final drawing displayed on the tube may then be stored in memory, or a permanent hard-copy drawing may be made using an output device such as the plotter shown in Fig. 16.43. Thus, by means of programmed data supplied to the computer, the designer or drafter is able to secure from a plotter an assembly drawing, Fig. 16.44, or a detail drawing complete with all required lines, lettering, dimensions, and so forth, Fig. 16.45.

Computer-aided design/computer aided manufacturing (CAD/CAM) refers to the integration of computers into the entire design-to-manufacturing process of a product or plant. In this method, the data are transmitted directly from the CAD system to the manufacturing plant

539

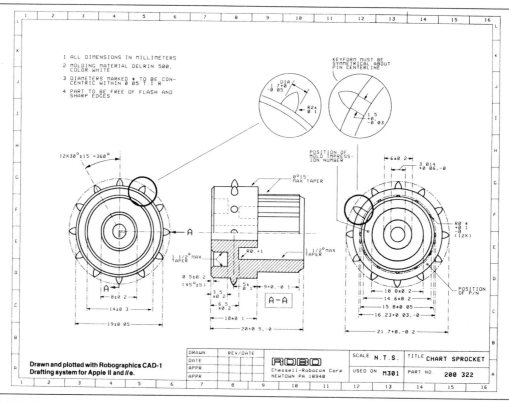

Fig. 16.45 Computer-Generated Drawing. *Courtesy of Chessell-Robocom Corp.*

(a)   (b)

Fig. 16.46 CAD/CAM System. *Courtesy of Control Data Corporation.*

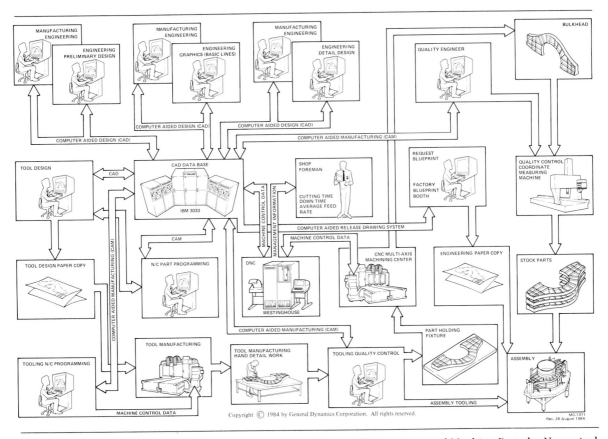

Fig. 16.47 Work Flow Diagram for Producing Detail and Assembly Drawings, and Machine Parts by Numerical Control. *Courtesy of Fort Worth Division of General Dynamics Corporation.*

to manage and control operations. The CAD/CAM system shown in Fig. 16.46 is used to cut seat patterns for Chrysler cars and trucks. A computer-controlled trim cutter in Chrysler's trim plant, (a), makes precision cuts in bolts of material according to directions given by the CAD terminal operator, (b), located in the company's corporate engineering complex. Before CAD/CAM the cutting of seat patterns was a painstaking manual job, requiring a worker to climb on top of layers of cloth and jigsaw through them following paper patterns the same way a seamstress follows a dress pattern.

The procedure for the design and preparation of engineering drawings and the production of machine parts by computer for the development of a new aircraft is shown in the work flow diagram, Fig. 16.47. Each of the operations from design to manufacture is systematically monitored and controlled by a CAD/CAM system.

The engineers and drafters must have a thorough understanding of the graphic language in order to prepare the correct input data for the computer and to evaluate the output of the plotter. The computer will do only what it is programmed to do. Regardless of how complex or automated the method of making a drawing becomes, the time-proven mode of graphic expression is indispensable for purposes of communication, specifications, records, and so on.

For additional information about computer-aided design and drafting, see Chapters 3 and 8.

541

DESIGN AND WORKING DRAWING PROBLEMS

DESIGN PROBLEMS

The following suggestions for project assignments are of a general and very broad nature, and it is expected that they will help generate many ideas for specific design projects. Much design work is undertaken to improve an existing product or system by utilization of new materials, new techniques, or new systems or procedures. In addition to the design of the product itself, another large amount of design work is essential for the tooling, production, and handling of the product. You are encouraged to discuss with your instructor any ideas you may have for a project.

1. Design new or improved playground, recreational, or sporting equipment. For example, a new child's toy could be both recreational and educational.
2. Design new or improved health equipment. For example, the physically handicapped have need for special equipment.
3. Design security or safety devices. Fire, theft, or poisonous gases are a threat to life and property.
4. Design devices and/or systems for waste handling. Home and factory waste disposal needs serious consideration.
5. Design new or improved educational equipment. Both teacher and student would welcome more efficient educational aids.
6. Design improvements in our land, sea, and air transportation systems. Vehicles, controls, highways, and airports need further refinement.
7. Design new or improved devices for material handling. A dispensing device for a powdered product is an example.
8. Improve the design of an existing device or system.
9. Design or redesign devices for improved portability.

Each solution to a design problem, whether prepared by an individual student or formulated by a group, should be in the form of a *report*, which should be typed or carefully lettered, assembled, and bound. Suitable folders or binders are usually available at the local school supply store. It is suggested that the report contain the following (or variations of them, as specified by your instructor).

1. A title sheet. The title of the design project should be placed in approximately the center of the sheet, and your name or the names of those in the group in the lower right-hand corner. The symbol PL should follow the name of the project leader.
2. Table of contents with page numbers.
3. Statement of the purpose of the project with appropriate comments.
4. Preliminary design sketches, with comments on advantages and disadvantages of each, leading to the final selection of the *best* solution. All work should be signed and dated.

5. An accurately made pictorial and/or assembly drawing(s), if more than one part is involved in the design.
6. Detail working drawings, freehand or mechanical as assigned. The 8.5″ × 11.0″ sheet size is preferred for convenient insertion in the report. Larger sizes may be bound in the report with appropriate folding.
7. A bibliography or credit for important sources of information, if applicable.

WORKING DRAWING PROBLEMS

The problems in Figs. 16.48–16.109 are presented to give you practice in making regular working drawings of the type used in industry. Many problems, especially those of the assemblies, offer an excellent opportunity for you to exercise your ability to redesign or improve on the existing design. Owing to the variations in sizes and in scales that may be used, you are required to select the sheet sizes and scales, when these are not specified, subject to the approval of the instructor. Standard sheet layouts are shown inside the back cover of this book. See also §2.62.

The statements for each problem are intentionally brief, so that the instructor may amplify or vary the requirements when making assignments. Many problems lend themselves to the preferred metric system or the acceptable complete decimal-inch system, while others may be more suitable for a combination of fractional and decimal dimensions. Either the preferred unidirectional or acceptable aligned dimensioning may be assigned.

Some tracings in ink may be assigned, but this is left to the discretion of the instructor.

It should be clearly understood that in problems presented in pictorial form, the placement of dimensions and finish marks cannot always be followed in the drawing. *The dimensions given are in most cases those needed to make the parts, but owing to the limitations of pictorial drawings they are not in all cases the dimensions that should be shown on the working drawing.* In the pictorial problems the rough and finished surfaces are shown, but finish marks are usually omitted. You should add all necessary finish marks and place all dimensions in the preferred places in the final drawings.

Each problem should be preceded by a thumbnail sketch or a complete technical sketch, fully dimensioned. Any of the title blocks shown inside the back cover of this book may be used, with modification if desired, or you may design the title block if so assigned by the instructor.

Since many of the problems in this chapter are of a general nature, they can also be solved on most computer graphics systems. If a system is available, the instructor may choose to assign specific problems to be completed by this method.

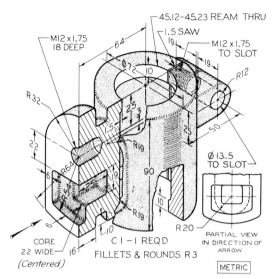

Fig. 16.48 Table Bracket. Make detail drawing. Use Size B or A3 sheet.

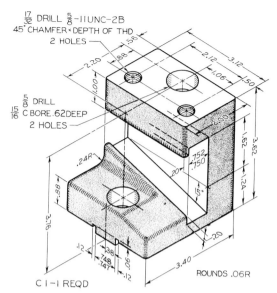

Fig. 16.49 RH Tool Post. Make detail drawing. Use Size B or A3 sheet. If assigned, convert dimensions to metric system.

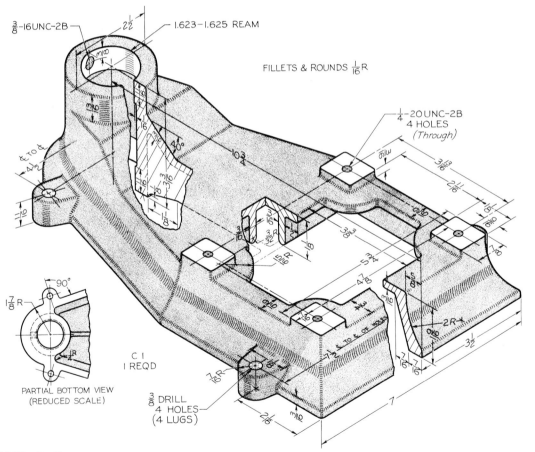

Fig. 16.50 Drill Press Base. Make detail drawing. Use Size C or A2 sheet. Use unidirectional metric or decimal-inch dimensions.

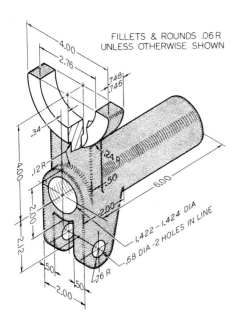

FILLETS & ROUNDS .06 R
UNLESS OTHERWISE SHOWN

4.00
2.76
.748
.746
.34
.24 R
.50
.12 R
4.00
6.00
2.00
2.00
1.422–1.424 DIA
2.12
.68 DIA – 2 HOLES IN LINE
.50
.76 R
.50
2.00

Fig. 16.51 Shifter Fork. Make detail drawing. Use Size B or A3 sheet. If assigned, convert dimensions to metric system.

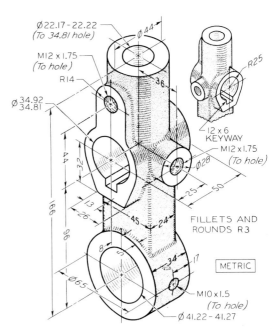

Ø 22.17–22.22
(To 34.81 hole)
M12 x 1.75
(To hole)
R14
Ø 34.92 / 34.81
Ø 44
36
R25
32
12 x 6 KEYWAY
M12 x 1.75
(To hole)
Ø 28
166
13
26
96
45
24
25
50
FILLETS AND ROUNDS R3
8
34
17
METRIC
Ø 65
M10 x 1.5
(To hole)
Ø 41.22–41.27

Fig. 16.52 Idler Arm. Make detail drawing. Use Size B or A3 sheet.

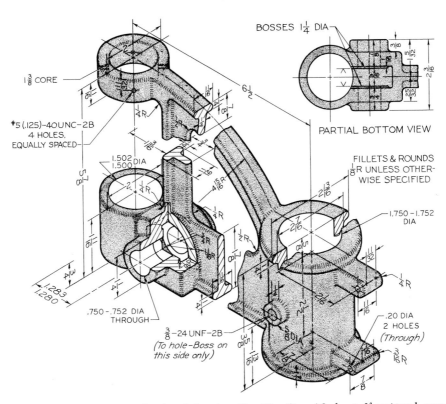

BOSSES 1¼ DIA
1⅜ CORE
*5 (.125)–40UNC–2B
4 HOLES,
EQUALLY SPACED
6½
1.502 / 1.500 DIA
2 – ¼ R
PARTIAL BOTTOM VIEW
FILLETS & ROUNDS ⅛ R UNLESS OTHER-WISE SPECIFIED
1.750–1.752 DIA
.750 –.752 DIA
THROUGH
1.283 / 1.280
¾ – 24 UNF–2B
(To hole–Boss on this side only)
.20 DIA
2 HOLES
(Through)

Fig. 16.53 Drill Press Bracket. Make detail drawing. Use Size C or A2 sheet. If assigned, convert dimensions to decimal inches or redesign the part with metric dimensions.

545

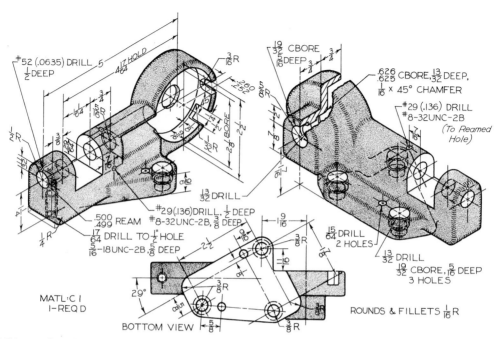

Fig. 16.54   Dial Holder. Make detail drawing. Use Size C or A2 sheet. If assigned, convert dimensions to decimal inches or redesign the part with metric dimensions.

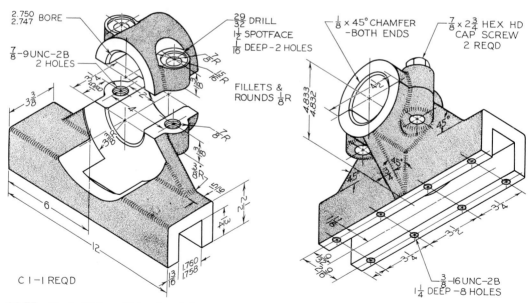

Fig. 16.55   Rack Slide. Make detail drawings. Draw half size on Size B or A3 sheet. If assigned, convert dimensions to decimal inches or redesign the part with metric dimensions.

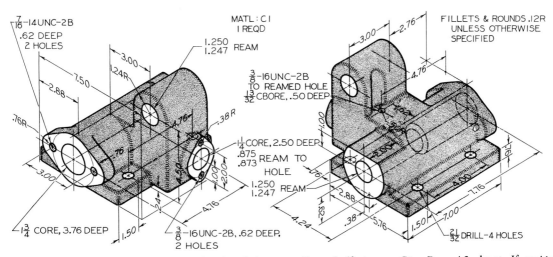

Fig. 16.56 Automatic Stop Box. Make detail drawing. Draw half size on Size B or A3 sheet. If assigned, redesign the part with metric dimensions.

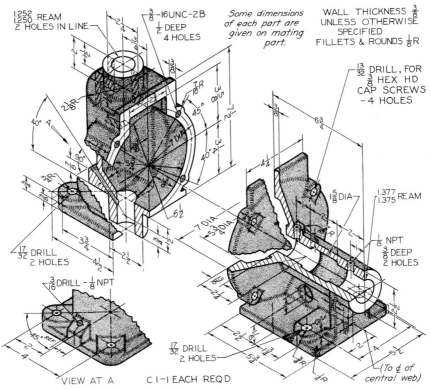

Fig. 16.57 Conveyor Housing. Make detail drawings. Draw half size on Size C or A2 sheets. If assigned, convert dimensions to decimal inches or redesign the parts with metric dimensions.

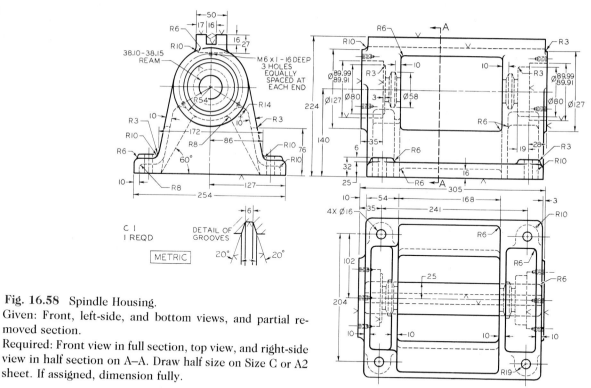

Fig. 16.58 Spindle Housing.

Given: Front, left-side, and bottom views, and partial removed section.

Required: Front view in full section, top view, and right-side view in half section on A–A. Draw half size on Size C or A2 sheet. If assigned, dimension fully.

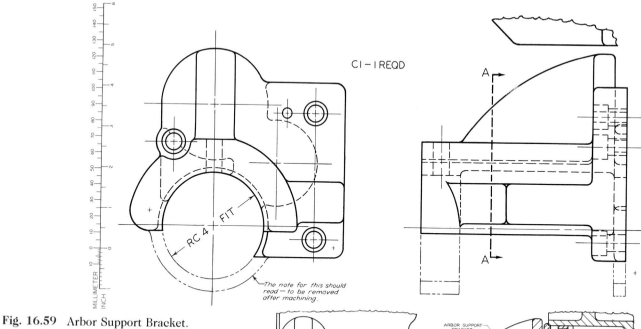

Fig. 16.59 Arbor Support Bracket.

Given: Front and right-side views.

Required: Front, left-side, and bottom views, and a detail section A–A. Use American National Standard tables for indicated fits and if required convert to metric values (see Appendixes 5–14). If assigned, dimension in the metric or decimal-inch system.

548

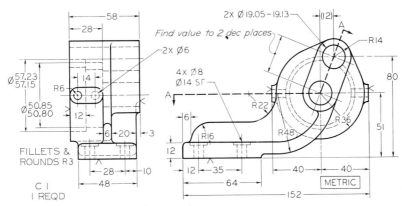

Fig. 16.60 Pump Bracket for a Thread Milling Machine.
Given: Front and left-side views.
Required: Front and right-side views, and top view in section on A–A. Draw full size on Size B or A3 sheet. If assigned, dimension fully.

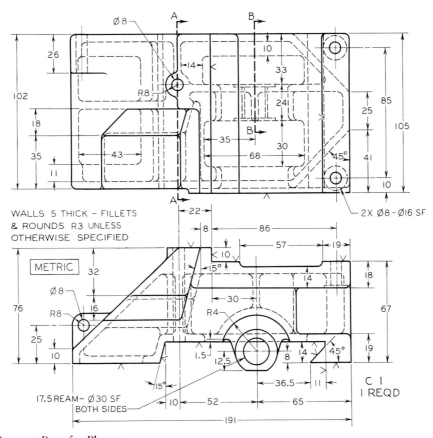

Fig. 16.61 Support Base for Planer.
Given: Front and top views.
Required: Front and top views, left-side view in full section A–A, and removed section B–B. Draw full size on Size C or A2 sheet. If assigned, dimension fully.

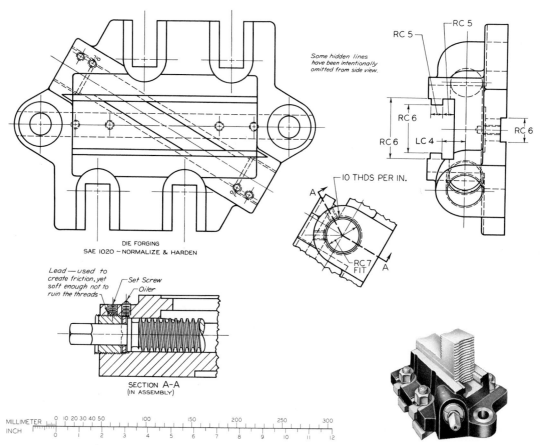

Some hidden lines have been intentionally omitted from side view.

RC 5
RC 5
RC 5
RC 6
RC 6
RC 6
LC 4

10 THDS PER IN.

A

A

RC 7 FIT

DIE FORGING
SAE 1020 — NORMALIZE & HARDEN

Lead — used to create friction, yet soft enough not to ruin the threads
Set Screw
Oiler

SECTION A-A
(IN ASSEMBLY)

MILLIMETER   0 10 20 30 40 50   100   150   200   250   300
INCH   0   1   2   3   4   5   6   7   8   9   10   11   12

Fig. 16.62   Jaw Base for Chuck Jaw.
Given: Top, right-side, and partial auxiliary views.
Required: Top, left-side (beside top), front, and partial auxiliary views complete with dimensions, if assigned. Use metric or decimal-inch dimensions. Use American National Standard tables for indicated fits or convert for metric values. See Appendixes 5–14.

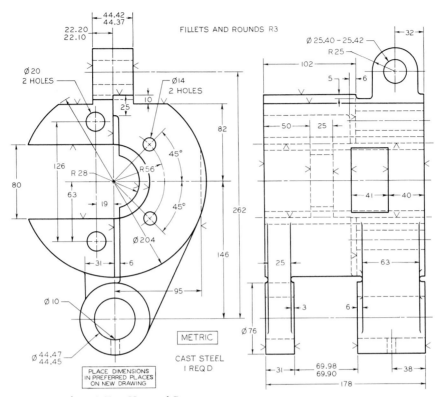

Fig. 16.63 Fixture Base for 60-Ton Vertical Press.

Given: Front and right-side views.

Required: Revolve front view 90° clockwise; then add top and left-side views. Draw half size on Size C or A2 sheet. If assigned, complete with dimensions.

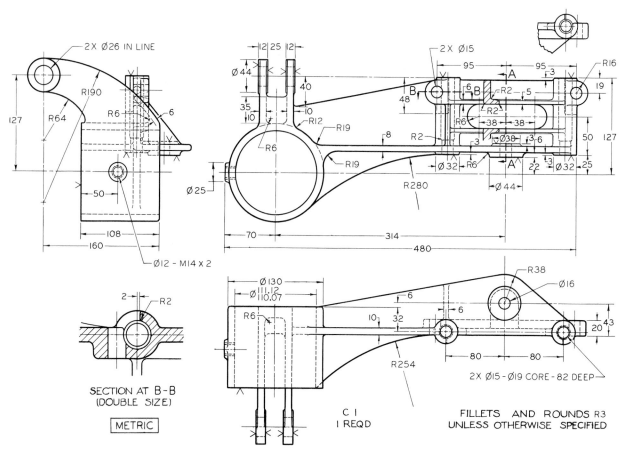

Fig. 16.64 Bracket.

Given: Front, left-side, and bottom views, and partial removed section.

Required: Make detail drawing. Draw front, top, and right-side views, and removed sections A–A and B–B. Draw half size on size C or A2 sheet. Draw section B–B full size. If assigned, complete with dimensions.

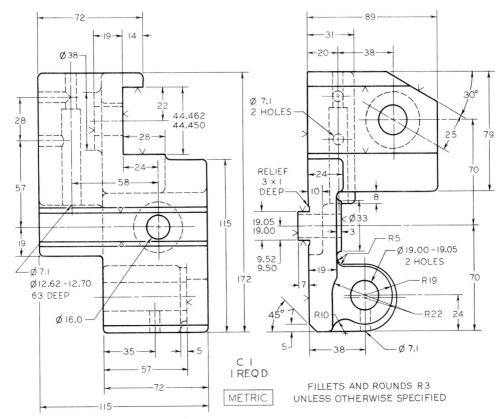

Fig. 16.65 Roller Rest Bracket for Automatic Screw Machine.
Given: Front and left-side views.
Required: Revolve front view 90° clockwise; then add top and left-side views. Draw half size on Size C or A2 sheet. If assigned, complete with dimensions.

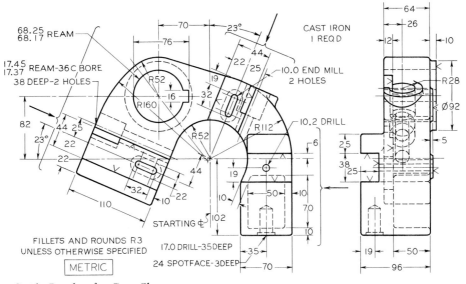

Fig. 16.66 Guide Bracket for Gear Shaper.
Given: Front and right-side views.
Required: Front view, a partial right-side view, and two partial auxiliary views taken in direction of arrows. Draw half size on Size C or A2 sheet. If assigned, complete with unidirectional dimensions.

553

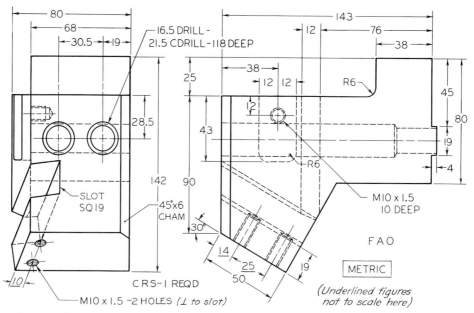

Fig. 16.67 Rear Tool Post.

Given: Front and left-side views.

Required: Take left-side view as new top view; add front and left-side views, approx. 215 mm apart, a primary auxiliary view, then a secondary view taken so as to show true end view of 19 mm slot. Complete all views, except show only necessary hidden lines in auxiliary views. Draw full size on Size C or A2 sheet. If assigned, complete with dimensions.

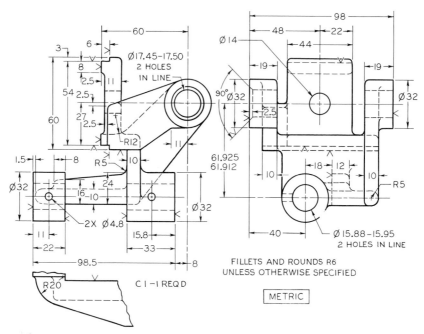

Fig. 16.68 Bearing for a Worm Gear.

Given: Front and right-side views.

Required: Front, top, and left-side views. Draw full size on Size C or A2 sheet. If assigned, complete with dimensions.

554

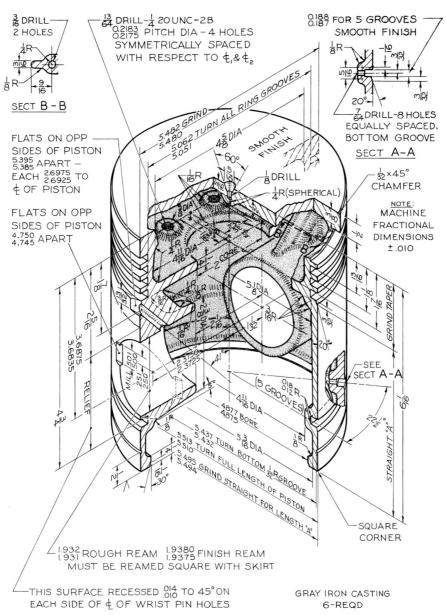

Fig. 16.69 Caterpillar Tractor Piston. Make detail drawing. Draw full size on Size C or A2 sheet. If assigned, use unidirectional decimal-inch system, converting all fractions to two-place decimal dimensions, or convert all dimensions to metric.

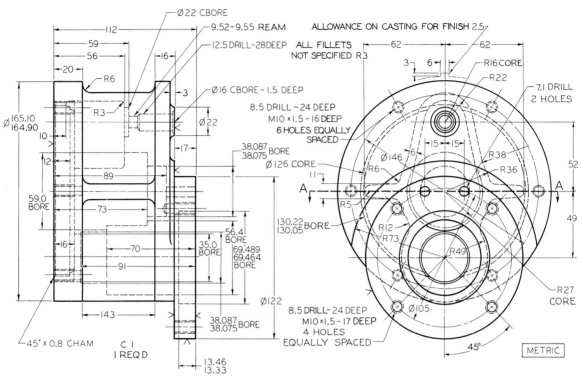

Fig. 16.70 Generator Drive Housing.

Given: Front and left-side views.

Required: Front view, right-side view in full section, and top view in full section on A–A. Draw full size on Size C or A2 sheet. If assigned, complete with dimensions.

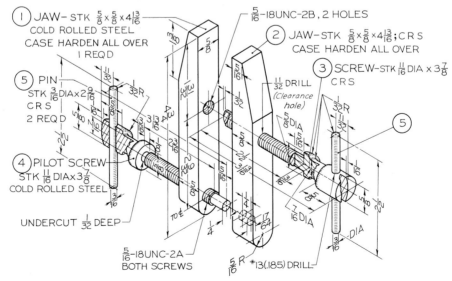

Fig. 16.71 Machinist's Clamp.   Draw details and assembly. If assigned, use unidirectional two-place decimal-inch dimensions or redesign for metric dimensions.

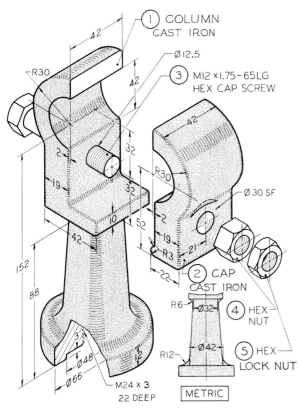

Fig. 16.72 Hand Rail Column. (1) Draw details. If assigned, complete with dimensions. (2) Draw assembly.

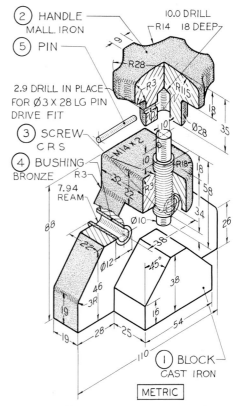

Fig. 16.73 Drill Jig. (1) Draw details. If assigned, complete with dimensions. (2) Draw assembly.

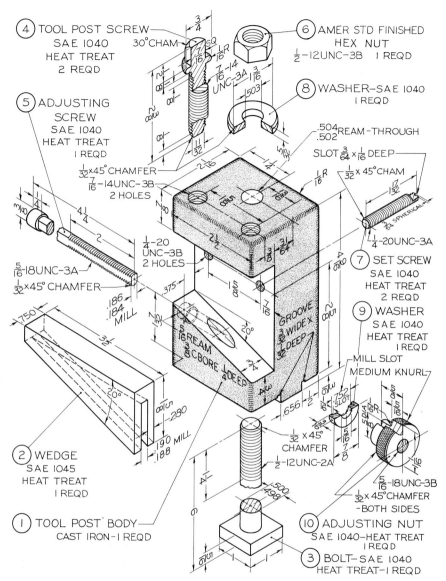

Fig. 16.74   Tool Post.   (1) Draw details. (2) Draw assembly. If assigned, use unidirectional two-place decimals for all fractional dimensions or redesign for all metric dimensions.

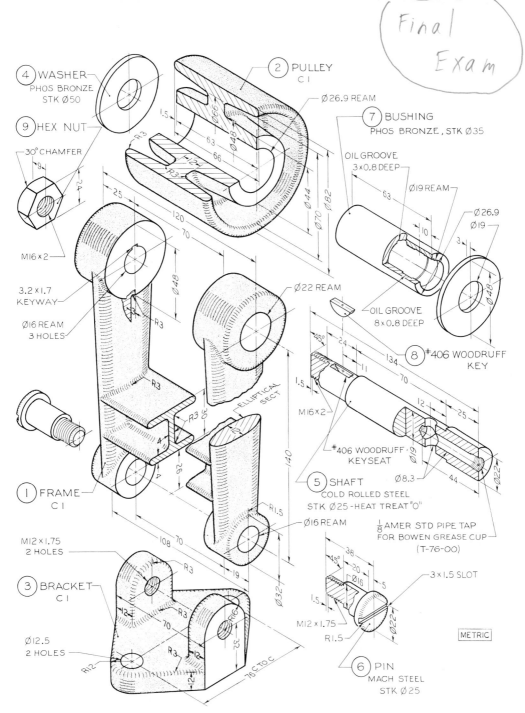

Fig. 16.75 Belt Tightener. (1) Draw details. (2) Draw assembly. It is assumed that the parts are to be made in quantity and they are to be dimensioned for interchangeability on the detail drawings. Use tables in Appendixes 11–14 for limit values. Design as follows.

a. Bushing fit in pulley: Locational interference fit.
b. Shaft fit in bushing: Free running fit.
c. Shaft fits in frame: Sliding fit.
d. Pin fit in frame: Free running fit.

e. Pulley hub length plus washers fit in frame: Allowance 0.13 and tolerances 0.10.
f. Make bushing 0.25 mm shorter than pulley hub.
g. Bracket fit in frame: Same as e above.

559

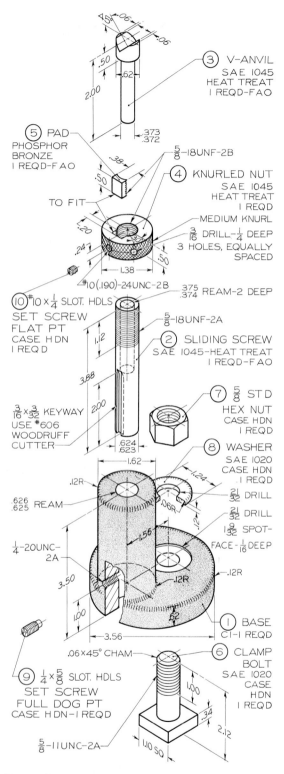

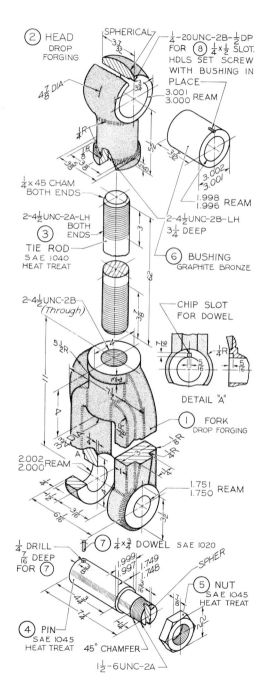

Fig. 16.76 Milling Jack. (1) Draw details. (2) Draw assembly. If assigned, convert dimensions to metric or decimal-inch system.

Fig. 16.77 Connecting Bar. (1) Draw details. (2) Draw assembly. If assigned, convert dimensions to metric or decimal-inch system.

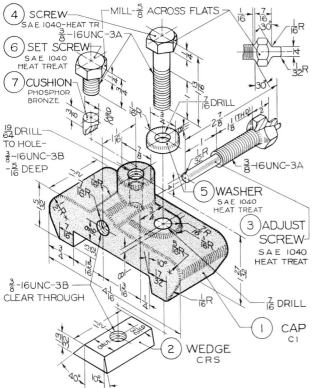

Fig. 16.78 Clamp Stop. (1) Draw details. (2) Draw assembly. If assigned, convert dimensions to decimal-inch system or redesign for metric dimensions.

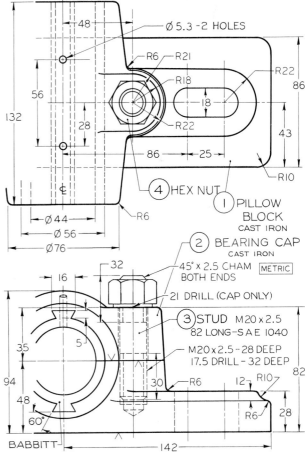

Fig. 16.79 Pillow Block Bearing. (1) Draw details. (2) Draw assembly. If assigned, complete with dimensions.

561

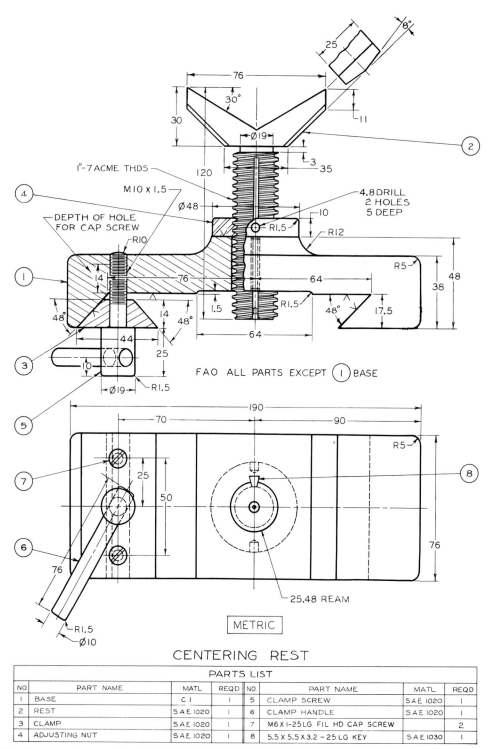

CENTERING REST

PARTS LIST							
NO.	PART NAME	MATL	REQD	NO.	PART NAME	MATL	REQD
1	BASE	C I	1	5	CLAMP SCREW	S A E 1020	1
2	REST	S A E 1020	1	6	CLAMP HANDLE	S A E 1020	1
3	CLAMP	S A E 1020	1	7	M6 X 1-25 LG FIL HD CAP SCREW		2
4	ADJUSTING NUT	S A E 1020	1	8	5.5 X 5.5 X 3.2 – 25 LG KEY	S A E 1030	1

Fig. 16.80   Centering Rest.   (1) Draw details. (2) Draw assembly. If assigned, complete with dimensions.

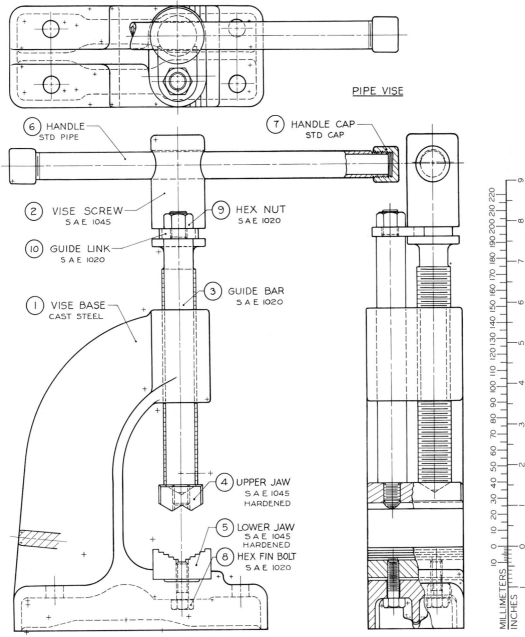

PIPE VISE

⑥ HANDLE
STD PIPE

⑦ HANDLE CAP
STD CAP

② VISE SCREW
S A E 1045

⑨ HEX NUT
S A E 1020

⑩ GUIDE LINK
S A E 1020

③ GUIDE BAR
S A E 1020

① VISE BASE
CAST STEEL

④ UPPER JAW
S A E 1045
HARDENED

⑤ LOWER JAW
S A E 1045
HARDENED

⑧ HEX FIN BOLT
S A E 1020

MILLIMETERS
INCHES

Fig. 16.81 Pipe Vise. (1) Draw details. (2) Draw assembly. To obtain dimensions, take distances directly from figure with dividers; then set dividers on printed scale and read measurements in millimeters or decimal inches as assigned. All threads are general-purpose metric threads (see Appendix 15) or Unified coarse threads except the American National Standard pipe threads on handle and handle caps.

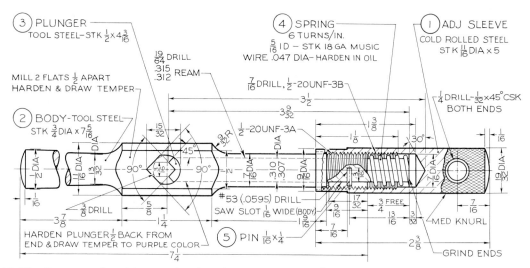

Fig. 16.82 Tap Wrench. (1) Draw details. (2) Draw assembly. If assigned, use unidirectional two-place decimals for all fractional dimensions or redesign for metric dimensions.

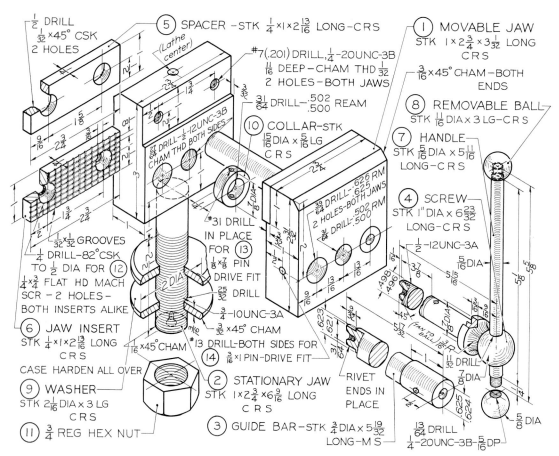

Fig. 16.83 Machinist's Vise. (1) Draw details. (2) Draw assembly. If assigned, use unidirectional two-place decimals for all fractional dimensions or redesign for metric dimensions.

564

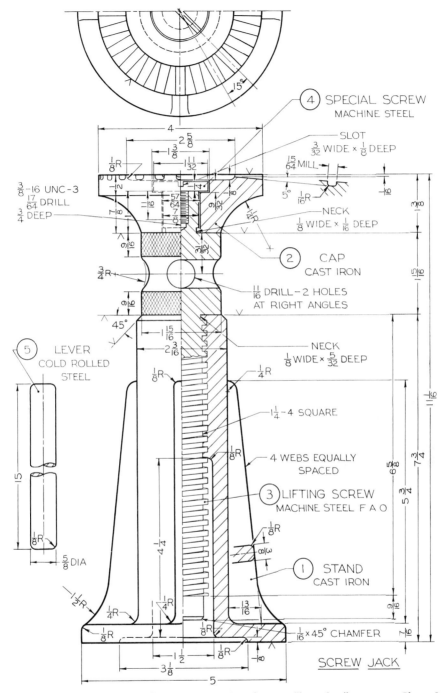

Fig. 16.84 Screw Jack. (1) Draw details. See Fig. 16.23, showing "boxed-in" views on Sheet Layout C–678 or A2–678 (see inside back cover). (2) Draw assembly. If assigned, convert dimensions to decimal inches or redesign for metric dimensions.

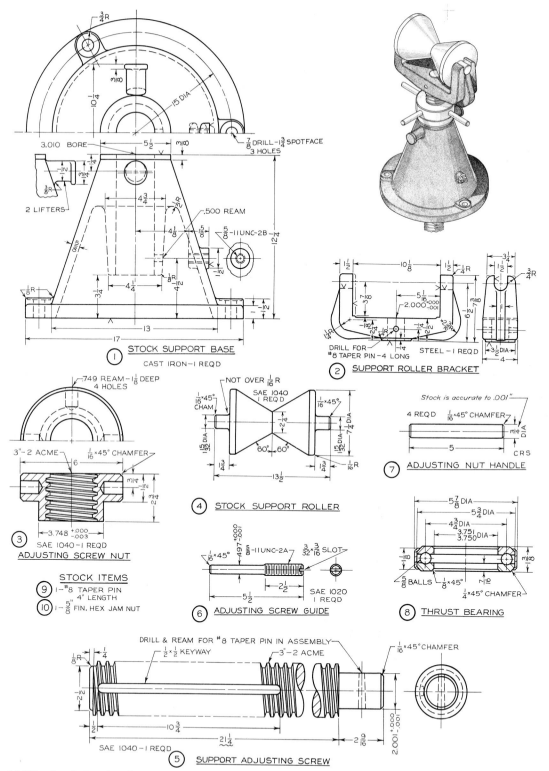

Fig. 16.85 Stock Bracket for Cold Saw Machine. (1) Draw details. (2) Draw assembly. If assigned, use unidirectional decimal dimensions or redesign for metric dimensions.

566

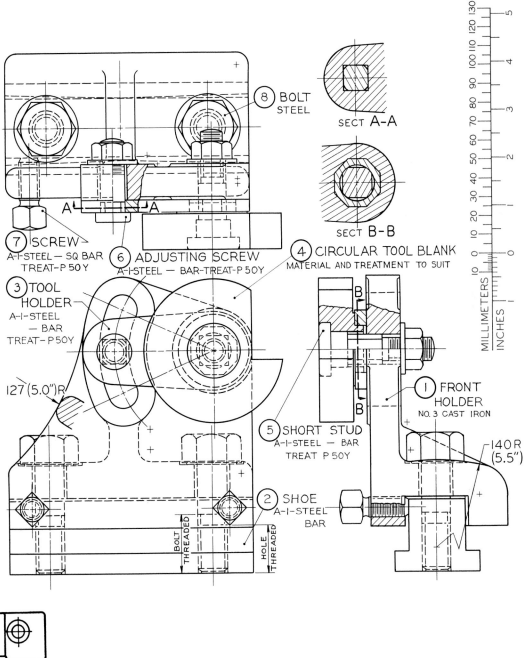

SECT A-A

SECT B-B

⑧ BOLT
STEEL

⑦ SCREW
A-I-STEEL — SQ BAR
TREAT-P 50 Y

⑥ ADJUSTING SCREW
A-I-STEEL — BAR-TREAT-P 50Y

④ CIRCULAR TOOL BLANK
MATERIAL AND TREATMENT TO SUIT

③ TOOL
HOLDER
A-I-STEEL
— BAR
TREAT- P 50Y

127 (5.0")R

⑤ SHORT STUD
A-I-STEEL — BAR
TREAT P 50Y

① FRONT
HOLDER
NO. 3 CAST IRON

140R
(5.5")

② SHOE
A-I-STEEL
BAR

BOLT THREADED

HOLE THREADED

MILLIMETERS
INCHES

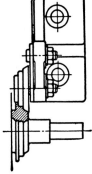

Fig. 16.86 Front Circular Forming Cutter Holder.   (1) Draw details. (2) Draw assembly.
To obtain dimensions, take distances directly from figure with dividers and set dividers
on printed scale. Use metric or decimal-inch dimensions as assigned.

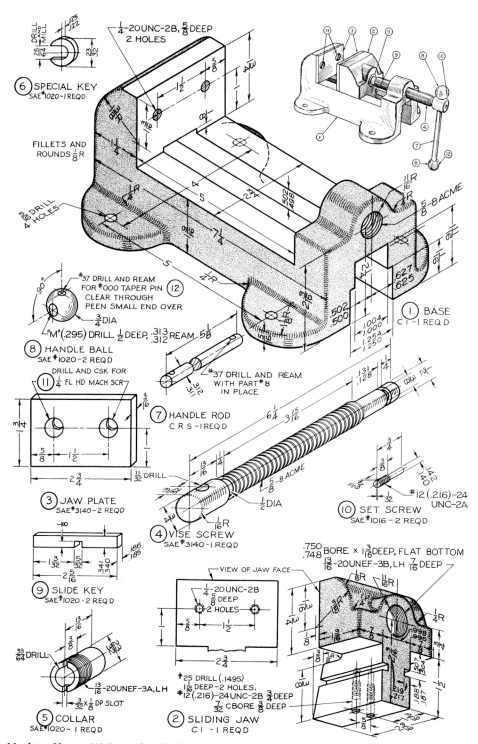

Fig. 16.87   Machine Vise. (1) Draw details. (2) Draw assembly. If assigned, convert dimensions to the decimal-inch system or redesign with metric dimensions.

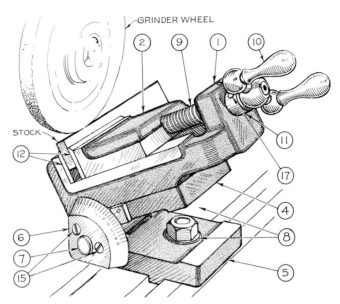

Fig. 16.88 Grinder Vise. See Figs. 16.89 and 16.90.

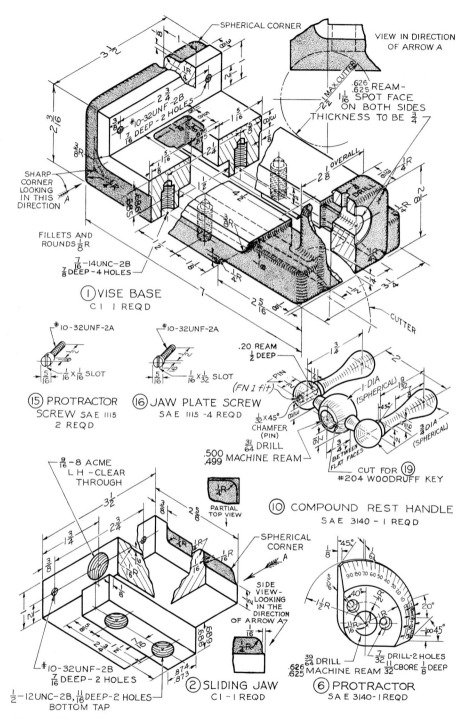

Fig. 16.89   Grinder Vise   (Continued). (1) Draw details. (2) Draw assembly. See Figs. 16.88 and 16.90. If assigned, convert dimensions to decimal inches or redesign with metric dimensions.

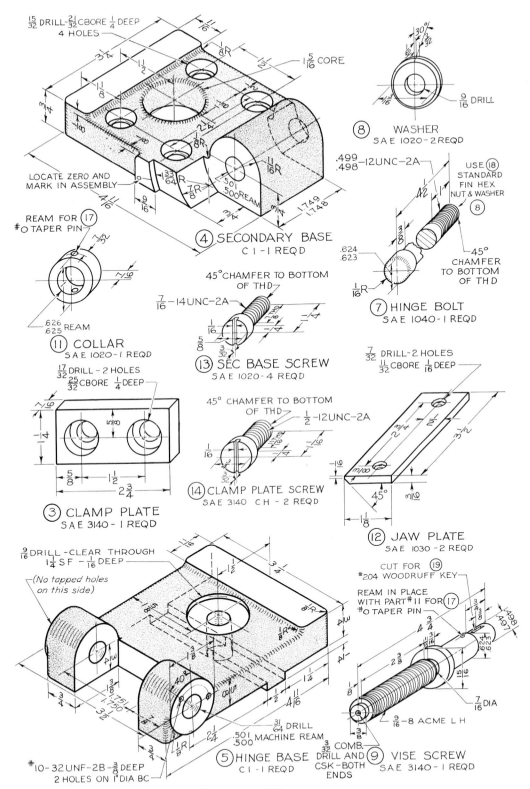

Fig. 16.90 Grinder Vise (Continued). See Fig. 16.89 for instructions.

571

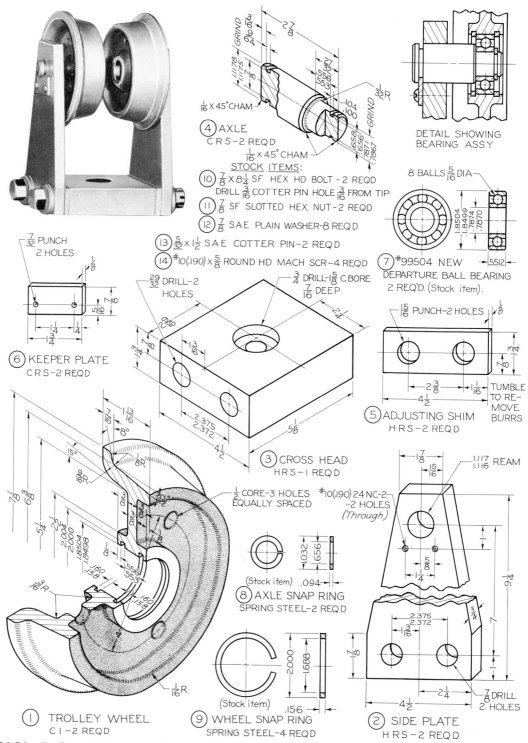

Fig. 16.91 Trolley. (1) Draw details, omitting parts 7–14. (2) Draw assembly. If assigned, convert dimensions to decimal inches or redesign for metric dimensions.

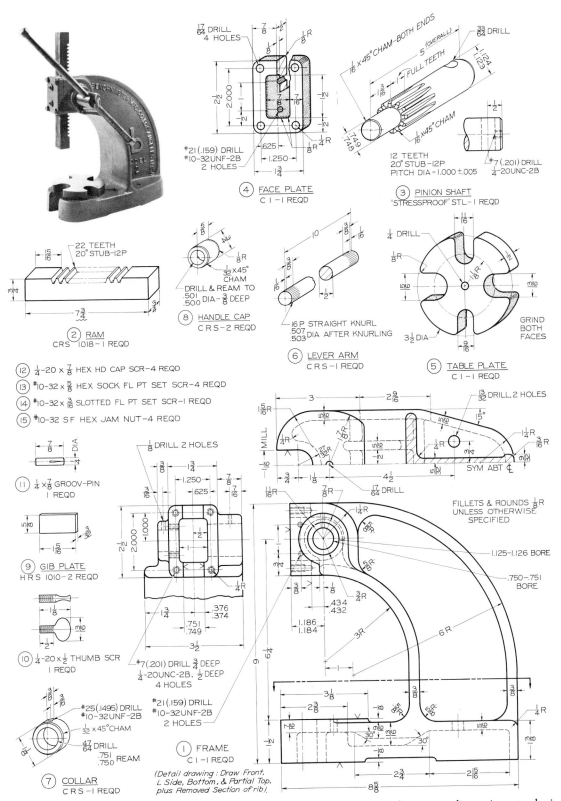

Fig. 16.92 Arbor Press. (1) Draw details. (2) Draw assembly. If assigned, convert dimensions to decimal inches or redesign for metric dimensions.

573

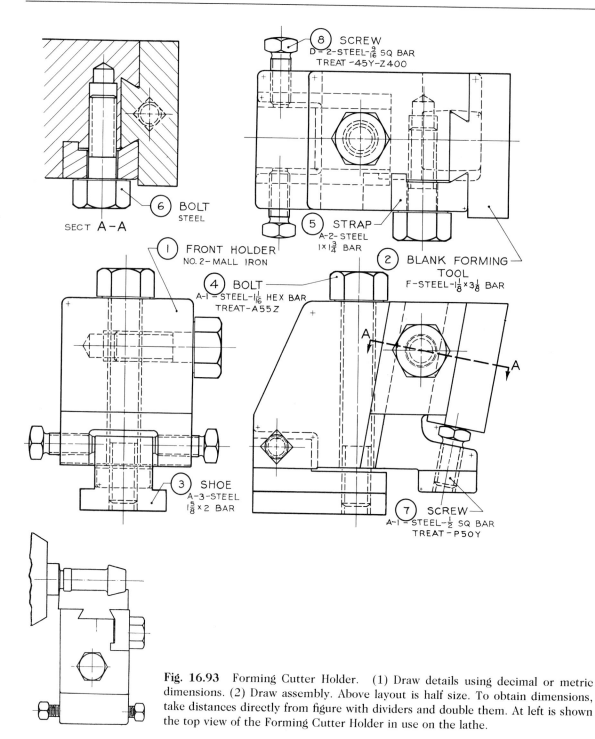

Fig. 16.93 Forming Cutter Holder. (1) Draw details using decimal or metric dimensions. (2) Draw assembly. Above layout is half size. To obtain dimensions, take distances directly from figure with dividers and double them. At left is shown the top view of the Forming Cutter Holder in use on the lathe.

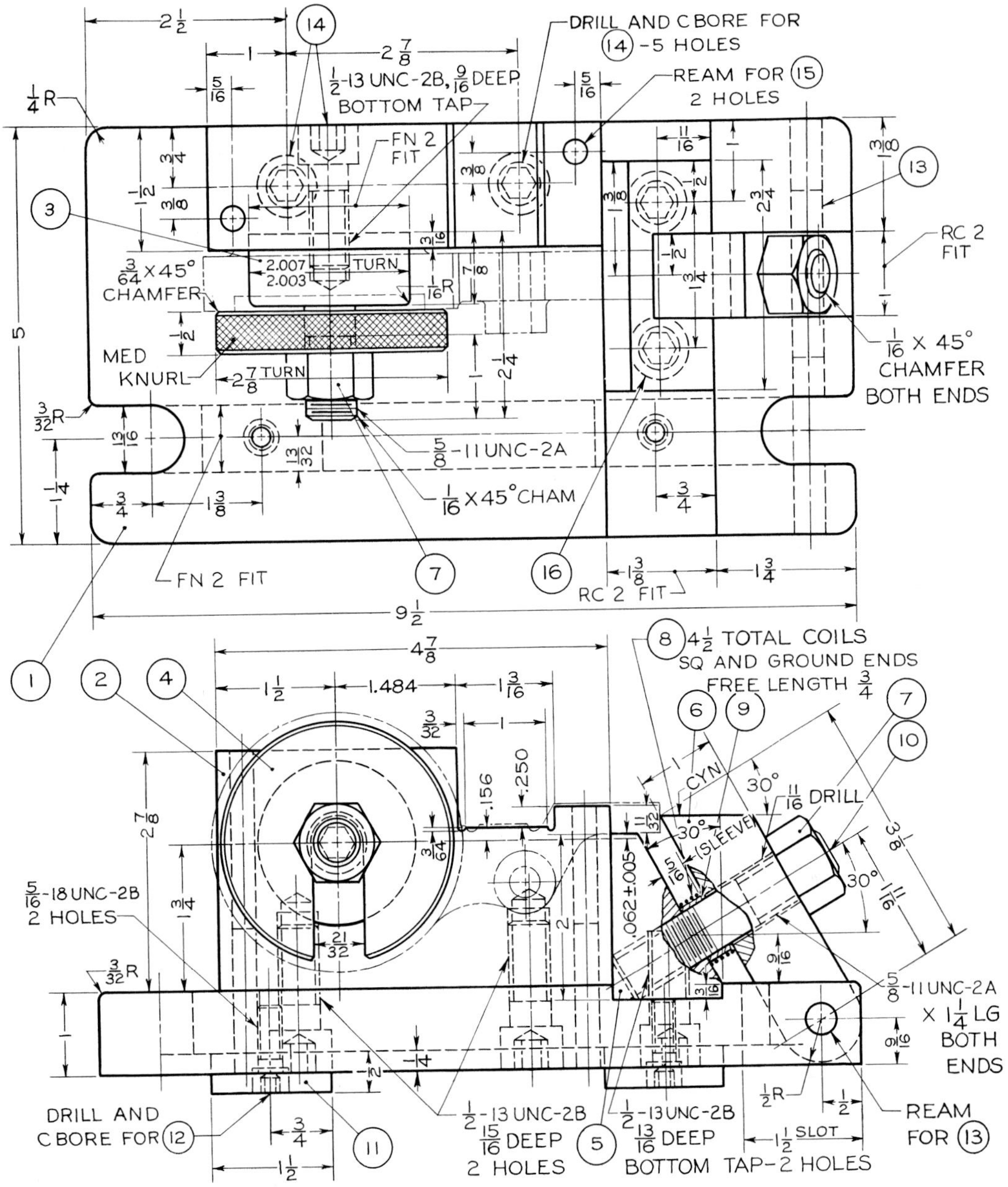

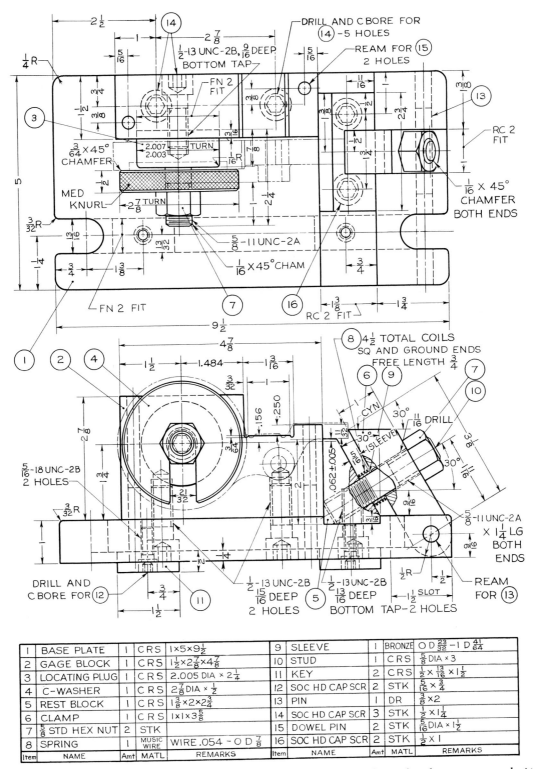

Item	NAME	Amt	MATL	REMARKS	Item	NAME	Amt	MATL	REMARKS
1	BASE PLATE	1	CRS	$1\times5\times9\frac{1}{2}$	9	SLEEVE	1	BRONZE	OD $\frac{23}{32}$ − ID $\frac{41}{64}$
2	GAGE BLOCK	1	CRS	$1\frac{1}{2}\times2\frac{7}{8}\times4\frac{7}{8}$	10	STUD	1	CRS	$\frac{5}{8}$ DIA × 3
3	LOCATING PLUG	1	CRS	2.005 DIA × $2\frac{1}{4}$	11	KEY	2	CRS	$\frac{1}{2}\times\frac{13}{16}\times1\frac{1}{2}$
4	C-WASHER	1	CRS	$2\frac{7}{8}$ DIA × $\frac{1}{2}$	12	SOC HD CAP SCR	2	STK	$\frac{5}{16}\times\frac{3}{4}$
5	REST BLOCK	1	CRS	$1\frac{3}{8}\times2\times2\frac{3}{4}$	13	PIN	1	DR	$\frac{3}{8}\times2$
6	CLAMP	1	CRS	$1\times1\times3\frac{5}{8}$	14	SOC HD CAP SCR	3	STK	$\frac{1}{2}\times1\frac{1}{4}$
7	$\frac{5}{8}$ STD HEX NUT	2	STK		15	DOWEL PIN	2	STK	$\frac{5}{16}$ DIA × $1\frac{1}{2}$
8	SPRING	1	MUSIC WIRE	WIRE .054 − OD $\frac{7}{8}$	16	SOC HD CAP SCR	2	STK	$\frac{1}{2}\times1$

Fig. 16.94 Milling Fixture for Clutch Arm. (1) Draw details using the decimal-inch system or redesign for metric dimensions, if assigned. (2) Draw assembly.

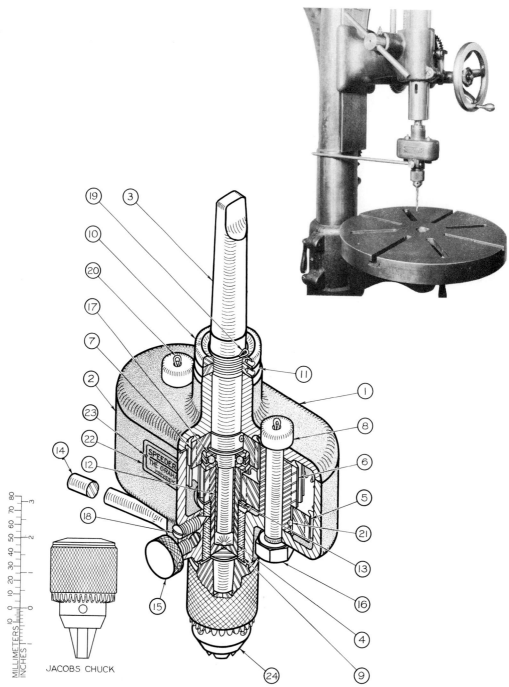

Fig. 16.95   Drill Speeder.   See Figs. 16.96 and 16.97.

JACOBS CHUCK

MILLIMETERS
INCHES

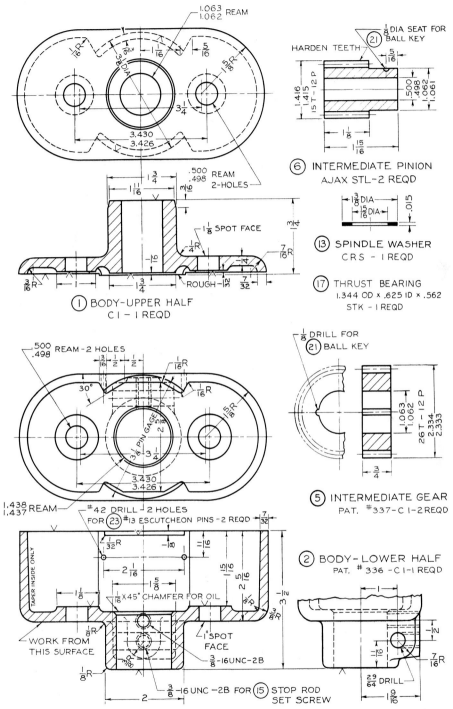

Fig. 16.96 Drill Speeder (Continued). (1) Draw details. (2) Draw assembly. See Fig. 16.95. If assigned, convert dimensions to decimal inches or redesign with metric dimensions.

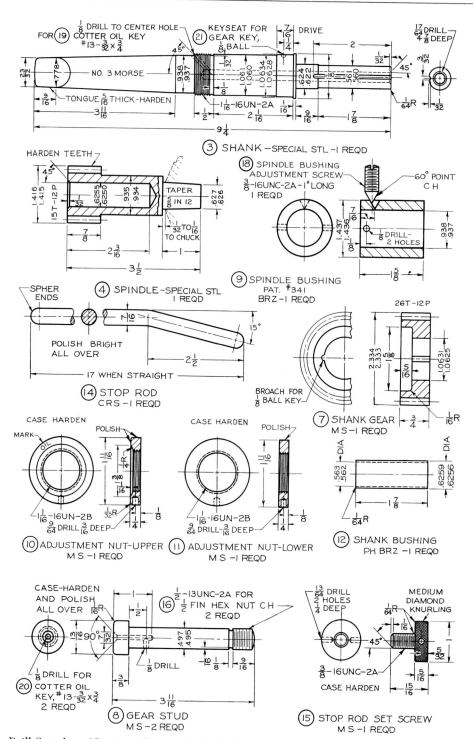

Fig. 16.97   Drill Speeder   (Continued). See Fig. 16.96 for instructions.

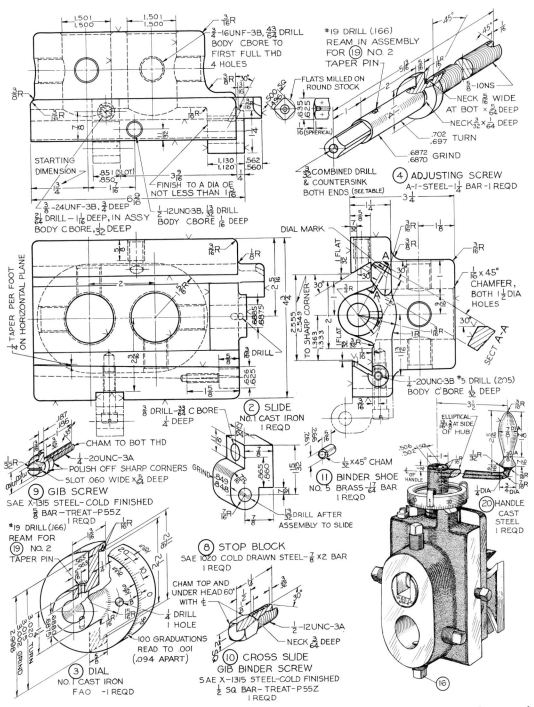

Fig. 16.98 Vertical Slide Tool. (1) Draw details. If assigned, convert dimensions to decimal inches or redesign for metric system. (2) Draw assembly. For part 2: Take given top view as front view in the new drawing; then add top and right-side views. See also Fig. 16.99. If assigned, use unidirectional dimensions.

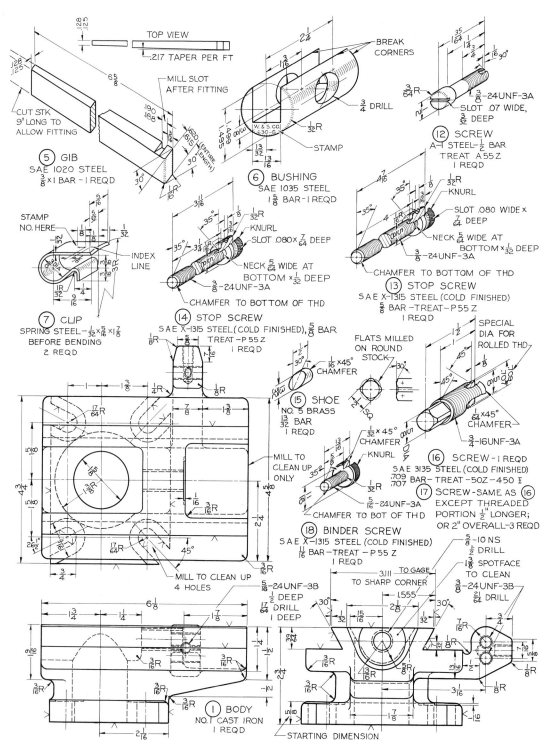

Fig. 16.99 Vertical Slide Tool (Continued). See Fig. 16.98 for instructions. For part 1: Take top view as front view in the new drawing; then add top and right-side views.

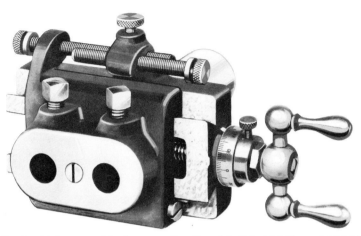

Fig. 16.100 Slide Tool. Make assembly drawing. See Figs. 16.102, 16.103, and 16.104.

PARTS LIST	NO. OF SHEETS 2	SHEET NO. 1	MACHINE NO. M-219
NAME NO. 4 SLIDE TOOL (SPECIFY SIZE OF SHANK REQ'D.)		LOT NUMBER
		NO. OF PIECES

TOTAL ON MACH.	NO. PCS.	NAME OF PART	PART NO.	CAST FROM PART NO.	TRACING NO.	MATERIAL	ROUGH WEIGHT PER PC.	DIA.	LENGTH	MILL	PART USED ON	NO.REQ. FINISH
	1	Body	219-12		D-17417	A-3-S D F						
	1	Slide	219-6		D-19255	A-3-S D F					219-12	
	1	Nut	219-9		E-19256	#10 BZ					219-6	
	1	Gib	219-1001		C-11129	S A E 1020					219-6	
	1	Slide Screw	219-1002		C-11129	A-3-S					219-12	
	1	Dial Bush.	219-1003		C-11129	A-1-S					219-1002	
	1	Dial Nut	219-1004		C-11129	A-1-S					219-1002	
	1	Handle	219-1011		E-18270	(Buy from Cincinnati Ball Crank Co.)					219-1002	
	1	Stop Screw (Short)	219-1012		E-51950	A-1-S					219-6	
	1	Stop Screw (Long)	219-1013		E-51951	A-1-S					219-6	
	1	Binder Shoe	219-1015		E-51952	#5 Brass					219-6	
	1	Handle Screw	219-1016		E-62322	X-1315 C.F.					219-1011	
	1	Binder Screw	219-1017		E-63927	A-1-S					219-6	
	1	Dial	219-1018		E-39461	A-1-S					219-1002	
	2	Gib Screw	219-1019		E-52777	A-1-S		$\frac{1}{4}$-20	1		219-6	
	1	Binder Screw	280-1010		E-24962	A-1-S					219-1018	
	2	Tool Clamp Screws	683-F-1002		E-19110	D-2-S					219-6	
	1	Fill Hd Cap Scr	1-A			A-1-S		$\frac{3}{8}$	1 $\frac{3}{8}$		219-6 219-9	
	1	Key	No.404 Woodruff								219-1002	

Fig. 16.101 Slide Tool Parts List.

581

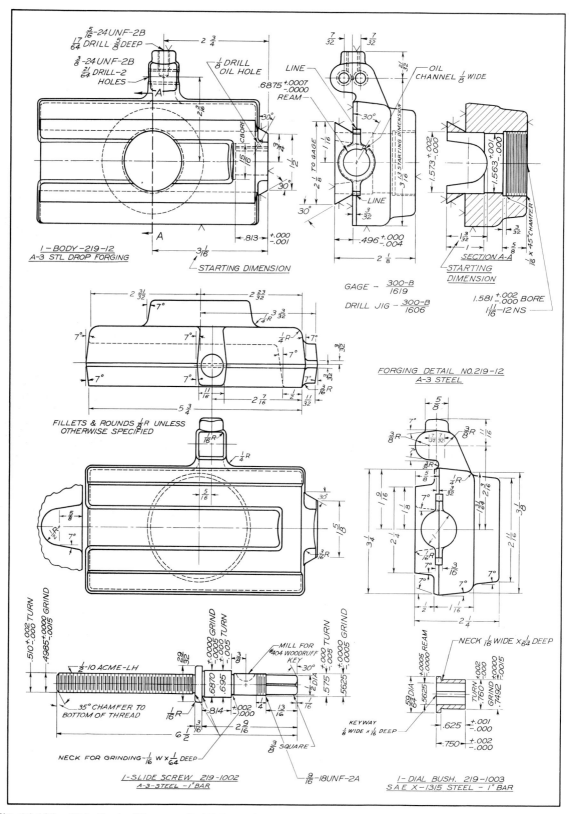

Fig. 16.102 Slide Tool   (Continued). (1) Draw details using decimal-inch dimensions or redesign with metric dimensions, if assigned. (2) Draw assembly. See Fig. 16.100.

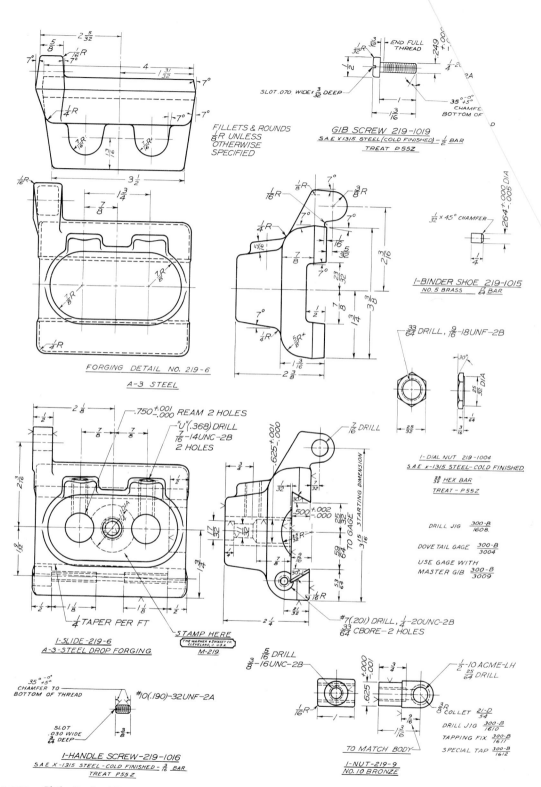

Fig. 16.103   Slide Tool   (Continued). See Fig. 16.102 for instructions.

583

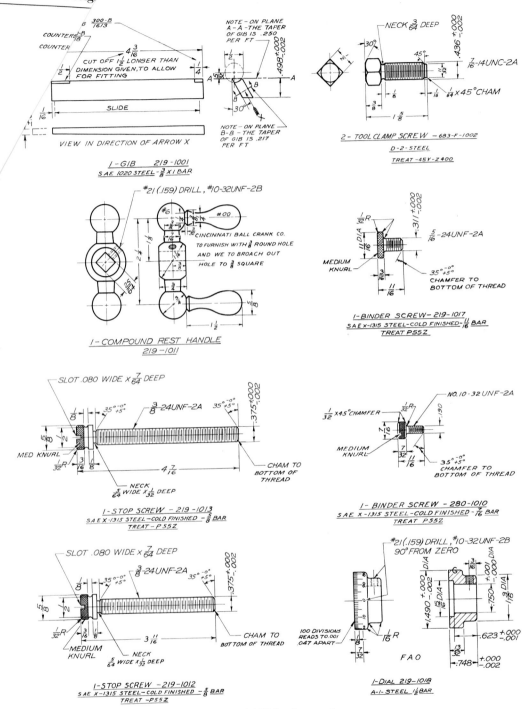

Fig. 16.104   Slide Tool   (Continued). See Fig. 16.102 for instructions.

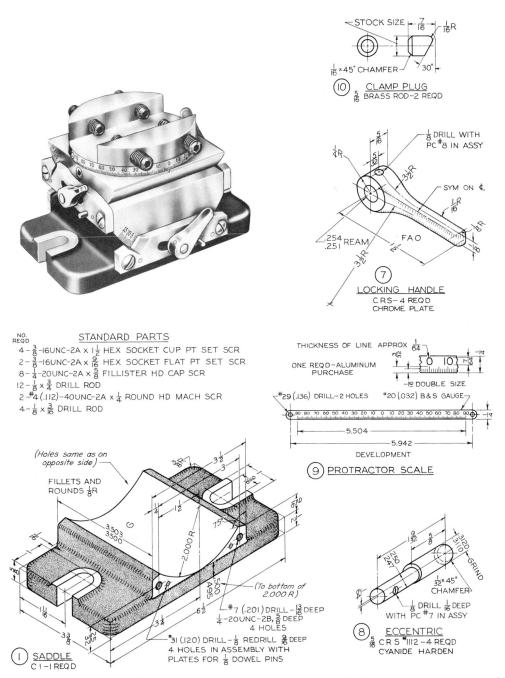

STOCK SIZE

$\frac{1}{16} \times 45°$ CHAMFER   30°

(10) $\frac{5}{16}$ CLAMP PLUG
BRASS ROD–2 REQD

$\frac{1}{8}$ DRILL WITH
PC #8 IN ASSY

SYM ON ₵

.254 REAM
.251   FAO

(7) LOCKING HANDLE
C R S– 4 REQD
CHROME PLATE

NO.
REQD   STANDARD PARTS
4 – $\frac{3}{8}$–16UNC–2A x $1\frac{1}{2}$ HEX SOCKET CUP PT SET SCR
2 – $\frac{3}{8}$–16UNC–2A x $\frac{9}{16}$ HEX SOCKET FLAT PT SET SCR
8 – $\frac{1}{4}$–20UNC–2A x $\frac{5}{8}$ FILLISTER HD CAP SCR
12 – $\frac{1}{8}$ x $\frac{3}{4}$ DRILL ROD
2 – #4 (.112)–40UNC–2A x $\frac{1}{4}$ ROUND HD MACH SCR
4 – $\frac{1}{8}$ x $\frac{3}{16}$ DRILL ROD

THICKNESS OF LINE APPROX $\frac{1}{64}$

ONE REQD–ALUMINUM
PURCHASE

DOUBLE SIZE

#29 (.136) DRILL–2 HOLES   #20 (.032) B & S GAUGE

5.504

5.942
DEVELOPMENT

(9) PROTRACTOR SCALE

(Holes same as on
opposite side)

FILLETS AND
ROUNDS $\frac{1}{8}$R

3.503
3.500

2.000 R

75°

(To bottom of
2.000 R)

#7 (.201) DRILL – $\frac{13}{16}$ DEEP
$\frac{1}{4}$–20UNC–2B, $\frac{5}{8}$ DEEP
4 HOLES

#31 (.120) DRILL – $\frac{1}{8}$ REDRILL $\frac{9}{8}$ DEEP
4 HOLES IN ASSEMBLY WITH
PLATES FOR $\frac{1}{8}$ DOWEL PINS

(1) SADDLE
C I – I REQD

$\frac{1}{32} \times 45°$
CHAMFER

$\frac{1}{8}$ DRILL $\frac{1}{8}$ DEEP
WITH PC #7 IN ASSY

(8) ECCENTRIC
$\frac{5}{16}$ C R S #1112–4 REQD
CYANIDE HARDEN

Fig. 16.105 "Any-Angle" Tool Vise. (1) Draw details using decimal-inch dimensions or redesign with metric dimensions, if assigned. (2) Draw assembly. See also Fig. 16.106.

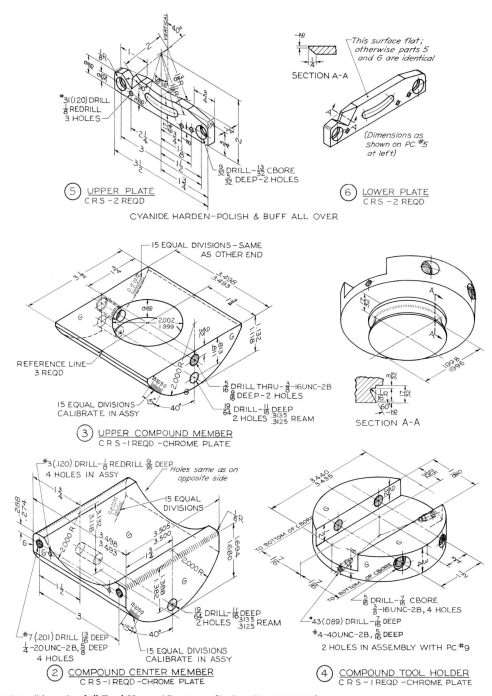

Fig. 16.106   "Any-Angle" Tool Vise   (Continued). See Fig. 16.105 for instructions.

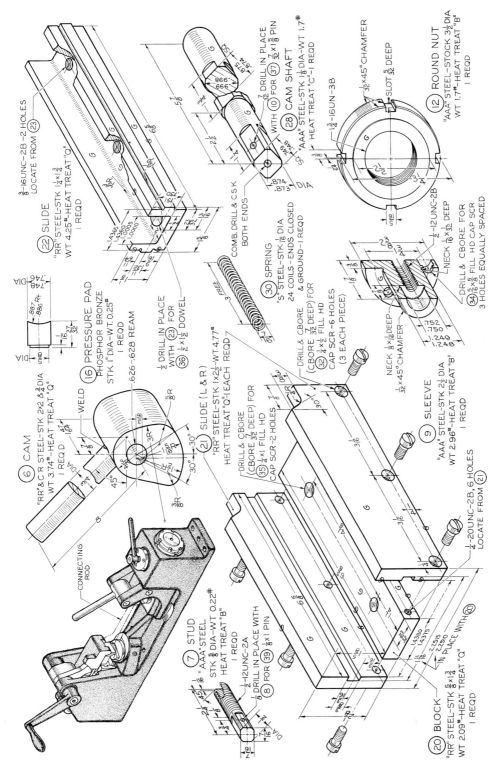

Fig. 16.107 Fixture for Centering Connecting Rod. (1) Draw details using decimal-inch dimensions or redesign with metric dimensions, if assigned. (2) Draw assembly. See also Figs. 16.108 and 16.109.

587

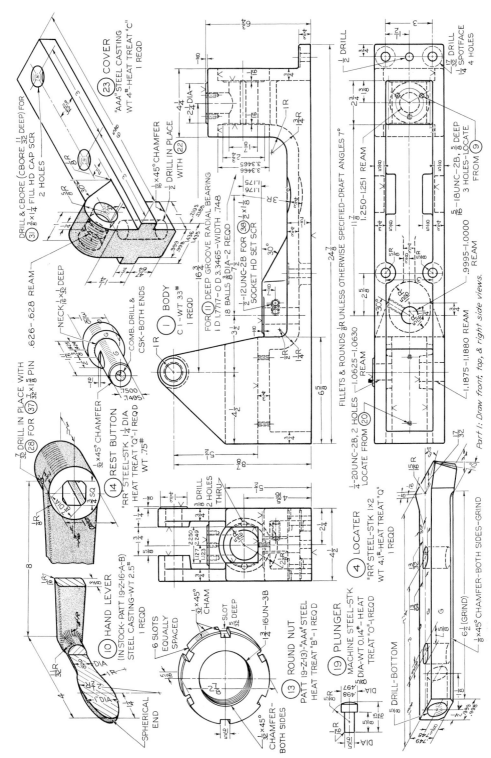

Part I: Draw front, top, & right side views. See Fig. 16.107 for instructions.

Fig. 16.108   Fixture for Centering Connecting Rod   (Continued). See Fig. 16.107 for instructions.

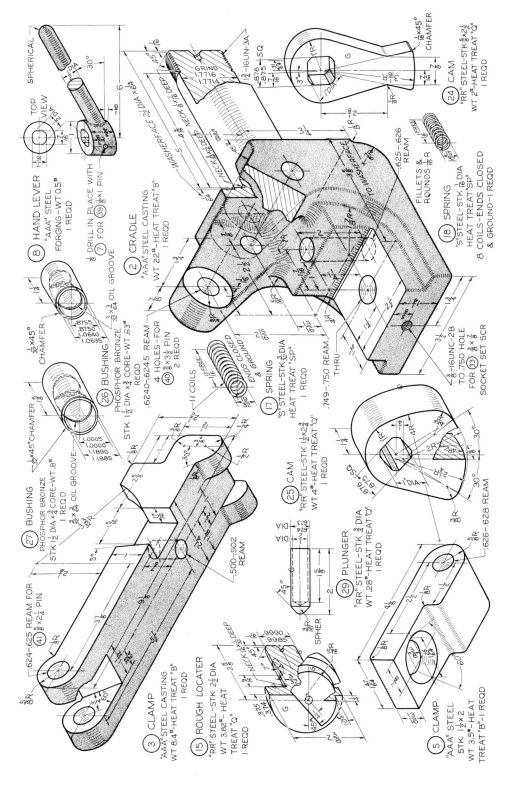

Fig. 16.109 Fixture for Centering Connecting Rod (Continued). See Fig. 16.107 for instructions.

589

CHAPTER 17

Reproduction and Control of Drawings

An essential part of designers' or drafters' education is a thorough knowledge of reproduction techniques and processes. Specifically, they should be familiar with the various processes available for the reproduction of drawings: *blueprint, diazo, microfilm, microfiche,* etc.

Because the average engineering drawing requires a considerable economic investment, adequate control and protection of the original drawing are mandatory. Such items as drawing numbers, §16.13, methods of filing, microfilming, security files, print making and distribution, drawing changes, and retrieval of drawings are all important. A proper drawing-control system will enable those in charge of drawings (1) to know the location and status of the drawing at all times; (2) to minimize the damage to original drawings from the handling required for revisions, printing, and so on; and (3) to provide distribution of prints to proper persons.

Those organizations with large computer-aided design (CAD) systems (see Chapters 3 and 8) use computer storage of finished drawings. In addition, digitized drawing information about frequently required components and elements, such as standard bolts, nuts, screws, pins, and piping valves, is stored in the computer for recall and placement on drawings as needed. CAD systems of this type are very expensive to purchase and maintain, but they have obvious advantages for production, storage, control, and recall of drawings. Most small and medium-size companies and a substantial number of large firms may not be able to justify the acquisition of high-capacity CAD systems and thus will continue to use the conventional methods of reproduction, storage, retrieval, and control of drawings described in this chapter. But CAD systems are becoming more compact and less expensive, and as advancing technology makes such systems more affordable, it is expected that many companies will make the change to computer graphics systems.

17.1 Storage of Drawings

Drawings may be stored flat in large flat-drawer files or hung vertically in cabinets especially designed for the purpose. Exceptionally large drawings are often rolled and stored in tubes in racks or cabinets. Prints are often folded and stored in standard office file cases. Proper control procedures will enable the user of the drawing to find it in the file, to return it to its proper place, and

591

to know where the drawing is when not in the file.

17.2   Reproduction of Drawings

After the drawings of a machine or structure have been completed, it is usually necessary to supply copies to many different persons and firms. Obviously, therefore, some means of exact, rapid, and economical reproduction must be used.

17.3   Blueprint Process

Of the several processes in use for reproduction, the *blueprint* process is the oldest and is still occasionally used for making prints of large architectural and structural drawings. It is essentially a photographic process in which the original drawing is the negative.

Blueprint papers are made by applying a coating of a solution of potassium ferricyanide and ferric ammonium citrate, a chemical preparation sensitive to light. They are available in various speeds and in rolls of various widths, or may be supplied in sheets of a specified size. The coated side of fresh paper is a light greenish yellow color. It will gradually turn to a greyish blue color, if not kept away from light, and may eventually be rendered useless.

After the paper has been exposed a sufficient length of time, it is subjected to a developing bath or a fixing bath, or to a fixing bath only, according to the method employed. Blueprint machines are available in noncontinuous types in which cut sheets are fed through the machine for exposure only and then washed in a separate washer. The continuous blueprint machine, Fig. 17.1, combines exposure, washing, and drying in one continuous operation.

Although best results are obtained when the original tracing is drawn in ink on cloth, vellum, or film, excellent prints may be made from penciled drawings or tracings if the tracing paper, cloth, or film is of good quality and if the drafter has made all required lines and lettering jet black.

Notations and corrections can be made on blueprints with any alkaline solution of sufficient strength to destroy the blue compound, for in-

Fig. 17.1   Automatic Blueprinting Machine. *Courtesy of Bruning Division of AM International.*

stance, with a 1.5 percent solution of caustic soda.

The blueprint process, which was the common method used for the reproduction of drawings for many years, has now been replaced to a large extent by other more convenient and efficient processes.

17.4   Vandyke Prints and Blue-Line Blueprints

A *negative Vandyke print* is composed of white lines on a dark brown background made by printing, in the same manner as for blueprinting, on a special thin Vandyke paper from an original pencil or ink tracing. This negative Vandyke is then used as an original to make positive blueprints or "blue-line blueprints." In this way the Vandyke print replaces the original tracing as the negative, and the original is not subjected to wear each time a run of prints is made. The blue-line blueprints have blue lines on white backgrounds and are often preferred because they can be easily marked on with an ordinary pencil or pen. They have the disadvantage of soiling easily in the shop.

17.5   Diazo-Moist Prints

A black-and-white print, composed of nearly black lines on a white background, may be made from ordinary pencil or ink tracings by exposure in the same manner as for blueprints, directly on special blackprint paper, cloth, or film.

Fig. 17.2   Blu-Ray 454 Whiteprinter. *Courtesy of Blu-Ray, Inc.*

Fig. 17.3   Ozalid Whiteprinter. *Courtesy of Ozalid Corporation.*

Exposure may be made in a blueprint machine or any machine using light in a similar way. However, the prints are not washed as in blueprinting, but must be fed through a special developer that dampens the coated side of the paper with a developing solution.

Colored-line prints in red, brown, or blue lines on white backgrounds may be made on the same machine simply by using the appropriate paper in each case.

These prints, together with diazo-dry prints, §17.6, have largely replaced the more cumbersome blueprint process.

17.6 The Diazo-Dry Process

The *diazo-dry* process is based on the sensitivity to light of certain dyestuff intermediates that have the characteristic of decomposing into colorless substances if exposed to ultraviolet light, and of reacting with coupling components to form an azo dyestuff upon exposure to ammonia vapors. It is a contact method of reproduction, and depends on the transmission of light through the original for the reproduction of positive prints. The subject matter may be pen or pencil lines, typewritten or printed matter, or any opaque image. There is no negative step involved; positives are used to obtain positive

prints. Sensitized materials can be handled under normal indoor illumination.

The diazo whiteprint method of reproduction consists of two simple steps—exposure and dry development by means of ammonia vapors. Exposure is made in a printer equipped with a source of ultraviolet light, a mercury vapor lamp, fluorescent lamp, or carbon arc. The light emitted by these light sources brings about a photochemical decomposition of the light-sensitive yellow coating of the paper except in those places where the surface is protected by the opaque lines of the original. The exposed print is developed dry in a few seconds in a dry-developing machine by the alkaline medium produced by ammonia vapors.

A popular combination printer (exposer) and developer, the Blue-Ray 454 Whiteprinter, is shown in Fig. 17.2. The tracing and the sensitized paper are fed into the machine, and when they emerge, the print is practically dry and ready for use.

Another exposer and developer combined in one machine is the Ozalid Whiteprinter, Fig. 17.3. Two operations are involved: (1) the tracing and the sensitized paper are fed into the printer slot for exposure to light, as shown in Fig. 17.4 (a), and (2) the paper is then fed through

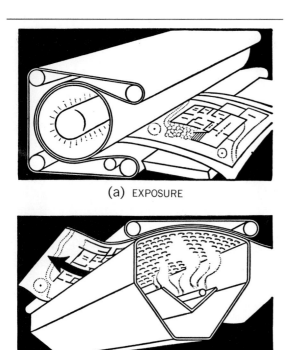

(a) EXPOSURE

(b) AMMONIA DEVELOPMENT

Fig. 17.4 Exposure and Development. *Courtesy of Ozalid Corporation.*

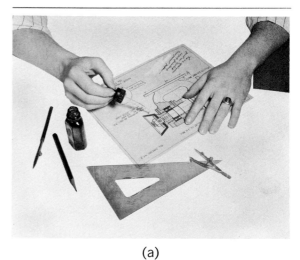

(a)

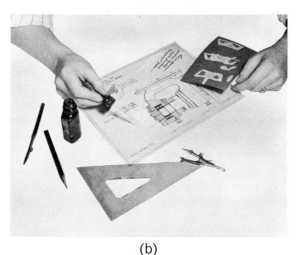

(b)

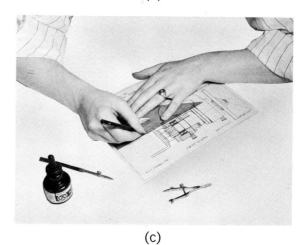

(c)

the developer slot for exposure to ammonia vapors, as shown at (b). If it is desired to remove the ammonia odor completely, the print is then fed through the printer with the back of the sheet next to the warm glass surrounding the light.

Ozalid prints may have black, blue, brown, or red lines on white backgrounds, according to which paper is used. All have the advantage of being easily marked on with pencil, pen, or crayon.

Ozalid "intermediates" are made in the same manner as regular prints, but on special translucent paper, cloth, or foil (transparent cellulose acetates). These are used in place of the original to produce regular prints. They may be used to save wear on the original or to permit changes to be made. Changes may be made by painting out parts with correction solution, Fig. 17.5 (a)

Fig. 17.5 Changing a Drawing. *Courtesy of Ozalid Corporation.*

Fig. **17.6**   Bruning Diazo-Dry Whiteprinter. *Courtesy of Bruning Division of AM International.*

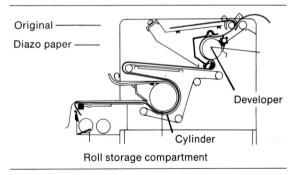

Fig. **17.7**   Diagram of Bruning Diazo-Dry Process. *Courtesy of Bruning Division of AM International.*

and (b), then drawing the new lines or lettering directly on the intermediate in pencil or ink, (c). Special masking and cut-out techniques can also be used.

Several types of Ozalid foil intermediates are available, including matte surfaces to facilitate pen and pencil additions. By means of foils, many new procedures are possible, such as a

composite print in which a wiring system is superimposed over a drawing.

The Bruning Whiteprinter, Fig. 17.6, is another medium-volume machine also equipped with fluorescent exposure lamps rather than a high-intensity mercury vapor light source. The use of fluorescent lamps allow the operator to switch on for copying and switch off when done. The machine is also capable of automatic separation. A diagram illustrating the Bruning diazo process is shown in Fig. 17.7.

The Ozalid Series 2000, Fig. 17.8, and the Printfold 825, Fig. 17.9, are both fully automatic

Fig. **17.8**   Ozalid Series 2000 High-Volume Automatic Diazo Printer. *Courtesy of Ozalid Corporation.*

Fig. 17.9 Printfold 825 Automated Whiteprint Machine. *Courtesy of Printfold Company.*

whiteprint machines designed for those operations that require high-volume productivity.

17.7 Thermo-Fax Process

The Thermo-Fax copy system is an infrared heat process in which the print paper is sensitive to heat instead of light. Copies are dry and have black lines on a light background. The original drawing need not be on transparent paper. The Thermo-Fax machine is easy to operate and produces a copy in a few seconds, but the process is rather expensive if several copies of a drawing are required. Copies are semipermanent and may fade unless stored in a cool, dark area.

17.8 Verifax Process

The Verifax process is a dye-transfer process. The original is exposed in the Verifax machine to produce a light-activated matrix, which is then placed in contact with the copy paper. The

two are then separated, with the print transferred to the copy paper. Additional copies may be had from the same matrix. The Verifax process is sensitive to all types of copy, but it is a wet process and relatively slow.

17.9 Xerography

Xerox prints are positive prints with black lines on a white background. A selenium-coated and electrostatically charged plate is used. A special camera is used to project the original onto the plate; hence, reduced or enlarged reproductions are possible. A negatively charged plastic powder is spread across the plate and adheres to the positively charged areas of the image. The powder is then transferred to paper by means of a positive electric charge and is baked onto the surface to produce the final print. Full-size prints or reductions can be made inexpensively and quickly in the fully automated Xerox or other similar copy machines. The process is dry and sensitive to all types of copy. The Xerox process is used also to produce mats for the Multilith offset duplicating method, §17.11.

Recent application of xerography includes volume print making from original drawings or from microfilms, §17.16. In addition, the portable Xerox Telecopier, Fig. 17.10, can receive or send $8\frac{1}{2}'' \times 11''$ documents over standard telephone lines in the office or in the field. After the telephone circuit is established, the document is fed into the sending machine. The copy is read and translated into signals for the receiving machine, which reproduces the document.

17.10 Digital Image Processing

Modern digital techniques have made possible the direct production of drawings on a laser printer from a variety of input sources, including computers and electronic video equipment.

In addition to drawings produced with the assistance of a CAD program, Chapters 3 and 8, conventional hand-produced drawings may be

Fig. 17.10   Xerox Portable Telecopier. *Courtesy of Xerox Corporation.*

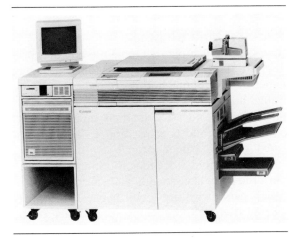

Fig. 17.11   Canon Color Laser Copier 500. *Courtesy of Canon USA, Inc.*

stored in computer memory through the use of digital scanning techniques or by manually digitizing the drawing.

Color laser copiers can reproduce drawings in four colors, with black lettering. The Canon Color Laser Copier 500, Fig. 17.11, is available with optional built-in computer processing unit and monitor, video player, and film projector to permit convenient viewing and editing of drawings.

17.11   Offset Printing

The Photolith, Multilith, and Planograph methods are generally known as *offset printing*. A camera is used to reproduce the original, enlarged or reduced if necessary, upon an aluminum or zinc sheet. This master plate is then mounted on a rotary drum that revolves in contact with a rubber roller that picks up the ink from the image and transfers it to the paper. The prints are excellent positive reproductions.

17.12   Photographic Contact Prints

Either transparent or opaque drawings may be reproduced the same size as the original by means of *contact printing*. The original is pressed tightly against a sheet of special photographic paper, either by mechanical spring pressure or by means of suction as in the "vacuum printer." The paper is exposed by the action of the transmitted or reflected light, and the print is developed in the manner of a photograph in a dark or semidark room.

Excellent duplicate tracings can be made on paper, on film, and on either opaque map cloth or transparent tracing cloth through this process. Poor pencil drawings or tracings can be duplicated and improved by intensifying the lines so as to be much better than the original. Pencil drawings can be transformed into "ink-like" tracings. Also, by this process, a reproduction can be made directly from an existing print.

The Eastman Kodak Company has developed a Kodagraph Autopositive paper by means of

Fig. 17.12   Visual Graphics Total Camera II. *Courtesy of Visual Graphics Corporation.*

which an excellent positive print can be made directly from either a transparent or an opaque original on an ordinary blueprint machine, ammonia developing machine, Copyflex machine, or any similar machine, without use of a darkroom or costly photographic equipment.

17.13   Photostats

The Visual Graphics Total Camera II, Fig. 17.12, is essentially a highly specialized camera. It is a completely modular daylight graphic reproduction system that includes a precision black-and-white stat camera with self-contained lighting, automatic focusing and processing, and programmable memory and microprocessor control. Various plug-in modular components are also available for special-purpose photo reproductions, such as enlargements or reductions, line or halftone stats, offset plates, reverses,

dropouts, color prints and film transparencies, and other special effects for a multitude of applications.

A photostat print may be the same size as or larger or smaller than the original, while photographic contact prints, §17.12, must be the same size. The original may be transparent or opaque. It is simply fastened in place, the camera is adjusted to obtain the desired type and size of print, and the print is made, developed, and dried in the machine (no darkroom is required). For black-and-white stats, the result is a negative print with white lines on a near-black background. A positive print having near-black lines on a white background is made by photostating the negative print.

17.14   Duplicate Tracings on Cloth

Specially prepared tracing cloth or film is available upon which a drawing may be reproduced from a negative Vandyke print of the original. Exposure of the duplicate tracing cloth may be made in a regular whiteprint machine or in any machine employing light in the same way.

Excellent duplicate tracings can also be produced with special materials on any of the several types of photographic print machines, as described in §17.12.

17.15   Line Etching

Line etching is a photographic method of reproduction. The drawing, in black lines on white paper or on tracing cloth or film, is placed in a frame behind a glass and photographed. This photographic negative is then mounted on a pane of glass and is printed on a sheet of planished zinc or copper. After the print has been specially treated to render the lines acid resistant, the plate is washed in a nitric acid solution, which eats away the metal between the lines, leaving them standing above the surface of the plate like type. The plate is then mounted upon a hardwood base, which can be used in any printing press as are blocks of type.

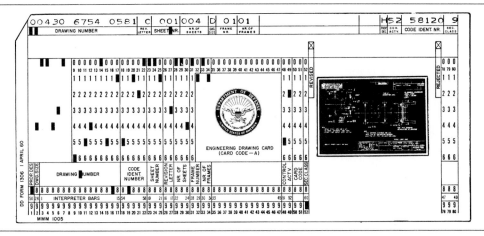

Fig. 17.13   A Standard Aperture Card. *Courtesy of C. S. Barber.*

17.16 Microfilm and Microfiche

A microfilm is a photographic image of information, records, or drawings that is stored on film at a greatly reduced scale.

Microfilm has come into considerable use, especially where large numbers of drawings are involved, to furnish complete duplication of all drawings in a small filing space. Microfilm also provides duplicate copies for security reasons. Enlargement copies can be made from the film through photographic, electrostatic, or photocopy methods. The films may be 16, 35, or 70 mm and may be produced in rolls, jacketed film, or individually mounted on aperture cards, Fig. 17.13.

The individual cards may be viewed in a reader and, if desired, a full-size print may be had from the reader-printer unit in Fig. 17.14.

Since the aperture card is essentially a standard data card used for electronic computers, mechanized equipment is capable of sorting, filing, and retrieving the cards. Duplicate aperture cards and reproductions of the drawings to the size desired may be produced on specialized equipment. A typical mechanized engineering data handling system is shown in Fig. 17.15.

Fig. 17.14   Microfilm Reader-Printer. *Courtesy of 3M Co.*

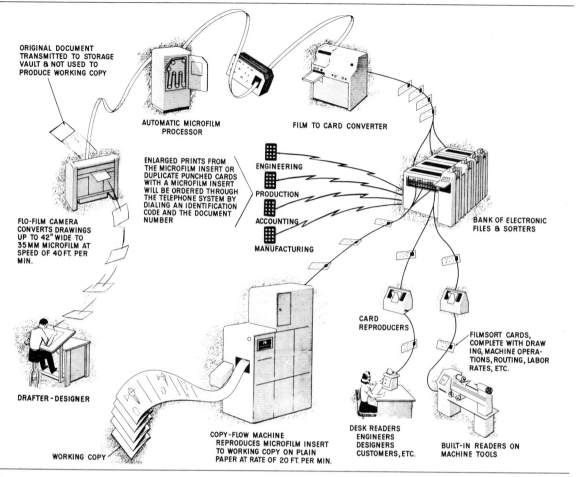

Fig. 17.15 A Mechanized Engineering Data Handling System. *Courtesy of United Technology Center, Sunnyvale, CA.*

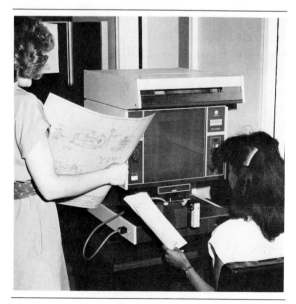

A *microfiche* is a cardlike film containing many rows of images of records or drawings. Card sizes used for storage are $3'' \times 5''$, $4'' \times 6''$, and $5'' \times 8''$. A typical $4'' \times 6''$ microfiche will contain the equivalent of 270 pages of information. The individual cards may be viewed on a reader and, if desired, a full-size copy may be made by using a reader-printer, Fig. 17.16.

Computer-output microfilm (COM) refers to a process used to produce drawings and records on microfilm, with the aid of a computer. A COM unit will produce a microfilm from database information converted to an image on a high-resolution screen that is then photographed. The main advantages of COM are storage capability and speed.

Fig. 17.16 Microfilm-Microfiche Reader-Printer.

C H A P T E R   1 8

Axonometric Projection

As described in Chapter 7, multiview drawing makes it possible to represent accurately the most complex forms of a design by showing a series of exterior views and sections. This type of representation has two limitations, however: its execution requires a thorough understanding of the principles of multiview projection, and its reading requires a definite exercise of the constructive imagination.

Frequently, it is necessary to prepare drawings for the presentation of a design idea that are accurate and scientifically correct and can be easily understood by persons without technical training. Such drawings show several faces of an object at once, approximately as they appear to the observer. This type of drawing is called a *pictorial drawing* [ANSI Y14.4–1957 (R1987)]. Since pictorial drawing shows only the appearances of parts or devices, it is not satisfactory for completely describing complex or detailed forms.

Pictorial drawing enables the person without technical training to visualize the design represented. It also enables the designer to visualize the successive stages of the design and to develop it in a satisfactory manner.

Various types of pictorial drawing are used extensively in catalogs, in general sales literature, and also in technical work,* to supplement and amplify multiview drawings. For example, pictorial drawing is used in Patent Office drawings, in piping diagrams, in machine, structural, and architectural designs, and in furniture design.

18.1   Methods of Projection

The four principal types of projection are illustrated in Fig. 18.1, and all except the regular multiview projection, (a), are pictorial types since they show several sides of the object in a single view. In all cases the views, or projections, are formed by the piercing points in the plane of projection of an infinite number of visual rays or projectors.

In both multiview projection, (a), and *axonometric projection,* (b), the observer is considered to be at infinity, and the visual rays are parallel to each other and perpendicular to the

---

*Practically all the pictorial drawings in this book were drawn by the methods described in Chapters 18–20. See especially Figs. 7.55–7.114 for examples.

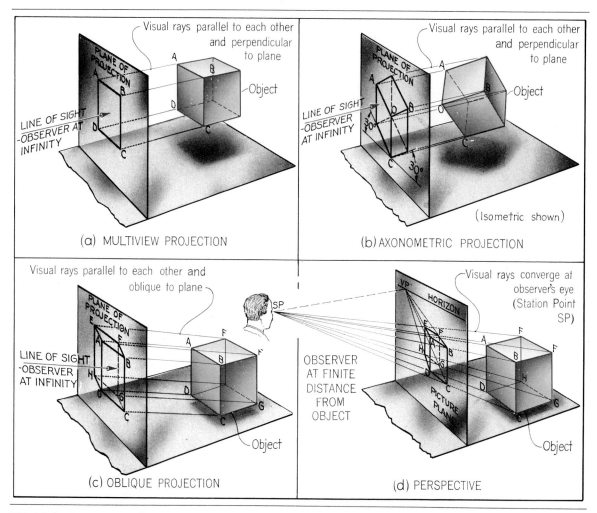

Fig. 18.1   Four Types of Projection.

---

plane of projection. Therefore, both are classified as *orthographic projections,* §1.10.

In *oblique projection,* (c), the observer is considered to be at infinity, and the visual rays are parallel to each other but oblique to the plane of projection. See Chapter 19.

In *perspective,* (d), the observer is considered to be at a finite distance from the object, and the visual rays extend from the observer's eye, or the station point (SP), to all points of the object to form a "cone of rays." See Chapter 20.

18.2   Types of Axonometric Projection

The distinguishing feature of axonometric projection, as compared to multiview projection, is the inclined position of the object with respect to the plane of projection. Since the principal edges and surfaces of the object are inclined to the plane of projection, the lengths of the lines, the sizes of the angles, and the general proportions of the object vary with the infinite number of possible positions in which the object may be

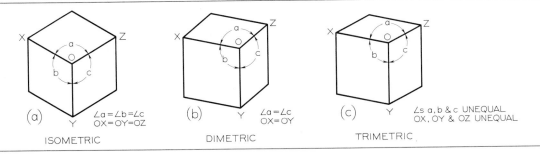

Fig. 18.2 Axonometric Projections.

placed with respect to the plane of projection. Three of these are shown in Fig. 18.2.

In these cases the edges of the cube are inclined to the plane of projection and are therefore foreshortened. See Fig. 7.20 (c). The degree of foreshortening of any line depends on its angle with the plane of projection; the greater the angle, the greater the foreshortening. If the degree of foreshortening is determined for each of the three edges of the cube that meet at one corner, scales can be easily constructed for measuring along these edges or any other edges parallel to them. See Figs. 18.41 (a) and 18.46.

It is customary to consider three edges of the cube that meet at the corner nearest the observer as the *axonometric axes*. In Fig. 18.1 (b), the axonometric axes, or simply the *axes*, are OA, OB, and OC. As shown in Fig. 18.2, axonometric projections are classified as (a) *isometric projection*, (b) *dimetric projection,* and (c) *trimetric projection,* depending on the number of scales of reduction required.

Isometric Projection

18.3   The Isometric Method of Projection

To produce an isometric projection (isometric means "equal measure"), it is necessary to place the object so that its principal edges, or axes, make equal angles with the plane of projection and are therefore foreshortened equally. See Fig. 6.21. In this position the edges of a cube would be projected equally and would make equal angles with each other (120°), as shown in Fig. 18.2 (a).

In Fig. 18.3 (a) is shown a multiview drawing of a cube. At (b) the cube is shown revolved through 45° about an imaginary vertical axis. Now an auxiliary view in the direction of the arrow will show the cube diagonal ZW as a point, and the cube appears as a true isometric projection. However, instead of the auxiliary view at

(b) being drawn, the cube may be further revolved as shown at (c), this time the cube being tilted forward about an imaginary horizontal axis until the three edges OX, OY, and OZ make equal angles with the frontal plane of projection and are therefore foreshortened equally. Here again, a diagonal of the cube, in this case OT, appears as a point in the isometric view. The front view thus obtained is a true isometric projection. In this projection the twelve edges of the cube make angles of about 35° 16′ with the frontal plane of projection. The lengths of their projections are equal to the lengths of the edges multiplied by $\sqrt{\frac{2}{3}}$, or by 0.816, approximately. Thus the projected lengths are about 80 percent of the true lengths, or still more roughly, about three-

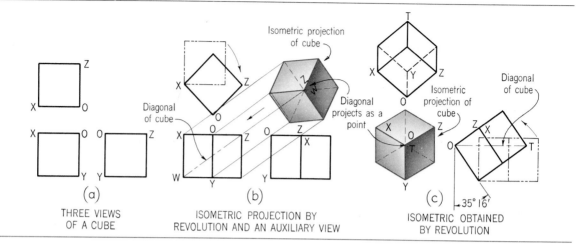

Fig. 18.3 Isometric Projection.

fourths of the true lengths. The projections of the axes OX, OY, and OZ make angles of 120° with each other and are called the *isometric axes*. Any line parallel to one of these is called an *isometric line;* a line that is not parallel is called a *nonisometric line*. It should be noted that the angles in the isometric projection of the cube are either 120° or 60° and that all are projections of 90° angles. In an isometric projection of a cube, the faces of the cube, and any planes parallel to them, are called *isometric planes.*

18.4   The Isometric Scale

A correct isometric projection may be drawn with the use of a special isometric scale, prepared on a strip of paper or cardboard, Fig. 18.4. All distances in the isometric scale are $\sqrt{\frac{2}{3}}$ times true size, or approximately 80 percent of true size. The use of the isometric scale is illustrated in Fig. 18.5 (a). A scale of $9'' = 1'-0$, or $\frac{3}{4}$-size scale (or metric equivalent), could be used to approximate the isometric scale.

18.5   Isometric Drawing

When a drawing is prepared with an isometric scale, or otherwise as the object is actually *projected* on a plane of projection, it is an *iso-*

metric projection, as illustrated in Fig. 18.5 (a). When it is prepared with an ordinary scale, it is an *isometric drawing,* illustrated at (b). The isometric drawing, (b), is about 25 percent larger than the isometric projection (a), but the pictorial value is obviously the same in both.

Since the isometric projection is foreshortened and an isometric drawing is full-scale size, it is usually advantageous to make an isometric drawing rather than an isometric projection. The drawing is much easier to execute and, for all practical purposes, is just as satisfactory as the isometric projection.

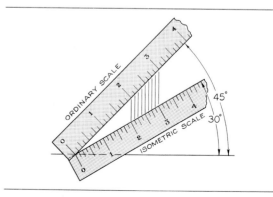

Fig. 18.4 Isometric Scale.

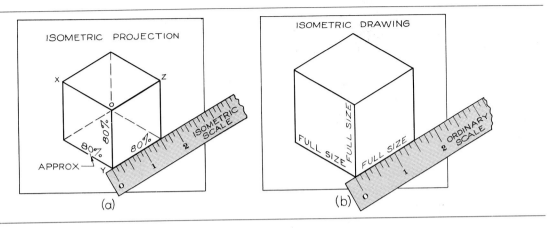

Fig. 18.5 Isometric and Ordinary Scales.

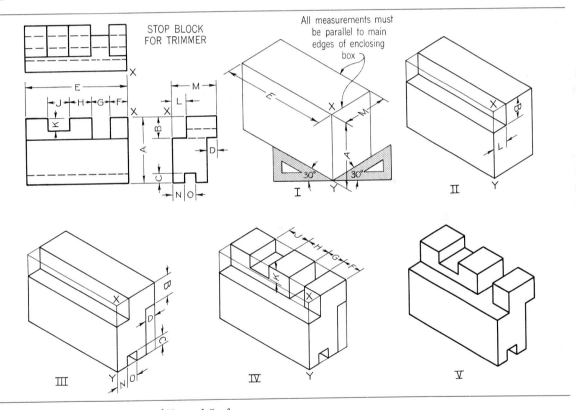

Fig. 18.6 Isometric Drawing of Normal Surfaces.

18.6   Steps in Making an Isometric Drawing

The steps in constructing an isometric drawing of an object composed only of normal surfaces, §7.19, are illustrated in Fig. 18.6. Notice that all

measurements are made parallel to the main edges of the enclosing box, that is, parallel to the isometric axes. No measurement along a diagonal (nonisometric line) on any surface or through the object can be set off directly with

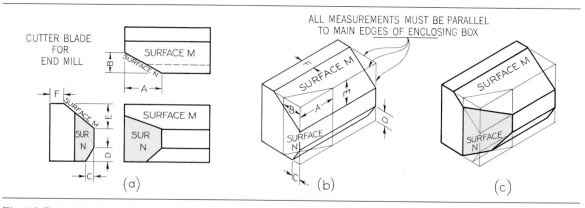

Fig. 18.7   Inclined Surfaces in Isometric.

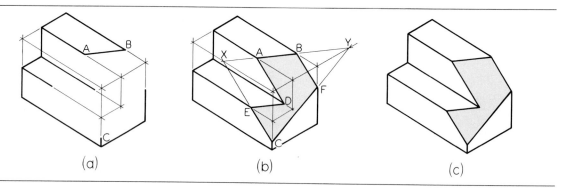

Fig. 18.8   Oblique Surfaces in Isometric.

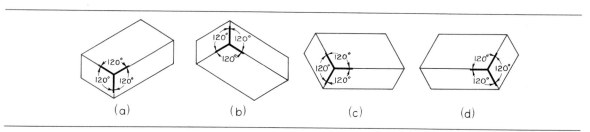

Fig. 18.9   Positions of Isometric Axes.

the scale. The object may be drawn in the same position by beginning at the corner Y, or any other corner, instead of at the corner X.

The method of constructing an isometric drawing of an object composed partly of inclined surfaces (and oblique edges) is shown in Fig. 18.7. Notice that inclined surfaces are located by *offset* or *coordinate measurements* along the isometric lines. For example, dimensions E and F are set off to locate the inclined surface M, and dimensions A and B are used to locate surface N.

For sketching in isometric, see §§6.12–6.14.

18.7 Oblique Surfaces in Isometric

Oblique surfaces in isometric may be drawn by establishing the intersections of the oblique surface with the isometric planes. For example, in Fig. 18.8 (a), the oblique plane is known to contain points A, B, and C. To establish the plane, (b), line AB is extended to X and Y, which are in the same isometric planes as C. Lines XC and YC locate points E and F. Finally AD and ED are drawn, using the rule of parallelism of lines. The completed drawing is shown at (c).

18.8 Other Positions of the Isometric Axes

The isometric axes may be placed in any desired position according to the requirements of the problem, as shown in Fig. 18.9, but the angle between the axes must remain 120°. The choice

of the directions of the axes is determined by the position from which the object is usually viewed, Fig. 18.10, or by the position that best describes the shape of the object. If possible, both requirements should be met.

If the object is characterized by considerable length, the long axis may be placed horizontally for best effect, as shown in Fig. 18.11.

18.9 Offset Location Measurements

The method of locating one point with respect to another is illustrated in Figs. 18.12 and 18.13. In each case, after the main block has been

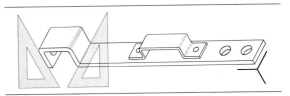

Fig. 18.11 Long Axis Horizontal.

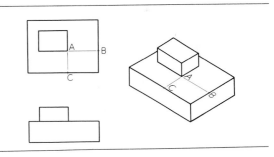

Fig. 18.12 Offset Location Measurements.

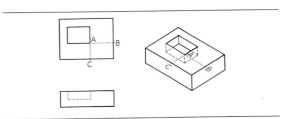

Fig. 18.13 Offset Location Measurements.

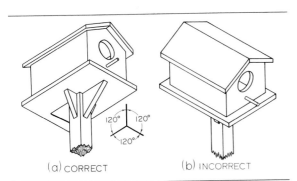

(a) CORRECT   (b) INCORRECT

Fig. 18.10 An Object Naturally Viewed from Below.

607

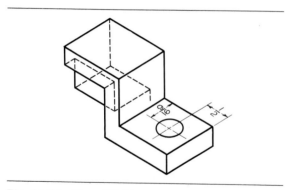

Fig. 18.14   Use of Hidden Lines.

drawn, the offset lines CA and BA in the multi-view drawing are drawn full size in the isometric drawing, thus locating corner A of the small block or rectangular recess. These measurements are called *offset measurements,* and since they are parallel to certain edges of the main block in the multiview drawings, they will be parallel, respectively, to the same edges in the isometric drawings, §7.25.

18.10   Hidden Lines

The use of hidden lines in isometric drawing is governed by the same rules as in all other types of projection: *Hidden lines are omitted unless they are needed to make the drawing clear.* A case in which hidden lines are needed is illustrated in Fig. 18.14, in which a projecting part cannot be clearly shown without the use of hidden lines.

18.11   Center Lines

The use of center lines in isometric drawing is governed by the same rules as in multiview drawing: *Center lines are drawn if they are needed to indicate symmetry or if they are needed for dimensioning,* Fig. 18.14. In general, center lines should be used sparingly and omitted in cases of doubt. The use of too many center lines may produce a confusion of lines, which diminishes the clearness of the drawing. Examples in which center lines are not needed are shown in Figs. 18.10 and 18.11. Examples in which they are needed are seen in Figs. 18.14 and 18.39 (a).

18.12   Box Construction

Objects of rectangular shape may be more easily drawn by means of *box construction,* which consists simply in imagining the object to be enclosed in a rectangular box whose sides coincide with the main faces of the object. For example, in Fig. 18.15 the object shown in two views is imagined to be enclosed in a construction box. This box is then drawn lightly with construction lines, as shown at I, the irregular features are then constructed as shown at II, and finally, as shown at III, the required lines are made heavy.

18.13   Nonisometric Lines

Since the only lines of an object that are drawn true length in an isometric drawing are the isometric axes or lines parallel to them, *nonisometric* lines cannot be set off directly with the

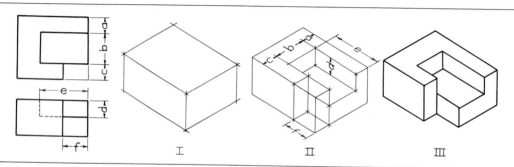

Fig. 18.15   Box Construction.

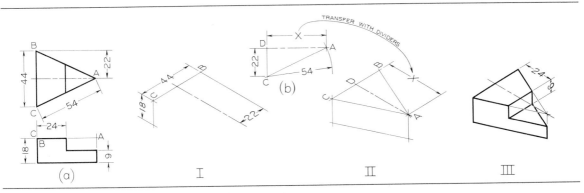

Fig. 18.16   Nonisometric Lines   (metric dimensions).

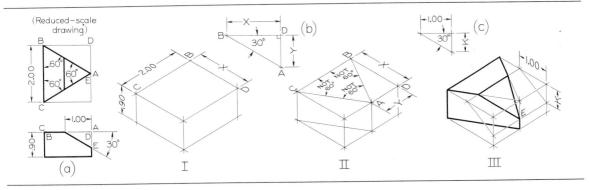

Fig. 18.17   Angles in Isometric.

scale. For example, in Fig. 18.16 (a), the inclined lines BA and CA are shown in their true lengths 54 mm in the top view, but since they are not parallel to the isometric axes, they will not be true length in the isometric. Such lines are drawn in isometric by means of box construction and offset measurements. First, as shown at I, the measurements 44 mm, 18 mm, and 22 mm can be set off directly since they are made along isometric lines. The nonisometric 54 mm dimension cannot be set off directly, but if one-half of the given top view is constructed full size to scale as shown at (b), the dimension X can be determined. This dimension is parallel to an isometric axis and can be transferred with dividers to the isometric at II. The dimensions 24 mm and 9 mm are parallel to isometric lines and can be set off directly, as shown at III.

To realize the fact that nonisometric lines will not be true length in the isometric drawing,

set your dividers on BA of II and then compare with BA on the given top view at (a). Do the same for line CA. It will be seen that BA is shorter and CA is longer in the isometric than the corresponding lines in the given views.

18.14   Angles in Isometric

As shown in §7.26, angles project true size only when the plane of the angle is parallel to the plane of projection. An angle may project larger or smaller than true size, depending on its position. Since in isometric the various surfaces of the object are usually inclined to the plane of projection, it follows that angles generally will not be projected true size. For example, in the multiview drawing in Fig. 18.17 (a), none of the three 60° angles will be 60° in the isometric drawing. To realize this fact, measure each angle in the isometric of II with the protractor and

609

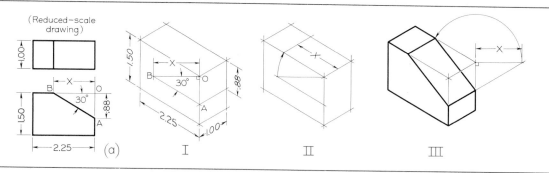

Fig. 18.18 Angle in Isometric.

note the number of degrees compared to the true 60°. No two angles are the same; two are smaller and one is larger than 60°.

As shown in I, the enclosing box can be drawn from the given dimensions, except for dimension X, which is not given. To find dimension X, draw triangle BDA from the top view full size, as shown at (b). Transfer dimension X to the isometric in I, to complete the enclosing box.

In order to locate point A in II, dimension Y must be used, but this is not given in the top view, (a). Dimension Y is found by the same construction, (b), and then transferred to the isometric, as shown. The completed isometric is shown at III where point E is located by using dimension K, as shown.

Thus, in order to set off angles in isometric, the regular protractor cannot be used.* *Angular measurements must be converted to linear measurements along isometric lines.*

In Fig. 18.18 (a) are two views of an object to be drawn in isometric. Point A can easily be located in the isometric, step I, by measuring .88″ down from point O. However, in the given drawing at (a) the location of point B depends on the 30° angle, and to locate B in the isometric linear dimension X must be known. This distance can be found graphically by drawing the

right triangle BOA attached to the isometric, as shown. The distance X is then transferred to the isometric with the compass or dividers, as shown at II. Actually, the triangle could be attached in several different positions. One of these is shown at III.

When angles are given in degrees, it is necessary to convert the angular measurements into linear measurements. This is best done by drawing a right triangle separately, as in Fig. 18.17 (b), or attached to the isometric, as in Fig. 18.18.

18.15 Irregular Objects

If the general shape of an object does not conform somewhat to a rectangular pattern, as shown in Fig. 18.19, it may be drawn as shown at (a) by using the box construction discussed previously. Various points of the triangular base are located by means of offsets a and b along the edges of the bottom of the construction box. The vertex is located by means of offsets OA and OB on the top of the construction box.

However, it is not necessary to draw the complete construction box. If only the bottom of the box is drawn, as shown at (b), the triangular base can be constructed as before. The orthographic projection of the vertex O′ on the base can then be located by offsets O′A and O′B, as shown, and from this point, the vertical center line O′O can be erected, using measurement C.

---

*Isometric protractors for setting off angles on isometric surfaces are available from drafting supplies dealers.

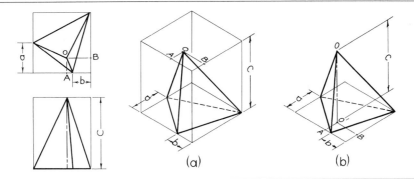

Fig. 18.19   Irregular Object in Isometric.

An irregular object may be drawn by means of a series of sections, as illustrated in Fig. 18.20. The edge views of a series of imaginary cutting planes are shown in the top and front views of the multiview drawing at (a). At I the various sections are constructed in isometric, and at II the object is completed by drawing lines through the corners of the sections. In the isometric at I, all height dimensions are taken from the front view at (a), and all depth dimensions from the top view.

18.16   Curves in Isometric

Curves may be drawn in isometric by means of a series of offset measurements similar to those discussed in §18.9. In Fig. 18.21 any desired

number of points, such as A, B, and C, are selected at random along the curve in the given top view at (a). Enough points should be chosen to fix accurately the path of the curve; the more points used, the greater the accuracy. Offset grid lines are then drawn from each point parallel to the isometric axes.

As shown at I, offset measurements a and b are laid off in the isometric to locate point A on the curve. Points B, C, and D are located in a similar manner, as shown at II. A light freehand curve is sketched smoothly through the points as shown at III. Points A′, B′, C′, and D′ are located directly under points A, B, C, and D, as shown at IV, by drawing vertical lines downward, making all equal to dimension c, the height of the block. A light freehand curve is then drawn

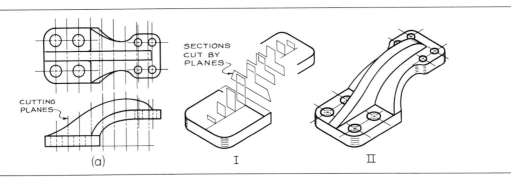

Fig. 18.20   Use of Sections in Isometric.

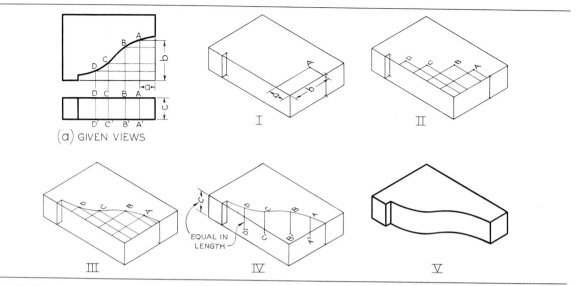

Fig. 18.21   Curves in Isometric.

through the points. The final curve is heavied in with the aid of the irregular curve, §2.54, and all straight lines are darkened to complete the isometric at V.

18.17   True Ellipses in Isometric

As shown in §§5.51, 6.13, and 7.30, if a circle lies in a plane that is not parallel to the plane of projection, the circle will be projected as a true ellipse. The ellipse can be constructed by the method of offsets, §18.16. As shown in Fig. 18.22 (a), draw parallel lines, spaced at random, across the circle; then transfer these lines to the isometric, as shown at (b), with the aid of the dividers. To locate points in the lower ellipse, transfer points of the upper ellipse down a distance equal to the height d of the block and draw the ellipse, part of which will be hidden, through these points. Draw the final ellipses with the aid of the irregular curve, §2.54.

A variation of the method of offsets, which provides eight points on the ellipse, is illustrated at (c) and (d). If more points are desired, parallel lines, as at (a), can be added. As shown at (c), circumscribe a square around the given circle,

and draw diagonals. Through the points of intersection of the diagonals and the circle, draw another square, as shown. Draw this construction in the isometric, as shown at (d), transferring distances a and b with the dividers.

A similar method that provides twelve points on the ellipse is shown at (e). The given circle is divided into twelve equal parts, using the 30° × 60° triangle, Fig. 2.23. Lines parallel to the sides of the square are drawn through these points. The entire construction is then drawn in isometric, and the ellipse is drawn through the points of intersection.

When the center lines shown in the top view at (a) are drawn in isometric, (b), they become the *conjugate diameters* of the ellipse. The ellipse can then be constructed on the conjugate diameters by the methods of Figs. 5.51 and 5.52 (b).

When the 45° diagonals at (c) are drawn in isometric at (d), they coincide with the major and minor axes of the ellipse, respectively. Note that the minor axis is equal in length to the sides of the inscribed square at (c). The ellipse can be constructed on the major and minor axes by any of the methods in §§5.48 and 5.51.

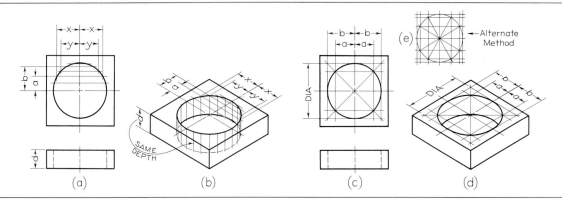

Fig. 18.22   True Isometric Ellipse Construction.

Remember the rule: *The major axis of the ellipse is always at right angles to the center line of the cylinder, and the minor axis is at right angles to the major axis and coincides with the center line.*

Accurate ellipses may be drawn with the aid of ellipse guides, §§5.56 and 18.21, or with a special *ellipsograph.*

If the curve lies in a nonisometric plane, not all offset measurements can be applied directly. For example, in Fig. 18.23 (a) the elliptical face shown in the auxiliary view lies in an inclined nonisometric plane. The cylinder is enclosed in a construction box, and the box is then drawn in isometric, as shown at I. The base is drawn by the method of offsets, as shown in Fig. 18.22. The inclined ellipse is constructed by locating a number of points on the ellipse in the isometric and drawing the final curve by means of the irregular curve, §2.54.

Measurements a, b, c, and so on, are parallel to an isometric axis and can be set off in the

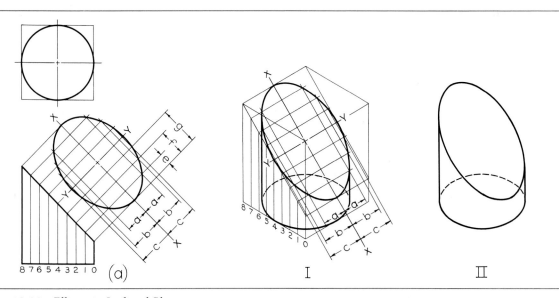

Fig. 18.23   Ellipse in Inclined Plane.

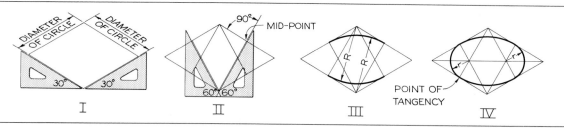

Fig. 18.24   Steps in Drawing Four-Center Ellipse.

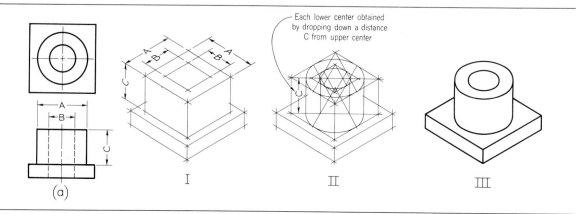

Fig. 18.25   Isometric Drawing of a Bearing.

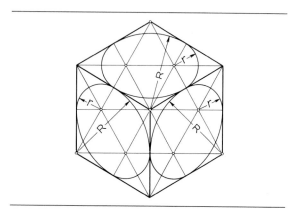

Fig. 18.26   Four-Center Ellipses.

isometric at I on each side of the center line X–X, as shown. Measurements e, f, g, and so on, are not parallel to any isometric axis and cannot be set off directly in isometric. However, when these measurements are projected to the front view and down to the base, as shown at (a), they can then be set off along the lower edge of the

614

construction box, as shown at I. The completed isometric is shown at II.

The ellipse may also be drawn with the aid of an appropriate ellipse template selected to fit the major and minor axes established along X–X and Y–Y, respectively. See Fig. 5.55.

18.18   Approximate Four-Center Ellipse

An approximate ellipse is sufficiently accurate for nearly all isometric drawings. The method commonly used, called the *four-center ellipse*, is illustrated in Figs. 18.24, 18.25, and 18.26. It can be used only for ellipses in isometric planes.

To apply this method, Fig. 18.24, draw, or conceive to be drawn, a square around the given circle in the multiview drawing; then

I.   Draw the isometric of the square, which is an equilateral parallelogram whose sides are equal to the diameter of the circle.

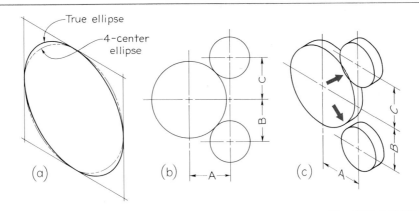

Fig. 18.27 Faults of Four-Center Ellipse.

II. Erect perpendicular bisectors to each side, using the 30° × 60° triangle as shown. These perpendiculars will intersect at four points, which will be centers for the four circular arcs.

III. Draw the two large arcs, with radius R, from the intersections of the perpendiculars in the two closest corners of the parallelogram, as shown.

IV. Draw the two small arcs, with radius r, from the intersections of the perpendiculars within the parallelogram, to complete the ellipse. As a check on the accurate location of these centers, a long diagonal of the parallelogram may be drawn, as shown. The midpoints of the sides of the parallelogram are points of tangency for the four arcs.

A typical drawing with cylindrical shapes is illustrated in Fig. 18.25. Note that the centers of the larger ellipse cannot be used for the smaller ellipse, though the ellipses represent concentric circles. Each ellipse has its own parallelogram and its own centers. Observe also that the centers of the lower ellipse are obtained by projecting the centers of the upper large ellipse down a distance equal to the height of the cylinder.

The construction of the four-center ellipse on the three visible faces of a cube is shown in Fig. 18.26, a study of which shows that all diagonals are horizontal or 60° with horizontal; hence, the entire construction is made with the T-square and 30° × 60° triangle.

Actually the four-center ellipse deviates considerably from the true ellipse. As shown in Fig. 18.27 (a), the four-center ellipse is somewhat shorter and "fatter" than the true ellipse. In constructions where tangencies or intersections with the four-center ellipse occur in the zones of error, the four-center ellipse is unsatisfactory, as shown at (b) and (c).

For a much closer approximation to the true ellipse, the Orth four-center ellipse, Fig. 18.28, which requires only one more step than the

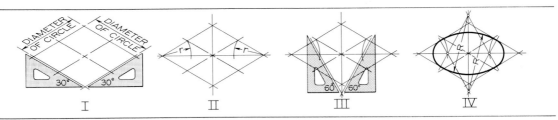

Fig. 18.28 Orth Four-Center Ellipse. *Courtesy of Professor H. D. Orth.*

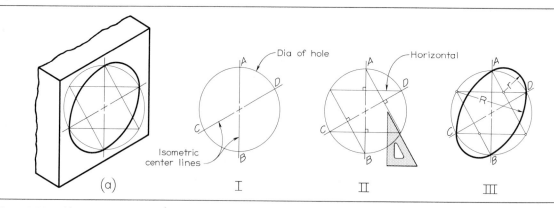

Fig. 18.29   Alternate Four-Center Ellipse.

regular four-center ellipse, will be found sufficiently accurate for almost any problem.

When it is more convenient to start with the isometric center lines of a hole or cylinder in drawing the ellipse, rather than the enclosing parallelogram, the *alternate four-center ellipse* is recommended, Fig. 18.29. A completely constructed ellipse is shown at (a), and the steps followed are shown at the right in the figure.

I.   Draw the isometric center lines. From the center, draw a construction circle equal to the actual diameter of the hole or cylinder. The circle will intersect the center lines at four points A, B, C, and D.

II.   From the two intersection points on one center line, erect perpendiculars to the other center line; then from the two intersection points on the other center line, erect perpendiculars to the first center line.

III.   With the intersections of the perpendiculars as centers, draw two small arcs and two large arcs, as shown.

NOTE   The above steps are exactly the same as for the regular four-center ellipse of Fig. 18.24 except for the use of the isometric center lines instead of the enclosing parallelogram.

18.19   Screw Threads in Isometric

Parallel partial ellipses spaced equal to the symbolic thread pitch, Fig. 15.9 (a), are used to represent the crests only of a screw thread in isometric, Fig. 18.30. The ellipses may be drawn by the four-center method of §18.18, or with the ellipse template, which is much more convenient, §§5.56 and 18.21.

18.20   Arcs in Isometric

The four-center ellipse construction is used in drawing circular arcs in isometric, as shown in Fig. 18.31. At (a) the complete construction is shown. However, it is not necessary to draw the complete constructions for arcs, as shown at (b) and (c). In each case the radius R is set off from the construction corner; then at each point, perpendiculars to the lines are erected, their intersection being the center of the arc. Note that the

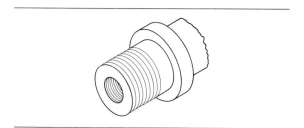

Fig. 18.30   Screw Threads in Isometric.

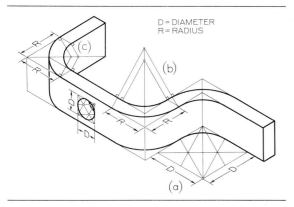

Fig. 18.31 Arcs in Isometric.

R distances are equal in both cases, (b) and (c), but that the actual radii used are quite different.

If a truer elliptic arc is required, the Orth construction, Fig. 18.28, can be used. Or a true elliptic arc may be drawn by the method of offsets, §18.17, or with the aid of an ellipse guide, §18.21.

18.21   Ellipse Guides

One of the principal time-consuming elements in pictorial drawing is the construction of ellipses. A wide variety of ellipse guides, or templates, is available for ellipses of various sizes and proportions. See §5.56. They are not available in every possible size, of course, and it may be necessary to "use the fudge factor," such as leaning the pencil or pen when inscribing the ellipse, or shifting the template slightly for drawing each quadrant of the ellipse.

The design of the ellipse template, Fig. 18.32, combines the angles, scales, and ellipses on the same instrument. The ellipses are provided with markings to coincide with the isometric center lines of the holes—a convenient feature in isometric drawing.

18.22   Intersections

To draw the elliptical intersection of a cylindrical hole in an oblique plane in isometric, Fig. 18.33, draw the ellipse in the isometric plane on

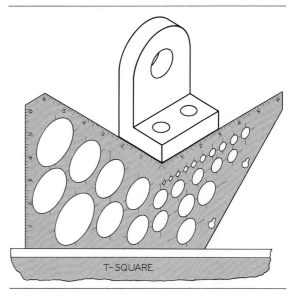

Fig. 18.32 Instrumaster Isometric Template.

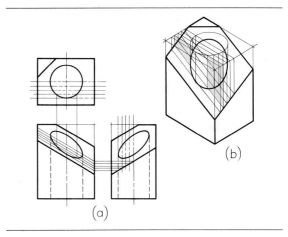

Fig. 18.33 Oblique Plane and Cylinder.

top of the construction box, (b); then project points down to the oblique plane as shown. It will be seen that the construction for each point forms a trapezoid, which is produced by a slicing plane parallel to a lateral surface of the block.

To draw the curve of intersection between two cylinders, Fig. 18.34, pass a series of imaginary cutting planes through the cylinders parallel to their axes, as shown. Each plane will cut

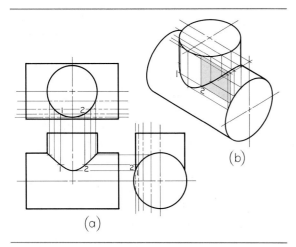

Fig. 18.34   Intersection of Cylinders.

elements on both cylinders that intersect at points on the curve of intersection, as shown at (b). As many points should be plotted as necessary to assure a smooth curve. For most accurate work, the ends of the cylinders should be drawn by the Orth construction, or with ellipse guides, or by one of the true-ellipse constructions.

18.23   The Sphere in Isometric

The isometric drawing of any curved surface is evidently the envelope of all lines that can be drawn on that surface. For the sphere, the great

circles (circles cut by any plane through the center) may be selected as the lines on the surface. Since all great circles, except those that are perpendicular or parallel to the plane of projection, are shown as ellipses having equal major axes, it follows that their envelope is a circle whose diameter is the major axis of the ellipses.

In Fig. 18.35 (a) two views of a sphere enclosed in a construction cube are shown. The cube is drawn at I, together with the isometric of a great circle that lies in a plane parallel to one face of the cube. Actually, the ellipse need not be drawn, for only the points on the diagonal located by measurements a are needed. These points establish the ends of the major axis from which the radius R of the sphere is determined. The resulting drawing shown at II is an *isometric drawing,* and its diameter is, therefore, $\sqrt{\frac{3}{2}}$ times the actual diameter of the sphere. The *isometric projection* of the sphere is simply a circle whose diameter is equal to the true diameter of the sphere, as shown at III.

18.24   Isometric Sectioning

In drawing objects characterized by open or irregular interior shapes, isometric sectioning is as appropriate as in multiview drawing. An *isometric full section* is shown in Fig. 18.36. In such cases it is usually best to draw the cut sur-

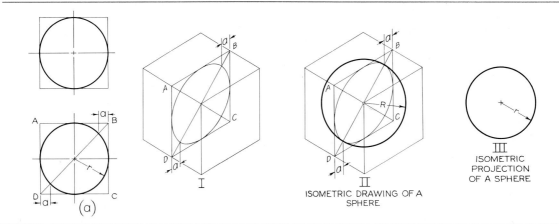

Fig. 18.35   Isometric of a Sphere.

face first and then to draw the portion of the object that lies behind the cutting plane. Other examples of isometric full sections are shown in Figs. 7.92, 9.11 (b), and 9.12 (d).

An *isometric half section* is shown in Fig. 18.37. The simplest procedure in this case is to make an isometric drawing of the entire object

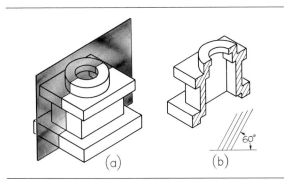

Fig. 18.36   Isometric Full Section.

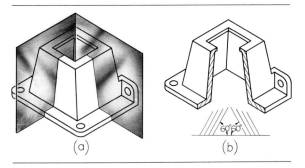

Fig. 18.37   Isometric Half Section.

and then the cut surfaces. Since only a quarter of the object is removed in a half section, the resulting pictorial drawing is more useful than full sections in describing both exterior and interior shapes together. Other typical isometric half sections are shown in Figs. 9.13, 9.41, and 9.42.

Isometric broken-out sections are also sometimes used. Examples are shown in Figs. 7.99, 9.51, and 16.48.

Section lining in isometric drawing is similar to that in multiview drawing. Section lining at an angle of 60° with horizontal, Figs. 18.36 and 18.37, is recommended, but the direction should be changed if at this angle the lines would be parallel to a prominent visible line bounding the cut surface, or to other adjacent lines of the drawing.

18.25   Isometric Dimensioning

Isometric dimensions are similar to ordinary dimensions used on multiview drawings but are expressed in pictorial form. Two methods of dimensioning are approved by ANSI, namely, the pictorial plane (aligned) system and the unidirectional system, Fig. 18.38. Note that *vertical lettering* is used for either system of dimensioning. Inclined lettering is not recommended for pictorial dimensioning. The method of drawing numerals and arrowheads for the two systems is shown at (a) and (b). For the 64 mm dimension in the aligned system at (a), the extension lines, dimension lines, and lettering are all drawn in

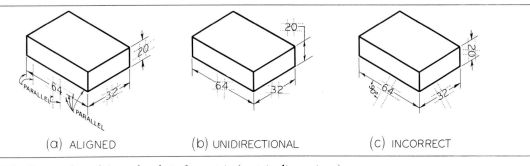

Fig. 18.38   Numerals and Arrowheads in Isometric (metric dimensions).

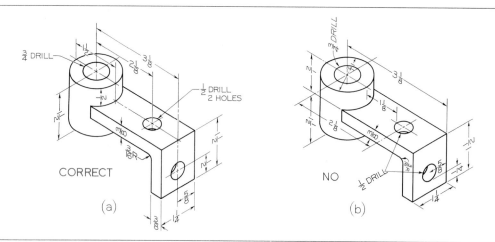

Fig. 18.39   Correct and Incorrect Isometric Dimensioning (Aligned System).

the isometric plane of one face of the object. The "horizontal" guide lines for the lettering are drawn parallel to the dimension line, and the "vertical" guide lines are drawn parallel to the extension lines. The barbs of the arrowheads should line up parallel to the extension lines.

For the 64 mm dimension in the unidirectional system at (b), the extension lines and dimension lines are all drawn in the isometric plane of one face of the object and the barbs of the arrowheads should line up parallel to the extension lines, all exactly the same as at (a). However, the lettering for the dimensions is vertical and reads from the bottom of the drawing. This simpler system of dimensioning is often used on pictorials for production purposes.

As shown at (c), the vertical guide lines for the letters should not be perpendicular to the dimension lines. The example at (c) is incorrect because the 64 mm and 32 mm dimensions are lettered neither in the plane of the corresponding dimension and extension lines nor in a vertical position to read from the bottom of the drawing. The 20 mm dimension is awkward to read because of its position.

Correct and incorrect practice in isometric dimensioning using the aligned system of dimensioning is shown in Fig. 18.39. At (b) the 3⅛″ dimension runs to a wrong extension line at the right, and consequently the dimension does not lie in an isometric plane. Near the left side, a

number of lines cross one another unnecessarily and terminate on the wrong lines. The upper ½″ drill hole is located from the edge of the cylinder when it should be dimensioned from its center line. Study these two drawings carefully to discover additional mistakes at (b).

The dimensioning methods described apply equally to fractional, decimal, and metric dimensions.

Many examples of isometric dimensioning are given in the problems at the ends of Chapters 7, 9, 10, and 16, and you should study these to find samples of almost any special case you may encounter.

18.26   Exploded Assemblies

Exploded assemblies are often used in design presentations, catalogs, sales literature, and in the shop, to show all the parts of an assembly and how they fit together. They may be drawn by any of the pictorial methods, including isometric, Fig. 18.40. Other isometric exploded assemblies are shown in Chapter 16.

18.27   Piping Diagrams

Isometric and oblique drawings are well suited for representation of piping layouts, as well as for all other structural work to be represented pictorially.

620

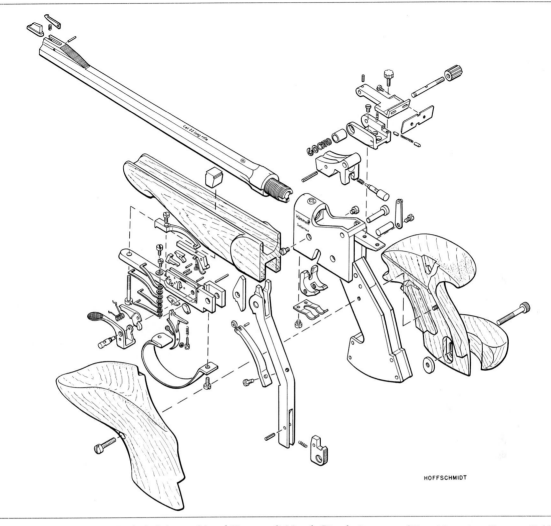

Fig. 18.40   Isometric Exploded Assembly of Hammerli Match Pistol. *Courtesy of* True *Magazine, Fawcett Publications.*

Dimetric Projection

18.28   The Dimetric Method of Projection

A *dimetric projection* is an axonometric projection of an object so placed that two of its axes make equal angles with the plane of projection and the third axis makes either a smaller or a greater angle. Hence, the two axes making equal angles with the plane of projection are fore-shortened equally, while the third axis is foreshortened in a different ratio.

Generally, the object is so placed that one axis will be projected in a vertical position. However, if the relative positions of the axes have been determined, the projection may be drawn in any revolved position, as in isometric drawing. See §18.8.

621

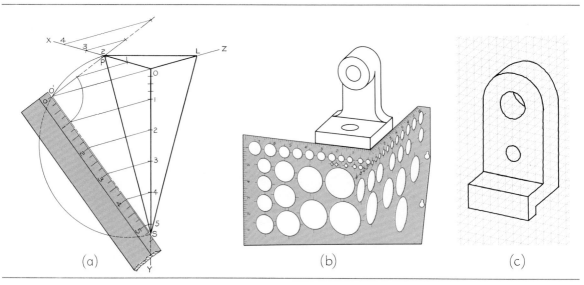

Fig. 18.41   Dimetric Projection.

The angles between the *projection of the axes* must not be confused with the angles which the *axes themselves* make with the plane projection.

The positions of the axes may be assumed such that any two angles between the axes are equal and over 90°, and the scales determined graphically, as shown in Fig. 18.41 (a), in which OP, OL, and OS are the projections of the axes or converging edges of a cube. In this case, angle POS = angle LOS. Lines PL, LS, and SP are the lines of intersection of the plane of projection with the three visible faces of the cube. From descriptive geometry we know that since line LO is perpendicular to the plane POS, in space, its projection LO is perpendicular to PS, the intersection of the plane POS and the plane of projection. Similarly, OP is perpendicular to SL, and OS is perpendicular to PL.

If the triangle POS is revolved about the line PS as an axis into the plane of projection, it will be shown in its true size and shape as PO'S. If regular full-size scales are marked along the lines O'P and O'S, and the triangle is counterrevolved to its original position, the dimetric scales may be laid off on the axes OP and OS, as shown.

In order to avoid the preparation of special scales, use can be made of available scales on the architects' scale by assuming the scales and calculating the positions of the axes, as follows:

$$\cos a = -\frac{\sqrt{2h^2v^2 - v^4}}{2hv}$$

where a is one of the two equal angles between the projections of the axes, h is one of the two equal scales, and v is the third scale.

Examples are shown in the upper row of Fig. 18.42, in which the assumed scales, shown encircled, are taken from the architects' scale. One of these three positions of the axes will be found suitable for almost any practical drawing.

The Instrumaster Dimetric Template, Fig. 18.41 (b), has angles of approximately 11° and 39° with horizontal, which provides a picture similar to that in Fig. 18.42 (III). In addition, the template has ellipses corresponding to the axes and accurate scales along the edges.

For other information on drawing of ellipses, see §18.32.

The Instrumaster Dimetric Graph paper, Fig. 18.41 (c), can be used to sketch in dimetrics as easily as to sketch isometrics on isometric paper. The grid lines slope in conformity to the angles on the Dimetric Template at (b) and, when

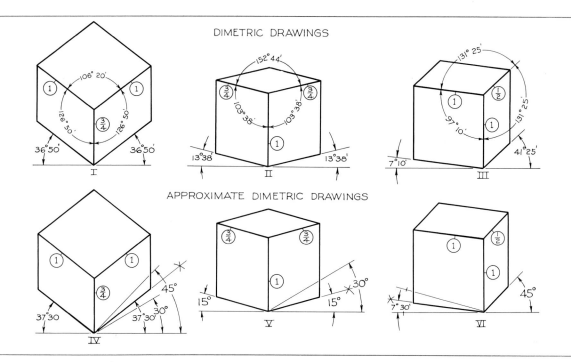

Fig. 18.42 Angles of Axes Determined by Assumed Scales.

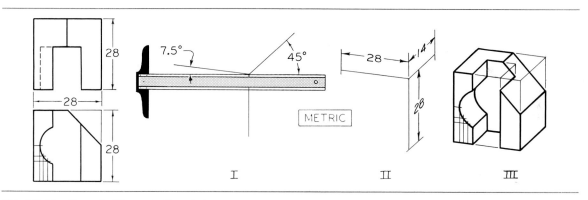

Fig. 18.43 Steps in Dimetric Drawing.

printed on vellum, the grid lines do not reproduce on prints.

18.29 Approximate Dimetric Drawing

Approximate dimetric drawings, which closely resemble true dimetrics, can be constructed by substituting for the true angles shown in the upper half of Fig. 18.42, angles that can be obtained with the ordinary triangles and compass, as shown in the lower half of the figure. The resulting drawings will be sufficiently accurate for all practical purposes.

The procedure in preparing an approximate dimetric drawing, using the position of VI in Fig. 18.42, is shown in Fig. 18.43. The offset method of drawing a curve is shown in the figure. Other methods for drawing ellipses are the same as in trimetric drawing, §18.32.

623

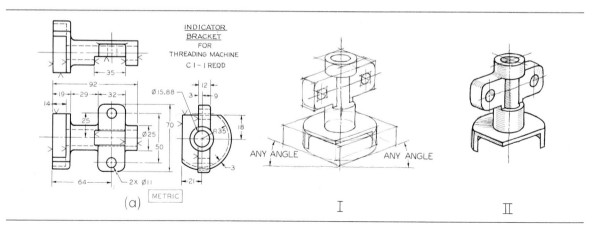

Fig. 18.44   Steps in Dimetric Sketching.

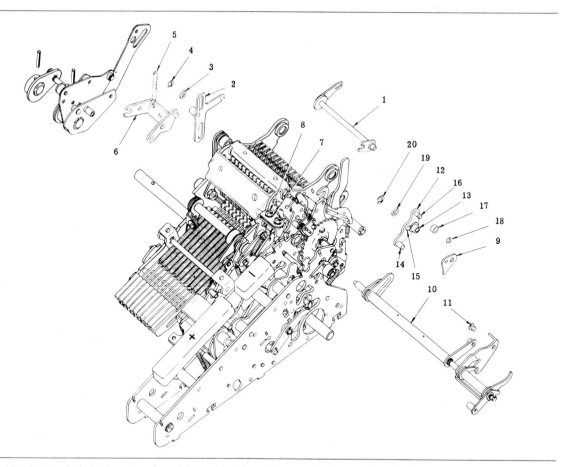

Fig. 18.45   Exploded Dimetric of an Adding Machine. *Courtesy of Victor Adding Machine Co.*

The steps in making a dimetric sketch, using a position similar to that in Fig. 18.42 (V), are shown in Fig. 18.44. The two angles are equal and about 20° with horizontal for the most pleasing effect.

An exploded approximate dimetric drawing of an adding machine is shown in Fig. 18.45. The dimetric axes used are those in Fig. 18.42 (IV). Pictorials such as this are often used in service manuals.

Trimetric Projection

18.30 The Trimetric Method of Projection

A *trimetric projection* is an axonometric projection of an object so placed that no two axes make equal angles with the plane of projection. In other words, each of the three axes and the lines parallel to them, respectively, have different ratios of foreshortening when projected to the plane of projection. If the three axes are assumed in any position on paper such that none of the angles is less than 90°, and if neither an isometric nor a dimetric position is deliberately arranged, the result will be a trimetric projection.

18.31 Trimetric Scales

Since the three axes are foreshortened differently, three different trimetric scales must be prepared and used. The scales are determined as shown in Fig. 18.46 (a), the method being the same as explained for the dimetric scales in §18.28. As shown at (a), any two of the three triangular faces can be revolved into the plane of projection to show the true lengths of the three axes. In the revolved position, the regular scale is used to set off inches or fractions thereof. When the axes have been counterrevolved to their original positions, the scales will be cor-

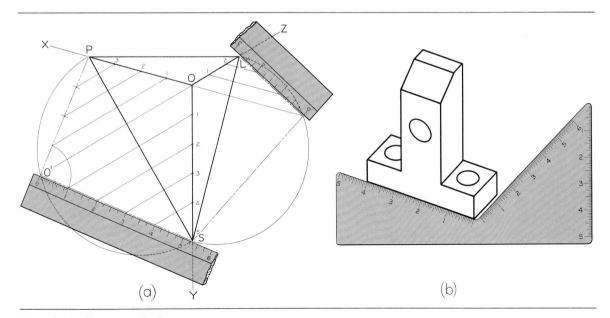

Fig. 18.46   Trimetric Scales.

rectly foreshortened, as shown. These dimensions should be transferred to the edges of three thin cards and marked OX, OZ, and OY for easy reference.

A special trimetric angle may be prepared from Bristol Board or plastic, as shown at (b). Perhaps six or seven such guides, using angles for a variety of positions of the axes, would be sufficient for all practical requirements.*

18.32 Trimetric Ellipses

The trimetric center lines of a hole, or on the end of a cylinder, become the conjugate diameters of the ellipse when drawn in trimetric. The ellipse may be drawn on the conjugate diameters by the methods of Fig. 5.51 or 5.52 (b). Or the major and minor axes may be determined from the conjugate diameters, Fig. 5.53 (c), and the ellipse constructed on them by any of the methods of Figs. 5.48–5.50 and 5.52 (a), or with the aid of an ellipse guide, Fig. 5.55.

One advantage of trimetric projection is the infinite number of positions of the object available. The angles and scales can be handled without too much difficulty, as shown in §18.31. However, the infinite variety of ellipses has been a discouraging factor.

In drawing any axonometric ellipse, keep the following in mind.

1. On the drawing, the major axis is always perpendicular to the center line, or axis, of the cylinder.

2. The minor axis is always perpendicular to the major axis; that is, on the paper it coincides with the axis of the cylinder.

3. The length of the major axis is equal to the actual diameter of the cylinder.

Thus we know at once the directions of both the major and minor axes, and the length of the major axis. *We do not know the length of the*

*Plastic templates of this type are available from drafting supplies dealers.

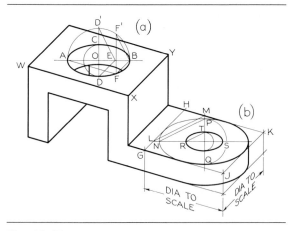

Fig. 18.47 Ellipses in Trimetric. *Method (b) courtesy of Professor H. E. Grant.*

minor axis. If we can find it, we can easily construct the ellipse with the aid of an ellipse guide or any of a number of ellipse constructions mentioned earlier.

In Fig. 18.47 (a), center O is located as desired, and horizontal and vertical construction lines that will contain the major and minor axes are drawn through O. Note that the major axis will be on the horizontal line perpendicular to the axis of the hole, and the minor axis will be perpendicular to it, or vertical.

Set the compass for the actual radius of the hole and draw the semicircle, as shown, to establish the ends A and B of the major axis. Draw AF and BF parallel to the axonometric edges WX and YX, respectively, to locate F, which lies on the ellipse. Draw a vertical line through F to intersect the semicircle at F′ and join F′ to B as shown. From D′ where the minor axis, extended, intersects the semicircle, draw D′E and ED parallel to F′B and BF, respectively. Point D is one end of the minor axis. From center O, strike arc DC to locate C, the other end of the minor axis. On these axes, a true ellipse can be constructed, or drawn with the aid of an ellipse guide. A simple method for finding the "angle" of ellipse guide to use is shown in Fig. 5.55 (c). If an ellipse guide is not available, an approximate four-center el-

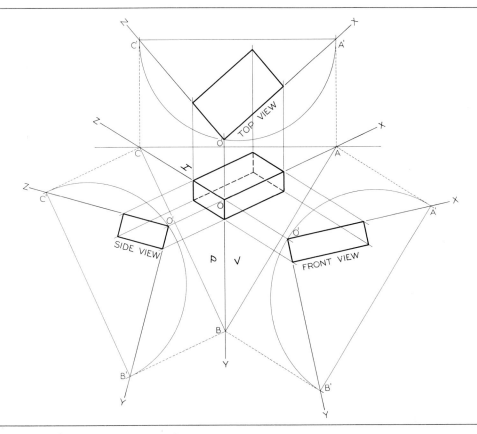

Fig. 18.48   Views from an Axonometric Projection.

lipse, Fig. 5.56, will be found satisfactory in most cases.

In constructions where the enclosing parallelogram for an ellipse is available or easily constructed, the major and minor axes can be readily determined as shown in Fig. 18.47 (b). The directions of both axes, and the length of the major axis, are known. Extend the axes to intersect the sides of the parallelogram at L and M, and join the points with a straight line. From one end N of the major axis, draw a line NP parallel to LM. The point P is one end of the minor axis. To find one end T of the minor axis of the smaller ellipse, it is only necessary to draw RT parallel to LM or NP.

The method of constructing an ellipse on an oblique plane in trimetric is similar to that shown for isometric in Fig. 18.33.

18.33   Axonometric Projection by the Method of Intersections

Instead of constructing axonometric projections with the aid of specially prepared scales, as explained in the preceding paragraphs, an axonometric projection can be obtained directly by projection from two orthographic views of the object. This method is called the *method of intersections;* it was developed by Profs. L. Eckhart and T. Schmid of the Vienna College of Engineering and was published in 1937.

Fig. 18.48   To understand this method, let us assume that the axonometric projection of a rectangular object is given, and it is required to find the three orthographic projections: the top view, front view, and side view.

627

Assume that the object is placed so that its principal edges coincide with the coordinate axes, and assume that the plane of projection (the plane on which the axonometric projection is drawn) intersects the three coordinate planes in the triangle ABC. From descriptive geometry, we know that lines BC, CA, and AB will be perpendicular, respectively, to axes OX, OY, and OZ. Any one of the three points A, B, or C may be assumed anywhere on one of the axes, and the triangle ABC drawn.

To find the true size and shape of the top view, revolve the triangular portion of the horizontal plane AOC, which is in front of the plane of projection, about its base CA, into the plane of projection. In this case, the triangle is revolved *inward* to the plane of projection through the smallest angle made with it. The triangle will then be shown in its true size and shape, and the top view of the object can be drawn in the triangle by projection from the axonometric projection, as shown, since all width dimensions remain the same. In the figure, the base CA of the triangle has been moved upward to C'A' so that the revolved position of the triangle will not overlap its projection.

In the same manner, the true sizes and shapes of the front view and side view can be found, as shown.

It is evident that if the three orthographic projections, or in most cases any two of them, are given in their relative positions, as shown in Fig. 18.48, the directions of the projections could be reversed so that the intersections of the projecting lines would determine the required axonometric projection.

In order to draw an axonometric projection by the method of intersections, it is well to make a sketch, Fig. 18.49, of the desired general appearance of the projection. Even if the object is a complicated one, this sketch need not be complete, but may be only a sketch of an enclosing box. Draw the projections of the coordinate axes OX, OY, and OZ, parallel to the principal edges of the object as shown in the sketch, and the triangle ABC to represent the intersection of the three coordinate planes with the plane of projection.

Revolve the triangle ABO about its base AB as the axis into the plane of projection. Line OA will revolve to O'A, and this line, or one parallel to it, must be used as the base line of the front view of the object. The projecting lines from the front view to the axonometric must be drawn parallel to the projection of the unrevolved Z-axis, as indicated in the figure.

Similarly, revolve the triangle COB about its base CB as the axis into the plane of projection. Line CO will revolve to CO″, and this line, or one parallel to it, must be used as the base line of the side view. The direction of the projecting lines must be parallel to the projection of the unrevolved X-axis, as shown.

Draw the front-view base line at a convenient location, but parallel to O'X, and with it as the base, draw the front view of the object. Draw the side-view base line also at a convenient location, but parallel to O″C, and with it as the base, draw the side view of the object, as shown. From the corners of the front view, draw projecting lines parallel to OZ, and from the corners of the side view, draw projecting lines parallel to OX. The intersections of these two sets of projecting lines determine the desired axonometric projection. It will be an isometric, a dimetric, or a trimetric projection, depending on the form of the sketch used as the basis for the projections, §18.2. If the sketch is drawn so that the three angles formed by the three coordinate axes are equal, the resulting projection will be an isometric projection; if two of the three angles are equal, the resulting projection will be a dimetric projection; and if no two of the three angles are equal, the resulting projection will be a trimetric projection.

In order to place the desired projection on a specific location on the drawing, Fig. 18.49, select the desired projection P of the point 1, for example, and draw two projecting lines PR and PS to intersect the two base lines and thereby to determine the locations of the two views on their base lines.

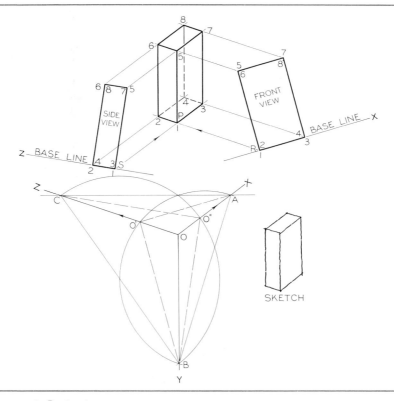

Fig. 18.49   Axonometric Projection.

Another example of this method of axonometric projection is shown in Fig. 18.50. In this case, it was deemed necessary only to draw a sketch of the plan or base of the object in the desired position, as shown. The axes are then drawn with OX and OZ parallel, respectively, to the sides of the sketch plan, and the remaining axis OY is assumed in a vertical position. The triangles COB and AOB are revolved, and the two base lines drawn parallel to O″C and O′A as shown. Point P, the lower front corner of the axonometric drawing, was then chosen at a convenient place, and projecting lines drawn toward the base lines parallel to axes OX and OZ to locate the positions of the views on the base lines. The views are drawn on the base lines or cut apart from another drawing and fastened in place with drafting tape.

To draw the elliptical projection of the circle, assume any points, such as A, on the circle in both front and side views. Note that point A is the same altitude d above the base line in both views. The axonometric projection of point A is found simply by drawing the projecting lines from the two views. The major and minor axes may be easily found by projecting in this manner or by methods shown in Fig. 18.47, and the true ellipse drawn by any of the methods of Figs. 5.48–5.50 and 5.52 (a), or with the aid of an ellipse guide, §§5.56 and 18.21. Or an approximate ellipse, which is satisfactory for most drawings, may be used, Fig. 5.56.

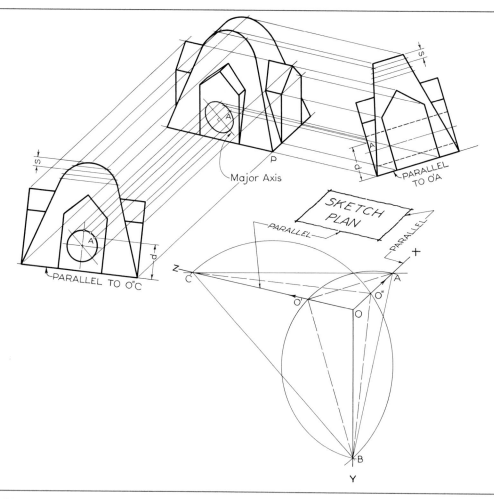

Fig. 18.50 Axonometric Projection.

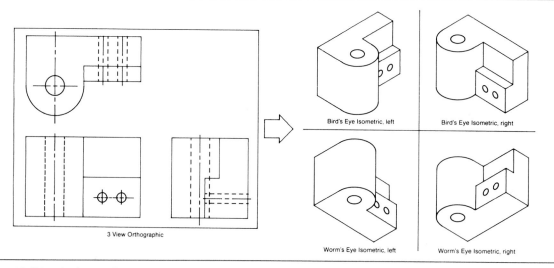

3 View Orthographic

Bird's Eye Isometric, left

Bird's Eye Isometric, right

Worm's Eye Isometric, left

Worm's Eye Isometric, right

Fig. 18.51 Orthographic to Isometric Conversion. The Auto-trol Orthographic to Axonometric Package (OTAP) system can be used to convert an orthographic drawing to axonometric. *Courtesy of Auto-trol Technology Corporation.*

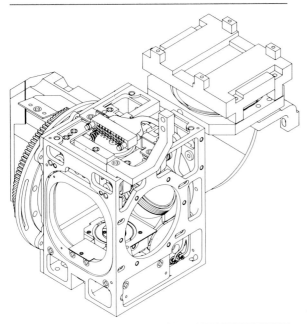

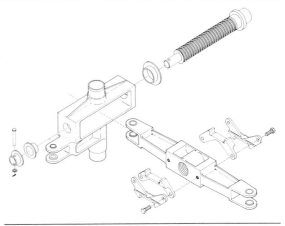

Fig. 18.53   Exploded Assembly Drawing Produced from Engineering Data by Using the Auto-trol Orthographic to Axonometric Package (OTAP) System. *Courtesy of Auto-trol Technology Corporation.*

Fig. 18.52   Isometric Assembly Drawing Produced by Using the Computervision Designer System. *Courtesy of Computervision Corporation, a subsidiary of Prime Computer, Inc.*

18.34   Computer Graphics

Complex pictorial illustrations can be among the most time-consuming tasks faced by the drafter. Modern computer graphics software, however, provides relief to the knowledgeable user. Orthographic-to-axonometric programs are available that can easily convert a multiview drawing to an isometric with the user's choice of alignment, Fig. 18.51. Isometric assembly drawings, Fig. 18.52, exploded assembly drawings, Fig. 18.53, and pictorial sections can also be readily generated.

AXONOMETRIC PROBLEMS

A large number of problems to be drawn axonometrically are given in Figs. 18.54–18.58. The earlier isometric sketches may be drawn on isometric paper, §6.14; later sketches should be made on plain drawing paper. On drawings to be executed with instruments, show all construction lines required in the solutions.

For additional problems, see Figs. 7.52–7.54, 19.23–19.26, and 20.34.

Since many of the problems in this chapter are of a general nature, they can also be solved on most computer graphics systems. If a system is available, the instructor may choose to assign specific problems to be completed by this method.

Axonometric problems in convenient form for solution may be found in *Technical Drawing Problems,* Series 1, by Giesecke, Mitchell, Spencer, Hill, Dygdon, and Novak; *Technical Drawing Problems,* Series 2, by Spencer, Hill, Dygdon, and Novak; and *Technical Drawing Problems* Series 3, by Spencer, Hill, Dygdon, and Novak, all designed to accompany this text and published by Macmillan Publishing Company.

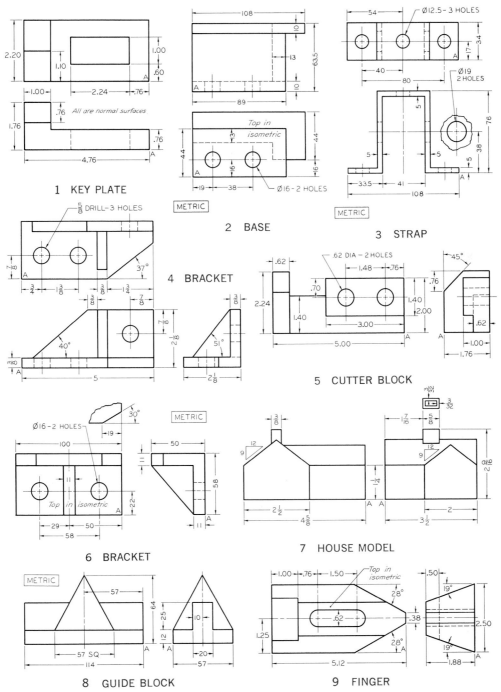

1 KEY PLATE

2 BASE

3 STRAP

4 BRACKET

5 CUTTER BLOCK

6 BRACKET

7 HOUSE MODEL

8 GUIDE BLOCK

9 FINGER

Fig. 18.54 (1) Make freehand isometric sketches. (2) Make isometric drawings with instruments on Layout A–2 or A4–2 (adjusted). (3) Make dimetric drawings with instruments, using Layout A–2 or A4–2 (adjusted), and position assigned from Fig. 18.42. (d) Make trimetric drawings, using instruments, with axes chosen to show the objects to best advantage. If dimensions are required, study §18.25.

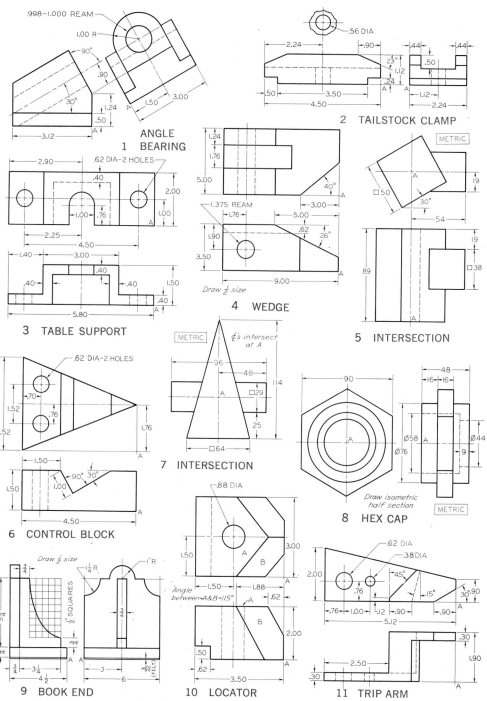

Fig. 18.55 (1) Make freehand isometric sketches. (2) Make isometric drawings with instruments on Layout A–2 or A4–2 (adjusted). (3) Make dimetric drawings with instruments, using Layout A–2 or A4–2 (adjusted), and position assigned from Fig. 18.42. (4) Make trimetric drawings, using instruments, with axes chosen to show the objects to best advantage. If dimensions are required, study §18.25.

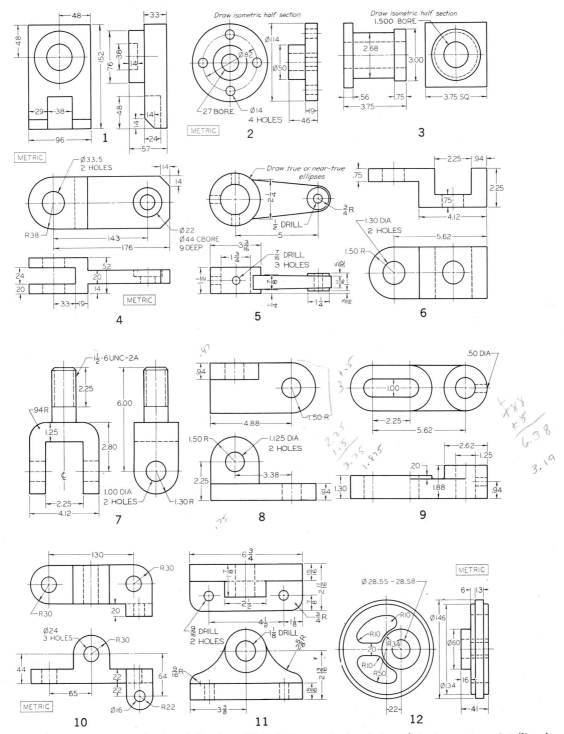

Fig. 18.56 (1) Make isometric freehand sketches. (2) Make isometric drawings with instruments, using Size A or A4 sheet or Size B or A3 sheet, as assigned. (3) Make dimetric drawings with instruments, using Size A or A4 sheet or Size B or A3 sheet, as assigned, and position assigned from Fig. 18.42. (4) Make trimetric drawings, using instruments, with axes chosen to show the objects to best advantage. If dimensions are required, study §18.25.

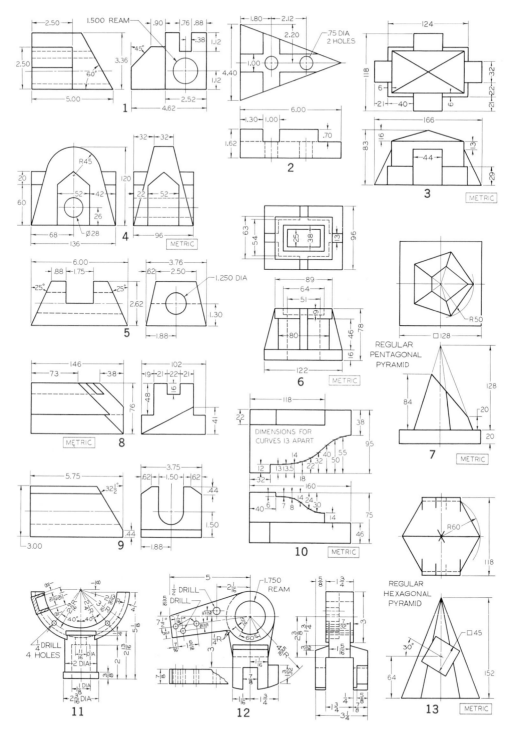

Fig. 18.57 (1) Make isometric freehand sketches. (2) Make isometric drawings with instruments, using Size A or A4 sheet or Size B or A3 sheet, as assigned. (3) Make dimetric drawings with instruments, using Size A or A4 sheet or Size B or A3 sheet, as assigned, and position assigned from Fig. 18.42. (4) Make trimetric drawings, using instruments, with axes chosen to show the objects to best advantage. If dimensions are required, study §18.25.

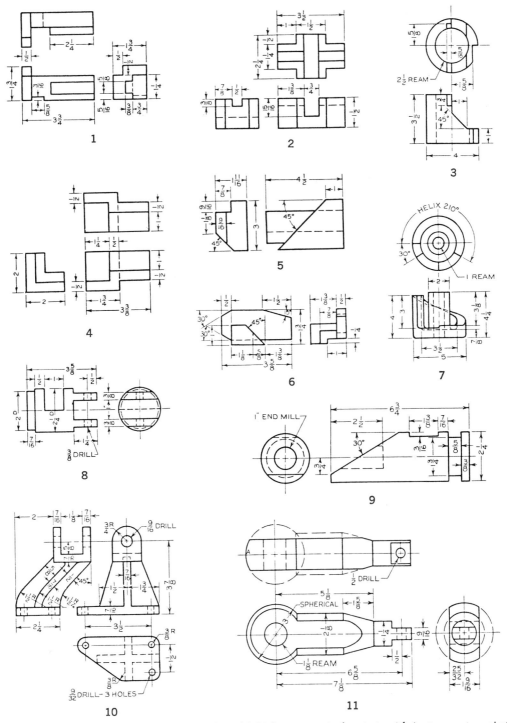

Fig. 18.58 (1) Make isometric freehand sketches. (2) Make isometric drawings with instruments, using Size A or A4 sheet or Size B or A3 sheet, as assigned. (3) Make dimetric drawings with instruments, using Size A or A4 sheet or Size B or A3 sheet, as assigned, and position assigned from Fig. 18.42. (4) Make trimetric drawings, using instruments, with axes chosen to show the objects to best advantage. If dimensions are required, study §18.25.

637

Oblique Projection

If the observer is considered to be stationed at an infinite distance from the object, Fig. 18.1 (c), and looking toward the object so that the projectors are parallel to each other and oblique to the plane of projection, the resulting drawing is an *oblique projection*. As a rule, the object is placed with one of its principal faces parallel to the plane of projection. This is equivalent to holding the object in the hand and viewing it approximately as shown in Fig. 6.27.

19.1 Oblique and Other Projections Compared

A comparison of oblique projection and orthographic projection is shown in Fig. 19.1. The front face A'B'C'D' in the oblique projection is identical with the front view, or orthographic projection, $A^VB^VC^VD^V$. Thus, if an object is placed with one of its faces parallel to the plane of projection, that face will be projected true size and shape in oblique projection as well as in orthographic or multiview projection. This is the reason why oblique projection is preferable to axonometric projection in representing certain objects pictorially. Note that surfaces of the object that are not parallel to the plane of projection will not project in true size and shape. For example, surface ABFE on the object (a square) projects as a parallelogram A'B'F'E' in the oblique projection.

In axonometric projection, circles on the object nearly always lie in surfaces inclined to the plane of projection and project as ellipses. In oblique projection, the object may be positioned so that those surfaces are parallel to the plane of projection, in which case the circles will project as true circles, and can be easily drawn with the compass.

A comparison of the oblique and orthographic projections of a cylindrical object is shown in Fig. 19.2. In both cases, the circular shapes project as true circles. Note that although the observer, looking in the direction of the oblique arrow, does see these shapes as ellipses, the drawing, or projection, represents not what is seen but what is projected on the plane of projection. This curious situation is peculiar to oblique projection.

Observe that the axis AB of the cylinder projects as a point A^VB^V in the orthographic projection, since the line of sight is parallel to AB. But in the oblique projection, the axis projects as a line A'B'. The more nearly the direction of sight approaches the perpendicular with respect to the plane of projection—that is, the larger the

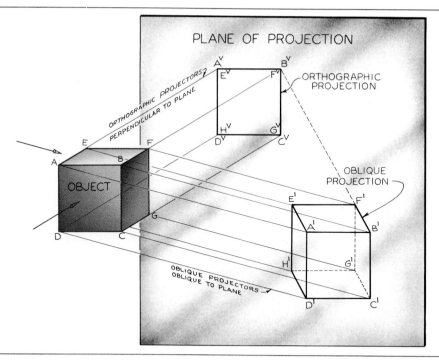

Fig. 19.1   Comparison of Oblique and Orthographic Projections.

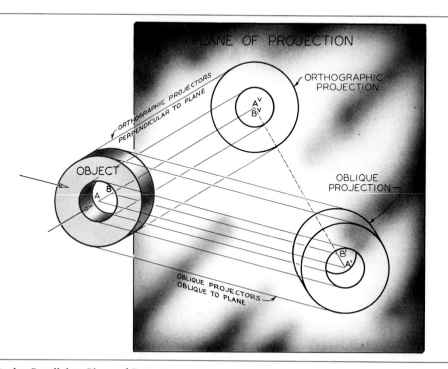

Fig. 19.2   Circles Parallel to Plane of Projection.

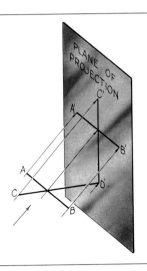

Fig. 19.3   Lengths of Projections.

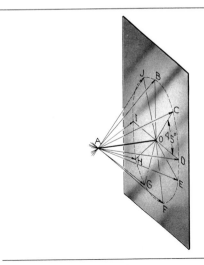

Fig. 19.4   Directions of Projections.

angle between the projectors and the plane—the closer the oblique projection moves toward the orthographic projection, and the shorter A'B' becomes.

19.2   Directions of Projectors

In Fig. 19.3, the projectors make an angle of 45° with the plane of projection; hence, the line CD', which is perpendicular to the plane, projects true length at C'D'. If the projectors make a greater angle with the plane of projection, the oblique projection is shorter, and if the projectors make a smaller angle with the plane of projection, the oblique projection is longer. Theoretically, CD' could project in any length from zero to infinity. However, the line AB is parallel to the plane and will project in true length regardless of the angle the projectors make with the plane of projection.

In Fig. 19.1, the lines AE, BF, CG, and DH are perpendicular to the plane of projection, and project as parallel inclined lines A'E', B'F', C'G', and D'H' in the oblique projection. These lines on the drawing are called the *receding lines*. As we have seen, they may be any length, from zero to infinity, depending on the direction of the line of sight. Our next concern is: What angle do these lines make on paper with respect to horizontal?

In Fig. 19.4, the line AO is perpendicular to the plane of projection, and all the projectors make angles of 45° with it; therefore, all the oblique projections BO, CO, DO, and so on, are equal in length to the line AO. It can be seen from the figure that the projectors may be selected in any one of an infinite number of directions and yet maintain any desired angle with the plane of projection. It is also evident that the directions of the projections BO, CO, DO, and so on, are independent of the angles the projectors make with the plane of projection. Ordinarily, this inclination of the projection is 45° (CO in the figure), 30°, or 60° with horizontal, since these angles may be easily drawn with the triangles.

19.3   Angles of Receding Lines

The receding lines may be drawn at any convenient angle. Some typical drawings with the receding lines in various directions are shown in Fig.

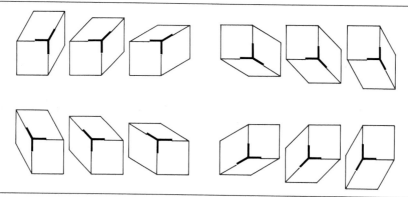

Fig. 19.5 Variation in Direction of Receding Axis.

19.5. The angle that should be used in an oblique drawing depends on the shape of the object and the location of its significant features. For example, in Fig. 19.6 (a) a large angle was used in order to obtain a better view of the rectangular recess on the top, while at (b) a small angle was chosen to show a similar feature on the side.

19.4 Length of Receding Lines

Since the eye is accustomed to seeing objects with all receding parallel lines appearing to converge, an oblique projection presents an unnatural appearance, with more or less serious distortion depending on the object shown. For example, the object shown in Fig. 19.7 (a) is a cube, the receding lines being full length, but the receding lines appear to be too long and to diverge toward the rear of the block. A striking example of the unnatural appearance of an oblique drawing when compared with the natu-

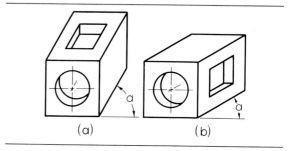

Fig. 19.6 Angle of Receding Axis.

ral appearance of a perspective is shown in Fig. 19.8. This example points up one of the chief limitations of oblique projection: objects characterized by great length should not be drawn in oblique with the long dimension perpendicular to the plane of projection.

The appearance of distortion may be materially lessened by decreasing the length of the receding lines (remember, we established in

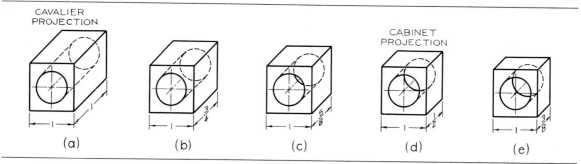

Fig. 19.7 Foreshortening of Receding Lines.

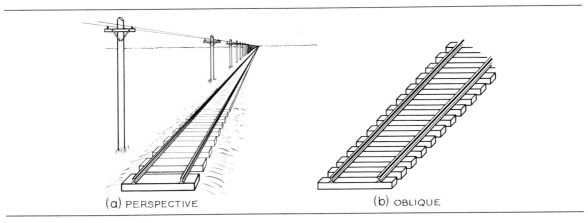

Fig. 19.8   Unnatural Appearance of Oblique Drawing.

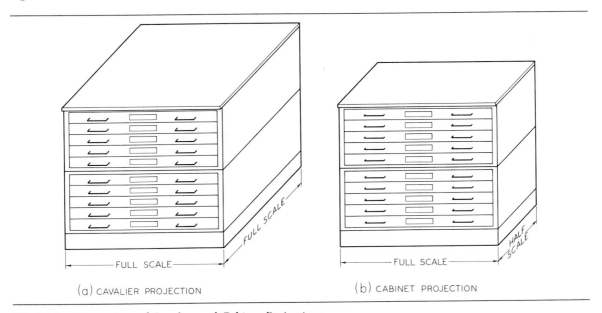

Fig. 19.9   Comparison of Cavalier and Cabinet Projections.

§19.2 that they could be any length). In Fig. 19.7 a cube is shown in five oblique drawings with varying degrees of foreshortening of the receding lines. The range of scales chosen is sufficient for almost all problems, and most of the scales are available on the architects', engineers', or metric scales.

When the receding lines are true length—that is, when the projectors make an angle of 45° with the plane of projection—the oblique drawing is called a *cavalier projection,* Fig. 19.7 (a). Cavalier projections originated in the drawing of medieval fortifications and were made on horizontal planes of projection. On these fortifications the central portion was higher than the rest, and it was called *cavalier* because of its dominating and commanding position.

When the receding lines are drawn to half size, as at (d), the drawing is commonly known as a *cabinet projection.* The term is attributed to the early use of this type of oblique drawing in the furniture industries. A comparison of cavalier projection and cabinet projection is shown in Fig. 19.9.

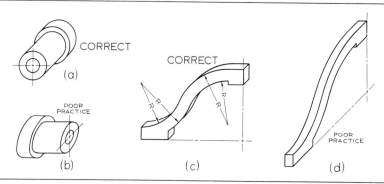

Fig. 19.10   Essential Contours Parallel to Plane of Projection.

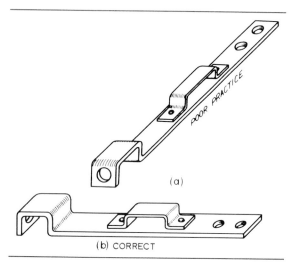

Fig. 19.11   Long Axis Parallel to Plane of Projection.

19.5   Choice of Position

The face of an object showing the essential contours should generally be placed parallel to the plane of projection, Fig. 19.10. If this is done, distortion will be kept at a minimum and labor reduced. For example, at (a) and (c) the circles and circular arcs are shown in their true shapes and may be quickly drawn with the compass, while at (b) and (d) these curves are not shown in their true shapes and must be plotted as free curves or in the form of ellipses.

The longest dimension of an object should generally be placed parallel to the plane of projection, as shown in Fig. 19.11 (b).

19.6   Steps in Oblique Drawing

The steps in drawing a cavalier drawing of a rectangular object are shown in Fig. 19.12. As shown in step I, draw the axes OX and OY per-

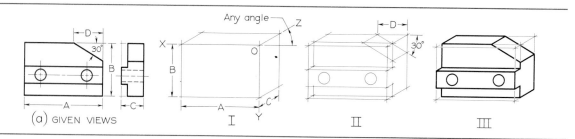

Fig. 19.12   Steps in Oblique Drawing—Box Construction.

pendicular to each other and the receding axis OZ at any desired angle with horizontal. On these axes, construct an enclosing box, using the overall dimensions of the object.

As shown at II, block in the various shapes in detail, and as indicated at III, heavy in all final lines.

Many objects most adaptable to oblique representation are composed of cylindrical shapes built on axes or center lines. In such cases, the oblique drawing is best constructed on the projected center lines, as shown in Fig. 19.13.

The object is positioned so that the circles shown in the given top view are parallel to the plane of projection and, hence, can be readily drawn with the compass in their true shapes. The general procedure is to draw the center-line skeleton, as shown in steps I and II, and then to build the drawing on these center lines.

It is very important to construct all points of tangency, as shown in step IV, especially if the drawing is to be inked. For a review of tangencies, see §§5.33–5.41. The final cavalier drawing is shown in step V.

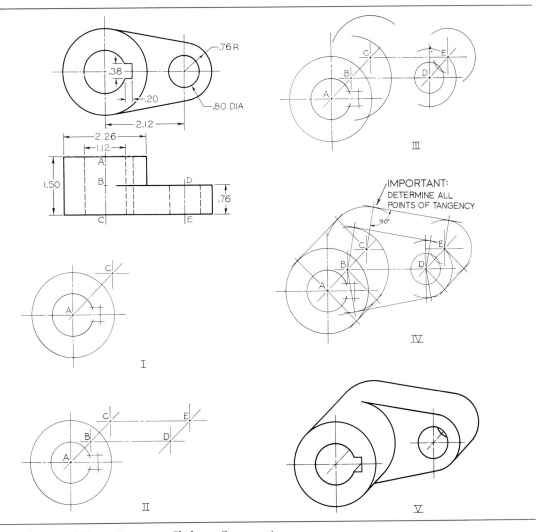

Fig. 19.13   Steps in Oblique Drawing—Skeleton Construction.

19.7 Four-Center Ellipse

It is not always possible to place an object so that all its significant contours are parallel to the plane of projection. For example, the object

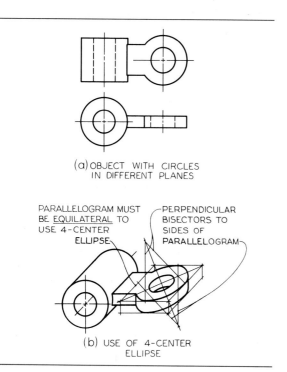

(a) OBJECT WITH CIRCLES IN DIFFERENT PLANES

PARALLELOGRAM MUST BE <u>EQUILATERAL</u> TO USE 4-CENTER ELLIPSE

PERPENDICULAR BISECTORS TO SIDES OF PARALLELOGRAM

(b) USE OF 4-CENTER ELLIPSE

Fig. 19.14 Circles and Arcs Not Parallel to Plane of Projection.

shown in Fig. 19.14 (a) has two sets of circular contours in different planes, and both cannot be placed parallel to the plane of projection.

In the oblique drawing at (b), the regular four-center method of Fig. 18.24 was used to construct ellipses representing circular curves not parallel to the plane of projection. This method can be used only in cavalier drawing in which case the enclosing parallelogram is equilateral—that is, the receding axis is drawn to full scale. The method is the same as in isometric: erect perpendicular bisectors to the four sides of the parallelogram; their intersections will be centers for the four circular arcs. If the angle of the receding lines is other than 30° with horizontal, as in this case, the centers of the two large arcs will not fall in the corners of the parallelogram.

The regular four-center method is not convenient in oblique drawing unless the receding lines make 30° with horizontal so that the perpendicular bisectors may be drawn easily with the 30° × 60° triangle and the T-square, parallel rule, or drafting machine without the necessity of first finding the midpoints of the sides. A more convenient method is the alternate four-center ellipse drawn on the two center lines, as shown in Fig. 19.15. This is the same method as used in isometric, Fig. 18.29, but in oblique drawing

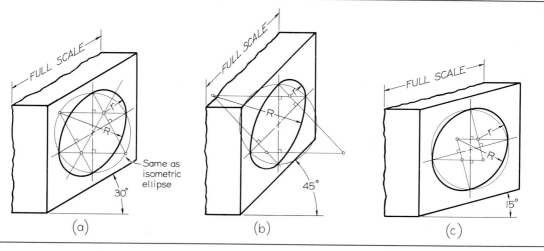

Fig. 19.15 Alternate Four-Center Ellipse.

646

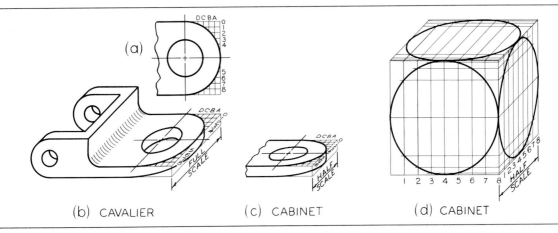

Fig. 19.16   Use of Offset Measurements.

it varies slightly in appearance according to the different angles of the receding lines.

First, draw the two center lines. Then, from the center, draw a construction circle equal in diameter to the actual hole or cylinder. The circle will intersect each center line at two points. From the two points on one center line, erect perpendiculars to the other center line, then, from the two points on the other center line, erect perpendiculars to the first center line. From the intersections of the perpendiculars, draw four circular arcs, as shown.

It must be remembered that the four-center ellipse can be inscribed only in an *equilateral* parallelogram; hence, it cannot be used in any oblique drawing in which the receding axis is foreshortened. Its use is limited, therefore, to cavalier drawing.

19.8   Offset Measurements

Circles, circular arcs, and other curved or irregular lines may be drawn by means of offset measurements, as shown in Fig. 19.16. The offsets are first drawn on the multiview drawing of the curve, as shown at (a), and these are transferred to the oblique drawing, as shown at (b). In this case, the receding axis is full scale, and therefore all offsets can be drawn full scale. The four-center ellipse could be used, but the method here is more accurate. The final curve is drawn with the aid of the irregular curve, §2.54.

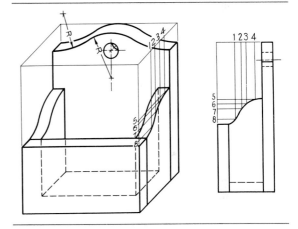

Fig. 19.17   Use of Offset Measurements.

If the oblique drawing is a cabinet drawing, as shown at (c), or any oblique drawing in which the receding axis is drawn to a reduced scale, the offset measurements parallel to the receding axis must be drawn to the same reduced scale. In this case, there is no choice of methods, since the four-center ellipse could not be used. A method of drawing ellipses in a cabinet drawing of a cube is shown at (d).

As shown in Fig. 19.17, a free curve may be drawn in oblique by means of offset measurements. This figure also illustrates a case in which hidden lines are used to make the drawing clearer.

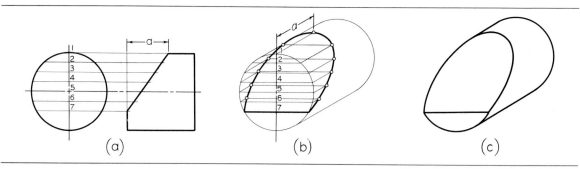

Fig. 19.18   Use of Offset Measurements.

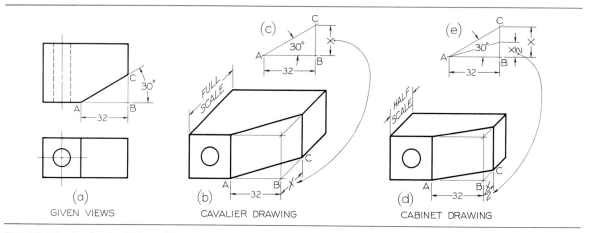

Fig. 19.19   Angles in Oblique Projection.

The use of offset measurements in drawing an ellipse in a plane inclined to the plane of projection is shown in Fig. 19.18. At (a) a number of parallel lines are drawn to represent imaginary cutting planes. Each plane will cut a rectangular surface between the front end of the cylinder and the inclined surface. These rectangles are drawn in oblique, as shown at (b), and the curve is drawn through corner points, as indicated. The final cavalier drawing is shown at (c).

19.9   Angles in Oblique Projection

If an angle that is specified in degrees lies in a receding plane, it is necessary to convert the angle into linear measurements in order to draw

the angle in oblique. For example, in Fig. 19.19 (a) an angle of 30° is given. In order to draw the angle in oblique, we need to know dimensions AB and BC. The distance AB is given as $1\frac{1}{4}''$ and can be set off directly in the cavalier drawing, as shown at (b). Distance BC is not known, but can easily be found by constructing the right triangle ABC at (c) from the given dimensions in the top view at (a). The length BC is then transferred with the dividers to the cavalier drawing, as shown.

In cabinet drawing, it must be remembered that *all receding dimensions* must be reduced to half size. Thus, in the cabinet drawing at (d), the distance BC must be half the side BC of the right triangle at (e), as shown.

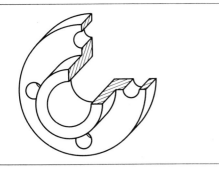

Fig. 19.20 Oblique Half Section.

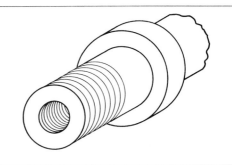

Fig. 19.21 Screw Threads in Oblique.

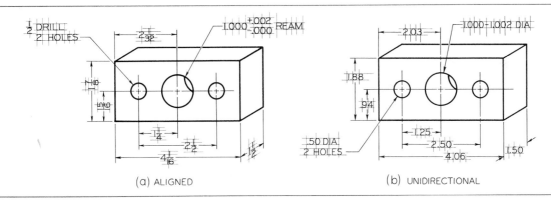

(a) ALIGNED   (b) UNIDIRECTIONAL

Fig. 19.22 Oblique Dimensioning.

19.10 Oblique Sections

Sections are often useful in oblique drawing, especially in the representation of interior shapes. An *oblique half section* is shown in Fig. 19.20. Other examples are shown in Figs. 9.41, 9.42, and 9.45. *Oblique full sections,* in which the plane passes completely through the object, are seldom used because they do not show enough of the exterior shapes. In general, all the types of sections discussed in §18.24 for isometric drawing may be applied equally to oblique drawing.

19.11 Screw Threads in Oblique

Parallel partial circles spaced equal to the symbolic thread pitch, Fig. 15.17 (b), are used to represent the crests only of a screw thread in a

cavalier oblique, Fig. 19.21. For cabinet oblique the space would be one-half of the symbolic pitch. If the thread is so positioned to require ellipses, they may be drawn by the four-center method of §19.7.

19.12 Oblique Dimensioning

An oblique drawing may be dimensioned in a similar manner to that described in §18.25 for isometric drawing, as shown in Fig. 19.22. The general principles of dimensioning, as outlined in Chapter 13, must be followed. As shown in the figure, all dimension lines, extension lines, and arrowheads must lie in the planes of the object to which they apply. The dimension figures also will lie in the plane when the aligned dimensioning system is used as shown at (a). For

649

the unidirectional system of dimensioning, (b), all dimension figures are set horizontal and read from the bottom of the drawing. This simpler system is often used on pictorials for production purposes. *Vertical lettering* should be used for all pictorial dimensioning.

Dimensions should be placed outside the outlines of the drawing except when greater clearness or directness of application results from placing the dimensions directly on the view. The dimensioning methods described apply equally to fractional, decimal, and metric dimensions. For many other examples of oblique dimensioning, see Figs. 7.61, 7.65, 7.66, and others on following pages.

19.13   Oblique Sketching

Methods of sketching in oblique on plain paper are illustrated in Fig. 6.27. Ordinary graph paper is very useful in oblique sketching, Fig. 6.28. The height and width proportions can be easily controlled by simply counting the squares. A very pleasing depth proportion can be obtained by sketching the receding lines at 45° diagonally through the squares and through half as many squares as the actual depth would indicate.

19.14   Computer Graphics

Using computer graphics, the drafter can easily create an oblique drawing that will provide the desired amount of foreshortening along the receding axis as well as the preferred direction of the axis. CAD programs also permit curves and circular features, which are not parallel to the frontal plane, to be readily shown on the drawing. Oblique sections, §19.10, and repetitive features such as screw threads, §19.11, may be quickly and accurately depicted.

OBLIQUE PROJECTION PROBLEMS

A large number of problems to be drawn in oblique—either cavalier or cabinet—are given in Figs. 19.23–19.26. They may be drawn freehand, §6.15, using graph paper or plain drawing paper as assigned by the instructor, or they may be drawn with instruments. In the latter case, all construction lines should be shown on the completed drawing.

Many additional problems suitable for oblique projection will be found in Figs. 7.52–7.54, 10.27–10.29, 18.54–18.57, and 20.34.

Since many of the problems in this chapter are of a general nature, they can also be solved on most computer graphics systems. If a system is available, the instructor may choose to assign specific problems to be completed by this method.

Problems in convenient form for solution may be found in *Technical Drawing Problems,* Series 1, by Giesecke, Mitchell, Spencer, Hill, Dygdon, and Novak; *Technical Drawing Problems,* Series 2, by Spencer, Hill, Dygdon, and Novak; and *Technical Drawing Problems,* Series 3, by Spencer, Hill, Dygdon, and Novak; all designed to accompany this text and published by Macmillan Publishing Company.

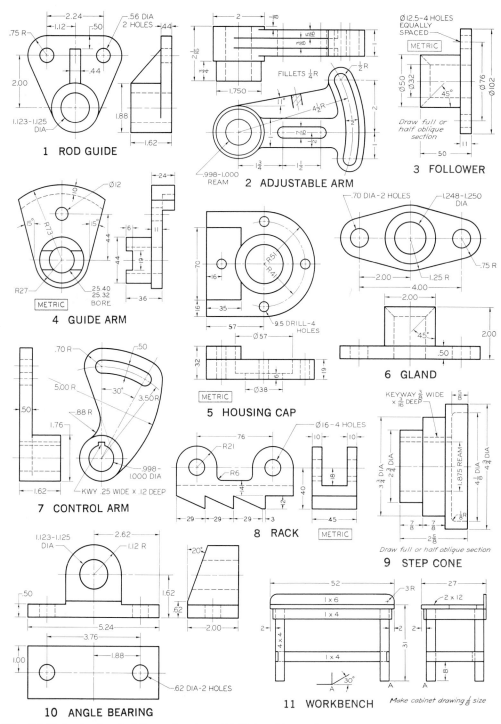

Fig. 19.23 (1) Make freehand oblique sketches. (2) Make oblique drawings with instruments, using Size A or A4 sheet, or Size B or A3 sheet, as assigned. If dimensions are required, study §19.12.

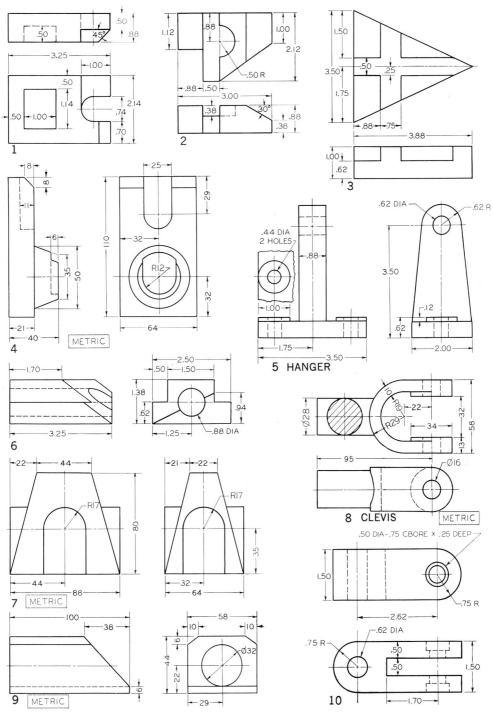

Fig. 19.24 Make oblique drawings with instruments, using Size A or A4 sheet, or Size B or A3 sheet, as assigned. If dimensions are required, study §19.12.

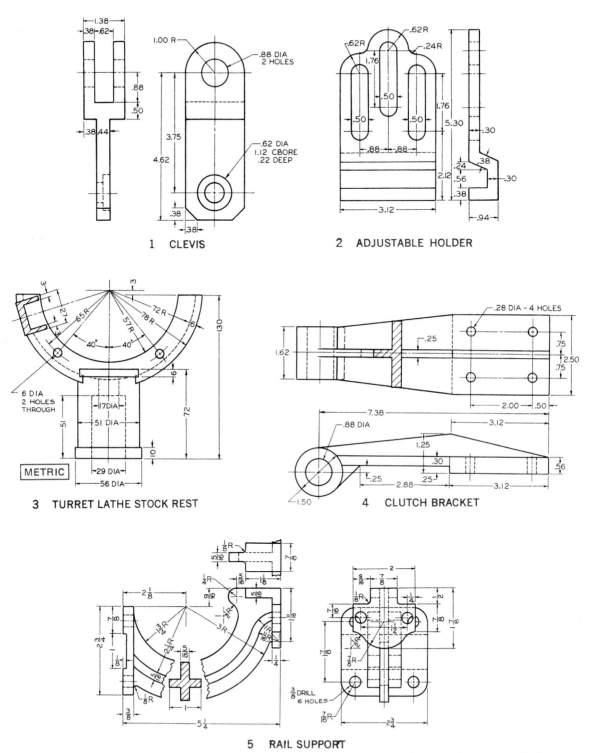

1 CLEVIS

2 ADJUSTABLE HOLDER

3 TURRET LATHE STOCK REST

METRIC

4 CLUTCH BRACKET

5 RAIL SUPPORT

Fig. 19.25 Make oblique drawings with instruments, using Size A or A4 sheet, or Size B or A3 sheet, as assigned. If dimensions are required, study §19.12.

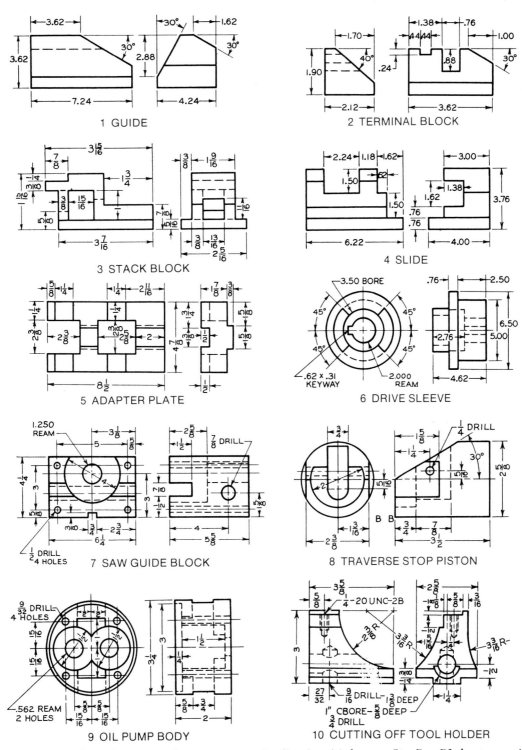

Fig. 19.26 Make oblique drawings with instruments, using Size A or A4 sheet, or Size B or B3 sheet, as assigned. If dimensions are required, study §19.12.

655

CHAPTER 20

Perspective

Perspective, or *central projection,* excels over all other types of projection in the pictorial representation of objects because it more closely approximates the view obtained by the human eye, Fig. 20.1. Geometrically, an ordinary photograph is a perspective. While perspective is of major importance to the architect, industrial designer, or illustrator, the engineer at one time or another is certain to be concerned with the pictorial representation of objects and should understand the basic principles of perspective. See ANSI/ASME Y14.4–1989.

20.1 General Principles

As explained in §1.10, a perspective involves four main elements: (1) the observer's eye, (2) the object being viewed, (3) the plane of projection, and (4) the projectors from the observer's eye to all points on the object. With rare exceptions, §20.6, the plane of projection is placed between the observer and the object, as shown in Fig. 1.10, and the collective piercing points in the plane of projection of all the projectors produce the perspective.

In Fig. 20.2, the observer is shown looking along a boulevard and through an imaginary plane of projection. This plane is called the *pic-*

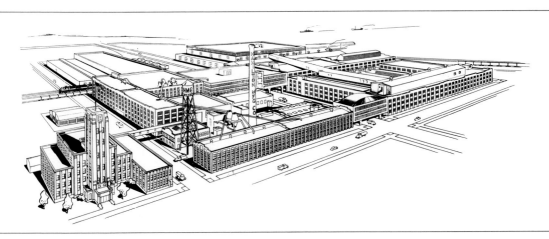

Fig. 20.1   Perspective of a Manufacturing Plant. *Courtesy of Hamilton Mfg. Co.*

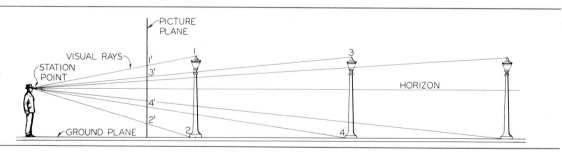

Fig. 20.2   Looking Through the Picture Plane.

ture plane or simply PP. The position of the observer's eye is called the *station point* or simply SP. The lines from SP to the various points in the scene are the projectors, or more properly in perspective, *visual rays.* The points where the visual rays pierce PP are the *perspectives* of the respective points. Collectively, these piercing points form the perspective of the object or the scene as viewed by the observer. The perspective thus obtained is shown in Fig. 20.3.

In Fig. 20.2, the perspective of lamp post 1–2 is shown at 1′–2′, on the picture plane; the perspective of lamp post 3–4 is shown at 3′–4′, and so on. Each succeeding lamp post, as it is farther from the observer, will be projected smaller than the one preceding. A lamp post at an infinite distance from the observer would appear as a point on the picture plane. A lamp post in front of the picture plane would be projected taller than it is, and a lamp post in the picture plane would be projected in true length. In the perspective, Fig. 20.3, the diminishing heights of the posts are apparent.

In Fig. 20.2, the line representing the *horizon* is the edge view of the *horizon plane,* which is parallel to the ground plane and passes through SP. In the perspective, Fig. 20.3, the horizon is the line of intersection of this plane with the picture plane and represents the eye level of the observer, or SP. Also, in Fig. 20.2, the *ground plane* is the edge view of the ground on which the object usually rests. In Fig. 20.3, the *ground line,* or GL, is the intersection of the ground plane with the picture plane.

In Fig. 20.3, it will be seen that lines that are parallel to each other but not parallel to the picture plane, such as curb lines, sidewalk lines, and lines along the tops and bottoms of the lamp posts, all converge toward a single point on the horizon. This point is called the *vanishing point,* or VP, of the lines. Thus, the first rule to learn in perspective is this: *All parallel lines that are not parallel to PP vanish at a single vanishing point, and if these lines are parallel to the ground, the vanishing point will be on the horizon.* Parallel lines that are also parallel to PP,

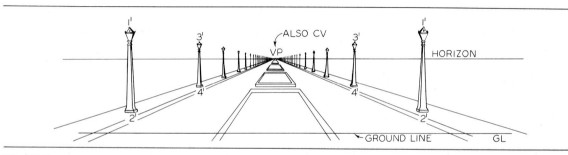

Fig. 20.3   A Perspective.

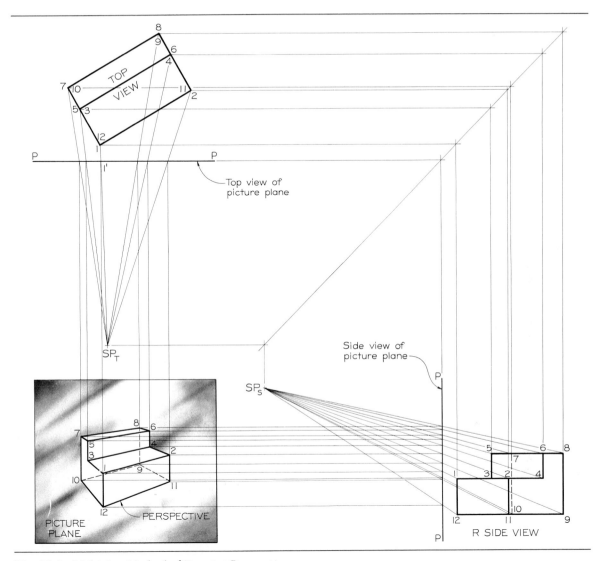

Fig. 20.4 Multiview Method of Drawing Perspective.

such as the lamp posts, remain parallel and do not converge toward a vanishing point.

20.2 Multiview Perspective

A perspective can be drawn by the ordinary methods of multiview projection, as shown in Fig. 20.4. In the upper portion of the drawing are shown the top view of the station point, the picture plane, the object, and the visual rays. In the

right-hand portion of the drawing are shown the right-side view of the same station point, picture plane, object, and visual rays. In the front view, the picture plane coincides with the plane of the paper, and the perspective is drawn on it. Note the method of projecting from the top view to the side view, which conforms to the usual multiview methods shown in Fig. 7.7.

To obtain the perspective of point 1, a visual ray is drawn in the top view from SP_T to point 1

659

on the object. From the intersection 1′ of this ray with the picture plane, a projection line is drawn downward till it meets a similar projection line from the side view. This intersection is the perspective of point 1, and the perspectives of all other points are found in a similar manner.

Observe that all parallel lines that are also parallel to the picture plane (the vertical lines) remain parallel and do not converge, whereas the other two sets of parallel lines converge toward vanishing points. As the vanishing points are not needed in the multiview construction of Fig. 20.4, they are not shown. But if the converging lines were extended, it would be found that they meet at two vanishing points (one for each set of parallel lines).

The perspective of any object may be constructed in this way, but if the object is placed at an angle with the picture plane, as is usually the case, the method is a bit cumbersome because of the necessity of constructing the side view in a revolved position. The revolved side view can be dispensed with, as shown in the following section.

20.3   The Setup for a Simple Perspective

The construction of a perspective of a simple form is shown in Fig. 20.5. The upper portion of the drawing, as in Fig. 20.4, shows the top views of SP, PP, and of the object. The lines SP–1, SP–2, SP–3, and SP–4 are the top views of the visual rays.

In the side view, a departure from Fig. 20.4 is made, in that a revolved side view is not required. All that is needed is any elevation view that will provide the necessary elevation or height measurements. If these dimensions are known, no view is required.

The perspective itself is drawn in the front-view position, the picture plane being considered as the plane of the paper on which the perspective is drawn. The ground line is the edge view of the ground plane or the intersection of the ground plane with the picture plane. The hori-

zon is a horizontal line in the picture plane that is the line of intersection of the horizon plane with the picture plane. Since the horizon plane passes through the observer's eye, or SP, the horizon is drawn at the level of the eye, that is, at the distance above the ground line representing, to scale, the altitude of the eye above the ground.

The *center of vision,* or CV, is the orthographic projection (or front view) of SP on the picture plane, and since the horizon is at eye level, CV will always be on the horizon, except in three-point perspective, §20.11. In Fig. 20.5, the top view of CV is CV′, found by dropping a perpendicular from SP to PP. The front view CV is found by projecting downward from CV′ to the horizon.

20.4   To Draw an Angular Perspective

Since objects are defined principally by edges that are straight lines, the drawing of a perspective resolves itself into drawing the *perspective of a line.* A drafter who can draw the perspective of a line can draw the perspective of any object, no matter how complex.

To draw the perspective of any horizontal straight line not parallel to PP—for example, the line 1–2 in Fig. 20.5—proceed as follows.

I.   *Find the piercing point in* PP *of the line.* In the top view, extend line 1–2 until it pierces PP at T; then project downward to the level of the line 1–2 projected horizontally from the side view. The point S is the piercing point of the line.

II.   *Find the vanishing point of the line.* The vanishing point of a line is the piercing point in PP of a line drawn through SP parallel to that line. Hence, the vanishing point VPR of the line 1–2 is found by drawing a line from SP parallel to that line and finding the top view of its piercing point O, and then projecting downward to the horizon. The line SP–O is actually a visual ray drawn toward the infinitely distant point on line 1–2 of the object, extended, and the vanishing point is the intersection of this visual ray

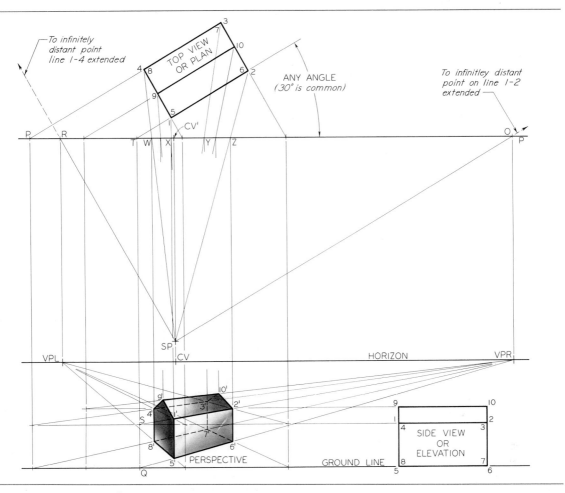

Fig. 20.5 Perspective of a Prism.

with the picture plane. The vanishing point is, then, the perspective of the infinitely distant point on the line extended.

III. *Join the piercing point and the vanishing point with a straight line.* The line S–VPR is the line joining these two points, and it is the perspective of a line of infinite length containing the required perspective of the line 1–2.

IV. *Locate the endpoints of the perspective of the line.* The endpoints 1′ and 2′ can be found by projecting down from the piercing points of the visual rays in PP, or by simply drawing the

perspectives of the remaining horizontal edges of the object. In practice, it is best to use both methods as a check on the accuracy of the construction. To locate the endpoints by projecting from the piercing points, draw visual rays from SP to the points 1 and 2 on the object in the top view. The top views of the piercing points are X and Z. Since the perspectives of points 1 and 2 must lie on the line S–VPR, project downward from X and Z to locate points 1′ and 2′.

After the perspectives of the horizontal edges have been drawn, the vertical edges and inclined edges can be drawn, as shown, to complete the

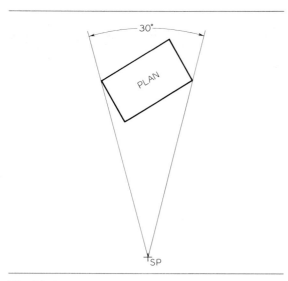

Fig. 20.6   Distance from Station Point to Object.

perspective of the object. Note that *vertical heights can be measured only in the picture plane*. If the front vertical edge 1–5 of the object was actually in PP—that is, if the object was situated with the front edge in PP—the vertical height could be set off directly full size. If the vertical edge is behind PP, a plane of the object, such as surface 1–2–5–6 can be extended forward until it intersects PP in line TQ. The line TQ is called a *measuring line,* and the true height SQ of line 1–5 can be set off with a scale or projected from the side view as shown.

If a large drawing board is not available, one vanishing point, such as VPR, may fall off the board. By using one vanishing point VPL, and projecting down from the piercing points in PP, vanishing point VPR may be eliminated. However, a valuable means of checking the accuracy of the construction will be lost.

20.5   Position of the Station Point
The center line of the cone of visual rays should be directed toward the approximate center, or center of interest, of the object. In two-point perspective, the type shown in Fig. 20.5, the loca-

tion of the station point (SP) in the plan view should be slightly to the left, not directly in front of the center of the object, and at such a distance that the object can be viewed at a glance without turning the head. This is accomplished if a cone of rays with its vertex at SP and a vertical angle of about 30° entirely encloses the object, as shown in Fig. 20.6.

In the perspective portion of Fig. 20.5, SP does not appear because the station point is in front of the picture plane. However, the orthographic projection CV of SP in the picture plane does show the height of the station point with respect to the ground plane. Since the horizon is at eye level, it also shows the altitude of the station point. Therefore, in the perspective portion of the drawing, the horizon is drawn a distance above the ground line at which it is desired to assume the station point. For most small and medium-size objects, such as machine parts or furniture, the station point is best assumed slightly above the top of the object. Large objects, such as buildings, are usually viewed from a station point about the altitude of the eye above the ground, or about 5′–6″.

20.6   Location of the Picture Plane
In general, the picture plane is placed in front of the object, as in Fig. 20.7 (b) and (c). However, it may be placed behind the object, as shown at (a), and it may even be placed behind SP, as shown at (d), in which event the perspective is reversed, as is the case of a camera. Of course, the usual position of the picture plane is between SP and the object. The perspectives in Fig. 20.7 differ in size but not in proportion. As shown at (b) and (c), with the picture plane between SP and the object, the farther that plane is from the object, the smaller the perspective will be. This distance may be assumed, therefore, with the thought of controlling the scale of the perspective. In practice, however, the object is usually assumed with the front corner in the picture plane to facilitate vertical measurements. See Fig. 20.12.

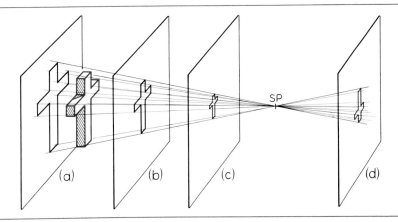

Fig. 20.7   Location of Picture Plane.

20.7   Position of the Object with Respect to the Horizon

To compare the elevation of the object with that of the horizon is equivalent to referring it to the level of the eye or SP, because the horizon is on a level with the eye, except in three-point perspective, §20.11. The differences in effect produced by placing the object on, above, or below the horizon are shown in Fig. 20.8.

If the object is placed above the horizon, it is above the level of the eye, or above SP, and will appear as seen from below. Likewise, if the object is below the horizon, it will appear as seen from above.

20.8   The Three Types of Perspectives

Perspective drawings are classified according to the number of vanishing points required, which in turn depends on the position of the object with respect to the picture plane.

If the object is situated with one face parallel to the plane of projection, only one vanishing point is required, and the result is a *one-point perspective* or *parallel perspective*, §20.9.

If the object is situated at an angle with the picture plane but with vertical edges parallel to the picture plane, two vanishing points are re-

quired, and the result is a *two-point perspective* or an *angular perspective*. This is the most common type of perspective drawing and is the one described in §20.4. See also §20.10.

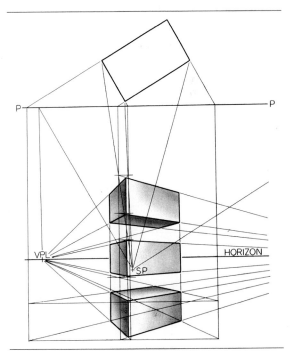

Fig. 20.8   Object and Horizon.

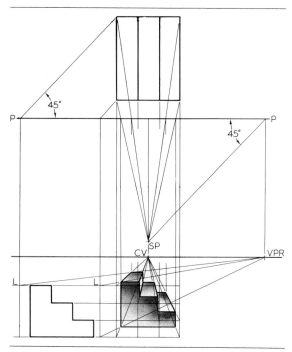

Fig. 20.9   One-Point Perspective.

If the object is situated so that no system of parallel edges is parallel to the picture plane, three vanishing points are necessary, and the result is a *three-point perspective,* §20.11.

20.9   One-Point Perspective

In one-point perspective, Fig. 18.1 (d), the object is placed so that two sets of its principal edges are parallel to PP, and the third set is perpendicular to PP. This third set of parallel lines will converge toward a single vanishing point in perspective, as shown.

In Fig. 20.9, the view shows the object with one face parallel to the picture plane. If desired, this face could be placed *in* the picture plane. The piercing points of the eight edges perpendicular to PP are found by extending them to PP and then projecting downward to the level of the lines as projected across from the elevation view.

To find the VP of these lines, a visual ray is drawn from SP parallel to them (the same as in step II of §20.4), and it is found that the *vanish-*

Fig. 20.10   One-Point Perspective. *Courtesy of Eaton Manufacturing Company*

ing point of all lines perpendicular to PP is in CV. By connecting the eight piercing points with the vanishing point CV, the indefinite perspectives of the eight edges are obtained.

To cut off on these lines the definite lengths of the edges of the object, horizontal lines are drawn from the ends of one of the edges in the top view and at any desired angle with PP, 45° for example, as shown. The piercing points and the vanishing point VPR of these lines are found, and the perspectives of the lines drawn. The intersections of these with the perspectives of the corresponding edges of the object determine the lengths of the receding edges. The perspective of the object may then be completed as shown.

One of the most common uses of parallel perspective is in the representation of interiors of buildings, as illustrated in Fig. 20.10.

An adaptation of one-point perspective, which is simple and effective in representing machine parts, is shown in Fig. 20.11. The front surface of the cylinder is placed in PP, and all circular shapes are parallel to PP; hence, these shapes will be projected as circles and circular arcs in the perspective. SP is located in front and to one side of the object, and the horizon is placed well above the ground line. The single vanishing point is on the horizon in CV.

The two circles and the keyway in the front surface of the object will be drawn true shape because they lie in PP. The circles are drawn with the compass on center O'. To locate R', the perspective center of the large arc, draw visual ray SP–R; then, from its intersection X with PP, project down to the center line of the large cylinder, as shown.

To find the radius T'W' at the right end of the perspective, draw visual rays SP–T and SP–W, and from their intersections with PP, project down to T' and W' on the horizontal center line of the hole.

20.10   Two-Point Perspective

In two-point perspective, the object is placed so that one set of parallel edges is vertical and has no vanishing point, while the two other sets each

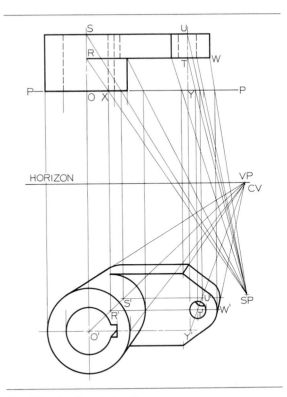

Fig. 20.11   One-Point Perspective.

have vanishing points. This is the most common type and is the method discussed in §20.4. It is suitable especially for representing buildings in the architectural drawing, or large structures in civil engineering, such as dams or bridges.

The perspective drawing of a small building is shown in Fig. 20.12. It is common practice (1) to assume a vertical edge of an object in PP so that direct measurements may be made on it and (2) to place the object so that its faces make unequal angles with PP; for example, one angle may be 30° and the other 60°. In practical work, complete multiview drawings are usually available, and the plane and elevation may be fastened in position, used in the construction of the perspective, and later removed.

Since the front corner AB lies in PP, its perspective A'B' may be drawn full size by projecting downward from the plan and across from the

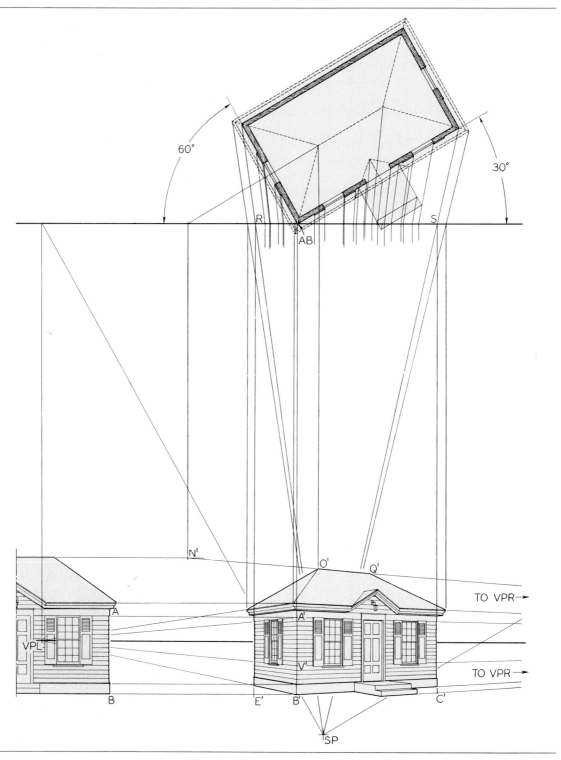

Fig. 20.12   Perspective Drawing of a Small Building.

elevation. The lengths of the receding lines from this corner are cut off by vertical lines SC′ and RE′ drawn from the intersections S and R, respectively, of the visual rays to these points of the object. The perspectives of the tops of the windows and the door are determined by the lines A′–VPR and A′–VPL, and their widths and lateral spacings are determined by projecting downward from the intersections with PP of the respective visual rays. The bottom lines of the windows are determined by the lines V′–VPR and V′–VPL.

The perspective of the line containing the ridge of the roof is found by joining N′, the point where the ridge line pierces the picture plane, and VPR. The ridge ends O′ and Q′ are found by projecting downward from the intersections of the visual rays with PP, or by drawing the perspectives of any two lines intersecting at the points. The perspective of the roof is completed by joining the points O′ and Q′ to the end of the eaves.

20.11   Three-Point Perspective

In three-point perspective, the object is placed so that none of its principal edges is parallel to the picture plane (PP); therefore, each of the three sets of parallel edges will have a separate vanishing point (VP), Fig. 20.13. The picture plane is assumed approximately perpendicular to the center line of the cone of rays.

In this figure, think of the paper as the picture plane, with the object behind the paper and placed so that all its edges make an angle with the picture plane. If a point CV is chosen, it will be the orthographic projection of your eye, or the station point, on the picture plane. The vanishing points P, Q, and R are found by conceiving lines to be drawn from a station point in space parallel to the principal axes of the object and finding their piercing points in the picture planes. It will be recalled that the basic rule for finding the vanishing point of a line in any type of perspective is to draw a visual ray, or line, from the station point parallel to the edge of the object whose vanishing point is required and

finding the piercing point of this ray in the picture plane. Since the object is rectangular, these lines to the vanishing points are at right angles to each other in space exactly as the axes are in axonometric projection, Fig. 18.48. The lines PQ, QR, and RP are perpendicular, respectively, to CV–R, CV–P, and CV–Q and are the *vanishing traces*, or horizon lines, of planes through SP parallel to the principal faces of the object.

The imaginary corner O is assumed in the picture plane and may coincide with CV; but as a rule the front corner is placed at one side near CV, thus determining how nearly the observer is assumed to be directly in front of this corner.

In this method the perspective is drawn directly from measurements and not projected from views. The dimensions of the object are given by the three views, and these will be set off on *measuring lines* GO, EO, and OF. See §20.13. The measuring lines EO and OF are drawn parallel to the vanishing trace PQ, and the measuring line GO is drawn parallel to RQ. These measuring lines are actually the lines of intersection of principal surfaces of the object, extended, with PP. Since these lines are in PP, true measurements of the object can be set off along them.

Three *measuring points* M_1, M_2, and M_3 are used in conjunction with the *measuring lines*. To find M_1, revolve triangle CV–R–Q about RQ as an axis. Since it is a right triangle, it can be easily constructed true size with the aid of a semicircle, as shown. With R as center, and R–SP_1 as radius, strike arc SP_1–M_1, as shown. M_1 is the measuring point for the measuring line GO. Measuring points M_2 and M_3 are found in a similar manner.

Height dimensions, taken from the given views, are set off full size or to any desired scale, along measuring line GO, at points 3, 2, and 1. From these points, lines are drawn to M_1, and heights on the perspective are the intersections of these lines with the perspective front corner OT of the object. Similarly, the true depth of the object is set off on measuring line EO from 0 to 5, and the true width is set off on measuring line OF from 0 to 8. Intermediate points can be constructed in a similar manner.

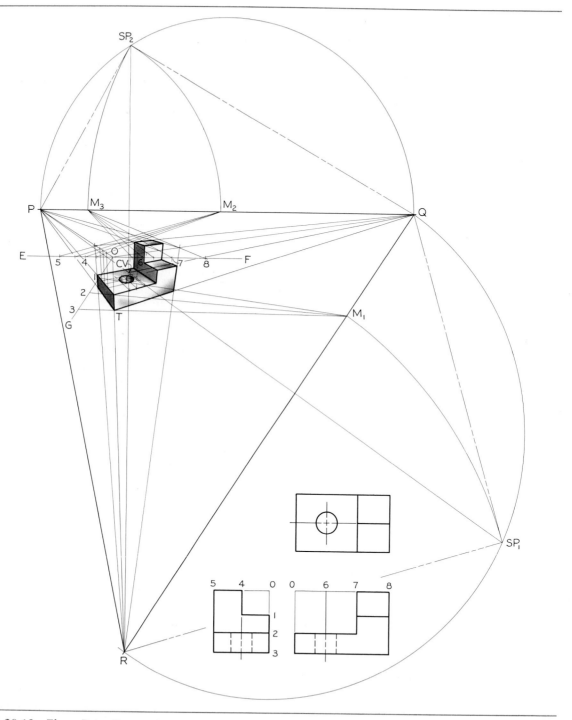

Fig. 20.13   Three-Point Perspective.

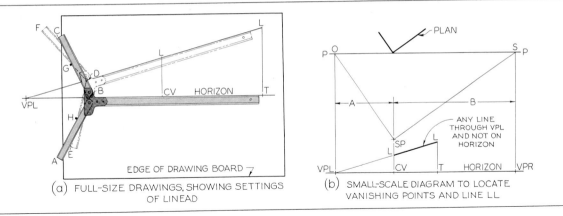

(a) FULL-SIZE DRAWINGS, SHOWING SETTINGS OF LINEAD

(b) SMALL-SCALE DIAGRAM TO LOCATE VANISHING POINTS AND LINE LL

Fig. 20.14 Perspective Linead.

20.12 The Perspective Linead and Template

PERSPECTIVE LINEAD *Fig. 20.14 (a)* The perspective linead consists of three straight-edged blades that can be clamped to each other at any desired angles. This instrument is convenient in drawing lines toward a vanishing point outside the limits of the drawing.

Before starting such a drawing, a small-scale diagram should be made, as indicated at (b), in which the relative positions of the object, PP, and SP are assumed, and the distances of the vanishing points from CV determined. Draw any line LL through a vanishing point as shown; then on the full-size drawing, assume CV and locate LL, as shown at (a).

To set the linead, clamp the blades in any convenient position; set the edge of the long blade along the horizon, and draw the lines BA and BC along the short blades. Then set the edge of the long blade along the line LL, and draw the lines DE and DF to intersect the lines first drawn at points G and H. Set pins at these points. If the linead is moved so that the short blades touch the pins, all lines drawn along the edge of the long blade will pass through VPL. This method is based on the principle that an angle inscribed in a circle is measured by half the arc it subtends.

TEMPLATE *Fig. 20.15* A template of thin wood or heavy cardboard, cut in the form of a circular arc, may be used instead of a perspective linead. If the template is attached to the drawing board so that the inaccessible VP is at the center of the circular arc, and the T-square is moved so that the head remains in contact with the template, lines drawn along the edge of the blade will, if extended, pass through the inaccessible VP.

If the edge of the blade does not pass through the center of the head, the lines drawn will be tangent to a circle whose center is at VP and

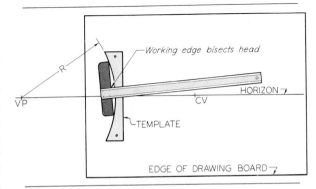

Fig. 20.15 Perspective Template.

Fig. 20.16   Klok Perspective Drawing Board. *Courtesy of Modulux Division, USG Co.*

whose radius is equal to the distance from the center of the head to the edge of the blade.

THE KLOK PERSPECTIVE DRAWING BOARD   Fig. 20.16 A specially designed board used for perspective drawing employs the preceding template idea in a ready-made form. Several useful scales are printed on the board for easy reference.

20.13   Measurements in Perspective

As explained in §20.1, all lines in PP are shown in their true lengths, and all lines behind PP are foreshortened.

Let it be required to draw the perspective of a line of telephone poles, Fig. 20.17. Let OB be the line of intersection of PP with the vertical plane containing the poles. In this line, the height AB of a pole is set off directly to the scale desired, and the heights of the perspectives of all poles are determined by drawing lines from A and B to VPR.

To locate the bottoms of the poles along the line B–VPR, set off along PP the distances 0–1, 1–2, 2–3, . . ., equal to the distance from pole to pole; draw the lines 1–1, 2–2, 3–3, . . ., forming a series of isosceles triangles 0–1–1, 0–2–2, 0–3–3, ... The lines 1–1, 2–2, 3–3, . . ., are parallel to each other, and, therefore, have a common vanishing point MP, which is found in the usual manner by drawing from SP a line SP–T parallel to the lines 1–1, 2–2, 3–3, . . ., and finding its piercing point MP (*measuring point*) in PP.

Since the line SP–X is parallel to the line of poles 1–2–3, . . ., the triangle SP–X–T is an isosceles triangle, and T is the top view of MP. The point T may be determined by setting off the distance X–T equal to SP–X or simply by drawing the arc SP–T with center at X and radius SP–X.

Having the measuring point MP, find the piercing points in PP of the lines 1–1, 2–2, 3–3, . . ., and draw their perspectives as shown. Since these lines are horizontal lines, their piercing points fall in a horizontal line BZ in PP, at the bottom of the drawing. Along BZ the true distances between the poles are set off; hence, BZ is called a *measuring line*. The intersections 1', 2', 3', . . ., of the perspectives of the lines 1–1, 2–2, 3–3, . . ., with the line B–VPR determine the spacing of the poles.

It will be seen that only a few measurements may be made along the measuring line BZ within the limits of the drawing. For additional measurements, the *diagonal method* of spacing may be employed, as shown. Since all diagonals from the bottom of each pole to the top of the succeeding pole are parallel, they have a common vanishing point VPI, which may be found as explained in §20.14. It is evident that the diagonal method exclusively may be used in the solution of this problem.

The method of direct measurements may also be applied to lines inclined to PP and to the ground plane, as illustrated in Fig. 20.18 for the line XE, which pierces PP at X. If the end of the house is conceived to be revolved about a vertical axis XO into PP, the line XE will be shown in its true length and inclination at XY. This line XY may be used as the measuring line for XE; it remains only to find the corresponding measur-

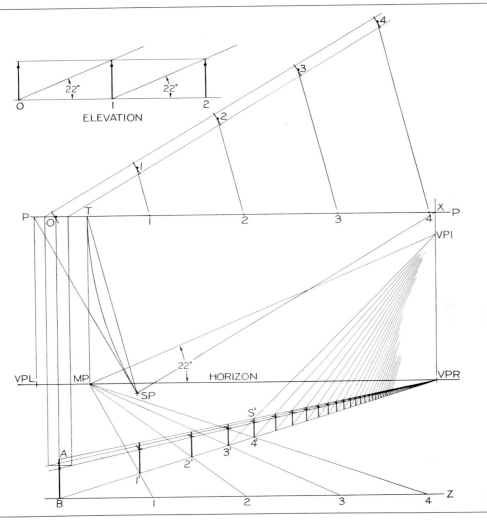

Fig. 20.17 Measurement of Vertical and Horizontal Lines.

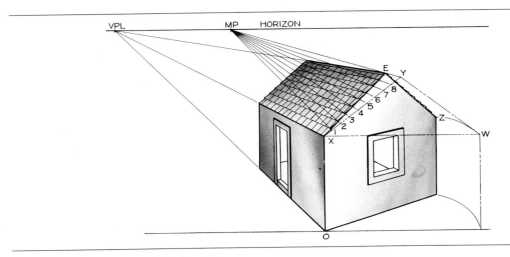

Fig. 20.18 Measurement of Inclined Lines.

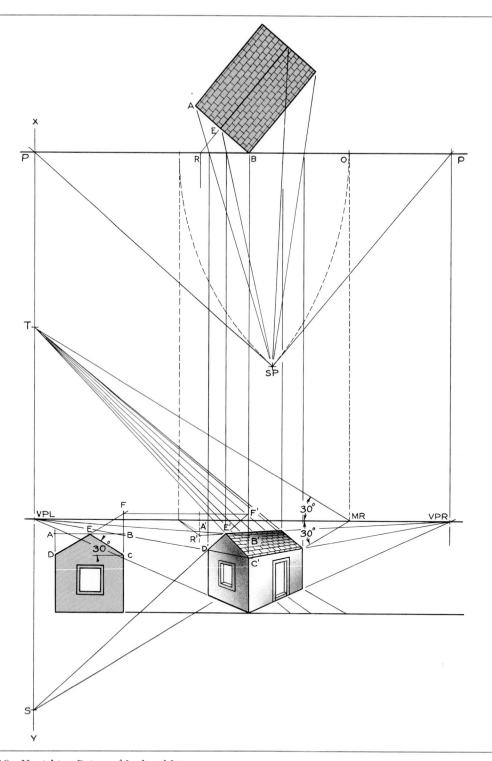

Fig. 20.19   Vanishing Points of Inclined Lines.

ing point MP. The line YE is the horizontal base of an isosceles triangle having its vertex at X, and a line drawn parallel to it through SP will determine MP, as described for Fig. 20.17.

20.14 Vanishing Points of Inclined Lines

The vanishing point of an inclined line is determined, as for all other lines, by finding the piercing point in PP of a line drawn from SP parallel to the given line.

In Fig. 20.19 is shown the perspective of a small building. The vanishing point of the inclined roof line C′E′ can be determined as follows: If a plane is conceived to be passed through the station point and parallel to the end of the house (plan view), it would intersect PP in the line XY, through VPL, and perpendicular to the horizon. Since the line drawn from SP parallel to C′E′ (in space) is in the plane SP–X–Y, it will pierce PP at some point T in XY. To find the point T, conceive the plane SP–X–Y revolved about the line XY as an axis into PP. The point SP will then fall on the horizon at a point shown by O in the top view and by MR in the front view. From the point MR draw the revolved position of the line SP–T (now MR–T) making an angle of 30° with the horizon and thus determining the point T, which is the vanishing point of the line C′E′ and of all lines parallel to that line. The vanishing point S of the line D′E′ is evidently in the line XY, because D′E′ is in the same vertical plane as the line C′E′. The vanishing point S is as far below the horizon as T is above the horizon, because the line E′D′ slopes downward at the same angle at which the line C′E′ slopes upward.

The perspectives of inclined lines can generally be found without finding the vanishing points, by finding the perspectives of the endpoints and joining them. The *perspective of any point may be determined by finding the perspectives of any two lines intersecting at the point.* Obviously, it would be best to use horizontal lines, parallel, respectively, to systems of lines whose vanishing points are already avail-

able. For example, in Fig. 20.19, to find the perspective of the inclined line EC, the point E′ is the intersection of the horizontal lines R′–VPR and B′–VPL. The point C′ is already established, since it is in PP; but if it were not in PP, it could be easily found in the same manner. The perspective of the inclined line EC is, therefore, the line joining the perspectives of the endpoints E′ and C′.

20.15 Curves and Circles in Perspective

If a circle is parallel to PP, its perspective is a circle. If the circle is inclined to PP, its perspective may be any one of the conic sections, in which the base of the cone is the given circle, the vertex is SP, and the cutting plane is PP. But since the center line of the cone of rays should be approximately perpendicular to the picture plane, the perspective will generally be an ellipse. The ellipse may be constructed by means of lines intersecting the circle, as shown in Fig. 20.20. The radial lines in the elevation view at the left can be easily drawn with the 45° and 30° × 60° triangles. A convenient method for determining the perspective of any plane curve is shown in Fig. 20.21.

20.16 The Perspective Plan Method

A perspective may be drawn by drawing first the perspective of the plan of the object, as shown in Fig. 20.22 (a), then the vertical lines, (b), and finally the connecting lines, (c). However, in drawing complicated structures, the superimposition of the perspective on the perspective plan causes a confusion of lines. For this reason, the perspective of the plan from which the location of vertical lines is determined is drawn either above or below its normal location. A suggestion of the range of possible positions of the perspective plan is given in Fig. 20.23; use of the perspective plan below the perspective is shown in Fig. 20.24.

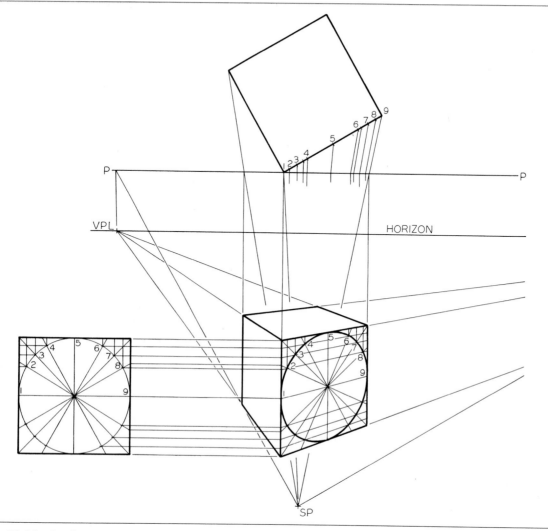

Fig. 20.20   Circles in Perspective.

The chief advantages of the perspective plan method over the ordinary plan method are that the vertical lines of the perspective can be spaced more accurately and that a considerable portion of the construction can be made above or below the perspective drawing, so that a confusion of lines on the required perspective is avoided.

When the perspective plan method is used, the ordinary plan view can be omitted and

measuring points used to determine distances along horizontal edges in the perspective.

20.17   Perspective Diagram

The spacing of vanishing points and measuring points may be determined graphically or may be calculated. In Fig. 20.25 a simple diagram of the plan layout shows the position of the object, the picture plane, the station point, and the con-

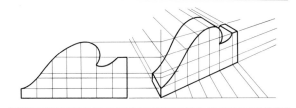

Fig. 20.21   Curves in Perspective.

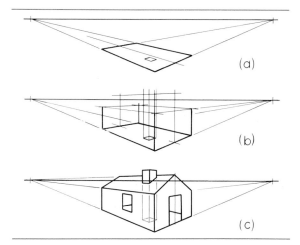

(a)

(b)

(c)

Fig. 20.22   Building upon the Perspective Plan.

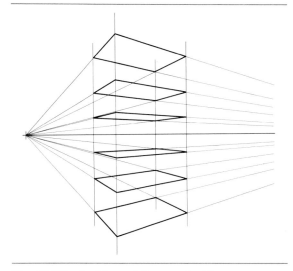

Fig. 20.23   Positions of Perspective Plan.

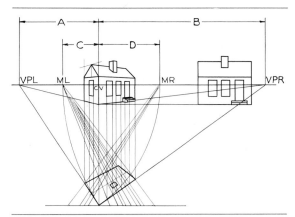

Fig. 20.24   Perspective Plan Method.

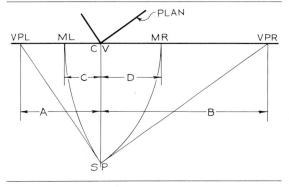

Fig. 20.25   Perspective Diagram.

structions for finding the vanishing points and measuring points for the problem in Fig. 20.24. As indicated in the figure, the complete plan need not be drawn. The diagram should be drawn to any small convenient scale, and vanishing points and measuring points set off in the perspective to the larger scale desired.

In practice, structures are usually considered in one of a limited number of simple positions with reference to the picture plane, such as 30° × 60°, 45° × 45°, and 20° × 70°. Therefore, a table of measurements for locating vanishing points and measuring points may be easily prepared, to avoid the necessity of a special construction for each drawing.

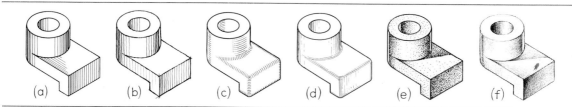

Fig. 20.26   Methods of Shading.

20.18   Shading

The effect of light can be utilized advantageously in describing the shapes of objects and in the finish and embellishment of such drawings as display drawings, patent drawings, and industrial pictorial drawings. Ordinary working drawings are not shaded.

Since the purpose of an industrial pictorial drawing is to show clearly the shape and not to be artistic, the shading should be simple and limited to producing a clear picture. Some of the common types of shading are shown in Fig.

20.26. Pencil or ink lines are drawn mechanically at (a) or freehand at (b). Two methods of shading fillets and rounds are shown at (c) and (d). Shading produced with pen dots is shown at (e), and pencil "tone" shading is shown at (f). Pencil shading applied to pictorial drawings on tracing paper may be reproduced with good results by making a whiteprint or a blueprint.

Examples of line shading on pictorial drawings often used in industrial sales literature are shown in Figs. 20.26, 20.27, and 20.28.

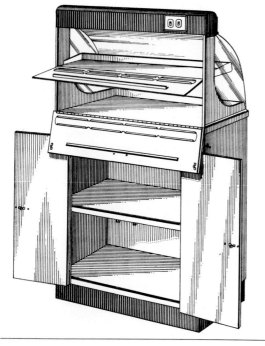

Fig. 20.27   Surface Shading Applied to Pictorial Drawing of Display Case.

Fig. 20.28   A Line-Shaded Drawing of an Adjustable Support for Grinding. *Courtesy of A. M. Byers Co.*

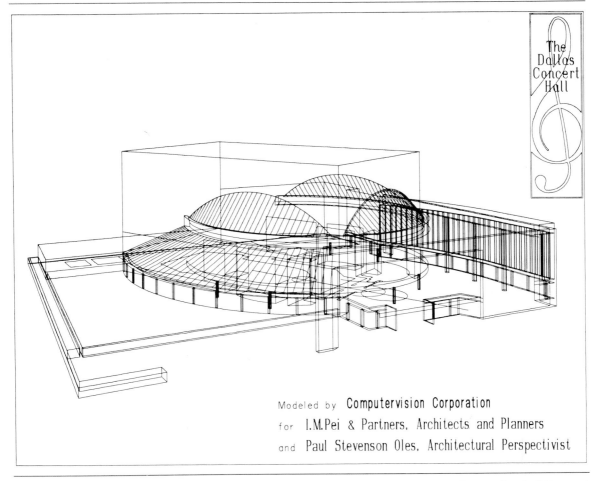

Modeled by **Computervision Corporation**

for I.M.Pei & Partners, Architects and Planners

and Paul Stevenson Oles, Architectural Perspectivist

Fig. 20.29 Perspective Drawing Produced by Using the Computervision Designer System for Building and Management (BDM). *Courtesy of Computervision Corporation, a subsidiary of Prime Computer, Inc.*

20.19   Computer Graphics

Perspective drawings, which provide pictorials most resembling photographs, are also the most time-consuming types of pictorials to draw. CAD programs are available that will produce either wireframe, Fig. 20.29, or solid perspective representations, with user selection of viewing distance, focal point, z-axis convergence, and arc resolution scale. Historically, perspectives have seen far greater application in architectural than in engineering drawing. Now the availability of these computer graphics routines makes perspective drawing a viable alternative for the drafter wishing to employ a pictorial representation of an object.

PERSPECTIVE PROBLEMS

Layouts for perspective problems are given in Figs. 20.30–20.33. These are to be drawn on a size B or A3 sheet of paper, vellum, or film, with the student's name, date, class, and other information lettered below the border as specified by the instructor.

Additional problems for perspective drawings on size B or A3 paper are given in Fig. 20.34. The student is to determine the arrangement on the sheet, so as to produce the most effective perspective in each case. Problems in which the student selects both sheet size and scale are given in Fig. 20.35.

Since many of the problems in this chapter are of a general nature, they can also be solved on most computer graphics systems. If a system is available, the instructor may choose to assign specific problems to be completed by this method.

In addition to these problems, many suitable problems for perspective will be found among the axonometric and oblique problems at the ends of Chapters 18 and 19.

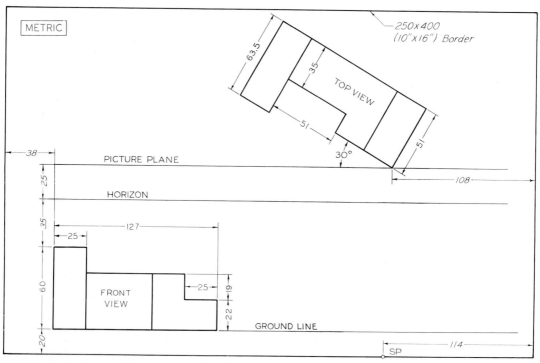

Fig. 20.30   Draw views and perspective. Omit dimensions. Use Size B or A3 sheet.

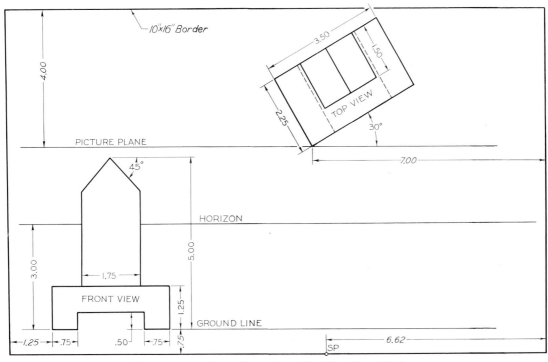

Fig. 20.31 Draw views and perspective. Omit dimensions. Use Size B or A3 sheet.

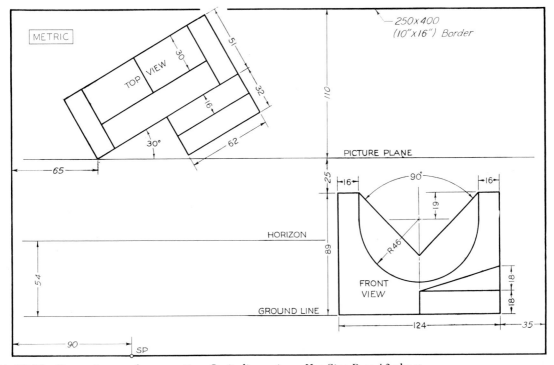

Fig. 20.32 Draw views and perspective. Omit dimensions. Use Size B or A3 sheet.

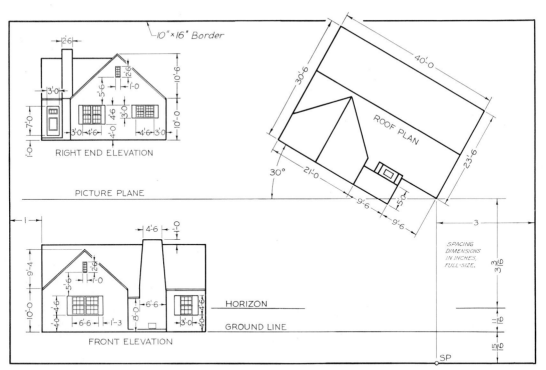

Fig. 20.33   Draw front elevation, plan, and perspective. Omit dimensions. Scale: $\frac{1}{8}'' = 1'-0$. Use Size B or A3 sheet.

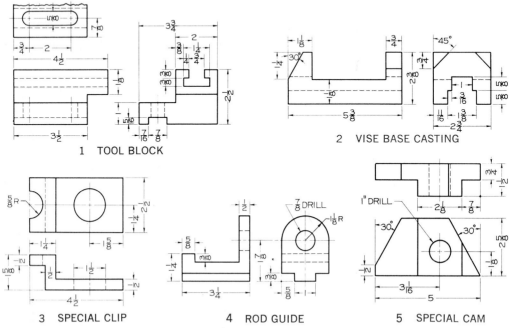

1   TOOL BLOCK

2   VISE BASE CASTING

3   SPECIAL CLIP

4   ROD GUIDE

5   SPECIAL CAM

Fig. 20.34   Draw side or front elevation, plan, and perspective of assigned problem. Omit dimensions. Use Size B or A3 sheet.

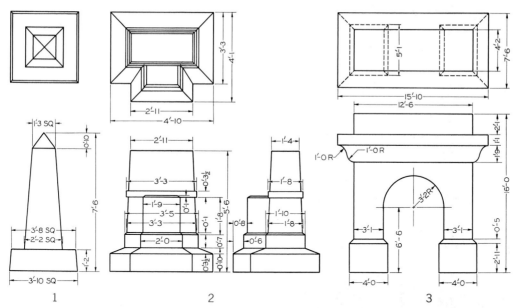

Fig. 20.35 Draw side or front elevation, plan, and perspective of assigned problem. Omit dimensions. Select sheet size and scale.

Intersections and Developments

A machine part or structure often consists of a number of geometric shapes so arranged as to produce the required form. Easily represented geometric forms frequently meet in lines of intersection that require knowledge and skill to produce in multiview projection. Accurate representation of the intersecting surfaces therefore becomes very important since precision fit is necessary for function and appearance. The development of these surfaces, such as those found in sheet-metal fabrication, is a flat pattern that represents the unfolded or unrolled surface of the form. The resulting plane figure gives the true size of each area of the form so connected that when fabricated the desired part or structure is produced.

21.1 Surfaces

A *surface* is a geometric magnitude having two dimensions. A surface may be generated by a line, called the *generatrix* of the surface. Any position of the generatrix is an *element* of the surface, Fig. 7.31 (a).

A *ruled surface* is one that may be generated by a straight line and may be a *plane*, a *single-curved surface*, or a *warped surface*.

A *plane* is a ruled surface that may be generated by a straight line one point of which moves along another straight line, while the generatrix remains parallel to its original position. Many of the geometric solids are bounded by plane surfaces, Fig. 5.7.

A *single-curved surface* is a developable ruled surface; that is, it can be unrolled to coincide with a plane. Any two adjacent positions of the generatrix lie in the same plane. Examples are the cylinder and the cone, Fig. 5.7.

A *warped surface* is a ruled surface that is not developable, Fig. 21.1. No two adjacent positions of the generatrix lie in the same plane. Many exterior surfaces on an airplane or automobile are warped surfaces.

A *double-curved surface* may be generated only by a curved line and has no straight-line elements. Such a surface, generated by revolving a curved line about a straight line in the plane of the curve, is called a *double-curved surface of revolution*. Common examples are the *sphere*, *torus*, *ellipsoid*, Fig. 5.7, and the *hyperboloid*, Fig. 21.1 (d).

A *developable surface* is one that may be unfolded or unrolled so as to coincide with a plane, §21.4. Surfaces composed of single-curved surfaces, or of planes, or of combinations of these types, are developable. Warped surfaces and double-curved surfaces are not developable. They may be developed approximately by dividing them into sections and substituting for each section a developable surface, that is, a plane or a single-curved surface. If the material used is sufficiently pliable, the flat sheets may be

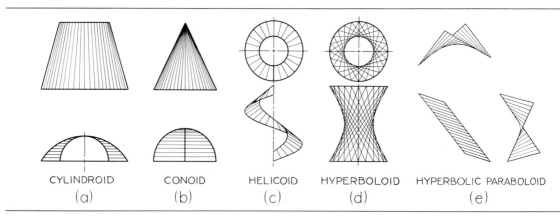

Fig. 21.1 Warped Surfaces.

stretched, pressed, stamped, spun, or otherwise forced to assume the desired shape. Nondevelopable surfaces are often produced by a combination of developable surfaces, which are then formed slightly to produce the required shape.

21.2   Solids

Solids bounded by plane surfaces are *polyhedra,* the most common of which are the pyramid and prism, Fig. 5.7. Convex solids whose faces are all equal regular polygons are *regular polyhedra.*

The simple regular polyhedra are the *tetrahedron, cube, octahedron, dodecahedron,* and *icosahedron,* known as the five *Platonic solids.*

Plane surfaces that bound polyhedra are *faces* of the solids. Lines of intersection of faces are *edges* of the solids.

A solid generated by revolving a plane figure about an axis in the plane of the figure is a *solid of revolution.*

Solids bounded by warped surfaces have no group name. The most common example of such solids is the screw thread.

Intersections and Developments of Planes and Solids

21.3   Principles of Intersections

The principles involved in intersections of planes and solids have their practical application in the cutting of openings in roof surfaces for flues and stacks and in wall surfaces for pipes and chutes, and so on, and in the building of sheet-metal structures (tanks, boilers, etc.).

In such cases, the problem is generally one of determining the true size and shape of the intersection of a plane and one of the more common geometric solids. The intersection of a

plane and a solid is the locus of the points of intersection of the elements of the solid with the plane. For solids bounded by plane surfaces, it is necessary only to find the points of intersection of the edges of the solid with the plane and to join these points, in consecutive order, with straight lines. For solids bounded by curved surfaces, it is necessary to find the points of intersection of several elements of the solid with the plane and to trace a smooth curve through these points. The curve of intersection of a plane and

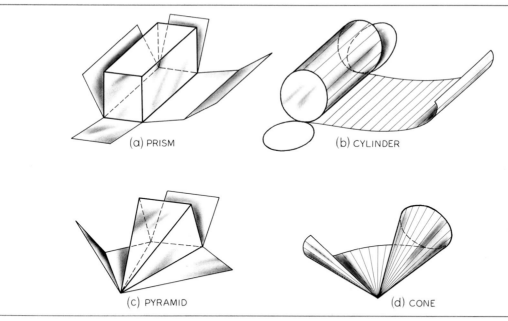

(a) PRISM

(b) CYLINDER

(c) PYRAMID

(d) CONE

Fig. 21.2   Development of Surfaces.

a circular cone is a *conic section*. The various conic sections are defined and illustrated in §5.47 and Fig. 5.46.

21.4   Developments

The *development* of a surface is that surface laid out on a plane, Fig. 21.2. Practical applications of developments occur in sheet-metal work, stone cutting, pattern making, packaging, and package design.

Single-curved surfaces and the surfaces of polyhedra can be developed. Warped surfaces and double-curved surfaces can be developed only approximately. See §21.1.

In sheet-metal layout, extra material must be provided for laps or seams. If the material is heavy, the thickness may be a factor, and the crowding of metal in bends must be considered. See §13.39. The drafter must also take stock sizes into account and should make layouts so as to economize in the use of material and of labor. In preparing developments, it is best to put the seam at the shortest edge and to attach the bases at edges where they match, to economize in soldering, welding, or riveting.

It is common practice to draw development layouts with the *inside surfaces up*. In this way, all fold lines and other markings are related directly to inside measurements, which are the important dimensions in all ducts, pipes, tanks, and other vessels, and in this position they are also convenient for use in the fabricating shop.

21.5   Hems and Joints for Sheet Metal and Other Materials

A wide variety of hems and joints are used in the fabrication of sheet-metal developments and other materials, Fig. 21.3. Hems are used to eliminate the raw edge as well as to stiffen the material. Joints and seams may be made for sheet metal by bending, welding, riveting, and soldering and by glueing and stapling for package materials.

Sufficient material as required for hems and joints must be added to the layout or development. The amount of allowance depends on the

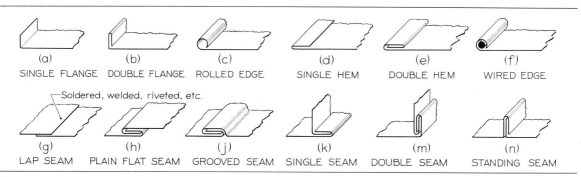

Fig. 21.3   Sheet-Metal Hems and Joints.

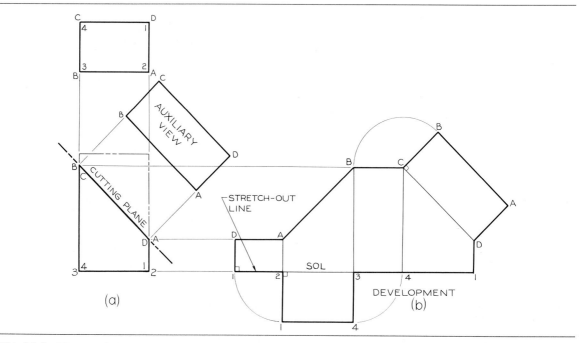

Fig. 21.4   Plane and Prism.

thickness of the material and the production equipment; therefore, no specific dimensions for allowances are given in this chapter. See §13.39.

21.6   To Find the Intersection of a Plane and a Prism and the Development of the Prism

INTERSECTION   *Fig. 21.4 (a)*   The true size and shape of the intersection is shown in the auxiliary view. See Chapter 10. The length AB is the

same as AB in the front view, and the width AD is the same as AD in the top view.

DEVELOPMENT   *Fig. 21.4 (b)*   On the straight line 1–1, called the *stretchout line* (SOL), set off the widths of the faces 1–2, 2–3, . . ., taken from the top view. At the division points, erect perpendiculars to 1–1, and set off on each the length of the respective edge, taken from the front view. The lengths can be projected across from the front view, as shown. Join the points thus found

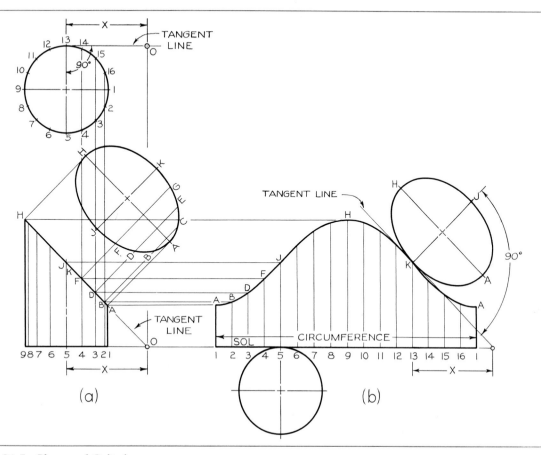

Fig. 21.5   Plane and Cylinder.

by straight lines to complete the development of the lateral surface. Attach to this development the lower base and the upper base, or auxiliary view, to obtain the development of the entire surface of the frustum of the prism.

21.7   To Find the Intersection of a Plane and a Cylinder and the Development of the Cylinder

INTERSECTION   *Fig. 21.5 (a)*   The intersection is an ellipse whose points are the piercing points in the secant plane of the elements of the cylinder. In spacing the elements, it is best, though not necessary, to divide the circumference of the base into *equal* parts and to draw an element at

each division point. In the auxiliary view, the widths BC, DE, . . ., are taken from the top view at 2–16, 3–15, . . ., respectively, and the curve is traced through the points thus determined, with the aid of the irregular curve, §2.54.

The major axis AH and the minor axis JK are shown true length in the front view and the top view, respectively. Therefore, the ellipse may also be constructed as explained in §§5.48–5.51 or with the aid of an ellipse template, §5.56.

DEVELOPMENT   *Fig. 21.5 (b)*   The base of the cylinder develops into a straight line 1–1, the stretchout line (SOL), equal to the circumference of the base, whose length may be determined by calculation (πd), by setting off with

687

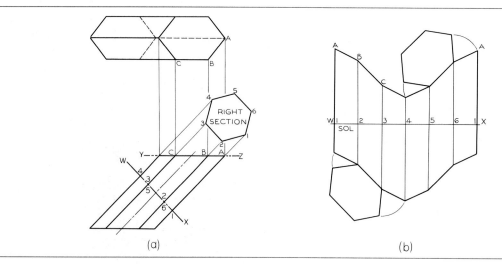

Fig. 21.6   Plane and Oblique Prism.

the bow dividers, or by rectifying the arcs of the base 1–2, 2–3, . . ., §5.45. Divide the stretchout line into the same number of equal parts as the circumference of the base, and draw an element through each division perpendicular to the line. Set off on each element its length, projected from the front view, as shown; then trace a smooth curve through the points A, B, D, . . ., §2.54, and attach the bases.

21.8   To Find the Intersection of a Plane and an Oblique Prism and the Development of the Prism

INTERSECTION   *Fig. 21.6 (a)* The right section cut by the plane WX is a regular hexagon, as shown in the auxiliary view; the oblique section, cut by the horizontal plane YZ, is shown in the top view.

DEVELOPMENT   *Fig. 21.6 (b)* The right section develops into the straight line WX, the stretchout line (SOL). Set off, on the stretchout line, the widths of the faces 1–2, 2–3, . . ., taken from the auxiliary view, and draw a line through each division perpendicular to the line. Set off, from the stretchout line, the lengths of the respective edges measured from WX in the front view. Join the points A, B, C, . . ., with straight lines, and

attach the bases, which are shown in their true sizes in the top view.

21.9   To Find the Intersection of a Plane and an Oblique Cylinder

INTERSECTION   *Fig. 21.7 (a)* The right section cut by the plane WX is a circle, shown in the auxiliary view. The intersection of the horizontal plane YZ with the cylinder is an ellipse shown in the top view, whose points are found as explained for the auxiliary view in Fig. 21.5 (a), §21.7. The major axis AH is shown true length in the top view, and the minor axis JK is equal to the diameter of the cylinder. Therefore, the ellipse may be constructed as explained in §§5.48–5.51, or with the aid of an ellipse template, §5.56.

DEVELOPMENT   *Fig. 21.7 (b)* The cylinder may be considered as a prism having an infinite number of edges; therefore, the development is found in a manner similar to that of the oblique prism shown in Fig. 21.6.

The circle of the right section cut by plane WX develops into a straight line 1–1, the stretchout line (SOL), equal in length to the circumference of the circle (πd). Divide the stretchout line into the same number of equal parts as the circumference of the circle as shown in the auxil-

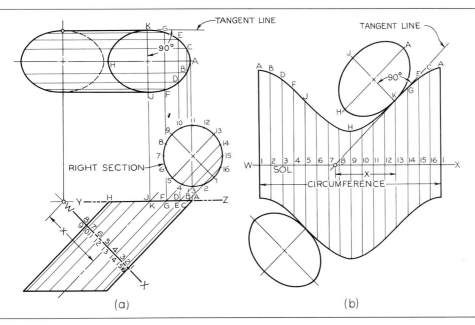

Fig. 21.7   Plane and Oblique Circular Cylinder.

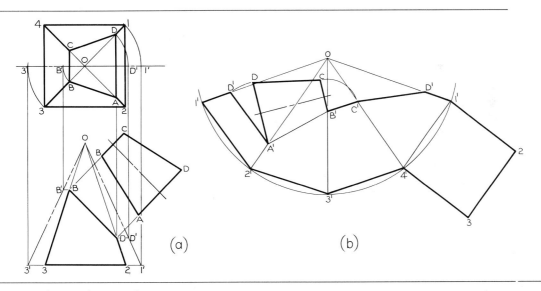

Fig. 21.8   Plane and Pyramid.

iary view, and draw elements through these points perpendicular to the line. Set off on each element its length, taken from the front view with dividers, as shown; then trace a smooth curve through the points A, B, D, . . ., §2.54, and attach the bases.

21.10   To Find the Intersection of a Plane and a Pyramid and to Develop the Resulting Truncated Pyramid

INTERSECTION   *Fig. 21.8 (a)*   The intersection is a trapezoid whose vertices are the points in which the edges of the pyramid pierce the secant plane. In the auxiliary view, the altitude of the trapezoid is projected from the front view, and the widths AD and BC are transferred from the top view with dividers.

DEVELOPMENT *Fig. 21.8 (b)* With O in the development as center and O–1′ in the front view (the true length of one of the edges) as radius, draw the arc 1′–2′–3′. . . . Inscribe the cords 1′–2′, 2′–3′, . . ., equal, respectively, to the sides of the base, as shown in the top view. Draw the lines 1′–O, 2′–O, . . ., and set off the true lengths of the lines OD′, OA′, OB′, . . ., respectively, taken from the true lengths in the front view, §11.10.

To complete the development, join the points D′, A′, B′, . . ., by straight lines, and attach the bases to their corresponding edges. To transfer an irregular figure, such as the trapezoid shown here, refer to §§5.28 and 5.29.

21.11 To Find the Intersection of a Plane and a Cone and to Develop the Lateral Surface of the Cone

INTERSECTION *Fig. 21.9 (a)* The intersection is an ellipse. If a series of horizontal cutting planes is passed perpendicular to the axis, as shown, each plane will cut a circle from the cone that will show in true size and shape in the top view. Points in which these circles intersect the original secant plane are points on the ellipse. Since the secant plane is shown edgewise in the front view, all of these piercing points may be found in that view and projected to the others, as shown.

Fig. 21.9 (b) This method is most suitable when a development also is required, since it utilizes elements that are also needed in the development. The piercing points of these elements in the secant plane are points on the intersection. Divide the base into any number of equal parts, and draw an element at each division point. These elements pierce the secant plane in points A, B, C, ... The top views of

these points are found by projecting upward from the front view, as shown. In the auxiliary view, the widths BL, CK, . . ., are taken from the top view. The ellipse is then drawn with the aid of the irregular curve, §2.54.

The major axis of the ellipse, shown in the auxiliary view, is equal to AG in the front view. The minor axis MN bisects the major axis and is equal to the minor axis of the ellipse in the top view. With these axes, the ellipse may also be constructed as explained in §§5.48–5.51 or with the aid of an ellipse template, §5.56.

DEVELOPMENT *Fig. 21.9 (c)* The cone may be considered as a pyramid having an infinite number of edges; hence, the development is found in a manner similar to that explained for the pyramid in §21.10. The base of the cone develops into a circular arc, with the slant height of the cone as its radius and the circumference of the base as its length, §5.45. The lengths of the elements in the development are taken from the element O–7 or O–1 in the front view, (b). Instead of our finding the true circumference of the base, the vertical angle 1–O–1 in the development can be set off equal to $\frac{r}{s}$ 360° (where r is the radius of the base and s is the slant height of the cone).

21.12 To Find the Development of a Hood and Flue

Fig. 21.10 Since the hood is a conical surface, it may be developed as described in §21.11. The two end sections of the elbow are cylindrical surfaces and may be developed as described in §21.7. The two middle sections of the elbow are cylindrical surfaces, but since their bases are not perpendicular to the axes, they will not develop into straight lines. They will be developed in a manner similar to that for an oblique cylinder, §21.9, Fig. 21.7 (b). If the auxiliary planes AB and DC are passed perpendicular to the axes, they will cut right sections from the cylinders, which will develop into the straight lines AB and CD in the developments.

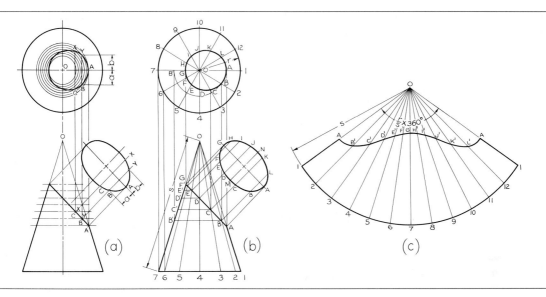

Fig. 21.9 Plane and Cone.

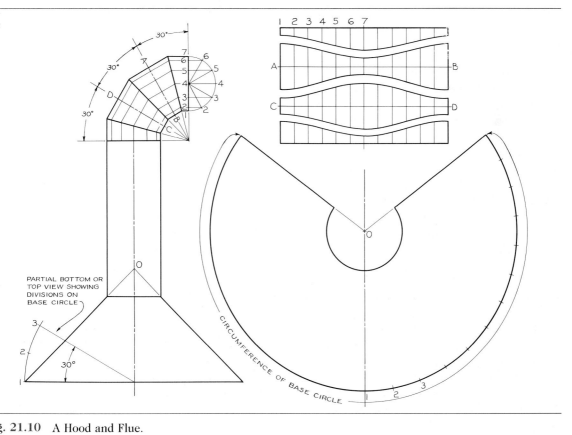

Fig. 21.10 A Hood and Flue.

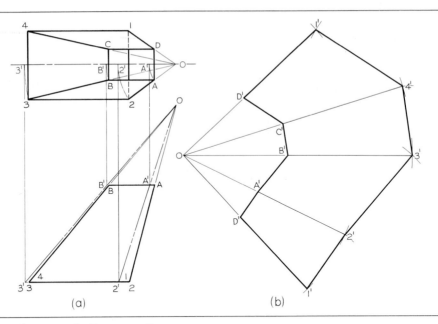

Fig. 21.11   Development of a Transition Piece.

If the developments are arranged as shown in Fig. 21.10, the elbow can be constructed from a rectangular sheet of metal without wasting material. The patterns are shown separated after cutting. Before cutting, the adjacent curves coincided.

21.13 To Find the Development of a Truncated Oblique Rectangular Pyramid

Fig. 21.11   None of the four lateral surfaces is shown in the multiview drawing in true size and shape. Using the method of §11.10, revolve each edge until it appears in true length in the front view, as shown. Thus, O–2 revolves to O–2', O–3 revolves to O–3', and so on. These true lengths are transferred from the front view to the development with the compass, as shown. Notice that true lengths OD', OA', OB', . . ., are found and transferred. The true lengths of the edges of the bases are given in the top view and are transferred directly to the development.

21.14 Triangulation

Triangulation is simply a method of dividing a surface into a number of triangles and transferring them to the development. A triangle is said to be "indestructible," because if its sides are of given lengths, it can be only one shape. A triangle can be easily transferred by transferring the sides with the aid of the compass, §5.28.

21.15 To Find the Development of an Oblique Cone by Triangulation

Fig. 21.12   Divide the base, in the top view, into any number of equal parts, and draw an element at each division point. Find the true length of each element, §11.10. If the divisions of the base are comparatively small, the lengths of the chords may be set off in the development as representing the lengths of the respective subtending arcs. In the development, set off O–1' equal to O–1 in the front view where it is shown true length. With 1' in the development as center, and the chord 1–2 taken from the top view as radius, strike an arc at 2'. With O as center,

692

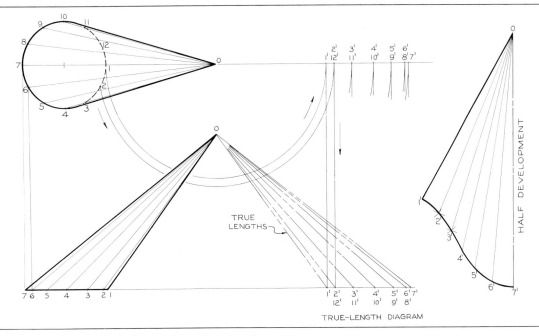

Fig. 21.12 Development of an Oblique Cone by Triangulation.

and O–2′, the true-length of the element O–2 from the "true-length" diagram, as radius, draw the arc at 2′. The intersection of these arcs is a point on the development of the base of the cone. The points 3′, 4′, . . ., in the curve are found in a similar manner, and the curve is traced through these points with the aid of the irregular curve, §2.54.

Since the development is symmetrical about element O–7′, it is necessary to lay out only half the development, as shown.

21.16   Transition Pieces

Fig. 21.13 A *transition piece* is one that connects two differently shaped, differently sized, or skewed-position openings. In most cases, transition pieces are composed of plane surfaces and conical surfaces, the latter being developed by triangulation. Triangulation can also be used to develop, approximately, certain warped surfaces. Transition pieces are used extensively in air conditioning, heating, ventilating, and similar construction.

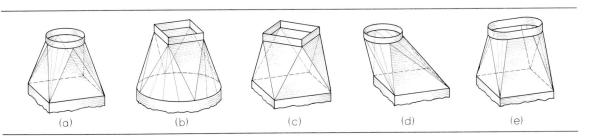

Fig. 21.13 Transition Pieces.

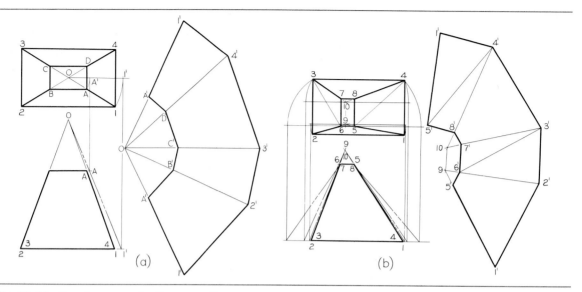

Fig. 21.14   Development of a Transition Piece.

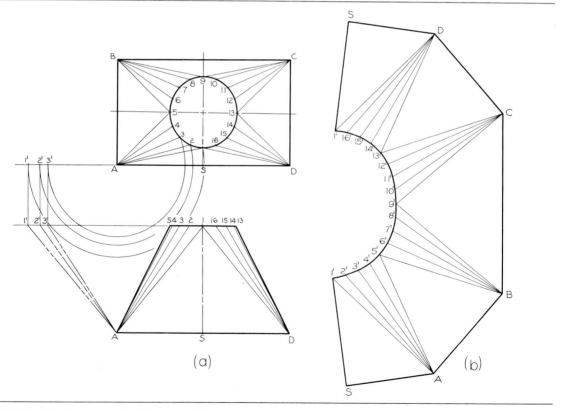

Fig. 21.15   Development of a Transition Piece.

21.17 To Find the Development of a Transition Piece Connecting Rectangular Pipes on the Same Axis

The transition piece is a frustum of a pyramid, Fig. 21.14 (a). Find the vertex O of the pyramid by extending its edges to their intersection. Find the true lengths of the edges by any one of the methods explained in §11.10. The development can then be found as explained in §21.10.

If the transition piece is not a frustum of a pyramid, as in Fig. 21.14 (b), it can best be developed by triangulation, §21.14, as shown for the faces 1–5–8–4 and 2–6–7–3, or by extending the sides to form triangles, as shown for faces 1–2–6–5 and 3–4–8–7, and then finding the true lengths of the sides of the triangles, §11.10, and setting them off as shown.

As a check on the development, lines parallel on the surface must also be parallel on the development; for example, 8'–5' must be parallel to 4'–1' on the development.

21.18 To Find the Development of a Transition Piece Connecting a Circular Pipe and a Rectangular Pipe on the Same Axis

Fig. 21.15 The transition piece is composed of four isosceles triangles and four conical surfaces. The seam is along line S–1. Begin the development on the line 1'–S, and draw the right triangle 1'–S–A, whose base SA is equal to half the side AD and whose hypotenuse A–1' is equal to the true length of side A–1.

The conical surfaces are developed by triangulation as explained in §§21.14 and 21.15.

21.19 To Find the Development of a Transition Piece Connecting Two Cylindrical Pipes on Different Axes

Fig. 21.16 The transition piece is a frustum of a cone, the vertex of which may be found by

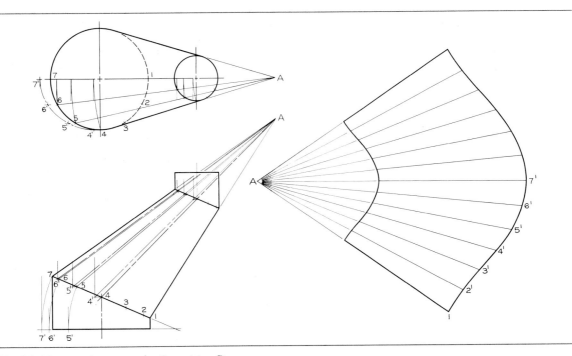

Fig. 21.16 Development of a Transition Piece.

extending the contour elements to their intersection A.

The development can be found by triangulation, as explained in §§21.14 and 21.15. The sides of each triangle are the true lengths of two adjacent elements of the cone, and the base is the true length of the curve of the base of the cone between the two elements. This curve is not shown in its true length in either view, and the plane of the base of the frustum must therefore be revolved until it is horizontal in order to find the distance from the foot of one element to the foot of the next. When the plane of the base is thus revolved, the foot of any element, such as 7, revolves to 7′, and the curve 6′–7′ (top view) is the true length of the curve of the base between the elements 6 and 7. In practice, the chord distances between these points are generally used to approximate the curved distances.

After the conical surface has been developed, the true lengths of the elements on the truncated section of the cone are set off from the vertex A of the development to secure points on the upper curve of the development.

If the transition piece is not a frustum of a cone, its development is found by another variation of triangulation, as shown in Fig. 21.17. The circular intersection with the large vertical pipe is shown true size in the top view, and the circular intersection with the small inclined pipe is shown true size in the auxiliary view. Since both intersections are true circles, and the planes containing them are not parallel, the lateral surface of the transition piece is a warped

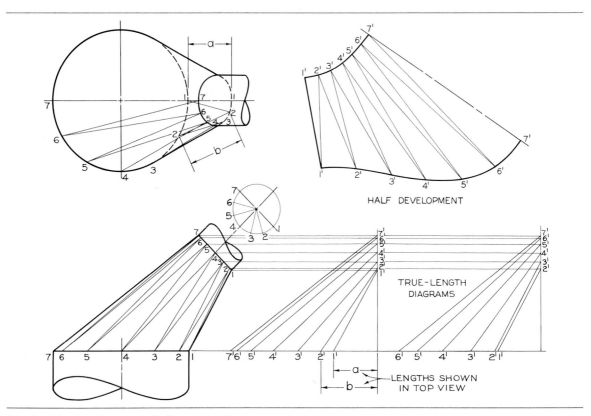

Fig. 21.17   Development of a Transition Piece.

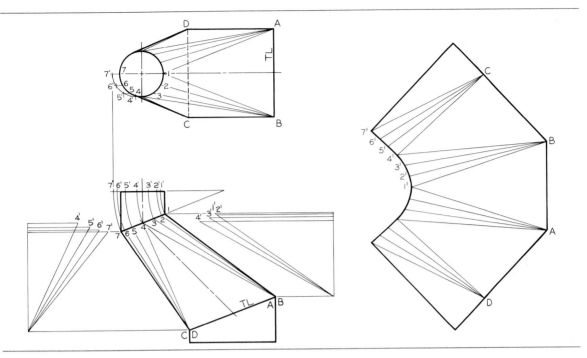

Fig. 21.18   Development of a Transition Piece.

surface and not conical (single-curved). It is theoretically nondevelopable but may be approximately developed by considering it to be made up of plane triangles, alternate ones of which are inverted, as shown in the development. The true lengths of the sides of the triangles are found by the method of Fig. 11.10 (d), but in a systematic manner so as to form true-length diagrams, as shown in Fig. 21.17.

21.20   To Find the Development of a Transition Piece Connecting a Square Pipe and a Cylindrical Pipe on Different Axes

Fig. 21.18   The development of the transition piece is made up of five plane triangular surfaces and four triangular conical surfaces similar to those in Fig. 21.15. The development is made in a similar manner to those described in §§21.15 and 21.18.

21.21   To Find the Intersection of a Plane and a Sphere and to Find the Approximate Development of the Sphere

INTERSECTION   *Fig. 21.19 (a)*   The intersection of a plane and a sphere is a circle, as shown in the top views in Fig. 21.19, the diameter of the circle depending on where the plane is passed. Any circle cut by a plane through the center of the sphere is called a *great circle*. If a plane passes through the center and perpendicular to the axis, the resulting great circle is called the *equator*. If a plane contains the axis, it will cut a great circle called a *meridian*.

DEVELOPMENT   *Fig. 21.19 (a)*   The surface of a sphere is a double-curved surface and is not developable, §21.1. The surface may be developed approximately by dividing it into a series of zones and substituting for each zone a frustum of a right-circular cone. The development of the

697

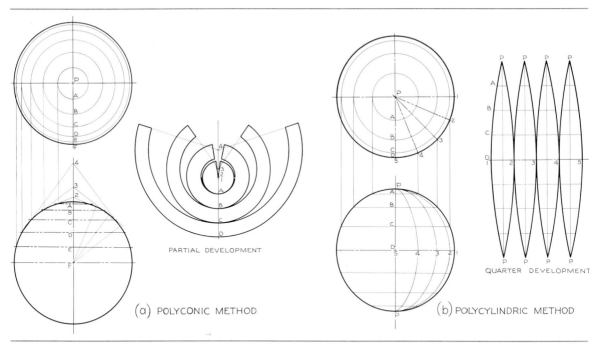

Fig. 21.19   Approximate Development of a Sphere.

conical surfaces is an approximate development of the spherical surface. If the conical surfaces are inscribed within the sphere, the development will be smaller than the spherical surface, while if the conical surfaces are circumscribed about the sphere, the development will be larger. If the conical surfaces are partly within and partly without the sphere, as indicated in the figure, the resulting development very closely approximates the spherical surface.

This method of developing a spherical surface is the *polyconic* method. It is used on government maps of the United States.

Fig. 21.19 (b)   Another method of making an approximate development of the double-curved surface of a sphere is to divide the surface into equal sections with meridian planes and substitute cylindrical surfaces for the spherical sections. The cylindrical surfaces may be inscribed within the sphere, or circumscribed about it, or located partly within and partly without. The development of the series of cylindrical surfaces is an approximate development of the spherical surface. This method is the *polycylindric* method, sometimes designated as the *gore* method.

Intersections and Developments of Solids

21.22   Principles of Intersections

Intersections of solids are generally regarded as in the province of descriptive geometry. For information on the more complicated intersec- tions the student is referred to any standard text on that subject. However, most of the intersec- tions encountered in drafting practice do not re- quire a knowledge of descriptive geometry, and

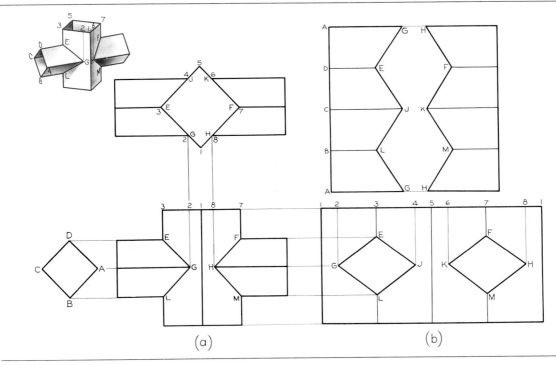

Fig. 21.20 Two Prisms at Right Angles to Each Other.

some of the more common solutions may be found in the paragraphs that follow.

An intersection of two solids is referred to as a *figure of intersection*. Two plane surfaces intersect in a straight line; hence, if two solids that are composed of plane surfaces intersect, the figure of intersection will be composed of straight lines, as shown in Figs. 21.20–21.23. The method generally consists of finding the piercing points of the edges of one solid in the surfaces of the other solid and joining these points with straight lines.

If curved surfaces intersect, or if curved surfaces and plane surfaces intersect, the figure of intersection will be composed of curves, as shown in Figs. 21.5, 21.9, and 21.24–21.29. The method generally consists of finding the piercing points of *elements* of one solid in the surfaces of the other. A smooth curve is then traced through these points, with the aid of the irregular curve, §2.54.

21.23   To Find the Intersection and Developments of Two Prisms

INTERSECTION   *Fig. 21.20 (a)*   The points in which the edges A, B, C, and D of the horizontal prism pierce the vertical prism are vertices of the intersection. The edges D and B of the horizontal prism intersect the edges 3 and 7 of the vertical prism at the points E, F, L, and M. The edges A and C of the horizontal prism intersect the faces of the vertical prism at the points G, H, J, and K. The intersection is completed by joining these points in order by straight lines.

DEVELOPMENTS   *Fig. 21.20 (b)*   To develop the lateral surface of the horizontal prism, set off on the vertical stretchout line A–A the widths of the faces AB, BC, . . ., taken from the end view, and draw the edges through these points, as shown. Set off, from the stretchout line, the lengths of the edges AG, BL, . . ., taken from the front

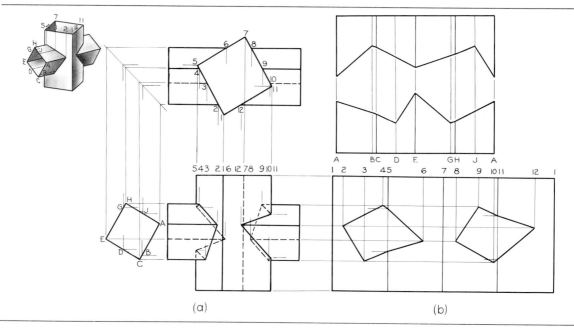

Fig. 21.21   Two Prisms at Right Angles to Each Other.

view or from the top view, and join the points G, L, J, . . ., by straight lines.

To develop the lateral surface of the vertical prism, set off on the stretchout line 1–1 the widths of the faces 1–2, 3–5, . . ., taken from the top view, and draw the edges through these points, as shown. Set off on the stretchout line the distances 1–2, 5–4, 5–6, and 1–8, taken from the top view, and draw the intermediate elements parallel to the principal edges. Take the lengths of the principal edges and of the intermediate elements from the front view, and join the points E, G, L, . . ., in order with straight lines, to complete the development.

21.24   To Find the Intersection and Developments of Two Prisms

INTERSECTION   *Fig. 21.21 (a)*   The points in which the edges ACEH of the horizontal prism pierce the surfaces of the vertical prism are found in the top view and are projected down-

ward to the corresponding edges ACEH in the front view. The points in which the edges 5 and 11 of the vertical prism pierce the surfaces of the horizontal prism are found in the left-side view at G, D, J, and B and are projected horizontally to the front view, intersecting the corresponding edges as shown. The intersection is completed by joining these points in order by straight lines.

DEVELOPMENTS   *Fig. 21.21 (b)*   The lateral surfaces of the two prisms are developed as explained in §21.23. True lengths of all lateral edges and lines parallel to them are shown in the front view of Fig. 21.21 at (a).

21.25   To Find the Intersection and Developments of Two Prisms

INTERSECTION   *Fig. 21.22 (a)*   The points in which edges 1–2–3–4 of the inclined prism pierce the surfaces of the vertical prism are vertices of the intersection. These points, found in the top

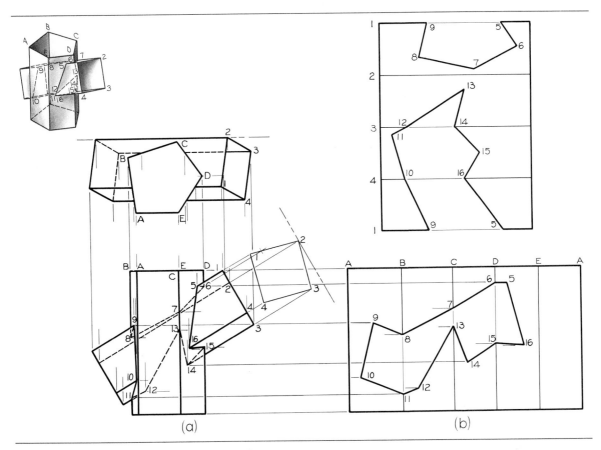

Fig. 21.22   Two Prisms Oblique to Each Other.

view, are projected downward to the corresponding edges 1–2–3–4 in the front view, as shown. The intersection is completed by joining these points in order by straight lines.

DEVELOPMENTS   *Fig. 21.22 (b)* The lateral surfaces of the two prisms are developed as explained in §21.23. True lengths of all edges of both prisms are shown in the front view of Fig. 21.22 at (a).

21.26   To Find the Intersection and the Developments of Two Prisms

Fig. 21.23 In this case the edges of the oblique prism are oblique to the planes of projection, and in the front and top views none of the edges is

shown true length, §7.24, and none of the faces is shown true size, §7.23. Furthermore, none of the angles, including the angle of inclination, is shown true size, §7.26. Therefore, it is necessary to draw a secondary auxiliary view, §10.19, to obtain the true size and shape of the right section of the oblique prism.

The direction of sight, indicated by arrow A, is assumed perpendicular to the end face 1–2–3, that is, parallel to the principal edges of the prism. The primary auxiliary view, taken in the direction of arrow B, shows the true lengths of the edges, the true inclination of the prism with respect to the horizontal, and, incidentally, the true length and inclination of arrow A. In the secondary auxiliary view, arrow A is shown as a point, and the end face 1–2–3 is shown in its true size.

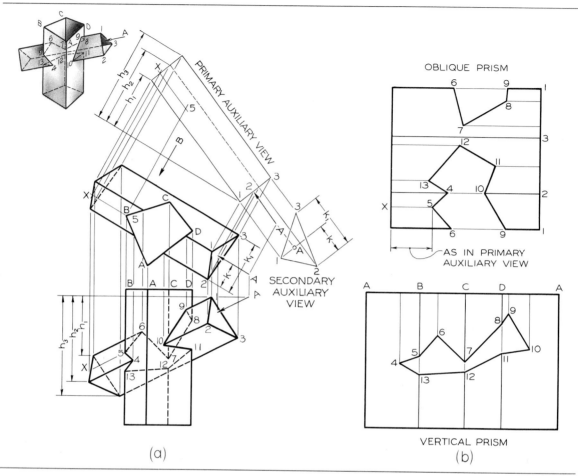

Fig. 21.23   Two Prisms Oblique to Each Other.

INTERSECTION   *Fig. 21.23 (a)*   The points in which the edges 1–2–3 of the oblique prism pierce the surfaces of the vertical prism are vertices of the intersection, found first in the top view and then projected downward to the front view.

DEVELOPMENTS   *Fig. 21.23 (b)*   The lateral surfaces of the two prisms are developed as explained in §21.23. True lengths of the edges of the vertical prism are shown in the front view. True lengths of the edges of the oblique prism can be shown in the primary auxiliary view; true

lengths to the vertices of the intersection may be found in this view, as shown for line X–5.

21.27   To Find the Intersection and Developments of Two Cylinders

INTERSECTION   *Fig. 21.24 (a)*   Assume a series of elements (preferably equally spaced) on the horizontal cylinder, numbered 1, 2, 3, . . ., in the side view, and draw their top and front views. Their points of intersection with the surface of the vertical cylinder are shown in the top view at A, B, C, . . ., and may be found in the front

702

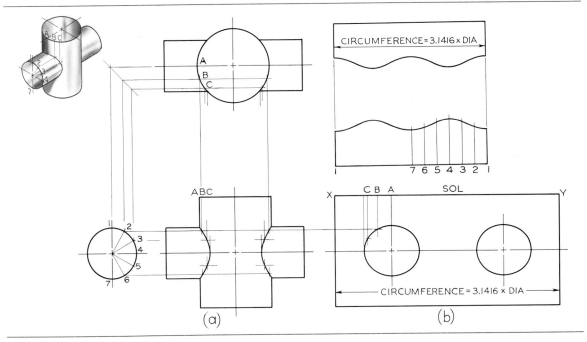

Fig. 21.24   Two Cylinders at Right Angles to Each Other.

view by projecting downward to their intersections with the corresponding elements 1, 2, 3, . . ., in the front view. When a sufficient number of points have been found to determine the intersection, the curve is traced through the points with the aid of the irregular curve, §2.54. See also Fig. 7.38 (c).

DEVELOPMENTS   *Fig. 21.24 (b)*   The lateral surfaces of the two cylinders are developed as explained in §21.7. True lengths of all elements of both cylinders are shown in the front view. Since both cylinders have bases at right angles to the center lines, the circles will develop as straight lines, and the developments will be rectangular, as shown. The length XY of the stretch-out line for the development of the vertical cylinder, is equal to the circumference of the cylinder, or πd, and the length 1–1 of the stretch-out line for the development of the horizontal cylinder is determined in the same way. Those

elements of the large cylinder that pierce the small cylinder can be identified in the top view as elements A, B, C, ... When these are drawn in the development, the points of intersections are found at their intersections with the corresponding elements of the horizontal cylinder taken from the front view, thus determining one of the figures of intersection, as shown in Fig. 21.24 (b).

21.28   To Find the Intersection and Developments of Two Cylinders

INTERSECTION   *Fig. 21.25 (a)*   A revolved right section of the inclined cylinder is divided into a number of equal parts, and an element is drawn at each of the division points, 1, 2, 3, ... The points of intersection of these elements with the surface of the vertical cylinder are shown in the top view at B, C, D, . . ., and are found in the front view by projecting downward to intersect

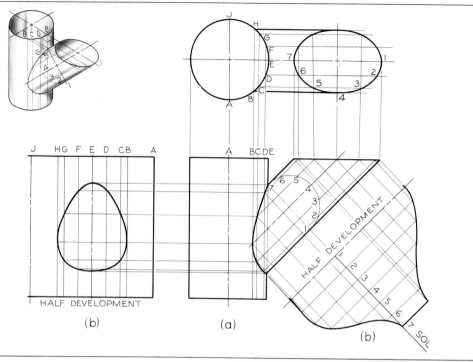

Fig. 21.25   Two Cylinders Oblique to Each Other.

the corresponding elements 1, 2, 3, ... The curve is traced through these points with the aid of the irregular curve, §2.54

DEVELOPMENTS   *Fig. 21.25 (b)* The lateral surfaces of the two cylinders are developed as explained in §§21.7 and 21.9. True lengths of all elements of both cylinders are shown in the front view.

21.29   To Find the Intersection and Developments of a Prism and a Cone

INTERSECTION   *Fig. 21.26 (a)* Points in which the edges of the prism intersect the surface of the cone are shown in the side view at A, C, and F. Intermediate points such as B, D, E, and G are piercing points of any lines on the lateral surface of the prism parallel to the edges. Through all of

the piercing points in the side view, elements of the cone are drawn and then drawn in the top and front views. The intersections of the elements of the cone with the edges of the prism (and lines along the prism drawn parallel thereto) are points of the intersections. The figures of intersections are traced through these points with the aid of the irregular curve, §2.54.

The elements 6, 5, 4, . . ., in the side view of the cone may be regarded as the edge views of cutting planes that cut these elements on the cone and edges or elements on the prism. The intersection of corresponding edges or elements on the two solids are points on the figure of intersection.

Another method of finding the figure of intersection is to pass a series of horizontal parallel planes through the solids in the manner of Fig. 21.9 (a). The plane will cut circles on the cone and straight lines on the prism, and their inter-

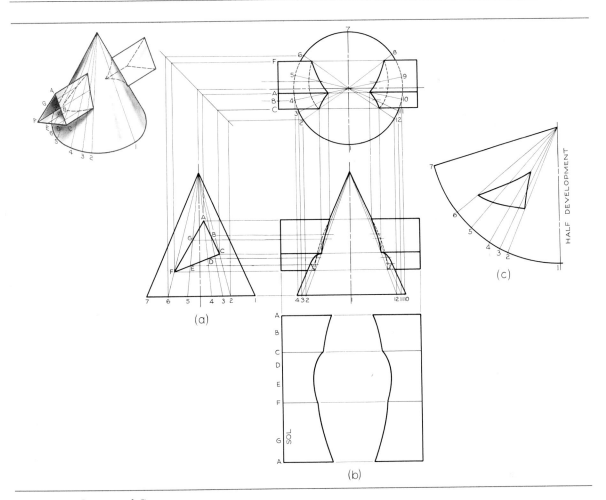

Fig. 21.26   Prism and Cone.

sections will be points on the figure of intersection. See also Fig. 21.27 (b).

DEVELOPMENTS   *Fig. 21.26 (b)* The lateral surface of the prism is developed as explained in §21.23. True lengths of all edges and lines parallel thereto are shown in both the front and top views.

The lateral surface of the cone, Fig. 21.26 (c), is developed as explained in §21.11. True lengths of elements from the vertex to points on the intersections are found as shown in Fig. 11.10 (a).

21.30   To Find the Intersection of a Prism and a Cone with Edges of Prism Parallel to Axis of Cone

Fig. 21.27 (a) Since the lateral surfaces of the prism are parallel to the axis of the cone, the figure of intersection will be composed of a series of hyperbolas, §§5.47 and 5.60. If a series of planes is assumed containing the axis of the cone, each plane will contain edges of the prism, or will cut lines parallel to them along the prism, and will cut elements on the cone that intersect these at points on the figure of intersection.

705

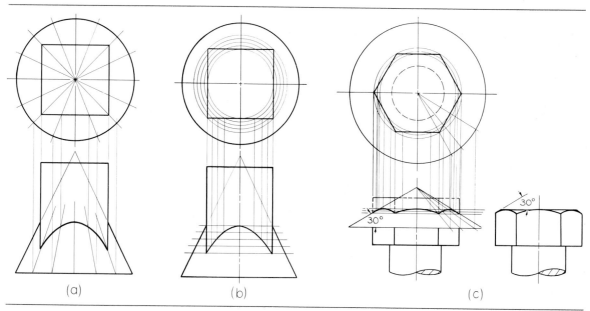

Fig. 21.27   Prisms and Cones.

Fig. 21.27 (b) The intersection is the same as at (a), but it is found in a different manner. Here a series of parallel planes perpendicular to the axis of the cone cut circles of varying diameters on the cone. These circles are shown true size in the top view, where the piercing points of these circles in the vertical plane surfaces of the prism are also shown. The front views of these piercing points are found by projecting downward to the corresponding cutting-plane lines.

Fig. 21.27 (c) The chamfer of an ordinary hexagon bolt head or hexagon nut is actually a conical surface that intersects the six vertical sides of a hexagonal prism to form hyperbolas. At (c) the methods of both (a) and (b) are shown to illustrate how points may be found by either method.

In machine drawings of bolts and nuts, these hyperbolic curves are approximated by means of circular arcs, as shown in Fig. 15.27.

21.31   To Find the Intersections and the Developments of a Cylinder and a Cone

INTERSECTIONS *Fig. 21.28 (a)* Points in which elements of the cylinder (preferably equally spaced to facilitate the development) intersect the surface of the cone are shown in the side view at A, B, C, ... The elements of the cylinder are shown here as points. Elements of the cone are then drawn from the vertex through each of these points, and then drawn in their correct locations in the top and front views. The intersections of these elements with the elements A, B, C, . . ., of the cylinder are points on the figures of intersection. The curves are then traced through these points with the aid of the irregular curve, §2.54.

As explained in §21.29, the elements 5, 6, 7, . . ., in the side view of Fig. 21.28 (a) could be regarded as edge views of cutting planes that cut elements from both the cone and the cylinder,

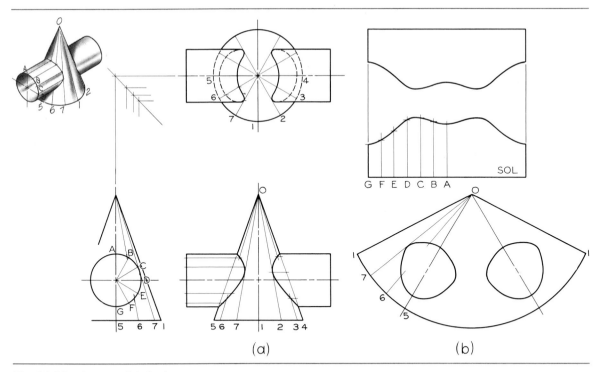

Fig. 21.28   Cone and Cylinder.

the elements meeting at points on the figure of intersection. Or a series of horizontal parallel planes can be passed through the solids that will cut circles from the cone and elements from the cylinder that intersect at points on the figure of intersection.

DEVELOPMENTS   *Fig. 21.28 (b)*   The lateral surface of the cylinder is developed as explained in §21.7. True lengths of all elements are shown in both the front and top views. The lateral surface of the cone is developed as explained in §21.11. True lengths of elements from the vertex to points on the intersections are found as shown in Fig. 11.10 (a).

21.32   To Find the Intersection of a Cylinder and a Sphere
Fig. 21.29   Horizontal planes 1, 2, 3, . . ., which appear edgewise in the front and side views, cut elements A, B, C, . . ., from the cylinder and cir-

cular arcs 1', 2', 3', . . ., from the sphere. The intersections of the elements with the arcs produced by the corresponding planes are points on the figure of intersection. Join the points with a smooth curve, §2.54.

21.33   Computer Graphics
Determining the intersections and developments of planes and solids that contain plane, curved, or warped surfaces can be very time consuming if done by traditional methods, Figs. 21.23, 21.25, and 21.28. Complex surfacing software is available that will permit the user to graphically model and mathematically define a wide variety of geometric surfaces and to determine their intersections with planes or with other surfaces. Once a solution has been determined, additional software is available that can provide computer-driven machining of the desired surface to produce a required part.

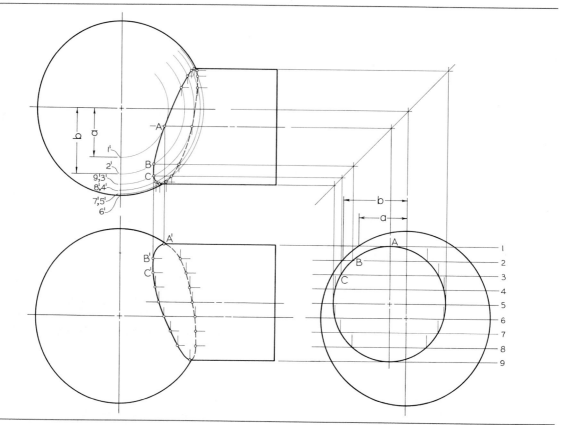

Fig. 21.29   Intersection of Sphere and Cylinder.

INTERSECTION AND DEVELOPMENT PROBLEMS

A wide selection of intersection and development problems is provided in Figs. 21.30–21.37. These problems are designed to fit size B (11.0″ × 17.0″) or A3 (297 mm × 420 mm) sheets. Dimensions should be included on the given views. The student is cautioned to take special pains to obtain accuracy in these drawings and to draw smooth curves as required.

Since many of the problems in this chapter are of a general nature, they can also be solved on most computer graphics systems. If a system is available, the instructor may choose to assign specific problems to be completed by this method.

Problems in convenient form for solution may be found in *Technical Drawing Problems,* Series 1, by Giesecke, Mitchell, Spencer, Hill, Dygdon, and Novak; *Technical Drawing Problems,* Series 2, by Spencer, Hill, Dygdon, and Novak; and *Technical Drawing Problems,* Series 3, by Spencer, Hill, Dygdon, and Novak, all designed to accompany this text and published by Macmillan Publishing Company.

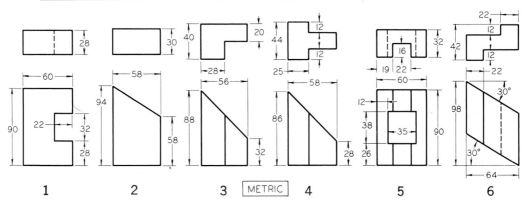

Fig. 21.30 Draw given views and develop lateral surface (Layout A3–3 or B–3).

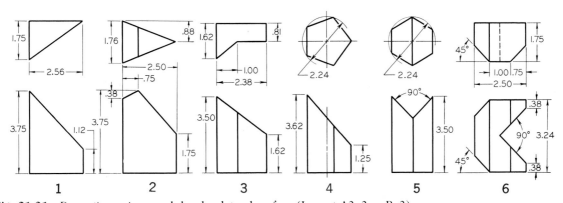

Fig. 21.31 Draw given views and develop lateral surface (Layout A3–3 or B–3).

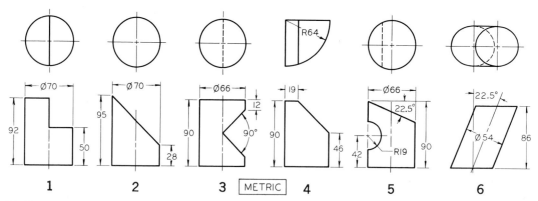

Fig. 21.32   Draw given views and develop lateral surface (Layout A3–3 or B–3).

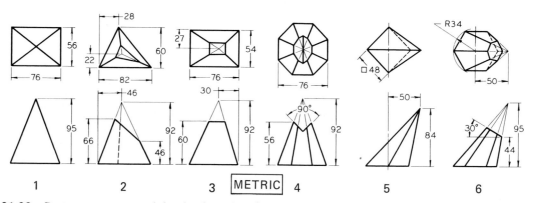

Fig. 21.33   Draw given views and develop lateral surface (Layout A3–3 or B–3).

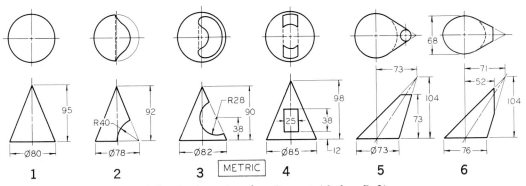

Fig. 21.34 Draw given views and develop lateral surface (Layout A3–3 or B–3).

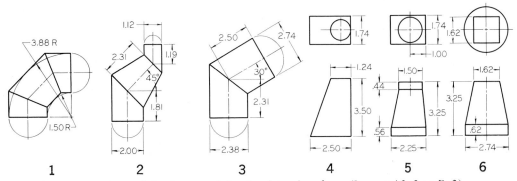

Fig. 21.35 Draw given views of the forms and develop lateral surfaces (Layout A3–3 or B–3).

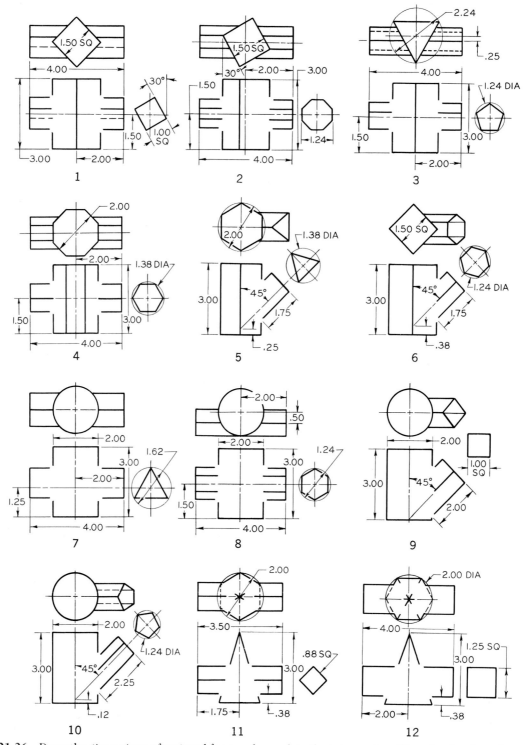

Fig. 21.36 Draw the given views of assigned form and complete the intersection. Then develop lateral surfaces (Layout A3–3 or B–3).

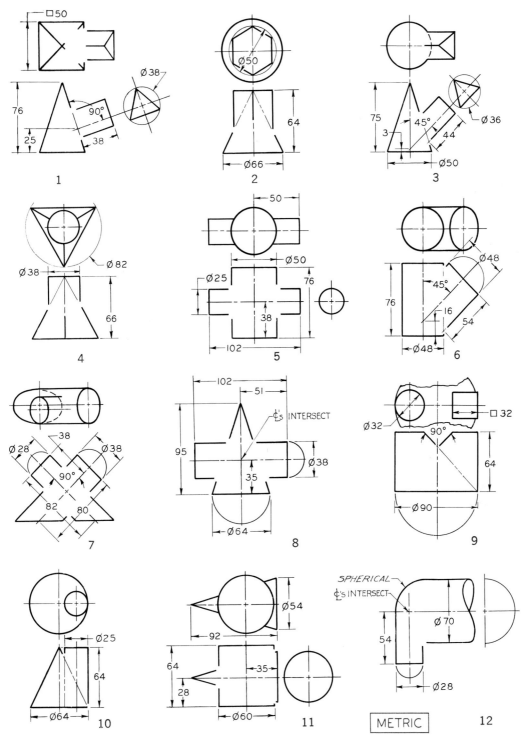

Fig. 21.37 Draw the given views of assigned form and complete the intersection. Then develop lateral surfaces (Layout A3–3 or B–3).

CHAPTER 22

Gearing and Cams

BY B. LEIGHTON WELLMAN*
REVISED BY WALTER D. CIBULSKIS†

In machine design and drafting, various devices are used to transmit power and motion from one machine member to another. *Gears, pulleys and belts, chains and sprockets, cams, linkages,* and other devices are common. In the layout of these machines, and in making detail drawings of these elements, certain standards must be followed, and a knowledge of the function of these devices is important. Some of the more important devices are described in this chapter.

22.1 Gears*

Gears, Fig. 22.1, are used to transmit power and rotating or reciprocating motion from one machine part to another. They may be classified according to the position of the shafts that they connect. Parallel shafts, for example, may be connected by *spur gears, helical gears,* or *herringbone gears.* Intersecting shafts may be connected by *bevel gears* having either straight, skew, or spiral teeth. Nonparallel, nonintersecting shafts may be connected by *helical gears, hypoid gears,* or a *worm* and *worm gear.* A spur gear meshed with a *rack* will convert rotary motion to reciprocating motion.

22.2 Spur Gears

The *friction wheels* shown in Fig. 22.2 (a) will transmit motion and power from one shaft to

*See ANSI Y14.7.1–1971 (R1988) and Y14.7.2–1978 (R1989) for additional information.

another parallel shaft. However, friction wheels are subject to slipping, and excessive pressure is required between the wheels to obtain the necessary frictional force; therefore, they are usually used for low-power applications, such as phonograph turntable drives. If teeth of the proper shape are provided on the cylindrical surfaces, the resulting spur gears, Fig. 22.2 (b), will transmit the same motion and power without slipping and with greatly reduced bearing pressures.

If a friction wheel of diameter D turns at n rpm (revolutions per minute), the linear velocity, v, of a point on its periphery will be πDn, since $\pi D = c$ (circumference).

EXAMPLE:

Let

$$D = 3 \text{ in.}$$
$$n = 100 \text{ rpm}$$

*Professor Emeritus of Mechanical Engineering, Worcester Polytechnic Institute.
†Professor, Department of Occupational Education, Chicago State University.

Fig. 22.1   An Assortment of Gears. *Courtesy Charles Bond Co.*

Then

$$v = \pi Dn = \pi(3)(100) \text{ in./min}$$
$$= 942 \text{ in./min}$$

or

$$\frac{942}{12} = 78.5 \text{ ft/min}$$

But the pitch circles of a pair of mating spur gears correspond exactly to the outside diameters of the friction wheels, and since the gears turn in contact without slipping, they must have the same linear velocity at the pitch line. Therefore,

$$\pi D_G n_G = \pi D_P n_P$$

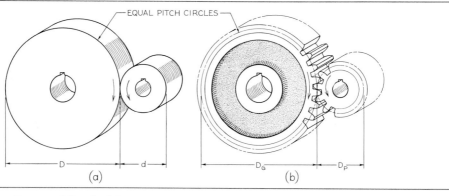

Fig. 22.2   Friction Gears and Toothed Gears.

or

$$\frac{D_G}{D_P} = \frac{n_P}{n_G} = m_G$$

where D_G = pitch diameter of larger gear (called *gear*)
D_P = pitch diameter of smaller gear (called *pinion*)
n_G = rpm of gear
n_P = rpm of pinion
m_G = gear ratio

The gear ratio is also expressed as n_P/n_G or D_G/D_P.

EXAMPLE:
Let

$$D_G = 27'' \quad \text{and} \quad D_P = 9''$$

Then

Gear ratio = m_G = 27 : 9
= 3 : 1 (read as 3 to 1)

EXAMPLE: For the same gear pair, let

$$n_P = 1725 \text{ rpm}$$

Find n_G.

$$\frac{n_P}{n_G} = \frac{D_G}{D_P}, \quad \frac{n_G}{n_P} = \frac{D_P}{D_G}$$

$$n_G = n_P \frac{D_P}{D_G} = 1725 \cdot \frac{9}{27}$$

$$= \frac{1725}{3}$$

$$= 575 \text{ rpm}$$

The teeth on mating gears must be of equal width and spacing; hence, the number of teeth on each gear, N, is directly proportional to its pitch diameter, or

$$\frac{N_G}{N_P} = \frac{D_G}{D_P} = \frac{n_P}{n_G} = m_G$$

22.3   Spur Gear Definitions and Formulas*

Proportions and shapes of gear teeth are well standardized, and the terms illustrated and defined in Fig. 22.3 are common to all spur gears. The dimensions relating to tooth height are for full-depth $14\frac{1}{2}°$ or 20° involute teeth.

22.4   The Shape of the Tooth

If gears are to operate smoothly with a minimum of noise and vibration, the curved surface of the tooth profile must be of a definite geometric form. The most common form in use today is the *involute profile.*

In the involute system, the form of the tooth depends basically on the *pressure angle,* which is ordinarily $14\frac{1}{2}°$ or 20°. This pressure angle determines the size of the *base circle;* from this the involute curve is generated.

The base circle is constructed in the following manner. At any point on the pitch circle, such as point P (the pitch point), Fig. 22.4, a line is drawn tangent to the pitch circle; a second line is drawn through P at the required pressure angle ($14\frac{1}{2}°$ is frequently approximated at 15° on the drawing). This line is called the *line of contact.* A line is then drawn perpendicular to the line of contact from the center, C. The base circle is then drawn with radius CJ tangent to the line of contact at J.

If the exact shape of the tooth is desired, the portion of the profile from the base circle to the addendum circle can be drawn as the involute of the base circle. In Fig. 22.5, the tooth profile from A to O is an involute of the base circle. The method of construction is shown in Fig. 5.64 (d) and (e). That part of the profile which is below the base circle, line OB, is drawn as a radial line (a straight line drawn from the gear center) that terminates in the fillet at the root circle. The fillet should be equal in radius to one and one-half times the clearance.

*See also AGMA 112.05–1976/ANSI B6.14–1976, AGMA 201.02–1968 (R1974)/ANSI B6.1–1968, AGMA 207.06–1977/ANSI B6.7–1977 and other AGMA (American Gear Manufacturers Association) standards for more details.

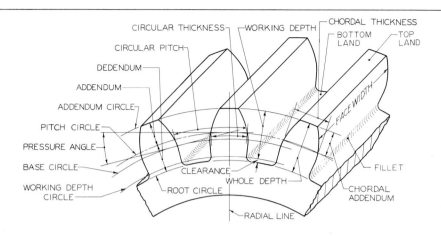

Term	Symbol	Definition	Formula
Addendum	a	Radial distance from pitch circle to top of tooth.	$a = 1/P$
Base circle		Circle from which involute profile is generated.	
Chordal addendum	a_c	Radial distance from top of a tooth to the chord of pitch circle.	$a_c = a + \frac{1}{2}D[1 - \cos(90°/N)]$
Chordal thickness	t_c	Thickness of a tooth measured along a chord of pitch circle.	$t_c = D \sin(90°/N)$
Circular pitch	p	Distance measured along pitch circle from a point on one tooth to corresponding point on the adjacent tooth; includes one tooth and one space.	$p = \pi D/N$ $p = \pi/P$
Circular thickness	t	Thickness of a tooth measured along the pitch circle; equal to one-half the circular pitch.	$t = p/2 = \pi/2P$
Clearance	c	Distance between top of a tooth and bottom of mating space; equal to the dedendum minus the addendum.	$c = b - a = 0.157/P$
Dedendum	b	Radial distance from pitch circle to bottom of tooth space.	$b = 1.157/P$
Diametral pitch	P	A ratio equal to number of teeth on the gear per inch of pitch diameter.	$P = N/D$
Number of teeth	N_G or N_P	Number of teeth on the gear or pinion.	$N = P \times D$
Outside diameter	D_o	Diameter of addendum circle; equal to pitch diameter plus twice the addendum.	$D_o = D + 2a = (N + 2)/P$
Pitch circle		An imaginary circle that corresponds to circumference of the friction gear from which the spur gear is derived.	
Pitch diameter	D_G or D_P	Diameter of pitch circle of gear or pinion.	$D = N/P$
Pressure angle	ϕ	Angle that determines direction of pressure between contacting teeth and designates shape of involute teeth—e.g., $14\frac{1}{2}°$ involute; also determines size of base circle.	
Root diameter	D_R	Diameter of the root circle; equal to pitch diameter minus twice the dedendum.	$D_R = D - 2b =$ $(N - 2.314)/P$
Whole depth	h_t	Total height of the tooth; equal to the addendum plus the dedendum.	$h_t = a + b = 2.157/P$
Working depth	h_k	Distance a tooth projects into mating space; equal to twice the addendum.	$h_k = 2a = 2/P$

Fig. 22.3 Spur Gear Nomenclature.

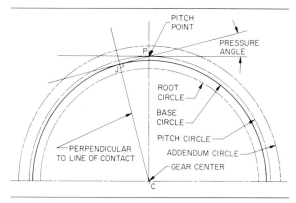

Fig. 22.4 Construction of a Base Circle.

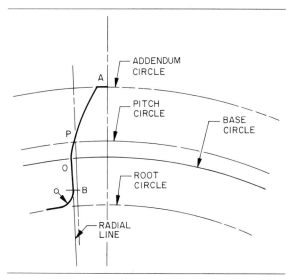

Fig. 22.5 The Involute Profile.

For display drawings and more rapid construction, the involute curve can be closely approximated with two circular arcs, as shown in Fig. 22.6. This method, originally devised by Grant,* employs a table of arc radii, an involute odontograph, for gears having various numbers

*G. B. Grant, "Teeth of Gears," 1891. Grant's tabulated radii, which have been used for many years, were based on a pressure angle of 15° and were empirically shortened to correct for interference at the tooth tips. The new radii given in Fig. 22.6 have been computed for both $14\frac{1}{2}$° and 20° to more closely approximate the true involute profile. To distinguish clearly the two tables, the new values are called *Wellman's involute odontograph.*

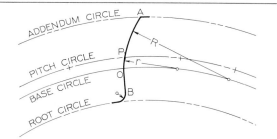

Divide radii given in table by Diametral Pitch.

No. of Teeth (N)	$14\frac{1}{2}$°		20°	
	R (in.)	r (in.)	R (in.)	r (in.)
12	2.87	0.79	3.21	1.31
13	3.02	0.88	3.40	1.45
14	3.17	0.97	3.58	1.60
15	3.31	1.06	3.76	1.75
16	3.46	1.16	3.94	1.90
17	3.60	1.26	4.12	2.05
18	3.74	1.36	4.30	2.20
19	3.88	1.46	4.48	2.35
20	4.02	1.56	4.66	2.51
21	4.16	1.66	4.84	2.66
22	4.29	1.77	5.02	2.82
23	4.43	1.87	5.20	2.98
24	4.57	1.98	5.37	3.14
25	4.70	2.08	5.55	3.29
26	4.84	2.19	5.73	3.45
27	4.97	2.30	5.90	3.61
28	5.11	2.41	6.08	3.77
29	5.24	2.52	6.25	3.93
30	5.37	2.63	6.43	4.10
31	5.51	2.74	6.60	4.26
32	5.64	2.85	6.78	4.42
33	5.77	2.96	6.95	4.58
34	5.90	3.07	7.13	4.74
35	6.03	3.18	7.30	4.91
36	6.17	3.29	7.47	5.07
37–39	6.36	3.46	7.82	5.32
40–44	6.82	3.86	8.52	5.90
45–50	7.50	4.46	9.48	6.76
51–60	8.40	5.28	10.84	7.92
61–72	9.76	6.54	12.76	9.68
73–90	11.42	8.14	15.32	11.96
91–120	0.118N		0.156N	
121–180	0.122N		0.165N	
Over 180	0.125N		0.171N	

Fig. 22.6 Wellman's Involute Odontograph.

of teeth. The base circle is drawn as described above, and the spacing of the teeth is set off along the pitch circle. Then the face of the tooth from P to A is drawn with the *face radius* R, and the portion of the flank from P to O is drawn with the *flank radius* r. Both arcs are drawn from centers located on the base circle. The table gives the correct face and flank radii for gears of *one* diametral pitch. For other pitches, the values in the table must be divided by the diametral pitch. For gears with more than 90 teeth, use a single radius (let R = r) computed from the appropriate formula given in Fig. 22.6, and then divide by diametral pitch. Below the base circle, the flank of the tooth is completed with a radial line OB and a fillet.

EXAMPLE: Suppose that $N = 20$ and $P = 4$ (the diametral pitch), for a $14\frac{1}{2}°$ involute tooth. The values from the table are $R = 4.02$, and $r = 1.56$. These must be divided by P; then $R = 4.02/4 = 1.005''$ and $r = 1.56/4 = 0.39''$. (Note that these values are for a full-size drawing. If the drawing is reduced or enlarged, the values must be adjusted accordingly.) Templates are also available for drawing gear tooth profiles and can increase the speed of drawing considerably.

The teeth may be spaced around the periphery by laying off equal angles, Fig. 22.7. The number of spaces should be 2N, twice the number of teeth, to make the space between teeth

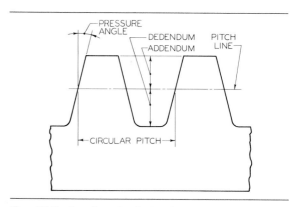

Fig. 22.8   Involute Rack Teeth.

equal to the tooth thickness at the pitch circle. In the example, since $N = 20$, $a = 360°/2N = 360°/40 = 9°$, the angle subtended by each tooth and each space.

When the pressure angle is 20° and the height of the tooth is reduced, the teeth are called *stub* teeth. Stub teeth are drawn in the same manner except that $a = 0.8/P$, $b = 1/P$, and the pressure angle is made 20°. The major advantage of stub teeth is that they are stronger than the $14\frac{1}{2}°$ standard full-depth teeth.

When the gear teeth are formed on a flat surface, the result is a rack, Fig. 22.8. In the involute system, the sides of rack teeth are straight and are inclined at an angle equal to the pressure angle. To mesh with a gear, the linear pitch of the rack must be the same as the circular pitch of the gear, and the rack teeth must have the same height proportions as the gear teeth.

22.5 Working Drawings of Spur Gears

A typical working drawing of a spur gear is shown in Fig. 22.9. Since the teeth are cut to standard shape with special cutters, it is not necessary to show individual teeth on the drawing. Instead, the addendum and root circles are drawn as phantom lines, Fig. 2.14, and the pitch circle as a center line. Thus, the drawing actually shows only a gear blank—a gear complete except for teeth. Since the machining of the blank and the

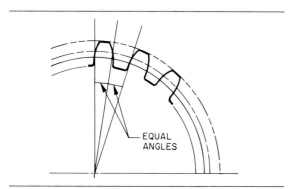

Fig. 22.7   Spacing Gear Teeth.

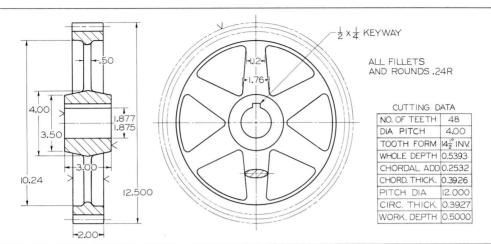

Fig. 22.9   Working Drawing of a Spur Gear.

The cutting data table in the figure reads:

CUTTING DATA	
NO. OF TEETH	48
DIA PITCH	4.00
TOOTH FORM	$14\frac{1}{2}°$ INV.
WHOLE DEPTH	0.5393
CHORDAL ADD	0.2532
CHORD. THICK.	0.3926
PITCH DIA	12.000
CIRC. THICK.	0.3927
WORK. DEPTH	0.5000

cutting of the teeth are separate operations in the shop, the necessary dimensions are arranged in two groups: the blank dimensions are shown on the views, and the cutting data are given in a note or table.

Before laying out the working drawing, the drafter must calculate the gear dimensions. For example, if the gear must have 48 teeth of 4 diametral pitch, with $14\frac{1}{2}°$ full-depth involute profile, as in Fig. 22.9, then the following items should be calculated in this order: pitch diameter, addendum, dedendum, outside diameter, root diameter, whole depth, chordal thickness, and chordal addendum. The dimensions shown in the figure are the minimum requirements for the spur gear. The chordal addendum and chordal thickness are given to aid in checking the finished gear. Other special data may be given in the table, according to the degree of precision required and the method of manufacture.

22.6   Design of Spur Gears

The design of spur gears normally begins with selection of pitch diameters to suit the required speed ratio, center distance, and space limitations. The size of the teeth—that is, the diametral pitch—depends on the gear speeds, gear materials, horsepower to be transmitted, and the selected tooth form. The complete analysis and design of precision gears is complex and beyond the scope of this textbook, but the chart in Fig. 22.10 will quickly yield a suitable diametral pitch for ordinary cut spur gears.*

EXAMPLE: Determine the diametral pitch and face width for a 5″ pitch diameter (G10400 carbon steel) pinion having $14\frac{1}{2}″$ full-depth teeth that must transmit 10 hp at 200 rpm.

Calculate the pitch-line velocity V as follows.

$$
\begin{aligned}
V &= \frac{\pi}{12} D_P n_P \\
&= 0.262 \times 5 \times 200 = 262 \text{ fpm}
\end{aligned}
$$

On the chart, draw a straight line from 25,000 psi on the working stress scale through 262 fpm on the velocity scale to intersect the pivot line at X. From X, draw a second straight line through 10 on the horsepower scale to intersect the Q scale at Y.

*The chart is based on the Lewis equation, with the Barth velocity modification, and assumes gear-face width equal to three times the circular pitch.

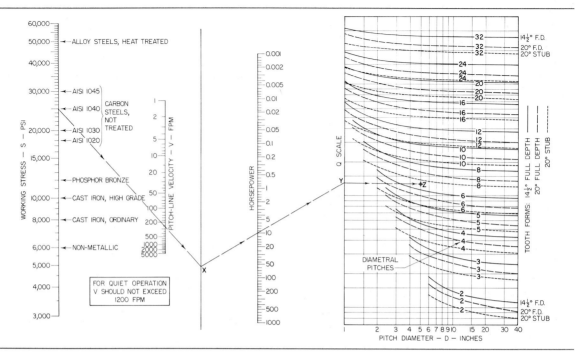

Fig. 22.10   Chart for Design of Cut Spur Gears.

From this point, enter the graph of pitches and go to the ordinate for 5″ pitch diameter. The junction point Z falls slightly above the curve for $14\frac{1}{2}°$ full-depth teeth of 6 diametral pitch; hence, choose 6 diametral pitch for the gears.

For good proportions the face width of a spur gear should be about three times the circular pitch, and this proportion is incorporated in the chart. Therefore, for 6 diametral pitch the circular pitch is $\pi/6$ or .5236″, and the face width is $3 \times .5236$ or 1.5708″. This width should be rounded off to $1\frac{5}{8}″$.

When pinion and gear are of the same material, the smaller gear is the weaker, and the design should be based on the pinion. When the materials are different, the chart should be used to determine the horsepower capacity of each gear.

22.7   Worm Gears*

Worm gears are used to transmit power between nonintersecting shafts that are at right angles to each other. The *worm* is a screw having a thread of the same shape as a rack tooth. The *worm wheel* is similar to a helical gear that has been cut to conform to the shape of the worm for more contact. A large speed ratio is obtainable with worm gearing, since a *single-thread worm* in one revolution advances the worm wheel only one tooth and a space.

In Fig. 22.11, a worm and a worm wheel are shown engaged. The section taken through the center of the worm and perpendicular to the axis of the worm wheel shows that the worm section is identical to a rack and that the wheel section is identical to a spur gear. Consequently, in this plane the height proportions of thread and gear

*See ANSI/AGMA 374.07–1977 for more details.

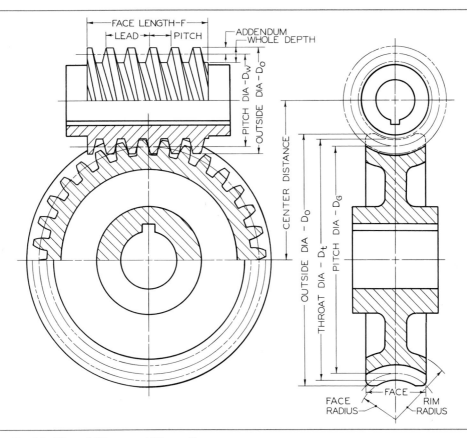

Fig. 22.11   Double Thread Worm and Worm Gear.

teeth are the same as for a spur gear of corresponding pitch.

Pitch (p)   The axial pitch of the worm is the distance from a point on one thread to the corresponding point on the next thread measured parallel to the worm axis. The pitch of the worm must exactly equal the circular pitch of the gear.

Lead (L)   The lead is the distance that the thread advances axially in one turn. The lead is always a multiple of the pitch. Thus, for a single-thread worm, the lead equals the pitch; for a double-thread worm, the lead is twice the pitch, and so on.

Lead angle (λ)   The lead angle is the angle between a tangent to the helix at the pitch diam-

eter and a plane perpendicular to the axis of the worm. The lead angle can be calculated from the formula

$$\tan \lambda = \frac{L}{\pi D_W}$$

where D_W = pitch diameter of the worm.

The speed ratio of worm gears depends only on the *number of threads* on the worm and the *number of teeth* on the gear. Therefore,

$$m_G = \frac{N_G}{N_W}$$

where N_G = number of teeth on the gear
N_W = number of threads on the worm

723

For $14\frac{1}{2}''$ standard involute teeth and single-thread or double-thread worms, the following proportions are the recommended practice of the AGMA (American Gear Manufacturers' Association). All formulas are expressed in terms of circular pitch p instead of diametral pitch, P. It is easier to machine the worm and the hob used to cut the gear if the circular pitch has an even rational value such as $\frac{5}{8}''$.

For the worm:

Pitch diameter	$D_W = 2.4p + 1.1^*$
Whole depth	$h_t = 0.686p$
Outside diameter	$D_o = D_W + 0.636p$
Face length	$F = p(4.5 + N_G/50)$

For the gear:

Pitch diameter	$D_G = p(N_G/\pi)$
Throat diameter	$D_t = D_G + 0.636p$
Outside diameter	$D_o = D_t + 0.4775p$
Face radius	$R_f = \frac{1}{2}D_W - 0.318p$
Rim radius	$R_r = \frac{1}{2}D_W + p$
Face width	$F = 2.38p + 0.25$
Center distance	$C = \frac{1}{2}(D_G + D_W)$

*Recommended value but may be varied.

22.8 Working Drawings of Worm Gears

In an assembly drawing, the engaged worm and gear can be shown as in Fig. 22.11, but it is customary to omit the gear teeth and represent the gear blank conventionally as shown in the lower half of the circular view. On detail drawings, the worm and gear are usually drawn separately as shown in Figs. 22.12 and 22.13. Although dimensioning of these parts is again largely dependent on the method of production, it is standard practice to dimension the blanks on the views and to give the cutting data in tabular form, as shown. Note that those dimensions that closely affect the engagement of the gear and worm have been given as three-place decimal or limit dimensions; other dimensions, such as rim radius, face lengths, and gear outside diameter have been rounded to convenient two-place decimal values.

22.9 Bevel Gears

Bevel gears are used to transmit power between shafts whose axes intersect. The analogous friction drive would consist of a pair of cones having a common apex at the point of intersection of the axes, as in Fig. 22.14. The axes may intersect at any angle, but axes at right angles are most common, and this is the only case that will be considered here. Bevel gear teeth have the same

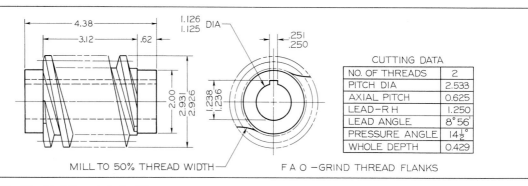

CUTTING DATA	
NO. OF THREADS	2
PITCH DIA	2.533
AXIAL PITCH	0.625
LEAD–RH	1.250
LEAD ANGLE	8°56'
PRESSURE ANGLE	$14\frac{1}{2}°$
WHOLE DEPTH	0.429

Fig. 22.12 Working Drawing of a Worm.

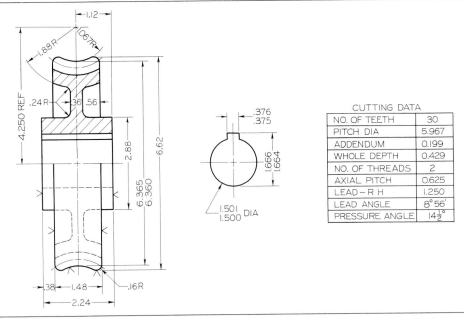

CUTTING DATA	
NO. OF TEETH	30
PITCH DIA	5.967
ADDENDUM	0.199
WHOLE DEPTH	0.429
NO. OF THREADS	2
AXIAL PITCH	0.625
LEAD – R H	1.250
LEAD ANGLE	8° 56'
PRESSURE ANGLE	$14\frac{1}{2}$°

Fig. 22.13 Working Drawing of a Worm Gear.

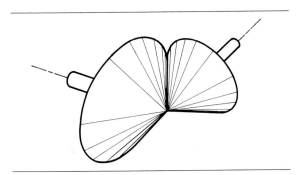

Fig. 22.14 Friction Cones.

involute shape as teeth on spur gears, but the bevel gear teeth are tapered toward the cone apex; hence, the height and width of a bevel gear tooth vary as the distance from the cone apex. Whereas spur gears are interchangeable—a spur gear of given pitch will run properly with any other spur gear of the same pitch and tooth form—this is not true of bevel gears, which must be designed in pairs and will run only with each other.

The speed ratio of bevel gears can be calculated from the same formulas given for spur gears in §22.2.

22.10 Bevel Gear Definitions and Formulas

The design of bevel gears is very similar to that of spur gears;* therefore, many of the spur gear terms are applied with slight modification to bevel gears. The *pitch diameter D* of a bevel gear is the diameter of the base of the pitch cone, and the *circular pitch p* of the teeth is measured along this circle. The *diametral pitch P* is also based on this circle; hence, the relationship of these three terms is the same as for spur gears.

The important dimensions and angles of a bevel gear are illustrated in Fig. 22.15. The *pitch*

*In the simplified formulas given here, tooth proportions are assumed equal on both gears, but in modern practice bevel gears are often designed with unequal addendums and unequal tooth thicknesses to balance the strength of gear and pinion. See ANSI/AGMA 208.03–1979.

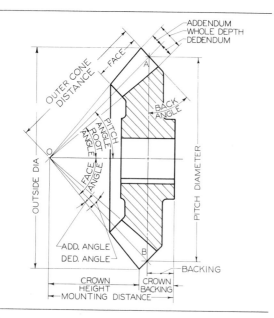

Fig. 22.15 Bevel Gear Nomenclature.

cone is shown as the triangle AOB, and the pitch angle is $\frac{1}{2}$ angle AOB. The root and face angle lines do not actually converge at point O, but are often drawn as if they do for simplicity. It can be seen from Fig. 22.16 that the pitch angle of each gear depends on the relative diameters of the gears. When the shafts are at right angles, the sum of the pitch angles for the two mating gears equals 90°.

Therefore, the pitch angles, Γ (gamma), are determined from the following equations:

$$\tan \Gamma_G = \frac{D_G}{D_P} = \frac{N_G}{N_P}$$

and

$$\tan \Gamma_P = \frac{D_P}{D_G} = \frac{N_P}{N_G}$$

Other terms are defined in Table 22.1.

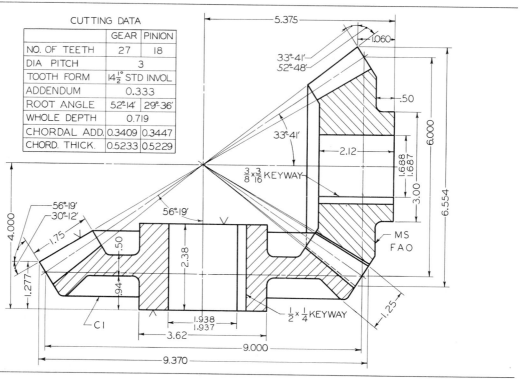

CUTTING DATA		
	GEAR	PINION
NO. OF TEETH	27	18
DIA PITCH	3	
TOOTH FORM	$14\frac{1}{2}°$ STD INVOL	
ADDENDUM	0.333	
ROOT ANGLE	52°14'	29°36'
WHOLE DEPTH	0.719	
CHORDAL ADD.	0.3409	0.3447
CHORD. THICK.	0.5233	0.5229

Fig. 22.16 Working Drawing of Bevel Gears.

Table 22.1 *Bevel Gear Terms, Definitions, and Formulas*

Term	Symbol	Definition	Formula
Addendum	a	Distance from pitch cone to top of tooth; measured at large end of tooth.	$a = 1/P$
Addendum angle	α	Angle subtended by addendum; same for gear and pinion.	$\tan \alpha = a/A$
Back angle		Usually equal to the pitch angle.	
Backing	Y	Distance from base of pitch cone to rear of hub.	
Chordal addendum	a_c ⎫	For bevel gears the formulas given for spur gears can be used if D is replaced by $D/\cos \Gamma$ and N is replaced by $N \cos \Gamma$.	
Chordal thickness	t_c ⎭		
Crown backing	Z	More practical than backing for shop use; dimension Z given on drawings instead of Y.	$Z = Y + a \sin \Gamma$
Crown height	X	Distance, parallel to gear axis, from cone apex to crown of the gear.	$X = \tfrac{1}{2}D_o/\tan \Gamma_o$
Dedendum	b	Distance from pitch cone to bottom of tooth; measured at large end of tooth.	$b = 1.188/P$
Dedendum angle	δ	Angle subtended by dedendum; same for gear and pinion.	$\tan \delta = b/A$
Face angle	Γ_o	Angle between top of teeth and the gear axis.	$\Gamma_o = \Gamma + \alpha$
Face width	F	Should not exceed $\tfrac{1}{3}A$ or $10/P$, whichever is smaller.	
Mounting distance	M	A dimension used primarily for inspection and assembly purposes.	$M = Y + \tfrac{1}{2}D/\tan \Gamma$
Outer cone distance	A	Slant height of pitch cone; same for gear and pinion.	$A = D/2 \sin \Gamma$
Outside diameter	D_o	Diameter of outside or crown circle of the gear.	$D_o = D + 2a \cos \Gamma$
Pitch diameter	D_G or D_P	Diameter of base of pitch cone of gear or pinion.	$D_G = N_G/P$ $D_P = N_P/P$
Root angle	Γ_R	Angle between the root of the teeth and the gear axis.	$\Gamma_R = \Gamma - \delta$

22.11 Working Drawings of Bevel Gears

As in the case of spur gears, a working drawing of a bevel gear gives only the dimensions of the gear blank. The necessary data for cutting the teeth are given in a note or table. A single sectional view, Fig. 22.16, usually will provide all necessary information. If a second view is required, only the gear blank is drawn, and the tooth profiles are omitted. Two gears are shown

727

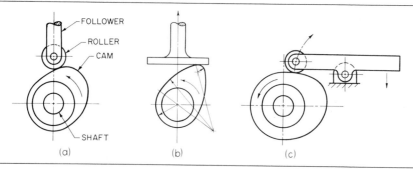

Fig. 22.17   Disk Cams.

in their operating relationship. On detail drawings, each gear is usually drawn separately, Fig. 22.15, and fully dimensioned. Proper placing of the gear-blank dimensions is largely dependent on the manufacturing methods used in producing the gear, but the scheme shown is commonly followed.

22.12   Cams

Cams provide a simple means for obtaining unusual and irregular motions that would be difficult to produce otherwise. Figure 22.17 (a) illustrates the basic principle of the cam. A shaft rotating at uniform speed carries an irregularly shaped disk called the *cam;* a reciprocating member, called the *follower,* presses a small roller against the curved surface of the cam. Rotation of the cam thus causes the follower to reciprocate with a definite cyclic motion according to the shape of the cam profile. The roller is held in contact with the cam by gravity or a spring. The problem of the drafter is to construct the cam profile necessary to obtain the desired motion of the follower.

An automobile valve cam that operates a flat-faced follower is shown at (b). At (c) is shown a disk cam with the roller follower attached to a linkage to transmit motion to another part of the device.

22.13   Displacement Diagrams

Since the motion of the follower is of primary importance, its rate of speed and its various positions should be carefully planned in a displacement diagram before the cam profile is constructed. A displacement diagram, Fig. 22.18, is a curve showing the displacement of the follower as ordinates erected on a base line that represents one revolution of the cam. The follower displacement should be drawn to scale, but any convenient length can be used to represent the 360° of cam rotation.

The motion of the follower as it rises or falls depends on the shape of the curves in the displacement diagram. In this diagram, four commonly employed types of curves are shown. If a straight line is used, such as the dashed line AD in the figure, the follower will move with a uniform velocity, but it will be forced to start and stop very abruptly, producing high acceleration and unnecessary force on the follower and other members. This straight-line motion can be modified as shown in the curve ABCD, where arcs have been introduced at the beginning and at the end of the period.

The curve shown at EF is one that gives harmonic motion to the follower. To construct this curve, a semicircle is drawn whose diameter is equal to the desired rise. The circumference of the semicircle is divided into equal arcs, the

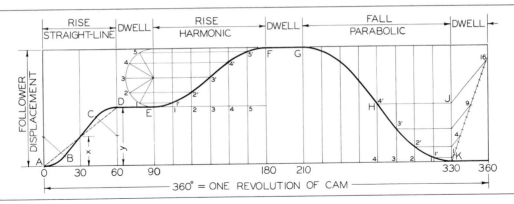

Fig. 22.18 Displacement Diagram with Typical Curves.

number of divisions being the same as the number of horizontal divisions. Points on the curve are then found by projecting horizontally from the divisions on the semicircle to the corresponding ordinates.

The parabolic curve shown at GHK gives the follower constantly accelerated and decelerated motion. The half of the curve from G to H is exactly the reverse of the half from H to K. To construct the curve HK, the vertical height from K to J is divided into distances proportional to 1^2, 2^2, 3^2, . . ., or 1, 4, 9, . . ., the number of such divisions being the same as the number of horizontal divisions. See §§5.16 and 5.58. Points on the curve are found by projecting horizontally from the divisions on the line JK to the corresponding ordinates. The parabolic curve and curves of higher polynomial equations produce smoother follower motion than those of the other curves discussed.

22.14 The Cam Profile

The general procedure in constructing a cam profile is shown in Fig. 22.19. The disk cam rotating counterclockwise on its shaft raises and lowers the roller follower that is constrained to move along the straight line AB. The displacement diagram at the bottom of the figure shows the desired follower motion.

With the follower in its lowest or initial position, the center of the roller is at A, and OA is the radius of the *base circle*. The diameter of the base circle is determined from design parameters, which are not discussed here.

Since the cam must remain stationary while it is being drawn, an equivalent rotative effect is obtained by imagining that the cam stands still while the follower rotates about the cam in the opposite direction. Therefore, the base circle is divided into twelve equiangular divisions corresponding to the divisions used in the displacement diagram. These divisions begin at zero and are numbered in an opposite direction to the cam rotation.

The points 1, 2, . . ., on the follower axis AB indicate successive positions of the center of the roller and are located by transferring ordinates such as x and y from the displacement diagram. Thus, when the cam has rotated 60°, the follower roller must rise a distance x to position 2, and after 90° of rotation, a distance y to position 3, and so on.

It should now be observed that while the center of the roller moved from its initial position A to position 2, for example, the cam rotated 60° *counterclockwise*. Therefore, point 2 must be revolved *clockwise* about the cam center O to the corresponding 60° tangent line to establish point 2′. In this position, the complete follower

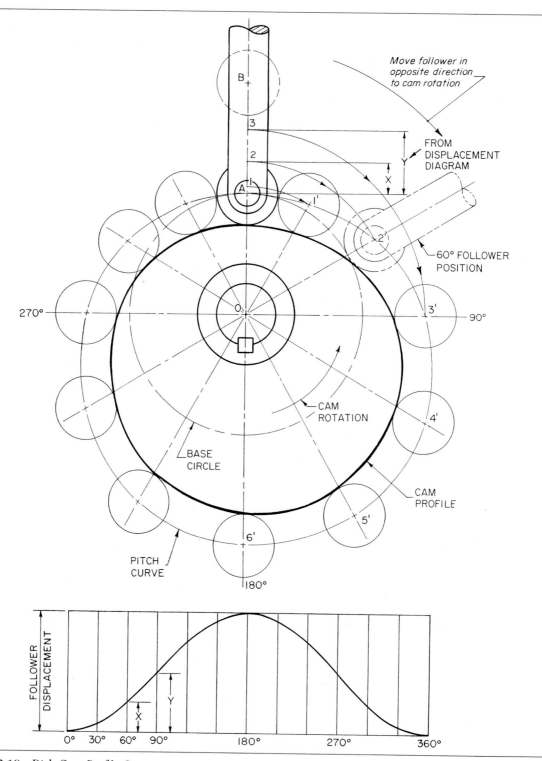

Move follower in opposite direction to cam rotation

B

3

2

FROM DISPLACEMENT DIAGRAM

Y

X

A

1'

1'

60° FOLLOWER POSITION

2'

270°   O   3'   90°

CAM ROTATION

4'

BASE CIRCLE

CAM PROFILE

5'

6'

PITCH CURVE

180°

FOLLOWER DISPLACEMENT

Y

X

0°   30°   60°   90°   180°   270°   360°

Fig. 22.19   Disk Cam Profile Construction.

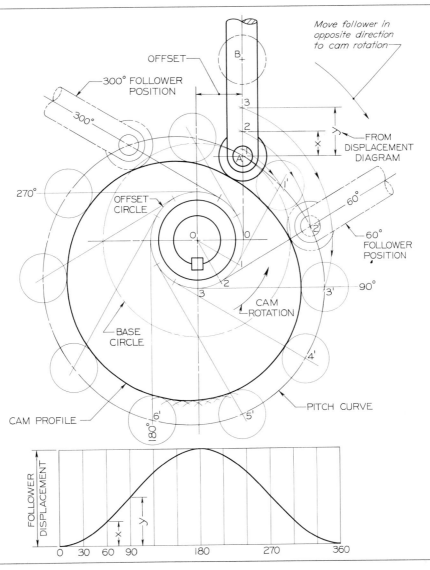

Fig. 22.20   Disk Cam Profile with Offset Follower Construction.

would appear as shown by the phantom outline. Points 1', 2', 3', ..., represent consecutive positions of the roller center, and a smooth curve drawn through these points is called the *pitch curve.* To obtain the actual cam profile, the roller must be drawn in a number of positions, and the cam profile drawn tangent to the roller circles, as shown. The best results are obtained by first drawing the pitch curve very carefully, and then drawing several closely spaced roller circles with centers on the pitch curve as shown between points 5' and 6'.

If the follower is offset as shown in Fig. 22.20, an offset circle is drawn with center O and radius equal to the amount of offset. As the cam turns, the extended center line of the follower will always be tangent to this offset circle. The equiangular spaces are stepped off on the offset circle, and tangent lines are then drawn from each point on the offset circle, as shown.

731

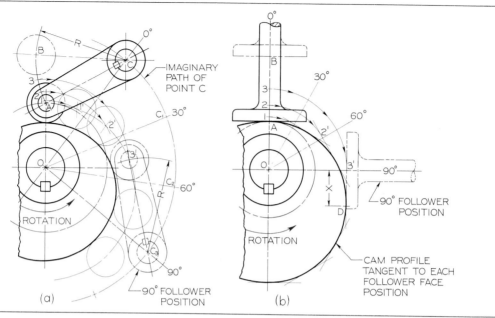

Fig. 22.21   Pivoted and Flat-Faced Followers.

If the roller is on a pivoted arm, as shown in Fig. 22.21 (a), then the displacement of the roller center is along the circular arc AB. The height of the displacement diagram (not shown) should be made equal to the rectified length of arc AB. Ordinates from the diagram are then transferred to arc AB to locate the roller positions 1, 2, 3, ... As the follower is revolved about the cam, pivot point C moves in a circular path of radius OC to the consecutive positions C_1, C_2, ... Length AC is constant for all follower positions; hence, from each new position of point C, the follower arc of radius R is drawn as shown at the 90° position. The roller centers 1, 2, 3, . . . , are now revolved about the cam center O to intersect the follower arcs at 1′, 2′, 3′, ... After the pitch curve is completed, the cam profile is drawn tangent to the roller circles.

The construction for a flat-faced follower is shown at (b). The initial point of contact is at A, and points 1, 2, 3, . . . , represent consecutive positions of the follower face. Then for the 90° position, point 3 must be revolved 90°, as shown, to position 3′, and the flat face of the follower is drawn through point 3′ at right angles to the cam radius. When this procedure has been repeated for each position, the cam profile will be enveloped by a series of straight lines, and the cam profile is drawn inside and tangent to these lines. Note that the point of contact, initially at A, changes as the follower rises: at 90°, for example, contact is at D, a distance X to the right of the follower axis.

22.15   Cylindrical Cams

When the follower movement is in a plane parallel to the cam shaft, some form of cylindrical cam must be employed. In Fig. 22.22, for ex-

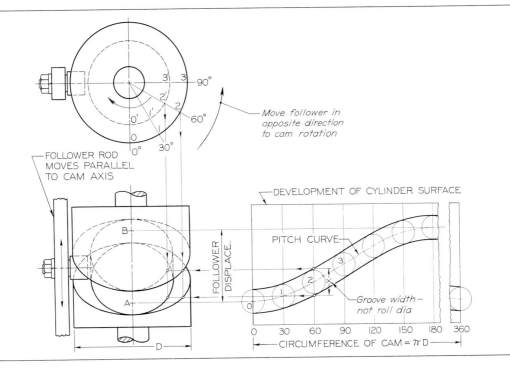

Fig. 22.22 Cylindrical Cam.

ample, the follower rod moves vertically parallel to the cam axis as the attached roller follows the groove in the rotating cam cylinder.

A cylindrical cam of diameter D is required to lift the follower rod a distance AB with harmonic motion in 180° of cam motion and to return the rod in the remaining 180° with the same motion. The displacement diagram is drawn first and conveniently placed directly opposite the front view. The 360° length of the diagram must be made equal to πD so that the resulting curves will be a true development of the outer surface of the cam cylinder. The pitch curve is drawn to represent the required motion, and a series of roller circles is then drawn to establish the sides of the groove tangent to these circles. This completes the development of the outer cylinder and actually provides all information needed for

making the cam; hence, it is not uncommon to omit the curves in the front view.

To complete the front view, points on the curves are projected horizontally from the development. For example, at 60° in the development, the width of the groove measured parallel to the cam axis is X, a distance slightly greater than the actual roller diameter. This width X is projected to the front view directly below point 2, the corresponding 60° position in the top view, to establish two points on the outer curves. The inner curves for the bottom of the groove can be established in the same manner, except that the groove width X is located below point 2′, which lies on the inner diameter. The inner curves are only approximate because the width at the bottom of the groove is actually slightly greater than X, but the exact bottom width can only be determined by drawing a second development for the inner cylinder.

22.16   Other Drive Devices

When the distance between drive and driven shafts makes the use of gears impractical, devices such as belt and pulley or chain and sprocket drives are often employed. Layout and detailing standards vary somewhat from one company to another, but design procedures and specification methods may be obtained from almost any good textbook or handbook on mechanical design.

22.17   Computer Graphics

Once the designer has determined the particular requirements of a gearset, CAD software is available that will permit the user to quickly generate drawings of single gears or two gears in mesh, Fig. 22.23, varying the pressure angle and number of teeth as desired. Features such as pitch circle center line and root and outer phantom circles can be shown or omitted from the drawing as desired.

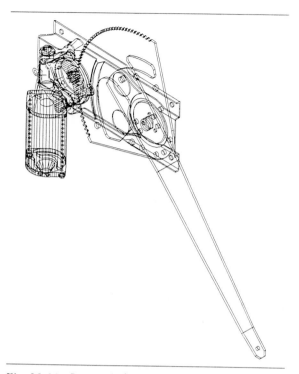

Fig. 22.23   Pictorial of Geared Window Crank Assembly. *Courtesy of CADKEY, Inc.*

PROBLEMS ON GEARING AND CAMS

The following problems will provide practice in laying out and making working drawings of the common types of gears and cams. Where paper sizes are not given, the student is to select both scale and sheet layout. If assigned, convert the design layouts to metric dimensions.

GEARING

Prob. 22.1 A 12-tooth 1DP pinion engages a 15-tooth gear. Make a full-size drawing of a segment of each gear showing how the teeth mesh. Construct the $14\frac{1}{2}°$ involute teeth exactly, noting any points where the teeth appear to interfere. Label gear ratio m_G on drawing. Include dimensions, notes, and cutting data as assigned by your instructor.

Alternate Assignment: Use 13 or 14 teeth, or any number of teeth as assigned by your instructor.

Prob. 22.2 The same as Prob. 22.1, but use 20° stub teeth.

Prob. 22.3 The same as Prob. 22.1, but use a rack in place of the 12-tooth pinion.

Prob. 22.4 Make a display drawing of the pinion in Fig. 22.24. Show two views, drawing the teeth by the odontograph method shown in Fig. 22.6. Draw double size. Omit dimensions.

Prob. 22.5 Make a pictorial display of the intermediate pinion shown in Fig. 16.96 (part 6). Show the gear as an oblique half section, §19.10, similar to Fig. 22.24. Draw four-times size, reducing the 30° receding lines by one-half as in cabinet projection, §19.4. Draw the teeth by the odontograph method.

Prob. 22.6 A spur gear has 60 teeth of 5 diametral pitch. The face width is 1.50". The shaft is 1.19" in diameter. Make the hub 2.00" long

and 2.25" in diameter. Calculate accurately all dimensions, and make a working drawing of the gear. Show six spokes, each 1.12" wide at the hub, tapering to .75" wide at the rim and .50" thick. Use your own judgment for any dimensions not given. Draw half size on Layout A–3 or A4–3 (adjusted) or full size on Layout B–3 or A3–3.

Prob. 22.7 Make a working drawing of the intermediate pinion shown in Fig. 16.96 (part 6). Check the gear dimensions by calculation.

Prob. 22.8 The same as Prob. 22.7, but use the intermediate gear shown in Fig. 16.96 (part 5).

Prob. 22.9 The same as Prob. 22.7, but use the pinion shown in Fig. 22.24.

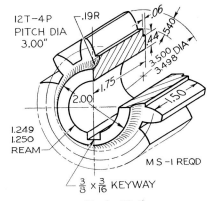

Fig. 22.24 Pinion (Prob. 22.4).

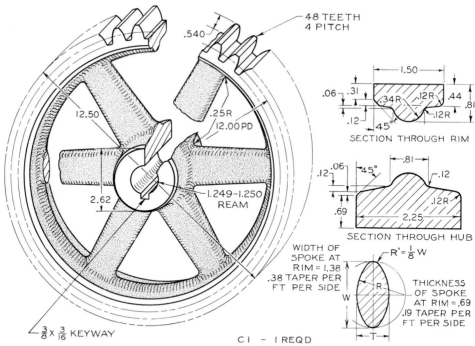

Fig. 22.25 Spur Gear (Prob. 22.10).

Prob. 22.10 The same as Prob. 22.7, but use the spur gear shown in Fig. 22.25.

Prob. 22.11 A pair of bevel gears have teeth of 4 diametral pitch. The pinion has 13 teeth, the gear 25 teeth. The face width is 1.12″. The pinion shaft is .94″ in diameter, and the gear shaft is 1.19″ in diameter. Calculate accurately all dimensions, and make a working drawing showing the gears engaged as in Fig. 22.16. Make the hub diameters approximately twice the shaft diameters. Select key sizes from Appendix 21. The backing for the pinion must be .62″, for the gear 1.25″. Use your own judgment for any dimensions not given.

Prob. 22.12 Make a working drawing of the pinion in Prob. 22.11.

Prob. 22.13 Make a working drawing of the gear in Prob. 22.11.

Prob. 22.14 The same as Prob. 22.12, but show two views of the pinion.

Prob. 22.15 The same as Prob. 22.11, but use 5 diametral pitch and 30 and 15 teeth. The face is 1.00″. Shafts: pinion, 1.00″ diameter; gear, 1.50″ diameter. Backing: pinion, .50″; gear, 1.00″.

Prob. 22.16 The same as Prob. 22.11, but use 4 diametral pitch, both gears 20 teeth. Select the correct face width. Shafts: 1.38″ diameter. Backing: 0.75″.

Prob. 22.17 Fig. 22.26 (a) shows a countershaft end used on a trough conveyor and (b) shows the layout for an assembly drawing of a similar unit. Using a scale of half size, make an assembly drawing of the complete unit (given layout dimensions are full size). The following full-size dimensions are sufficient to establish the position and general outline of the parts; the minor detail dimensions, web and rib shapes,

736

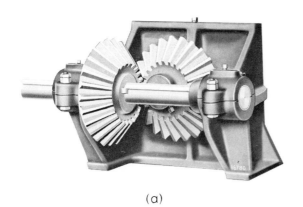

(a)

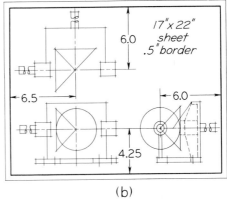

17"x22"
sheet
.5"border

6.0

6.5

6.0

4.25

(b)

Fig. 22.26 Countershaft End (Prob. 22.17). *Courtesy of Link-Belt Co.*

and fillets and rounds should be designed by the student with the help of the photograph.

The gears are identical in size and are similar in shape to the one in Fig. 22.15. There are 24 teeth of 3 diametral pitch in each gear. Face width, 1.50". Shafts, 2.00" diameter, extending 7.00" beyond left bearing, 4.00" beyond rear (break shafts as shown). Hub diameters, 2.75". Backing, 1.00". Hub lengths, 3.25". Front gear is held by a square gib key and a .38" set screw. Collar on front shaft next to right bearing is .75" thick, 2.25" outside diameter, with a .38" set screw.

On the main casting, the split bearings are 2.75" diameter, 3.00" long, and 10.00" apart. Each bearing cap is held by two $\frac{1}{2}$" bolts, 3.50" apart, center to center. Oil holes have a $\frac{1}{4}$" pipe tap for plug or grease cup. Shaft center lines are 6.50" above bottom surface, 8.00" from rear surface of casting. Main casting is 11.50" high, and its base is 16.00" long, 8.75" wide, and .75" thick. All webs, ribs, and walls are uniformly .50" thick. In the base are eight holes (not shown in the photograph) for $\frac{1}{2}$" bolts, two outside each end, and two inside each end.

Proceed by first blocking in the gears in each view, and then block in the principal main casting dimensions. Fill in details only after principal dimensions are clearly established.

Prob. 22.18 The worm and worm gear shown in Fig. 22.11 have a circular pitch of .62", and the gear has 32 teeth of $14\frac{1}{2}°$ involute form. The worm is double threaded. Make an assembly drawing similar to Fig. 22.11. Draw the teeth on the gear by the odontograph method of Fig. 22.6. Calculate dimensions accurately, and use AGMA proportions. Shafts: worm, 1.12" diameter; gear, 1.62" diameter.

Prob. 22.19 Make a working drawing of the worm in Prob. 22.18.

Prob. 22.20 Make a working drawing of the gear in Prob. 22.18.

Prob. 22.21 The same as Prob. 22.19, but the worm is single thread.

Prob. 22.22 A single-thread worm has a lead of .75". The worm gear has 28 teeth of standard form. Make a working drawing of the worm. The shaft is 1.25" in diameter.

Prob. 22.23 Make a working drawing of the gear in Prob. 22.22. The shaft is 1.50" in diameter.

737

CAMS

Prob. 22.24   Fig. 22.27 (a). Draw the displacement diagram, and determine the cam profile that will give the radial roller follower this motion: up 1.50″ in 120°, dwell 60°, down in 90°, dwell 90°. Motions are to be unmodified straight line and of uniform velocity. The roller is .75″ in diameter, and the base circle is 3.00″ in diameter. Note that the follower has zero offset. The cam rotates clockwise.

Prob. 22.25   The same as Prob. 22.24, except that the straight-line motions are to be modified by arcs whose radii are equal to one-half the rise of the follower.

Prob. 22.26   The same as Prob. 22.24, except that the upward motion is to be harmonic and the downward motion parabolic.

Prob. 22.27   The same as Prob. 22.26, except that the follower is offset 1.00″ to the left of the cam center line.

Prob. 22.28   Fig. 22.27 (b). Draw the displacement diagram, and determine the cam profile that will give the flat-faced follower this motion: dwell 30°, up 1.50″ on a parabolic curve in 180°,

dwell 30°, down with harmonic motion in 120°. The base circle is 3.00″ in diameter. After completing the cam profile, determine the necessary width of face of the follower by finding the position of the follower where the point of contact with the cam is farthest from the follower axis. The cam rotates counterclockwise.

Prob. 22.29   Fig. 22.27 (c). Draw the displacement diagram, and determine the cam profile that will swing the pivoted follower through an angle of 30° with the same motion as prescribed in Prob. 22.28. The radius of the follower arm is 2.75″, and in its lowest position the center of the roller is directly over the center of the cam. The base circle is 2.50″ in diameter, and the roller is .75″ in diameter. The cam rotates counterclockwise.

Prob. 22.30   The same as Prob. 22.29, except that the motion is to be that given in Probs. 22.24 and 22.25.

Prob. 22.31   The same as Prob. 22.29, except that the motion is to be that given in Probs. 22.24 and 22.26.

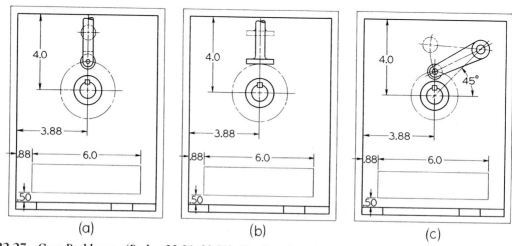

(a)   (b)   (c)

Fig. 22.27   Cam Problems   (Probs. 22.24–22.31). Use Size A or A4 sheet.

Prob. 22.32 Using an arrangement like Fig. 22.22, construct a *half* development, and complete the front view for the following cylindrical cam. Cam, 2.50″ diameter, 2.50″ high. Roller, .50″ diameter; cam groove, .38″ deep. Cam shaft, .62″ diameter; follower rod, .62″ wide, .31″ thick. Motion: up 1.50″ with harmonic motion in 180°, down with same motion in remaining 180°. Cam rotates counterclockwise. Assume lowest position of the follower at the front center of cam. Use Layout A–3 or A4–3 (adjusted).

Prob. 22.33 The same as Prob. 22.32, but construct a *full* development for the following motion: up $1\frac{1}{2}$″ with parabolic motion in 120°, dwell 90°, down with harmonic motion in 150°. Use Layout B–3.

Electronic Diagrams

Working drawings for the fabrication of electrical machinery, switching devices, chassis for electronic equipment, cabinets, housings, and other "mechanical" elements associated with electrical equipment are based on the same principles as given in Chapter 16, "Design and Working Drawings." A working drawing of a chassis for an electronic device is shown in Figure 23.1. Because of the complexity of modern electronics, however, a considerable number of graphic symbols have been developed for describing the connections and functions of such circuits and systems.

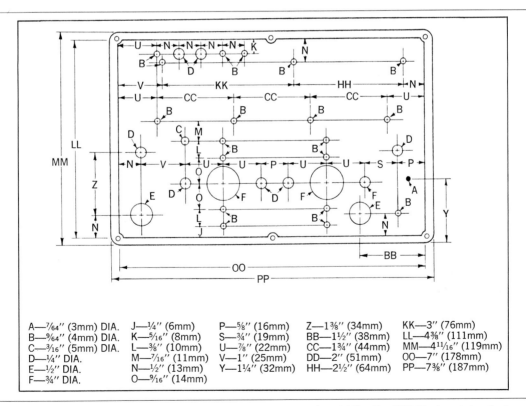

A—⁷⁄₆₄″ (3mm) DIA.	J—¼″ (6mm)	P—⅝″ (16mm)	Z—1⅜″ (34mm)	KK—3″ (76mm)
B—⁹⁄₆₄″ (4mm) DIA.	K—⁵⁄₁₆″ (8mm)	S—¾″ (19mm)	BB—1½″ (38mm)	LL—4⅜″ (111mm)
C—³⁄₁₆″ (5mm) DIA.	L—⅜″ (10mm)	U—⅞″ (22mm)	CC—1¾″ (44mm)	MM—4¹¹⁄₁₆″ (119mm)
D—¼″ DIA.	M—⁷⁄₁₆″ (11mm)	V—1″ (25mm)	DD—2″ (51mm)	OO—7″ (178mm)
E—½″ DIA.	N—½″ (13mm)	Y—1¼″ (32mm)	HH—2½″ (64mm)	PP—7⅜″ (187mm)
F—¾″ DIA.	O—⁹⁄₁₆″ (14mm)			

Fig. 23.1   Working Drawing for Chassis Fabrication (top plate of Bud Mfg. Co. CU247 chassis box). *Courtesy of American Radio Relay League.*

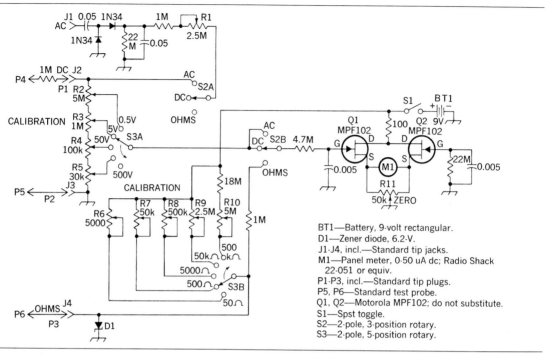

Fig. 23.2 Schematic Diagram of a FET (field-effect transistor) VOM (volt-ohm-milliammeter). *Courtesy of American Radio Relay League.*

23.1 Sources of Standardized Symbols

The symbols approved by the American National Standards Institute (ANSI) are published in "Graphic Symbols for Electrical and Electronic Diagrams," ANSI Y32.2–1975/IEEE 315–1975. Selected portions of this standard are shown in Appendix 36. In addition, modern printed circuitry techniques, which are used extensively in today's electronic equipment, require specially prepared diagrams or drawings (§23.15).

Recommendations for the preparation of electrical and electronic diagrams using the American National Standard symbols appear in "Electrical and Electronics diagrams," ANSI Y14.15. Figure 23.2 is a typical schematic diagram. Much of the material in this chapter is extracted or adapted from that standard.

In addition to the ANSI standards, other standards should be referred to if applicable or required. The U.S. government has developed a series of standards to be used by military contractors, including the following:

MIL–STD–275   Printed Wiring for Electronic Equipment

MIL–STD–429   Printed Wiring and Printed Circuits—Terms and Definitions

MIL–STD–454   Standard General Requirements for Electronic Equipment

MIL–STD–681   Identification Coding and Application of Hook Up and Lead Wire

MIL–STD–1495   Multilayer Printed Wiring Boards for Electronic Equipment

In other applications, adherence to standards developed by Underwriters' Laboratory (UL) may be required, particularly where safety considerations may dictate a special printed circuit board layout or component spacing.

23.2 Types of Electronic Diagrams

The following types of diagrams are commonly used in the electronics industry. Definitions are quoted from ANSI Y14.15.

Schematic diagrams. The type of drawing used by those concerned with the design, construction, and maintenance of electronic equipment, defined as

> A diagram which shows, by means of graphic symbols, the electrical connections and functions of a specific circuit arrangement. The schematic diagram facilitates tracing the circuit and its functions without regard to the actual physical size, shape, or location of the component devices or parts.

Single-line or one-line diagram. Frequently this type of diagram is called a Block Diagram; it is used to simplify the representation of complex equipment by using blocks or rectangles to depict stages, units, or groups of components in a system. ANSI defines it as

> A diagram which shows, by means of single lines and graphic symbols, the course of an electric circuit or system of circuits and the component devices or parts used therein.

Connection or wiring diagram. Such a diagram is used to represent the wiring between component devices in electrical or electronic equipment. It is defined as follows.

> A diagram which shows the connections of an installation or its component devices or parts. It may cover internal or external connections, or both, and contains such detail as is needed to make or trace connections that are involved. The Connection Diagram usu-

ally shows general physical arrangement of the component devices or parts.

Interconnection diagram. This diagram is similar to the Connection Diagram; both serve to supplement Schematic Diagrams. It is described as follows.

> A form of Connection or Wiring Diagram which shows only external connections between unit assemblies or equipment. The internal connections of the unit assemblies or equipment are usually omitted.

23.3 Drawing Size, Format, and Title

The sizes and formats for drawings of electrical and electronic diagrams should conform to American National Standard Y14.1. See §2.63.

The title of a drawing of one of the types defined in §23.2 should include the standardized wording in the definition and a description of the application, for example,

SINGLE–LINE DIAGRAM—AM–FM RECEIVER
SCHEMATIC DIAGRAM—
       250 WATT AMATEUR TRANSCEIVER

It is often useful to include some basic wiring information on schematic diagrams, such as connection details of coils and switches. The title of such a combined form is not altered, however, but describes the primary function of the diagram.

23.4 Line Conventions and Lettering

As in any drawing intended for reproductions, line thicknesses and letter sizes should be selected according to the amount of reduction or enlargement involved so that legibility will not be adversely affected.

ANSI recommends a line of medium thickness for general use on electrical diagrams. A thin line may be used for brackets, leader lines,

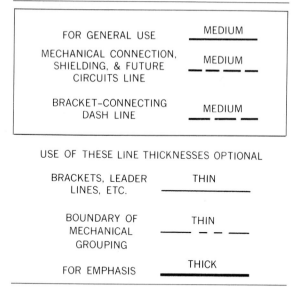

Fig. 23.3 Line Conventions for Electronic Diagrams [ANSI Y14.15].

and so on. To emphasize special features such as main signal paths, a thicker line may be used to provide the desired contrast. For recommended line thickness and lettering, refer to ANSI Y14.2M. See also §§2.11 and 4.6–4.26.

Line conventions for electrical diagrams are shown in Fig. 23.3.

23.5 Conventional Practices in Electronic Diagrams

SYMBOLS As stated earlier, symbols used should conform to ANSI Y32.2–1975/IEEE 315–1975 or to some other nationally approved standard. If no symbol is available, a special symbol may be devised (or a standard symbol augmented), provided an explanatory note is included.

SIZE OF SYMBOLS It is recommended in ANSI Y14.15 that symbols be drawn roughly 1.5 times the size of the samples shown in ANSI Y32.2–1975/IEEE 315–1975 (Appendix 36), pro-

vided anticipated reduction is no greater than 2.5 to 1. Following this recommendation, circular tube envelopes will be about 22 mm or .88″ to 25 mm or 1.00″ in diameter and circular envelopes for semiconductors will be about 16 mm or .62″ to 19 mm or .75″ in diameter, although the envelope for semiconductors may be omitted if no confusion results. Some textbooks on this subject contain much more detailed recommendations about symbol sizes, as well as spacing and arrangements.

SWITCHES AND RELAYS These in general should be shown in the "normal" position—with no operating force or applied energy. If exceptions are necessary, as for switches that may operate in several positions with no applied force, an explanatory note should describe the conditions shown.

ABBREVIATIONS As in other fields of technical drawing, abbreviations on electrical diagrams should conform to ANSI Y1.1. See Appendix 4.

GROUPING OF PARTS When certain parts or components are naturally grouped as in separately obtained subassemblies or assembled components, such as relays, tuned circuit transformers, hermetically sealed units, and printed circuit boards, this grouping may be indicated by means of an enclosing dashed-line "box" as in Fig. 23.6 or by extra space from adjacent circuitry.

23.6 Single-Line Diagrams

The single-line diagram is intended to describe the basic functions of a circuit or system. As such, it normally includes little of the detailed information and individual-parts identification of the schematic diagram. The individual lines connecting the symbols may represent single conductors or multiple conductors. The emphasis is on the function of each stage of a device rather than on the composition of the stage. A single-line diagram as used in electronics and communications is shown in Fig. 23.4.

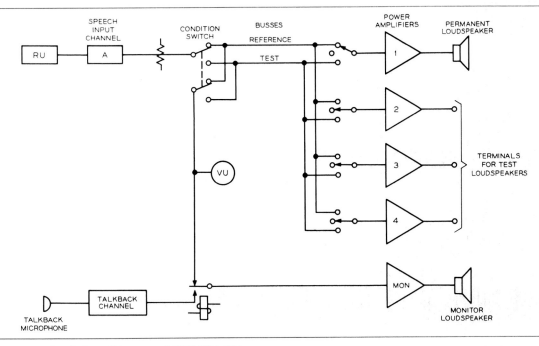

Fig. 23.4   Single-Line Diagram   [ANSI Y14.15].

Layout of a single-line diagram involves essentially the same principles and procedures (except for lesser detail) suggested for schematic diagrams, §23.7.

23.7   Schematic Diagrams

A convenient procedure for making a schematic (or single-line) diagram is to begin with a free-hand sketch on grid paper, Fig. 23.5. It is frequently desirable to make the final drawing, Fig. 23.6, on tracing paper or film placed over similar grid paper. Note that some details are rearranged in Fig. 23.6 to improve the final diagram.

The use of templates, such as those shown in Fig. 23.7, is a convenient time-saver in drawing symbols.

Computer-aided drafting techniques, as described in Chapter 8, may also be used to produce complicated schematic drawings effi-

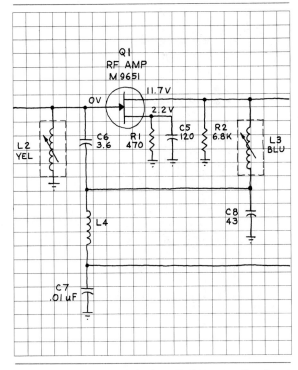

Fig. 23.5   A Portion of a Preliminary Freehand Sketch of a Schematic Diagram.

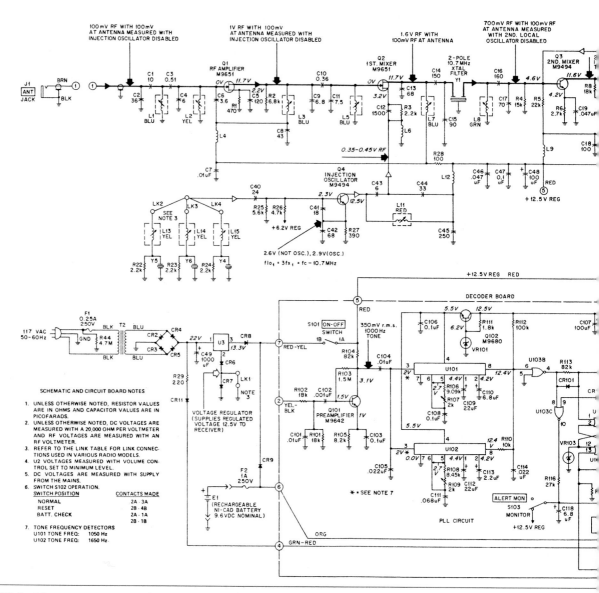

Fig. 23.6 Maintenance-Type Schematic Diagram of FM Weather Monitor Radio. *Copyright Motorola, Inc., 1980. All rights reserved. Reprinted with permission of Motorola, Inc.*

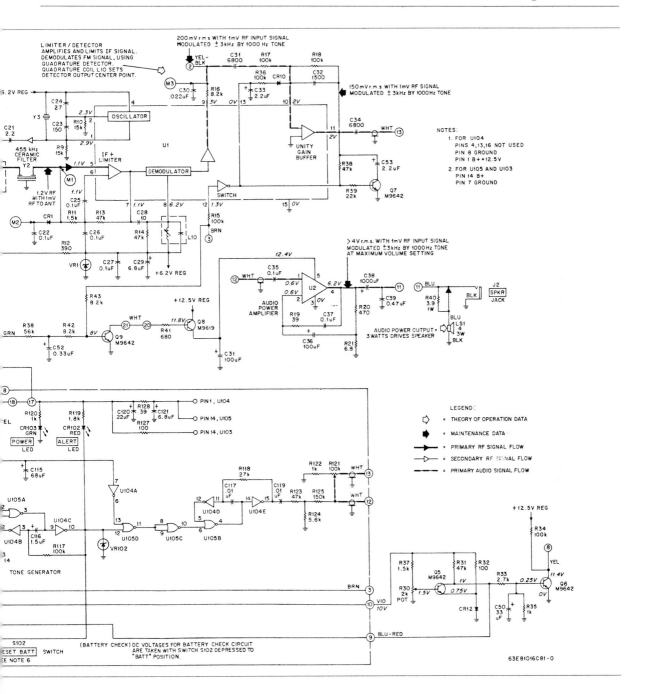

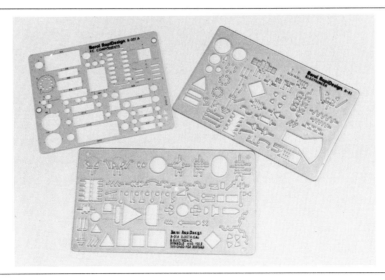

Fig. 23.7   Electronic Symbol Templates.

ciently. Using a graphics workstation, Fig. 23.8, and appropriate software, the drafter/designer can lay out and revise a schematic on the terminal's video display and then have a drawing, such as the example shown in Fig. 23.9, produced on a computer-driven plotter.

The various parts or symbols should be arranged so as to provide a pleasing balance between blank areas and lines. It is very important that sufficient blank spaces be provided adjacent to symbols for insertion of reference designations and notes. Exceptionally large spaces will, of course, give an unbalanced effect and should be avoided unless it is necessary to provide for possible later circuit additions.

SIGNAL PATH   It is standard practice to arrange schematic and single-line diagrams so that the signal or transmission path from input to output proceeds from left to right and from top to bottom if it is necessary to draw the diagram in successive "layers." Supplementary circuits, such as a power supply and an oscillator circuit, are usually drawn below the main circuit.

Stages of an electronic device are groups of components, usually associated with an electronic tube or transistor or other semiconductor, which together perform one function of the device.

For example, the sketch in Fig. 23.5 shows the RF (radio frequency) amplifier stage of the FM Weather Monitor Radio whose schematic is shown in Fig. 23.6. Some other stages that may be seen in Fig. 23.6 are the first and second mixer stages associated with transistors Q2 and Q3, the limiter/detector stage containing integrated circuit U1, and finally the audio output stage, U2, and speaker. The power supply, which rectifies alternating current through diodes CR2–CR5 to produce direct current voltage, regulated through integrated circuit U3, is shown at the lower left, along with a rechargeable battery backup power supply, E1.

Also shown in the lower portion of the diagram is the circuitry contained on a separate tone decoder printed circuit board (within the dashed lines).

The experienced electronic drafter is able to visualize fairly accurately the space required by the circuitry involved in each stage. The drafter generally keeps the semiconductor symbols or tube envelope ovals in horizontal lines and

Fig. 23.8 Prime Computer Graphics Workstation. *Courtesy of Prime Computer, Inc., Natick, MA. Copyright 1984.*

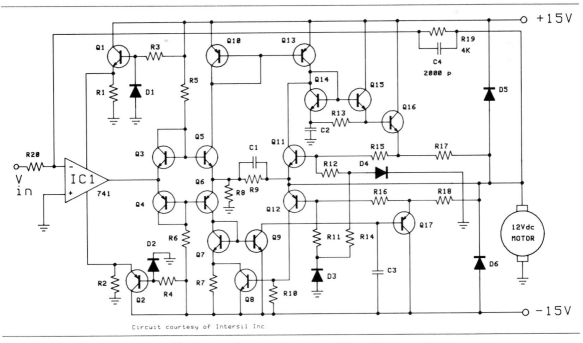

Fig. 23.9 Computer-Generated Schematic Diagram. *Courtesy of Chessel-Robocom Corp.*

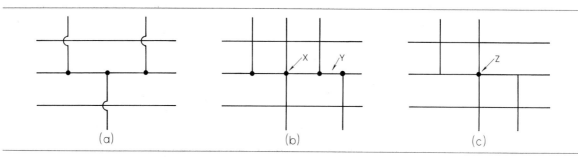

Fig. 23.10   Crossovers.

groups the associated circuitry in reasonably symmetrical arrangement about them. Thus, each stage is fairly well confined to an assigned area of the drawing. Note again that *the signal path from input to output follows the general path from left to right and top to bottom.* The input is at the upper left and the output at the lower right in Fig. 23.6. Note that the signal path is clearly designated by a heavier than normal line. This is typical of maintenance-type schematic diagrams, which are drawn so as to be easily read and interpreted by service technicians. Supplementary circuitry (the power supply) is in the lower portion of the diagram.

Connecting lines (for conductors) are drawn horizontally or vertically, for the most part, minimizing bends and crossovers. Long interconnecting lines should be avoided. *Interrupted paths* may be used in place of long, awkward interconnecting lines or where a diagram occupies more than one sheet. When parallel connecting lines must be drawn close together, the spacing between lines should not be less than .06″ after reduction. As a further visual aid, parallel lines should be grouped with consideration of function, and with at least double spacing between groups.

Crossovers are not usually completely avoidable. When they are necessary, they should be handled with care. The looped crossovers of Fig. 23.10 (a) are not American National Standard but are still employed when there is possibility of confusion. A simpler practice recognized by ANSI is shown at (b). Connection of more than three lines at one point, shown at X, is not rec-

ommended and can usually be avoided by moving or "staggering" one or more lines as at Y.

The ANSI recommended practice [ANSI Y14.15] is shown at (c). In this system it is understood that termination of a line signifies a connection. If more than three lines come together, as at Z, the dot symbol becomes necessary. This is avoidable, however, by staggering the connecting lines.

Interrupted paths, either for a single line or groups of lines, may be used where desirable for overall simplification of a diagram, Figs. 23.11 and 23.12. These must be carefully labeled with letters, numbers, abbreviations, or other identification so that their destinations are unmistakable. Grouped lines are also bracketed as well as labeled, Fig. 23.12. When convenient, interrupted grouped lines may be connected by dashed lines, as in Fig. 23.13.

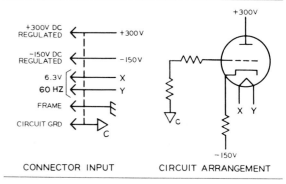

CONNECTOR INPUT   CIRCUIT ARRANGEMENT

Fig. 23.11   Identification of Interrupted Lines [ANSI Y14.15].

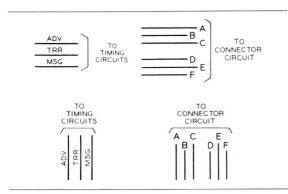

Fig. 23.12   Typical Arrangement of Line Identifications and Destinations   [ANSI Y14.15].

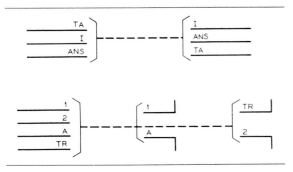

Fig. 23.13   Interrupted Lines Interconnected by Dash Lines   [ANSI Y14.15].

23.8 Terminals

Terminal circles need not be shown unless they are needed for clarity and identification. Switches S101 and S102 of Fig. 23.6 illustrate the use of terminal symbols and their identification.

When actual physical markings appear on or near terminals of a component, these may be shown on the electrical diagram. Otherwise, for convenient reference it is well to assign arbitrary numbers or letters to the terminals. These designations must then be explained by a simple diagram that associates the arbitrarily assigned numbers or letters with the actual physical arrangement of terminals on the component, Figs. 23.14, 23.15, and 23.16 (b). Figure 23.16 (a) is a pictorial explanation of (b).

Terminals or leads are frequently identified by colors or symbols, which should be indicated on the diagram. In Fig. 23.6, color coding is shown for transformer T2, and the colors of the insulated wire leads attached to the decoder board subchassis are noted near each terminal.

Terminal identification for rotary variable resistors follows practices similar to the foregoing, except that in addition it is frequently desirable to indicate the direction of rotation. This is normally described as clockwise or counterclockwise as viewed from the knob or actuator end of the control. Any contrary arrangement must be explained by a note or diagram. The letters CW placed near the terminal adjacent to the movable contact identify the extreme clockwise position

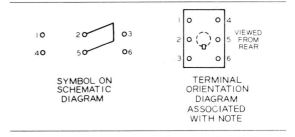

Fig. 23.14   Terminal Identification—Toggle Switch [ANSI Y14.15].

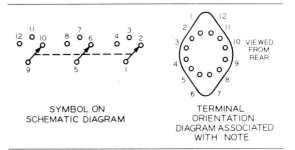

Fig. 23.15   Terminal Identification—Rotary Switch [ANSI Y14.15].

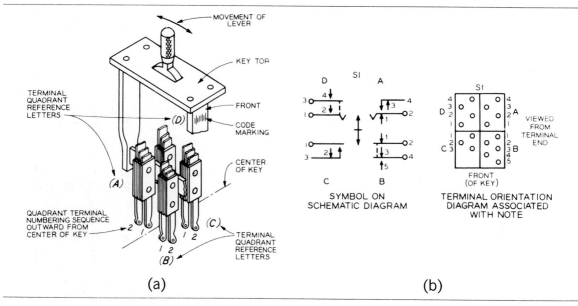

Fig. 23.16   Terminal Identification—Lever-Type Key   [ANSI Y14.15].

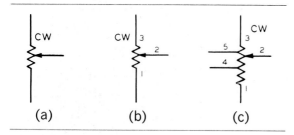

Fig. 23.17   Terminal Identification—Variable Resistor   [ANSI Y14.15].

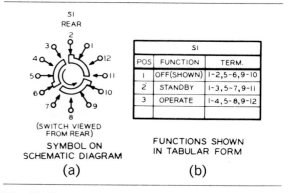

	S1	
POS	FUNCTION	TERM.
1	OFF(SHOWN)	1-2,5-6,9-10
2	STANDBY	1-3,5-7,9-11
3	OPERATE	1-4,5-8,9-12

FUNCTIONS SHOWN
IN TABULAR FORM

(a)   (b)

Fig. 23.19   Position-Function Relationships for Rotary Switches—Tabular Method Only   [ANSI Y14.15].

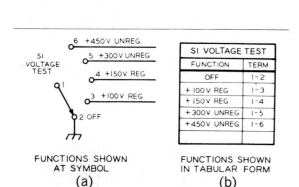

SI VOLTAGE TEST	
FUNCTION	TERM.
OFF	1-2
+100V REG	1-3
+150V REG	1-4
+300V UNREG	1-5
+450V UNREG	1-6

FUNCTIONS SHOWN
AT SYMBOL
(a)

FUNCTIONS SHOWN
IN TABULAR FORM
(b)

Fig. 23.18   Position-Function Relationships for Rotary Switches—Optional Methods   [ANSI Y14.15].

of the movable terminal, Fig. 23.17. When terminals are numbered, number 2 is assigned to the movable contact, (b). Additional fixed taps are assigned sequential numbers as at (c).

Switch position, as it affects function, should also be shown on a schematic diagram. Depending on complexity, the designation may range from the simple ON–OFF of switch S101 in Fig. 23.6 to the functional labeling of the terminals in Fig. 23.18 (a) or to the tabular form of func-

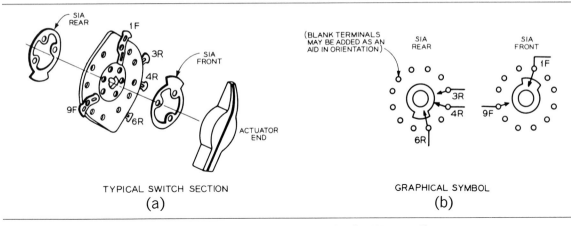

TYPICAL SWITCH SECTION

(a)

GRAPHICAL SYMBOL

(b)

Fig. 23.20   Development of a Graphical Symbol—Rotary Switch   [ANSI Y14.15].

tion listing, Fig. 23.19 (b). The tabular forms indicate by dashes which terminals are connected for each position. For example, in Fig. 23.19 (b), position 2 connects terminal 1 to terminal 3, terminal 5 to terminal 7, and terminal 9 to terminal 11.

23.9   Division of Parts

For clarity, sections of multielement parts may be drawn separately in a schematic diagram, with the subdivisions indicated by suffix letters. Thus, for the rotary switch of Fig. 23.20, the first portion S1A of switch S1 is identified as S1A FRONT and S1A REAR, viewed as usual from the actuator end. Note in Fig. 23.6 that sections of integrated circuit U104 are designated as U104A, U104B, U104C, U104D, and U104E, and are separated from each other by other components on the drawing.

Another example of suffixing is shown in Fig. 23.21. Here the two piezoelectric crystals A and B are enclosed in one unit, Y1. They are then referred to as Y1A and Y1B.

Another method of identifying parts of components that are functionally separated on the diagram is with the PART OF prefix, Fig. 23.22 (a). If the portion is indicated incomplete by a short-break line, as at (b), the words PART OF may be omitted.

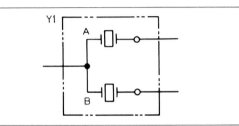

Fig. 23.21   Identification of Parts by Suffix Letters [ANSI Y14.15].

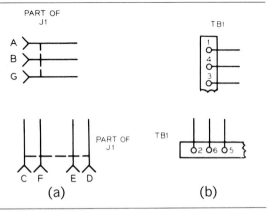

(a)   (b)

Fig. 23.22   Identification of Portions of Items   [ANSI Y14.15]

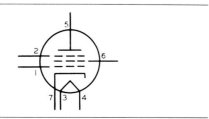

Fig. 23.23   Terminal Identification for Electron Tube Pins   [ANSI Y14.15].

23.10   Electron Tube Pin Identification

Electron tube pins are conventionally numbered clockwise from the tube base key or other point of reference, with the tube viewed from the bottom. In the recommended method of pin identification on a diagram, the corresponding numbers are shown immediately outside the tube envelope and adjacent to the connecting line, Fig. 23.23.

23.11   Reference Designations

An essential part of any electrical diagram is the identification of each separately replaceable part, shown with appropriate combinations of letters and numbers. In addition, portions not separately replaceable are frequently identified, as in previous examples. It should be noted that in electronics and communications, the recognized form is the appropriate letter followed by a number of the same size and on the same line, with no separating hyphen or space. This in turn is followed by any suffix needed, again with no hyphen.

The assignment of numbers for each category (resistances, capacitances, inductances, etc.) should start in the upper left corner of the diagram and proceed from left to right, then top to bottom, ending at the lower right corner. If in revision some parts are deleted, the remaining components should not be renumbered. The numbers not used should be listed in a table, such as Fig. 23.24. If the circuit contains many parts, it may also be helpful to include, as shown, the highest numbers used in each category.

Electron tubes and semiconductors such as transistors are identified by the reference designation and by the type number. It is frequently desirable also to include the tube function below the type number. All this information should be placed near the tube symbol—above it, if convenient, Fig. 23.25. Additional examples of transistor reference designations are shown in Figs. 23.2 and 23.6, although sometimes the transistor types are identified on a separate sheet.

23.12   Numerical Values

Along with reference designations, numerical values for resistance, capacitance, and inductance should be shown, preferably in the form

HIGHEST REFERENCE DESIGNATIONS	
R65	C35
REFERENCE DESIGNATIONS NOT USED	
R7, R9 R60, R62	C11, C14 C19, C23

Fig. 23.24   Highest and Omitted Reference Designations   [ANSI Y14.15].

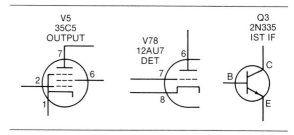

Fig. 23.25   Reference Designation, Type Number, and Function for Electron Tubes and Semiconductors [ANSI Y14.15].

Multiplier	Prefix	Symbol	
		Method 1	Method 2
10^{12}	tera	T	T
10^9	giga	G	G
10^6 (1,000,000)	mega	M	M
10^3 (1000)	kilo	k	K
10^{-3} (.001)	milli	m	MILLI
10^{-6} (.000001)	micro	μ	U
10^{-9}	nano	n	N
10^{-12}	pico	p	P
10^{-15}	femto	f	F
10^{-18}	atto	a	A

(a)   MULTIPLIERS

Range in Ohms	Express as	Example
Less than 1000	ohms	.031 470
1000 to 99,999	ohms or kilohms	1800 15,853 10 k 22 K
100,000 to 999,999	kilohms or megohms	380 k .38 M
1,000,000 or more	megohms	3.3 M

(b)   RESISTANCE

Range in Picofarads	Express as	Example
Less than 10,000	picofarads	152.4 pF 4700 PF
10,000 or more	microfarads	.015 μF 30 UF

(c)   CAPACITANCE

Fig. 23.26   Numerical Values for Components.

employing the fewest ciphers. This may be done by a judicious combination of the multipliers shown in Fig. 23.26 (a) and the symbols in (b) and (c). Note that commas are not used in four-digit numbers: 4700 is recommended, not 4,700.

INDUCTANCE   According to magnitude, inductance may be expressed in henries (H), milli-henries (mH or MILLI H), or microhenries (μH or UH). Following the principle of minimizing the number of ciphers, 2 μH is preferable to .002 mH, and 5 mH is better than either .005H or 5000 UH.

NOTE: Much repetition of units of measurement may be avoided by notes such as the following: UNLESS OTHERWISE SPECIFIED, RESIST-ANCE VALUES ARE IN OHMS AND CAPACI-TANCE VALUES ARE IN MICROFARADS.

PART VALUE PLACEMENT   For clarity, numerical values should be immediately adjacent to the symbol. Suggested forms are shown in Fig. 23.27.

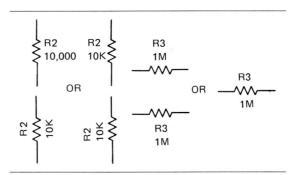

Fig. 23.27   Methods of Reference Designation and Part Value Placement   [ANSI Y14.15].

23.13   Functional Identification and Other Information

It sometimes contributes to readability of a diagram to include special functional identification of certain parts or stages. In particular, where functional designations (TUNER, OUTPUT, etc.) are to be shown on a panel or chassis surface, they should also be shown on the electrical diagram in an appropriate place.

755

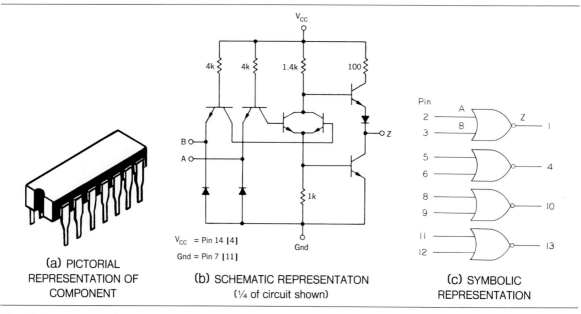

Fig. 23.28   A Typical Dual In-Line Package Integrated Circuit Representation. *(b) Courtesy of Motorola, Inc.*

Test points may be identified by the words TEST POINT on the drawing. If several are to be shown, the abbreviated form TP1, TP2, or M1, M2, and so on may be used, Fig. 23.6.

Still other information that may be shown on schematic diagrams, if desired, includes the following.

DC resistance of transformer windings and coils.

Critical input or output impedance values.

Voltage or current wave shapes at selected points.

Wiring requirements for critical ground points, shielding pairing, and so on.

Power or voltage ratings of parts.

Indication of operational controls or circuit functions.

23.14   Integrated Circuits

Modern technological advancements have resulted in the development of the integrated circuit, a semiconductor wafer or "chip" that has been processed to produce a microminiature replacement for discrete components such as transistors, diodes, resistors, capacitors, and connecting wiring.

Integrated circuits are widely used in consumer products such as pocket calculators, personal computers, radios, and television sets as well as in mainframe computers, microprocessors, and many other industrial applications.

Fig. 23.28   A typical integrated circuit, housed in a "DIP" (dual in-line package) less than an inch in length, (a), contains circuitry replacing all the discrete components shown in (b). To simplify the representation of such circuitry, a schematic symbol is used, (c).

23.15   Printed Circuits

In the interests of miniaturization and mass production, printed circuit boards have come into wide usage in the electronics industry, replacing expensive and tedious hand-wiring manufacturing methods.

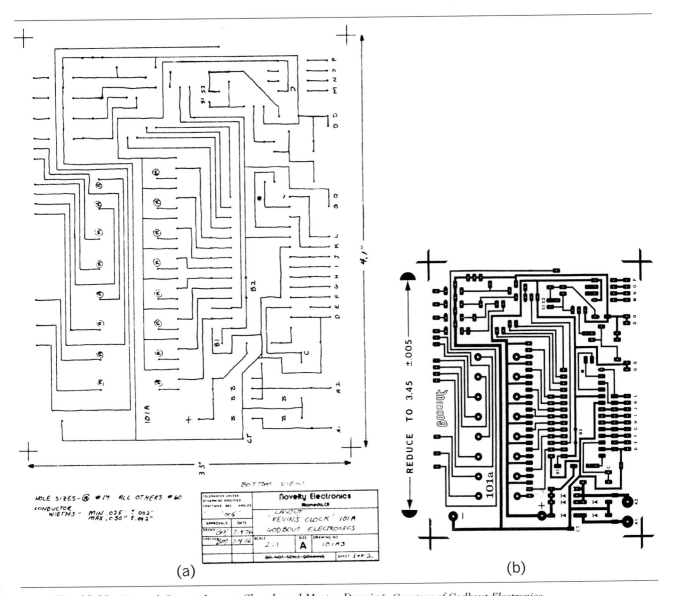

(a)   (b)

Fig. 23.29   Printed Circuit Layout Sketch and Master Drawing. *Courtesy of Godbout Electronics.*

A variety of procedures have been developed for fabricating these boards. Their name is derived from an early method of production, depositing (printing) a pattern of conductive ink on a base of insulating material. The method now most commonly used consists of etching a wiring pattern on copper-coated phenolic, glass-polyester, or glass-epoxy material.

Once the electronic designer has decided on the physical arrangement of components to be mounted on the printed circuit board, a freehand layout sketch is made, as seen in Fig. 23.29 (a).

The layout shown was made double size on tracing paper over a .1″-square-grid graph-paper guide.

Several methods may be used to make the finished artwork from which the production master is derived, Fig. 23.29 (b). These include pen-and-ink drafting, the application of dry-transfer material, or preprinted self-adhesive material. The pen-and-ink method is tedious and may be less accurate than the other two methods since it requires drawing all conductor paths by hand on vellum or film.

Use of preprinted self-adhesive symbols and tapes is the preferred method, providing an accurate, fast, and easily made master, such as the example in Fig. 23.30. It is important to maintain close tolerances in the production of printed circuit artwork, so that the various electronic components' leads will fit properly in the spaces provided for them on the finished board.

Fig. 23.30   Printed Circuit Master Made with Self-Adhesive Material. *Courtesy of W. H. Brady Co.*

23.16   Computer Graphics

The production of printed circuit artwork has been greatly simplified through the use of computer-aided drafting (CAD). Sophisticated programs such as P-CAD's Master Designer have been developed for personal computers that aid the drafter or designer in reducing the physical size of the board and minimizing the need for crossover jumper wires.

Double-sided printed circuit boards are common, and many compact designs require multilayer boards, in which the conductive paths are sandwiched between insulating layers. These board layouts, Fig. 23.31, require precise registration of the artwork for each layer to insure proper alignment of corresponding features, such as holes that are to be "plated through" from side to side.

A CAD component layout program such as Master Placer allows the designer to create the best possible parts placement in the shortest possible time, Fig. 23.32.

Increased use is being made of microminiature surface-mount components, which have

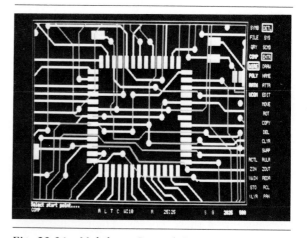

Fig. 23.31   Multilayer Printed Circuit Board Layout Produced with P-CAD Software. *Courtesy of CADAM, an IBM company.*

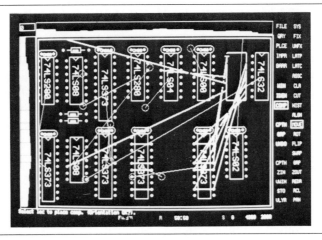

Fig. 23.32 Computer-Assisted Component Layout. *Courtesy of CADAM, an IBM Company.*

short solder tab connections instead of wire leads and so do not require drilled mounting holes. The small size of these parts requires the high degree of accuracy in board layout and design that CAD systems can easily provide.

Interactive CAD systems, Chapter 8, are capable of providing full-size or scaled artwork along with other data required to set up automated soldering and drilling equipment required to produce a finished printed circuit board.

ELECTRONIC DIAGRAM PROBLEMS

The following exercises in the construction of schematic diagrams are designed for Layout A–3 or A4–3 (adjusted). If freehand sketches are assigned, 8.5″ × 11.0″ grid paper with .25″ grid squares is suggested. For a mechanical drawing, tracing paper or film is convenient, the drawing being preceded by a sketch on grid paper to work out the details of arrangement and spacing. Refer to Appendix 36 for standard graphical symbols.

Prob. 23.1   The circuit shown in Fig. 23.33 is suitable for classroom or laboratory demonstrations. After a local radio station is tuned with a regular 9-volt battery in place, other devices such as photoelectric cells and homemade batteries are substituted.

Make a freehand or mechanical schematic diagram of the circuit, approximately twice the size of Fig. 23.33. Use standard symbols and show component values and reference designations, following recommended practices (Layout A–3).

Components List

C1	10-365 PF variable capacitor
D1	Germanium diode
L1	Loopstick antenna coil with center tap
T1	Output transformer: 2500 ohm primary, 3–4 ohm secondary
Q1	NPN type RF transistor (*B = base, C = collector, E = emitter*)
Q2	PNP type audio transistor

Prob. 23.2   The unit shown in Fig. 23.34 is used to calibrate the tuning dials of short-wave communications receivers by providing a check signal every 100 kilohertz throughout the commonly used bands.

Components List

Capacitors	
C1	1000 PF
C2	7-45 PF trimmer
C3	10 PF
Crystal	
Y1	100 KHz
Resistors (ohms)	
R1	56K
R2	56K
Integrated Circuit	
U1	HEP 570 (quad 2-input gate)

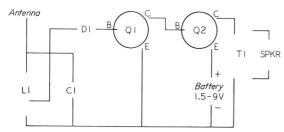

Fig. 23.33   Circuit Demonstrating the Generation of Electrical Energy   (Prob. 21.1).

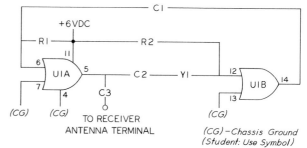

Fig. 23.34   Crystal Calibrator   (Prob. 23.2).

Reproduce the schematic diagram approximately double size, freehand or mechanically, as assigned. Add standard symbols and include reference designations, component values, and other data following recommended practices (Layout A–3 or A4–3 adjusted).

Prob. 23.3 The circuit of Fig. 23.35 is that of a Variable Frequency Oscillator (VFO) that may be used with transmitters operating in the 80-meter amateur radio ("ham") band.

Make a freehand or mechanical schematic diagram, approximately twice the size of Fig. 23.35. Use standard symbols and include reference designations and component values, following recommended practices (Layout A–3 or A4–3 adjusted).

Components List

Capacitors
C1	100 PF trimmer
C2	50 PF
C3	150 PF variable
C4	470 PF
C5, C6	1000 PF
C7, C8	.1 UF

Resistors (ohms)
R1, R2, R3	22K
R4	270

Coils
L1	6 UH ceramic or air core
L2	1 MH RF choke

Semiconductors
D1	1N914 silicon diode
D2	6 V, 1 W zener diode
Q1	40673 dual-gate MOSFET transistor

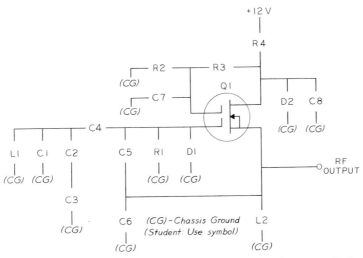

Fig. 23.35   Variable Frequency Oscillator   (Prob. 23.3). *Circuit courtesy of American Radio Relay League.*

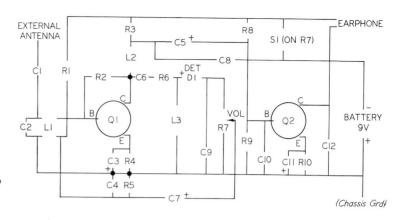

Fig. 23.36 Transistor Pocket Radio (Prob. 23.4).

(Chassis Grd)

Components List

Capacitors
C1	10 PF
C2	10-365 PF variable
C3	90 UF
C4	.005 UF
C5	5 UF
C6	.001 UF
C7	5 UF
C8	.01 UF
C9	.005 UF
C10	.005 UF
C11	90 UF
C12	.005 UF

Coils
L1	Loopstick antenna (same symbol as magnetic core transformer)
L2	Magnetic core, .5 MILLI H
L3	Magnetic core, 1.0 MILLI H

Resistors (ohms)
R1	47K
R2	82K
R3	2200
R4	1000
R5	10K
R6	82
R7	10K variable (potentiometer)
R8	82K
R9	5600
R10	3900

Transistors
(*E = emitter, B = base, C = collector*)
Q1	PNP type, RF-AF Amplifier
Q2	PNP type, audio amplifier

Miscellaneous
S1	On-off switch (mounted on *R7*)
D1	Diode detector

Prob. 23.4 Make a freehand or mechanical schematic diagram of the circuit shown in Fig. 23.36 approximately double size. Use standard symbols and show reference designations and component values, following recommended practices (Layout A–3 or A4–3 adjusted).

Note: By rearranging slightly a few connecting lines, you can eliminate the three dot symbols.

Prob. 23.5 Figures 23.37 (a) and (b) are pictorial wiring diagrams of a transistorized code practice oscillator. The unit provides for practicing International Morse Code either by listening

to an audible tone over headphones or speaker or by watching a flashing light.

Study the illustrations and the components list carefully, making preliminary sketches of connections. Then make a complete schematic diagram of the circuit following recommended practices and including component values and reference designations (Layout A–3 or A4–3 adjusted).

Note: Terminal strips TS–1 and TS–2 and the battery holder are for mechanical function and convenience in wiring. They should *not* be shown in a schematic diagram.

762

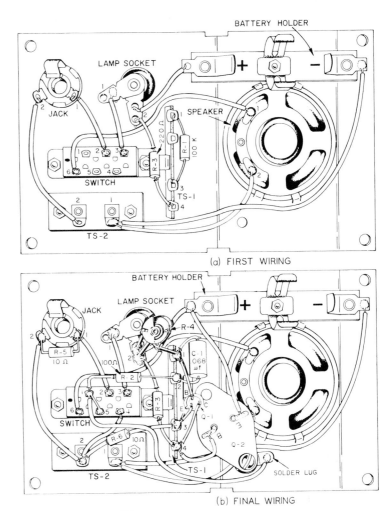

Fig. 23.37 Code Practice Oscillator (Prob. 23.5).

Components List

Capacitors
  C1 .068 UF

Resistors (ohms)
  R1 100K
  R2 100
  R3 220
  R4 50K variable (potentiometer)
  R5 10
  R6 10

Transistors
  (*E = emitter, B = base, C = collector*)
  Q1 NPN type, oscillator
  Q2 PNP type, amplifier

Miscellaneous
  Battery "C" Type, $1\frac{1}{2}$ volt
  Battery holder

Jack 2-conductor (for headphones)
Key Telegraph key (not shown—connected between terminals 1 and 2 of TS–2)
Lamp
Lamp socket
Speaker
Switch 3-position slide switch
Connections
  Position one (OFF-LIGHT)—terminals 1–5, 3–4
  Position two (SPEAKER)—terminals 1–6, 2–4
  Position three (PHONES)—terminals 2–5
Terminal strips TS–1 and TS–2

CHAPTER 24

Structural Drawing

BY E. I. FIESENHEISER*

Structural drawing consists of the preparation of design and working drawings for buildings, bridges, domes, tanks, towers, and other structures. It comprises a very large field for the drafter. Although the basic principles are the same as for making drawings of machines and machine parts, certain of the methods of representation are different in structural drafting.

Ordinarily the civil engineer determines the form and shape of a structure as well as the sizes of the main members to be used. The drafter then makes the detail drawings, frequently under the engineer's supervision. Thus, the drafter is the engineer's first assistant and has the opportunity to learn something about design. In many cases, the drafter's position is a stepping-stone to greater responsibility.

24.1   Structure and Materials

A spider web is one of nature's fine examples of a structure. The web is made of many parts connected to form a unit strong enough to support the spider. Most of our structures, Figs. 24.1 and 24.2, also have many interconnecting parts, such as beams, girders, and columns, that form a framework strong enough to support loads.

The materials most commonly used in construction are wood, glass, structural steel, aluminum, concrete (plain, reinforced, and prestressed), structural clay products, and stone masonry.

24.2   Wood Construction

Many different types of wood are used as structural timber, including ash, birch, cedar, cypress, Douglas fir, elm, oak, pine, poplar, redwood, and spruce. Information concerning the strength properties of the various types and grades can be obtained from *Design Values for Wood Construction,* published by the National Forest Products Association, Washington, DC, in 1982. For design of different members and connections, the student should refer to the NFPA's *National Design Specifications, Wood Construction,* 1982.

Wood is commonly used for sills, columns, studs, joists, rafters, purlins, trusses, and sheathing in homes and other buildings. Typical drawings of wood structures and construction details may be obtained from the various publications of the National Forest Products Association. Common methods of fastening timber members include nails, screws, lag screws, drift bolts, bolts, steel plates, and various special timber connectors. Ordinarily, a structural timber is cut so that the wood fibers, or grain, run parallel to

---

*Professor Emeritus of Civil Engineering, Illinois Institute of Technology.

Fig. 24.1   Model of Lamella Dome of Light-Weight Standard Steel Trusses. *Courtesy of Department of Architecture, Illinois Institute of Technology.*

Fig. 24.2   Model of Football Stadium of Post-Tensioned Concrete. *Courtesy of Department of Architecture, Illinois Institute of Technology.*

766

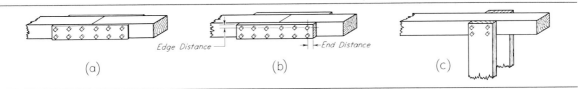

Fig. 24.3 Typical Bolted Joints.

the length. The strength resistance of wood is not the same in a direction perpendicular to the grain as it is parallel to the grain. Therefore, in designing connections, the direction of the force to be transmitted must be taken into account. Also, proper spacing, edge distance, and end distance must be maintained for screws, bolts, and other connectors.

Typical bolted joints are shown in Fig. 24.3. To transmit the forces, either steel or wood splice plates may be used, as shown at (a) and (b). A detail without splice plates is shown at (c). It should be realized that each type of connection requires a different design.

The use of *split-ring metal connectors*, Fig. 24.4, is now common. Figure 24.5 shows a drawing of the left half of a timber roof truss in which this type of connector is used. Only the left half is drawn, since the truss is symmetrical. By referring to the views of the top and bottom chords, the student may visualize the relative positions of the connecting members of the structure. In drawings of trusses, the view of the top chord is simply an auxiliary view. However, it is customary to show the lower chord by means of a section taken just above the lower chord with the *line of sight downward*. The lower chord, therefore, is shown in first-angle projection, §7.38.

Ordinarily, wood is surfaced or dressed. Because of the wood loss in surfacing, a nominal 1″ thickness is reduced to a minimum of $\frac{3}{4}$″ for dressed thickness if dry, and $\frac{25}{32}$″ if the wood is green. For nominal dimensions from 2″ to 4″, the dressed sizes, dry, are less in each case by $\frac{1}{2}$″, or by $\frac{7}{16}$″ if green. For 5″, 6″, and 7″ dimensions, the dressed-size reduction is $\frac{1}{2}$″ for dry wood and $\frac{3}{8}$″ for green wood. For 8″ and greater sizes, the reduction is $\frac{3}{4}$″ for dry and $\frac{1}{2}$″ for green wood. Lumber is considered "green" if its moisture content exceeds 19 percent. The symbols S2S and S2E indicate, respectively, surfacing of two sides and two edges. If all four faces are to be surfaced, the symbol S4S is used. The working drawings must show whether standard dressed, standard rough, or special sizes are to be used.

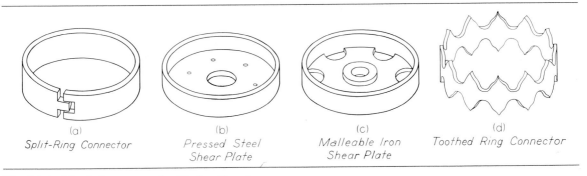

(a)	(b)	(c)	(d)
Split-Ring Connector	*Pressed Steel Shear Plate*	*Malleable Iron Shear Plate*	*Toothed Ring Connector*

Fig. 24.4 Metal Connectors for Timber Structures.

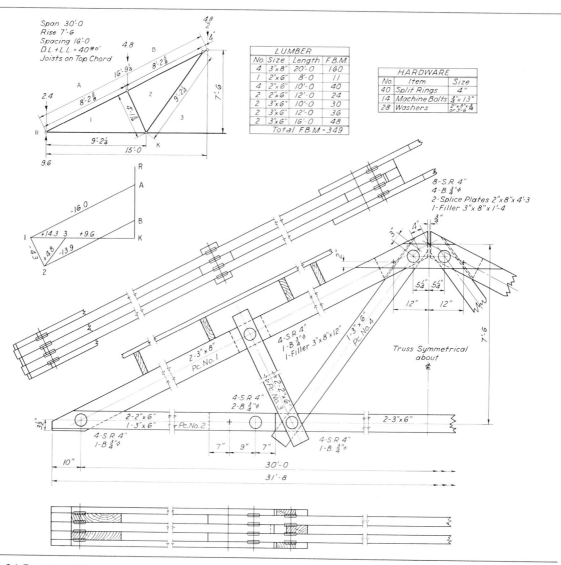

Fig. 24.5 A Roof Truss. *Based on a design by Timber Engineering Company.*

24.3 Metal Ring Connectors

If properly installed by skilled workers in wood of proper grade and moisture content, the metal ring connector is a very satisfactory and useful device. The method consists of using either a toothed ring, called an *alligator*, or a *split ring*, Fig. 24.4. If the toothed ring, (d), is used, it is placed between the two members to be connected, and these are drawn together by tightening the bolt so that the teeth of the ring are forced into the two members, and the ring thus assists in transmitting stress from one member to the other. If the split ring is used, a groove is cut into each of the two members to be connected, the ring is placed in the grooves, and the two members are held together by means of a bolt, as shown in Fig. 24.6. The open joint of the ring should be in a direction at right angles to that of the stress, so that as the stress is applied the ring is deformed slightly and transmits the

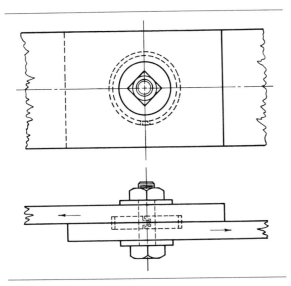

Fig. 24.6   Method of Installing Split-Ring Connectors.

pressure to the wood within the ring as well as to that without. With this connection, the tensile and shearing strengths of wood are developed to a higher degree than by other methods of connection, and it is possible to use timber in tension much more economically.

It is standard practice to show ring connectors by solid lines in order to save time, as shown in Fig. 24.5.

24.4   Structural Steel Drafting

Structural steel drawings are ordinarily of two types: *engineering design drawings* made in the design engineer's office and *manufacturing drawings* usually made in the office of the steel fabricator. For a more complete treatment of this subject, consult *Detailing for Steel Construction,* American Institute of Steel Construction, Chicago, IL.

Design drawings are concerned primarily with showing clearly the overall dimensions of the structure, such as the locations of columns, beams, angles, and other structural shapes, and the listing of the sizes of these members. It is necessary to show a certain amount of detail, usually in the form of typical cross sections, special connections required, various notes, and so on. In the case of a building floor, for example, a *floor plan* is drawn, showing the steel columns in cross section and the beam or girder framing by the use of single heavy lines, as shown in Fig. 24.7. Members framing from column to column,

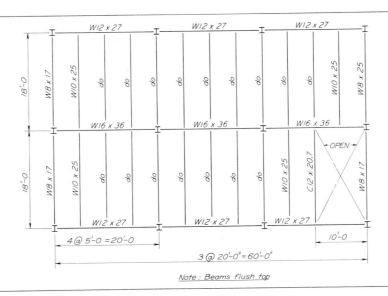

Fig. 24.7   Typical Steel Floor Design Plan.

providing end support for other beams, are called *girders,* while smaller beams framing between girders are called *filler beams.* The designer's plans are sent to the steel fabricator who is to furnish the steel for the job. From these plans the fabricator makes the necessary detailed shop drawings and erection plans. However, before shop work is begun, the fabricator's drawings are sent to the design engineer for final checking, as the engineer has the authority to make any changes necessary to conform to the required strength and safety of connections. As soon as the fabricator has received the shop drawings approved by the engineer, the shop work may be carried out.

Shop drawings consist of detail drawings of all parts of the entire structure, showing exactly how the parts are to be made. See Figs. 24.11–24.15 and Fig. 24.17. Essentially, such drawings show all dimensions necessary for fabrication calculated to the nearest $\frac{1}{16}''$, the location of all holes for connections, details of connection parts, and the required sizes of all material. In addition, fabrication or construction methods may be specified by appropriate notes on the detail drawings whenever such items are not covered by separate specifications. Obviously, fabrication and shop methods, as well as suitable field construction methods, must be fully understood by the detailer. The design of details and connections is an important part of the engineering of the structure and should not be neglected, since the connections of the various members must be adequate to transmit the forces in these members. Connection details should be drawn to a scale sufficiently large to show them clearly without crowding, although overall lengths of members need not be drawn to scale. All dimensions should be shown, since detail drawings should never be scaled in the shop or in the field by the workers making the piece. An adequate system of piece marking should be employed. Each piece that is separately handled should have its own piece mark, and this piece mark should be shown wherever the member appears on the drawings. This mark also is painted on the member in the shop and

later serves as a shipping mark and erection mark in the field for final assembly of the member in the structure.

Erection plans, ordinarily made by the steel fabricator, are essentially skeleton assembly drawings showing the relationship of the various members or parts to be fitted together in the final structure in the field. In all cases, the piece marks of the individual members are shown on the erection plans. Only sufficient detail to enable the complete assembly of the various members by skilled workers is required, because the detail drawings, already made, fully describe each member and its connections. In most cases, line diagrams in which members are represented simply by heavy straight lines are adequate, although when complex assemblies are required, these should be shown in greater detail. Assembly views should be drawn to scale but, like the detail shop drawings, are never scaled by workers to obtain dimensions. Appropriate notes on these drawings may be used to indicate how the structure is to be assembled. An erection plan of roof steel framing, which is an addition to an existing building, is shown in Fig. 24.8. This industrial structure houses a pulp mill for the manufacture of paper roofing products. New steel is shown by full lines, whereas existing roof members are shown by dashed lines. Connections to existing steel members are shown in sectional views, and timber framing for the support of large roof ventilators is also shown.

24.5   Structural Steel Shapes

Although aluminum and magnesium shapes are available for use in construction, only steel shapes are within the scope of this chapter. Structural steel is available in many standard shapes that are formed at the mill by rolling steel billets under high temperatures. The shapes available are square, flat, or round bars; plates; equal-leg and unequal-leg angles; American Standard and miscellaneous channels; S, W, M, and HP shapes for use as beams, columns, and bearing piles; structural tees, mill cut from W, S, or M shapes; pipe, in standard, strong, and extra-

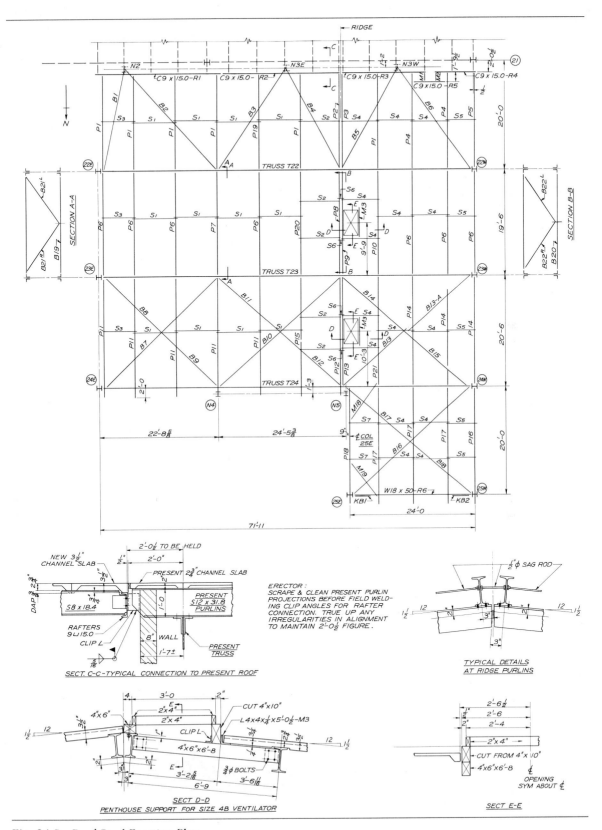

Fig. 24.8   Roof Steel Erection Plan.

771

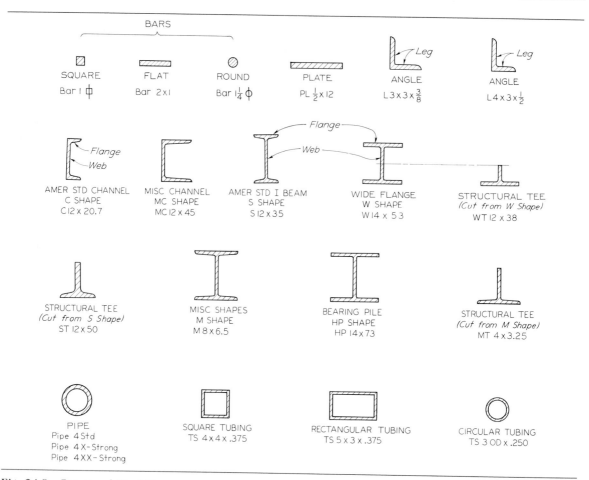

Fig. 24.9 Structural Steel Shapes.

strong weights; and tubing, in square, rectangular, or circular cross sections. Figure 24.9 shows some typical cross sections of these shapes with the conventional manner of designation or billing in each case. The correct designations must be used on both the design and the detailed drawings in every case where the structural member appears. As a typical example, a wide flange section of 14″ nominal depth, weighing 53 lb per ft and having an overall length of 26′–2⅛″, would be designated as W14 × 53 × 26′–2⅛.

The use of templates by the drafter, such as that shown in Fig. 24.10, is a convenient time-saver in drawing structural symbols. The sym-

bols may be drawn in pencil or inked with a technical fountain pen.

24.6 Scales for Detailing

Details should be drawn to a scale of ¾″ = 1′–0, or 1″ = 1′–0, using the architects' scale, §2.29. Overall lengths of members, however, need not be drawn to scale.

24.7 Specifications

Detailed information concerning dimensions, weights, and properties of rolled steel structural

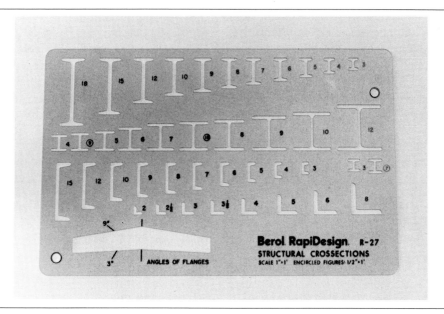

Fig. 24.10   Structural Cross-sections Template.

shapes, rolling mill practice, and miscellaneous data for designing and estimating may be found in the *Manual of Steel Construction* (AISC). The information contained in this manual is essential to structural steel drafting.

Currently, 15 different types or grades of structural steel are available. These differ in chemical composition and physical properties. Each is controlled in its manufacture by a separate ASTM (American Society for Testing Materials) specification, and there are considerable variations in the strengths and costs of the various grades. The design engineer must be aware of the various physical properties of steels, such as strengths, ductilities, corrosion resistance, and costs, in order to make an economical selection of the grade of steel. The types used, in any case, must be specified by ASTM designation on the drawings. The grade most commonly used now is ASTM A36, the number 36 specifying that the guaranteed minimum yield strength is 36 ksi (36,000 lb/in.²).

Main members of a steel structure must be joined together to form the complete structure in the field. In the modern steel fabricating shop the most common method of forming suitable connections is by welding of connection material to the main members, with provision of open holes for insertion of field bolts to join with other members of the assembly. The connection material, consisting, for example, of angles or connection plates, may also be attached to the main members by riveting, or by bolting with either ordinary or high-strength steel bolts in either bearing-type or friction-type connections. Formerly, shop riveting and field riveting were used exclusively in structural work. Although still used to some extent, the riveted connection tends to be replaced by welding or high-strength bolting. High-strength bolts must be field-tightened by a definite amount of torque applied to the wrench.

24.8   Riveting

Structural rivets are made of soft carbon steel and are available in diameters ranging from $\frac{1}{2}''$ to $1\frac{1}{4}''$. Rivets driven in the shop are called *shop rivets,* and those driven in the field (at the construction site) are called *field rivets.* Rivets are usually of the button-head type, Fig. 15.40 (a), and are driven hot, into holes $\frac{1}{16}''$ larger than the rivet diameter. The length of a rivet is the thickness (grip) of the parts being connected, plus the length of the shank necessary to form the driven head and to fill the hole. Excess shank length will produce capped heads, whereas lengths too short will not permit the formation of a full head. Shop rivets are ordinarily driven by large riveting machines that are part of the permanent shop equipment.

Field rivets are heated in a forge at the construction site. When properly heated, the rivet is held firmly in the hole with a dolly bar while the pneumatic hammer or rivet gun forms a round smooth head.

A shop drawing for a riveted roof truss is shown in Fig. 24.11. Only the left half is drawn, since the truss is symmetrical about the center line. The use of the gage lines of the members should be noted. Gage lines should be located as closely as possible to the centroidal axes of the members, and at the joints where members intersect, these gage lines should intersect at a single point to avoid unnecessary moment stresses due to eccentricities. It is noted that the locations of the open holes indicate where the field splices are to be, namely, at the center line of the truss ($\mathcal{C}$) for the top chord and at $5'-2\frac{5}{16}$ from the center line for the bottom chord.

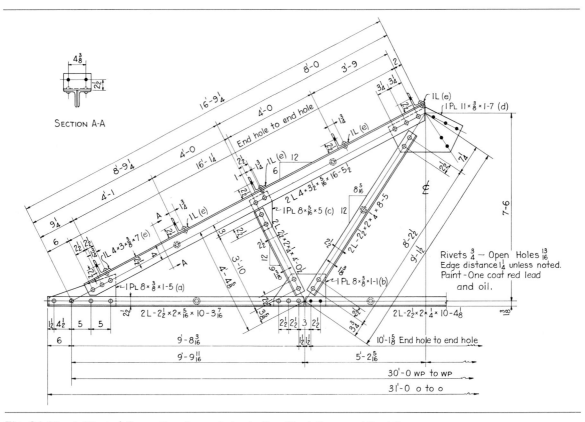

Fig. 24.11   A Riveted Truss. *Based on a design by Fort Worth Structural Steel Company.*

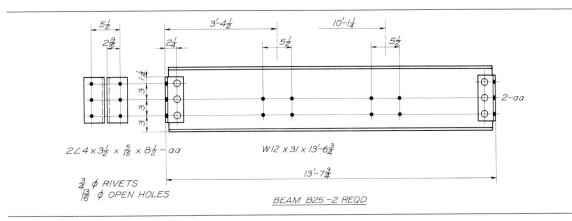

Fig. 24.12   Floor Beam Shop Drawing.

24.9   Frame Beam Connections

Because of their common usage, the American Institute of Steel Construction recommends certain standard connections for attaching beams to other members. These connections are ordinarily adequate to transmit the end forces that beams carry. However, the drafter should know the strength of these connections and use them only when they are sufficient. For complete information, see the AISC *Manual of Steel Construction*. This manual shows details of standard framed beam connections using $\frac{3}{4}''$, $\frac{7}{8}''$, and 1″ diameter bolted or riveted connections, and allowable loads for both friction-type and bearing-type connections. The fasteners may have a regular or a staggered arrangement. The holes may be either standard round or oversized slotted ones.

The use of angles for standard two-angle framed beam connections in a typical detail drawing of a floor beam is shown in Fig. 24.12. This drawing illustrates several important features: shop rivets are shown as open circles on shop drawings, whereas holes for field rivets or bolts are blacked in solid, Fig. 15.42, *gage lines* (lines passing through rivets or holes like center lines) are always shown, and it is desirable to line up holes or rivets on these lines where pos-

sible, rather than to "break the gage." It is necessary to locate the gage line of an angle for each leg in all cases, unless it has already been shown for the identical angle elsewhere on the drawing. The edge distance, from end rivet or hole to the end of the angle, must be given at one end, the billed length of the piece being worked out to provide the necessary edge distance at the other end. It is not necessary that the beam extend the full length of the distance back-to-back of end angles. In this case, as is customary, it is shown "set back" at both ends, the length of beam called for being 1″ less than the $13'-7\frac{3}{4}$ distance. Below the drawing, the mark B25 is the piece or shipping mark that appears on the erection plan and is to be painted on the member in the shop for identification. The end connection angles are fully detailed at the left end of the beam and are given the assembly mark **aa**. Therefore, at the right end where these same angles are again used, only the assembly mark **2–aa**, to indicate two angles, is given. The figure $10'-1\frac{1}{4}$ is called an *extension figure*, as this is the distance from the back of the left-end angles to the center of the group of four holes. Note that this dimension is on the same horizontal line as the $3'-4\frac{1}{2}$ figure just to the left. It is customary, in giving feet and

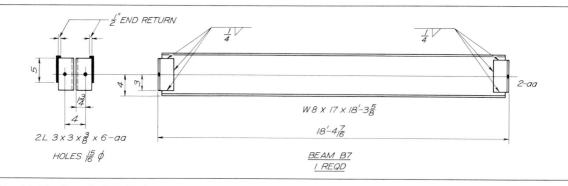

Fig. 24.13   Detail of Welded Beam.

inches, to give the foot mark (') after dimensions in feet, but not to give the (") mark, designating inches.

24.10   Welding

The use of welding as a method of connecting steel members of buildings and bridges is common. Most steel fabricators have riveting, bolting, and welding equipment available, although a few are equipped to handle only welded fabrication. The metal-arc process is used, energy being supplied through an electrode to unite both the metal of the electrode and the parent or base metal. Electrodes may be either bare or coated, although most welding today is done with coated electrodes. Of all types of welds, the *fillet weld* is most common in structural steel fabrication. For additional information, the student should refer to Chapter 27 and to the specifications of the AISC *Manual of Steel Construction*. When a structure is to be welded, it should be designed throughout for this method of fabrication, as it is not possible to obtain the maximum economy of construction by merely substituting welding for riveting. The designations of welds by the use of standard symbols, Fig. 27.3, have greatly simplified the making of shop drawings.

A beam with end connection angles shop welded to the beam web is shown in Fig. 24.13.

The outstanding legs of the angles are to be welded to the connecting columns in the field, as shown in the end view. This view pertains only to field erection. Open holes in the outstanding legs are for bolts, to facilitate positioning.

A shop drawing of diagonal bracing between two columns is shown in Fig. 24.14. Here the diagonal angle members are shop welded to gusset plates that are to be bolted to the column flanges as a permanent installation in the field.

Figure 24.15 is the complete shop drawing for a symmetrical welded roof truss. Because it is symmetrical, it is only necessary to draw the left half of the structure; symmetry is shown on the drawing by the note Sym. abt. ℄. The clip angles marked aa are to be used for the attachment of roof purlins to the truss. In this structure the only plate material needed is the small gusset plate marked pb, most of the connections of web members to chords being made by simply fillet welding the angles against the webs of the chords. The chords are of structural tee (WT) type. The top chords are joined at the ridge by butt welding, which is also used at the end of the truss where the chords join. Note the use of welding symbols, the marking of members, and the listing of all sizes and quantities in the Bill of Material. The drawing is uncluttered and clear, with considerable important information conveyed under the heading General Notes.

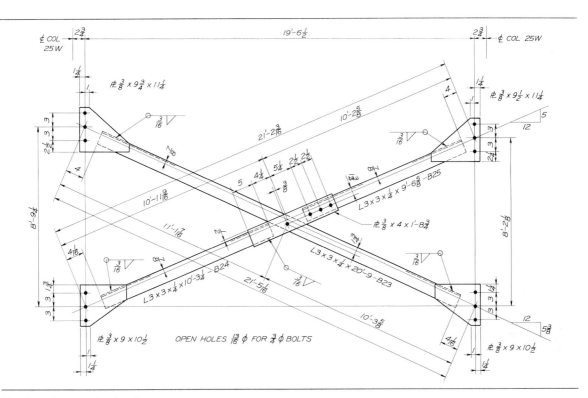

Fig. 24.14   Details of Column Bracing.

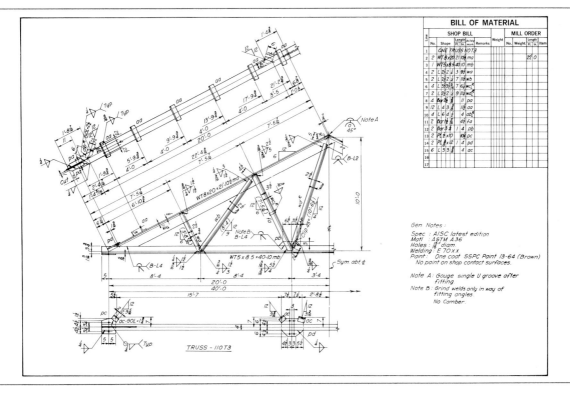

Fig. 24.15   A Welded Roof Truss. *Courtesy of American Institute of Steel Construction.*

24.11   High-Strength Bolting for Structural Joints

The two basic types of high-strength steel bolts in common use are known as ASTM A325 and A490. The type A449 is similar in physical properties to A325, except that ordinary rather than special nuts may be used with this type. These bolts are heat treated by quenching and tempering. The A325 bolt is made of medium-carbon steel, whereas the A490 is of alloy steel. Metric high-strength structural bolts, nuts, and washers are available. Consult a manufacturer's catalog or Appendix 18 for specifications. At the time of installation, such bolts are tightened a prescribed amount either by the "turn of the nut method" or by use of a calibrated torque wrench.

The use of a high-strength steel bolt to transmit a force from the center plate into the two outside plates is shown in Fig. 24.16. When fully tightened, the connecting parts are held together by friction, thus preventing joint slip. Fatigue failures due to impact or stress reversals are then minimized. The slip resistance depends not only on the amount of clamping force but also on the nature of the contact surfaces. Figure 24.16 shows hardened washers under both head and nut. Whether one or two washers are needed, or none, depends on the method of tightening used, the yield stress of the material being joined, and whether the joint is to be of "friction" or of "bearing" type. In the bearing-type connection, no allowance is made for friction due to clamping action, and the strength of the bolt shank bearing against the material is relied on. Numerous spec-

ifications govern accepted practice. These, together with installation and inspection procedures, design examples, and reference tables, are well covered in *High Strength Bolting for Structural Joints,* Bethlehem Steel Corporation, Bethlehem, PA.

Figure 24.17 represents the complete shop drawing of a steel column of two-floor height, the floors being indicated by reference lines Ⓑ, ①, and ② of the drawing. Here, high-strength bolts are used to attach connection material to the W12 × 96 column shaft. The top views at levels ① and ② show the outstanding legs of the angles in sectional plan views, with open holes shown by the round dots located by dimensions taken from the column center lines. The open holes are for the insertion of field bolts in beam connections when the entire assembly is erected.

Faces Ⓐ, Ⓑ, Ⓒ, and Ⓓ are designated. Viewing a column from above, we designate the faces in that order counterclockwise, with faces Ⓐ and Ⓒ designating the flanges. With regard to web faces Ⓑ and Ⓓ, only face Ⓑ needs to be drawn in the usual case. However, when the connections on face Ⓓ are exceptionally complicated this face also should be drawn as an additional elevation view.

Note the manner of showing the shop bolts and their designation "HSB" (high-strength bolts).

At the top of this column, splice connection material is provided by Bars (b), for attachment of the shaft above. All material for this column is listed in the Bill of Material that is a part of the drawing.

24.12   Calculation of Dimensions

Perhaps the most important part of the structural drafter's work is the accurate calculation of dimensions. If there are incorrect dimensions on the drawings, they will result in serious errors and misfits when members are assembled in the field. Correction of such errors not only entails considerable expense but often delays the completion of the work. Since in ordinary steel work, dimensions are given to the closest $\frac{1}{16}''$, consid-

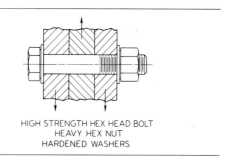

HIGH STRENGTH HEX HEAD BOLT
HEAVY HEX NUT
HARDENED WASHERS

Fig. 24.16   High-Tensile-Strength Steel Bolt.

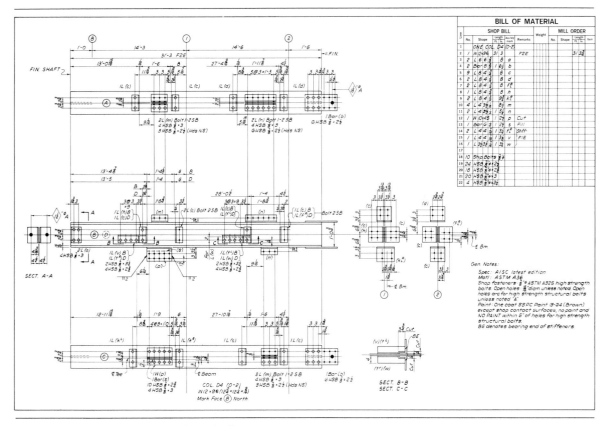

Fig. 24.17 Steel Column Detail—Bolted Connections. *Courtesy of American Institute of Steel Construction.*

erable precision is demanded. For skewed members, such as those in a truss, some trigonometry is involved. The dimensions may be easily and accurately determined with trigonometry and a small handheld electronic calculator that features natural functions. For example, to calculate the dimensions for the truss of Fig. 24.18 that has point B at midpoint of AC and BD perpendicular to top chord AC, proceed as follows:

For dimensions in feet plus inches with inch fractions to closest $\frac{1}{16}''$, convert feet and inches to inches and decimals and reconvert final results to feet, inches, and fractions.

Bevel = angle tangent × 12; sixteenths = inch decimals × 16.

$$\tan \theta = \frac{90''}{180''} = 0.5$$

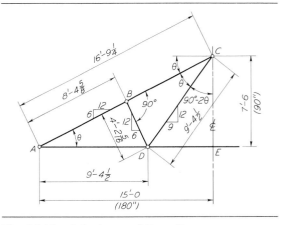

Fig. 24.18 Calculation of Truss Dimensions.

and

$$\theta = \text{arc tan } 0.5 = 26.565°$$

Bevel (AC and BD):

$$0.5 \times 12'' = 6''$$

Bevel (CD):

$$90° - 2\theta = 36.87°$$
$$12'' \times \text{tan } 36.87° = 12'' \times 0.75 = 9.00''$$
$$\text{Length AC} = \frac{180''}{\cos \theta} = 201.25'' = 16' - 9\tfrac{1}{4}$$
$$\text{Length BD} = \frac{201.25''}{2} (\tan \theta) = 50.3125''$$
$$= 4' - 2\frac{5}{16}$$
$$\text{Length AD} = \frac{201.25''}{2 \cos \theta} = 112.5 = 9' - 4\tfrac{1}{2}$$
$$\text{Length CD} = \text{AD}$$

(Note in the value 50.3125″ for the length of BD that 50″ = 4 × 12 + 2, therefore, 4′–2; .3125″ × 16 = 5 sixteenths. Hence, the total length of BD is 4′–2$\frac{5}{16}$.)

24.13   Concrete Construction

Concrete is a building material made by mixing sand (fine aggregate) and gravel (coarse aggregate) or other fine and coarse aggregates with *portland cement* and water. The strength of concrete varies with the quality and relative quantities of the materials, with the manner of mixing, placing, and curing, and with the age of the concrete. The compressive strength of concrete depends on the mix design, but has been manufactured to develop an ultimate strength at 28 days as high as 7000 psi. The tensile strength of the material is limited to about one-tenth the compressive strength. A new generation of concrete used in today's highrise construction is "high-strength" concrete with compressive strength up to 17,000 psi. Portland cement is a controlled, manufactured product as compared

to natural cements found in some localities. It derives its name from its color, which resembles that of a famous building stone found on the island of Portland in southern England.

Since the tensile strength of *plain concrete* is very limited, the usefulness of concrete as a building material can be materially improved by embedding steel reinforcing bars in it in such a way that the steel resists the tension, and the concrete mainly the compression. In this way the two materials act together in resisting forces and flexure. Concrete combined with steel in this way is called *reinforced concrete*. When the steel is pretensioned before the application of the superimposed load, thus producing an interior force within the member, the material is called *prestressed concrete*. The steel in prestressed concrete is of a flexible but very strong type.

24.14   Reinforced Concrete Drawings

The design of the reinforcing for a reinforced concrete structure and the preparation of the corresponding drawings are complicated. In order to simplify this work and to secure uniformity in the many engineering offices, the American Concrete Institute, Detroit, MI, has prepared a *Manual of Standard Practice for Detailing Reinforced Concrete Structures*.

It is recommended in the manual that two sets of drawings be prepared, an *engineering drawing* and a *placing drawing*. The engineering drawing is prepared by the engineer who designs the structure, and the placing drawing is prepared by the manufacturer who fabricates the reinforcing steel. The engineering drawing is to show the general arrangement of the structure, the sizes and reinforcements of the several members, and such other information as may be necessary for the correct interpretation of the designer's ideas. The drawing is also used for making forms with precise dimensions before placing the reinforcing bars and casting concrete. The placing drawing is to show the sizes

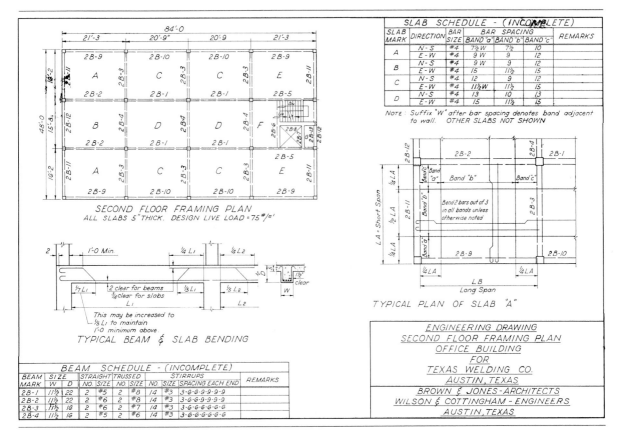

Fig. 24.19   Engineering Drawing for a Two-Way Slab and Beam Floor.

and shapes of the several rods, stirrups, hoops, ties, and so forth, and to arrange them in tabular forms for ready reference by the building contractor. The method of preparing an engineering drawing for a two-way slab and beam floor of a multistory building is illustrated in Fig. 24.19. For methods of preparing placing drawings, the student should consult the manual referred to earlier.

The design drawing for a reinforced concrete pier, one of the supporting members for a highway bridge, is shown in Fig. 24.20. Note that the steel bars, even though embedded, are shown by full lines and that concrete is always stippled in cross section. Unlike shop drawings for structural steel, concrete drawings are ordinarily made to scale in both directions. Usually, a scale

of $\frac{1}{4}''$ to the foot is adequate, although when the structure is complicated, scales of $\frac{3}{8}''$ or $\frac{1}{2}''$ to the foot may be used. An effort should be made to avoid a cluttered appearance, which is usually the result of crowding the drawing with many notes. Cluttering can be avoided largely by using tables and schedules for listing of bar sizes and other necessary data and by covering many important points in a single set of notes, as shown in Fig. 24.20.

24.15   Structural Clay Products

Brick and tile, which are manufactured clay products, have been in use for centuries and comprise some of the best known forms of building construction. Brick and tile units are made

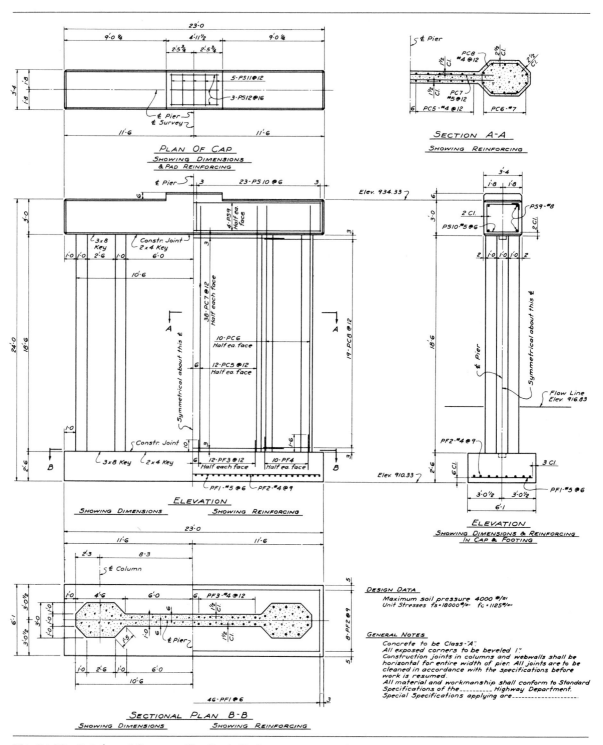

Fig. 24.20 Reinforced Concrete Pier Deck Girder. *From* Manual of Standard Practice for Detailing Reinforced Concrete Structures, *ACI 315-80 (American Concrete Institute, Detroit, MI, 48219).*

from many different types of clay and in many different shapes, forms, and colors. Ordinarily, they are built into masonry forms by the skilled brick or tile mason, who places the units one at a time in a soft mortar. After the mortar hardens, it becomes an integral part of the structure. Typical mortars contain sand, lime, portland cement, and water. Although the compressive and tensile strengths of the clay units themselves are considerable, the overall strength of the structure is limited by the strength of the mortar joints. As with concrete, therefore, the result is a structure of high compressive strength and relatively low tensile strength. Similarly, it is possible to reinforce brick and tile masonry by embedding steel rods within the members, thus adding greatly to their tension resistance and strength. When this is done, the material is called *reinforced brick* or *tile masonry* (RBM).

Information concerning the manufacture, the weight and strength properties, and the various uses and applications of structural clay products can be obtained from the Brick Institute of America handbooks *Principles of Brick Engineering* and *Principles of Tile Engineering*. These references will be found invaluable to both the designer and the drafter concerned with designs in this material.

Bricks are made of various sizes, the $2\frac{1}{4}'' \times 3\frac{3}{4}'' \times 8''$ building brick being the most common. Thickness of mortar joints usually varies from $\frac{1}{4}''$ to $\frac{3}{4}''$, with $\frac{3}{8}''$ and $\frac{1}{2}''$ most common. Of the several methods of bonding brick, Fig. 24.21, the following are the most common: *running bond*—all face brick are stretchers and are generally bonded to the backing by metal ties; *American bond*—the face brick are laid alternatively, five courses of stretchers and one course of headers; *Flemish bond*—the face brick are laid with alternate stretchers and headers in every course; *English bond*—the face brick are laid alternatively, one course of stretchers and one course of headers. In modern work, these standard methods of bonding are frequently modified to produce various artistic effects. Typical brick lintel arches are shown in Fig. 24.22.

In addition to its use as a basic building material, tile is used extensively in fireproofing structural steel members. Most building codes require that the steel members be enclosed in concrete or masonry so that fire will not cause collapse of the structure. Hollow tile units, being light and relatively inexpensive, are well adapted to this usage.

24.16   Stone Construction

Natural stone, used in masonry construction—most commonly today for ornamental facing—is generally limestone, marble, sandstone, or granite, Fig. 24.21.

Ashlar (or *ashler*) masonry is formed of stones cut accurately to rectangular faces and laid in regular courses or at random with thin mortar joints.

Rubble masonry is formed of stones of irregular shapes and laid in courses or at random with mortar joints of varying thickness.

Manufactured stone is concrete made of fine aggregate for the facing, and coarse aggregate for the backing. The fine aggregate consists of screenings of limestone, marble, sandstone, or granite, so that the manufactured stone presents an appearance similar to that of natural stone. Manufactured stone is made of any desired shape, with or without architectural ornament.

Architectural terra cotta is a hard-burned clay product and is used primarily for architectural decoration and for wall facing and wall coping.

Brick, stone, tile, and terra cotta are combined in many different ways in masonry construction. A few examples are shown in Fig. 24.21.

24.17   Computer Graphics

Structural drawings utilize a variety of standard symbols and methods of representation. A number of CAD symbol libraries are available that will allow the user to quickly generate drawings of structural steel shapes, Fig. 24.23, connecting members, and steel erection plans. Once the plans have been created, other programs are available that will automatically calculate areas and perform finite element analysis on the design.

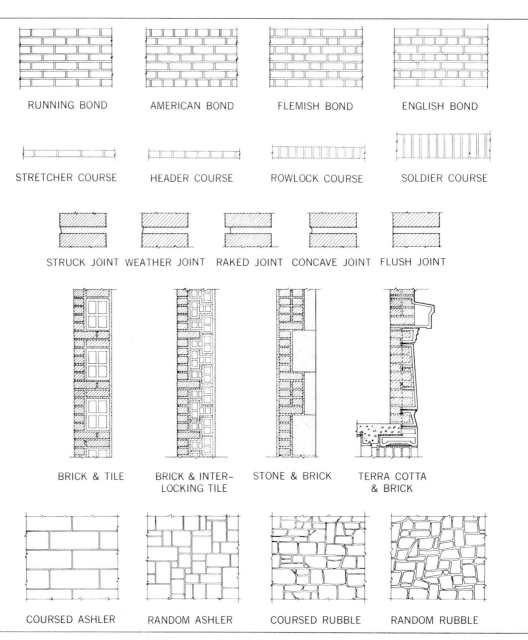

RUNNING BOND   AMERICAN BOND   FLEMISH BOND   ENGLISH BOND

STRETCHER COURSE   HEADER COURSE   ROWLOCK COURSE   SOLDIER COURSE

STRUCK JOINT   WEATHER JOINT   RAKED JOINT   CONCAVE JOINT   FLUSH JOINT

BRICK & TILE   BRICK & INTER-LOCKING TILE   STONE & BRICK   TERRA COTTA & BRICK

COURSED ASHLER   RANDOM ASHLER   COURSED RUBBLE   RANDOM RUBBLE

Fig. 24.21   Methods of Laying and Bonding Brick, Tile, and Stone.

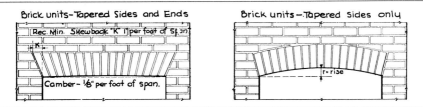

Fig. 24.22 Typical Lintels. *From* Brick and Tile Engineering, *courtesy of Structural Clay Products Institute.*

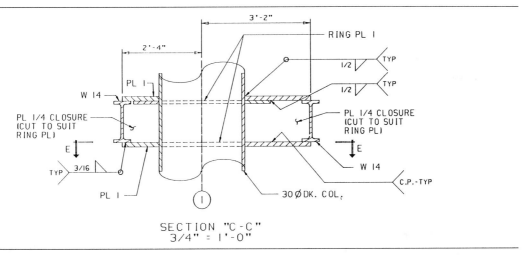

Fig. 24.23 CAD-Generated Structural Steel Detail Drawing. *Courtesy of Computervision Corporation, a subsidiary of Prime Computer, Inc.*

STRUCTURAL DRAWING PROBLEMS

The following problems are intended to afford practice in drawing and dimensioning simple structures and in illustrating methods of construction.

Prob. 24.1 Calculate point-to-point lengths (the distances between centers of joints) of the web members of the truss of Fig. 24.5. Make a detailed drawing of web member piece No. 4.

Prob. 24.2 Make a complete detail of the top chord member of the truss shown in Fig. 24.5 for a 30° angle of inclination.

Prob. 24.3 Assuming riveted construction, with rivets of $\frac{3}{4}''$ or 20 mm diameter, make a complete shop drawing for a typical filler beam of Fig. 24.7. Detail the same beam for welded construction. Consult the AISC *Manual of Steel Construction.*

Prob. 24.4 Assuming the column size to be W8 × 31, detail the W16 × 36 girder at the center of the drawing of Fig. 24.7.

Prob. 24.5 Referring to Fig. 24.15, make a complete detail for a truss of the same length, but change the height from 10′–0 to 6′–8. Use angle members of the same cross-section sizes

as those shown, but of different lengths, as needed.

Prob. 24.6 Referring to Fig. 24.14, make a similar bracing detail, changing the distance between column centers from 20′–0 to 18′–6.

Prob. 24.7 Draw cross sections through panel D of Fig. 24.19, in both directions. Include the supporting beams in each cross section, and show all dimensions, size and spacing of reinforcing steel, and dimensions to locate the ends of bars and the points of bend for the bent bars. Also show and locate the stirrups in these views.

Prob. 24.8 Detail the brickwork surrounding a window frame for an opening 4′–0$\frac{7}{8}$ wide by 6′–9 high. Use type of curved arch lintel similar to that of Fig. 24.22. Assume standard-size building brick with $\frac{1}{2}''$ mortar joints.

Prob. 24.9 Consult *Principles of Tile Engineering*, and draw a cross-sectional view through a 12″ wall of composite brick and tile construction.

CHAPTER 25

Topographic Drawing and Mapping

BY E. I. FIESENHEISER*

In previous chapters you were introduced to the methods and techniques used in drawing manufactured objects. *Topographic drawing* and *mapping* have to do with the representation of portions of the earth's surface—mainly its natural features—to a convenient scale. On such drawings the relative positions of natural features, with respect to certain definitely located points, are shown. Since the shape of the earth is spherical, any representation on a plane, such as a piece of paper, is necessarily somewhat distorted. See §21.21. In drawing large areas, therefore, some method of projection must be used that results in a minimum of distortion. In such work, the positions of the control or reference points are usually defined by spherical coordinates of latitude and longitude (meridians and parallels), which are shown as reference lines on the drawing. In drawing small areas to a relatively large scale, the distortion due to earth curvature is so slight as to be unnoticeable and may therefore be entirely neglected. Orthographic projection, as used in technical drawings, in which the line of sight is assumed to be perpendicular to the plane of the map, is the method most commonly used in topographic representation.

25.1 Definitions

The purpose for which a map is to be used determines what features should be represented, what scale should be used, and what detail should be included. Certain commonly used symbols are employed in representing natural features and man-made objects. Also, certain terms and various types of maps are common. Therefore, the following definitions are essential to an understanding of the subject.

A *plat* is a map, usually of a small area, plotted directly from a land survey. It does not ordinarily show relative ground elevations and is drawn for some specific purpose, such as the calculation of areas, the location of property lines, or the location of a building project. Usually it contains a *traverse*. See Fig. 25.1.

A *traverse* consists of a series of intersecting straight lines of accurately measured lengths. At the points of intersection, the deflection angles between adjacent lines are measured and recorded. Starting at one point, therefore, we can calculate rectangular coordinates of the other intersection points by trigonometry. A *closed traverse* thus becomes a closed polygon and provides a method for checking the accuracy of the work.

---

*Professor Emeritus of Civil Engineering, Illinois Institute of Technology.

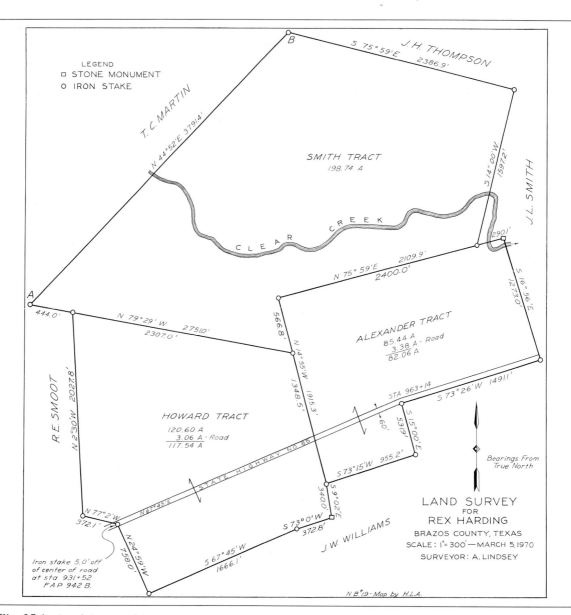

Fig. 25.1   Land Survey Plat.

The land survey plat of Fig. 25.1 illustrates a closed traverse.

Elevations are vertical distances above a common *datum* or reference plane or point. The elevation of a point on the surface of the ground is usually determined by differential leveling from some other point of known elevation. Commonly, elevations are referenced to mean sea level.

A *profile* is a line contained in a vertical plane, and it depicts the relative elevations of various points along the line. For example, if a

788

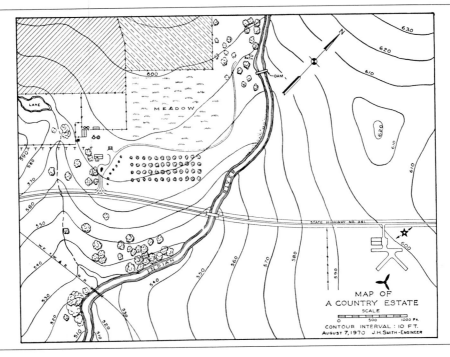

Fig. 25.2   A Topographic Map.

vertical section were to be cut into the earth, the top line of this section would represent the ground profile.

Contours are lines drawn on a map to locate, in the plan view, points of equal ground elevation. On a single contour line, therefore, all points have the same elevation.

Hatchures are short, parallel, or slightly divergent lines drawn in the direction of the slope. They are closely spaced on steep slopes and converge toward the tops of ridges and hills. Hatchures are shade lines to show relief. See Appendix 33.

Monuments are special installations of stone or concrete to mark the locations of points accurately determined by precise surveying. It is intended that monuments be permanent or nearly so, and they are usually tied in by references to nearby natural features, such as trees and large boulders.

Cartography is the science or art of map making.

Topographic maps depict (1) water, including seas, lakes, ponds, rivers, streams, canals, and swamps; (2) *relief* or elevations of mountains, hills, valleys, cliffs, and the like; (3) *culture,* or human constructions, such as towns, cities, roads, railroads, airfields, and boundaries. See Figs. 25.2 and 25.12.

Hydrographic maps convey information concerning bodies of water, such as shoreline locations; relative elevations of points of lake, stream, or ocean beds; and sounding depths.

Cadastral maps are accurately drawn maps of cities and towns, showing property lines and other features that control property ownership.

Military maps contain information of military importance in the area represented.

Nautical maps and charts show navigational features and aids, such as locations of buoys, shoals, lighthouses and beacons, and sounding depths.

Aeronautical maps and charts show prominent landmarks, towers, beacons, and elevations for the use of air navigators.

Engineering maps are made for special projects as an aid to locations and construction. See Figs. 25.6 and 25.9–25.11.

Landscape maps are used in planning installations of trees, shrubbery, drives, and other garden features in the artistic design of area improvements. See Fig. 25.8.

25.2   Sources of Information

The basis of all maps and topographic drawing is the *survey*. Several surveying methods are used to obtain the information necessary for the making of a map.

Short distances are ordinarily measured by steel tape, with stakes being driven to mark the points between field measurements. Distances may also be determined by measuring aerial photographs, when the scale of the photograph is known. An instrumental method, known as the *stadia* method, is also widely used in map making. The *stadia transit* is an optical instrument used in conjunction with a special *stadia rod*. By sighting on the rod and using the necessary conversion factor, the instrument reading can easily be converted to distance.

Recent developments in the art of surveying have revolutionized distance measurement by the use of electronic instruments. To measure to a distant point, the surveyor directs the instrument's aiming head toward the point, where a passive reflector or prism has been set. The instrument generates either a modulated infrared light signal, focused into a narrow beam, or a laser beam, aimed directly at the reflector. When the reflector bounces the beam back to the aiming head, the beam's travel time is electronically measured and directly converted into the distance to the point. A major advantage of the electronic measurement over the distance-taping method is that moving traffic does not have to be stopped while measurements are being taken.

Such time-saving instruments are capable of distance measurements up to 4 miles, depending on the quality of the instrument. Accuracy varies from .01 to .03 ft, more than sufficient for topographic surveying.

By use of a *compass,* the bearing of a line, which is the angle between the line and the *magnetic north,* may be read on the compass. When all bearings of the lines of a *traverse* have been determined, the angles between the lines are easily computed by addition or subtraction. The correct method of listing bearings requires that they be referenced either to the north or to the south. A bearing, therefore, is either north, and east or west; or it is south, and east or west. See N 44°52′E for traverse line AB of Fig. 25.1. Compass readings are not to be regarded as of high accuracy, since magnetic north and true north are not quite the same; also local magnetism may affect the position of the compass needle.

When accurate measurement of angles is desired, the *transit* is the instrument usually employed. This optical instrument may be set up directly over a point, then sighted successively on two other points, after which the deflection angle in a horizontal plane can be read on the transit. This instrument is also used for the measurement of vertical angles.

For accurate orientation of lines on a map, the angle of a line with the *true north* is often desired. For this purpose the transit may again be used for sighting on a star (usually *Polaris,* the North Star), or on the sun. From instrument readings, the true angle of a line may then be calculated.

The *level,* also an optical instrument, is equipped with a telescope for sighting long distances. This instrument is commonly used to determine differences in elevation in the field, which is called the process of *differential leveling.* When this accurate instrument is leveled, the line of sight of its telescope is horizontal. A level rod, graduated in feet and decimals of feet, may be held on various points. Instrument readings of the rod then serve to determine the differences in elevations of the points.

Photogrammetry is now widely used for map surveying. This method utilizes actual photographs of the Earth's surface and of manufactured objects on the Earth. Aerial photogrammetry, via aircraft or satellite, is used for such activities as governmental and commercial surveying, explorations, and property valuation. It has the great advantage of being easy to use in difficult terrain having steep slopes, where ground surveying would be difficult or nearly impossible. The utilization of aerial photographs is called *aerial photogrammetry*, whereas that utilizing photographs taken from ground stations with the axis of the camera lens nearly horizontal is known as *terrestrial photogrammetry*. By combining the results of both types of observations, it is possible not only to determine the relative positions of objects in a horizontal plane, but also to determine relative elevations. Thus, the science of photogrammetry can be the basis of contour mapping as well as plan mapping. Generally, aerial photographs are used in matched groups to form a *mosaic*. To form a satisfactory mosaic map, the group photographs must overlap slightly. A distinct advantage of photogrammetry is that a large area can be mapped from a single clear photograph. The method may be used in connection with ground surveying by photographing *control points* already located on the ground by precise surveying.

For large land developments and large construction projects, more modern methods for producing topographic maps have evolved. Some state highway departments and most engineering firms who specialize in surveying and mapping now make use of aerial photography with computers, terrain digitizers, stereoplotters, and various new photolaboratory techniques. Investment in this expensive equipment can be justified when the volume of work is great and sufficient hours can be saved through its use, as is frequently the case. See §25.8 for topographic maps of the United States.

For additional information, which is beyond the scope of this chapter, the student may refer to *Elements of Photogrammetry* by P. R. Wolf (McGraw-Hill); *Elements of Cartography* by A. Robinson and R. Sale (John Wiley); and *Manual of Surveying Instructions for the Survey of the Public Lands of the United States*, prepared and published by the Bureau of Land Management (U.S. Government Printing Office, Washington, DC).

25.3   Contours

Although contours have already been defined, their uses and characteristics and the methods of plotting them require further discussion.

A *contour interval* is the vertical distance between horizontal planes passing through successive contours. For example, in Fig. 25.3 the contour interval is 10′. The contour interval should not change on any one map. It is customary to show every fifth contour by a line heavier than those representing intermediate contours.

If extended far enough, every contour line will close. At streams, contours form V s pointing upstream. Should successive contours be evenly spaced, this means that the ground slopes uniformly, whereas closely spaced contours indicate steep slopes.

Locations of points on contour lines are determined by *interpolation*. In Fig. 25.3, the locations and elevations of seven control points are determined, and contour lines will be drawn on the assumption that the slope of the surface of the ground is uniform between station A and the six adjacent stations. To draw the contour lines, a contour interval of 10′ was adopted, and the locations of the points of intersection of the contour lines with the straight lines, joining the point A and the six adjacent points, was calculated as follows.

The horizontal distance between stations A and B is 740′. The difference in elevation of those stations is 61′. The difference in elevation of station A and contour 300 is 9′; therefore, contour 300 crosses the line AB at a distance from station A of $\frac{9}{61}$ of 740, or 109.1′. Contour 290 crosses the line AB at a distance from contour 300 of $\frac{10}{61}$ of 740, or 121.3′. This 121.3′ distance between contour lines is constant along the line AB and can be set off without further calculation.

In the same way, points in which the contours cross the other lines of the survey can be

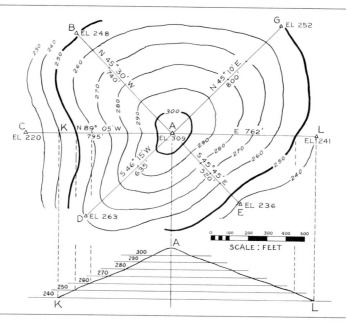

Fig. 25.3   Contours Determined from Control Points.

interpolated. After this process is completed, the several contour lines can be drawn through points of equal elevation, as shown.

After contours have been plotted, it is easy to construct a profile of the ground line in any direction. In Fig. 25.3, the profile of line KAL is shown in the lower or front view. It is customary, as shown here, to draw the profile to a larger vertical scale than that of the plan in order to emphasize the varying slopes.

Contour lines may also be plotted by use of the recorded elevations of points on the ground surface, as in Fig. 25.4 (a). This figure illustrates a *checkerboard survey,* in which lines are drawn at right angles to each other, dividing the survey into 100 ft squares, and where elevations have been determined at the corners of the squares. The contour interval is taken as 2', and the slope of the ground between adjacent stations is assumed to be uniform.

The points where the contour lines cross the survey lines can be located approximately by inspection, accurately by the graphical method shown in Fig. 5.14, or by the numerical method explained for Fig. 25.3.

The points of intersection of contour lines with survey lines may also be found by constructing a profile of each line of the survey, as shown for line 1 in Fig. 25.4 (b). Horizontal lines are drawn at elevations at which it is desired to show contours. The points in which the profile line intersects these horizontal lines indicate the elevations of points in which corresponding contour lines cross the survey line 1 and, therefore, can be projected upward, as shown, to locate these points.

It is obvious that the profile of any line can be constructed from the contour map by the converse of the process just described.

25.4   Symbols

Various natural and manufactured features are designated by special symbols. A reference list of the most commonly used map symbols is given in Appendix 33. The student should refer to this list for identification of the symbols used in the figures of this chapter.

The use of templates by the drafter, such as that shown in Fig. 25.5, is a convenient time-

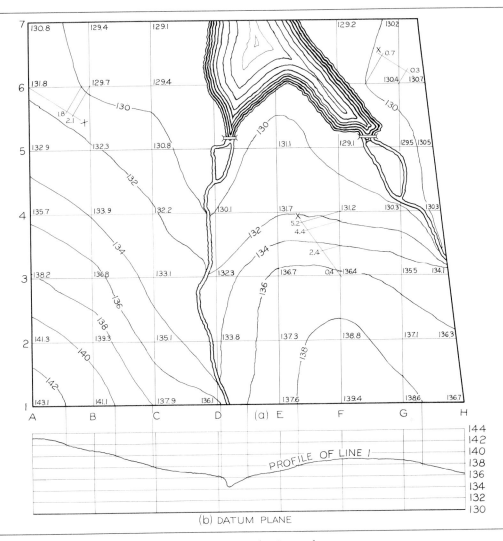

Fig. 25.4   Contours Determined from Readings at Regular Intervals.

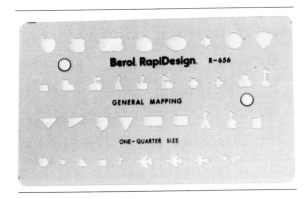

Fig. 25.5   General Mapping Template.

saver in drawing mapping symbols. The symbols may be drawn in pencil or with a technical fountain pen.

25.5   City Maps

The special use of a map determines what features are to be shown. Maps of city areas may be put to many uses, some of which will be described. Figure 25.6, a city plan for location of a new road construction, shows only those features of importance to the location and construc-

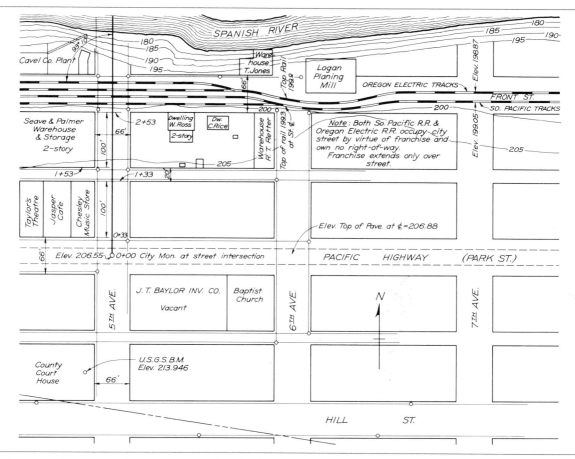

Fig. 25.6   A City Plan for Location of a New Road Project.

tion of the road. The transit line starts at the center-line intersection of Park St. and 5th Ave., and it is marked as station 0 + 00. From here it extends north over the railroad yard to cross the river. Features near the transit line, such as buildings, are shown and identified by name. Street widths are important and are shown. Contour lines between the railroad yard and the river indicate the steeply sloping terrain.

Maps perform an important function for those who plan the layout of lots and streets. For example, Fig. 25.7 (a) shows an original layout of these features for a new residential area. An examination of the contours will show that this layout is not satisfactory, since the directions of the streets do not fit the natural ground slopes.

Streets should be arranged so that the subdivision can be entered from a low point and so that a maximum number of lots will be above street grade. The layout at (b) is a decided improvement, for in it the streets curve to fit the topography and the entrance is located at a low point.

Maps have a definite use in landscape planning. Figure 25.8 is a landscape map showing a proposed layout of lots and trees for beautification of a new subdivision.

25.6   Structure Location Plans

When a construction project is contemplated in a particular area, the construction should be located most advantageously to fit the topography

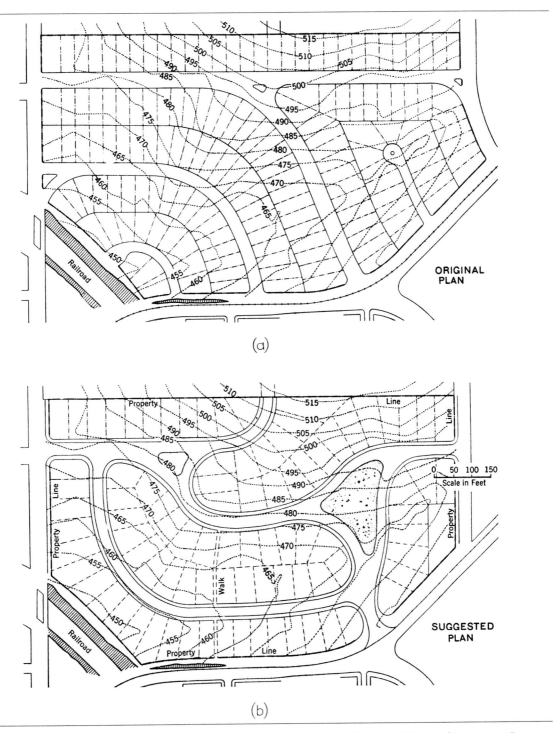

(a)

(b)

Fig. 25.7   Adjustment of Streets to Topography. *From* Land Subdivision, *ASCE Manual No. 16 of Engineering Practice.*

Fig. 25.8   Land Subdivision Showing Use of Culs-de-Sac (U-shaped streets), Stamford, Connecticut. *From* Land Subdivision, *ASCE Manual No. 16 of Engineering Practice.*

796

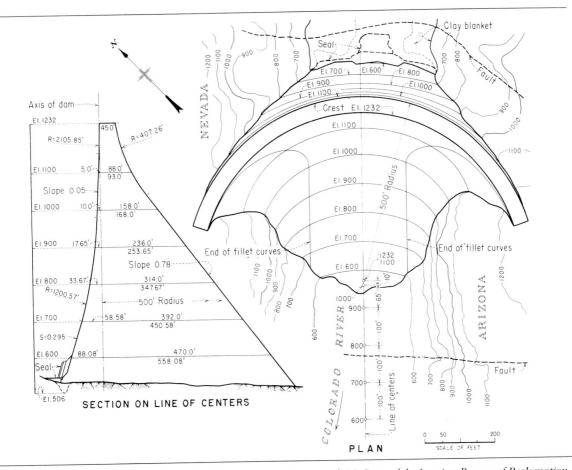

Fig. 25.9   Plan for Hoover Dam. *From* Treatise on Dams. *Courtesy of U.S. Dept. of the Interior, Bureau of Reclamation.*

of the area. Thus, a location plan on paper is of great advantage. A project location plan for a dam is shown in Fig. 25.9. This map shows the important natural features, contours, a plan view of the structure, and a cross section through Hoover Dam.

Although many drawings, perhaps hundreds, comprise the complete detail drawings for a large bridge, one of the most important early drawings is a general arrangement plan and elevation in the form of a line diagram. As an example, Fig. 25.10 shows a plan and elevation of a large bridge structure.

25.7   Highway Plans

Before highway construction starts, it is necessary first to plan the location and to arrange both the horizontal and vertical alignment most advantageously. Commonly, both plan and profile are drawn on the same sheet, as in Fig. 25.11. In the plan at the top of the figure, the topography, consisting of such features as trees, fences, and farmhouses along the right of way, is shown. The transit line, locating the center line of the new road, is drawn with stations spotted at 100′ distances apart. Data for laying in the horizontal curves in the field have been calcu-

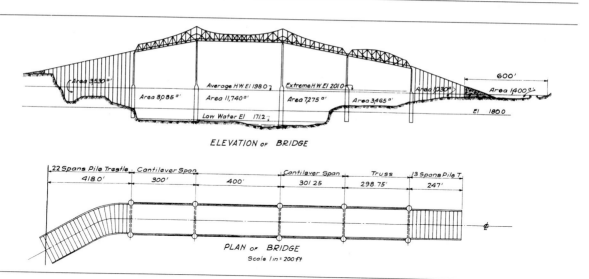

Fig. 25.10   Plan and Elevation of a Bridge Structure.

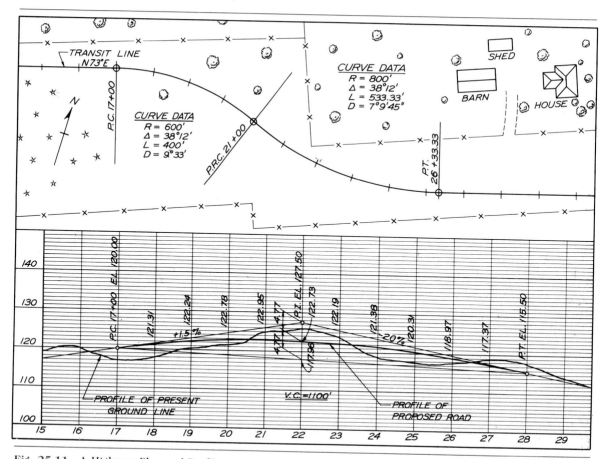

Fig. 25.11   A Highway Plan and Profile.

lated and are listed. The point of curve (P.C.) at station 17 + 00 is the point at which the line begins to curve with a 600′ radius, for a curve length of 400′. The central angle is 38°12′, and the degree of curve (angle subtended by a 100′ chord) is shown as 9°33′. The reverse curve of 800′ radius begins at station 21 + 00. Note also the north point and the bearing N73°E of the transit line.

The vertical alignment is shown in profile below the plan. Note the listing of station numbers below the profile. The scale of this view is larger in a vertical than in a horizontal direction in order to show the elevations clearly. The existing ground profile along the center line of the road is shown, as well as the profile of the proposed vertical alignment. The symbol P.I. denotes the point of intersection of the grade lines, and grade slopes are given in percentages. A 1 percent grade would rise vertically 1′ in each 100′ of horizontal distance. Station 22 + 00, therefore, is the intersection point of an upgrade of 1.5 percent and a downgrade of −2.0 percent.

To provide a smooth transition between these grades, a vertical curve (V.C.) of 1100′ length is used. This curve is parabolic and tangent to grade at stations 17 + 00 and 28 + 00. The straight line joining these points has an elevation 117.96′ directly below the P.I. At this point the parabolic curve must pass through the midpoint of the vertical distance, at a height of 4.77′ below the P.I. Ordinates to parabolas, measured from tangents, are proportional to the squares of the horizontal distances from the points of tangency. Therefore, it is possible to calculate the elevations of points along the curve by first determining the grade elevations, then subtracting the parabolic curve ordinates. The final profile elevations are given in the figure. Calculations are given in Table 25.1.

25.8 United States Maps

Maps of the United States, prepared by the U.S. Coast and Geodetic Survey and the U.S. Geological Survey, are excellent examples of topographic mapping. A small section of a typical

Table 25.1 *Calculation of Vertical Curve Elevations*

Station	Tangent Elevations	Ordinate	Curve Elevations
18	121.50	.19[a]	121.31
19	123.00	.76	122.24
20	124.50	1.72	122.78
21	126.00	3.05	122.95
22	127.50	4.77	122.73
23	125.50	3.31	122.19
24	123.50	2.12	121.38
25	121.50	1.19	120.31
26	119.50	.53	118.97
27	117.50	.13	117.37

[a] $\dfrac{(100)^2 4.77′}{(500)^2} = .19′$.

U.S. Geological Survey map is shown in Fig. 25.12. Such maps cover large areas, the largest scale used being 1 : 62,500, very nearly one mile to the inch. The contour interval in this example is 20′. To so small a scale, it would be impossible to show clearly any small features—vegetation and fences, for example. Therefore, these maps can show only the main features of the terrain, such as contours, roads, railroads, rivers, lakes, and streams. These maps are very reliable, as they are based on precise surveying.

The U.S. Department of Interior Geological Survey has prepared a series of standard topographic maps that cover the United States, Puerto Rico, Guam, American Samoa, and the Virgin Islands. Each unit of survey (map) is a *quadrangle* bounded by parallels of latitude and meridians of longitude. Quadrangles covering $7\frac{1}{2}$ minutes of latitude and longitude are published at a scale of 1 : 24,000 (1″ = 2000′). Quadrangles covering 15 minutes of latitude and longitude are published at a scale of 1 : 62,500 (1″ = 1 mile, approx.).

Maps (27″ × 41″) showing the status of various phases of mapping and the areas covered by aerial photography in the United States at a scale

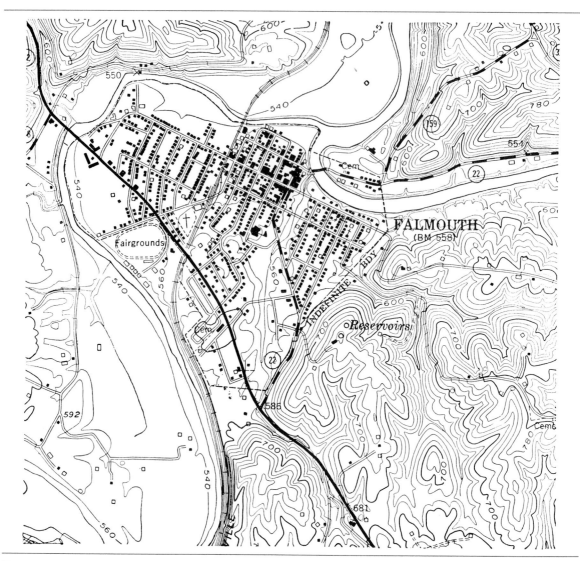

Fig. 25.12   A Typical Map of the U.S. Geological Survey. *Courtesy of U.S. Geological Survey.*

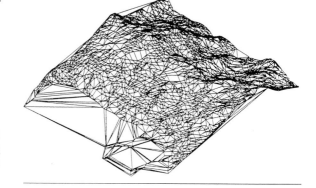

Fig. 25.13   Three-Dimensional Digital Terrain Model Created by Computek, Inc., of Taipei, Taiwan, Using CADL (CADKEY Advanced Design Language). *Courtesy of CADKEY, Inc.*

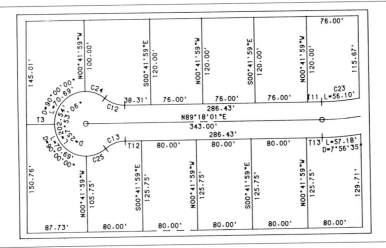

Fig. 25.14 CAD-Produced Tract Map. *Courtesy of California Computer Products, Inc.*

of 1 : 5,000,000 (1″ = 80 miles, approx.) are available on request to the U.S. Geological Survey, Reston, VA 22092.

25.9 Computer Graphics

Various CAD programs allow the user to create contour maps easily by entering random site data. Point descriptions, contour lines, grid lines, and features such as bodies of water or structures can be depicted, and topographic pictorial drawings produced, Fig. 25.13. Planners are able to lay out subdivision plots with programs that will also provide the building contractor with earthwork calculations for individual lots or the entire site. Street intersections and features such as cul-de-sacs can be automatically generated, Fig. 25.14.

TOPOGRAPHIC DRAWING PROBLEMS

The following problems are given to afford practice in topographic drawing. The drawings are designed for a Size B or A3 sheet. The position and arrangement of the titles should conform approximately to that of Fig. 25.1.

Prob. 25.1 Draw symbols of six of the common natural surface features (streams, lakes, etc.) and six of the common development features (roads, buildings, etc.) shown in Appendix 33.

Prob. 25.2 Draw, to assigned horizontal and vertical scales, profiles of any three of the six lines shown in Fig. 25.3.

Prob. 25.3 Assuming the slope of the ground to be uniform and assuming a horizontal scale of $1'' = 200'$ and a contour interval, §25.3, of $5'$, plot, by interpolation, the contours of Fig. 25.3.

Prob. 25.4 Using the elevations shown in Fig. 25.4 (a) and a contour interval of $1'$, plot the contours to any convenient horizontal and vertical scales, and draw profiles of lines 3 and 5 and of any two lines perpendicular to them. Check, graphically, the points in which the contours cross these lines.

Prob. 25.5 Using a contour interval of $1'$ and a horizontal scale of $1'' = 100'$, plot the contours from the elevations given, Fig. 25.15, at $100'$ stations; check, graphically, the points in which the contours cross one of the horizontal lines and one of the vertical lines, using a vertical scale of $1'' = 10'$; sketch, approximately, the drainage channels.

Prob. 25.6 Draw a plat of the survey shown in Fig. 25.1 to as large a scale as practicable. Use

144.7	139.2	143.1	144.6	144.3	143.5	142.2
142.5	138.0	139.0	141.3	142.7	139.3	139.1
140.7	137.5	136.1	138.6	138.0	136.1	137.2
138.8	136.5	135.0	136.2	135.7	135.9	136.1
139.1	136.4	134.6	133.5	133.7	134.1	135.8
135.3	134.5	133.0	132.7	132.0	131.9	132.3
135.9	134.0	132.7	131.3	130.8	129.6	131.5

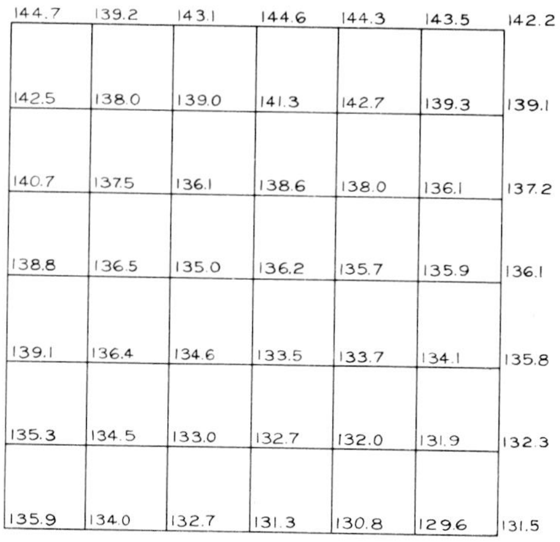

Fig. 25.15 To Draw Contours (Prob. 25.5).

an engineers' scale to set off distances and a protractor to set off bearings; if the drawing is accurate, the plat will close.

Prob. 25.7 Draw a topographic map of a country estate similar to that shown in Fig. 25.2.

Prob. 25.8 Calculate profile elevations for a vertical curve $800'$ long to join grades of $+3.00$ percent and -3.00 percent. Assume grade elevations at points of tangency to be $100.00'$.

CHAPTER 26

Piping Drawing

Pipe is used for transporting liquids and gases and for structural elements such as columns and handrails. The choice of the type of pipe is determined by the purpose for which it is to be used.

Pipe is made of aluminum, brass, clay, concrete (concrete made with ordinary aggregates and in combination with other materials), copper, glass, iron, lead, plastics, rubber, wood, and other materials or combinations of them. Cast-iron, steel, wrought iron, brass, copper, and lead pipes are most commonly used for transporting water, steam, oil, or gases.

26.1 Steel and Wrought-Iron Pipe

Steel or wrought-iron pipe is in common use for water, steam, oil, and gas. Up to the early 1930s it was available in only three weights known as "standard," "extra strong," and "double extra strong." At that time increasing pressures and temperatures, particularly for steam service, made the availability of more diversified wall thicknesses desirable. The American National Standards Institute Sectional Committee B36 has developed dimensions for ten different schedules of pipe. See Appendix 38. This table shows that dimensions are established for nominal sizes from $\frac{1}{8}''$ to $24''$ and that dimensions have not been established for all schedules. In the different schedules, the outside diameters (OD) are maintained for each nominal size to facilitate threading and the uniform use of fittings and valves.

Certain of the schedule dimensions correspond to the dimensions of "standard" and "extra strong" pipe. These are shown in **boldface**

type in the appendix. The schedule dimensions so shown for Schedules 30 and 40 correspond to standard pipe, and those for Schedule 80 correspond to extra strong pipe. There are no schedule dimensions corresponding to "double extra strong" pipe. Pipe corresponding to all the established schedule dimensions is not always commercially available and should be investigated before specifying pipe on drawings. Generally, Schedules 40, 80, and 160 are readily available; others may or may not be.

Note that the actual outside diameter of pipe in nominal sizes $\frac{1}{8}''$ to $12''$ inclusive is larger than the nominal size, whereas the outside diameter of pipe in nominal sizes $14''$ and larger corresponds to the nominal size. This pipe in nominal sizes $14''$ and larger is commonly referred to as OD pipe.

Pipe is available as welded or as seamless pipe. Welded pipe is available in Schedules 40 and 80 in the smaller sizes. Lap-welded pipe is made in sizes up to and including $2''$. Butt-welded pipe is available as furnace-welded material, where a formed length is heated in a furnace and then welded, in sizes up to and including 3″. Butt-welded pipe is also available as continuous welded pipe, where the finished pipe is continuously heated, formed, and welded from a roll of strip steel, in sizes up to and including 4″. Seamless pipe is made in both small and large sizes.

Many of today's applications require the use of alloys to withstand the pressure-temperature conditions without having to be excessively thick. Numerous alloys in both ferritic and austenitic material are available. Reference should be made to the various specifications of the American Society for Testing Materials (ASTM) for these alloys and for dimensional tolerances.

Steel pipe can be obtained as *black pipe* or as *galvanized pipe*. Galvanized pipe is not always available from the original producer, due to lack of facilities at the pipe mill.

Steel or wrought-iron pipe is available in lengths up to about 40′ in the small sizes, the length decreasing with increasing size and wall thickness.

26.2   Cast-Iron Pipe

Cast-iron (C.I.) pipe is generally used for water or gas service and as soil pipe. For water and gas pipe, it is available in sizes from 3″ to 60″ inclusive and in standard lengths of 12′. Various wall thicknesses, to satisfy the internal pressure requirements, can be secured. The dimensions and the pressure ratings of the various classes are shown in Appendix 39.

Generally speaking, water and gas pipes are connected with *bell and spigot* joints, Fig. 26.1 (a), or *flanged* joints, (b), although other types of joints, (c), are also used.

As soil pipe, cast-iron pipe is available in sizes 2″ to 15″ inclusive, in standard lengths of 5′, and in *service* and *extra heavy* weights. Soil pipe is generally connected with bell and spigot joints, but soil pipe with threaded ends is available in sizes up to 12″.

In using cast-iron pipe, the designer must consider both the internal pressure and the external loading due to fill and other loadings, such as roads and tracks. Cast-iron pipe is brittle, and settlement can cause fracture unless sufficient flexibility is provided in the joints. For this reason, flange joints are not usually employed for buried pipes unless adequately supported.

26.3   Seamless Brass and Copper Pipe

Pipe made of brass and copper, and having approximately the same dimensions as "standard" and "extra strong" steel pipe, is available. Such pipe is suitable for plumbing, including supply, soil, waste drain, and vent lines. It is also particularly suitable for process work where formation of scale and oxidation of steel pipe are objectionable. Brass and copper pipe is available in straight lengths up to 12′.

Brass pipe is generally known as *red brass pipe,* which is an alloy of approximately 85 percent copper and 15 percent zinc. Copper pipe is

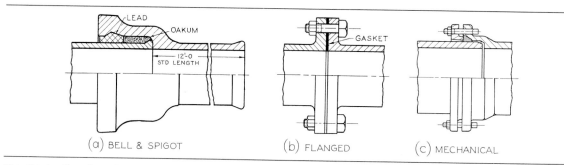

Fig. 26.1   Cast-Iron Pipe Joints.

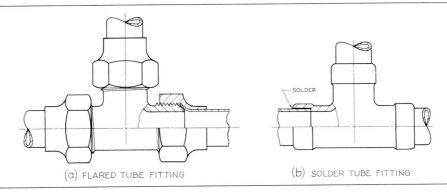

(a) FLARED TUBE FITTING   (b) SOLDER TUBE FITTING

Fig. 26.2   Copper Pipe Fittings.

practically pure copper with less than 0.1 percent of alloying elements.

Brass and copper pipe should be joined with fittings of copper-base alloy in order to avoid galvanic action resulting in corrosion. Where screwed joints are used, fittings similar to cast or malleable iron fittings, Fig. 26.5, are available. Flanged fittings of brass and copper are of different dimensions than those made of ferrous material. Reference should be made to dimensional standards published by the American National Standards Institute [ANSI B16.24–1979] for dimensions of such fittings.

26.4   Copper Tubing

Where nonferrous construction in sizes below 2″ is used, copper tubing is frequently employed. Such tubing is suitable for process work, as mentioned in §26.3, for plumbing (particularly supply lines for hot and cold water), and for heating systems (particularly radiant heating).

Copper tubing is made as *hard temper* and as *soft tubing*. Hard temper tubing is much stiffer than soft tubing and should be used where rigidity is desired. Soft tubing can be easily bent and, therefore, is generally used where bending during assembly is required. Neither hard- nor soft-temper tubing has the rigidity of iron or steel pipe and, consequently, must be supported at much more frequent intervals than the latter. Where multiple runs of parallel tubes are em-

ployed for long distances (20′ or more), it is common practice to use soft tubing and to lay the parallel runs in a trough construction, thus obtaining continuous support.

Copper tubing joints are usually made with *flared* joints, Fig. 26.2 (a), or with *solder* joints, (b). There are several types of flared joints, but the basic design of making a metal-to-metal joint is common to all. Fittings, such as tees, elbows, and couplings, are available for flared joints.

Solder joints are also known as *capillary* joints because the annular space between the tube and the fitting is so small that the molten solder is drawn into the space by capillary action. The solder may be introduced through a hole in the fitting, Fig. 26.3, through the outer end of the annular space, Fig. 26.2 (b), or fittings having a factory-assembled ring of solder in the fitting can be purchased. Solder joints may be made with soft solder (usually 50–50 or 60–40

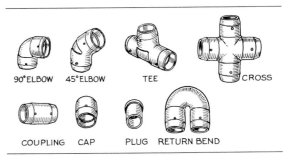

90°ELBOW   45°ELBOW   TEE   CROSS

COUPLING   CAP   PLUG   RETURN BEND

Fig. 26.3   Solder Fittings.

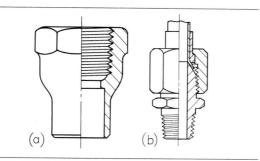

Fig. 26.4   Adapters—Copper Tube to Threaded Pipe.

tin and lead) or with silver solder. This latter material has a higher melting point than does soft solder, makes a stronger joint, and is suitable for higher operating temperatures.

Copper pipe or tubing has an upper operating temperature limit of 406 °F. If solder fittings are used, the upper temperature limit will be dependent on the softening point of the solder rather than the limit of temperature of the base material.

Copper tubing can be connected to threaded pipe or fittings by means of *adapters*. These adapters are available with either male or female pipe threads and with either flared or solder connections for the tubing. Two types of such adapters are shown in Fig. 26.4.

Copper tubing is available in straight lengths up to 20′ or in coils of 60′ for soft-temper material. Hard-temper material is available in straight lengths only, since it cannot successfully be coiled. Installation costs of coiled material are lower than for straight material, because of the

fewer number of joints to be made. Copper tubing is available in both OD and nominal sizes. American Society for Testing Materials Specifications B88 and B251 give details of dimensions of such tubing.

26.5   Special Pipes

Pipe and tubing of other materials, such as aluminum and stainless steel, are also available. Pipe and tubing of plastics are being increasingly used as these materials undergo development. Service for which these miscellaneous materials are suitable varies widely because of the physical properties and temperature-pressure limitations of the materials. Regardless of the material used in a piping system, the procedure followed in the design of the system and the creation of the necessary drawings remains basically the same.

26.6   Pipe Fittings

Pipe fittings are used to join adjacent lengths of pipe and frequently also to provide changes of direction, to provide branch connections at different angles, or to effect a change in size. They are made of cast iron, malleable iron, cast or forged steel, nonferrous alloys, and other materials for special applications. They can be obtained in various weights that should be matched to the pipe with which they are to be used. Ferrous fittings are made for threaded, welded, or flanged joints. Nonferrous fittings are made for threaded, solder, flared, or flanged joints. The common types of fittings for threaded joints are shown in Fig. 26.5, for welded joints in Fig. 26.6,

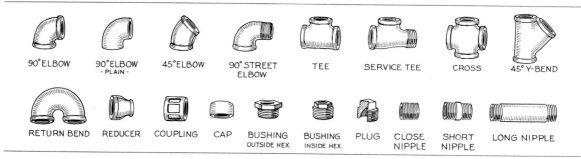

Fig. 26.5   Screwed Fittings.

for flanged joints in Fig. 26.7, and for solder joints in Fig. 26.3.

Where both or all ends of a fitting are the same nominal size, the fitting is designated by the nominal size and the description—for example, a 2″ *screwed tee*. Where two or more ends of a fitting are not the same nominal size, the fitting is designated as a *reducing fitting*, and the

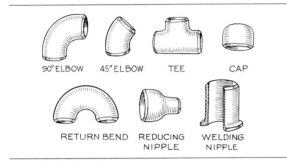

Fig. 26.6 Butt-Welded Fittings.

dimensions of the run precede those of the branches, and the dimension of the larger opening precedes that of the smaller opening—for example, a 2″ × 1½″ × 1″ *screwed reducing tee*. See Fig. 26.8 for typical designations.

The threads of screwed fittings conform to the pipe thread with which they are to be used, either male or female as the case may be. See §26.7.

Dimensions of 125 lb cast-iron screwed fittings, 250 lb cast-iron screwed fittings, 125 lb cast-iron flanged fittings, and 250 lb cast-iron flanged fittings are shown in Appendixes 40–43 and 45. See §26.7 for reference to steel fittings.

26.7 Pipe Joints

The joints between pipes, fittings, and valves may be *screwed, flanged, welded,* or for nonferrous materials, joints may be *soldered.*

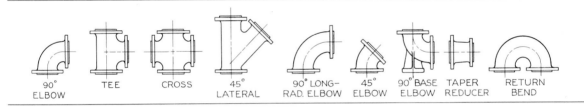

Fig. 26.7 Flanged Fittings.

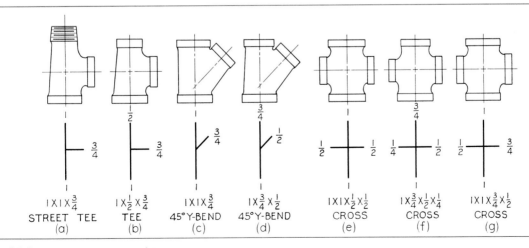

Fig. 26.8 Designating Sizes of Fittings.

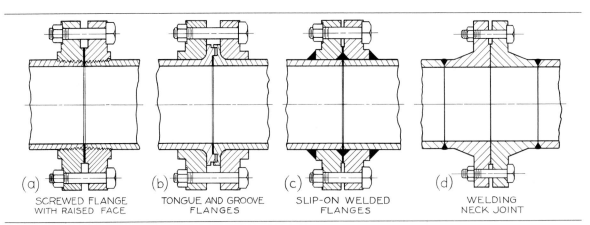

(a)	(b)	(c)	(d)
SCREWED FLANGE WITH RAISED FACE	TONGUE AND GROOVE FLANGES	SLIP-ON WELDED FLANGES	WELDING NECK JOINT

Fig. 26.9   Special Types of Flanged Joints.

The American National Standard pipe threads are illustrated in Figs. 15.20–15.22, and tabular dimensions are shown in Appendix 38. The threads of the American Petroleum Institute (API) differ somewhat from the American National Standard pipe threads. Refer to the API Standards for these differences.

Threaded joints can be made up tightly by simply screwing the cleaned threads together. However, it is common practice to use *pipe compound* in making such joints, as this provides lubrication for the threads and enables them to be screwed together more tightly. It also serves to seal irregularities, thus providing a tighter joint. Such material should be applied to the male thread only, to avoid forcing it into the pipe where contamination or obstruction may result.

Flanged joints are made by bolting two flanges together with a resilient *gasket* between the flange faces. Flanges may be attached to the pipe, fitting, or appliance by means of a screwed joint, by welding, by lapping the pipe, or by being cast integrally with the pipe, fitting, or appliance.

The faces of the flanges between which the gasket is placed have different standard *facings*, such as *flat face*, $\frac{1}{16}''$ *raised face*, $\frac{1}{4}''$ *raised face*, *male and female*, *tongue and groove*, and *ring joints*. Flat face and $\frac{1}{16}''$ raised face are standard for cast-iron flanges in the 125 lb and 250 lb classes, respectively. The other types of facing are standard for steel flanges.

The number and size of the bolts joining these flanges vary with the size and the working pressure of the joint. Bolting for Class 125 cast-iron and Class 250 cast-iron flanges is shown in Appendixes 43 and 46, respectively.

For dimensions of the various flange facings and for flange and bolting dimensions of the various sizes and pressure standards of steel flanges, refer to the American National Standard for Steel Pipe Flanges and Flanged Fittings [ANSI/ASME B16.5–1988], which is too voluminous to be included here. Some typical types of flanged joints are shown in Fig. 26.9.

Piping construction employing welded joints is in almost universal use today, particularly for higher pressure and temperature conditions. Such joints may either be *socket welded* or *butt welded*, Fig. 26.10. Socket-welded joints are limited to use in small sizes. The contours of the butt-welded joints shown at (b) and (c) are those shown in American National Standard B16.25–1986.

26.8 Valves

Valves are used to stop or to regulate the flow of fluids in a pipe line. The more common types are *gate valves*, *globe valves*, and *check valves*. Other types of valves, such as *pressure-reducing valves* and *safety valves*, are special devices used to maintain automatically a desired lower

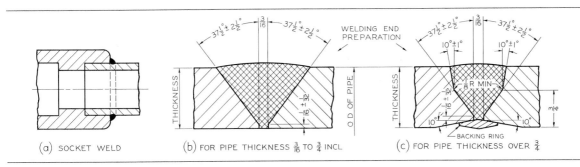

Fig. 26.10 Welded Joints.

pressure on the downstream side of the valve or to prevent automatically undesirable overpressure, respectively.

26.9 Globe Valves

Globe valves have approximately spherical bodies with the seating surface at either a right or an acute angle to the center line of the pipe, Fig. 26.11 (a). In such a valve the flowing fluid must make abrupt turns in the body, thus resulting in considerably higher pressure loss than for a gate valve.

Globe valves are commonly used where close regulation of flow is desired, because they lend themselves better to this type of regulation and are less subject to cutting action in throttling service than gate valves.

Valves of the *inside screw* and *outside screw and yoke* (O S & Y) types are available, §26.11. Angle valves and needle valves are special designs of the general class of globe valves, Fig. 26.11 (c).

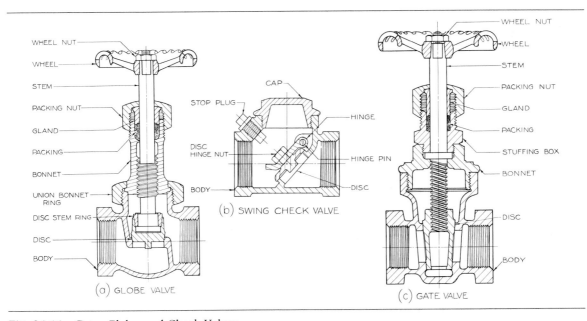

Fig. 26.11 Gate, Globe, and Check Valves.

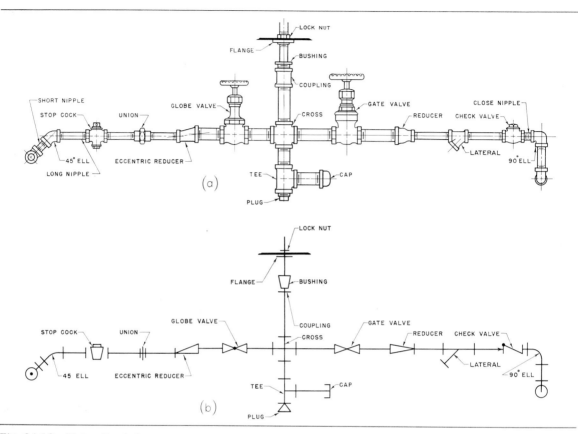

Fig. 26.12   Piping Symbols.

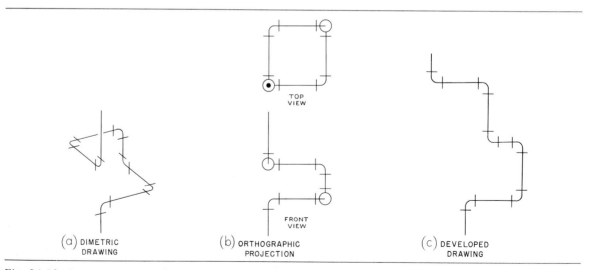

Fig. 26.13   Representations of Pipe Expansion Joint.

26.10 Check Valves

Check valves are used to limit the flow of fluids to one direction only. The disk may be hinged so as to swing partially out of the stream, Fig. 26.11 (b), or it may be guided in such a manner that it can rise vertically from its seat. The two types are called *swing checks* and *lift checks,* respectively.

26.11 Gate Valves

Gate valves have full-sized straightway openings that offer small resistance to the flow of fluids. The gate, or disk, may rise on the stem (*inside screw* type), Fig. 26.11 (c), or the gate may rise with the stem, which in turn rises out of the body (*rising stem,* or *outside screw and yoke—O S & Y*—type). Inside screw type valves are employed in the smaller sizes and lower pressures.

Seating may be on nonparallel seats, in which case the disk is solid and wedge shaped. There is also a type of gate valve employing parallel seats. In this type two disks are hung loosely on the stem and are free of the seats until an adjusting wedge reaches a lug at the closed position of the valve, when further movement of the stem causes the wedge to spread the disks and form a tight joint on the parallel seats. Such valves are used only on low-pressure and low-temperature services.

26.12 American National Standard Code for Pressure Piping

The American National Standards Institute has adopted an American National Standard Code for Pressure Piping [ANSI/ASME B31.1–1989]. This compilation of recommended practices and minimum safety standards covers various types of piping, such as power piping, industrial gas and air piping, oil refinery piping, oil transportation piping, refrigerating piping, chemical industry process piping, and gas transmission and distribution piping.

26.13 Piping Drawings

To simplify the preparation of working drawings of piping systems, the set of symbols shown in Appendix 34 has been developed to represent the various pipe fittings and valves in common use. An application of these symbols in a piping drawing is shown in Fig. 26.12 (b). The use of templates to draw piping symbols is convenient and timesaving.

Drawings of piping systems may be made as *single-line* drawings, Figs. 26.12 (b) and 26.13–26.15, or as *double-line* drawings, Figs. 26.12

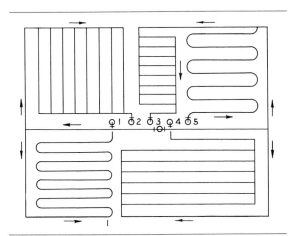

Fig. 26.14 Pipe Grids and Serpentine Coils for a Panel Radiant Heating System.

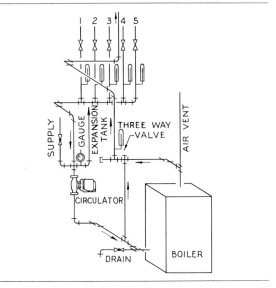

Fig. 26.15 Schematic Drawing of Piping Connecting Boiler to Heating Coils.

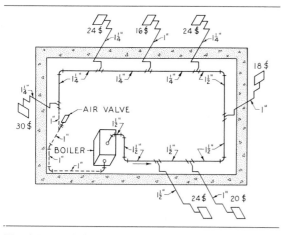

Fig. 26.16 A One-Pipe Steam Heating System.

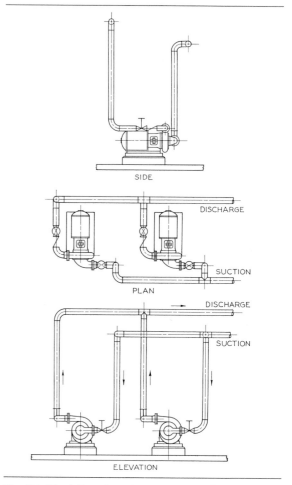

Fig. 26.17 A Two-Line Piping Drawing for a Pumping Plant.

(a), 26.17, and 26.18. Either type of drawing may be made as multiview projections, Figs. 26.13 (b), 26.17, and 26.19; as axonometric projections, Figs. 26.13 (a) and 26.20; or as oblique projections, Figs. 26.15 and 26.16. The oblique projection in Fig. 26.16 is a modified form of oblique projection generally used in representing the piping arrangement for heating systems. In these cases, the pipe mains are shown in plan and the risers in oblique projection in various directions so as to make the representation as clear as possible.

In most installations, some pipes are vertical and some are horizontal. If the vertical pipes are assumed to be revolved into the horizontal plane or the horizontal pipes revolved into the vertical plane by turning some of the fittings, Fig. 26.13 (c), the entire installation can be shown in one plane. Such a drawing is a *developed piping drawing*.

In complicated systems where a large amount of piping of various sizes is run in close proximity, and where clearances are important, the use of double-line multiview drawings, made accurately to scale, is desirable. The use of such drawings, which show the relative positions of component parts in all views, greatly reduces the probability of interferences when the piping is erected and is almost a necessity where piping components are prefabricated in a shop and sent to the job in finished dimensions. Such method of fabrication is universal in large systems using large piping, most piping $2\frac{1}{2}''$ and larger being shop fabricated. See Figs. 26.17 and 26.18.

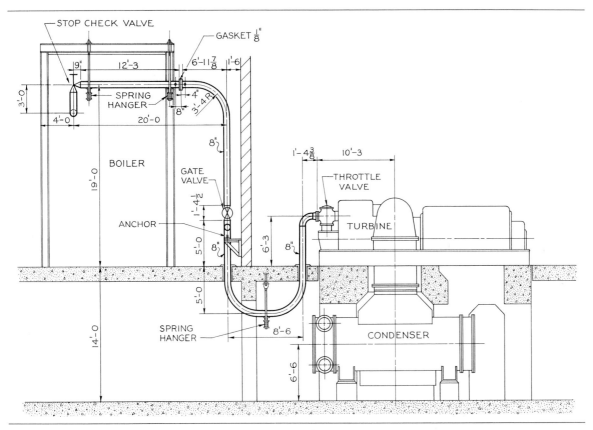

Fig. 26.18   A Dimensioned Piping Drawing—Side View of Steam Piping.

26.14 Dimensioning

In dimensioning a piping drawing, distances from center to center (c to c), center to end (c to e), or end to end (e to e) of fittings or valves and the lengths of all straight runs of pipe should be given, Fig. 26.18. Fully dimensioned single-line drawings need not be drawn to scale. Allowances in pipe lengths for makeup in fittings and valves must be made in preparing a bill of materials. All double-line drawings should have center lines if they are to be dimensioned. The size of the pipe for each run is shown by a numeral or by a note at the side of the pipe, with a leader when necessary.

26.15 Computer Graphics

Computer-aided drafting techniques, as described in Chapters 3 and 8, can produce complicated piping drawings efficiently. Using a CAD system with appropriate software, the drafter/designer can create a piping drawing on the terminal's video display and then have a drawing, such as shown in Fig. 26.19, produced on a computer-driven plotter.

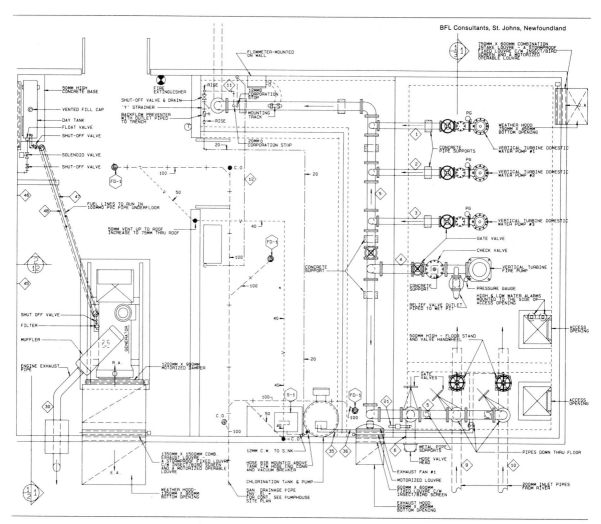

Fig. 26.19   Piping System Drawing Produced with the VersaCAD Advanced Drafting Software Package. *Courtesy of VersaCAD.*

PIPING DRAWING PROBLEMS

The drawings for Probs. 26.1–26.4 are to be three times as large as the corresponding illustrations in the book.

Prob. 26.1 Make a double-line drawing, similar to Fig. 26.12 (a), showing the following fittings: a union, a 45° Y-bend, an eccentric reducer, a globe valve, a tee, a stopcock, and a 45° ell. Use $\frac{1}{2}''$ and $1''$ wrought-steel pipe and 125 lb C.I. screwed fittings.

Prob. 26.2 Make a single-line drawing, similar to Fig. 26.12 (b), showing the following fittings: a 45° ell, a union, a 45° Y-bend, an eccentric reducer, a tee, a reducer, a gate valve, a plug, a cap, and a cross.

Prob. 26.3 Make a single-line drawing of the system of pipe coils and grids shown in Fig. 26.14; show, by their respective standard symbols, the elbows and tees that must be used to connect pipes meeting at right angles if welding is not used to make the joints.

Prob. 26.4 Make an oblique projection, similar to that shown in Fig. 26.15, of the one-pipe steam heating system shown in Fig. 26.16. Show the pipes by single lines, the fittings by their standard symbols, and the boiler and radiators as parallelepipeds.

Prob. 26.5 Make a single-line isometric drawing of the piping layout shown in Fig. 26.20. Use

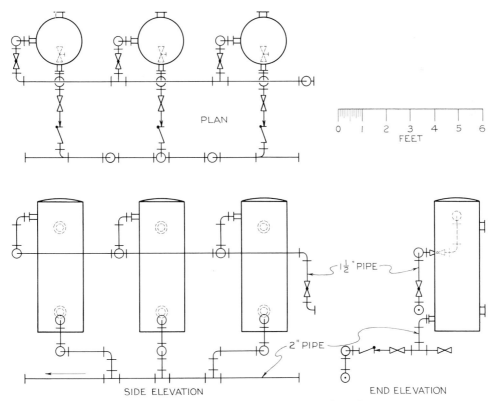

Fig. 26.20 Pipe Layout for Battery of Air Receivers (Probs. 26.5 and 26.7).

a scale of $\frac{3}{4}'' = 1'-0$, and a Size C or A2 sheet. (Similar to isometric layout shown in Fig. 26.21.)

Prob. 26.6   Make a single-line multiview drawing of the piping layout shown in Fig. 26.21. Use a scale of $1'' = 1'-0$ and a Size C or A2 sheet. (Similar to piping layout shown in Fig. 26.20.)

Prob. 26.7   Make a double-line multiview drawing of the piping layout shown in Fig. 26.20, to

a scale selected by the student. (Similar to two-line piping drawing shown in Fig. 26.17.)

Prob. 26.8   Make a double-line multiview drawing of the piping layout shown in Fig. 26.21, to a scale selected by the student. Use Schedule 80 wrought-steel pipe throughout, with Class 250 C.I. flanged fittings where pipe is larger than 2'', and Class 250 C.I. screwed fittings where pipe is 2'' and smaller. (Similar to two-line piping drawing shown in Fig. 26.17.)

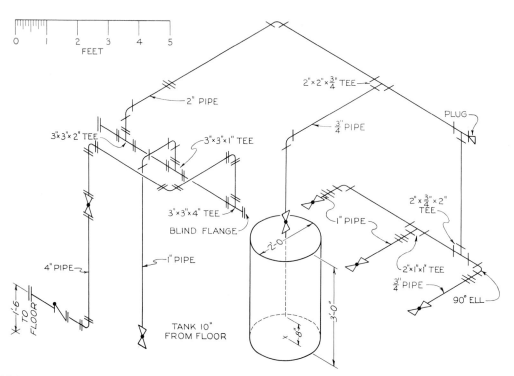

Fig. 26.21   Isometric Pipe Layout   (Probs. 26.6 and 26.8).

C H A P T E R   2 7

Welding Representation

Welding is often the method of choice for fastening parts together permanently rather than bolts, screws, rivets, or other fasteners. It is also widely used in fabricating machine parts or other structures that formerly would have been formed by casting or forging, and it is used to a considerable extent in the erection of structural steel frames for buildings, ships, and other structures.

27.1 Importance of the Drawing

Since welding is used so extensively, and for so large a variety of purposes, it is essential to have an accurate method of showing on the working drawings of machines or structures the exact types, sizes, and locations of welds desired by the designer. The old practice of simply lettering a note on the drawing, "To be welded throughout" or "To be completely welded," which actually shifted responsibility for welding control to the welding shop, not only is dangerous but also may be unnecessarily expensive because shops will usually "play safe" by welding more than necessary.

To provide a means for placing complete welding information on the drawing in a simple manner, a system of welding symbols was developed by the American Welding Society and originally published in 1947 under the title "Standard Welding Symbols." These symbols were adopted by the American National Standards Institute and are currently published as ANSI/AWS A2.4–1986. The text and illustrations of this chapter are based primarily on this standard.

A typical welding drawing is shown in Fig. 27.1. It is an assembly drawing in the sense that

it is composed of a number of separate pieces fastened together as a unit. The welds themselves are not drawn, but are clearly and completely indicated by the welding symbols. The joints are all shown as they would appear before welding. Dimensions are given to show the sizes of the individual pieces to be cut from stock. Each component piece is identified by encircled numbers and by specifications in the parts list, as shown.

27.2 Welding Processes

Three of the principal methods of welding are the oxyacetylene method, generally known as *gas welding*, the electric-arc method, generally known as *arc welding*, and electric-resistance welding, generally called *resistance welding*.

Gas welding was originated in 1895, when the French chemist Le Châtelier discovered that the combustion of acetylene gas with oxygen produced a flame of very high temperature—high enough to melt metals. This discovery was soon followed by the development of practical methods of producing and transporting oxygen and acetylene, and by the construction of suitable torches and welding rods.

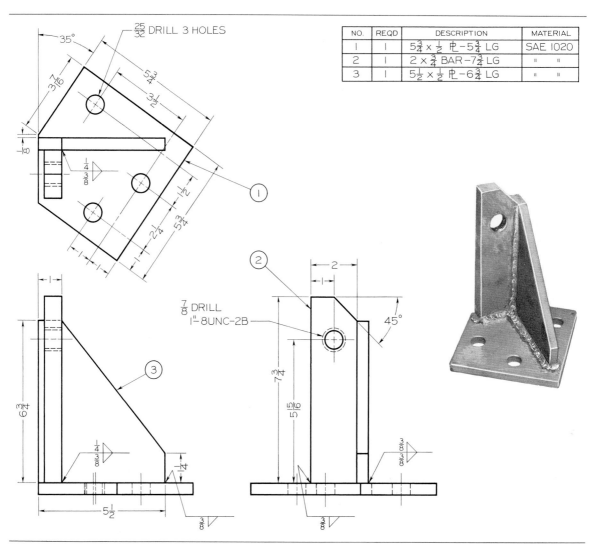

Fig. 27.1 Welding Drawing. *Courtesy of The Lincoln Electric Co.*

In arc welding, the heat of an electric arc is used to fuse the metals that are to be welded or cut. The first arc welding was done in 1881 by De Meritens in France, and for a number of years the development of arc welding was very slow. During World War I, the U.S. Navy used welding to a limited extent for repairing machinery. Since that time, as the result of extensive research, basic improvements have been made in the manufacture of electrodes and in the me-

chanical equipment used for welding, so that now arc and gas welding are important construction processes in industry.

In resistance welding, two pieces of metal are held together under some pressure, and a large amount of electric current is passed through the parts. The resistance of the metals to the passage of the current causes great heating at the junction of the two pieces, resulting in the welding of the metals.

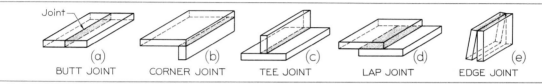

Fig. 27.2   The Basic Types of Welded Joints.

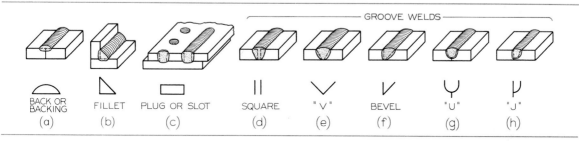

Fig. 27.3   Basic Welds and Symbols.

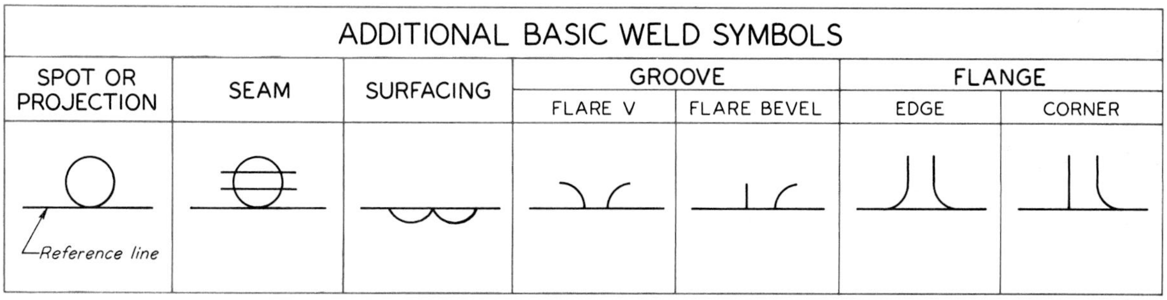

ADDITIONAL BASIC WELD SYMBOLS						
SPOT OR PROJECTION	SEAM	SURFACING	GROOVE		FLANGE	
			FLARE V	FLARE BEVEL	EDGE	CORNER
(Reference line)						

Fig. 27.4   Additional Basic Weld Symbols.

27.3   Types of Welded Joints

Fig. 27.2   There are five basic types of welded joints, classified according to the positions of the parts being joined. A number of different types of welds are applicable to each type of joint, depending on the thickness of metal, the strength of joint required, and other considerations.

27.4   Types of Welds

Fig. 27.3   The four types of arc and gas welds are the *back* or *backing*, (a); the *fillet*, (b); the *plug* or *slot*, (c); and *groove*. The groove welds are further classified as square, V, bevel, U, and

J, shown in (d)–(h). More than one type of weld may be applied to a single joint. For example, a V weld may be on one side and a back weld on the other side. Frequently, the same type of weld is used on opposite sides, forming such welds as a double-V, a double-U, or a double-J.

The four basic resistance welds are the *spot weld, projection weld, seam weld,* and *flash* or *upset weld.* Except for the flash or upset weld, the corresponding symbols for these welds and additional basic weld symbols for surfacing, groove, and flange joints are given in Fig. 27.4. See Fig. 27.25 (d) for the use of the square groove weld symbol for the flash or upset resist-

SUPPLEMENTARY SYMBOLS						
WELD ALL AROUND	FIELD WELD	MELT-THRU	BACKING OR SPACER MATL	CONTOUR		
				FLUSH	CONVEX	CONCAVE

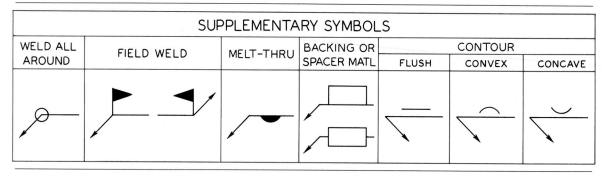

Fig. 27.5   Supplementary Symbols.

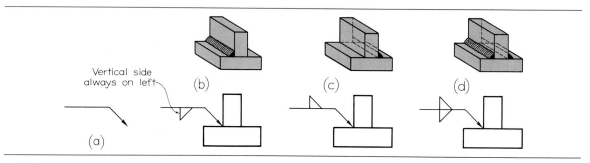

Fig. 27.6   Welding Symbols.

ance weld. Supplementary symbols are represented in Fig. 27.5.

27.5   Welding Symbols

Fig. 27.6   The basic element of the symbol is the "bent" arrow, (a). The arrow points to the joint where the weld is to be made, (b), and attached to the reference line, or shank, of the arrow is the weld symbol for the desired weld. The symbol would be one of those illustrated in Figs. 27.3 and 27.4. In this case a fillet weld symbol has been used.

The weld symbol is placed below the reference line if the weld is to be on the *arrow side* of the joint as at (b), or above the reference line if the weld is to be on the *other side* of the joint as at (c). If the weld is to be on *both* the arrow side and the other side of the joint, weld symbols

are placed on both sides of the reference line, (d). This rule for placement of the weld symbol is followed for all the arc or gas weld symbols.

When a joint is represented by a single line on a drawing, as in the top and side views of Fig. 27.7 (b), the *arrow side* of the joint is regarded as the "near" side to the reader of the drawing, according to the usual conventions of drafting.

For the plug, slot, seam, and projection welding symbols, the arrow points to the outer surface of one of the members at the center line of the weld. In such cases, the arrow side of the joint is the one to which the arrow points, or the "near" side to the reader. See Figs. 27.20 and 27.24.

Note that for all fillet or groove symbols, the *vertical side of the symbol is always drawn on the left,* Fig. 27.6 (b).

For best results, the welding symbols should be drawn mechanically, but in certain cases

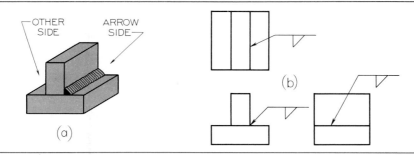

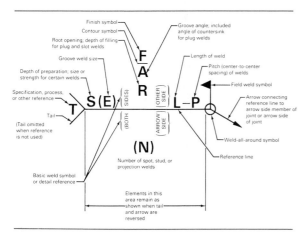

Fig. 27.7 Arrow Side and Other Side.

Fig. 27.8 The Standard Locations of the Elements of the Welding Symbol [ANSI/AWS A2.4–1986].

where necessary, the symbols may be drawn freehand.

The complete welding symbol, enlarged, is shown in Fig. 27.8. See Appendix 32 for additional welding symbol details. In practice, some companies will need to use only a simple symbol composed of the minimum elements, the arrow and the weld symbol; others will make use of additional components.

Reference to a specification, process, or other supplementary information is indicated by any desired symbol in the tail of the arrow, Fig. 27.9 (a). Otherwise a general note may be placed on the drawing, such as

UNLESS OTHERWISE INDICATED, MAKE ALL WELDS PER SPECIFICATION NO. XXX.

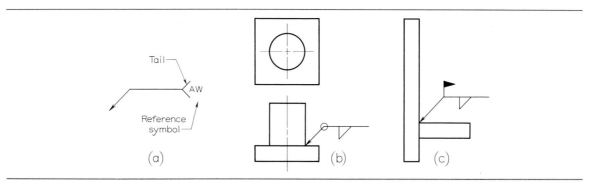

Fig. 27.9 Welding Symbols.

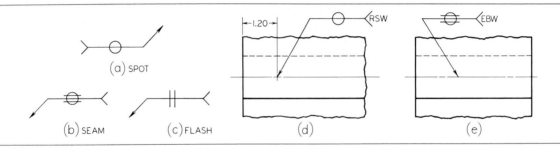

Fig. 27.10   Spot, Seam, and Flash Welding Symbols.

If no reference is indicated in the symbol, the tail may be omitted.

To avoid repeating the same information on many welding symbols on a drawing, general notes may be used, such as

FILLET WELDS $\frac{5}{16}''$ UNLESS OTHERWISE

INDICATED

or

ROOT OPENINGS FOR ALL GROOVE WELDS $\frac{3}{16}''$

UNLESS OTHERWISE INDICATED

Welds extending completely around a joint are indicated by an open circle around the elbow of the arrow, Fig. 27.9 (b). *When the weld-all-around symbol is not used, the welding symbol is understood to apply between abrupt changes in direction of the weld, unless otherwise shown.* A short vertical staff with a solid trian-gular flag at the elbow of the arrow indicates a weld to be made "in the field" (on the site) rather than in the fabrication shop, as shown at (c).

Spot, seam, flash, or upset symbols usually do not have *arrow-side* or *other-side* significance, and are simply centered on the reference line of the arrow, Fig. 27.10 (a), (b), and (c). Spot and seam symbols are shown on the drawing as indicated at (d) and (e). Note the required process must be specified in the tail of the symbol (RSW = resistance spot weld, EBW = electron beam weld).

For bevel- or J-groove welds, the arrow should point with a definite change of direction, or break, *toward the member* that is to be beveled or grooved, as shown in Fig. 27.11 (a) and (b). In this case, the upper member is grooved. The break is omitted if the location of the bevel or groove is obvious.

Lettering for the symbols should be placed to read from the bottom or from the right side of the drawing in accordance with the aligned sys-

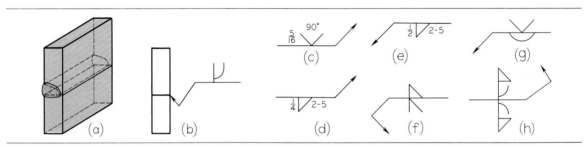

Fig. 27.11   Welding Symbols.

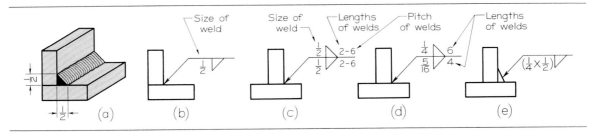

Fig. 27.12   Dimensioning of Fillet Welds.

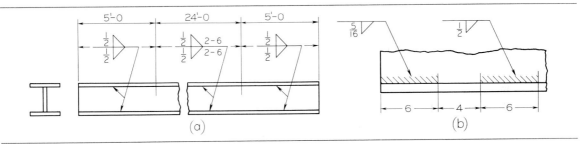

Fig. 27.13   Lengths of Fillet Welds.

tem, as shown at (c), (d), and (e). Dimensions may be indicated on a drawing in the fractional, decimal-inch, or metric system.

When a joint has more than one weld, the combined symbols are used, as at (f), (g), and (h).

27.6 Fillet Welds

Fig. 27.12   The usual fillet weld has equal legs, (a). The size of the weld is the length of one leg and is indicated, (b), by a dimension figure (fraction, decimal-inch, or metric) at the left of the weld symbol. For fillet welds on both sides of a joint, the dimensions may be indicated on one side or both sides of the reference line, (c). The lengths of the welds and the pitch (center-to-center spacing of welds) are indicated as shown. When the welds on opposite sides are different in size, the sizes are given as shown at (d). If a fillet weld has unequal legs, the weld orientation

is shown on the drawing, if necessary, and the lengths of the legs given in parentheses to the left of the weld symbol, as at (e). If a general note is given on the drawing, such as

ALL FILLET WELDS $\frac{5}{16}$″ UNLESS OTHERWISE NOTED

the size dimensions are omitted from the symbols.

No length dimension is needed for a weld that extends the full distance between abrupt changes of direction. For each abrupt change in direction, an additional arrow is added to the symbol, except when the weld-all-around symbol is used.

Fig. 27.13   Lengths of fillet welds may be indicated by symbols in conjunction with dimension lines, (a). The extent of fillet welding may be shown graphically by means of section lining, (b), if desired.

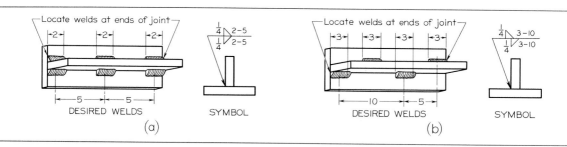

Fig. 27.14   Intermittent Welds.

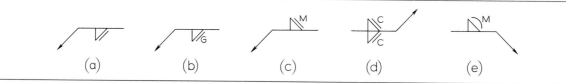

Fig. 27.15   Surface Contour of Fillet Welds.

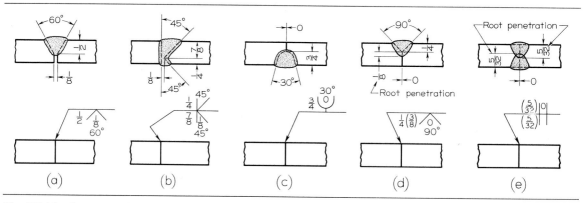

Fig. 27.16   Groove Welds.

Fig. 27.14   Chain intermittent fillet welding is indicated as shown in (a). If the welds are staggered, the weld symbols are staggered, (b).

Fig. 27.15   Unfinished flat-faced fillet welds are indicated by adding the flush symbol, Fig. 27.5, to the weld symbol, as shown at (a). If fillet welds are to be made flat-faced by mechanical means, the flush-contour symbol and the user's standard finish symbol are added to the weld symbol, (b), (c), and (d). These finish symbols indicate the method of finishing (C = chipping, G = grinding, M = machining, R = rolling, H = hammer-

ing) and not the degree of finish. If fillet welds are to be finished to a convex contour, the convex-contour symbol is added, together with the finish symbol, as shown at (e).

27.7   Groove Welds

Fig. 27.16   Various groove welds are shown above, and the corresponding symbolic representations below. The sizes of the groove welds (depth of the V, bevel, U, or J) are indicated on the left of the weld symbol. For example, at (a) the size of the V-weld is $\frac{1}{2}''$, at (b) the sizes are

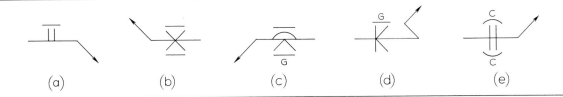

Fig. 27.17 Surface Contour of Groove Welds.

$\frac{1}{4}''$ and $\frac{7}{8}''$, at (c) the size is $\frac{3}{4}''$, and at (d) the size is $\frac{1}{4}''$. For the symbol at (d), the size is followed by $\frac{1}{8}''$, which is the additional "root penetration" of the weld. At (e), the root penetration is $\frac{5}{32}''$ from zero, or from the outside of the members. Note the overlap of the root penetration in this case.

The root opening or space between members, when not covered by a company standard, is shown within the weld symbol. At (a) and (b) the root openings are $\frac{1}{8}''$. At (c), (d), and (e) the openings are zero.

The groove angles, when not covered by a company standard, appear just outside the openings of the weld symbols, (a)–(d).

A general note may be used on the drawing to avoid repeating the symbols, such as

ALL V-GROOVE WELDS TO HAVE 60° GROOVE
ANGLE UNLESS OTHERWISE SHOWN

However, when the dimensions of one or both of two opposite welds differ from the general note, both welds should be completely dimensioned.

When single-groove or symmetrical double-groove welds extend completely through, the size need not be added to the welding symbol. For example, in Fig. 27.16 (a), if the V-groove extended entirely through the joint, the depth or size would be simply the thickness of the stock and would not need to be indicated in the welding symbol.

Fig. 27.17 When groove welds are to be approximately flush without finishing, the flush-contour symbol, Fig. 27.5, is added to the weld symbols, (a) and (b). If the welds are to be machined, the flush-contour symbol and the user's standard finish symbol are added to the weld symbol, (c) and (d). These finish symbols indicate the method of finishing (C = chipping, G = grinding, M = machining), and not the degree of finish. If a groove weld is to be finished with a convex contour, the convex-contour and finish symbols are added, as at (e).

27.8 Back or Backing Welds

Fig. 27.18 The bead-type welds used as back or backing welds on single-groove welds are indicated, (a), by a back or backing symbol opposite the groove weld symbol. Dimensions of

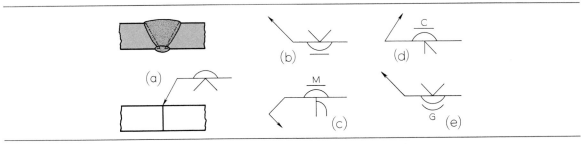

Fig. 27.18 Back or Backing Weld Symbols.

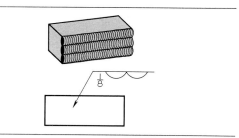

Fig. 27.19   Surface Weld Symbol.

such back or backing welds are not shown on the symbol, but may be shown, if necessary, directly on the drawing.

When back or backing welds are to be approximately flush without machining, the flush-contour symbol is included in the weld symbols, as shown at (b). If they are to be machined, the user's finish symbol is added, (c) and (d). If the welds are to be finished with a convex contour, the convex-contour symbol and the finish symbol are included in the weld symbol, as at (e).

27.9   Surface Welds

Fig. 27.19 The surface weld symbol is used to indicate a surface to be built up by welding, whether by single- or multiple-pass bead-type welds. Since this symbol does not indicate a welded joint, there is no arrow-side or other-side significance; hence, the symbol is always drawn

below the reference line. The minimum height of the weld deposit is indicated at the left of the weld symbol, as shown, except where no specific height is required. When a specific area of a surface is to be built up, the dimensions of the area are given directly on the drawing.

27.10   Plug and Slot Welds

As shown in Fig. 27.3, the same symbol is used for plug welds and slot welds. A hole or a slot is made in one member to receive the weld, as shown in Fig. 27.20 (a) and (d). If the hole or slot is in the arrow-side member, the weld symbol is placed below the reference line, (b) and (c); if in the other-side member, the weld symbol is placed above the line, (e) and (f).

The size of a plug weld, which is the smallest diameter of the hole if countersunk, is placed at the left of the weld symbol, as shown at (b) and (c). If the included angle of countersink of plug welds is in accordance with the user's standard, it is omitted from the welding symbol; otherwise, it is indicated adjacent to the weld symbol, as shown at (b) and (c).

A plug weld is understood to fill the depth of the hole unless the depth of plug is indicated by a number, in inches or metric units, inside the weld symbol, Fig. 27.21 (a). The pitch (center-to-center spacing) of plug welds is shown in inches or metric units at the right of the weld symbol, as shown at (b). If the weld is to be ap-

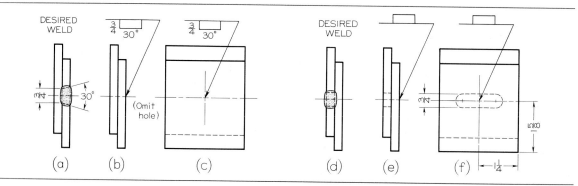

Fig. 27.20   Plug and Slot Welds.

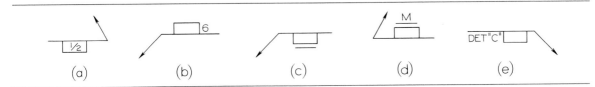

Fig. 27.21   Plug Welds.

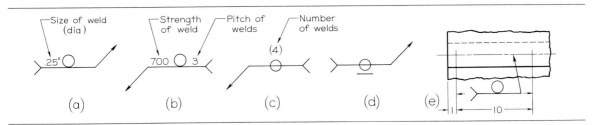

Fig. 27.22   Spot Welds.

proximately flush without finishing, the flush-contour symbol is added, (c). If the weld is to be made flush by mechanical means, a finish symbol is added, as at (d). The flush-contour and finish symbols are used in the same manner for slot welds and for plug welds.

The depth of filling of slot welds is indicated in the same manner as for plug welds, (a). The size and location dimensions of slot welds cannot be shown on the welding symbol, and must be shown directly on the drawing, Fig. 27.20 (f), or by a detail with a reference to it on the welding symbol, as shown in Fig. 27.21 (e).

27.11   Spot Welds

Fig. 27.22   The spot weld symbol, with the required welding process indicated in the tail, may or may not have arrow-side or other-side significance. Dimensions shall be shown on the same side of the reference line as the symbol, or on either side when the symbol is centered on the reference line and no arrow-side or other-side significance is intended.

The size of a spot weld is its diameter. This value, expressed in customary or metric units,

may be shown at the left of the weld symbol on either side of the reference line, as shown at (a). If it is desired to indicate the minimum acceptable shear strength in pounds per spot, instead of the size of the weld, this value is placed at the left of the weld symbol, as shown at (b). The pitch (center-to-center spacing) is indicated in appropriate units at the right of the weld symbol, (b). In this case the spot welds are 3″ apart. If a joint requires a certain number of spot welds, the number is given in parentheses above or below the symbol, as at (c). If the exposed surface of one member is to be flush, the flush-contour symbol is added, above the symbol if it is the other-side member, and below if it is the arrow-side member, as shown at (d). The use of the welding symbol in conjunction with ordinary dimensions is shown at (e).

27.12   Seam Welds

Fig. 27.23   The seam weld symbol, with the welding process indicated in the tail, may or may not have arrow-side or other-side significance. Dimensions (customary or metric units) shall be shown on the same side of the reference line as

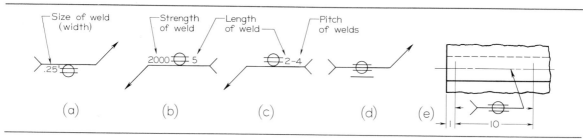

Fig. 27.23　Seam Welds.

the symbol or on either side when the symbol is centered on the reference line and no arrow-side or other-side significance is intended.

The size of a seam weld is its width. This value, expressed in customary or metric units, may be shown at the left of the weld symbol, on either side of the reference line, as shown at (a). If it is desired to indicate the minimum acceptable shear strength in pounds per linear inch, instead of the size of the weld, this value is placed at the left of the weld symbol, as shown at (b). The length of a seam weld may be shown in appropriate units at the right of the weld symbol, (b). In this case, the seam weld is 5″ long. If the weld extends the full distance between abrupt changes of direction, no length dimension in the symbol is given.

The pitch (center-to-center spacing) of intermittent seam welding is the distance between centers of lengths of welding. The pitch, in customary or metric units, is shown at the right of

the length figure, as shown at (c). In this case, the welds are 2″ long and spaced 4″ center to center.

When the exposed surface of one member is to be flush, the flush-contour symbol is added, above the symbol if it is the other-side member and below if it is the arrow-side member, as shown at (d). The use of the welding symbol in conjunction with ordinary dimensions is shown at (e).

27.13　Projection Welds

Fig. 27.24　In projection welding, one member is embossed in preparation for the weld, as shown at (a). When welded, the joint appears in section as at (b). The weld symbols, in this case, are placed below the reference lines, as shown at (c), to indicate that the arrow-side member is the one that is embossed. The weld symbols would be placed above the lines if the other member were embossed.

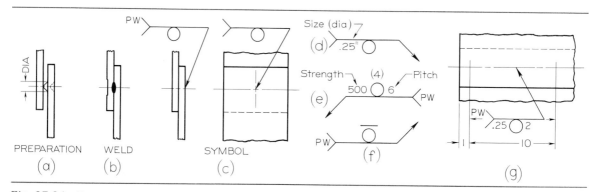

Fig. 27.24　Projection Welds.

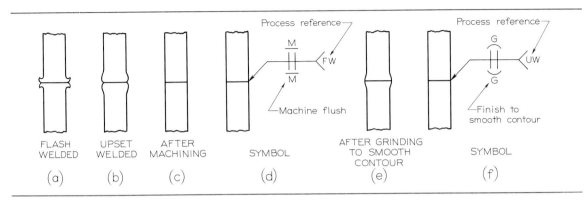

Fig. 27.25   Flash and Upset Welds.

Projection welds are dimensioned (customary or metric units) by either size or strength. The size is the diameter of the weld. This value is shown, to the left of the weld symbol, as shown at (d). If it is desired to indicate the minimum acceptable shear strength in pounds per weld, the value is placed at the left of the weld symbol, as at (e). The pitch (center-to-center spacing) is indicated in appropriate units at the right of the weld symbol, as at (e). In this case, the welds are spaced 6″ (152 mm) apart. If the joint requires a definite number of welds, the number is given in parentheses, as shown at (e). If the exposed surface of one member is to be flush, the flush-contour symbol is added, as shown at (f). The use of the welding symbol in conjunction with ordinary dimensions is shown at (g). The welding process reference is required in the tail of the symbol.

27.14   Flash and Upset Welds

Flash and upset weld symbols have no arrow-side or other-side significance, but the supplementary symbols do. A flash-welded joint is shown in Fig. 27.25 (a), and an upset welded joint at (b). The joint after machining flush is shown at (c). The complete symbol at (d) includes the weld symbol together with the flush-contour and machining symbols.

If the joint is ground to smooth contours, (e), the resulting welding drawing and symbol would be constructed as shown at (f), which includes

convex-contour and grind symbols. At either (d) or (f), the joint may be finished on only one side, if desired, by indicating the contour and machining symbols on the appropriate side of the reference line. The dimensions of flash and upset welds are not shown on the welding symbol. Note that the process reference for flash welding (**FW**) or upset welding (**UW**) must be placed in the tail of the symbol.

27.15   Welding Templates

To simplify the drawing of welding symbols, and to speed up the drafting, welding templates are available. The template shown in Fig. 27.26 has all the forms needed for drawing the arrow, the

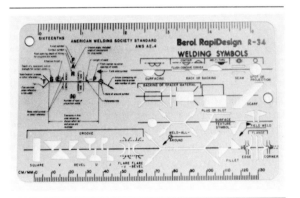

Fig. 27.26   Welding Template. *Courtesy of Berol USA, Div. of Berol Corporation.*

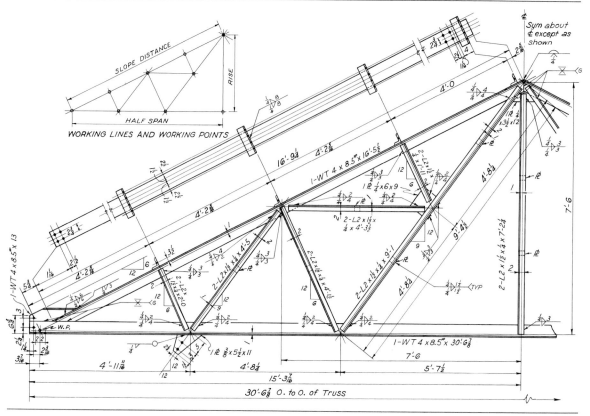

Fig. 27.27 A Welded Truss.

weld symbols, and the supplementary symbols, as well as an illustration of the complete composite welding symbol for quick reference. The symbols may be drawn in pencil or ink. For the latter, the technical fountain pen is recommended, Fig. 2.57.

27.16 Welding Applications

A typical example of welding fabrication for machine parts is shown in Fig. 27.1. In many cases, especially when only one or a few identical parts are required, it is cheaper to produce by welding than to make patterns and sand castings and do the necessary machining. Thus, welding is particularly adaptable to custom-built constructions.

Welding is also suitable for large structures that are difficult or impossible to fabricate entirely in the shop and is coming into greater use

for steel structures, such as building frames, bridges, and ships. A welded beam is shown in Fig. 24.13, and a welded assembly of diagonal bracing between two columns is shown in Fig. 24.14. A welded truss is shown in Fig. 27.27. It is easier to place members in such a welded truss so that their center-of-gravity axes coincide with the working lines of the truss than is the case in a riveted truss. The student should compare this welded truss with the riveted truss of Fig. 24.11.

27.17 Computer Graphics

Welding symbols libraries available to the CAD user, Fig. 27.28, allow for rapid application of accurate, uniform symbols that are in compliance with AWS standards, Fig. 27.29. In addition to standard symbols, many CAD programs permit the operator to create custom symbols as required.

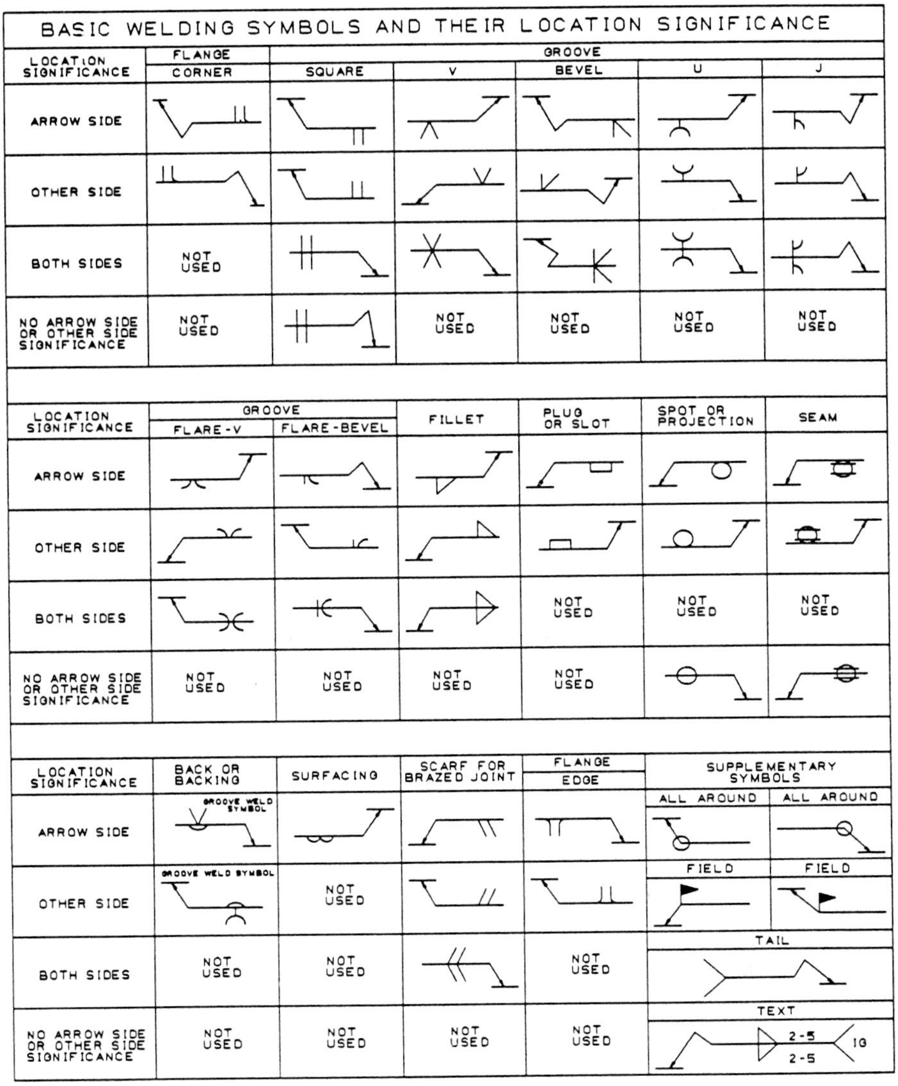

BASIC WELDING SYMBOLS AND THEIR LOCATION SIGNIFICANCE

LOCATION SIGNIFICANCE	FLANGE CORNER	SQUARE	GROOVE — V	GROOVE — BEVEL	GROOVE — U	GROOVE — J
ARROW SIDE	(symbol)	(symbol)	(symbol)	(symbol)	(symbol)	(symbol)
OTHER SIDE	(symbol)	(symbol)	(symbol)	(symbol)	(symbol)	(symbol)
BOTH SIDES	NOT USED	(symbol)	(symbol)	(symbol)	(symbol)	(symbol)
NO ARROW SIDE OR OTHER SIDE SIGNIFICANCE	NOT USED	(symbol)	NOT USED	NOT USED	NOT USED	NOT USED

LOCATION SIGNIFICANCE	GROOVE FLARE-V	GROOVE FLARE-BEVEL	FILLET	PLUG OR SLOT	SPOT OR PROJECTION	SEAM
ARROW SIDE	(symbol)	(symbol)	(symbol)	(symbol)	(symbol)	(symbol)
OTHER SIDE	(symbol)	(symbol)	(symbol)	(symbol)	(symbol)	(symbol)
BOTH SIDES	(symbol)	(symbol)	(symbol)	NOT USED	NOT USED	NOT USED
NO ARROW SIDE OR OTHER SIDE SIGNIFICANCE	NOT USED	NOT USED	NOT USED	NOT USED	(symbol)	(symbol)

LOCATION SIGNIFICANCE	BACK OR BACKING	SURFACING	SCARF FOR BRAZED JOINT	FLANGE EDGE	SUPPLEMENTARY SYMBOLS	
ARROW SIDE	GROOVE WELD SYMBOL (symbol)	(symbol)	(symbol)	(symbol)	ALL AROUND (symbol)	ALL AROUND (symbol)
OTHER SIDE	GROOVE WELD SYMBOL (symbol)	NOT USED	(symbol)	(symbol)	FIELD (symbol)	FIELD (symbol)
BOTH SIDES	NOT USED	NOT USED	(symbol)	NOT USED	TAIL (symbol)	
NO ARROW SIDE OR OTHER SIDE SIGNIFICANCE	NOT USED	NOT USED	NOT USED	NOT USED	TEXT 2-5 2-5 IG	

Fig. 27.28 Computervision Production Drafting CADDS 4× Welding Symbol Library. *Courtesy of Computervision Corporation, a subsidiary of Prime Computer, Inc.*

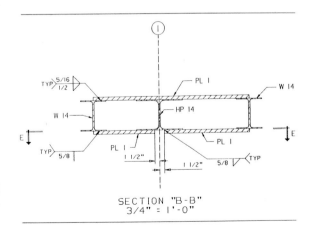

Fig. 27.29 CAD-Generated Welded Structural Detail Drawing. *Courtesy of Computervision Corporation, a subsidiary of Prime Computer, Inc.*

WELDING DRAWING PROBLEMS

The following problems are given to familiarize the student with some applications of welding symbols to machine construction and to steel structures.

Prob. 27.1 Fig. 7.60. Change to a welded part. Make working drawing, using appropriate welding symbols.

Prob. 27.2 Fig. 7.63. Same instructions as for Prob. 27.1.

Prob. 27.3 Fig. 7.68. Same instructions as for Prob. 27.1.

Prob. 27.4 Fig. 7.79. Same instructions as for Prob. 27.1.

Prob. 27.5 Fig. 7.86. Same instructions as for Prob. 27.1.

Prob. 27.6 Fig. 7.98. Same instructions as for Prob. 27.1.

Prob. 27.7 Fig. 7.107. Same instructions as for Prob. 27.1.

Prob. 27.8 Fig. 7.111. Same instructions as for Prob. 27.1.

Prob. 27.9 Fig. 7.116. Same instructions as for Prob. 27.1.

Prob. 27.10 Make a half-size drawing of the joint at the center of the lower chord of the truss in Fig. 27.27 where the chord is supported by two vertical angles. The chord is a structural tee, cut from an $8 \times 5\frac{1}{4}$, 17 lb wide-flange shape. Draw the front and side views, and show the working lines, the two angles, the structural tee, and all welding symbols.

Prob. 27.11 Make a half-size front view, showing the welding symbols, of any joint of the truss in Fig. 27.27 in which three or four members meet.

Prob. 27.12 Draw half-size front, top, and left-side views of the end joint of the truss in Fig. 27.27 showing the welding symbols.

C H A P T E R 2 8

Graphs and Diagrams

BY E. J. MYSIAK*

In previous chapters we have seen how graphical representation is used instead of words to describe the size, shape, material, and fabrication methods for the manufacture of actual objects. Graphical representation is also used extensively to present and analyze data and to solve technical problems. A pictorial or graphical presentation is much more impressive and easier to understand than a numerical tabulation or a verbal description. These graphical descriptions are synonymously termed *graphs, charts,* or *diagrams.*

The term *chart* has two meanings. It is associated with maps and also describes any of the forms of graphical presentation described in this chapter. Therefore, the term *chart* includes all graphs and diagrams. A *graph* is a special form of *chart* with data plotted on some type of grid. A *diagram* is a *chart* without the use of a grid.

Tabulated data in Fig. 28.1 (a), showing the average weekly earnings of United States manufacturing workers for the years 1974–1984, are presented as a line graph at (b) and a bar graph at (c). The greater effectiveness of graphical representation is evident.

28.1 Uses of Graphical Representation

Graphs and diagrams can be classified into two broad categories depending on whether their application or use is for technical purposes or for popular appeal.

Technically trained personnel communicate by means of graphs similar to Fig. 28.1 (b). When graphs are used to present data to lay personnel, a form similar to Fig. 28.1 (c) is preferred.

Scientific or technical personnel use graphs and diagrams to (a) present results of experimental investigations, (b) represent phenomena that follow natural laws, (c) represent equations for further computational purposes, and (d) derive equations to represent empirical test data. The forms of graphical representation most commonly used are

1. Rectangular coordinate line charts.
2. Semilogarithmic coordinate line charts.
3. Logarithmic coordinate line charts.

*Engineering Manager, Phoenix Company of Chicago, Wood Dale, IL.

833

AVERAGE WEEKLY EARNINGS OF U.S. MANUFACTURING WORKERS

Year	Current Dollars	1977 Dollars
1974	176.80	217.20
1975	190.79	214.85
1976	209.32	222.92
1977	228.90	228.90
1978	249.27	231.66
1979	268.94	224.64
1980	288.62	212.64
1981	318.00	212.00
1982	330.65	207.96
1983	354.08	216.03
1984	372.91	221.05

(a)

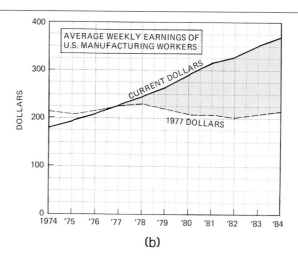

(b)

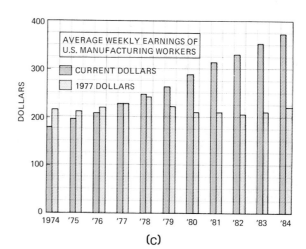

(c)

Fig. 28.1   Comparison of Tabulated and Graphical Presentations.

4. Trilinear coordinate line charts.
5. Polar coordinate line charts.
6. Nomographs or alignment charts.
7. Volume charts.
8. Rectangular coordinate distribution charts.
9. Flowcharts.

Often information and ideas must be presented in a way understandable to the layperson in order to have the information or idea ac-

cepted. For popular appeal, graphs and diagrams frequently used are

1. Rectangular coordinate line charts.
2. Bar or column charts.
3. Rectangular coordinate surface or area charts.
4. Pie charts.
5. Volume (map) charts.
6. Flowcharts.
7. Map distribution charts.

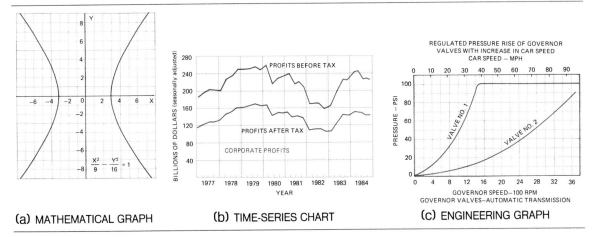

(a) MATHEMATICAL GRAPH   **(b) TIME-SERIES CHART**   **(c) ENGINEERING GRAPH**

Fig. 28.2   Rectangular Line Chart Classification.

28.2   Rectangular Coordinate Line Graphs

The rectangular coordinate line graph is the type in which values of two related variables are plotted on coordinate paper, and the points, joined together successively, form a continuous line or "curve."

The following are some of the purposes for which a line graph can be used to advantage.

1. Comparison of a large number of plotted values in a compact space.

2. Comparison of the relative movements (trends) of more than one set of data on the same graph. There should not be more than two or three curves on the same graph, and there should be some definite relationship among them.

3. Interpolation of intermediate values.

4. Representation of movement or overall trend (relative change) of a series of values rather than the difference between values (absolute amounts).

Line graphs are *not* well suited for (1) presenting relatively few plotted values in a series, (2) emphasizing changes or difference in absolute amounts, or (3) showing extreme or irregular movement of data.

Rectangular line graphs may be classified as (1) *mathematical graphs*, (2) *time-series charts*, or (3) *engineering graphs*. Any of these may have one or more curves on the same graph. If the values plotted along the axes are pure numbers (positive and negative), showing the relationship of an equation, the plot is commonly called a mathematical graph, Fig. 28.2 (a). When one of the variables is any unit of time, the chart is known as a time-series chart, (b). This is one of the most common forms, since time is frequently one of the variables. Line, bar, or surface chart forms may be used for time-series charts, line charts being the most widely used in engineering practice. The plotting of values of any two related physical variables on a rectangular coordinate grid is referred to as an engineering chart, graph, or diagram, (c).

Line curves are generally presented for any of three types of relationships.

1. Observed relationships, usually plotted with observed data points connected by straight, irregular lines, Fig. 28.3 (a).

2. Empirical relationships, (b), normally reflecting the author's interpretation of his series of observations, represented as smooth curves or straight lines fitted to the data by eye or by formulas chosen empirically.

835

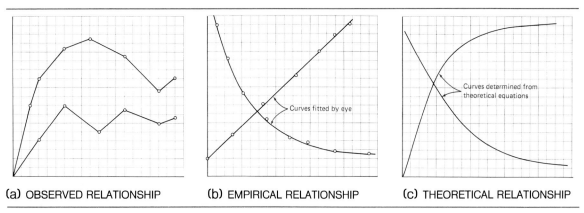

(a) OBSERVED RELATIONSHIP   (b) EMPIRICAL RELATIONSHIP   (c) THEORETICAL RELATIONSHIP

Fig. 28.3 Curve Fitting.

3. Theoretical relationships, (c), in which the curves are smooth and without point designations, though observed values may be plotted to compare them with a theoretical curve if desired. The curve thus drawn is based on theoretical considerations only, in which a theoretical function (equation) is used to compute values for the curve.

28.3 Design and Layout of Rectangular Coordinate Line Graphs

The steps in drawing a typical coordinate line graph are shown in Fig. 28.4.

I. a. Compute and/or assemble data in a convenient arrangement.

b. Select the type of graph and coordinate paper most suitable, §28.4.

c. Determine the size of the paper and locate the axes.

d. Determine the variable for each axis and choose the appropriate scales, §28.5. Letter the unit values along the axes, §28.5.

II. Plot the points representing the data and draw the curve or curves, §28.6.

III. Identify the curves by lettering names or symbols, §28.6. Letter the title, §28.7. Ink in the graph, if desired.

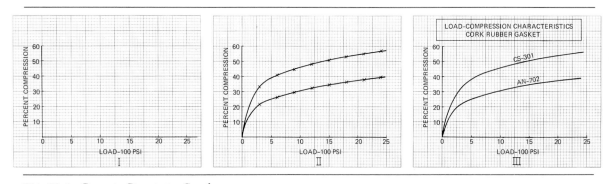

Fig. 28.4 Steps in Drawing a Graph.

The completed graph, which includes the curves, captions, and designations, should have a balanced arrangement relative to the axes. Much of the foregoing procedure is also applicable to the other forms of charts, graphs, and diagrams discussed in this chapter.

28.4 Grids and Composition

To simplify the plotting of values along the perpendicular axes and to eliminate the use of a special scale to locate them, *coordinate paper,* or "graph paper," ruled with grids, is generally used and can be purchased already printed. Alternatively, the grids can be drawn on blank paper.

Printed coordinate papers are available in various sizes and spacings of grids, $8\frac{1}{2}'' \times 11''$ being the most common paper size. The spacing of grid lines may be $\frac{1}{10}''$, $\frac{1}{20}''$, or multiples of $\frac{1}{16}''$. A spacing of $\frac{1}{8}''$ to $\frac{1}{4}''$ is preferred. Closely spaced coordinate ruling is generally avoided for publications, charts reduced in size, and charts used for slides. Much of engineering graphical analysis, however, requires (1) close study, (2) interpolation, and (3) only one copy, with possibly a few prints that can be readily prepared with little effort. Therefore, engineering graphs are usually plotted on the closely spaced, printed coordinate paper. Printed papers can be obtained in several colors of lines and in various weights and grades. A thin, translucent paper may also be used when prints are required. A special nonreproducible-grid coordinate paper is available for use when reproductions without visible grids are desired.

Scale values and captions should be placed outside the grid axes, if possible. Since printed papers do not have sufficient margins to accommodate the axes and nomenclature, the axes should be placed far enough inside the grid area to permit sufficient space for axes and lettering, as shown in Fig. 28.5 (a) and (b). As much of the remaining grid space as possible should be used for the curve—that is, the scale should be such as to spread the curve out over the available space. A title block (and tabular data, if any) should be placed in an open space on the chart, as shown. If only one copy of the chart is required, tabular data may be placed on the back of the graph or on a separate sheet.

Charts prepared for printed publications, conferences, or projection (slides) generally do not require accurate or detailed interpolation and should emphasize the major facts presented. For such graphs, coordinate grids drawn or

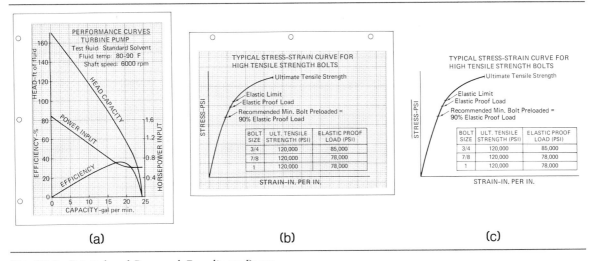

Fig. 28.5   Printed and Prepared Coordinate Paper.

Fig. 28.6   Vertical Rulings—Specially Prepared Grids.

traced on blank paper, cloth, or film have definite advantages when compared to printed paper. The charts in Fig. 28.5 (b) and (c) show the same information plotted on printed coordinate paper and plain paper, respectively. The specially prepared sheet should have as few grid rulings as necessary—or none, as at (c)—to allow a clear interpretation of the curve. For ease of reading, lettering is not placed on grid lines. The title and other data can be lettered in open areas, completely free of grid lines.

The layout of specially prepared grids is restricted by the overall paper size required, or space limitations for slides and other considerations. Space is first provided for margins and for axes nomenclature; the remaining space is then divided into the number of grid spaces needed for the range of values to be plotted. Another important consideration of composition that affects the spacing of grids is the slope or trend of the curve, as discussed in §28.5.

Since independent variable values are generally placed along the horizontal axis, especially in time-series charts, vertical rulings can be made for each value plotted, if uniformly spaced, Fig. 28.6 (a). If there are many values to plot, intermediate values can be designated by *ticks* on the curves or along the axis, (b) and (c), with the grid rulings omitted. The horizontal lines are generally spaced according to the available space and the range of values.

The weight of the grid rulings should be thick enough to guide the eye in reading the values,

but thin enough to provide contrast and emphasize the curve. The thickness of the lines generally should decrease as the number of rulings increases. As a general rule, as few rulings should be used as possible, but if a large number of rulings is necessary, major divisions should be drawn heavier than the subdivision rulings, for ease of reading.

28.5   Scales and Scale Designation

The choice of scale is the most important factor of composition and curve significance. Rectangular coordinate line graphs have values of the two related variables plotted with reference to two mutually perpendicular coordinate axes, meeting at a zero point or origin, Fig. 28.7 (a). The horizontal axis, normally designated as an *x*-axis, is called the *abscissa*. The vertical axis is denoted a *y*-axis and is called the *ordinate*. It is common practice to place independent values along the abscissa and the dependent values along the ordinate. For example, if in an experiment at certain time intervals, related values are observed, recorded, or determined, the amount of these values is dependent on the time intervals (independent or controlled) chosen. The values increase from the point of origin toward the right on the *x*-axis and upward on the *y*-axis.

Mathematical graphs, (b), quite often contain positive and negative values, which necessitates the division of the coordinate field into four

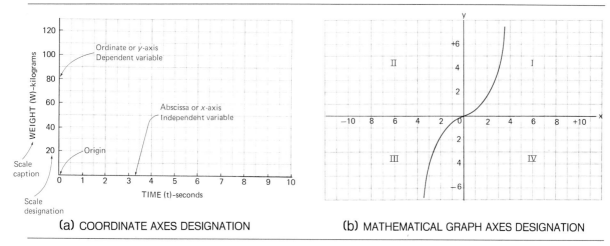

(a) COORDINATE AXES DESIGNATION   (b) MATHEMATICAL GRAPH AXES DESIGNATION

Fig. 28.7   Axes Designation.

quadrants, numbered counterclockwise as shown. Positive values increase toward the right on the x-axis and upward on the y-axis, from the origin. Negative values increase (negatively) to the left on the x-axis and downward on the y-axis.

Generally, a full range of values is desirable, beginning at zero and extending slightly beyond the largest value, to avoid crowding. The available coordinate area should be used as completely as possible. However, certain circumstances require special consideration to avoid wasted space. For example, if the values to be plotted along one of the axes do not range near zero, a "break" in the grid may be shown, as in Fig. 28.8 (a). However, when relative amount of change is required, as it is in Fig. 28.5 (a), the axes or grid should not be broken, and the zero line should not be omitted. If the absolute amount is the important consideration, the zero line may be omitted, as in Fig. 28.8 (b), (c), and (d). Time designations of years naturally are fixed and have no relation to zero.

If a few given values to be plotted are widely separated in amount from the others, the total range may be very great, and when this is com-

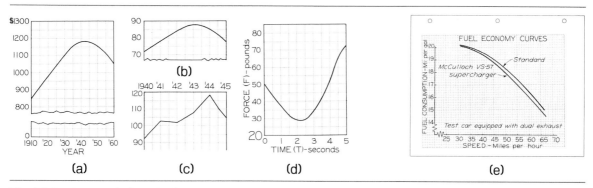

Fig. 28.8   Axes Scale "Breaks."

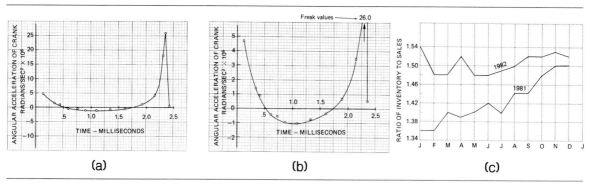

Fig. 28.9 "Freak" Values and Combined Curves.

pressed to fit on the sheet, the resulting curve will tend to be "flat," as shown in Fig. 28.9 (a). It is best to arrange for such values to fall off the sheet and to indicate them as "freak" values. The curve may then be drawn much more satisfactorily, as shown at (b).

A convenient manner in which to show related curves having the same units along the abscissa, but different ordinate units, is to place one or more sets of ordinate units along the left margin and another set of ordinate values along the right margin, as shown in Fig. 28.5 (a), using the same rulings. Multiple scales are also sometimes established along the abscissa, such as for time units of months covering multiple years, as in Fig. 28.6 (b). A more compact arrangement

for the curve is shown in Fig. 28.9 (c), where the purpose is to compare the inventory/sales ratios for 1981 and 1982 on a monthly basis.

The choice of scales deserves careful consideration, since it has a controlling influence on the depicted rate of change of the dependent variable. The *slope* of the curve (trend) should be chosen to represent a true picture of the data or a correct impression of the trend.

The slope of a curve is affected by the spacing of the rulings and their designations. A slope or trend can be made to appear "steeper" by increasing the ordinate scale or decreasing the abscissa scale, Fig. 28.10 (a), and "flatter" by increasing the abscissa scale or decreasing the ordinate scale, (b). As shown in Fig. 28.11, a

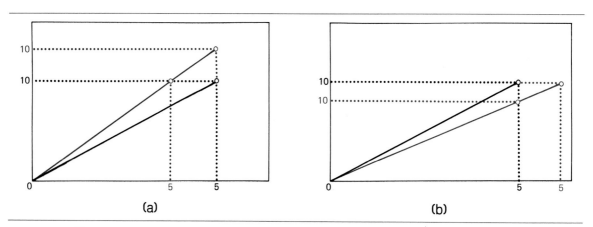

Fig. 28.10 Slopes.

840

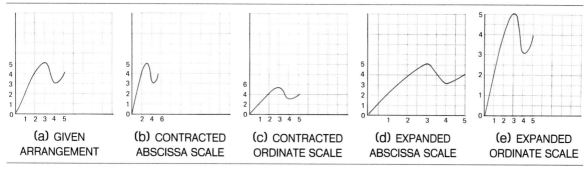

Fig. 28.11   Effects of Scale Designation.

variety of slopes or shapes can be obtained by expanding or contracting the scales. A deciding factor is the impression desired to be conveyed graphically.

Normally, an angle greater than 40° with the horizontal gives an impression of a significant rise or increase of ordinate values, while an angle of 10° or less suggests an insignificant trend, Fig. 28.12 (a) and (b). *The slope chosen should emphasize the significance of the data plotted.* Some relationships are customarily presented in a conventional shape, as shown at (c) and (d). In this case, an expanded abscissa scale, as shown at (d), should be avoided.

Scale designations should be placed outside the axes, where they can be shown clearly. Ab-

scissa nomenclature is placed along the axis so that it can be read from the bottom of the graph. Ordinate values are generally lettered so that they can also be read from the bottom; but ordinate captions, if lengthy, are lettered to be read from the right. The values can be shown on both the right and left sides of the graph, or along the top and bottom, if necessary for clearness, as when the graph is exceptionally wide or tall or when the rulings are closely spaced and hard to follow. When the major interest (e.g., maximum or minimum values) is situated at the right, the ordinate designations may be placed along the right, Fig. 28.31 (e). This arrangement also encourages reading the chart first and then the scale magnitudes.

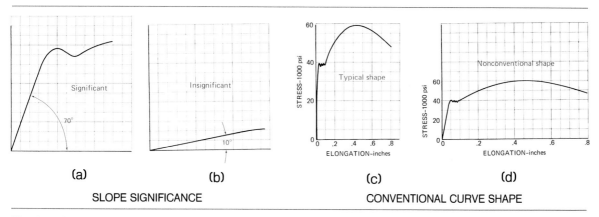

Fig. 28.12   Curve Shapes.

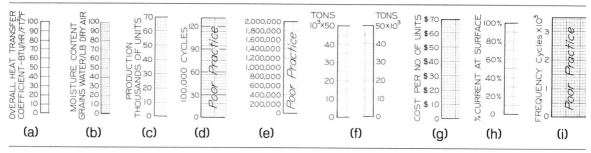

Fig. 28.13   Scale Designations.

When grid rulings are specially prepared on blank paper, every major division ruling should have its value designated, Fig. 28.13 (a). The labeled divisions should not be closer than .25″ and rarely more than 1″ apart. Intermediate values (rulings or ticks) should not be identified and should be spaced no closer than .05″. If the rulings are numerous and close together, as on printed graph paper, only the major values are noted, (b) and (c). The assigned values should be consistent with the minor divisions. For example, major divisions designated as 0, 5, 10, . . ., should not have 2 or 4 minor intervals, since resulting values of 1.25, 2.5, 3.75, . . ., are undesirable. Similarly, odd-numbered major divisions of 3, 5, 7, . . ., or multiples of odd numbers with an even number of minor divisions, should be avoided, as (d) indicates. The numbers, if three digits or smaller, can be fully given. If the numbers are larger than three digits, (e), dropping the ciphers is recommended, if the omission is indicated in the scale caption, as at (e). Values are shortened to even hundreds, thousands, or millions, in preference to tens of thousands, for example. Graphs for technical use can have the values shortened by indicating the shortened number times some power of 10, as at (f). In special cases, such as when giving values in dollars or percent, the symbols may be given adjacent to the numbers, as at (g) and (h).

Designations other than numbers usually require additional space; therefore, standard abbreviations should be used when available, Fig. 28.14. These abscissa values may be lettered vertically, as in the center at (a) and (b), or inclined, as at the right in (c), to fit the designations along the axes.

Scale captions (or titles) should be placed along the scales so that they can be read from the bottom for the abscissa, and from the right for the ordinate. Captions include the name of the variable, symbol (if any), units of measurement, and any explanation of digit omission in the values. If space permits, the designations are lettered completely, but if necessary, standard abbreviations may be used for the units of measurement. Notations such as shown in Fig. 28.13 (i) should be avoided, since it is not clear whether the values shown are to be multiplied by the power of 10 or already have been. Short captions may be placed above the values, Fig. 28.13 (f), especially when graphs are prepared for projection slides, since reading from the right is difficult.

28.6   Points, Curves, and Curve Designations

In mathematical and popular-appeal graphs, curves without designated points are commonly used, since the purpose is to emphasize the general significance of the curves. On graphs prepared from observed data, as in laboratory experiments, points are usually designated by various symbols, Fig. 28.15. If more than one curve is plotted on the same grid, a combination of these symbols may be used, one type for each curve, although labels are preferable if clear. The

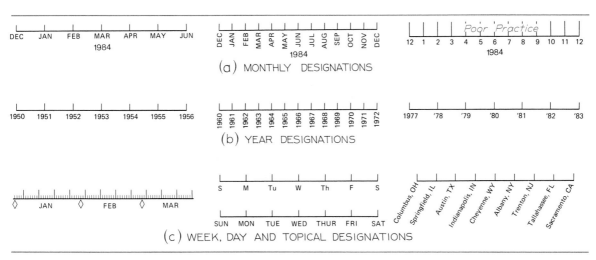

Fig. 28.14   Nonnumerical Designations.

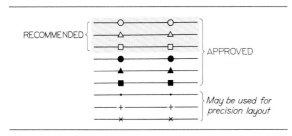

Fig. 28.15   Point Symbols.

use of open-point symbols is recommended, except in cases of "scatter" diagrams, Fig. 28.37, where the filled-in points are more visible. In general, filled-in symbols should be used only when more than three curves are plotted on the same graph and a different identification is required for each curve. The curve should not be drawn through the point symbols as they may be needed for reference later for additional information.

When several curves are to be plotted on the same grid, they can be distinguished by the use of various types of lines, Fig. 28.16 (a). However, solid lines are used for the curves wherever pos-

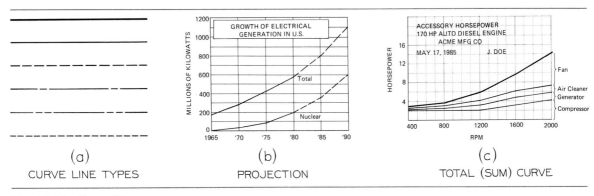

Fig. 28.16   Curve Lines.

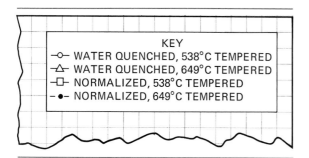

Fig. 28.17   Keys.

sible, while the dashed line is commonly used for *projections* (estimated values, such as future expectations), as shown at (b). The curve should be heavier in weight than the grid rulings, but a difference in weight can also be made between various curves to emphasize a preferred curve or a total value curve (sum of two or more curves), as shown at (c). A *key,* or *legend,* should be placed in an isolated portion of the grid, preferably enclosed by a border, to denote point symbols or line types that are used for the curves. If the grid lines are drawn on blank paper, a space should be left vacant for this information, Fig. 28.17. Keys may be placed off the grids below the title, if space permits. However, it is preferable to designate curves with labels, if possible, rather than letters, numbers, or keys, Fig. 28.2 (c). Colored lines are very effective for distinguishing the various curves on a grid, but they may not be suitable for multiple copies.

28.7   Titles

Titles for a graph may be placed on or off the grid surface. If placed on the grid, white space should be left for the title block, but if printed coordinate paper is used, a heavy border should enclose the title block. If further emphasis is desired, the title may be underlined. The contents of title blocks vary according to method of presentation. Typical title blocks include title, subtitle, institution or company, date of preparation, and name of the author, Fig. 28.16 (c).

Some relationships may be given an appropriate name. For example, a number of curves showing the performance of an engine are commonly entitled "Performance Characteristics." If two variables plotted do not have a suitable title, "Dependent variable (name) vs. independent variable (name)" will suffice. For example,

GOVERNOR PRESSURE vs. SPEED

Notes, when required, may be placed under the title for general information, Fig. 28.18 (a); labeled adjacent to the curve, (b), or along the curve, Fig. 28.20 (a); or referred to by means of reference symbols, Fig. 28.18 (c) and (d).

Any chart can be made more effective, whether it is drawn on blank or printed paper, if it is inked. For reproduction purposes, as for slides or for publications, inking is necessary.

28.8   Semilogarithmic Coordinate Line Charts

In a semilog chart, also known as a *rate-of-change* or *ratio chart,* two variables are plotted on semilogarithmic coordinate paper to form a continuous straight line or curve. Semilog paper contains uniformly spaced vertical rulings and logarithmically spaced horizontal rulings.

Semilog charts have the same advantages as rectangular coordinate line charts (arithmetic charts), §28.2. When rectangular coordinate line charts give a false impression of the trend of a curve, the semilog charts will be more effective in revealing whether the rate of change is increasing, decreasing, or constant. Semilog charts are also useful in the derivation of empirical equations.

Semilog charts, like rectangular coordinate line graphs, are not recommended for presenting only a few plotted values in a series, for emphasizing change in absolute amounts, or for showing extreme or irregular movement or trend of data.

In Fig. 28.19 (a) and (b), data are plotted on rectangular coordinate grids (arithmetic) and on semilogarithmic coordinate grids, respectively.

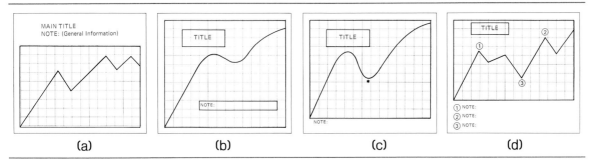

Fig. 28.18 Notes.

The same data, which produce curves on the arithmetic graph, produce straight lines on the semilog grid. The straight lines permit an easier analysis of the trend or movements of the variables. If the logarithms of the ordinate values are plotted on a rectangular coordinate grid, instead of the actual values, straight lines will result on the arithmetic graph, as shown at (c). The straight lines produced on semilog grid provide a simple means of deriving empirical equations. Straight lines are not necessarily obtained on a semilog grid, but if they do occur, it means that

the rate of change is constant, Fig. 28.20 (a). Irregular curves can be compared to constant-rate scales individually or between a series of curves, as shown at (b).

28.9 Design and Layout of Semilog Charts

Semilog graphs are usually prepared on printed semilog coordinate paper. Graphs for publication, however, requiring fewer grid rulings, are plotted on specially prepared grid scales or

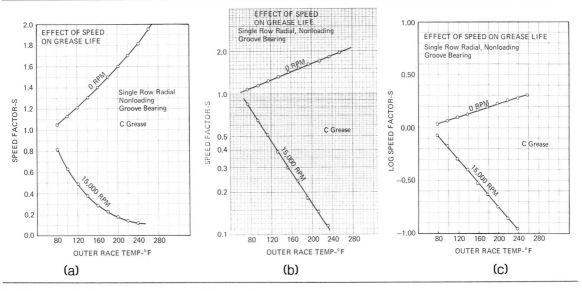

Fig. 28.19 Arithmetic and Semilogarithmic Plottings.

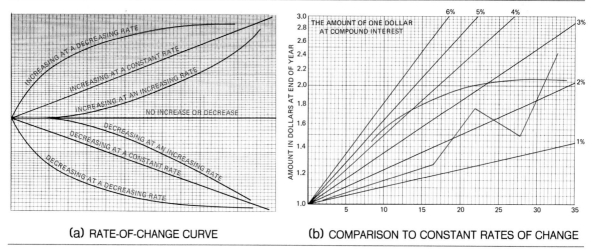

(a) RATE-OF-CHANGE CURVE   (b) COMPARISON TO CONSTANT RATES OF CHANGE

Fig. 28.20   Rates of Change.

traced from the original plot. As illustrated in Fig. 28.21 (a), the logarithmically divided scale is generally placed along the ordinate. The abscissa scale is uniformly divided into the required number of rulings for the independent variable. Logarithmic scales are nonuniform, beginning with the largest space at the bottom and decreasing in size for 10 divisions, which is known as one *log cycle*. The locations of major divisions and subdivisions for the semilog grid are determined by taking the logarithm to the base 10 (common logarithm) of the value to be plotted. As shown at (b), the major division of

one log cycle, from 1 to 10, ranges in value (logarithmically) from 0 to 1. If this logarithmic proportion is maintained, the log cycle can be laid out to any height, but the remaining divisions depend on the height chosen for one unit of log cycle. For example, if a length of 10″ for one log cycle is suitable, the number two division is at $3.01″ = (.301 \times 10″)$ from the origin of the cycle (number one division), and the number three is at 4.77″ from the origin. A convenient device for laying out logarithmic scales is a *log cycle sector* prepared on a separate sheet of paper, as shown at (c). The sector is folded along the length of

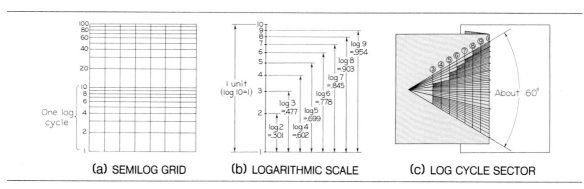

(a) SEMILOG GRID   (b) LOGARITHMIC SCALE   (c) LOG CYCLE SECTOR

Fig. 28.21   Semilog Chart Design.

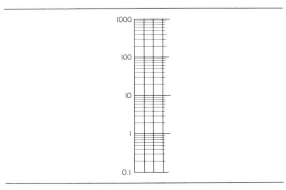

Fig. 28.22   Cycle Designation.

log cycle desired, and divisions are marked on the scale being prepared adjacent to the folded edge.

Printed semilog paper is available with as many as five log cycles on an 8.5″ × 11.0″ sheet. Nonreproducible grids for prints are also available.

The composition and layout techniques of the semilog graphs are similar to those of the rectangular coordinate graphs, discussed in §28.5.

As can be observed from the layout of a logarithmic cycle scale, the first designation (origin of cycle) is at log 1 = 0, and the last designation is at log 10 = 1. Log cycle designation must start at some power of 10 and end at the next power of 10 (e.g., 1 to 10, 10 to 100, .1 to 1.0, .01 to .1, or correspondingly 10^0 to 10^1, 10^1 to 10^2, 10^{-1} to 10^0, 10^{-2} to 10^{-1}), Fig. 28.22. As can also be observed from a log scale, the values plotted can be decreasingly small in quantity, approaching zero, but never reaching zero. Logarithms of numbers greater than one are positive, and of numbers between 0 and 1 are negative. As a positive variable quantity approaches zero, its logarithm becomes negatively infinite. *Semilog charts, therefore, never have a zero value on the logarithmic scale.* The cycle designation determines the number of cycles required as a result of observing the range of values to be plotted. Cycle designation also permits the plotting of an extensive range of values, since each cycle accommodates an entire range of one power of 10.

The techniques of scale and chart designation and captions are similar to those discussed for rectangular coordinate line graphs. Values shown along the logarithmic axis, however, can be designated in a number of ways, as shown in Fig. 28.23. The scale shown at (a) is preferred, since it is the simplest and will be understood by any person who makes frequent use of semilog charts.

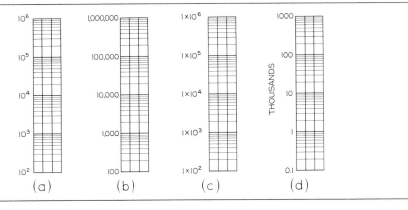

Fig. 28.23   Log Scale Designation.

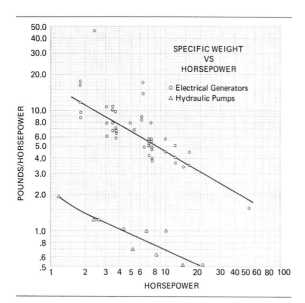

Fig. 28.24   Logarithmic Chart.

28.10   Logarithmic Coordinate Line Charts

Logarithmic charts have two variables plotted on a logarithmic coordinate grid to form a continuous line or a "curve." Printed logarithmic paper contains logarithmically spaced horizontal and vertical rulings. As in the case of semilog charts, paper containing as many as five cycles on an axis can be purchased. See §28.9.

Log charts are applicable for the comparison of a large number of plotted values in a compact space and for the comparison of the relative trends of several curves on the same chart. This form of graph is not the best form for presenting relatively few plotted values in a series or for emphasizing change in absolute amounts. The designation of log cycles, however, permits the plotting of very extensive ranges of values.

Logarithmic charts are used primarily to determine empirical equations by fitting a single straight line to a series of plotted points, Chapter 30. They are also used to obtain straight-line relationships when the data are suitable, as in Fig. 28.24. The design and layout of log charts are the same as for semilog charts, §28.9, and sim-

ilar in many respects to those for rectangular coordinate charts, §§28.3–28.7.

28.11   Trilinear Coordinate Line Charts

Trilinear charts have three related variables plotted on a coordinate paper in the form of an equilateral triangle. The points joined together successively form a continuous straight line or "curve."

Trilinear charts are particularly suited for the following uses.

1. Comparing three related variables relative to their total composition (100 percent).

2. Analyzing the composition structure by a combination of curves—for example, metallic microstructure of trinary alloys, Fig. 28.26 (a).

3. Emphasizing change in amount or differences between values.

The trilinear chart is *not* recommended for (1) emphasizing movement or trend of data or for (2) comparing three related but dissimilar physical quantities—for example, force, acceleration, and time.

Trilinear charts are most frequently applied in the metallurgical and chemical fields, because of the frequency of three variables in metallurgical and chemical composition. The basis of application is the geometric principle that the sum of the perpendiculars to the three sides from any point within an equilateral triangle is equal to the altitude of the triangle.

In an equilateral triangle ABC, Fig. 28.25 (a), the sum of the distances X–f, X–m, and X–n from the point X within the triangle is equal to the altitude A–r, B–s, or C–t of the triangle. For example, if the distances X–f, X–m, and X–n are, respectively, 50, 30, and 20 units, the altitude of the triangle is 100 units, and the point X will represent a quantity composed of 50, 30, and 20 parts of the three variables. At (b) is shown a chart for various freezing temperatures, with the mixture proportions by volume of water, metha-

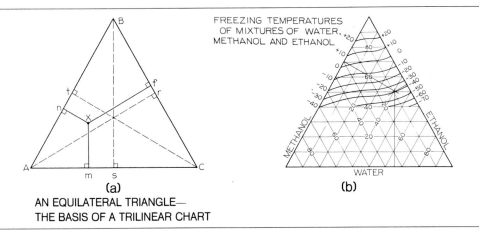

(a)

AN EQUILATERAL TRIANGLE—
THE BASIS OF A TRILINEAR CHART

(b)

Fig. 28.25   Trilinear Coordinate Line Charts.

nol, and ethanol required. For example, a freezing temperature of $-40°F$ can be established by mixing 50 parts of water with 10 parts of ethanol and 40 parts of methanol.

28.12   Design and Layout of Trilinear Charts

Trilinear charts are also usually drawn on printed grid paper. Charts for publications, requiring maximum clarity, can be plotted on grids drawn on plain paper or traced from an original

plotting on a printed sheet. The trilinear coordinate grid is an equilateral triangle, which must have the same space between rulings along each axis. Normally, each side of the triangle is divided into 10 divisions for percent plots, Fig. 28.26, with as many subdivisions as required. If only a portion of a coordinate field is to be used, only a portion of the grid need be shown, Fig. 28.27.

The scale designations can be placed to be read from the bottom of the sheet or can be tilted as shown in Fig. 28.25 (b), but for ease of read-

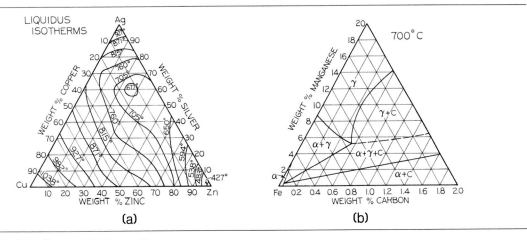

(a)

(b)

Fig. 28.26   Metallurgical Trilinear Charts.

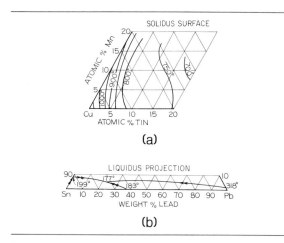

SOLIDUS SURFACE

(a)

LIQUIDUS PROJECTION

(b)

Fig. 28.27   Partial Trilinear Charts.

ing, all scale notations should be placed outside the grid area. Other techniques of scale designations are the same as for rectangular coordinate line graphs, §28.5.

Since individual points are generally not shown on trilinear charts, the curves are drawn as continuous lines.

Curves are frequently single-weight solid lines, but dashed lines are used for projected (anticipated) information, Fig. 28.26 (b). When the curve-fitting information varies within a range, shaded bands are used, Fig. 28.28.

When the chart indicates the composition of various parts or different parts, the symbols are placed within the appropriate areas—for example, α, β, α + γ, which are microstructure sym-

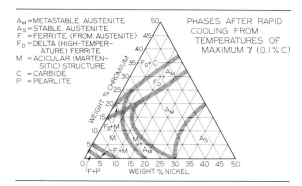

A_M = METASTABLE AUSTENITE
A_S = STABLE AUSTENITE
F = FERRITE (FROM AUSTENITE)
F_D = DELTA (HIGH-TEMPER-
ATURE) FERRITE
M = ACICULAR (MARTEN-
SITIC) STRUCTURE
C = CARBIDE
P = PEARLITE

PHASES AFTER RAPID
COOLING FROM
TEMPERATURES OF
MAXIMUM γ (0.1% C)

Fig. 28.28   Shaded-Band Curves.

bols, Fig. 28.26 (b). Curve designations are placed along the curves or in breaks in the curve—for example, the temperature values shown in Fig. 28.27 (a).

Since most of the grid area is utilized by the curves, the title, keys, and notes are placed off the grid, as shown in Fig. 28.28.

28.13   Polar Coordinate Line Charts

Polar charts have two variables, one a *linear* magnitude and the other an *angular* quantity, plotted on a polar coordinate grid with respect to a pole (origin) to form a continuous line or "curve."

Polar charts are particularly applicable to

1. Comparing two related variables, one being a linear magnitude (called a *radius vector*) and the second an angular value.

2. Indicating movement or trend or location with respect to a pole point.

Polar charts are *not* suited for (1) emphasizing changes in amounts or differences between values or (2) interpolating intermediate values.

As shown in Fig. 28.29 (a), the zero degree line is the horizontal right axis. To locate a point P, it is necessary to know the radius vector r, and an angle θ (e.g., 5, 70°). The point P could also be denoted as (5, 430°), (5, −290°), (−5, 250°), and (−5, −110°), for example. If we plot the equation r = a (no angular designation), we will obtain a circle with the center at the pole, as shown at (b). The value a is a constant value, which determines the relative size of the radius vector and the curve. The equation r = 2a cos θ produces a circle going through the pole point, with its center on the polar axis, (c). The plot of r = a sin 3θ produces a "three-leaved rose," as shown at (d). Polar charts also have many practical applications concerned with the magnitude of some values and their location with respect to a pole point. For example, Fig. 28.30 (a) and (b) illustrate stress charts from experimentation and for stress visualization, respectively. At (c) is shown a polar chart of a bearing load diagram,

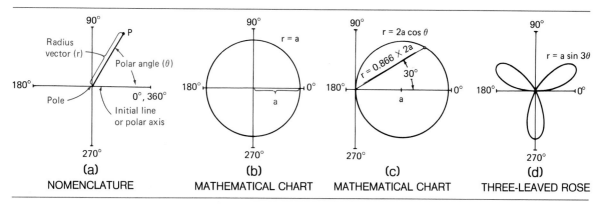

Fig. 28.29   Polar Coordinate Charts.

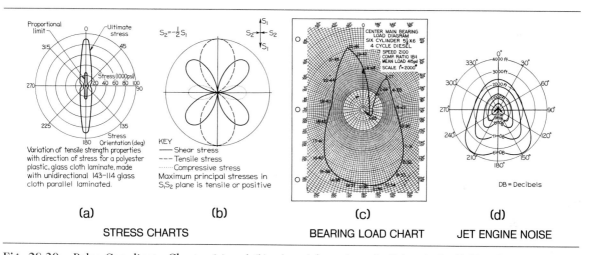

Fig. 28.30   Polar Coordinate Charts. *(a) and (b) adapted from charts by Robert L. Stedfield and F. W. Kinsman, respectively, with permission of* Machine Design. *(c) adapted from R. R. Slaymaker,* Bearing Lubrication Analysis, *copyright 1955 by John Wiley and Sons, with permission of the publisher and Clevite Corporation. (d) adapted from chart by G. S. Schairer, with permission of author and Society of Automotive Engineers.*

and the graph at (d) indicates the noise distribution from a jet engine.

28.14   Nomographs or Alignment Charts

Nomographs or alignment charts consist of straight or curved scales, arranged in various configurations so that a straight line drawn across the scales intersects them at values satisfying the equation represented.

Alignment charts can be used for analysis, but the predominant application is for computation. The design and layout of some common forms of nomographs are discussed in Chapter 29.

28.15   Bar or Column Charts

Bar charts are graphic representations of numerical values by lengths of bars, beginning at a base line, indicating the relationship between two or more related variables.

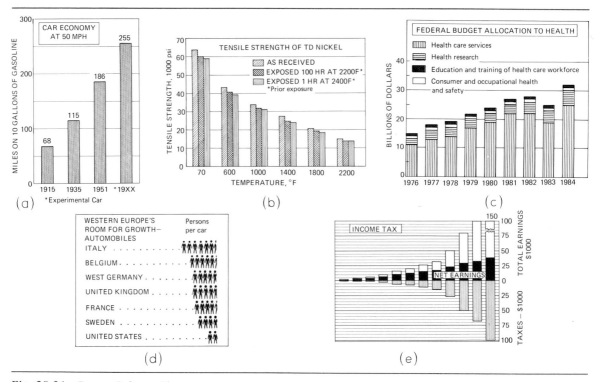

Fig. 28.31   Bar or Column Charts.

Bar charts are particularly suited for the following purposes.

1. Presentation for the nontechnical reader.
2. A simple comparison of two values along two axes.
3. Illustration of relatively few plotted values.
4. Representation of data for a total period of time in comparison to point data.

Bar or column charts are *not* recommended for (1) comparing several series of data or (2) plotting a comparatively large number of values.

The bar chart is effective for nontechnical use because it is most easily read and understood. Therefore, newspapers, magazines, and similar publications use it extensively. The bars may be placed horizontally or vertically; when they are placed vertically, the presentation is sometimes called a column chart.

Bar charts are plotted with reference to two mutually perpendicular coordinate axes, similar to those in rectangular coordinate line charts. The graph may be a simple bar chart with two related variables, Fig. 28.31 (a); a grouped bar chart (three or more related variables), (b); or a combined bar chart (three or more related variables), (c). Another type of bar graph employs pictorial symbols, composed to form bars, as shown at (d). Bar charts can also effectively indicate a "deviation" or difference between values, (e).

28.16   Design and Layout of Bar Charts

If only a few values are to be represented by vertical bars, the chart should be higher than wide, Fig. 28.31 (a). When a relatively large

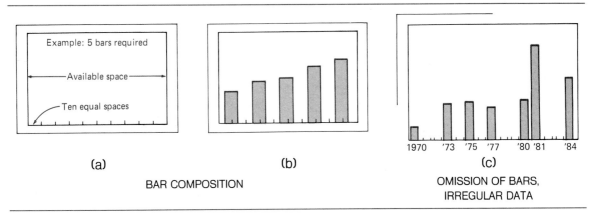

(a)   (b)   (c)

BAR COMPOSITION   OMISSION OF BARS,
                                    IRREGULAR DATA

Fig. 28.32   Bar Composition.

number of values are plotted, a chart wider than high is preferred, (b) and (c). Composition is dictated by the number of bars used, whether they are to be vertical or horizontal, and the available space.

A convenient method of spacing bars is to divide the available space into twice as many equal spaces as bars are required, Fig. 28.32 (a). Center the bars on every other division mark, beginning with the first division at each end, as shown at (b). When the series of data is incomplete, the missing bars should be indicated by the use of ticks, (c), indicating the lack of data. The bars should be spaced uniformly when the data used are distributed. When irregularities in data exist, the bars should be spaced accordingly, as shown.

Bar charts may be drawn on printed coordinate paper; however, clarity for popular use is promoted by the use of blank paper and the designation of only the major rulings perpendicular to the bars. If the values of bars are individually noted, the perpendicular rulings may be omitted completely.

Since bar charts are used extensively to show differences in values for given periods of time, the values or amounts should be proportionate to the heights or lengths of the bars. The zero or principal line of reference should never be omit-

ted. Normally, the full length of the bars should be shown to the scale chosen. When a few exceptionally large values exist, the columns may be broken as shown in Fig. 28.31 (e), with a notation included indicating the full value of the bar.

Scale designations are normally placed along the base line of the bars and adjacent to the rulings.

Standard abbreviations may be used for bar designations. The techniques of scale designations used for line curves, §28.5, are also applicable to bar charts. The values of the bars may be designated above the bars for simple and grouped bar charts, Fig. 28.31 (a). Spaces between bars should be wider than the bars when there are relatively few bars. Spaces between bars should be narrower than the bars when there are many bars, as at (b) and (c).

Bars are normally emphasized by shading or by filling in solid. Some of the common forms of shading are shown in the figure. Weight and spacing of shading depend on the amount of area to be shaded and the final size of the chart and are therefore a matter of judgment.

Bar designations may be placed across several bars when possible, as shown at (e). Other methods include the use of notes with leaders and arrowheads and the use of keys.

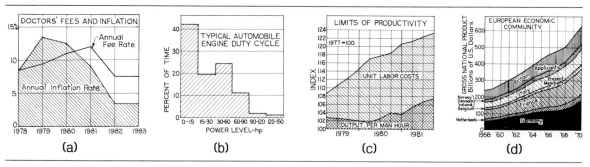

Fig. 28.33 Surface or Area Charts.

As in the case of line graphs, §28.7, titles for bar charts are placed where space permits and contain similar information.

28.17 Rectangular Coordinate Surface or Area Charts

A surface chart, or area chart, is a line or bar chart with ordinate values accentuated by shading the entire area between the outlining curve and the abscissa axis, Fig. 28.33 (a).

A surface or area chart can be a simple chart (called a staircase chart), (b); a multiple surface or "strata" chart with one surface or layer on top of another, (c); or a combined surface chart indicating the distribution of components in relation to their total, (d).

Surface charts are used effectively to

1. Accentuate or emphasize data that appear weak as a line chart.

2. Emphasize amount of ordinate values.

3. Represent the components of a total, usually expressed as a percent of a total, or compared to 100 percent.

4. Present a general picture.

Surface charts are *not* recommended for (1) emphasizing accurate reading of values of charts containing more than one curve or (2) showing line curves that intersect or cross one another.

A map of terrestrial or geographic features is a form of area or surface chart. It represents a graphic picture of area surfaces, or the relationship of their component parts to the total configuration.

28.18 Design and Layout of Surface Charts

Since surface charts are used as a general picture, and accurate reading of values is difficult, only major rulings need be shown. Techniques of grids and composition are the same as for line curves, §28.4. Printed grids generally should not be used, since the printed grid lines would interfere with the shading of areas.

Similarly to bar charts, surface charts represent a comparison of values from a zero line or base line; therefore, the ordinate scale should not be broken, nor should the zero (base) line be omitted. Other procedures and techniques of scale and scale designations are the same as for line curves.

The shading of surfaces is accomplished by the use of black (solid) and white areas, lines, and dots, Fig. 28.33. The darker shading tones should be used at the bottom of the chart, and progressively lighter tones should be used as each stratum is shaded proceeding upward, as shown at (c). The weight and spacing of lines and dots are dependent on the final size of the chart, and are a matter of judgment on the part

854

of the drafter. Projections, or extensions of a curve beyond present available data, can be distinguished by dashed outlines and lighter line shading.

Surfaces should be designated by placing the labels entirely on the surface, where possible, as at (c). Small surfaces can be denoted by a label with a leader and arrowhead, (d). The area of the labels should be clear of shading for ease in reading. Legends should not be used as a means of designation if direct labeling is possible.

The methods of chart designations are the same as for rectangular coordinate line graphs.

28.19 Pie Charts

Pie charts, or sector charts, are used to compare component parts in relation to their total by the use of circular areas.

Pie charts are effective for

1. Representing data on a percentage basis.
2. Presenting a general picture.
3. Showing relatively few plotted values.
4. Emphasizing amounts rather than the trend of data.

A pie chart is normally presented as a true circular area, Fig. 28.34 (a), or in pictorial form,

(b). Since most applications are concerned with monetary values, a disk or "coin" is commonly used for the circular area.

28.20 Design and Layout of Pie Charts

Grids are not used for pie charts. The circular area is drawn to a desired size, within a permissible space. The determination of the various sizes of the sectors is based on 360° being equivalent to 100 percent. Therefore, the following relationship exists.

$$\frac{100\%}{360°} = \frac{x\%}{\theta°}$$

or

$$\theta° = \frac{x\%(360°)}{100\%} = 3.6°(x\%)$$

and 1 percent is represented by 3.6°, 2 percent by 7.2°, and so on.

Sector values (percent and amount) are placed within the sectors where possible. If a sector is small, a note with a leader and arrowhead will suffice. Labels should be clear of any shading and lettered to read from the bottom of the chart, where possible, Fig. 28.34 (c) and (d). Another technique of sector designation is to

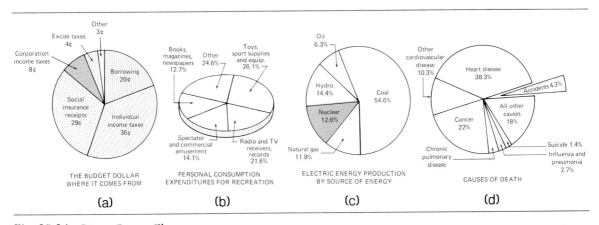

Fig. 28.34 Pie or Sector Charts.

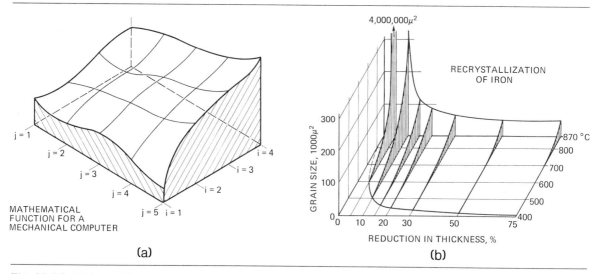

MATHEMATICAL
FUNCTION FOR A
MECHANICAL COMPUTER

(a)

(b)

Fig. 28.35 Volume Charts. *(a) adapted from drawing by Eugene W. Pike and Thomas R. Silverberg, with permission of Machine Design.*

shade the areas, (a). If one of the parts is to be emphasized and compared with the other parts, it can be shaded a different tone from the others, (c), or separated from the circular area, as shown at (d).

The title should be placed above or below the figure.

28.21 Volume Charts

A volume chart is the graphic representation of three related variables with respect to three mutually perpendicular axes in space. Volume charts are generally difficult to prepare, so they are not often used. The method of construction is not discussed in detail, but some of the forms are shown.

Figure 28.35 illustrates line volume charts plotted with respect to three axes (Cartesian coordinates), in isometric and oblique projections at (a) and (b), respectively. Bar graphs can be similarly presented in three dimensions on an isometric grid.

A combination of bars and maps may be used to represent related variables, Fig. 28.36 (a).

Topographic map construction also is a graphic representation of three related variables (two dimensions in a horizontal plane and one dimension in a vertical direction), all drawn in one plane of the drawing paper, as shown at (b).

28.22 Rectangular Coordinate Distribution Charts

When the data observed or obtained vary greatly, they can be plotted on a rectangular coordinate grid for the purpose of observing the distribution or areas of major concentration, with no attempt to fit a curve, Fig. 28.37. Charts of this nature are commonly called *scatter diagrams*.

28.23 Flowcharts

Flowcharts are predominantly schematic representations of the flow of a process—for example, manufacturing production processes and electric or hydraulic circuits, Fig. 28.38. Pictorial forms may be used as shown at (a). Schematic symbols are also used, if applicable, (b) and (c). Blocks with captions are used in the simplest form of flowchart, as illustrated at (d).

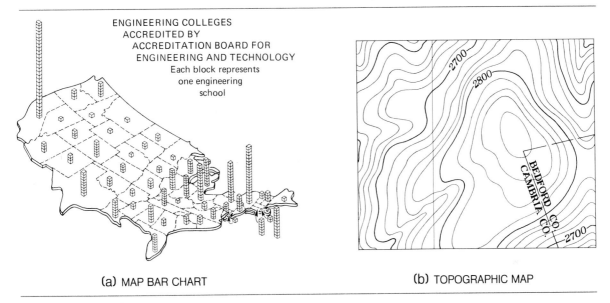

(a) MAP BAR CHART

(b) TOPOGRAPHIC MAP

Fig. 28.36 Map Charts.

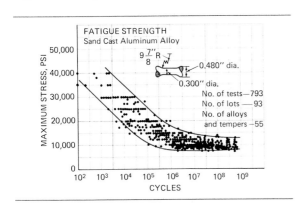

Fig. 28.37 "Scatter" Diagram. *Courtesy of Product Engineering*

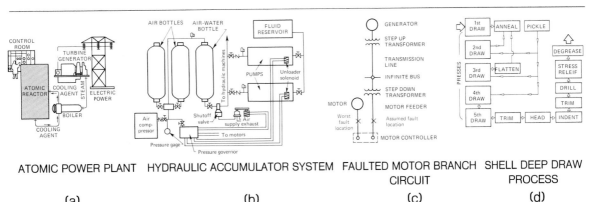

ATOMIC POWER PLANT   HYDRAULIC ACCUMULATOR SYSTEM   FAULTED MOTOR BRANCH CIRCUIT   SHELL DEEP DRAW PROCESS

(a)   (b)   (c)   (d)

Fig. 28.38 Flowcharts. *(b) adapted from chart by A. F. Welsh, with permission of* Machine Design.

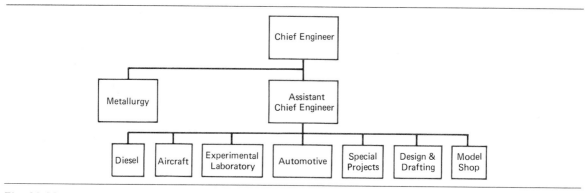

Fig. 28.39   Organization Chart.

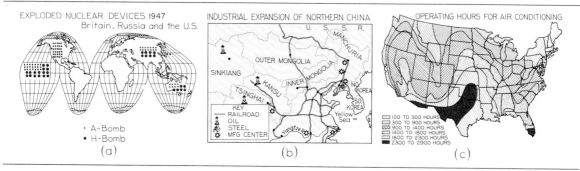

Fig. 28.40   Map Distribution Charts. *(a), (b), and (c) courtesy of* Look *Magazine,* Life *Magazine, and* Heating, Piping and Air Conditioning, *respectively.*

Organization charts are similar to flow-charts, except that they are usually representations of the arrangement of personnel and physical items of specific organizations, Fig. 28.39.

28.24   Map Distribution Charts

When it is desired to present data according to geographical distribution, maps are commonly used. Locations and emphasis of data can be shown by dots of various sizes, Fig. 28.40 (a); by the use of symbols, (b); by shading of areas, (c); and by the use of numbers or colors.

28.25   Computer Graphics

The graphic presentations discussed in this chapter have all been constructed using conventional drafting equipment. Graphs, charts, and

Fig. 28.41   Graphic Display Device. *Courtesy of AT&T.*

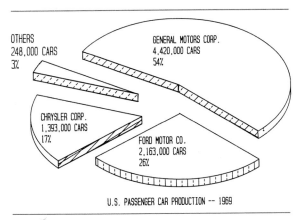

U.S. PASSENGER CAR PRODUCTION -- 1969

Fig. 28.42 Computer-Generated Pictorial Pie Chart.
Courtesy of Department of Engineering Graphics, Illinois Institute of Technology.

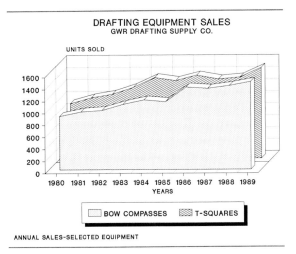

Fig. 28.44 Computer-Generated Pictorial Line Graph. *Courtesy of Department of Engineering Graphics, Illinois Institute of Technology.*

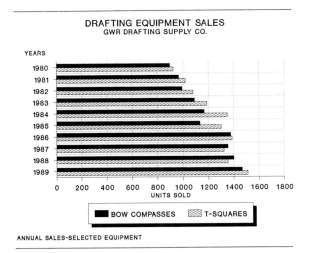

Fig. 28.43 Computer-Generated Horizontal Bar Graph. *Courtesy of Department of Engineering Graphics, Illinois Institute of Technology.*

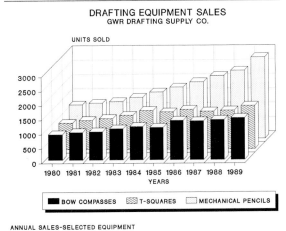

Fig. 28.45 Computer-Generated Pictorial Vertical Bar Graph. *Courtesy of Department of Engineering Graphics, Illinois Institute of Technology.*

diagrams can also be produced by using a computer graphics system. Graphics software packages intended for use with PCs make it possible to store information and data that can be accessed when needed to create drawings and other graphic images in black and white or color. Text material can be incorporated with the graphics so that reports can be quickly prepared, revised, and published. The images created may be viewed on a graphic display device or monitor, Fig. 28.41. A record or copy of these images may then be produced using an output device such as a dot-matrix or laser printer. Examples of graphs and charts produced in this manner are shown in Figs. 28.42–28.45. For additional information, see Chapters 3 and 8.

GRAPH PROBLEMS

Construct an appropriate form of graph for each set of data listed. The determination of graph form (line, bar, surface, etc.) is left to the discretion of the instructor or the student and should be based on the nature of the data or the form of presentation desired. In some cases, more than one curve or more than one series of bars are required.

Since many of the problems in this chapter are of a general nature, they can also be solved on most computer graphics systems. If a system is available, the instructor may choose to assign specific problems to be completed by this method.

Prob. 28.1 Medicine for Profit: A Healthy Market (revenues generated by for-profit acute-care hospitals)

Year	Gross Revenue (billions of dollars)
1969	0.75
1970	1.00
1971	1.25
1972	1.60
1973	2.00
1974	2.50
1975	3.00
1976	3.50
1977	4.00
1978	4.50
1979	5.25
1980	6.25
1981	7.30
1982	9.00
1983	10.50
1984	13.00

Prob. 28.2 Metal Hardness Comparison (Mohs' hardness scale)

Metal	Comparative Degree of Hardness (diamond = 10)
Lead	1.5
Tin	1.8
Cadmium	2.0
Zinc	2.5
Gold	2.5
Silver	2.7
Aluminum	2.9
Copper	3.0
Nickel	3.5
Platinum	4.3
Iron	4.5
Cobalt	5.5
Tungsten	7.5
Chromium	9.0
Diamond	10.0

Prob. 28.3 Effect of Accuracy of Gear Manufacture on Available Strength in Terms of Horsepower (60 teeth gear and 30 teeth pinion of 6 diametral pitch, 1.5″ face width, $14\frac{1}{2}°$ pressure angle; 500 Brinell Case hardness)

Pitch Line Velocity, ft/min	Horsepower			
	Perfect Gear	Aircraft Quality Gear	Accurate Quality Gear	Commercial Quality Gear
0	0	0	0	0
1000	140	100	75	50
2000	290	180	105	60
3000	Straight-line	250	120	70
4000	curve through	320	130	72
5000	two points	380	140	72

Noise limit: 2750 ft/min—accurate quality gear; 1400 ft/min—commercial quality gear.

Prob. 28.4 U.S. Population

Year	Population
1800	5,308,483
1810	7,239,881
1820	9,638,453
1830	12,866,020
1840	17,069,453
1850	23,191,876
1860	31,443,321
1870	38,558,371
1880	56,155,783
1890	62,947,714
1900	75,994,575
1910	96,977,266
1920	105,710,620
1930	122,775,046
1940	131,669,275
1950	151,325,798
1960	179,323,175
1970	203,302,031
1980	226,545,805
1990	249,000,000[a]
2000	267,000,000[a]

[a]Estimated.

Prob. 28.5 U.S. Pedestrian Deaths, 1975–1979

Age	Number of Deaths		Age	Number of Deaths	
	Female	Male		Female	Male
1	500	800	47	225	600
3	575	1000	50	225	600
5	600	1040	52	225	575
7	400	800	55	250	500
8	300	625	57	250	550
10	325	500	60	225	575
12	425	625	62	275	575
15	400	1100	65	250	500
17	375	1075	67	290	525
20	300	1000	70	300	525
22	275	900	72	350	475
25	200	750	75	340	450
27	225	700	77	300	475
30	200	600	80	300	400
32	175	500	82	225	375
35	150	450	85	150	300
37	160	475	87	100	200
40	175	475	90	50	150
42	200	500	92	25	50
45	175	500	95	0	0

Prob. 28.6 World Population

Year	Population (billions)
1950	2.51
1960	3.03
1970	3.63
1980	4.42
1990	5.28[a]
2000	6.20[a]

[a]Projections.

Prob. 28.7 Worldwide Growth of Engineering Periodicals

Years	Number of Periodicals
1921–1930	166
1931–1940	215
1941–1950	301
1951–1960	485
1961–1970	705
1971–1980	927
1981–1985	991

Prob. 28.8 Lumber Prices—Softwoods

Year	Index (1967 = 100)			
	Mar.	June	Sept.	Dec.
1980	362.4	328.6	349.4	355.7
1981	346.0	357.0	335.3	321.8
1982	318.9	328.1	320.9	322.7
1983	369.4	397.9	361.4	360.3

Prob. 28.9 U.S. Personal Consumption Expenditures

Year	Billions of Dollars		
	Durables	Nondurables	Services
1965	63.0	188.6	178.7
1970	85.2	265.7	270.8
1975	132.2	407.3	437.0
1980	214.7	668.8	784.5

Prob. 28.10 Industrial Energy Sources

Year	Percent of Total U.S. Industrial Use of Energy			
	Coal	Oil	Natural Gas	Electricity
1969	23	23	45	9
1971	20	24	46	10
1973	17	36	39	8
1975	16	36	38	10
1977	15	39	35	11
1979	14	41	33	12
1981	15	37	35	13
1983	12	40	35	13
1985	13	38	35	14
1987	13	39	34	14

Prob. 28.11 Psychological Analysis of Work Efficiency and Fatigue

Hours of Work	Relative Production Index	
	Heavy Work	Light Work
9–10 A.M.	100	96
10–11 A.M.	108	104
11–12 A.M.	104	104
12–1 P.M.	98	103
Lunch		
2–3 P.M.	103	100
3–4 P.M.	99	102
4–5 P.M.	98	101
After 8-hour day		
5–6 P.M.	91	94
6–7 P.M.	86	93
7–8 P.M.	68	83

Prob. 28.12 The World Economy—Industrial Production

Year	Index (1980 = 100)					
	United Kingdom	France	U.S.	West Germany	Italy	Japan
1976	94	90	86	89	80	77
1977	100	96	90	94	92	84
1978	101	94	94	95	85	85
1979	104	99	103	98	95	92
1980	106	102	103	102	102	100
1981	96	98	102	98	99	99
1982	98	97	98	97	98	102
1983	100	95	93	93	94	101
1984	105	99	108	99	95	112

Plot on semilog shows rate of change of index.

Prob. 28.13   Essential Qualities of a Successful Engineer (average estimate based on 1500 questionnaires from practicing engineers)

Quality	Percent
Character	41
Judgment	$17\frac{1}{2}$
Efficiency	$14\frac{1}{2}$
Understanding human nature	14
Technical knowledge	13
Total	100

Prob. 28.14   Automobile Accident Analysis

Type of Accident	Percent
Cross traffic (grade crossing, highway, railway)	21
Same direction	30
Head-on	21
Fixed object	11
Pedestrian	10
Miscellaneous	7
Total	100

Prob. 28.15   Comparison of Horsepower at the Rear Wheels (as shown by dynamometer tests)

Engine rpm	Horsepower at Rear Wheels			
	McCulloch Supercharged with Dual Exhausts	McCulloch Supercharged	Unsupercharged —Dual Exhausts	Unsupercharged
2000	77.0	73.0	64.5	59.5
2200	82.0	77.5	70.5	65.0
2400	88.0	83.5	75.0	69.0
2600	95.5	91.5	79.0	73.5
2800	105.0	99.0	82.5	76.5
3000	112.5	105.5	84.5	78.5
3200	117.0	109.5	85.5	79.5
3400	119.0	111.5	83.5	77.0
3600	118.5	111.5		

Prob. 28.16   How the World Uses Its Work Force: Employment by Economic Sector, 1980

Country	Percent of Workers			
	Agriculture[a]	Mining and Construction	Manufacturing	Services[b]
United States	3.6	7.2	22.1	67.1
Canada	5.4	7.7	19.7	67.2
Australia	6.5	9.1	19.9	64.5
Japan	10.1	10.1	25.0	54.8
France	8.7	9.3	25.8	56.2
Great Britain	2.7	8.3	28.4	60.6
Italy	14.2	11.2	26.9	47.7
Netherlands	6.0	9.6	21.3	63.1
Sweden	5.6	7.2	24.3	62.9

[a]Agriculture, forestry, hunting, and fishing.
[b]Transportation, communication, public utilities, trade, finance, public administration, private household services, and miscellaneous services.

Plot the data for Problems 28.17 and 28.18 on rectangular coordinate paper and on semilog paper.

Prob. 28.17   Rupture Strength of T. D. Nickel—
High-Temperature Alloy

Temperature (T),°F	100-hr Rupture Stress (s_r), psi × 1000 (log scale)
1200	24
1400	17
1600	12.5
1800	9
2000	6.5
2200	4.75
2400	3.5

Prob. 28.18   Nuisance Noise

Frequency (f), cps (log scale)	Octave Band Level, db			
	Hearing Loss Risk Region		Power Lawn Mower at 3 ft	5-hp Chainsaw at 3 ft
	Negligible	Serious		
53	104	122	84	93
106	93	113	93	103
220	87	107	94	103
425	85	105	90	111
850	85	105	84	112
1700	85	105	84	107
3400	85	105	82	104
6800	85	105	75	98

Plot the data for the following problems on rectangular coordinate paper and on logarithmic paper.

Prob. 28.19   Loss of Head for Water Flowing in Iron Pipes

Prob. 28.20   Material and Process Economics

Velocity (v), ft/sec	Loss of Head, ft/1000 ft			
	1″ Pipe	2″ Pipe	4″ Pipe	6″ Pipe
0	0	0	0	0
1	6.0	2.9	1.6	.7
2	23.5	9.5	4.2	2.4
3	50	20	8.2	4.7
4		34	13.5	7.7
5		51	20	11.4
6			28	15.5
7			37	20

Weight of Steel Forging (W), lb	Unit Cost (C) for Forging, $		
	Simple	Average	Complex
0.1	0.22		
0.2	0.38		
0.4	0.67		
1.0	1.4		
2.0	2.5	2.75	4.5
4.0	4.4	5.0	8.5
10.0	9.2	11.0	19.5
20.0	16.0	20.0	37.0
40.0	28.5	36.5	70.0
100	60.0	80.0	165.0

Prob. 28.21 Pressurized Square Tubing, Fig. 28.46.

Orientation, degrees	Stress Ratio (σ_1/P)	
	$D/a = 0.80$	$D/a = 0.86$
0	2.0	2.6
15	3.2	4.3
30	4.2	6.3
45	4.3	5.0
60	4.2	6.3
75	3.2	4.3
90	2.0	2.6

Data are given for one quadrant; quadrants are identical.

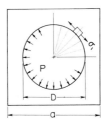

Fig. 28.46 Prob. 28.21.

Prob. 28.22 Variation of Modulus of Elasticity (*E*) with Direction in Copper

Orientation, degrees	$E \times 10^6$ psi	
	As Rolled	Annealed
0	20.0	9.5
15	18.5	11.0
30	16.5	14.5
45	15.0	17.5
60	16.5	14.5
75	18.5	11.0
90	20.0	9.5
105	18.5	11.0
120	16.5	14.5
135	15.0	17.5
150	16.5	14.5
165	18.5	11.0
180	20.0	9.5

Direction of rolling is from 0° to 180°.

Prob. 28.23 Light Distribution in a Vertical Plane for a Bulb Suspended from the Ceiling with the Filament at the Origin of the Polar Chart

Orientation, degrees	Candle Power
0	140
10	210
20	310
30	320
40	310
50	310
60	300
70	290
80	280
90	250
100	270
110	290
120	295
130	300
140	315
150	330
160	340
170	350
180	340

Prob. 28.24   Main Bearing Load Diagram (4000 rpm, no counterweight)

Orientation, degrees	Load 1000 lb	(1000 kg)	Orientation, degrees	Load 1000 lb	(1000 kg)
0	6.6	(3.0)	190	5.3	(2.4)
10	4.8	(2.2)	200	5.2	(2.4)
20	4.2	(1.9)	210	4.9	(2.2)
30	3.7	(1.7)	220	4.5	(2.1)
40	3.2	(1.5)	230	4.2	(1.9)
50	3.0	(1.4)	240	3.7	(1.7)
60	2.8	(1.3)	250	3.3	(1.5)
70	2.7	(1.2)	260	3.1	(1.4)
80	2.7	(1.2)	270	2.9	(1.3)
90	2.8	(1.3)	280	2.8	(1.3)
100	2.9	(1.3)	290	2.7	(1.2)
110	3.3	(1.5)	300	2.8	(1.3)
120	3.6	(1.6)	310	3.0	(1.4)
130	4.1	(1.9)	320	3.2	(1.5)
140	4.6	(2.1)	330	3.7	(1.7)
150	4.9	(2.2)	340	4.4	(2.0)
160	5.0	(2.3)	350	5.3	(2.4)
170	5.2	(2.4)	360	6.6	(3.0)
180	5.3	(2.4)			

Max. load = 6600 lb (3000 kg); mean load = 4530 lb (2059 kg).

CHAPTER 29

Alignment Charts

BY E. J. MYSIAK*

The engineer, in performing his design work, uses mathematics for calculations with equations, for deriving equations, and when solving particular equations. Graphical solutions of equations are performed with the use of alignment charts and coordinate axes graphs. Alignment charts are explained in this chapter. The derivation of equations is covered in Chapter 30, "Empirical Equations." Particular solutions of equations are explained in §§31.1 and 31.2.

29.1 Nomographs or Alignment Charts

A nomograph is a diagram or a combination of diagrams representing a mathematical law or equation. The Greek roots *nomos* (law) and *graphein* (to write) suggest this definition. The rectangular coordinate graphs discussed in the preceding chapter are the most common examples, showing graphically the relationship of two or more variables and their function. The term "nomograph," however, is more popularly applied to a combination of scales, arranged properly to represent mathematical laws (equations) for computational purposes. Some of the more common forms of nomographs are shown in Fig. 29.1.

Basically, a nomograph is used to solve a three-variable equation. A straight line (*isopleth*) joining known or given values of two of the variables intersects the scale of the third variable at a value that satisfies the equation represented, as at (a), (c), (f), (g), and (h). For this reason

they are also called *alignment charts*. Alignment charts are nomographs; however, nomographs are not necessarily alignment charts. A rectangular coordinate graph can be classified as a nomograph. Two or more such charts can sometimes be combined to solve an equation containing more than three variables, as at (b), (d), and (e). The forms shown have fixed scales and a movable alignment line. However, movable-scale nomographs can be designed with a fixed direction alignment line, the slide rule being an example of this form.

Although alignment charts require time to construct, they have considerable popular appeal for the following reasons.

1. They save time when it is necessary to make repeated calculations of certain numerical relationships (equations).

2. They enable one unskilled or lacking a background in mathematics to handle analytical solutions.

---

*Engineering Manager, Phoenix Company of Chicago, Wood Dale, IL.

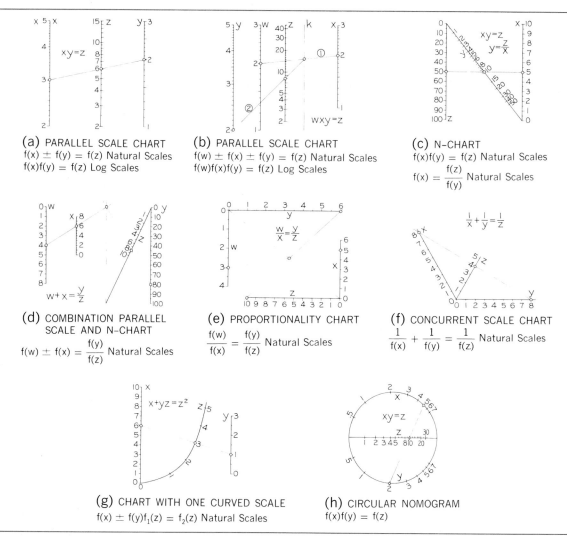

Fig. 29.1 Common Forms of Nomographs or Alignment Charts.

3. When constructed properly, they are limited to the scale values for which the equation is valid. This prevents the use of the equation with values that are not applicable.

The design of alignment charts by the use of plane geometry is described here. For the design of alignment charts by means of determinants (matrix algebra), more advanced textbooks should be consulted.

29.2   Functional Scales

A mathematical equation expresses the relationship of a group of *variables* and *constants*. For example, the equation

$$y = x^2 + 2x + 3$$

contains two variables, x and y, which are letter symbols designating quantities that may have several values in the equation. The variable x is expressed as the *function* $x^2 + 2x$; that is, the function of x is $f(x) = x^2 + 2x$. The function of y is $f(y) = y$. Other typical expressions of functions of a variable are

$$\frac{1}{x}, \quad \log x, \quad \sin x, \quad x^2 + \frac{3}{x^3}, \quad \text{and} \quad x - 1.$$

A *constant* is any quantity that always has the same value, such as the number 3 in the equation. The number 2 of $2x$ is a *constant coefficient*.

The common forms of nomographs are composed of scales, each representing only one variable; hence, scales are designed for each individual variable. To draw a scale to a certain length L, representing the function of a variable between definite limits, the difference between the extreme values is multiplied by a proportionality factor called the *functional modulus*, m.

$$L = m[f(x)_{\max} - f(x)_{\min}]$$
$$= m[f(x_2) - f(x_1)]$$

where x_2 and x_1 represent the values of the variable x corresponding, respectively, to the maximum and minimum values of the function, Fig. 29.2.

In this relationship a convenient modulus could be chosen to simplify construction of the scale and consequently determine the length, which is a preferred procedure by those experienced in scale layout.

It is sometimes difficult to choose a functional modulus that will result in a scale length long enough or short enough to fit the paper size to be used. Rearranging the equation to

$$m = \frac{L}{f(x_2) - f(x_1)}$$

the functional modulus can be defined as the length on the scale for a unit value of the function variable. The equation in this form with the determined range of values, $f(x_2)$ and $f(x_1)$, and an assumed scale length, L, may result in an inconvenient functional modulus, m. Since a cer-

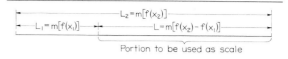

Fig. 29.2   Functional Scale Relationship.

tain convenient scale length is seldom necessary, the resulting functional modulus, m, can then be rounded off to a convenient value for scale layout, as explained in §29.5. The rounded-off functional modulus will result in a new scale length, which should be close enough to the desired length.

If any of the values for the function are negative, they must be treated algebraically as negative values. For example, if $f(x_1)$ is negative,

$$L = m[f(x_2) - f(-x_1)]$$
$$= m[f(x_2) + f(x_1)]$$

29.3   Scale Layout

As an example of functional scale construction, the equation of §29.2 is used, and the typical procedure for scale layout is illustrated in Fig. 29.3 for the variable x, $f(x) = x^2 + 2x + 3$. The consequence of including the constant in the function of the variable is to add $\frac{3}{8}''$ to the scale, which is not used. Therefore, the resulting scale in effect is for the terms with variables only, $f(x) = x^2 + 2x$.

29.4   Scale Modulus

When each term of a variable in a function contains a constant coefficient, it may be more convenient to plot values from the function after combining the constant coefficient with the functional modulus. For example, if $f(x) = 2x^2 + 2x$, with x ranging in value from .5 to 5.0,

$$L = m[f(x_2) - f(x_1)]$$
$$= m[2(25) + 2(5)] - [2(.25) + 2(.5)]$$
$$= m[60 - 1.5] = 58.5m$$

If the length $L = 10''$ is chosen, then $10'' = 58.5m$ and $m = .171$. The scale could be prepared from tabulated values for $2x^2 + 2x$. A

STEP 1. Assume or determine values of the variable and the limits of the scale.	x	0	1	2	3	4	5
STEP 2. Compute the values of the function, substituting the values of step 1.	$f(x) = x^2 + 2x + 3$	3	6	11	18	27	38

STEP 3. Choose a convenient functional modulus and determine total length of the scale.	$L = m[f(x_2) - f(x_1)]$ If $m = \dfrac{1}{8}$, $L = \dfrac{1}{8}[(x_2{}^2 + 2x_2 + 3) - (x_1{}^2 + 2x_1 + 3)]$ $= \dfrac{1}{8}[(5^2 + 2(5) + 3) - (3)] = \dfrac{35}{8} = 4.375''$

STEP 4. If scale length is suitable, multiply functional modulus by the function values. These are the measuring lengths for the graduations on the scale for the values of the function.	$\left(m = \dfrac{1}{8}\right)$ $mf(x) =$	$\dfrac{3}{8} =$ .375	$\dfrac{6}{8} =$ .750	$\dfrac{11}{8} =$ 1.375	$\dfrac{18}{8} =$ 2.250	$\dfrac{27}{8} =$ 3.375	$\dfrac{38}{8} =$ 4.750

STEP 5. Lay off the distances obtained along a chosen line.	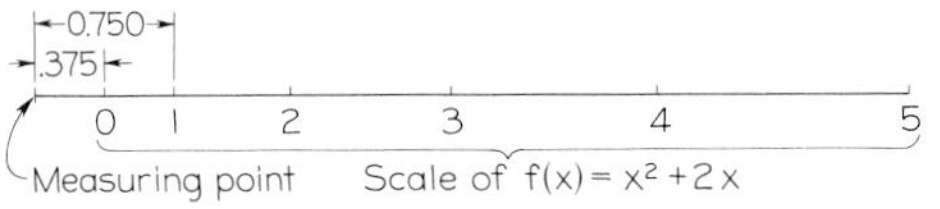
STEP 6. Denote the values of the variable (not the function) at the corresponding points determined.	

Fig. 29.3   Functional Scale Layout and Design.

more convenient method is to use the equation as

$$L = 2m[(x_2^2 + x_2) - (x_1^2 + x_1)]$$

A common factor has been withdrawn from the coefficients and combined with the functional modulus. The product of the functional modulus and the constant coefficient is called a *scale modulus* and is commonly designated by a capital letter M. The scale modulus would be $M = 2m = .342$, and the function of the variable used would be $x^2 + x$.

A further advantage in preparing this scale would be to choose a more convenient scale modulus, resulting in a length close to the length of $10''$ desired. If $M = .333 = \frac{1}{3}$ is chosen, then

$$L = .333[(25 + 5) - (.25 + .5)] = 9.74''$$

Since $M = \frac{1}{3} = \frac{10}{30}$, the 30 scale on an engineers' scale could be used, §29.5.

29.5   Engineers' Scale

The use of an engineers' scale, where appropriate, eliminates the tedious operation of multiplying each functional value by the scale modulus to obtain the measurement of the graduations. The scale graduations are laid out directly, using the function values on the engineers' scale. The functional or scale moduli are determined only to denote which scale to use and to determine the total scale length. If a definite scale length is not required, it is best to choose a scale modulus of 1, $\frac{1}{2}$, $\frac{1}{3}$, $\frac{1}{4}$, $\frac{1}{5}$, or $\frac{1}{6}$, or a multiple of these such as $\frac{10}{3}$, $\frac{1}{30}$, and $\frac{10}{30}$, that would permit the use of the 10, 20, 30, 40, 50, or 60 scales on the engineers' scale, §2.27.

The numbers 10, 20, 30, 40, and so on represent the number of subdivisions per inch on the scale. For example, the 30 scale, Fig. 29.4 (a), has 1 inch divided into 30 parts. The number 3 *on the engineers' scale* can be taken to represent .3, 3, 30, or 300, depending on the scale

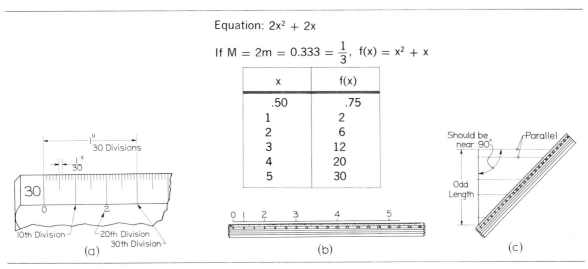

Equation: $2x^2 + 2x$

If $M = 2m = 0.333 = \frac{1}{3}$, $f(x) = x^2 + x$

x	f(x)
.50	.75
1	2
2	6
3	12
4	20
5	30

(a)

(b)

(c)

Fig. 29.4   Use of Engineers' Scale.

modulus used: $\frac{10}{3}$, $\frac{1}{3}$, $\frac{1}{30}$, or $\frac{1}{300}$, respectively. At (b) is shown the equation of §29.4. Since $M = \frac{1}{3}$, the figure at $1''$ represents 3 as a value of the function $(x^2 + x)$. Number 5 (a value of the variable x) would be at 30 on the engineers' scale, number 2 would be at 6, and so on. Notice that the function values are measured directly from the engineers' scale onto the scale line. When an odd scale modulus or length is necessary (occurs frequently for the middle scale of a three-scale alignment chart), the odd length can be laid off by the parallel-line method, (c). In this case the functional or scale moduli are not used for scale calibration. The engineers' scale is used to calibrate a line at an angle to the scale line, and it is then projected onto the scale line within the determined scale length. For accuracy in projection, the parallel lines should be as near $90°$ as possible to the scale line. In the example of §29.4, $f(x) = x^2 + x$, $M = 2m = .342$, $L = 10''$; the scale can be laid out for a length of $10''$. The scale modulus for the final scale can still be .342. On a line at an angle to the final scale line, the 30 engineers' scale $(M = .333)$, or any other appropriate scale, can be used to calibrate a scale with the functional values of .75, 2, 6, 12, 20, and 30 units, which is then projected onto the final $10''$ long scale line.

29.6   Approximate Subdivision of Nonuniform Spaces

After the major graduations are located along the scale, the scale should be subdivided into smaller divisions. If the spaces between the major divisions are nonuniform, an approximate method of subdivision is to use a *sector*, constructed on a separate sheet of translucent paper, tracing cloth, or film, as shown in Fig. 29.5. In use, the

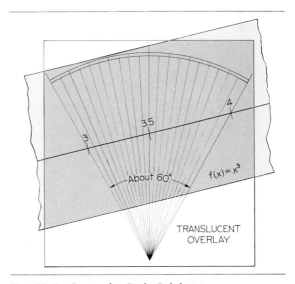

Fig. 29.5   Sector for Scale Subdivision.

Variable °C	Function 1.8°C + 32	M × Function $\frac{1}{50}$ (1.8°C + 32)	Variable °F	Function °F	M × Function $\frac{1}{50}$ (°F)
−40	−40	$-\dfrac{40}{50}$	−40	−40	$-\dfrac{40}{50}$
−30	−22	$-\dfrac{22}{50}$	−20	−20	$-\dfrac{20}{50}$
−20	− 4	$-\dfrac{4}{50}$	0	0	$-\dfrac{0}{50}$
−10	14	$\dfrac{14}{50}$	20	20	$\dfrac{20}{50}$
0	32	$\dfrac{32}{50}$	40	40	$\dfrac{40}{50}$
10	50	$\dfrac{50}{50}$	60	60	$\dfrac{60}{50}$
20	68	$\dfrac{68}{50}$	80	80	$\dfrac{80}{50}$
30	86	$\dfrac{86}{50}$	100	100	$\dfrac{100}{50}$
40	104	$\dfrac{104}{50}$	120	120	$\dfrac{120}{50}$
50	122	$\dfrac{122}{50}$	140	140	$\dfrac{140}{50}$
60	140	$\dfrac{140}{50}$	160	160	$\dfrac{160}{50}$
70	158	$\dfrac{158}{50}$	180	180	$\dfrac{180}{50}$
80	176	$\dfrac{176}{50}$	200	200	$\dfrac{200}{50}$
90	194	$\dfrac{194}{50}$	212	212	$\dfrac{212}{50}$
100	212	$\dfrac{212}{50}$			

If $M = m = \dfrac{1}{50}$, $L_{°c} = \dfrac{1}{50} [212 - (-40)] = \dfrac{1}{50} (252) = 5.04''$

When °C = −40°, °F = 1.8(−40) + 32 = −40°F; when °C = 100°, °F = 1.8(100) + 32 = 212°F

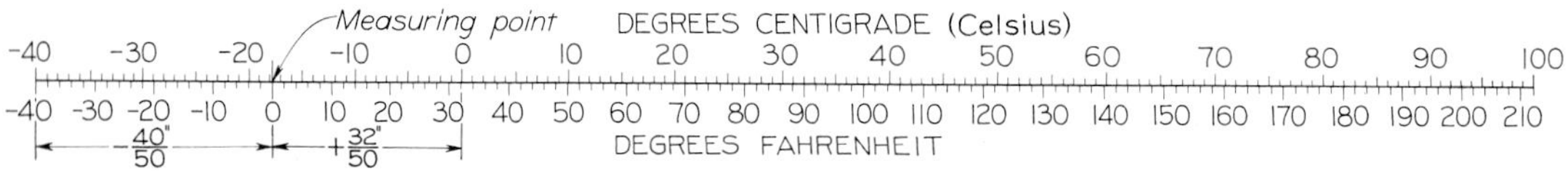

Fig. 29.6   Design and Layout of Temperature Conversion Chart.

sector is fitted to three previously located points on the scale, such as 3, 3.5, and 4 in the example. The subdivisions are located on the scale by piercing through the sector paper with a compass or divider needle point. If the scale is drawn on a transparent medium, the scale may be placed over the sector chart.

29.7   Conversion Scales

An equation with two variables can be arranged to equate the function of one variable to the function of the second variable. If functional scales are designed for each function of the two related variables and placed adjacent to one another, the resulting chart is known as a *conversion scale*. The procedure for designing each scale is the same as discussed in §§29.3 and 29.4, with the following exceptions: *the functional modulus must be the same for both scales; both scales are laid out from the same origin*, although it may not appear on the scale. The origin (measuring point) is at the zero value of the functions, not the variables. As an example, the classical problem of a temperature conversion scale is illustrated in Fig. 29.6. The equation for converting °C to °F is

$$°F = \tfrac{9}{5}°C + 32 = 1.8°C + 32$$

The first step is to separate the two variables, expressing their functions individually. Thus, $f(°F) = °F$; $f(°C) = 1.8°C + 32$. If °C is to range from $-40°C$ to $100°C$, the range of °F must be determined from the equation.

After tabulation of an appropriate distribution of intermediate values and the corresponding computed function values, the scale modulus multiplied by the function need not be determined if a convenient modulus is chosen for the use of the engineers' scale. If the Celsius scale is the first scale laid out, the measuring point can be located $\tfrac{40''}{50}$ from the left end. This point can be used to locate the remaining calibrations; negative values to the left, positive values to the right. Notice the measuring point (function value origin) is at $-17.8°C$ and $0°F$ after the variable

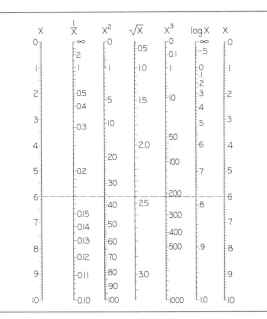

Fig. 29.7   Conversion Scales.

values have been added. For this particular conversion scale, the origin (zero value) for the function and the °F variable value are the same point.

Another convenient form of conversion scale is the series illustrated in Fig. 29.7. The scales are parallel, but not adjacent, and a horizontal line must be drawn at right angles to the scales to determine the various functions of x. Thus, if from 6 on the scale x, a horizontal line is drawn across to scale x^2, the value 36 is found. These scales are called *natural (functional) scales*. If the function is the log of a variable, it is a natural scale, but if it is a function laid out logarithmically, it is called a log scale, §29.10. In Fig. 29.7 only the first and the last scales are uniformly divided; the intermediate scales are nonuniform.

29.8   Natural Parallel-Scale Nomographs, $f(x) \pm f(y) = f(z)$

The construction of nomographs for equations with three variables, involving the sum or difference of two variables, is based on the following geometric derivation.

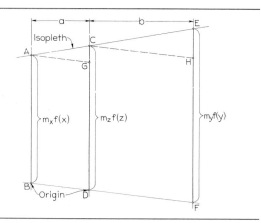

Fig. 29.8   Parallel-Scale Chart Geometric Derivation.

In Fig. 29.8, AB, CD, and EF represent three parallel functional scales. The origins are aligned at any angle, and an *isopleth* (index line) crosses the three scales in any direction. Lines AG and CH are drawn parallel to BF.

From similar triangles,

$$\frac{CD - GD}{a} = \frac{EF - HF}{b}$$

Correspondingly,

$$\frac{m_z f(z) - m_x f(x)}{a} = \frac{m_y f(y) - m_z f(z)}{b}$$

Collecting terms,

$$m_x f(x) + \frac{a}{b} m_y f(y) = \left(1 + \frac{a}{b}\right) m_z f(z)$$

If $f(x) \pm f(y) = f(z)$, the coefficients must be equal; therefore,

$$m_x = \frac{a}{b} m_y = \left(1 + \frac{a}{b}\right) m_z$$

or

$$\frac{m_x}{m_y} = \frac{a}{b} \tag{1}$$

Substituting equation (1) into the coefficient of $f(z)$,

$$m_x = \left(1 + \frac{m_x}{m_y}\right) m_z$$

or

$$m_z = \frac{m_x m_y}{m_x + m_y} \tag{2}$$

Equation (1) is the relationship for spacing the outside scales after choosing moduli for these scales. Equation (2) is used to determine the modulus of the middle scale. These two relationships are used on all parallel-scale nomographs. Note that the two variables on one side of the equality sign, $f(x)$ and $f(y)$, are the outside scales.

A few principles applicable to all parallel-scale charts are illustrated in Fig. 29.9. When positive values are laid out in one direction, negative values must be laid out in the opposite direction, (b). Note that a function with a negative sign has an inverted scale and the origins are in alignment. Another approach would be to rearrange this relationship to be $x = z + y$, x being the middle scale. In this case, no scale would be inverted, which is preferred.

As mentioned previously, changing the alignment of origins to any angle, including horizontal, does not affect the equation.

29.9   Design and Layout of Three-Variable Parallel-Scale Nomographs

The example to be used is the equation for the outside diameter of a 20° stub-tooth gear,

$$d_0 = d + 2a$$

where d_0 = outside diameter of pinion, inches; d = pitch diameter, 3.000″ to .250″ (for diametral pitches 4 to 54); and a = addendum, .250″ to .0393″ (for diametral pitches 4 to 54, respectively).

The first step is to determine or assume the range of values (maximum and minimum) for

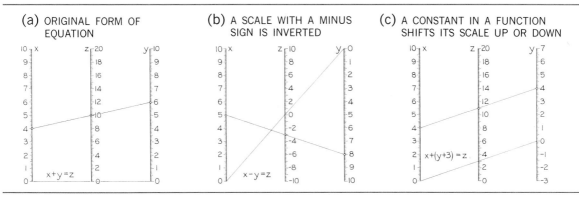

Fig. 29.9 Parallel-Scale Variations.

two of the functions of the equation. The maximum and the minimum values of the third function can then be determined. The computation must be performed, since the layout of the alignment chart is based on the alignment of the three scales within these specific maximum and minimum values.

The next step is to separate the three functions and determine an appropriate length (for $8\frac{1}{2}'' \times 11''$ paper size) and functional moduli for the two outside scales. See upper portion of Fig. 29.10 (a). The variables d and a were selected for the outside scales, since they are positive value scales. If d_0 was selected as an outside scale, one of the other variables would become negative and inverted on the alignment chart ($d_0 - d = 2a$ or $d_0 - 2a = d$). Remember, the two variables on one side of the equality sign are the outside scales.

If $m_d = 3$,

$$L_d = m_d(d_2 - d_1)$$
$$= 3(3.00 - .250) = 8.250''$$

If $m_a = 20$,

$$L_a = m_a(a_2 - a_1)$$
$$= 20(2)(.250 - .0393) = 8.428''$$

Correspondingly, the spacing of the outside scales is

$$\frac{a}{b} = \frac{m_d}{m_a} = \frac{3}{20} = \frac{.60}{4.00}$$

The functional modulus of the middle scale is

$$m_{d_0} = \frac{m_d m_a}{m_d + m_a} = \frac{(3)(20)}{3 + 20} = 2.61$$

The range of d_0 is found by substituting the maximum and minimum values into the equation.

$$L_{d_0} = m_{d_0}(d_{0_2} - d_{0_1})$$
$$= 2.61(3.500 - .329) = 8.277''$$

The scale modulus is found by multiplying the functional modulus by the constant coefficient of each variable function.

Note that *functional moduli are used to determine spacing and length of scales.* A convenient scale modulus, however, aids in the layout of the scale.

It is not necessary to compute the length of the middle scale, v_f, but its computation is desirable, since it serves as a check on the preceding work.

To construct the chart, three parallel lines are drawn, spaced proportionate to .60 :: 4.00, about 9″ to 10″ long, Fig. 29.10. *Caution:* One of the most frequently made mistakes is erroneous placing of the outside scales because of interchanging the proportionate distances. Since all the terms in the equation are positive, the three scales increase in value in the same direction,

Equation	Var.	Funct.	Variable Range	Funct. Modulus (m)	Scale Modulus (M)	Scale Length	Distance from Center Scale	Dir.	Scale Used
$d_0 = d + 2a$	d	d	3.00″–.250″	$3 = \dfrac{90}{30}$	$1 \times \dfrac{90}{30} = \dfrac{90}{30}$	8.250″	.60″	↑	30
	a	2a	.250″–.0393″	$20 = \dfrac{1000}{50}$	$2 \times \dfrac{1000}{50} = \dfrac{2000}{50}$	8.428″	4.00″	↑	50
	d_0	d_0	3.50″–.329″	2.61	$1 \times 2.61 = 2.61$	8.277″	—	↑	40

(a) DESIGN

d	$f(d) = d$	a	$f(a) = a$	d_0	$f(d_0) = d_0$
.25	.25	.039	.039	.329	.329
.50	.50	.05	.05	.50	.50
1.00	1.00	.10	.10	1.0	1.0
1.50	1.50	.15	.15	1.5	1.5
2.00	2.00	.20	.20	2.0	2.0
2.50	2.50	.25	.25	2.5	2.5
3.00	3.00			3.0	3.0
				3.5	3.5

(b) DATA

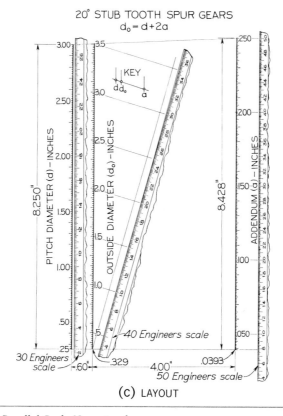

(c) LAYOUT

Fig. 29.10   Three-Variable Parallel-Scale Nomograph.

normally upward, as shown here. In order to align the scale origins, draw a base line across the three scales. Usually the base line is drawn in a horizontal direction for a convenience. With the 30 engineers' scale, lay out the d scale, with 1" representing 30. The major values of .50, 1.00, 1.50, and 2.00 are located by multiplying these values by 9 (since the scale modulus is 90/30) and measuring them at 4.50, 9.00, 13.50, and 18.00, respectively. The use of the engineers' scale eliminates the need for multiplying each functional value by the functional modulus. The a scale is laid out in a similar manner by using the 50 scale. Since the constant coefficient was accounted for in the functional modulus, it is not necessary to multiply each value by 2 or the functional modulus.

After the two outside scales are laid out for their entire range, a line joining their top ends determines the correct height of the middle scale. The calculated length of 8.277" should check with the layout, if outside scales were laid out correctly. The major scale divisions of the middle scale could be located by multiplying each functional value by the functional modulus and measuring this distance with a regular scale from the origin point. Since the modulus is odd, however, a more convenient method is to use a slightly larger engineers' scale (40 or 50) and locate the major divisions by the parallel-line method, as shown. From the equation, the values of d_0 range from .329 to 3.500, and their functional values are the same, since $f(d_0) = d_0$. With the 40 scale placed with 3.29 at the origin (1" = 4), the major values are set off with the engineers' scale. The limit values of the scale designations need not be exactly .329 and 3.500. The scale can be extended to round figures of .250" and 3.750", but the values of .329 and 3.500 must lie on the base line and top isopleth, respectively.

After the scales have been properly arranged, the subdivisions can be located with the aid of a sector chart, Fig. 29.5. Addition of title, key, and scale captions completes the chart.

As a final check, two or three arithmetic computations should be compared to the same data on the alignment chart.

For a second illustration, the example to be used is the equation for the velocity of a free-falling body,

$$v_f^2 = v_0^2 + 2gs$$

where v_f = final velocity, ft/sec; v_0 = original velocity, 10 to 50 ft/sec; g = constant acceleration due to gravity, 32.2 ft/sec^2; and s = displacement of body, 5' to 50'.

If $m_{v_0} = \frac{1}{300}$,

$$L_{v_0} = m_{v_0}(v_{02}^2 - v_{01}^2)$$
$$= \tfrac{1}{300}(2500 - 100) = 8''$$

If $m_s = \frac{1}{322}$,

$$L_s = m_s(2g)(s_2 - s_1)$$
$$= \tfrac{1}{322}(64.4)(50 - 5) = 9''$$

$m_s = \frac{1}{322}$ was chosen since 322 is a multiple divisor of 64.4 and conveniently results in a scale modulus of $M_s = \frac{10}{50}$. The use of the 50 engineers' scale is anticipated.

Correspondingly, the spacing of the outside scales is

$$\frac{a}{b} = \frac{m_{v_0}}{m_s} = \frac{\frac{1}{300}}{\frac{1}{322}} = \frac{322}{300} = \frac{3.22}{3.00}$$

Note the algebraic treatment of the reciprocal functional moduli. The functional modulus of the middle scale is

$$m_{v_f} = \frac{m_{v_0}m_s}{m_{v_0} + m_s} = \frac{(\frac{1}{300})(\frac{1}{322})}{(\frac{1}{300}) + (\frac{1}{322})} = \frac{1}{622}$$

The range of v_f is found by substituting the maximum and minimum values into the equation.

$$L_{v_f} = m_{v_f}(v_{f2}^2 - v_{f1}^2)$$
$$= \frac{1}{622}(5720 - 422) = 8.518''$$

The design, data, and layout for the alignment chart are shown in Fig. 29.11.

Equation	Var.	Funct.	Variable Range	Funct. Modulus (m)	Scale Modulus (M)	Scale Length	Distance from Center Scale	Dir.	Scale Used
	V_0	V_0^2	10–50 ft/sec	$\dfrac{1}{300}$	$1 \times \dfrac{1}{300}$	8″	3.22″	↑	30
$V_f^2 = V_0^2 + 2gs$	s	2gs	5–50 ft	$\dfrac{1}{322}$	$\dfrac{64.4}{322} = \dfrac{2}{10} = \dfrac{10}{50}$	9″	3.00″	↑	50
	V_f	V_f^2	20.5–75.7 ft/sec	$\dfrac{1}{622}$	$1 \times \dfrac{1}{622}$	8.518″	0	↑	40

(a) DESIGN

V_0	$f(V_0) = V_0^2$	s	$f(s) = s^*$	V_f	$f(V_f) = V_f^2$
10	100	5	5	20.5	422
15	225	10	10	25	625
20	400	15	15	30	900
25	625	20	20	35	1225
30	900	25	25	40	1600
35	1225	30	30	45	2025
40	1600	35	35	50	2500
45	2025	40	40	55	3025
50	2500	45	45	60	3600
		50	50	65	4225
				70	4900
				75	5625
				75.7	5720

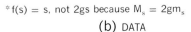
* $f(s) = s$, not 2gs because $M_s = 2gm_s$

(b) DATA

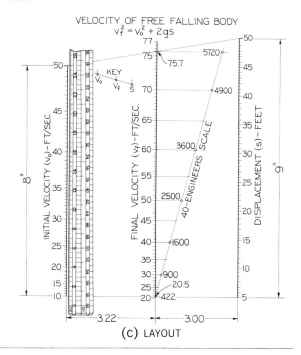

(c) LAYOUT

Fig. 29.11   Three-Variable Parallel-Scale Nomograph.

29.10 Logarithmic Parallel-Scale Nomographs, $f(x)f(y) = f(z)$ or $f(x) = f(z)/f(y)$

Many engineering equations appear in the form of the product of variables. This form can be transformed into the sum or difference of variables by taking logarithms of both sides of the equation:

or

$$f(x)f(y) = f(z)$$

or

$$\log f(x) + \log f(y) = \log f(z)$$

$$f(x) = \frac{f(z)}{f(y)}$$

or

$$\log f(x) = \log f(z) - \log f(y)$$

880

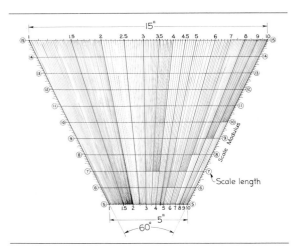

Fig. 29.12 Log Sector Chart.

Parallel-scale charts can be constructed for this form of equation by the use of logarithmic scales instead of natural scales. A convenient device for the construction of logarithmic scales is a log sector chart, Fig. 29.12. The use of a logarithmic sector eliminates the necessity of multiplying the logarithm of each value by the modulus to construct the scale. The use of the log sector chart is detailed in §29.11.

29.11 Design and Layout of Logarithmic Parallel-Scale Nomographs

The primary design of log scale nomographs is similar to natural scale nomographs. The majority of the calculations are tabulated in Fig. 29.13 (a), for the example equation:

$$P = VI$$

The first step is to compute the range of values for P, using the maximum and minimum values for V and I. Since $P = VI$,

$$\log P = \log V + \log I$$

V and I will be on the outside scales. The function of the variables for scale layout is the log of

the original function. The determination of scale and functional modulus, scale lengths, and spacing is similar to the form discussed in §29.9.

In the chart layout, (b), notice that the scale modulus is the length of one complete log cycle used. A portion of a cycle, or more than one cycle, may be used. Cycles longer than those on the sector can be constructed by doubling or tripling a smaller cycle length. Odd-length cycles can be conveniently determined by folding the sector at the appropriate subdivision between two even-length cycles—for example, the 1.875″ cycle for P is below the 2″ cycle and is extended lower until a cycle length of 1.875″ is obtained.

Similar to other parallel-scale alignment charts, the maximum and minimum values of the variables, as determined and calculated, must be on the top isopleth and base line, respectively.

For a second illustration, the majority of the calculations are tabulated in Fig. 29.14 (a) for the equation:

$$f = 4.42 \sqrt{F_0/W}$$

Since $f = 4.42\sqrt{F_0/W}$,

$$f^2 = 19.5 F_0/W$$

and

$$2 \log f = \log 19.5 + \log F_0 - \log W \quad (1)$$

or

$$f^2 W = 19.5 F_0$$

and

$$2 \log f + \log W = \log 19.5 + \log F_0 \quad (2)$$

or

$$W = 19.5 F_0/f^2$$

and

$$\log W = \log 19.5 + \log F_0 - 2 \log f \quad (3)$$

ELECTRICAL POWER — P=VI

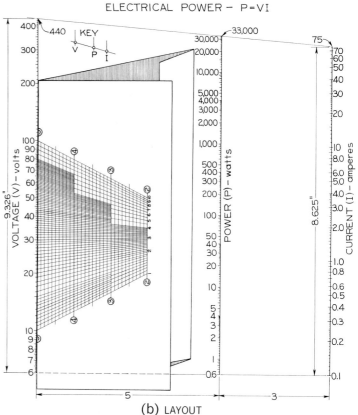

(b) LAYOUT

ELECTRICAL POWER

P = VI	P = power, watts
P_{min} = 6(.1)	V = voltage, 6–440 volts
= .6 watts	I = current,
P_{max} = 440(75)	.1–75 amperes
= 33,000 watts	

P = VI

$\log P = \log V + \log I$

If $m_V = 5$, $L_V = 5(\log 440 - \log 6)$
$\qquad = 5(2.6435 - .7783) = 9.326''$

If $m_I = 3$, $L_I = 3(\log 75 - \log .1)$
$\qquad = 3[1.8750 - (-1)] = 8.625''$

$m_P = \dfrac{5 \times 3}{5 + 3} = \dfrac{15}{8} = 1.875$

$L_P = 1.875 (\log 33,000 - \log .6)$
$\qquad = 1.875 [4.519 - (-.222)] = 8.89''$

$$\frac{a}{b} = \frac{m_V}{m_I} = \frac{5}{3}$$

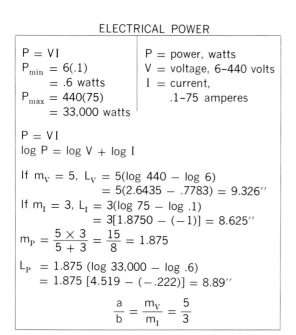

Var.	Funct.	m	M	Scale Length	Scale Dir.	Length of One Log Cycle
V	log V	5	5	9.326″	↑	5″
I	log I	3	3	8.625″	↑	3″
P	log P	1.875	1.875	8.890″	↑	1.875″

(a) DESIGN

Fig. 29.13   Logarithmic Parallel-Scale Nomograph.

Equation (2) was selected as the best arrangement, since all the scales for the variables will be positive. Equations (1) and (3) would have a negative or inverted scale ($-\log W$ or $-2 \log f$, respectively). W and f will be on the outside scales.

The scale for the variable W is one complete log cycle, since the range of values is from 0.1 to 1.0, and will be 8″ long. The 20″ cycle for f can be constructed by doubling lengths measured on the 10″ cycle, only a portion of a cycle being used (from 19.75 to 44.2). The 4.44″ cycle for F_0 is between 4″ and 5″ cycles, and is laid out almost twice, since F_0 ranges from 2 to 10 and from 10 to 100. Notice that tabulated values of the variables for scale layout are not necessary, since any variable to a power merely increases the log scale length required—for ex-

VIBRATION OF CYLINDRICAL SPRINGS
FIXED AT ONE END

$$f = 4.42\sqrt{\frac{F_0}{W}}$$

f = frequency of vibration, vibrations/second

$$f_{min} = 4.42\sqrt{\frac{2}{.1}}$$

$$= 19.75 \text{ vib/sec}$$

F_0 = force necessary to deflect spring 1″, 2–100 lb/in.

$$f_{max} = 4.42\sqrt{\frac{100}{1}}$$

$$= 44.2 \text{ vib/sec}$$

W = wt. of spring, .1–1.0 lb

$Wf^2 = 19.5 F_0$

$\log W + 2 \log f = \log 19.5 + \log F_0$

If $m_W = 8$, $L_W = 8(\log 1 - \log .1)$

$\qquad = 8[0 - (-1)] = 8″$

If $m_f = 10$, $L_f = 10(2 \log 44.2 - 2 \log 19.75)$

$\qquad = 20(1.6455 - 1.2955) = 7″$

$m_{F_0} = \dfrac{8 \times 10}{8 + 10} = \dfrac{80}{18} = 4.44$

$L_{F_0} = 4.44(\log 100 - \log 2)$

$\qquad = 4.44(2 - .301) = 7.55″$

$\dfrac{a}{b} = \dfrac{m_W}{m_f} = \dfrac{8}{10} = \dfrac{4}{5}$

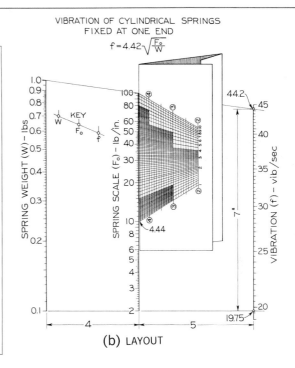

VIBRATION OF CYLINDRICAL SPRINGS
FIXED AT ONE END

$$f = 4.42\sqrt{\frac{F_0}{W}}$$

(b) LAYOUT

Var.	Funct.	m	M	Scale Length	Scale Dir.	Length of One Log Cycle
W	log W	8	8	8″	↑	8″
f	2 log f	10	20	7″	↑	20″
F_0	log F_0	4.44	4.44	7.55″	↑	4.44″

(a) DESIGN

Fig. 29.14   Logarithmic Parallel-Scale Nomograph.

ample, f^2 becomes $2 \log f$. The calibrations are laid out directly with the log sector chart.

Accounting for the term with a constant C ($\log 19.5$) may be questioned. Whether the form of the equation is $x + y = z + C$ or $xy = zC$, the constant C is accounted for by determining the range of values for all variables from the equation with the constant. Then the primary concern is to place the last scale in the proper position to accommodate the constant. This is automatically accomplished by placing the maximum and minimum values chosen or determined on the upper isopleth and base line, respectively.

29.12   Four-Variable Parallel-Scale Nomographs, $f(w) \pm f(x) \pm f(y) = f(z)$ or $f(w)f(x)f(y) = f(z)$ or $f(w)f(x) = f(z)f(y)$

Parallel-scale nomographs may be constructed for any of the four-variable equation forms shown.

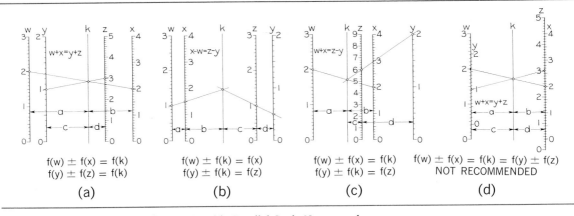

$$f(w) \pm f(x) = f(k)$$
$$f(y) \pm f(z) = f(k)$$

(a)

$$f(w) \pm f(k) = f(x)$$
$$f(y) \pm f(k) = f(z)$$

(b)

$$f(w) \pm f(x) = f(k)$$
$$f(y) \pm f(k) = f(z)$$

(c)

$$f(w) \pm f(x) = f(k) = f(y) \pm f(z)$$
NOT RECOMMENDED

(d)

Fig. 29.15   Arrangements of Four-Variable Parallel-Scale Nomographs.

The design of four-variable nomographs requires transforming the equation into two groups of two variables each and equating these two groups to a third term, $f(k)$.

$$f(w) \pm f(x) = f(z) \pm f(y) = f(k)$$

and

$$\log f(w) \pm \log f(x) = \log f(z) \pm \log f(y)$$
$$= \log f(k)$$

From the transformed equation, two separate three-variable parallel-scale nomographs are constructed. Both nomographs include the new term $f(k)$. The $f(k)$ is used to determine a scale and functional modulus, which *must be the same for both nomographs*. The scale for $f(k)$ is never calibrated, but is used as a *pivot scale* and must have the same direction as the other scales.

Several scale arrangements for four-variable nomographs are shown in Fig. 29.15.

29.13   Design and Layout of Four-Variable Parallel-Scale Nomographs

Since the method of design and layout is similar to the other forms discussed, the information for an example problem is tabulated and illustrated in Fig. 29.16 (a) and (b). As shown in the upper portion of the figure at (a), equation (1) could be used, but that would require an inverted scale; therefore, equation (2) is preferred. One three-variable nomograph will contain scales for H, k, and I. Scales I and k will be outside scales. The second three-variable nomograph will have scales for k, R, and t. Variables R and t will be outside scales. The k term will be an outside scale for one nomograph and the middle scale for the second nomograph. The functional and scale moduli for k, m_k, and M_k, calculated for one nomograph, *must be the same moduli values for the second nomograph*. The use of a log sector chart to lay out the logarithmic scales is the same as explained in §29.11. The scale moduli are the lengths of the log cycles used for direct calibration of complete, partial, or multiple log cycles, depending on the range of values for the variables. The constant value ($\log 2.39$) is disregarded, since it was used to determine the range of values for the variables and is automatically accounted for by placing maximum and minimum values on the upper isopleth and base line, respectively. Note that the functional modulus of a middle scale does not have to be the last modulus determined. A common error, to be especially avoided, is incorrect association of the scale spacings a, b, c, and d, with their respective scales.

JOULE'S LAW OF HEATING OF ELECTRICAL CONDUCTORS

$H = .239\, RI^2 t$

$\dfrac{H}{I^2} = .239\, Rt$

(1) $\log H - 2 \log I = \log k$
$= \log 2.39 + \log R + \log t$
(2) $\log H = \log k + 2 \log I$
$\log k = \log 2.39 + \log R + \log t$

H = heat, calories
R = resistance, 1–10 ohms
I = current, 5–25 amperes
t = time, 1–60 seconds
$H = 0.239(1)(25)(1)$
$= 5.975$ cal
$H = 0.239(10)(625)(60)$
$= 89,600$ cal

If $m_H = 2$, $L_H = 2(\log 89,600 - \log 5.975)$
$= 2(4.9525 - .7765) = 8.352''$
If $m_I = 6$, $L_I = 6(2 \log 25 - 2 \log 5)$
$= 6(2)(1.398 - .699) = 8.388''$

Since $f(H)$ is the middle scale, $m_H = \dfrac{m_I m_k}{m_I + m_k}$; $2 = \dfrac{6 m_k}{6 + m_k}$; $m_k = 3$

$\dfrac{a}{b} = \dfrac{m_I}{m_k} = \dfrac{6}{3} = \dfrac{2}{1}$

If $m_R = 8$, $L_R = 8(\log 10 - \log 1) = 8(1 - 0) = 8''$

Since $f(k)$ is the middle scale, $m_k = \dfrac{m_R m_t}{m_R + m_t}$; $3 = \dfrac{8 m_t}{8 + m_t}$; $m_t = 4.8$

$L_t = 4.8(\log 60 - \log 1) = 4.8(1.778 - 0) = 8.53''$

$\dfrac{c}{d} = \dfrac{m_R}{m_t} = \dfrac{8}{4.8} = \dfrac{2}{1.2}$

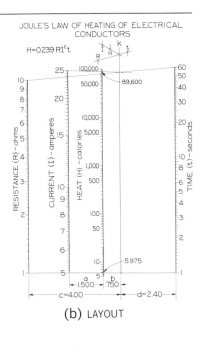

JOULE'S LAW OF HEATING OF ELECTRICAL CONDUCTORS

(b) LAYOUT

Equation	Var.	Funct.	Variable Range	m	M	Scale Length	Dir.	Distance from Center Scale	Length of One Log Cycle
log H = log k + 2 log I	H	log H	5.975–89,600	2	2	8.352″	↑	0	2″
	I	2 log I	5–25	6	12	8.388″	↑	1.500	12″
	k	log k	—	3	3	—	↑	.750	—
log k = log 2.39 + log R + log t	k	log k	—	3	3	—	↑	0	—
	R	log R	1–10	8	8	8.000″	↑	4.000	8″
	t	log t	1–60	4.8	4.8	8.530″	↑	2.400	4.8″

(a) DESIGN

Fig. 29.16   Four-Variable Parallel-Scale Nomograph.

29.14   Natural-Scale N-charts, $f(z)f(y) = f(x)$ or $f(z) = f(x)/f(y)$

Equations that are in the form of the product or division of two variables equaling a third variable may be constructed with natural (nonloga-rithmic) scales in the form of an N-chart (also called a Z-chart).

In Fig. 29.17, AB and CD represent two parallel scales for the variables that are the dividend and divisor in the division equation form. The

885

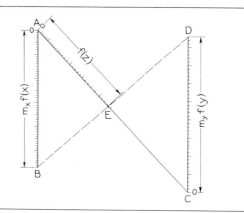

Fig. 29.17   N-chart Geometry.

term that represents the quotient appears on the diagonal scale AE.

The functional moduli and lengths of the two vertical scales are arbitrarily chosen, limited only by the paper size. *These scales must begin at zero* and extend in opposite directions. The diagonal scale connects the two zero points from A to C and is graduated from A to the maximum value desired. The distance between the vertical scales is any convenient distance.

The outside scales are graduated in the same manner as previously described in §§29.3 and 29.5. The diagonal scale may be calibrated analytically or graphically. The analytical method may be found in the texts listed in the bibliography. A simpler graphical method is discussed in §29.15.

29.15   Design and Layout of an N-chart

The calculations and layout of an example problem are shown in Fig. 29.18. The equation may be arranged in any of the following forms:

$$\frac{I}{b} = \frac{h^3}{12}; \quad \frac{I}{h^3} = \frac{b}{12}; \quad \text{or} \quad \frac{12I}{b} = h^3$$

If the variable to a power were one of the vertical scales, it would be a nonuniform functional scale, requiring calibration and computation. Since the diagonal scale is always nonuniform and calibrated, $f(h) = h^3$ was chosen for this scale. This corresponds to the equation form $12I/b = h^3$. The zero values of the function $12I$ and h^3 will coincide.

After reasonable scale lengths have been determined for $f(I)$ and $f(b)$, construct these scales parallel to each other and any convenient distance apart. Usually, a roughly square chart is satisfactory. Proper selection of moduli permits direct use of the engineers' scale. A base line for alignment is not required.

Draw a line connecting the zero values of the two vertical scales. To calibrate the diagonal scale for $f(h)$, select one convenient value of b (or I). This value should be so selected as to simplify the form of the equation for computation of the other values, when substituted into the equation. A value of $b = 6$ simplifies the equation to $I = h^3/2$. With this expression, determine the values of I for various values of h. Draw lines from 6 on the b scale to the values of I on the I scale to locate the corresponding values of h on the diagonal scale.

29.16   Large Value N-charts

When the range of values for an N-chart are large numbers and do not extend to zero, the chart may be constructed with an additional computation. As shown in Fig. 29.19, the scales of the chart are nonintersecting within the area of the paper. The conditions for an N-chart, however, are met, since the scales extended to their zero values would intersect. The lengths of the two vertical scales are determined for their large-number range of values. The two lengths (L) must be the same.

After computing new scale lengths from the zeros to the maximum values, the necessary horizontal distances b and c can be easily determined, since by similar triangles,

$$\frac{a}{d_1} = \frac{b}{d_2} = \frac{c}{d_3}$$

MOMENT OF INERTIA—RECTANGULAR CROSS SECTION

$$I = \frac{bh^3}{12}$$

I = moment of inertia, in.^{4}

b = width, 0–8″

$$h^3 = \frac{12 I}{b}$$

h = height, 0–12″

$$I = \frac{8(12)^3}{12} = 1152 \text{ in.}^4$$

If $m_b = 1$, $L_b = m_b(b_2 - b_1) = 1(8 - 0) = 8″$

If $m_I = \dfrac{1}{2400}$, $L_I = m_I 12(I_2 - I_1) = \dfrac{1}{2400} 12(1152 - 0)$

$$= \frac{1}{200} (1152) = 5.76″$$

Equation	Var.	Funct.	Variable Range	m	M	Scale Length	Dir.	Scale Used
	b	b	0–8″	1	1	8″	↑	10
$h^3 = \dfrac{12 I}{b}$	I	12 I	0–1152 in.^{4}	$\dfrac{1}{2400}$	$12m = \dfrac{12}{2400} = \dfrac{1}{200}$	5.76″	↓	20
	h	h^3	0–12″	—	—	—	↘	—

(a) DESIGN

(b) DATA

h	Values of I with b = 6″ I = h^3/2
1	.5
2	4.0
3	13.5
4	32.0
5	62.5
6	108.0
7	171.5
8	256.0
9	364.5
10	500.0
11	665.5
12	864.0

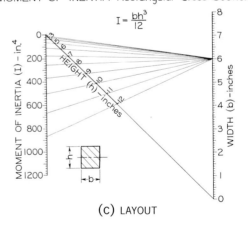

(c) LAYOUT

Fig. 29.18   N-chart.

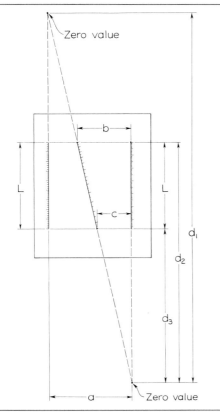

Fig. 29.19   Nonintersecting N-chart.

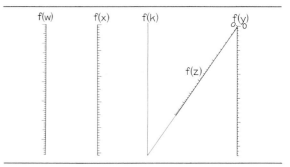

Fig. 29.20   Combination Chart.

properly represent the equation, as shown in Fig. 29.20. A fifth term, $f(k)$, equated to the equation, is used as an uncalibrated vertical outside scale for the parallel-scale nomograph and as a vertical scale for the N-chart, since

$$f(w) \pm f(x) = f(k) = \frac{f(y)}{f(z)} \quad [\text{or} = f(y)f(z)]$$

29.18   Design and Layout of a Four-Variable Combination Nomograph

The method of design and layout is similar to the other forms discussed; the given data for an example problem are tabulated and illustrated in Fig. 29.21. The equations are arranged as shown in step 2, since the $f(k)$ scale is to be an outside scale for the parallel-scale chart and a vertical scale for the N-chart. If the equation contains a product of two functions, either variable can be placed on the diagonal scale of the N-chart. If the equation contains a division of two functions, the variable in the denominator must be placed on the diagonal scale. The lengths of only the three vertical scales must be determined. The resulting functional moduli for the parallel-scale chart are used for spacing the scales, while the N-chart spacing is not definite and is arranged simply for convenience.

The 10 engineers' scale is used to calibrate the v_0 scale and the 20 scale is used to calibrate the v_f scale. The v_0 scale was designed for a

29.17   Combination Parallel-Scale and N-chart, $f(w) \pm f(x) = f(y)/f(z)$ [or $= f(y)f(z)$]

The arrangement of variables within an equation for a nomograph makes it necessary that all the scales have the same type of graduations, either logarithmic or natural. An equation with addition or subtraction of variables has natural scales, while an equation with a multiplication or division of variables suggests logarithmic scales. If an equation of four variables contains any combination of addition or subtraction with multiplication or division, the use of an N-chart (natural scales) for the multiplication or division terms in combination with a natural parallel-scale for the addition or subtraction terms, can

LINEAR MOTION—CONSTANT ACCELERATION

$$V_f^2 - V_0^2 = 2as$$

(1) $f(V_f) - f(V_0) = f(k)$

V_f = final velocity, 0–120 mph
V_0 = original velocity, 0–60 mph
a = acceleration, 0–2.2 ft/sec²
s = displacement, 0–1 mile

$$= f(a)f(s)$$

(2) $V_f^2 = k + V_0^2$; $2a = \dfrac{k}{s}$

$$L_{V_f} = m_{V_f}(V_{f_2}^2 - V_{f_1}^2) = m_{V_f}(14400 - 0)$$

If $m_{V_f} = .0005$, $L_{V_f} = 7.20''$

$$L_{V_0} = m_{V_0}(V_{0_2}^2 - V_{0_1}^2) = m_{V_0}(3600 - 0)$$

If $m_{V_0} = .001$, $L_{V_0} = 3.60''$

$$m_{V_f} = \frac{m_{V_0} m_k}{m_{V_0} + m_k}; \quad .0005 = \frac{.001\, m_k}{.001 + m_k}; \quad m_k = .001;$$

$$\frac{a}{b} = \frac{m_{V_0}}{m_k} = \frac{1}{1}$$

$$L_s = m_s(5280 - 0); \text{ if } m_s = .001, L_s = 5.28''$$

LINEAR MOTION—CONSTANT ACCELERATION

$V_f^2 - V_0^2 = 2as$

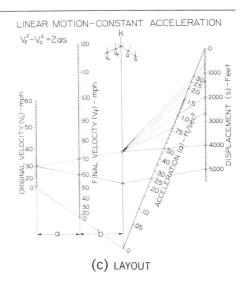

(c) LAYOUT

Equation	Variable	Function	m	M	Scale Length	Distance from Center Scale	Direction	Scale Used
	V_f	V_f^2	$.0005 = \dfrac{1}{2000}$	$\dfrac{1}{2000}$	7.20″	0	↑	20
$V_f^2 = k + V_0^2$	V_0	V_0^2	$.001 = \dfrac{1}{1000}$	$\dfrac{1}{1000}$	3.60″	1	↑	10
	k	k	$.001 = \dfrac{1}{1000}$	$\dfrac{1}{1000}$	—	1	↑	—
	k	k	$.001 = \dfrac{1}{1000}$	$\dfrac{1}{1000}$	—	—	↑	—
$2a = \dfrac{k}{s}$	s	s	$.001 = \dfrac{1}{1000}$	$\dfrac{1}{1000}$	5.28″	—	↓	10
	a	$2a$	—	—	—	—	↗	—

(a) DESIGN

V_f	$f(V_f) = V_f^2$	V_0	$f(V_0) = V_0^2$
0	0	0	0
10	100	10	100
20	400	20	400
30	900	30	900
40	1,600	40	1600
50	2,500	50	2500
60	3,600	60	3600
70	4,900		
80	6,400		
90	8,100		
100	10,000		
110	12,100		
120	14,400		

DATA FOR N-CHART DIAGONAL CALIBRATION

(b) DATA

a	s
.50	4000
.75	2670
1.00	2000
1.50	1333
2.00	1000
2.50	800
3.00	666

$k = V_f^2 - V_0^2$

when $V_f = 70$, $V_0 = 30$, $k = 70^2 - 30^2 = 4000$

when $k = 4000$, $s = \dfrac{4000}{2a} = \dfrac{2000}{a}$

when $V_f = 60$, $V_0 = 30$, $k = 2500$

when $k = 2500$, $a = .25$, $s = 5000$

Fig. 29.21   Combination Parallel Scale and N-chart.

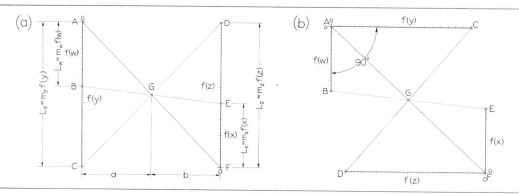

Fig. 29.22   Proportional Chart.

shorter length in order to keep the pivot point of the k scale on the paper (within reason) when using low v_0 values and large v_f values.

To calibrate the diagonal scale of the N-chart, a value must be located on the uncalibrated k scale as a pivot point ($k = 4000$). A second point ($k = 2500$) was found to locate the value of $a = .25$ ft/sec² only.

29.19   Proportional Charts, $f(w)/f(x) = f(y)/f(z)$

Equations with four variables in the form of a proportion, or in the form $f(w)f(z) = f(y)f(x)$, can be represented by a nomograph with natural scales, called a proportional chart. The geometry of its construction is shown in Fig. 29.22 (a).

From similar triangles,

$$\frac{AB}{EF} = \frac{a}{b} = \frac{AC}{DF}$$

If $AB = L_w = m_w f(w)$ and $EF = L_x = m_x f(x)$, and so on,

$$\frac{m_w f(w)}{m_x f(x)} = \frac{m_y f(y)}{m_z f(z)}$$

and since

$$\frac{f(w)}{f(x)} = \frac{f(y)}{f(z)}$$

then

$$\frac{m_w}{m_x} = \frac{m_y}{m_z}$$

The proportionality of the functional moduli is the only relationship necessary for the construction of the chart. After three moduli have been determined for the scale lengths desired, the fourth modulus must be in proportion to the other values. Many arrangements of scale are possible; the most convenient form is shown at (b). The uncalibrated diagonal line connects the zero values of the four scales and is used as an index or pivot line. The relationship of the reversed directions for the vertical and horizontal scales, as based on the geometry employed, should be noted. The $f(y)$ scale must be opposite the $f(z)$ scale, and in the reverse direction. The same is true of the $f(x)$ and $f(w)$ scales.

The design calculations and layout of an illustrative example of a proportional chart are shown in Fig. 29.23.

FIBER STRESS—HELICAL COMPRESSION AND EXTENSION SPRINGS

$$s = \frac{8PD}{\pi d^3}$$

$$\frac{s}{P} = \frac{8D}{\pi d^3}$$

s = fiber stress in torsion, 0–63,600 psi
P = axial load, 0–20,000 lb
D = mean diameter of spring, 0–10″
d = wire diameter, 0–2.0″

$$L_s = m_s[f(s_2) - f(s_1)] = m_s(63,600 - 0)$$
If $m_s = .0001$, $L_s = 6.36″$
$$L_P = m_P[f(P_2) - f(P_1)] = m_P(20,000 - 0)$$
If $m_P = .0004$, $L_P = 8.000″$
$$L_D = m_D[f(D_2) - f(D_1)] = m_D[8(10) - 8(0)] = 80\,m_D$$
If $m_D = .05$, $L_D = 4.00″$
$$\frac{m_s}{m_P} = \frac{m_D}{m_d}; \quad \frac{.0001}{.0004} = \frac{.05}{m_d}; \quad m_d = .20$$

$$L_d = m_d[f(d_2) - f(d_1)] = .2[\pi(2)^3 - \pi(0)^3] = 5.026$$

(a) DESIGN

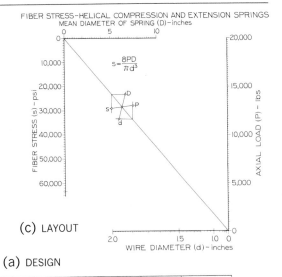

(c) LAYOUT

Variable	Function	m	M	Scale Length	Dir.	Scale Used
s	s	$.0001 = \dfrac{1}{10,000}$	$\dfrac{1}{10,000}$	6.36″	↓	10
P	P	$.0004 = \dfrac{4}{10000} = \dfrac{2}{5000}$	$\dfrac{2}{5000}$	8.00″	↑	50*
D	8D	$.05 = \dfrac{5}{100} = \dfrac{1}{20}$	$8m = \dfrac{8}{20} = \dfrac{2}{5}$	4.00″	→	50*
d	πd^3	$.2 = \dfrac{2}{10} = \dfrac{1}{5}$	$\dfrac{1}{5}$	5.026″	←	50

*Halved. (Actually $M = \frac{1}{25}$)

(b) DATA

f(s) = s	f(P) = P	f(D) = D	d	f(d) = πd^3
0	0	0	0	0
10,000	5,000	1	1	3.14
20,000	10,000	2	1.5	10.60
30,000	15,000	3	2.0	25.12
40,000	20,000	4		
50,000		5		
60,000		6		
63,600		7		
		8		
		9		
		10		

Fig. 29.23   Proportional Chart.

ALIGNMENT CHART PROBLEMS

The following problems provide laboratory experience in applying the methods described in this chapter and, in addition, introduce many typical engineering problems to the student. For convenience in assignment, the problems are classified according to type. Some problems have data also in the metric system; such data are shown in parentheses ( ).

Since many of the problems in this chapter are of a general nature, they can also be solved on most computer graphics systems. If a system is available, the instructor may choose to assign specific problems to be completed by this method.

Scale Layout. Construct scales for the following.

Prob. 29.1 $f(d) = \pi d$
d = diameter of circle,
    0″ to 6″ (0 to 152 mm)

Prob. 29.2 $f(x) = 2x - 3$ $x = 0$ to 15

Prob. 29.3 $f(x) = 2 \log x$ $x = 1$ to 10

Prob. 29.4 $f(x) = 3x$ $x = 1$ to 10

Prob. 29.5 $f(x) = \dfrac{1}{1 + x}$ $x = 1$ to 10

Two Variables: Conversion Scales. Construct two-variable conversion nomographs for the following.

Prob. 29.6 Area of a Circle.

$$A = \frac{\pi d^2}{4}$$

A = area of a circle, in.^{2} (mm^2)
d = diameter of a circle, 0″ to 6″
    (0 to 152 mm)

Prob. 29.7 Radian-Degrees Conversion.

$$r = \frac{N°}{57.3}$$

r = radians
$N°$ = number of degrees, 0° to 360°
    (1 revolution, 360° = 2π radians;
        1 radian = 57.3°)

Prob. 29.8 Velocity of Efflux (liquid flow through an orifice).

$$v = \sqrt{2gh}$$

v = velocity, ft/sec
h = height of liquid surface above orifice,
    0′ to 5′
g = acceleration due to gravity, constant
    = 32.2 ft/sec^2

Prob. 29.9 Major Diameter of Numbered Thread Sizes.

$$D = .013″N + .060″$$

D = major diameter of thread, in.
N = thread numbers, 0, 1, 2, 3, 4, 5, 6,
    8, 10, 12

Prob. 29.10 Hardness Conversion.

$$R_c = 88B^{.162} - 192$$

R_C = Rockwell-C hardness number
B = Brinell hardness number, 153 to 767

Prob. 29.11 Hardness Conversion.

$$R_B = \frac{(B - 47)}{0.0074B + .154}$$

R_B = Rockwell-B hardness number
B = Brinell hardness number, 100 to 352

Prob. 29.12 Coefficient of Friction Between Steel Wheels and Cast Iron Blocks—Rubbing Velocity.

$$f = \frac{.6}{\sqrt[3]{v}}$$

f = coefficient of friction
v = rubbing velocity, 0 to 88 ft/sec
   (0 to 27 m/sec)

Three Variables. Construct parallel-scale nomographs or N-charts for the following.

Prob. 29.13 Velocity of Free-Falling Body.

$$v_f = v_0 + gt$$

v_f = final velocity, ft/sec
v_0 = initial velocity, 0 to 50 ft/sec
 t = time, 0 to 20 sec
 g = acceleration due to gravity,
   constant = 32.2 ft/sec^2

Prob. 29.14 Rectangular Moment of Inertia for Circular Hollow Cylinder.

$$I = \frac{\pi(D^4 - d^4)}{64}$$

 I = rectangular moment of inertia, in.^{4} (mm^4)
D = outside diameter, .5" to 5.0"
   (13 to 127 mm)

d = inside diameter, .375" to 4.75"
   (9.5 to 120 mm)

Prob. 29.15 Weight of Steel Tube per Foot.

$$W = 2.65(d_0^2 - d_i^2)$$

W = weight per foot, lb
d_0 = outside diameter, 1" to 12"
d_i = inside diameter, $\frac{7}{8}$" to $11\frac{3}{4}$"

Prob. 29.16 Tube—Polar Radius of Gyration.

$$k_0 = .354\sqrt{D_1^2 + D_2^2}$$

k_0 = polar radius of gyration, in. (mm)
D_1 = outside diameter of tube, 1" to 12"
   (25.4 to 305 mm)
D_2 = inside diameter of tube, $\frac{3}{4}$" to $11\frac{1}{2}$"
   (19.1 to 292 mm)

Prob. 29.17 Differential Chain-Block Rise, Fig. 29.24.

$$H = \frac{\pi}{2}(D_a - D_b)$$

 H = rise, in. (mm) per one revolution of
   sheaves in upper block
D_a = large sheave diameter, upper block,
   4" to 12" (102 to 305 mm)
D_b = small sheave diameter, upper block,
   3" to 11" (76 to 280 mm)

Fig. 29.24 (Prob. 29.17).

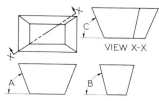

Fig. 29.25 (Prob. 29.18).

Prob. 29.18 Valley Slopes for Bin Hoppers, Fig. 29.25.

$$\cot^2 C = \cot^2 A + \cot^2 B$$

A = side angle, 30° to 75°
B = side angle, 30° to 75°
C = valley angle, degrees

Prob. 29.19 Ohm's Law.

$$V = IR$$

V = potential (voltage), volts
I = electric current, .1 to 10 amperes
R = resistance, 1 to 100 ohms

Prob. 29.20 Kinetic Energy.

$$E_k = \frac{Wv^2}{2g}$$

E_k = kinetic energy, ft-lb (kg-m)
W = weight of body in motion, 10 to 200 lb
(4.5 to 91 kg)
v = velocity of body, 5 to 100 ft/sec
(1.5 to 30.5 m/sec)
g = acceleration due to gravity = 32.2 ft/sec²
(9.8 m/sec²)

Prob. 29.21 Coefficient of Friction.

$$\mu = \frac{f}{N}$$

μ = coefficient of friction, .15 to .6
f = frictional force, lb
N = normal force (to plane of movement),
5 to 150 lb

Prob. 29.22 Torque-rpm-hp Relationship.

$$T = \frac{63025 \text{ hp}}{N}$$

T = torque, 1000 to 100,000 in.-lb
hp = horsepower, .02 to 3000 hp
N = shaft speed, rpm

Prob. 29.23 Pipe Flow Loss.

$$H = 6.2 \frac{V^{1.9}}{D^{1.16}}$$

H = head loss, ft per 1000' of pipe length
V = flow velocity, 1.0 to 50 ft/sec
D = internal pipe diameter, 1″ to 36″

Prob. 29.24 Internal Stress—Thick Cylinders (Lame's equation for brittle materials), Closed-End Cylinders.

$$s_t = p\left(\frac{1 + R^2}{R^2 - 1}\right)$$

s_t = working stress, tension at inner
radius, psi
p = working pressure in cylinder,
10 to 150 psi
$R = r_0/r_i$ = ratio of outside radius to inside
radius of cylinder, 1.2 to 5.0

Prob. 29.25 Energy of a Magnetic Field.

$$W = \tfrac{1}{2}LI^2$$

W = energy, joules
L = inductance, .01 to 4.0 henries
I = current, 1 to 30 amperes

Prob. 29.26 Discharge of Suppressed Weirs (Francis formula).

$$q = 3.33bh^{3/2}$$

q = discharge, ft³/sec
b = length of weir, 1' to 150'
h = head on weir, .1' to 15'

Prob. 29.27 Shaft Design.

$$s = 5.09 \frac{T}{d^3}$$

s = shear stress, psi
T = torque, 2500 to 90,000 in.-lb
D = shaft diameter, 1" to 10"

Four Variables. Construct the appropriate four-variable nomograph (parallel-scale, combination N-chart and parallel-scale nomographs, or proportional chart) for the following.

Prob. 29.28 Automotive Gear Ratios.

$$RS_c = DS_e$$

R = reduction ratios, 1 to 16
S_c = car speed, 0 to 120 mph
D = wheel diameter, in.
S_e = engine speed, 0 to 3500 rpm

Prob. 29.29 Electromagnetic Field Intensity.

$$H = \frac{2\pi NI}{10r}$$

H = field intensity, Oersteds
N = coil turns, 5 to 1000 turns
I = current, 1 to 20 amperes
r = coil radius, 1 to 10 cm

Prob. 29.30 Centrifugal Force.

$$F = .00034 \, N^2 rW$$

F = centrifugal force, lb
N = rotational speed, 10 to 10,000 rpm
r = radius of center of gravity, .1' to 5.0'
W = weight of rotating body, .5 to 500 lb

Prob. 29.31 Vibration Frequency of Spring.

$$n = \frac{761,500 \, d}{ND^2}$$

n = vibration per minute of spring

d = wire diameter, 0" to .1875"
   (0 to 4.75 mm)
D = mean diameter of spring, 0" to 1.0"
   (0 to 25.5 mm)
N = number of active coils, 4 to 50

Prob. 29.32 Section Moduli of Floor Beams.

$$s = \frac{12 \, WBL^2}{8(16,000)}$$

s = section modulus of beam or girder, in.^{2}
   (for maximum fiber stress in beam
   of 16,000 psi)
W = uniform load, 15 to 700 lb/ft^2
B = mean spacing of beams, 4' to 30'
L = span of beam, length of girder, 5' to 50'

Prob. 29.33 Orifice Flow.

$$R = 19.64 \, Cd^2\sqrt{h}$$

R = rate of flow, gal/min
C = orifice constant, .5 to 1.6
d = diameter of orifice, .06" to 1.0"
h = head of water, 3' to 100'

Prob. 29.34 Cylinder Bending Stress.

$$s = \frac{M}{\pi R^2 T}$$

s = maximum allowable stress, tensile or
   compressive, psi (kN/m^2)
M = bending moment, 100 to 100,000,000
   in.-lb (.115 to 115 000 kg-cm)
R = cylinder radius, 1" to 10"
   (2.54 to 25.4 cm)
T = cylinder wall thickness, .01" to 1.0"
   (0.25 to 25.4 cm)

Prob. 29.35 Maximum Deflection of Simple Steel Beam Under Uniform Load.

$$\Delta_{max} = \frac{5}{384}\left(\frac{1728 \, WL^4}{EI}\right)$$

Δ_{max} = maximum deflection, in.

W = total load per ft, 100 to 30,000 lb/ft

L = span in ft, 5' to 60'

$\ $ = modulus of elasticity

$\ $ = 29×10^6 lb/in.^{2} (constant value)

I = moment of inertia, 100 to 30,000 in.^{4}

Prob. 29.36 Critical Speed of End-Supported Bare Steel Shafts.

$$N_c = \frac{46.886(10)^5\sqrt{D_1^2 + D_2^2}}{L^2}$$

N_c = critical speed, rpm

L = shaft length, 20" to 60"
(508 to 1524 mm)

D_1 = outside diameter, 1" to 4"
(25.4 to 102 mm)

D_2 = inside diameter, .50" to 3.0"
(12.7 to 76.2 mm)

For aluminum, multiply N_c by 1.0026.
For magnesium, multiply N_c by .9879.

Prob. 29.37 Reynolds Number—Hydraulics.

$$R = \frac{7750\,VD}{v}$$

R = Reynolds number

V = velocity, 1 to 100 ft/sec

D = pipe inside diameter, .10" to 2.0"

v = kinematic viscosity, 5 to 500 centistokes

Prob. 29.38 Electrodeposit Design.

$$A = .167\,CBL$$

A = plating current, amperes

C = current density, 1 to 1000 amperes/ft^2

B = width of strip, 1" to 100"

L = length of immersed strip, 1' to 100'

Prob. 29.39 Steel Tape—Temperature Correction.

$$C_t = .0000065\,s(T - T_0)$$

C_t = temperature correction to measured
length, ft

s = measured length, ft

T = temperature at which measurements
are made, 0° to 120° Fahrenheit

T_0 = temperature at which tape is
standardized, 65° to 75° Fahrenheit
(68°F = Std)

Prob. 29.40 Moment of Inertia of Rectangular Bar.

$$I = \frac{M}{12}(b^2 + l^2)$$

I = moment of inertia of rectangular bar
about axis through its center and at right
angles to dimensions b and l, lb-ft^2

M = mass of bar, 2 to 10 lb

b = width of bar, .042' to .25'

l = length of bar, .50' to 2.0'

C H A P T E R 3 0

Empirical Equations

BY E. J. MYSIAK[*]

Empirical equations by definition are equations derived from experimental data or experience as distinguished from equations derived from logical reasoning or hypothesis (rational equations). At times, tabulated data or the analysis of a graph are inadequate, and an equation for the data is required. A graphical plot shows the trend of the data and the value of one variable relative to the corresponding second variable. The derivation of empirical equations is a more comprehensive method of analysis and can be used to calculate additional data not obtained in experimentation.

The derivations of equations are varied in methods. A basic procedure is to plot the data on rectangular, semilogarithmic, or logarithmic coordinate graph paper, in an attempt to obtain a straight line. If the plot results in a reasonably straight line on one of these papers, an approximate (empirical) equation can be derived by geometric and algebraic methods. Three of the more common methods of deriving the equation of a straight line are described in the following sections.

30.1 Data Plotting

The plot and graph-paper size should be large enough to suit the accuracy of the data. In most cases an 8.5″ × 11.0″ sheet will suffice. However, if the data contain three or four significant figures that cannot be accurately located, larger graph paper may be used. The major divisions of the rectangular grid should be subdivided into 10 parts for convenient plotting. See §28.5.

The standard practice of plotting the independent variable on the x-axis and the dependent variable on the y-axis is followed. The one exception to this practice may be when attempting to obtain a straight line on a semilogarithmic paper. If a straight line cannot be obtained when the y-axis is on the logarithmic grid, the data can be plotted on reversed axes. If a straight line is obtained, the notation of the equation is correspondingly reversed.

If the derived equation is to be of any value, the initial plotting must be done with care and accuracy. The points should be denoted carefully with small, sharp, light crosses or dots, which are better than circles, Fig. 30.1. Circles can be used later when preparing the graph for presentation.

---

[*]Engineering Manager, Phoenix Company of Chicago, Wood Dale, IL.

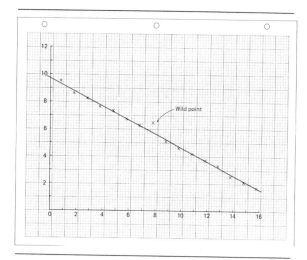

Fig. 30.1   Data Plotting and Curve Fitting.

30.2   Curve Fitting

After all the points of data are plotted, the straight line can be fitted by eye, Fig. 30.1. As many points as possible should be on the line. It is normal to have some points above and below the line, owing to deviations in reading instruments when obtaining the data. The straight line should be fitted with as many points above the

line as below the line to "balance out" the errors. The properly drawn straight line minimizes and accounts for the error of the data, resulting in an average line. Occasionally, a "wild" point will be plotted, which is at an abnormal distance from the general path of the curve. A point of this type may be due to a misread instrument or a mistake in recording the data and should be disregarded.

When the data are numerous and the deviations of instrument readings are greater, a "scatter" plotting will result, Fig. 30.2. The straight line is fitted by enveloping the maximum and minimum points with light straight lines. An average line is determined, which will be near the midpoints within the envelopment, depending on the density of the points plotted.

30.3   Limitations and Accuracy

The derived empirical equation is at best an approximate representation of the data. The equation is valid only within the range of the experiment or observed data. Extrapolation of the curve (extension beyond the known range) or calculations from the equation for extrapolated values should be treated with caution.

The equation derived cannot be more accurate than the data used. If the data are accurate to three significant figures, the equation must also be expressed to the same number of significant figures.

Errors are possible for a number of reasons. Variations of instrument accuracy (sensitivity) or the readings taken, which result in deviations of the points plotted from a straight line, are self-compensating in the graphical solution. This is accomplished by fitting the average line to the plotting and does not affect the validity of the data. "Wild" points, which occur infrequently due to the erroneous readings or data recording, are noticeable in a graphical method and can be disregarded. Errors due to faulty instruments, calibrations, or settings will result in all the data being high or low. Graphical solutions will not compensate for this situation. The instrument must be corrected.

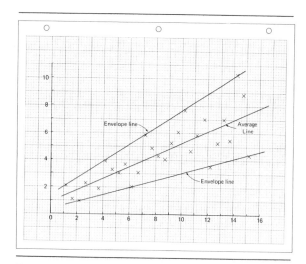

Fig. 30.2   Curve Fitting—"Scatter" Data.

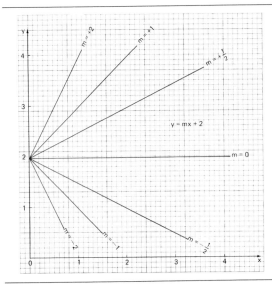

Fig. 30.3 Equation of Straight Lines on Rectangular Grid.

Torque (T), oz-in.	Current Input to Motor (I), amperes
10	6.6
20	8.1
30	9.6
40	11.2
50	12.6
60	14.3

(a) TABULATED DATA

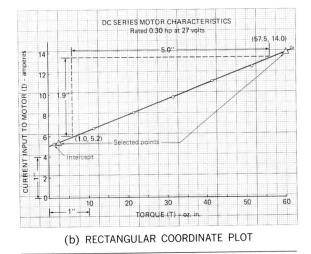

(b) RECTANGULAR COORDINATE PLOT

Fig. 30.4   Rectangular Coordinate Solution.

30.4   Empirical Equations—Solution by Rectangular Coordinates

The equation for a straight line on a rectangular coordinate grid is $y = mx + b$. As shown in Fig. 30.3, m represents the slope of the line (the tangent of the angle between the line and the x-axis) and b is the intercept on the y-axis (when $x = 0$). A negative slope is inclined downward to the right. A positive slope has an upward trend to the right. The intercept may be positive or negative.

In Fig. 30.4, the data presented at (a), plotted on rectangular coordinate paper, allow a straight line to be fitted to the points, as shown at (b). The procedure for derivation of the equation requires the determination of the slope of the line m and the I-intercept b when $T = 0$.

SLOPE-INTERCEPT METHOD   The slope-intercept method requires the use of a graphical plot. Since in a majority of cases the scale modulus on the y-axis does not equal the scale modulus on the x-axis, the trigonometric slope of the line

s must be divided by the ratio of the axes scale moduli r to determine the slope of the line.

$$m = \frac{s}{r} = \frac{\text{trig. slope} = \dfrac{y \text{ distance}}{x \text{ distance}}}{\dfrac{\text{in. per unit } y\text{-axis}}{\text{in. per unit } x\text{-axis}}}$$

$$= \frac{\dfrac{1.9}{5.0}}{\dfrac{.250}{.10}} = \frac{.380}{2.50} = .152$$

The slope of the line may also be obtained as follows.

$$m = \frac{I_2 - I_1}{T_2 - T_1}$$

$$= \frac{14.0 - 5.2}{57.5 - 1.0} = \frac{8.8}{56.5} = .156$$

The use of coordinates for slope determination will automatically result in the appropriate slope sign if treated algebraically. The difference in the slope values obtained from the two methods of calculation, though small, exists because the data used were different in each case. The first method required measurements of a y distance and the corresponding x distance. Selected coordinates on the straight line were used for the second method.

By observation, the intercept value (I value, when $T = 0$) is 5.05. The derived equation is

$$I = .152T + 5.05.$$

SELECTED POINTS METHOD   Since two unknowns must be determined, m and b, two equations solved simultaneously provide another method of equation derivation. Two points are selected; then data values on the line or points are established on the line. The points selected should be widely spread. The T and I values for these two points, substituted into the equation $I = mT + b$, will provide a solution.

Selected points: (57.5, 14.0) and (1.0, 5.2)

$I = mT + b$:

$$14.0 = 57.5m + b \qquad (1)$$
$$5.2 = 1.0m + b \qquad (2)$$

Subtracting: $8.8 = 56.5m$
$$m = .156$$

Substituting in (1):

$$14.0 = 57.5(.156) + b$$
$$b = 14.0 - 8.97 = 5.03$$

Equation is

$$I = .156T + 5.03$$

The solution of the simultaneous equations, algebraically treated, automatically denotes the sign of the slope.

METHOD OF AVERAGES   If a definite number of data values of y were added together, and an equal number of computed values of y (from a derived equation) were also added, a difference of zero ($\Sigma y_{data} - \Sigma y_{computed} = 0$) between the two sums would indicate close agreement. On this premise, if the equation $\Sigma y = m\Sigma x + nb$ is applied to two equal groups of data values of x and y, solution of two simultaneous equations will provide the solution for the slope and intercept values. The term n in the equation is the number of data values of x and y summed.

In the example used, six items of data are available; therefore, two groups of three are added together.

$$\Sigma I = m\Sigma T + nb$$

Sum of second
3 data items: $38.1 = 150m + 3b$ (1)
Sum of first 3
data items: $24.3 = 60m + 3b$ (2)

Subtracting: $13.8 = 90m$
$$m = .153$$

Substituting in (1):

$$38.1 = 150(.153) + 3b$$
$$b = \frac{15.15}{3} = 5.05$$

Equation is

$$I = .153T + 5.05$$

The method of averages is based on a numerical derivation from the data only. The graphical plot is required only to verify the form of the curve and need not be accurate. This method requires additional time, and care must be taken to prevent errors in calculation. Equation derivation by the slope-intercept method or selected points method is dependent on the accuracy of the representative straight line and not the data.

30.5 Computation of Residuals

In order to determine the best approximation of the three derived equations, *residuals* should be computed. With the use of the derived empirical equations, the *dependent variable* is computed, using the independent variable data values. A comparison of computed values with observed data values reveals a plus or minus difference (residual), Table 30.1. If the observed value is smaller than the computed value, the residual is commonly denoted as minus. The algebraic sum of the residuals for each equation suggests which is the best equation (the smallest sum, plus or minus). For the example problem used in §30.4, the equation derived by the method of averages seems to be the best approximation.

If the data are accurate and fit a straight line closely, any of the three methods will produce good results. If the data are approximate and still fit a straight line well, the selected points method will produce the best results. When the straight line must be extended an appreciable distance to obtain the intercept value, inaccuracy can be

expected, and the slope-intercept method is not a reliable method.

As a verification of the fact that the accuracy of derivation is dependent on the data accuracy in comparison to the method (graphical or numerical), the following correlation was performed.

In the method of averages, the first three and second three data values were added to obtain the equation $I = .153T + 5.05$. If the 1st, 3rd, and 5th data values, and the 2nd, 4th, and 6th data values were added instead, the solution of the simultaneous equations would result in the equation $I = .160T + 4.80$. A comparison of the two equations indicates a great difference and a poor correlation because of the data. The numerical method cannot be assumed to be automatically a more accurate method in comparison to the graphical or semigraphical methods of slope-intercept or selected points, respectively.

30.6 Empirical Equations— Semilog Coordinates, $y = b(10)^{mx}$ or $y = be^{mx}$

Data plotted on rectangular coordinate paper, which do not result in a straight line, may rectify to an approximate straight line graph on semi-

Table 30.1 *Computation of Residuals for Equations in §30.4*

Observed Data		Slope-Intercept $I = .152T + 5.05$		Selected Points $I = .156T + 5.03$		Method of Averages $I = .153T + 5.05$	
T	I	I_{C_1}	Residual	I_{C_2}	Residual	I_{C_3}	Residual
10	6.6	6.6	.0	6.6	.0	6.6	.0
20	8.1	8.1	.0	8.2	−.1	8.1	.0
30	9.6	9.6	.0	9.7	−.1	9.6	.0
40	11.2	11.1	+.1	11.3	−.1	11.2	.0
50	12.6	12.7	−.1	12.8	−.2	12.7	−.1
60	14.3	14.2	+.1	14.4	−.1	14.2	+.1
		Sum = +.1		Sum = −.6		Sum = .0	

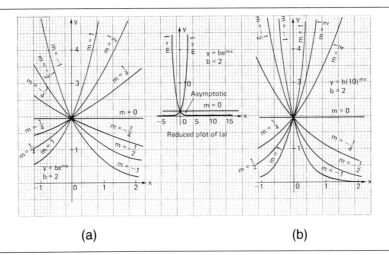

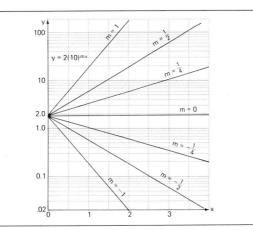

Fig. 30.5   Exponential Curves.

logarithmic coordinate grid, if the rectangular coordinate curve resembles an exponential curve, as shown in Fig. 30.5 (a) or (b). The base of the exponent may be either e ($e = 2.718$) or 10. An exponential curve intersects one of the axes at a steep angle and is asymptotic to the other axis. The same data used to plot some of the curves in Fig. 30.5 (b), when plotted on semi-log coordinate paper, rectify to straight lines as shown in Fig. 30.6.

Fig. 30.6   Semilog Plot.

Since semilog paper has logarithmic divisions along one of the axes (normally designated the y-axis), the equation for a straight line on this type of graph paper is

$$\log y = mx + \log b$$

or

$$\ln y = mx + \ln b$$

therefore,

$$y = b(10)^{mx} \quad \text{or} \quad y = be^{mx}$$

The derivation of the empirical equation requires the solution for the values b (y-axis intercept, when $x = 0$) and m (the slope of the straight line to the x-axis). The same three methods used for rectangular coordinate solutions, §30.4, are applicable.

An alternate method is to plot $\log y$ values on rectangular coordinate graph paper, if semilog coordinate paper is not readily available, Fig. 30.7.

When a rectangular coordinate plot results in an x-axis intercept and a curve appearing asymptotic to the y-axis, Fig. 30.8, the semilog plotting may rectify to a straight line if the logarithmic

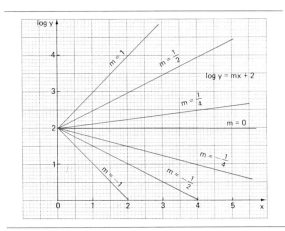

Fig. 30.7 Rectangular Grid Solution.

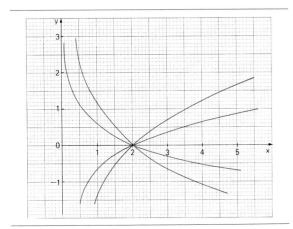

Fig. 30.8 Reverse Plot.

scale is placed on the *x*-axis. The equation becomes

$$x = a(10)^{my} \quad \text{or} \quad x = ae^{my}$$

The data for friction in thrust bearings, Fig. 30.9 (a), when plotted on rectangular coordinate paper, result in the curve shown at (b). The *y*-axis intercept and appearance of being asymptotic to the *x*-axis suggest the possibility of a straight-line plot on semilog coordinate paper, (c).

Friction coefficient factor (k)	Log k	Ratio (I/a)
1.00	—	.0
1.05	.021	.25
1.15	.061	.50
1.30	.114	.75
1.45	.161	1.00
1.65	.218	1.25
1.80	.255	1.50
2.00	.301	1.75
2.15	.332	2.00
2.35	.371	2.25
2.60	.415	2.50
2.80	.447	2.75
3.00	.477	3.00

(a) DATA

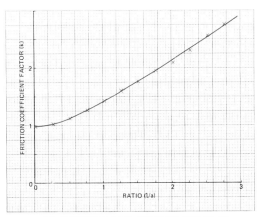

(b) RECTANGULAR COORDINATE PLOT

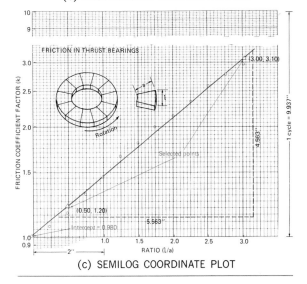

(c) SEMILOG COORDINATE PLOT

Fig. 30.9 Semilog Empirical Equation Derivation.

SLOPE-INTERCEPT METHOD

Intercept $= .980$

$$\text{Slope } (m) = \frac{s}{r}$$

$$= \frac{\dfrac{4.563''}{5.563''}}{\dfrac{9.937'' \text{ per unit cycle}}{2'' \text{ per unit}}} = \frac{.820}{4.97}$$

$$= .165$$

Notice that the scale moduli are the length of one log cycle on the log axis and the length per unit (one) on the rectangular coordinate axis. The log cycle should be measured, since printing variations occur. The $10''$ log cycle measured $9.937''$ on the semilog graph paper used in this example. The intercept in this example is $.980$.

ALTERNATE SLOPE DETERMINATION (COORDINATE METHOD)

$$\text{Slope } (m) = \frac{\log y_2 - \log y_1}{x_2 - x_1}$$

$$= \frac{\log 3.1 - \log 1.2}{3.00 - .5}$$

$$= \frac{.4125}{2.50} = .165$$

$$y = .980(10)^{.165x} \qquad \boxed{k = .980(10)^{.165(l/a)}}$$

If logarithms to the base e are preferred,

$$\text{Slope } (m) = \frac{\ln y_2 - \ln y_1}{x_2 - x_1}$$

$$= \frac{\ln 3.1 - \ln 1.2}{3.00 - .5}$$

$$= \frac{1.132 - .183}{2.50}$$

$$= \frac{.949}{2.50} = .380$$

$$y = .980e^{.380x} \qquad \boxed{k = .980e^{.380(l/a)}}$$

SELECTED POINTS METHOD
Selected points: $(.50, 1.2)$ and $(3.00, 3.10)$

$y = b(10)^{mx}$:

$$\log y = \log b + mx \log 10$$
$$\log 3.10 = \log b + m(3.00)$$
$$.492 = \log b + 3.00m \qquad (1)$$
$$\log 1.20 = \log b + m(.50)$$
$$.079 = \log b + .50m \qquad (2)$$

Subtracting (2) from (1) gives

$$.413 = 2.50m$$
$$m = .165$$

Substituting into equation (1):

$$.492 = \log b + 3.00(.165)$$
$$\log b = .492 - .495$$
$$= -.003 = 9.997 - 10$$
$$b = .993 \qquad \boxed{k = .993(10)^{.165(l/a)}}$$

The solution for logarithms to the base e is as follows.

$y = be^{mx}$:

$$\ln y = \ln b + mx \ln e$$
$$\ln 3.10 = \ln b + m(3.00) \ln e$$
$$1.132 = \ln b + 3.00m \qquad (1)$$

$$\ln 1.2 = \ln b + m(.50) \ln e$$
$$.183 = \ln b + .50m \qquad (2)$$

Subtracting (2) from (1) gives

$$.949 = 2.50m$$
$$m = .380$$

Substituting into equation (1):

$$1.132 = \ln b + 3.00(.380)$$
$$\ln b = 1.132 - 1.140$$
$$= -.008$$
$$b = .992$$

$$y = .992e^{.380x} \qquad \boxed{k = .992e^{.380(l/a)}}$$

METHOD OF AVERAGES

$$\Sigma \log y = n \log b + m\Sigma x \log 10$$

Table 30.2   Computation of Residuals for Equations in §30.6

Observed Data		Slope-Intercept $k = .980(10)^{.165(l/a)}$		Selected Points $k = .993(10)^{.165(l/a)}$		Method of Averages $k = .980(10)^{.168(l/a)}$	
l/a	k	k_{C_1}	Residual	k_{C_2}	Residual	k_{C_3}	Residual
.00	1.00	.98	$+.02$	.99	$+.01$	.98	$+.02$
.25	1.05	1.08	$-.03$	1.09	$-.04$	1.08	$-.03$
.50	1.15	1.19	$-.04$	1.20	$-.05$	1.19	$-.04$
.75	1.30	1.30	.00	1.32	$-.02$	1.31	$-.01$
1.00	1.45	1.42	$+.03$	1.45	.00	1.44	$+.01$
1.25	1.65	1.58	$+.07$	1.60	$+.05$	1.59	$+.06$
1.50	1.80	1.73	$+.07$	1.76	$+.04$	1.75	$+.05$
1.75	2.00	1.91	$+.09$	1.93	$+.07$	1.93	$+.07$
2.00	2.15	2.09	$+.06$	2.12	$+.03$	2.12	$+.03$
2.25	2.35	2.31	$+.04$	2.33	$+.02$	2.34	$+.01$
2.50	2.60	2.53	$+.07$	2.57	$+.03$	2.58	$+.02$
2.75	2.80	2.79	$+.01$	2.83	$-.03$	2.83	$-.03$
3.00	3.00	3.07	$-.07$	3.11	$-.11$	3.13	$-.13$
		Sum $= +.32$		Sum $= .00$		Sum $= +.03$	

Sum of last 6 terms:

$$2.343 = 6 \log b + m(14.25) \qquad (1)$$

Sum of first 6 terms:

$$.830 = 6 \log b + m(5.25) \qquad (2)$$

Subtracting (2) from (1) gives

$$1.513 = \qquad m(9.00)$$
$$m = .168$$

Substituting into equation (2):

$$.830 = 6 \log b + 5.25(.168)$$
$$\log b = \frac{.830 - .833}{6}$$
$$= -.009 = 9.991 - 10$$
$$b = .980 \qquad k = .980(10)^{.168(l/a)}$$

The solution for logarithms to the base e is as follows.

$$\Sigma \ln y = n \ln b + m \Sigma x \ln e$$

Sum of last 6 terms:

$$5.397 = 6 \ln b + m(14.25) \qquad (1)$$

Sum of first 6 terms:

$$1.911 = 6 \ln b + m(5.25) \qquad (2)$$

Subtracting (2) from (1) gives

$$3.486 = \qquad m(9.00)$$
$$m = .387$$

Substituting into equation (2):

$$1.911 = 6 \ln b + .387(5.25)$$
$$\ln b = \frac{1.911 - 2.032}{6}$$
$$= -.0202$$
$$b = .980$$
$$y = .980e^{.387x} \qquad k = .980e^{.387(l/a)}$$

Residuals for the equation form $y = b(10)^{mx}$ were computed for the best approximation and are tabulated in Table 30.2.

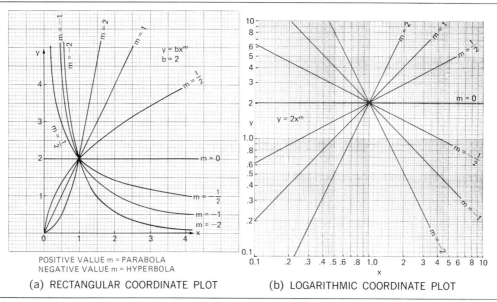

POSITIVE VALUE m = PARABOLA
NEGATIVE VALUE m = HYPERBOLA

(a) RECTANGULAR COORDINATE PLOT   (b) LOGARITHMIC COORDINATE PLOT

Fig. 30.10   Power Curves.

30.7   Empirical Equations— Logarithmic Coordinates, $y = bx^m$

Curves that plot as a parabola through the origin, or a hyperbola asymptotic to the x- and y-axes, are known as power curves, Fig. 30.10 (a), and can be rectified to a straight line by plotting the same data on logarithmic coordinate paper, as shown at (b). The equation for a straight line on logarithmic paper is

$$\log y = m \log x + \log b$$

or

$$y = bx^m$$

The derivation of an empirical equation requires the determination of the values for the slope m of the line and the intercept value b (y intercept when $x = 1$). The intercept is at the axis value of 1.0 since $\log 1.0 = 0$. The three methods of equation derivation used for rectangular and semilog coordinate solutions, §§30.4 and 30.6,

are applicable to logarithmic coordinate plottings.

As an example of equation derivation, the data for head discharge for a 1.00″ diameter circular orifice, Fig. 30.11 (a), were plotted on rectangular coordinate grid paper, (b). The resulting curve, a parabola through the origin, suggests a power curve. Plotting the same data on logarithmic graph paper yields a straight-line graph, as shown in (c).

SLOPE-INTERCEPT METHOD   In order to obtain the intercept value b on the y-axis, when $x = 1.0$, the straight line must be extended downward and to the left, since the y-axis values are only to .1. Three-cycle log paper was used; however, the intercept is still below the bottom cycle. The intercept can be measured, as shown. The use of four-cycle log paper would result in a less accurate plotting. This procedure is necessary only for the slope-intercept method.

(a) DATA

Discharge (D), ft³/sec	Head (H), ft³
4	1.15
5	1.75
6	2.5
7	3.4
8	4.5
9	5.5
10	6.9
15	15.5
20	27
25	43
30	63

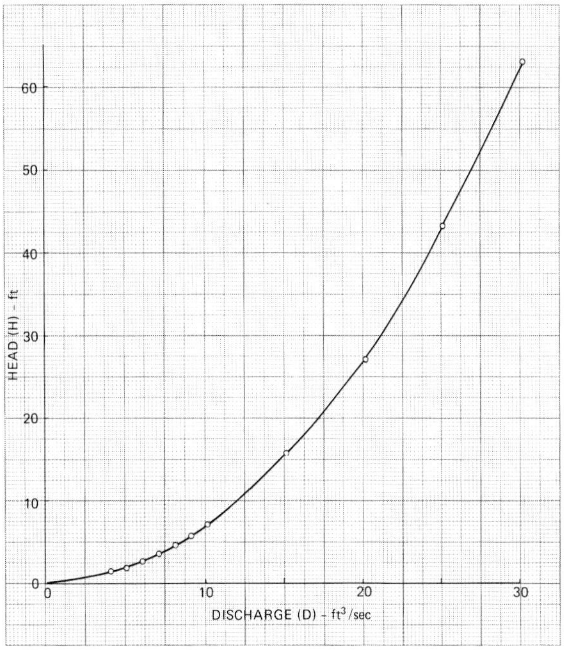

(b) RECTANGULAR COORDINATE PLOT

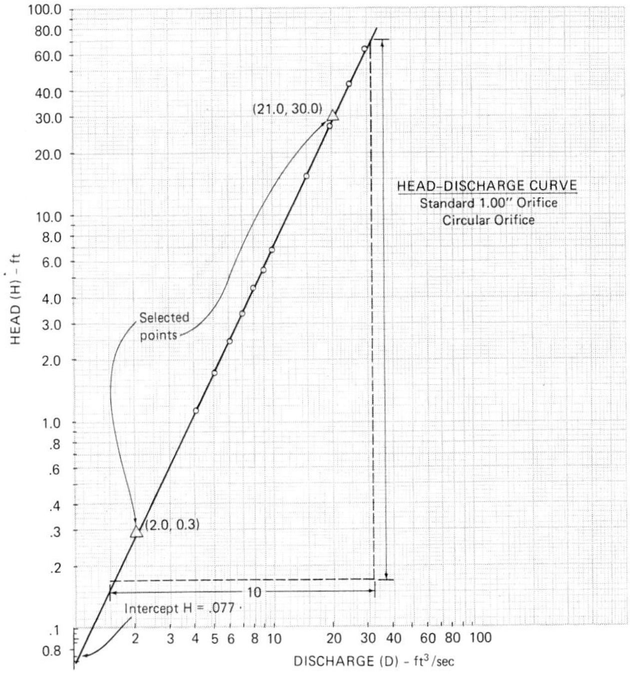

(c) LOGARITHMIC PLOT

Fig. 30.11   Logarithmic Empirical Equation Derivation.

Intercept $b = .077$,   when $D = 1.00$.

If the lengths of the log cycles are the same along both axes, the slope m is the tangent of the angle between the straight line and the x-axis.

$$\text{Slope } m = \frac{19.55}{10} = 1.955$$

Equation is

$$H = .077D^{1.955}$$

SELECTED POINTS METHOD
Selected points: $(2.0, .3)$ and $(21.0, 30.0)$

$y = bx^m$:

$$
\begin{aligned}
\log y &= \log b + m \log x \\
1.477 &= \log b + m(1.322) \quad (1) \\
-.523 &= \log b + m(.301) \quad (2)
\end{aligned}
$$

Subtracting: $2.000 = \qquad m(1.021)$
$$m = 1.960$$

Substituting in equation (1):

$$
\begin{aligned}
1.477 &= \log b + 1.960(1.322) \\
\log b &= 1.477 - 2.590 \\
&= -1.113 = 8.887 - 10 \\
b &= .0771 \qquad H = .0771D^{1.960}
\end{aligned}
$$

METHOD OF AVERAGES

$y = bx^m$:

$$
\begin{aligned}
\log y &= \log b + m \log x \\
\sum \log y &= n \log b + m \sum \log x
\end{aligned}
$$

Sum of second 5:

$$5.836 = 5 \log b + m(5.829) \quad (1)$$

Sum of first 5:

$$1.887 = 5 \log b + m(3.827) \quad (2)$$

Subtracting: $3.949 = \qquad m(2.002)$
$$m = 1.971$$

Table 30.3   *Computation of Residuals for Equation in §30.7*

Observed Data		Slope-Intercept $H = .077DE^{1.955}$		Selected Points $H = .0771D^{1.960}$		Method of Averages $H = .0739D^{1.971}$	
D	H	H_{C_1}	Residual	H_{C_2}	Residual	H_{C_3}	Residual
4	1.15	1.16	− .01	1.17	− .02	1.14	+ .01
5	1.75	1.80	− .05	1.81	− .06	1.77	− .02
6	2.50	2.56	− .06	2.59	− .09	2.53	− .03
7	3.40	3.47	− .07	3.51	− .11	3.44	− .04
8	4.50	4.52	− .02	4.57	− .07	4.47	+ .03
9	5.50	5.67	− .17	5.74	− .24	5.67	− .17
10	6.90	6.98	− .08	7.05	− .15	6.95	− .05
15	15.50	15.5	.00	15.7	− .20	15.5	.00
20	27	27.1	− .10	27.5	− .50	27.4	− .40
25	43	42	+ 1.0	42.8	+ .20	42.7	+ .30
30	63	60	+ 3.0	60.9	+ 2.1	60.6	+ 2.4
		Sum =	+ 3.44	Sum =	+ .86	Sum =	+ 2.03

INFLUENCE OF TOLERANCE ON COST OF MACHINING

Tolerance (T), in.	Relative Cost of Machining (C)	Log T	Log C
.001	5.00	−3.000	.699
.002	2.60	−2.699	.415
.003	1.70	−2.523	.231
.004	1.25	−2.398	.097
.005	1.00	−2.301	.000
.006	.90	−2.222	−.046
.007	.80	−2.155	−.097

(a) DATA

Substituting in equation (1):

$$5.836 = 5 \log b + (1.971)(5.829)$$
$$\log b = \frac{5.836 - 11.49}{5}$$
$$= -1.131 = 8.869 - 10$$
$$b = .0739 \qquad H = .0739D^{1.971}$$

To test for the best approximation, residuals were computed and tabulated in Table 30.3.

As a second example of equation derivation, the data for influence of tolerance on cost of machining, Fig. 30.12 (a), were plotted on rectangular coordinate grid paper, (b). The resulting curve, asymptotic to both axes, suggests a power curve. The same data plotted on logarithmic graph paper rectifies to a straight-line graph, as shown in (c).

SLOPE-INTERCEPT METHOD   In order to obtain the intercept value b on the y-axis, when $x = 1.0$, the straight line must be extended downward and to the right, since the x-axis values are only to .10. Point A was shifted to point B (corresponding value) on the upper portion of the graph paper to enable the extension of the curve on the same graph paper. This procedure is necessary only for the slope-intercept method.

Intercept $b = .0062$,   when $T = 1.00$.

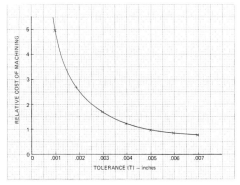

(b) RECTANGULAR COORDINATE PLOT

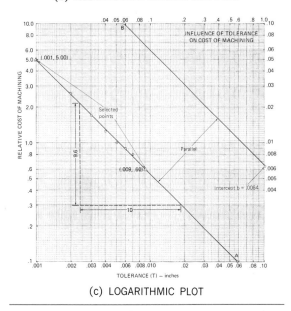

(c) LOGARITHMIC PLOT

Fig. 30.12   Logarithmic Empirical Equation Derivation.

909

Table 30.4 *Computation of Residuals for Equations in §30.7*

Observed Data		Slope-Intercept $C = .0062T^{-.960}$		Selected Points $C = .00637T^{-.965}$		Method of Averages $C = .00528T^{-.994}$	
T	C	C_{C_1}	Residual	C_{C_2}	Residual	C_{C_3}	Residual
.001	5.00	4.77	+.23	4.97	+.03	5.02	−.02
.002	2.60	2.43	+.17	2.55	+.05	2.54	+.06
.003	1.70	1.64	+.06	1.73	−.03	1.69	+.01
.004	1.25	1.25	.00	1.31	−.06	1.28	−.03
.005	1.01	1.01	−.01	1.06	−.06	1.02	−.02
.006	.85	.85	+.05	.89	+.01	.86	+.04
.007	.73	.73	+.07	.77	+.03	.73	+.07
			Sum = +.57		Sum = −.03		Sum = +.11

The lengths of the log cycles are usually the same along both axes; the slope m is the tangent of the angle between the straight line and the x-axis.

$$\text{Slope } m = \frac{-9.6}{10.0} = -.960$$

The sign of the slope must be included in the computation.

$$C = .0062T^{-.960}$$

SELECTED POINTS METHOD
Selected points: (.001, 5.00) and (.009, .60)

$y = bx^m$:
$$\log y = \log b + m \log x$$
$$.699 = \log b + m(-3.00) \qquad (1)$$
$$9.778 - 10 = \log b + m(7.954 - 10)$$

or
$$-.222 = \log b + m(-2.046) \qquad (2)$$

Subtracting: $.921 = \qquad m(-.954)$
$$m = -.965$$

Substituting in equation (1):

$$.699 = \log b + (-.965)(-3.000)$$
$$\log b = .699 - 2.895$$
$$= -2.196 = 7.804 - 10$$
$$b = .00637 \qquad \boxed{C = .0064T^{-.965}}$$

METHOD OF AVERAGES

$y = bx^m$:

$$\log y = \log b + m \log x$$
$$\Sigma \log y = n \log b + m\Sigma \log x$$

Sum of first 3:

$$1.345 = 3 \log b + m(-8.222) \qquad (1)$$

Sum of second 3:

$$.051 = 3 \log b + m(-6.921) \qquad (2)$$

Subtracting (2) from (1) gives

$$1.294 = \qquad m(-1.301)$$
$$m = -.994$$

Substituting in equation (1):

$$1.345 = 3 \log b + (-.994)(-8.222)$$
$$\log b = \frac{1.345 - 8.175}{3}$$
$$= -2.277 = 7.723 - 10$$
$$b = .00528 \qquad \boxed{C = .0053T^{-.994}}$$

To test for the best approximation, residuals were computed and tabulated in Table 30.4.

EMPIRICAL EQUATIONS PROBLEMS

Plot the data for the following problems on the appropriate form of graph paper. Determine the equation by the three methods included in the text, and analyze with residuals the best equation. Some problems have data also in the metric system; such data are shown in parentheses ( ).

Since many of the problems in this chapter are of a general nature, they can also be solved on most computer graphics systems. If a system is available, the instructor may choose to assign specific problems to be completed by this method.

RECTANGULAR COORDINATE SOLUTION

Prob. 30.1 Thermal Conductivity of Insulating Materials.

Prob. 30.2 Power at Different Altitudes for Compound Compression of Air.

Mean Temperature (T), °F (°C)	Thermal Conductivity (C), Btu/ft²/hr/°F/in.	
	Celotex	Rockwool Blanket
40 (4.4)	.35	.24
80 (26.7)	.355	.27
120 (48.9)	.36	.30
160 (71.1)	.365	.32
200 (93.3)	.372	.35
240 (115.6)	.380	.38

Altitude (A), ft above sea level	Power (P), indicated hp/100 ft³/min	
	80 lb gage	125 lb gage
1000	13.55	16.83
2000	13.30	16.50
3000	13.05	16.17
4000	12.8	15.83
5000	12.55	15.50
6000	12.30	15.17
7000	12.05	14.85
8000	11.80	14.50
9000	11.55	14.17
10,000	11.30	13.82

Prob. 30.3 Impact Breaking Strength of Gray Iron (specimen: cantilever beam $\frac{9}{16}''$ long, $\frac{1}{2}''$ wide; Class 20 gray cast iron).

Thickness of $\frac{1}{2}''$ Wide Specimen (t), in. (mm)	Fracture Energy (E), (scatter data)	
	ft-lb	(kg-m)
0 (.00)	0	(.00)
$\frac{1}{8}$ (3.175)	1–2	(.138–.277)
$\frac{1}{4}$ (6.350)	$2\frac{3}{8}$–$3\frac{7}{8}$	(.329–.536)
$\frac{3}{8}$ (9.525)	$3\frac{3}{4}$–$5\frac{1}{2}$	(.519–.760)
$\frac{1}{2}$ (12.70)	$5\frac{1}{2}$–7	(.760–.968)
$\frac{5}{8}$ (15.875)	7–$8\frac{5}{8}$	(.968–1.192)
$\frac{3}{4}$ (19.05)	$8\frac{1}{2}$–$10\frac{1}{4}$	(1.175–1.418)

Prob. 30.4 Gear Pump Efficiency (internal-external gear pump at 600 rpm).

Pressure (P), psi (kN/m^2)	Volumetric Efficiency (V), %
100 (6.90)	82.0
200 (1380)	75.0
300 (2070)	68.5
400 (2760)	61.5
500 (3450)	54.5
600 (4140)	47.5
700 (4830)	40.0
800 (5520)	33.0

Prob. 30.5 Thermal Aging of Adhesives.

Aging Time (T), yr	Shear Strength (s_s), 1000 psi (1000 kN/m^2)			
	Nitrile Phenolic		Epoxy Phenolic	
	At 250°F (121°C)	At 350°F (177°C)	At 250°F (121°C)	At 350°F (177°C)
0	2.20 (15.2)	1.60 (11.0)	2.08 (14.3)	1.95 (13.4)
$\frac{1}{4}$	2.18 (15.0)	1.50 (10.3)	1.93 (13.3)	1.70 (11.7)
$\frac{1}{2}$	2.13 (14.7)	1.40 (9.7)	1.80 (12.4)	1.45 (10.0)
$\frac{3}{4}$	2.10 (14.5)	1.30 (9.0)	1.65 (11.4)	1.20 (8.3)
1	2.05 (14.1)	1.20 (8.3)	1.52 (10.5)	.90 (6.2)
$1\frac{1}{4}$	2.00 (13.8)	1.10 (7.6)	1.38 (9.5)	.65 (4.5)
$1\frac{1}{2}$	1.97 (13.6)	1.00 (6.9)	1.25 (8.6)	.40 (2.8)
$1\frac{3}{4}$	1.93 (13.3)	.90 (6.2)	1.10 (7.6)	—
2	1.87 (12.9)	.80 (5.5)	.95 (6.6)	—
$2\frac{1}{4}$	1.83 (12.6)	.70 (4.8)	.80 (5.5)	—
$2\frac{1}{2}$	1.80 (12.4)	.60 (4.1)	.68 (4.7)	—

Prob. 30.6 Coefficient of Friction versus Feed Rate (material: SAE 4340 steel, 200 BHN; cutting speed: 542 rpm; tool: sintered carbide, $+0°$ rake angle).

Feed Rate (t), in./revolution	Coefficient of Friction (μ)
.00055	1.175
.0011	1.075
.00235	1.050
.00295	1.025
.00370	.975
.00620	.850

Prob. 30.7 Warmup Heat.

Temperature Rise ($T_{final} - T_{initial} = \Delta t$) °F (°C)	Warmup Heat (H), Btu/lb			
	Water	Paraffin Wax	Aluminum, Air, Nitrogen	Copper, Steel, Stainless
0 (-18)	0	0	0	0
50 (10)	50	35	10	5
100 (38)	100	70	25	10
150 (66)	150	105	35	15
200 (93)	200	140	47	20
250 (121)	—	175	58	25
300 (149)	—	210	70	30
350 (177)	—	245	83	35
400 (204)	—	280	95	40

SEMILOG COORDINATE SOLUTION

Prob. 30.8   Allowable Working Stresses for Compression Springs (chrome–vanadium wire, ASTM–A231, SAE 6150)

Wire Diameter (D), in. (mm)	Torsional Stress (S_t), 1000 psi (1000 kN/m^2)	
	Severe Service	Average Service
.03 (.0008)	91 (630)	121 (830)
.05 (.0013)	84 (580)	112 (770)
.07 (.0018)	79 (550)	105 (720)
.09 (.0023)	76 (520)	101 (700)
.11 (.0028)	73 (500)	97 (670)
.13 (.0033)	71 (490)	94 (650)
.15 (.0038)	69 (480)	92 (630)
.17 (.0043)	67 (460)	90 (620)
.19 (.0049)	66 (455)	87 (600)
.21 (.0053)	64 (440)	86 (590)
.23 (.0058)	63 (435)	84 (580)
.25 (.0064)	62 (430)	83 (570)

Prob. 30.9   Electrical Resistance of 95% Alumina Ceramic.

Temperature (T), °C	Electrical Resistance (R), megohms
400	240
450	120
500	60
550	31
600	15
650	7.5
700	3.7
750	2.0
800	.9
850	.47

Prob. 30.10   High-Temperature Alloy (René 41 strength).

Rupture Life (L) at 1800°F (982°C), hr	Stress (S) 1000 psi (1000 kN/m^2)	
20	14.3	(98.6)
40	12.9	(89.0)
60	12.0	(82.7)
80	11.5	(79.3)
100	11.0	(75.9)
200	9.7	(66.9)
400	8.3	(57.2)
600	7.5	(51.7)
800	6.9	(47.6)
1000	6.5	(44.8)

Prob. 30.11 K Factors for Preliminary Estimates of Helical and Spur Gear Size.

Brinell Hardness Number (BHN)	K Factor (scatter data)
200	200–420
250	240–560
300	350–680
350	460–840
400	575–1050
450	750–1380
500	900–1700
550	1050–2100
600	1250–2500

Prob. 30.12 Velocity of Sound in Air at 70°F (21°C).

Pressure (P), 1000 psi (1000 kN/m²)		Velocity (V), ft/sec (m/sec)	
.5	(3.5)	1150	(351)
1.0	(6.9)	1175	(358)
1.5	(10.3)	1215	(370)
2.0	(13.8)	1260	(384)
2.5	(17.2)	1310	(399)
3.0	(20.7)	1365	(416)
3.5	(24.1)	1420	(433)
4.0	(27.6)	1480	(451)
5.0	(34.5)	1610	(491)

Prob. 30.13 Bearing Lubrication Test (single row, radial, nonloading groove bearings, operating at 3000 rpm with negligible load and one-fourth standard pack of New Departure Code C grease).

Bearing Outer Race Temperature (T), °F (°C)		Base Grease Life (B), hr
80	(27)	40,000
100	(38)	28,500
120	(49)	21,000
140	(60)	15,000
160	(71)	10,750
180	(82)	7,600
200	(93)	5,500
220	(104)	4,000
240	(116)	2,850

LOG–LOG COORDINATE SOLUTION

Prob. 30.14 Power Requirements for Rotary Drilling Operations (300 hydraulic hp required to pump 80 lb mud through a $4\frac{1}{2}''$ full hole string).

Mud Circulation (G), gal/min (liter/min)		Standpipe Pressure (P), 1000 psi (1000 kN/m²)	
300	(1140)	17.2	(119.0)
400	(1510)	12.8	(88.3)
500	(1890)	10.2	(70.3)
600	(2270)	8.5	(58.6)
700	(2650)	7.3	(50.3)
800	(3030)	6.35	(43.8)
900	(3410)	5.7	(39.3)
1000	(3790)	5.1	(35.2)

Prob. 30.16 Thermalizing Heat-Treatment Process of Molybdenum Alloy (Mo–.5% Ti).

Distance from Hardened Surface (d), in. (mm)		Microhardness (H) (20-gram load)
0.0003	(.008)	1800
0.001	(.025)	950
0.002	(.051)	660
0.003	(.076)	540
0.004	(.102)	460
0.005	(.127)	410
0.006	(.152)	380
0.007	(.178)	350
0.008	(.203)	330
0.009	(.229)	310

Prob. 30.15 Gas Carburizing Heat-Treatment of Low Carbon Steel at 1700°F.

Carburizing Time (t), hr	Case Depth (D), in. (mm)
4	.047 (1.19)
8	.067 (1.70)
12	.084 (2.13)
16	.096 (2.44)
20	.110 (2.79)
24	.124 (3.15)
28	.136 (3.45)

Prob. 30.17 Glass Bead Filter Efficiency.

Rate of Flow (Q), gal/min (liter/min)		Flow Loss (p), in. Hg (mm Hg)	
5	(19)	.14	(3.5)
10	(38)	.5	(12.7)
15	(57)	1.1	(27.9)
20	(76)	1.9	(48.3)
25	(95)	2.9	(73.7)
30	(114)	4.0	(101.6)
35	(133)	5.4	(137.2)

Prob. 30.18 Fluorothene Dielectric Strength.

Thickness (t), mils (mm)		Dielectric Strength (E), volts/mil (volts/mm)	
13	(.33)	1400	(55 100)
19	(.48)	1200	(47 200)
25	(.64)	1000	(39 400)
40	(1.02)	800	(31 500)
50	(1.27)	700	(27 600)
70	(1.78)	600	(23 600)
90	(2.29)	500	(19 700)
120	(3.05)	450	(17 700)

CHAPTER 31

Graphical Mathematics

BY E. J. MYSIAK*

In general, mathematical problems may be solved *algebraically* (using numerical and mathematical symbols) or *graphically* (using drawing techniques). The algebraic method is predominantly a verbal approach in comparison to the visual methods of graphics; therefore, errors and discrepancies are more evident and subject to detection in a graphical presentation. The advantages of graphics are quite evident in any mathematics text, since most writers in the field of mathematics supplement their algebraic notations with graphical illustrations to illuminate their writings and improve the visualization of the solutions.

Naturally, the question of accuracy arises, but it must be remembered that measuring instruments are themselves graphical devices. The data obtained from an instrument reading, or from measurements of physical quantities, in many cases are recorded to three significant figures. Such a degree of accuracy is practical and readily substantiated by graphical methods of computation.

The graphical methods cannot be used exclusively, but neither can the algebraic methods be used to the fullest degree of effectiveness without the use of graphics. The engineer should be familiar with both methods in order to convey a clearer understanding of his analyses and designs.

Since equations are mathematical expressions, mathematical operations can be (1) computations with equations, (2) derivations of equations, and (3) solutions of particular equations. Graphic computations with the use of coordinate paper are included in this chapter as well as solutions of particular equations, algebra, and the calculus. The use of alignment charts for graphic computations and the specific topic of graphic empirical equation derivation are discussed in Chapters 29 and 30, respectively.

31.1 Rectangular Coordinate Algebraic Solution Graphs

Rectangular coordinate paper can be effectively used to visualize algebraic solutions. A common application is to solve two simultaneous poly-nomial equations with two unknowns. A graph of two linear equations is shown in Fig. 31.1 (a), indicating the solution as the intersection of the two lines. Linear equations are simple to solve algebraically; however, a graphical approach is

---

*Engineering Manager, Phoenix Company of Chicago, Wood Dale, IL.

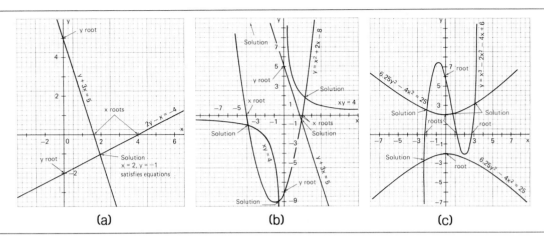

Fig. 31.1   Rectangular Coordinate Graphs—Algebraic Solutions.

advantageous in the solution of a quadratic equation for a parabola ($y = x^2 + 2x - 8$) with a linear equation for a straight line ($y + 3x = 5$) or an equation for an equilateral hyperbola ($xy = 4$), Fig. 31.1 (b). The graphical plot can also be used for further analysis. Note the linear equation and the equation for the equilateral hyperbola do not have a real simultaneous equation solution. The ease of visualization and solution is evident as the equations become more complex. Figure 31.1 (c) is a graphical solution for the cubic equation $y = x^3 - 2x^2 - 4x + 6$ and the equation for a hyperbola $6.25y^2 - 4x^2 = 25$. If the data were empirical, without an equation, the graphical solution would be necessary.

Another graphic application to algebraic problems is the determination of the roots for the equations. The roots of the equation are determined by the intersection of the curve and the corresponding axis. The linear equation of Fig. 31.1 (a) has one root per variable, as denoted. The highest power of the quadratic equation for the x term of Fig. 31.1 (b) being 2, there are two roots on the x-axis. The y term has only one root. The equilateral hyperbola, in the same graph, is asymptotic to both axes and therefore has no real roots. The symmetrical hyperbola of Fig. 31.1 (c) has one plus and one minus root. A cubic equation can have one or three real roots, depending on the location of the curve relative

to the axes. The cubic equation of Fig. 31.1 (c) has three roots.

A more convenient procedure to determine roots of an equation is to separate the equation and plot two separate curves. The quadratic equation of Fig. 31.1 (b), $y = x^2 + 2x - 8$, was separated to $y = x^2 = 8 - 2x$. The equation $y = x^2$ results in a symmetrical parabola, Fig. 31.2. The remaining portion of the quadratic equation, $y = 8 - 2x$, is a linear equation. The intersection of the parabola and the straight line

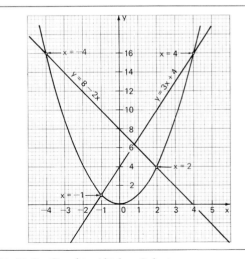

Fig. 31.2   Graphic Algebra Solutions.

918

provides the same root solution for the quadratic equation. The parabolic curve can be used with other linear portions of similar quadratic equations to provide the respective root solutions on the same plot.

31.2   The Graphical Calculus

If two variables are so related that the value of one of them depends on the value assigned the other, then the first variable is said to be a function of the second. For example, the area of a circle is a function of the radius. *The calculus is* that branch of mathematics pertaining to the change of values of functions due to finite changes in the variables involved. It is a method of analysis called *the differential calculus* when concerned with the determination of the *rate of change* of one variable of a function with respect to a related variable of the same function. A second principal operation, called *the integral calculus*, is the inverse of the differential calculus and is defined as a *process of summation* (finding the total change).

31.3   The Differential Calculus

Fundamentally, the differential calculus is a means of determining for a given function the limit of the ratio of change in the dependent variable to the corresponding change in the independent variable, as this change approaches zero. This limit is the *derivative* of the function with respect to the independent variable. The derivative of a function $y = f(x)$ may be denoted by dy/dx, $f'(x)$, y', or $D_x y$.

31.4   Geometric Interpretation of the Derivative

As illustrated in Fig. 31.3,

1. The symbol $f(x)$ is used to denote a function of a single variable x and is read "f of x."

2. The value of the function when x has the value x_0 may be denoted by $f(x_0)$ or $F(x_0)$.

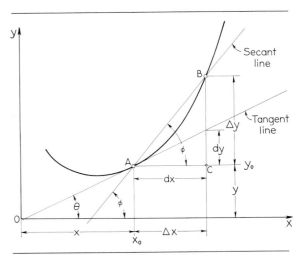

Fig. 31.3   Geometric Interpretation of the Derivative.

3. The increment Δy (read: "delta y") is the change in y produced by increasing x from x_0 to $x_0 + \Delta x$; therefore, $\Delta y = f(x_0 + \Delta x) - f(x_0)$.

4. The differential, dy, of y is the value that Δy would have if the curve coincided with its tangent. The differential, dx, of x is the same as Δx, when x is the independent variable.

5. The slope of the secant line is represented by the ratio $\Delta y/\Delta x$.

6. Ratio dy/dx represents the slope of the tangent line.

Draw a secant through A (coordinates x, y) and a neighboring point B on the curve (coordinates $x + \Delta x$, $y + \Delta y$). The slope of the secant line through A and B is

$$\frac{BC}{AC} = \frac{(y + \Delta y) - y}{(x + \Delta x) - x} = \frac{\Delta y}{\Delta x}$$
$$= \text{tangent of angle } BAC = \text{tangent } \phi$$

Let point B move along the curve and approach point A indefinitely. The secant line will revolve about point A, and its limiting position is the tangent line at A, when Δx, varying, approaches zero as a limit. The angle of inclination,

919

θ, of the tangent line at A becomes the limit of the angle ϕ when Δx approaches zero. Therefore,

$$\tan \phi = \tan \theta = \frac{dy}{dx}$$
$$= \text{slope of tangent line at } A$$

The result of this analysis is the following theorem.

The value of the derivative at any point of a curve is equal to the slope of the tangent line to the curve at that point.

Accordingly, graphical differentiation is a process of drawing tangents or the equivalent (chords parallel to tangents) to a curve at any point and determining the value of a differential (value of the slope of the tangent line to the curve or to a parallel chord, at the point selected). The value of the slope is the corresponding ordinate value on the derived curve.

31.5   Graphical Differentiation—The Slope Law

The slope of the curve at any point is the tangent of the angle (θ) between the x-axis of the graph and the tangent line to the curve at the selected point, Fig. 31.4. The slope is also the rise or fall of the tangent line in the y direction per one unit of travel in the x direction, the value of the slope being calculated in terms of the scale units of the x- and y-axes.

The *slope law* as applied to differentiation may be stated as follows.

The slope at any point on a given curve is equal to the ordinate of the corresponding point on the next lower derived curve.

The slope law is the graphical equivalent of differentiation of the calculus. The applications of the principles of the slope law are illustrated in the following examples.

In Fig. 31.5 (a), the slopes of OA and CD are constant and positive, indicated by the straight lines increasing in value upward (positively).

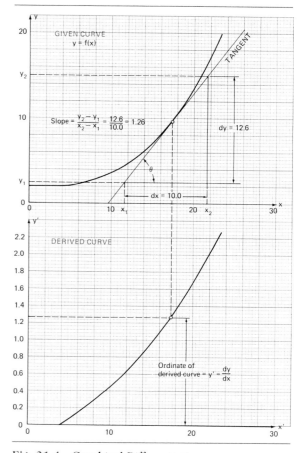

Fig. 31.4   Graphical Differentiation.

Curve AB has a constant zero slope; curve BC has a constant negative slope as shown by the straight line decreasing in value downward (negatively). The constant slopes produce level derivative curves, which are derived in a definite order and graphed in the same order in projection below the given curve. Units assigned to the axes are at any convenient scale.

At (b), any point on the curve has a slope equal to dy/dx. The corresponding point on the derived curve has an ordinate y' equal to the numerical value of dy/dx. Negative slopes have negative ordinate values on the derived curve. Zero slopes at B and D have zero values at B' and D' and are the x-axis on the derived curve.

Since the differential calculus is concerned with the determination of a rate of change, it is

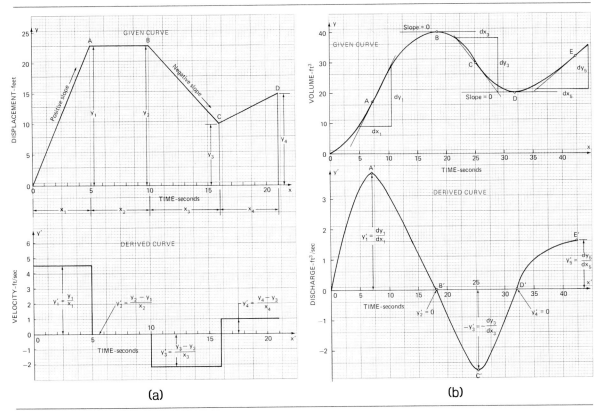

Fig. 31.5   Graphical Differentiation.

applicable to problems involving the change of a physical quantity as a second related quantity varies. The example at Fig. 31.5 (a) involves the displacement of a body with respect to time; the derived curve represents the velocity of the body at any instant. The second example, at (b), is a graph of the quantity of water discharged from a pipe plotted versus time. The ordinates of the derived curve represent $\frac{\text{quantity}}{\text{time}}$, which is the velocity of the water in the pipe at any given instant. The derived curve is one degree lower than the given curve. For example, if $y = x^3$, $dy/dx = 3x^2$. It is to be noted that the exponent of x in the derivative is one degree less or lower. If the derived curve $\frac{\text{velocity}}{\text{time}}$ were differentiated, an $\frac{\text{acceleration}}{\text{time}}$ curve would be derived, completing the family of $\frac{\text{displacement}}{\text{time}}$, $\frac{\text{velocity}}{\text{time}}$, and $\frac{\text{acceleration}}{\text{time}}$ curves.

31.6   Tangent-Line Constructions

The principal manipulation involved in graphical differentiation is the drawing of lines tangent to the given curve. Generally, it is more convenient to draw a tangent line in a given direction and then determine the point of tangency than it is to select a point of tangency and then determine a tangent line at this point.

Under certain conditions, the tangent line can be constructed geometrically. If a portion of the curve is assumed to be approximately circular, the point of tangency will lie on the perpendicular bisector of any chord, and the tangent line will be parallel to the chord, Fig. 31.6 (a).

If the curve approximates a parabola, ellipse, hyperbola, or circle, the tangent point is located at the intersection of the curve and a line drawn

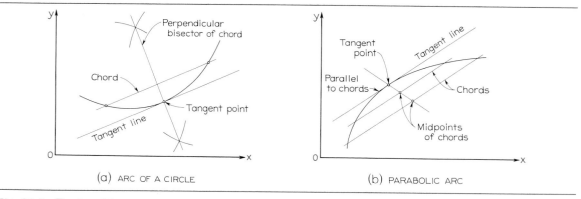

Fig. 31.6   Tangent Lines.

through the midpoint of two parallel chords, as shown at (b). The tangent line is drawn parallel to the chords.

If the curve is of unknown analytical form, difficulties arise in drawing the tangent. One method of drawing tangents is to mark a transparent strip at the edge with two dots no more than .25″ (6 mm) apart. If the dots lie on the curve, the edge of the strip approximately coincides with the tangent—the point of tangency is located midway between the dots, Fig. 31.7 (a).

A second method requires the use of a stainless steel or glass mirror, as shown at (b). The mirror is placed perpendicular to the graph surface and across the curve. The mirror is then moved until the curve and image coincide to form an unbroken line, as shown at the left at (b). A line drawn along the front edge of the stainless steel or the back edge of the glass mirror determines a normal to the tangent, as shown.

31.7   Graphical Differentiation Using Ray, String, or Funicular Polygons—The Chord Method

The method of differentiating described in §31.5 is semigraphical. The computed tangent values were transferred to the derived curve ordinate scale. A more convenient, entirely graphical,

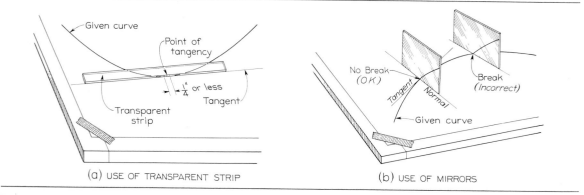

Fig. 31.7   Tangent Line Determination.

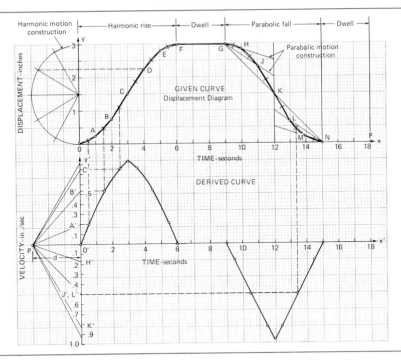

Fig. 31.8   Differentiation—Ray Polygon Method.

method using *ray, string, or funicular polygons* is described in the following example. In Fig. 31.8 a harmonic-rise and parabolic-fall motion for a cam is determined in the displacement diagram (given curve), and a derived curve is required. The given curve is divided into a number of short arcs, with equally spaced ordinates. Draw chords from *O* to *A*, *A* to *B*, and so on. The distance between ordinates along the *x*-axis is chosen small enough so that the chords between the two points are approximately parallel to a tangent line at the midpoint of the chosen arc.

Locate in projection below the given curve the origin *O'* of the derived curve and a *pole point P* at a convenient distance *d* from the origin point *O'* on the *x*-axis. To avoid "flat" derived curves, *d* is generally chosen at some multiple distance of the scale along the *x*-axis. The slope values of the tangent points determined on the given curve, and plotted on the ordinate scale of the derived curve, are computed as the

y value per *unit x* value. Slope $= \dfrac{y_2 - y_1}{x_2 - x_1}$ accounts for the axes scale designations, whatever they may be. If the pole distance, which is an *x* value, is any multiple of 1 (unit value), the ordinate values of the derived curve are reduced by a proportional amount, in order to transfer the same slope value from one graph to the other. If the pole distance is twice the unit *x* value, the ordinate scale of the derived curve is $\frac{1}{2}$ the scale unit of the given curve ordinates. The pole distance of the example problem is 3 units; therefore the same length for 1 unit on the given curve ordinate scale equals $\frac{1}{3}$ of a unit on the ordinate scale of the derived curve.

From the pole point *P*, draw *rays* parallel to the chords in the given curve. For example, *PA'* is parallel to *OA* and *PB'* is parallel to *AB*.

Locate the tangent points on the given curve for each subtended arc, as described in §31.6. Project each tangent point from the given curve

923

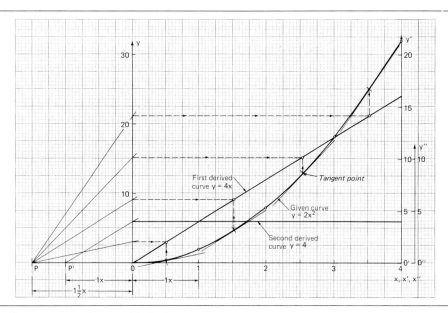

Fig. 31.9   Differentiation—Tangent Method.

downward to the derived curve until it intersects a horizontal line drawn from the appropriate points on the y'-axis of the derived curve. The process is repeated for all the tangent points and their corresponding points on the y'-axis. Construct the derived (derivative) curve by drawing a smooth curve through the located points. The computation of the slope values is not involved.

31.8   Graphical Differentiation— Ray Polygon—Tangent Method

Another method similar to the chord method consists in the determination of the tangent points and lines, without concern for the chords. In Fig. 31.9, a derived curve is obtained from a given curve $y = 2x^2$ by the tangent method. A ray is drawn from the pole point parallel to each tangent line, and the derived curve points are located by projecting the tangent points vertically from the given curve until the lines intersect the appropriate horizontal projection lines from the y'-axis of the derived curve.

The pole point P in this case is at a distance equal to $1\frac{1}{2}$ times the unit scale; therefore the ordinate scale for the first derived curve (y') is $\frac{2}{3}$ of the ordinate scale of the given curve.

The differentiating of the derived curve (successive differentiation) results in a second derived curve. The pole point selected, P', is at a distance equal to the unit scale of the x'-axis of the derived curve; therefore the ordinate scale for the second derived curve (y'') is equal to the ordinate scale of the first derived curve (y').

If clarity permits, the successively derived curves can be conveniently placed one on top of the other, since the abscissa scale is the same for all the curves. Each curve, however, requires a different ordinate scale, since a "flat" curve is to be avoided.

In the example used, the first derived curve is an inclined straight line, which is expected, since if $y = 2x^2$,

$$\text{1st derivative:} \quad \frac{dy}{dx} = 4x$$

924

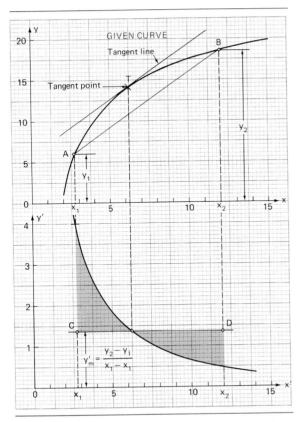

Fig. 31.10   Area Law.

chord and tangent are parallel, their slopes are equal. The slope of the tangent line at T is equal to the mean ordinate y'_m of T' in the derived curve

$$y'_m = \frac{y_2 - y_1}{x_2 - x_1}$$

and

$$y_2 - y_1 = y'_m(x_2 - x_1)$$

but

$$y'_m(x_2 - x_1) = \text{area of rectangle } CDx_2x_1$$

Therefore

$$y_2 - y_1 = \text{area of rectangle } CDx_2x_1$$

The law derived from this analysis, the *area law,* may be stated as follows.

> *The difference in the length of any two ordinates to a continuous curve equals the total net area between the corresponding ordinates of the next lower curve.*

The application of the area law as stated, which provides for dividing the given curve into short arcs, permits the determination of the derivative curve for the given curve. The law is also applicable to the process of integration, as discussed in §31.13.

The second derived curve appears as a line parallel to the x-axis, with a value of $y = 4$, which is also expected, since if $y = 4x$

2nd differential: $\dfrac{d^2y}{dx^2} = 4$

31.9   Area Law

Given a curve, Fig. 31.10 (upper portion), a tangent to the arc is to be constructed at T. A length of arc AB is chosen so that T is at the midpoint of the length of arc. A chord is constructed through A and B, and the tangent is drawn through T parallel to the chord. The coordinates of A are x_1, y_1, and of B are x_2, y_2. Since the

31.10   Graphical Differentiation—Area Law

As an example of the application of the area law, a practical problem in rate of interest may be used, as follows: An office building being erected costs $60,000 for the first story, $62,500 for the second, $65,000 for the third, and so forth. A basic cost of $400,000 is for the lot, plans, basement, and other preliminaries. The net annual income is $4000 per story. What number of stories will give the greatest rate of interest on the investment?

Initial computations are

Stories	Cost (1)	Income (2)	Percent Interest $\frac{(2)}{(1)} \times 100$
1	$460,000	$ 4,000	.87%
2	522,500	8,000	1.53
3	587,500	12,000	2.04
etc.			

In Fig. 31.11, the given curve is plotted using the computed coordinate values. The derived (derivative) curve is determined from the given curve by applying the area law.

Divide the given curve into a number of short arcs, with equally spaced ordinates y_1, y_2, ... The intervals between the ordinates along the x-axis should be chosen so that the chord inter-

cepting the two points on the arcs OA, AB, BC, ..., will be approximately parallel to the tangent lines at the midpoints of the subtended arcs.

Determine the length of each mean ordinate for the approximate interval by dividing the difference in length of the ordinates by the x distance between them. For example,

$$y'_m = \frac{y_2 - y_1}{x_2 - x_1}$$

The solution by this procedure is semigraphical.

If the intervals between ordinates are made equal, the length of the mean ordinate for each interval of the derived curve will be proportional to SA, TB, UC, ..., which eliminates the calculations involved in the semigraphical method.

The ray-polygon method of transferring ordinate values from one graph to the other also can be used.

Many of the derived curves determined are apt to appear "flat." To avoid a flat curve, the scale for the ordinates of the derived curve may be twice the scale of the ordinates of the given curve,

$$y'_m = 2\left(\frac{y_2 - y_1}{x_2 - x_1}\right)$$

The distance on the ordinate of the derived curve can be rapidly determined by applying the dividers to the given curve ordinate differences SA, TB, UC, ..., and placing twice these distances on the corresponding ordinates of the derived curve.

Horizontal lines for each interval and a smooth curve through the centers of horizontal lines are then drawn so that the triangular areas above and below the curve appear by eye to be approximately equal in area.

31.11   Practical Applications— Differentiation

The application of the calculus, especially by graphical methods, is based on the practicality and knowledge of the subject matter. Table 31.1

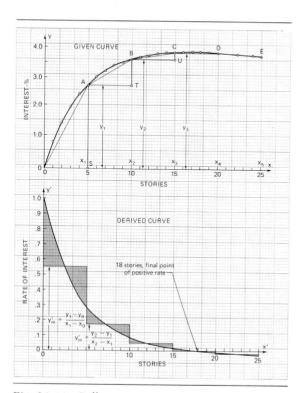

Fig. 31.11   Differentiation—Area Law.

Table 31.1 *Differentiation Applications*

Plotted Data		Derivative of Derived Curve
Independent	**Dependent**	
Time	Displacement	Velocity
Time	Velocity	Acceleration

Time	Amount of
	  Population
	  Inches
	  Volume
	  Temperature
		Rate of
		  Growth
		  Growth
		  Flow
		  Cooling, heating

Time	Energy or work	Power
Displacement	Energy or work	Force
Quantity	Cost	Rate of cost
Variable #2	Related variable #1	Rate of change of variable #1 as variable #2 changes

summarizes only a few of the practical applications for the differential calculus. The derived curve also shows maximum and minimum values for the derivative and if the rate is constant, variable, or zero.

31.12 The Integral Calculus

Integration is a process of summation, the integral calculus having been devised for the purpose of calculating the areas bounded by curves. If the given area is assumed to be divided into an infinite number of infinitesimal parts called *elements,* the sum of all these elements is the total area required. The integral sign ∫, the long S, was used by early writers and is still used in calculations to indicate "sum."

One of the most important applications of the integral calculus is the determination of a function from a given derivative, the inverse of differentiation. The process of determining such a function is *integration,* and the resulting function is called the *integral* of the given derivative or *integrand.* In many cases, however, graphical

integration is used to determine the area bounded by curves, in which case the expression of the function is not necessary.

Graphical solutions may be classified into three general groups: graphical, semigraphical, and mechanical methods. Each of these methods will be discussed.

31.13 Graphical Solution— Area Law

Since integration is the inverse of differentiation, the area law as analyzed for differentiation, §31.9, is applicable to integration, but in reverse. The area law as applied to integration may be stated as follows.

The area between any two ordinates of a given curve is equal to the difference in length between the corresponding ordinates of the next higher curve.

If an $\frac{acceleration}{time}$ curve were given, the integral of this derivative curve would be the $\frac{velocity}{time}$ curve. If the $\frac{velocity}{time}$ curve is integrated as shown in Fig.

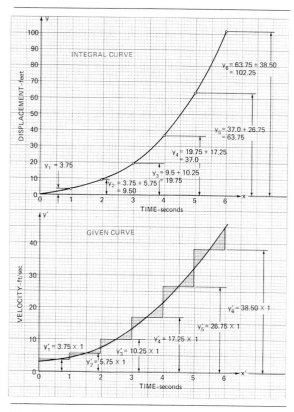

Fig. 31.12 Integration—Area Law.

mated between two consecutive ordinates by drawing a horizontal line that cuts the arc between the two ordinates, forming a pair of approximately equal "triangles," as shown in the lower part of the figure. It must be remembered that the area between any two ordinates on the given curve is calculated. In the example used, the distance between ordinates is one-second unit value; therefore, the area is equal to the mean ordinate value multiplied by 1. If the distance between ordinates is greater or less than unit value, this value must be considered in calculating the areas.

The points on the integral curve in this example may be obtained directly from observation of the given curve. The first ordinate value (y_1) is at the end of the first one-second interval and is equal to the mean ordinate (y_1') obtained from the given curve since $y_1' \times 1 = y_1'$. The second ordinate value (y_2) is for the end of the second one-second interval and is equal to the first mean ordinate value plus the second mean ordinate value ($y_2 = y_1' + y_2'$), and so on. If the scales of the ordinates for the given and integral curves are equal or proportional, these additions can be made graphically with the dividers.

31.14   Graphical Solution—Funicular Polygon Method

Another graphical method of integration is the funicular or string polygon method. If the lower-degree curve is given, Fig. 31.13, and the area between the curve ABCDEF and the x-axis is required (integration), the area obviously equals the sum of the rectangles ABKG, KLDC, and EFML.

Locate a pole point P at a distance d to the left of the origin O on the extended x-axis. See §31.15. Project the lines AB, CD, and EF to the y-axis to determine the points A', C', and E', respectively. Draw rays from the pole point to points A', C', and E'. From point G draw a straight line (string) parallel to ray PA' until it intersects the ordinate CKB, extended to the point of intersection at H. From point H draw a string parallel to ray PC' until it intersects the

31.12, the $\frac{\text{displacement}}{\text{time}}$ curve is obtained (the next higher curve).

In Fig. 31.12, the integral curve is drawn directly above the given curve, using the same scale along the x-axis as that on the x'-axis. The scale for the y-axis must provide the maximum value expected, and its length can be estimated. The value is equal to the total area under the given curve. For this example, the mean ordinate of the given curve appears to be about 17; therefore, the total area is approximately 102 square units (17 × 6). The ordinate scale for the integral curve is then selected to accommodate slightly more than this value.

The procedure for locating points on the integral curve is as follows: On the given derivative curve, uniformly (equally) spaced ordinates are established, and a mean ordinate is approxi-

928

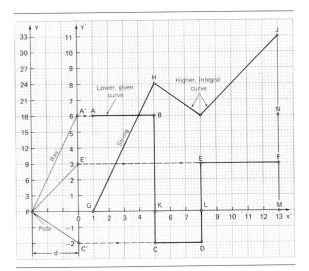

Fig. 31.13 Integration—Funicular Polygon.

ordinate *DLE*, extended to the point of intersection at *I*. From point *I* draw a string parallel to ray *PE'* until it intersects the ordinate *MF*, extended to the intersection at *J*.

The curve *GHIJ* is the higher-degree integral curve. Ordinate *KH* represents the area under line *AB*, ordinate *IL* represents the area under line *ABCD*, and ordinate *JM* represents the area under line *ABCDEF*.

From Fig. 31.13,

$$\text{Slope of ray } PA' = \frac{OA'}{PO} = \frac{OA'}{d}$$

$$\text{Slope of string } GH = \frac{KH}{GK}$$

Since these slopes were constructed equal,

$$\frac{OA'}{d} = \frac{KH}{GK}$$

or

$$OA' = d\left(\frac{KH}{GK}\right)$$

The ordinate *OA'* of the lower curve therefore equals the slope of the string or upper curve multiplied by a constant *d*, the pole distance.

It will be seen that this analysis is a verification of the slope law, §31.5. A rearrangement of the slope relationship verifies the area law, since *OA'* = *d*(*KH*/*GK*), *OA'*(*GK*) = *d*(*KH*) = area of rectangle *ABGK*. *KH* is the ordinate of the higher-degree integral curve.

31.15 Ordinate Scale of Integral Curve

The ordinate values placed along the *y*-axis of the next higher degree (integral) curve are dependent on the location of the pole point used to obtain strings and rays. The abscissa scale along the *x*-axis is kept the same as that of the original curve.

It is desirable to choose the unit scale of the integral curve ordinates a proper length so as to avoid "flat" integral curves. Larger units along the *y*-axis, Fig. 31.14, in comparison to the units along the *y'*-axis, will produce a steeper curve.

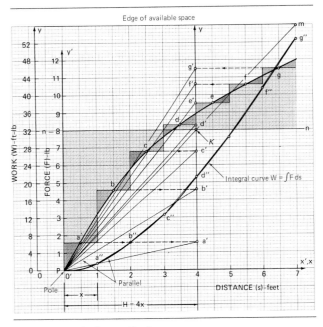

Fig. 31.14 Ordinate Scale.

929

The factor to consider is the choice of the pole distance (H). An even whole number multiple of the unit scale length along the abscissa will simplify the layout of the y-axis (ordinate) scale units.

In the figure, H was chosen equal to four times the unit length of the abscissa scale, to the left of the origin. The vertical (ordinate) scale units of the integral curve are, therefore, four times the unit scale of the ordinate of the given curve (based on areas under the curves).

The problem in Fig. 31.14 was arranged to obtain the best use of the available space, plus increased accuracy for the required integral. After the given curve is plotted within the available space, a line $0'-m$ is drawn from the origin to the desired endpoint of the integral curve. An average ordinate, line $n-n$, is estimated so that the two approximately triangular areas appear to be equal. An auxiliary ordinate axis for the integral curve is erected at the intersection of line $n-n$ and $0'-m$. Divide the given curve into small increments (area law method) and establish an average ordinate for each increment, a, b, c, and so on. Project these ordinates to the auxiliary ordinate axis, a to a', b to b', and so on. From the pole point, a ray is drawn to each of these points, Pa', Pb', Pc', and so on. Draw a segment $0'-a''$ across the first increment from the origin parallel to the corresponding ray Pa'. This segment coincides with $0'-a'$. From the end of this segment, a second segment $a''b''$ is drawn across the second increment parallel to the corresponding second ray Pb'. This process is continued for the remaining increments. A smooth curve is drawn through the endpoints of the segments.

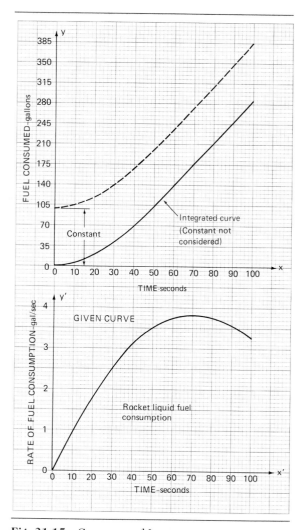

Fig. 31.15   Constants of Integration.

31.16   Constants of Integration

The integral curves in many cases begin at zero, but it cannot be assumed that this is always the case.

In Fig. 31.15, since the area under the given curve between any pair of ordinates gives only the difference between the lengths of the corresponding ordinates of the integral curve, the integration of the given curve gives only the shape of the integral curve. Its position with respect to

the x-axis is determined by initial conditions that must be known or assumed. Therefore, when a curve is integrated by the calculus and limits are not stated, a quantity C, called the *constant of integration,* must be added to the resulting equation to provide for initial conditions—that is, to fix the curve with respect to the coordinate axes.

The actual constant value can be determined by knowing one point on the integral curve or

Table 31.2 *Practical Applications—Integration*

Plotted Data		Integral of Integration Curve
Independent	Dependent	
Time	Acceleration	Velocity
Time	Velocity	Displacement
y values	x values	Area of X–Y plot enclosed by X-axis and/or Y-axis and the curve
Area	Pressure	Force
Displacement	Force	Work or energy
Time	Power	Work or energy
Volume	Pressure	Work or energy
Time	Force	Momentum
Velocity	Momentum	Impulse
Time	Rate based on time	Total change or cumulative change
Quantity #2	Rate of change of quantity #1 based on change of quantity #2	Max. or min. quantity #1 for values of quantity #2

GRAPHICAL MATHEMATICS PROBLEMS

The following problems provide laboratory experience in applying the methods described in this chapter and, in addition, introduce many typical engineering problems to the student. For convenience in assignment, the problems are classified according to type. Some problems have data also in the metric systems; such data are shown in parentheses ( ).

Since many of the problems in this chapter are of a general nature, they can also be solved on most computer graphics systems. If a system is available, the instructor may choose to assign specific problems to be completed by this method.

ALGEBRAIC SOLUTIONS

Roots and Simultaneous Equations Plot the following equations and determine the x and y roots (if any). Plot any two of the following equations and determine their solution (simultaneous), if any.

Prob. 31.1 $y = 4 + 3x$
Prob. 31.2 $y = 6 - 5x$
Prob. 31.3 $2y = 4 + 8x$
Prob. 31.4 $3y = 6 - 2x$
Prob. 31.5 $4y = 3x$
Prob. 31.6 $-y = 2x + 2$
Prob. 31.7 $y = 2x - 3$
Prob. 31.8 $y = x^2 + 3x - 4$

Prob. 31.9 $y = x^2 - 2x - 3$
Prob. 31.10 $y = x^2 + 4x + 3$
Prob. 31.11 $y = x^2 - 3x + 2$
Prob. 31.12 $y = x^3 + 6x^2 - 30x - 40$
Prob. 31.13 $y = x^3 + 4x^2 - 9x - 36$
Prob. 31.14 $x = y^3 + 2y^2 - 3y - 12$
Prob. 31.15 $x^2y = 16$
Prob. 31.16 $xy^2 = 15$
Prob. 31.17 $xy = 9$
Prob. 31.18 $2y^2 - 3x = 5$
Prob. 31.19 $\log y = 2x^3$
Prob. 31.20 $\log y = 5x^2$
Prob. 31.21 $x^2 + y^2 = 9$
Prob. 31.22 $4y^2 - 9x^2 = 16$

THE GRAPHICAL CALCULUS

In the following problems, plot the given data and perform the graphical differentiation or integration by any of the methods included in the text.

Differential Calculus

Prob. 31.23 Determine the acceleration (mph/sec) versus time curve for the following tests performed on an 8-cylinder automobile engine.

| Time (t), sec | Velocity (v), mph | |
	Without Supercharger	With Supercharger
0	0	0
5	28	35
10	45	55
15	54	67
20	61	75
25	67	81
30	71	86
35	75	90
47	—	97.1
57	86.7	—

Prob. 31.24 Determine the power curve for the following data. Power is the rate of doing work. (1 hp = 33,000 ft-lb/min = 44,750 J/min; 1 joule = 0.74 ft-lb).

Time (t), min	Work (W), ft-lb	(joules)
1	100,000	(136 000)
2	70,000	(95 000)
3	50,000	(68 000)
4	37,000	(50 000)
5	29,500	(40 000)
6	24,500	(33 000)
7	21,500	(29 000)
8	19,500	(26 500)

Prob. 31.25 From the data of tank content versus time for a rocket, determine the rate of fuel consumption, gal/sec. Plot fuel consumed versus time.

Time (t), sec	Liquid Fuel Tank Content (C), gal (liters)
0	940 (3560)
5	900 (3410)
10	810 (3070)
15	660 (2500)
20	430 (1630)
25	290 (1100)
30	195 (738)
35	130 (492)
40	75 (284)
45	40 (151)
50	20 (76)
55	5 (19)
60	0 (0)

Prob. 31.26 Plot the data for the recovery of hydrogen-embrittled mild steel, .505″ diameter rods, and determine the rate of cooling.

Time (t), hr	Temperature (T), °C
$\frac{1}{2}$	700
1	500
2	425
4	340
6	275
8	225
10	200
12	170
14	150
16	125
18	120
20	110
22	100
24	95

Prob. 31.27   Plot the velocity data for the ram of a shaper machine and determine the acceleration by graphical differentiation.

Time (t), sec	Velocity (v), ft/sec	(m/sec)
0	0	(0)
.25	.333	(0.102)
.50	.533	(0.163)
.75	.650	(0.198)
1.00	.733	(0.223)
1.25	.783	(0.239)
1.50	.800	(0.244)
1.75	.817	(0.249)
2.00	.800	(0.244)
2.25	.783	(0.239)
2.50	.733	(0.223)
2.75	.650	(0.198)
3.00	.533	(0.163)
3.25	.400	(0.122)
3.50	.200	(0.061)
3.75	−.133	(−0.041)
4.00	−.533	(−0.163)
4.25	−1.017	(−0.310)
4.50	−1.417	(−0.432)
4.75	−1.625	(−0.495)
4.85	−1.650	(−0.503)
5.00	−1.608	(−0.490)
5.25	−1.350	(−0.412)
5.50	−.917	(−0.280)
5.75	−.417	(−0.127)
6.00	0	(0)

Prob. 31.28   A manufacturer of machine parts can sell x parts per week at a price $p = 200 - 1.5x =$ (price per part). The cost of production $c = .5x^2 + 20x + 600$. Determine the total selling price and production cost for increments of 5 parts, for a total of 50 parts. Plot each curve, total selling price versus number of parts and production cost versus number of parts on the same sheet of graph paper.

Determine the "break-even" point, where selling price equals production costs. Determine the number of parts to be manufactured with the largest profit (profit = selling price minus production cost).

Determine the rate of production cost and the rate of the selling price.

Integral Calculus

Prob. 31.29   For the data of Prob. 31.23, determine the displacement versus time curves.

Prob. 31.30   Determine the work curve for the following data. Plot volume as abscissa and pressure as ordinate. The integral curve is the work curve, inch-pounds versus cubic inches (kg-cm versus cm³), and is equivalent to the area under the given curve. The units for work are inch-pounds (kg-cm).

Volume (V), in.³ (cm³)		Pressure (P), lb/in.² (kg/cm²)	
0	(0)	100	(7.03)
10	(164)	57	(4.01)
20	(328)	41	(2.88)
30	(492)	33	(2.32)
40	(656)	27.5	(1.93)
50	(819)	24	(1.69)
60	(983)	22	(1.55)
70	(1147)	21	(1.48)
80	(1311)	20	(1.41)
90	(1475)	20	(1.41)

Prob. 31.31 Plot the x and y coordinates given. The resulting shape is one-fourth the cross-sectional area of an aircraft fuel tank. After integrating the given curve to determine the area, compute the volume of the tank, if the length of the tank is 3 ft (0.914 m). Lay out a line EF to an appropriate scale representing the vertical height of the tank and calibrate it as a measuring stick to the nearest 10 gal (1 ft^3 = 7.48 gal; 1 gal = 3.785 liters).

x		y	
ft	(cm)	ft	(cm)
2.0	(60.96)	0	(0)
1.975	(60.20)	.25	(7.62)
1.890	(57.61)	.50	(15.24)
1.740	(53.04)	.75	(22.86)
1.500	(45.72)	1.00	(30.48)
1.120	(34.14)	1.25	(38.10)
.750	(22.86)	1.40	(42.67)
0	(0)	1.50	(45.72)

Prob. 31.32 Plot the velocity data for the ram of a shaper machine of Prob. 31.27, and determine the displacement of the ram by graphical integration.

Prob. 31.33 Plot the extrusion force data, with force as the ordinate axis. Integrate graphically the curve. The resultant integral is the work done by the punch, since $W = \int F ds$.

Impact Extrusion of Soft Copper	
Punch Stroke (s), mm	Extruding Force (F), tons
0	0
0.35	.75
0.60	1.25
0.70	1.50
1.00	2.75
1.30	5.25
1.40	6.50
1.45	7.25
1.50	8.25
1.70	10.25
1.80	10.75
2.10	11.75
2.25	11.90
2.50	12.10
2.65	11.95
3.05	11.75
3.60	11.25
3.80	10.75
4.25	10.50
4.35	10.25
4.50	10.50
4.65	10.75
4.80	11.50
4.85	11.00
4.90	0

Prob. 31.34 Plot the acceleration data for the explosion stroke of a piston of a 6-cylinder gas automobile engine (engine rpm: 2800; cylinder diameter: 3.38″ (85.7 mm); stroke: 5.50″ (139.7 mm); connecting rod length: 11.00″ (279.4 mm). Determine the velocity of the piston by graphical integration.

Prob. 31.35 Plot the piston force data (explosion stroke), with the force as the ordinate axis; data for the engine specified in Prob. 31.34. Integrate graphically the curve. The resultant integral is the work done by the piston, since $W = \int F ds$.

Crank Angle Position (θ), degrees	Acceleration (a), ft/sec² (cm/sec²)
0	24,630 (750 700)
15	23,500 (711 700)
30	19,600 (597 400)
45	13,980 (426 100)
60	7,420 (226 200)
75	753 (22 950)
90	5,160 (157 300)
105	9,500 (289 600)
120	12,300 (374 900)
135	13,830 (421 500)
150	14,460 (440 700)
165	14,730 (449 000)
180	14,770 (450 200)

Piston Stroke (s), in. (cm)	Piston Force (F), lb (newton)
0 (0)	834 (3 710)
.125 (0.318)	2,550 (11 340)
.438 (1.110)	2,005 (8 918)
.969 (2.461)	1,370 (6 094)
1.625 (4.128)	950 (4 226)
2.375 (6.033)	682 (3 034)
3.094 (7.858)	518 (2 304)
3.781 (9.604)	406 (1 806)
4.375 (11.113)	327 (1 455)
4.875 (12.383)	288 (1 281)
5.219 (13.256)	251 (1 117)
5.438 (13.811)	187 (832)
5.500 (13.970)	101 (449)

Appendix

1 Bibliography of American National Standards

American National Standards Institute, 1430 Broadway, New York, N.Y. 10018. For complete listing of standards, see ANSI catalog of American National Standards.

Abbreviations

Abbreviations for Use on Drawings and in Text, ANSI Y1.1–1972 (R1984)

Bolts, Screws, and Nuts

Bolts, Metric Heavy Hex, ANSI B18.2.3.6M–1979 (R1989)

Bolts, Metric Heavy Hex Structural, ANSI B18.2.3.7M–1979 (R1989)

Bolts, Metric Hex, ANSI B18.2.3.5M–1979 (R1989)

Bolts, Metric Round Head Short Square Neck, ANSI B18.5.2.1M–1981

Bolts, Metric Round Head Square Neck, ANSI/ASME B18.5.2.2M–1982

Hex Jam Nuts, Metric, ANSI B18.2.4.5M–1979

Hex Nuts, Heavy, Metric, ANSI B18.2.4.6M–1979

Hex Nuts, Slotted, Metric, ANSI B18.2.4.3M–1979 (R1989)

Hex Nuts, Style 1, Metric, ANSI B18.2.4.1M–1979 (R1989)

Hex Nuts, Style 2, Metric, ANSI B18.2.4.2M–1979 (R1989)

Hexagon Socket Flat Countersunk Head Cap Screws (Metric Series), ANSI/ASME B18.3.5M–1986

Mechanical Fasteners, Glossary of Terms, ANSI B18.12–1962 (R1981)

Miniature Screws, ANSI B18.11–1961 (R1983)

Nuts, Metric Hex Flange, ANSI B18.2.4.4M–1982

Plow Bolts, ANSI B18.9–1958 (R1977)

Round Head Bolts, Metric Round Head Short Square Neck, ANSI B18.5.2.1M–1981

Screws, Hexagon Socket Button Head Cap, Metric Series, ANSI/ASME B18.3.4M–1986

Screws, Hexagon Socket Head Shoulder, Metric Series, ANSI/ASME B18.3.3M–1986

Screws, Hexagon Socket Set, Metric Series, ANSI/ASME B18.3.6M–1986

Screws, Metric Formed Hex, ANSI B18.2.3.2M–1979 (R1989)

Screws, Metric Heavy Hex, ANSI B18.2.3.3M–1979 (R1989)

Screws, Metric Hex Cap, ANSI B18.2.3.1M–1979 (R1989)

Screws, Metric Hex Flange, ANSI/ASME B18.2.3.4M–1984

Screws, Metric Hex Lag, ANSI B18.2.3.8M–1981

Screws, Metric Machine, ANSI/ASME B18.6.7M–1985

Screws, Socket Head Cap, Metric Series, ANSI/ASME B18.3.1M–1986

Screws, Tapping and Metallic Drive, Inch Series, Thread Forming and Cutting, ANSI B18.6.4–1981

Slotted and Recessed Head Machine Screws and Machine Screw Nuts, ANSI B18.6.3–1972 (R1983)

Slotted Head Cap Screws, Square Head Set Screws, and Slotted Headless Set Screws, ANSI B18.6.2–1972 (R1983)

Socket Cap, Shoulder, and Set Screws (Inch Series) ANSI/ASME B18.3–1986

Square and Hex Bolts and Screws, Inch Series, ANSI B18.2.1–1981

Square and Hex Nuts (Inch Series) ANSI/ASME B18.2.2–1987

Track Bolts and Nuts, ANSI/ASME B18.10–1982

Wing Nuts, Thumb Screws, and Wing Screws, ANSI B18.17–1968 (R1983)

Wood Screws, Inch Series, ANSI B18.6.1–1981

Charts and Graphs

Illustrations for Publication and Projection, ANSI Y15.1M–1979 (R1986)

Process Charts, ANSI Y15.3M–1979 (R1986)

Time Series Charts, ANSI Y15.2M–1979 (R1986)

Dimensioning and Surface Finish

General Tolerances for Metric Dimensioned Products, ANSI B4.3–1978 (R1984)

Preferred Limits and Fits for Cylindrical Parts, ANSI B4.1–1967 (R1987)

Preferred Metric Limits and Fits, ANSI B4.2–1978 (R1984)

Surface Texture, ANSI/ASME B46.1–1985

Drafting Manual (Y14)

Sect. 1 Drawing Sheet Size and Format, ANSI Y14.1–1980 (R1987)

Sect. 2 Line Conventions and Lettering, ANSI Y14.2M–1979 (R1987)

Sect. 3 Multi and Sectional View Drawings, ANSI Y14.3–1975 (R1987)

Sect. 4 Pictorial Drawing, ANSI/ASME Y14.4–1989

Sect. 5 Dimensioning and Tolerancing, ANSI Y14.5M–1982 (R1988)

Sect. 6 Screw Thread Representation, ANSI Y14.6–1978 (R1987)

Sect. 6a Screw Thread Representation, Metric, ANSI Y14.6aM–1981 (R1987)

Sect. 7.1 Gear Drawing Standards—Part 1, for Spur, Helical, Double Helical, and Rack, ANSI Y14.7.1–1971 (R1988)

Sect. 7.2 Gear and Spline Drawing Standards—Part 2, Bevel and Hypoid Gears, ANSI Y14.7.2–1978 (R1989)

Sect. 13 Mechanical Spring Representation, ANSI Y14.13M–1981 (R1987)

Sect. 15 Electrical and Electronics Diagrams, ANSI Y14.15–1966 (R1988)

Sect. 15a Electrical and Electronics Diagrams—Supplement, ANSI Y14.15a–1971 (R1988)

Sect. 15b Electrical and Electronics Diagrams—Supplement, ANSI Y14.15b–1973 (R1988)

Sect. 17 Fluid Power Diagrams, ANSI Y14.17–1966 (R1987)

Sect. 34 Parts Lists, Data Lists, and Index Lists, ANSI Y14.34M–1982 (R1988)

Sect. 36 Surface Texture Symbols, ANSI Y14.36–1978 (R1987)

Gears

Design for Fine-Pitch Worm Gearings, ANSI/AGMA 374.04–1977

Gear Nomenclature—Terms, Definitions, Symbols, and Abbreviations, ANSI/AGMA 112.05–1976

Nomenclature of Gear-Tooth Failure Modes, ANSI/AGMA 110.04–1980

System for Straight Bevel Gears, ANSI/AGMA 208.03–1979

Tooth Proportions for Coarse-Pitch Involute Spur Gears, ANSI/AGMA 201.02

Tooth Proportions for Fine-Pitch Involute Spur and Helical Gears, ANSI/AGMA 207.06–1977

Graphic Symbols (Y32)

Fire-Protection Symbols for Architectural and Engineering Drawings, ANSI/NFPA 172–1986

Fire-Protection Symbols for Risk Analysis Diagrams, ANSI/NFPA 174–1986

Graphic Symbols for Electrical and Electronics Diagrams (Including 1986 Supplement) ANSI/IEEE 315–1975 (R1989)

Graphic Symbols for Electrical Wiring and Layout Diagrams Used in Architecture and Building Construction, ANSI Y32.9–1972 (R1989)

Graphic Symbols for Fluid Power Diagrams, ANSI Y32.10–1967 (R1987)

Graphic Symbols for Grid and Mapping Used in Cable Television Systems, ANSE/IEEE 623–1976 (R1989)

Graphic Symbols for Heat-Power Apparatus, ANSI Y32.2.6M–1950 (R1984)

Graphic Symbols for Heating, Ventilating and Air Conditioning, ANSI Y32.2.4–1949 (R1984)

Graphic Symbols for Logic Functions, ANSI/IEEE 91–1984

Graphic Symbols for Pipe Fittings, Valves, and Piping, ANSI/ASME Y32.2.3–1949 (R1988)

Graphic Symbols for Plumbing Fixtures for Diagrams Used in Architecture and Building Construction, ANSI Y32.4–1977 (R1987)

Graphic Symbols for Process Flow Diagrams in the Petroleum and Chemical Industries, ANSI Y32.11–1961 (R1985)

Graphic Symbols for Railroad Maps and Profiles, ANSI Y32.7–1972 (R1987)

Instrumentation Symbols and Identification, ANSI/ISA S5.1–1984

Reference Designations for Electrical and Electronics Parts and Equipment, ANSI/IEEE 200–1975 (R1989)

Symbols for Fire Fighting Operations, ANSI/NFPA 178–1986

Symbols for Mechanical and Acoustical Elements as Used in Schematic Diagrams, ANSI Y32.18–1972 (R1985)

Symbols for Welding and Nondestructive Testing, Including Brazing, ANSI/AWS A2.4–86

Keys and Pins

Clevis Pins and Cotter Pins, ANSI B18.8.1–1972 (R1983)

Hexagon Keys and Bits (Metric Series), ANSI B18.3.2M–1979 (R1986)

Keys and Keyseats, ANSI B17.1–1967 (R1973)

Pins-Taper Pins, Dowel Pins, Straight Pins, Grooved Pins and Spring Pins (Inch Series), ANSI B18.8.2–1978 (R1983)

Woodruff Keys and Keyseats, ANSI B17.2–1967 (R1978)

Piping

Bronze Pipe Flanges and Flanged Fittings, Class 150 and 300, ANSI B16.24–1979

Cast Bronze Threaded Fittings, Class 125 and 250, ANSI/ASME B16.15–1985

Cast Iron Pipe Flanged and Flanged Fittings, Class 25, 125, 250 and 800, ANSI/ASME B16.1–1989

Cast Iron Threaded Fittings, Class 125 and 250, ANSI/ASME B16.4–1985

Ductile Iron Pipe, Centrifugally Cast, in Metal Molds or Sand-Lined Molds for Water or Other Liquids, ANSI/AWWA C151/A21.51–1986

Factory-Made Wrought Steel Buttwelding Fittings, ANSI/ASME B16.9–1986

Ferrous Pipe Plugs, Bushings, and Locknuts with Pipe Threads, ANSI B16.14–1983

Flanged Ductile-Iron and Gray Iron Pipe with Threaded Flanges, ANSI/AWWA C115/A21.15–88

Malleable-Iron Threaded Fittings, Class 150 and 300, ANSI/ASME B16.3–1985

Pipe Flanges and Flanged Fittings, Steel Nickel Alloy and Other Special Alloys, ANSI/ASME B16.5–1988

Stainless Steel Pipe, ANSI/ASME B36.19M–1985

Welded and Seamless Wrought Steel Pipe, ANSI/ASME B36.10M–1985

Rivets

Large Rivets ($\frac{1}{2}$ Inch Nominal Diameter and Larger) ANSI B18.1.2–1972 (R1983)

Small Solid Rivets ($\frac{7}{16}$ Inch Nominal Diameter and Smaller) ANSI B18.1.1–1972 (R1983)

Small Solid Rivets, Metric, ANSI/ASME B18.1.3M–1983

Small Tools and Machine Tool Elements

Jig Bushings, ANSI B94.33–1974 (R1986)

Machine Tapers, ANSI B5.10–1981 (R1987)

Milling Cutters and End Mills, ANSI/ASME B94.19–1985

Reamers, ANSI B94.2–1983 (R1988)

T-Slots—Their Bolts, Nuts and Tongues, ANSI/ASME B5.1M–1985

Twist Drills, Straight Shank and Taper Shank Combined Drills and Countersinks, ANSI B94.11M–1979 (R1987)

Threads

Acme Screw Threads, ANSI/ASME B1.5–1988

Buttress Inch Screw Threads, ANSI B1.9–1973 (R1985)

Class 5 Interference-Fit Thread, ANSI/ASME B1.12–1987

Dryseal Pipe Threads (Inch), ANSI B1.20.3–1976 (R1982)

Dryseal Pipe Threads (Metric), ANSI B1.20.4–1976 (R1982)

Hose Coupling Screw Threads, ANSI/ASME B1.20.7–1966 (R1983)

Metric Screw Threads for Commercial Mechanical Fasteners—Boundary Profile Defined, ANSI B1.18M–1982 (R1987)

Metric Screw Threads—M Profile, ANSI/ASME B1.13M–1983 (R1989)

Metric Screw Threads—MJ Profile, ANSI B1.21M–1978

Nomenclature, Definitions and Letter Symbols for Screw Threads, ANSI/ASME B1.7M–1984

Pipe Threads, General Purpose (Inch), ANSI/ASME B1.20.1–1983

Stub Acme Threads, ANSI/ASME B1.8–1988

Unified Screw Threads (UN and UNR Thread Form), ANSI/ASME B1.1–1989

Unified Miniature Screw Threads, ANSI B1.10–1958 (R1988)

Washers

Lock Washers, ANSI B18.21.1–1972 (R1983)

Plain Washers, ANSI B18.22.1–1965 (R1981)

Plain Washers, Metric, ANSI B18.22M–1981

Miscellaneous

Knurling, ANSI/ASME B94.6–1984

Preferred Metric Sizes for Flat Metal Products, ANSI/ASME B32.3M–1984

Preferred Metric Equivalents of Inch Sizes for Tubular Metal Products Other Than Pipe, ANSI/ASME B32.6M–1984

Preferred Metric Sizes for Round, Square, Rectangle and Hexagon Metal Products, ANSI B32.4M–1980 (R1986)

Preferred Metric Sizes for Tubular Metal Products Other Than Pipe, ANSI B32.5–1977 (R1988)

Preferred Thickness for Uncoated Thin Flat Metals (Under 0.250 in.), ANSI B32.1–1952 (R1988)

Surface Texture (Surface Roughness, Waviness and Lay), ANSI/ASME B46.1–1985

2 Technical Terms

"The beginning of wisdom is to call things by their right names."
—Chinese Proverb

n *means a* noun; v *means a* verb

acme (n) Screw thread form, §§15.3 and 15.13

addendum (n) Radial distance from pitch circle to top of gear tooth.

allen screw (n) Special set screw or cap screw with hexagon socket in head, §15.31.

allowance (n) Minimum clearance between mating parts, §14.12.

alloy (n) Two or more metals in combination, usually a fine metal with a baser metal.

aluminum (n) A lightweight but relatively strong metal. Often alloyed with copper to increase hardness and strength.

angle iron (n) A structural shape whose section is a right angle, §12.20.

anneal (v) To heat and cool gradually, to reduce brittleness and increase ductility, §12.22.

arc-weld (v) To weld by electric arc. The work is usually the positive terminal.

babbitt (n) A soft alloy for bearings, mostly of tin with small amounts of copper and antimony.

bearing (n) A supporting member for a rotating shaft.

bevel (n) An inclined edge, not at right angle to joining surface.

bolt circle (n) A circular center line on a drawing, containing the centers of holes about a common center, §13.25.

bore (v) To enlarge a hole with a boring mill, Figs. 12.15 (b) and 12.18 (b).

boss (n) A cylindrical projection on a casting or a forging.

BOSS

brass (n) An alloy of copper and zinc.

braze (v) To join with hard solder of brass or zinc.

Brinell (n) A method of testing hardness of metal.

broach (n) A long cutting tool with a series of teeth that gradually increase in size which is forced through a hole or over a surface to produce a desired shape, §12.15.

bronze (n) An alloy of eight or nine parts of copper and one part of tin.

buff (v) To finish or polish on a buffing wheel composed of fabric with abrasive powders.

burnish (v) To finish or polish by pressure upon a smooth rolling or sliding tool.

burr (n) A jagged edge on metal resulting from punching or cutting.

bushing (n) A replaceable lining or sleeve for a bearing.

calipers (n) Instrument (of several types) for measuring diameters, §12.17.

cam (n) A rotating member for changing circular motion to reciprocating motion.

carburize (v) To heat a low-carbon steel to approximately 2000°F in contact with material which adds carbon to the surface of the steel, and to cool slowly in preparation for heat treatment, §12.22.

caseharden (v) To harden the outer surface of a carburized steel by heating and then quenching.

castellate (v) To form like a castle, as a castellated shaft or nut.

casting (n) A metal object produced by pouring molten metal into a mold, §12.3.

cast iron (n) Iron melted and poured into molds, §12.3.

center drill (n) A special drill to produce bearing holes in the ends of a workpiece to be mounted between centers. Also called a "combined drill and countersink," §13.35.

COMBINED DRILL
& C SINK

chamfer (n) A narrow inclined surface along the intersection of two surfaces.

CHAMFER

chase (v) To cut threads with an external cutting tool.

cheek (n) The middle portion of a three-piece flask used in molding, §12.3.

chill (*v*)   To harden the outer surface of cast iron by quick cooling, as in a metal mold.

chip (*v*)   to cut away metal with a cold chisel.

chuck (*n*)   A mechanism for holding a rotating tool or workpiece.

coin (*v*)   To form a part in one stamping operation.

cold-rolled steel (CRS) (*n*)   Open hearth or Bessemer steel containing 0.12–0.20% carbon that has been rolled while cold to produce a smooth, quite accurate stock.

collar (*n*)   A round flange or ring fitted on a shaft to prevent sliding.

COLLAR

colorharden (*v*)   Same as *caseharden,* except that it is done to a shallower depth, usually for appearance only.

cope (*n*)   The upper portion of a flask used in molding, §12.3.

core (*v*)   To form a hollow portion in a casting by using a dry-sand core or a green-sand core in a mold, §12.3.

coreprint (*n*)   A projection on a pattern which forms an opening in the sand to hold the end of a core, §12.3.

cotter pin (*n*)   A split pin used as a fastener, usually to prevent a nut from unscrewing, Fig. 15.31 (g) and (h) and Appendix 30.

counterbore (*v*)   To enlarge an end of a hole cylindrically with a *counterbore.* §12.16.

COUNTERBORE

countersink (*v*)   To enlarge an end of a hole conically, usually with a *countersink,* §12.16.

COUNTERSINK

crown (*n*)   A raised contour, as on the surface of a pulley.

cyanide (*v*)   To surface-harden steel by heating in contact with a cyanide salt, followed by quenching.

dedendum (*n*)   Distance from pitch circle to bottom of tooth space.

development (*n*)   Drawing of the surface of an object unfolded or rolled out on a plane.

diametral pitch (*n*)   Number of gear teeth per inch of pitch diameter.

die (*n*)   (1) Hardened metal piece shaped to cut or form a required shape in a sheet of metal by pressing it against a mating die. (2) Also used for cutting small male threads. In a sense is opposite to a tap.

die casting (*n*)   Process of forcing molten metal under pressure into metal dies or molds, producing a very accurate and smooth casting.

die stamping (*n*)   Process of cutting or forming a piece of sheet metal with a die.

dog (*n*)   A small auxiliary clamp for preventing work from rotating in relation to the face plate of a lathe.

dowel (*n*)   A cylindrical pin, commonly used to prevent sliding between two contacting flat surfaces.

DOWEL

draft (*n*)   The tapered shape of the parts of a pattern to permit it to be easily withdrawn from the sand or, on a forging, to permit it to be easily withdrawn from the dies, §§12.3 and 12.21.

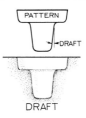

PATTERN

DRAFT

DRAFT

drag (*n*)   Lower portion of a flask used in molding, §12.3.

draw (*v*)   To stretch or otherwise to deform metal. Also to temper steel.

drill (*v*)   To cut a cylindrical hole with a drill. A *blind hole* does not go through the piece, §12.16.

drill press (*n*)   A machine for drilling and other hole-forming operations, §12.8.

drop forge (*v*)   To form a piece while hot between dies in a drop hammer or with great pressure, §12.21.

face (*v*)   To finish a surface at right angles, or nearly so, to the center line of rotation on a lathe.

FAO   Finish all over, §13.17.

feather key (*n*)   A flat key, which is partly sunk in a shaft and partly in a hub, permitting the hub to slide lengthwise of the shaft, §15.34.

file (*v*)   To finish or smooth with a file.

fillet (*n*) An interior rounded intersection between two surfaces, §§7.34 and 12.5.

fin (*n*) A thin extrusion of metal at the intersection of dies or sand molds.

fit (*n*) Degree of tightness or looseness between two mating parts, as a *loose fit*, a *snug fit*, or a *tight fit*. §§13.9 and 14.1–14.3.

fixture (*n*) A special device for holding the work in a machine tool, *but not for guiding the cutting tool.*

flange (*n*) A relatively thin rim around a piece.

FLANGE

flash (*n*) Same as *fin*.

flask (*n*) A box made of two or more parts for holding the sand in sand molding, §12.3.

flute (*n*) Groove, as on twist drills, reamers, and taps.

forge (*v*) To force metal while it is hot to take on a desired shape by hammering or pressing, §12.21.

galvanize (*v*) To cover a surface with a thin layer of molten alloy, composed mainly of zinc, to prevent rusting.

gasket (*n*) A thin piece of rubber, metal, or some other material, placed between surfaces to make a tight joint.

gate (*n*) The opening in a sand mold at the bottom of the *sprue* through which the molten metal passes to enter the cavity or mold, §12.3.

graduate (*v*) To set off accurate divisions on a scale or dial.

grind (*v*) To remove metal by means of an abrasive wheel, often made of carborundum. Use chiefly where accuracy is required, §12.14.

harden (*v*) To heat steel above a critical temperature and then quench in water or oil, §12.22.

heat-treat (*v*) To change the properties of metals by heating and then cooling, §12.22.

interchangeable (*adj.*) Refers to a part made to limit dimensions so that it will fit any mating part similarly manufactured, §14.1.

jig (*n*) A device *for guiding a tool* in cutting a piece. Usually it holds the work in position.

journal (*n*) Portion of a rotating shaft supported by a bearing.

kerf (*n*) Groove or cut made by a saw.

KERF

key (*n*) A small piece of metal sunk partly into both shaft and hub to prevent rotation, §15.34.

keyseat (*n*) A slot or recess in a shaft to hold a key, §15.34.

KEYSEAT

keyway (*n*) A slot in a hub or portion surrounding a shaft to receive a key. §15.34.

KEYWAY

knurl (*v*) To impress a pattern of dents in a turned surface with a knurling tool to produce a better hand grip, §13.37.

lap (*v*) To produce a very accurate finish by sliding contact with a *lap*, or piece of wood, leather, or soft metal impregnated with abrasive powder.

lathe (*n*) A machine used to shape metal or other materials by rotating against a tool, §12.7.

lug (*n*) An irregular projection of metal, but not round as in the case of a *boss*, usually with a hole in it for a bolt or screw.

malleable casting (*n*) A casting that has been made less brittle and tougher by annealing.

mill (*v*) To remove material by means of a rotating cutter on a milling machine, §12.9.

mold (*n*) The mass of sand or other material that forms the cavity into which molten metal is poured, §12.3.

MS (*n*) Machinery steel, sometimes called *mild steel* with a small percentage of carbon. Cannot be hardened.

neck (*v*) To cut a groove called a *neck* around a cylindrical piece.

NECK

normalize (*v*)   To heat steel above its critical temperature and then to cool it in air, §12.22.

pack-harden (*v*)   To *carburize,* then to *case-harden.* §12.22.

pad (*n*)   A slight projection, usually to provide a bearing surface around one or more holes.

PAD

pattern (*n*)   A model, usually of wood, used in forming a mold for a casting. In sheet metal work a pattern is called a *development.*

peen (*v*)   To hammer into shape with a ballpeen hammer.

pickle (*v*)   To clean forgings or castings in dilute sulphuric acid.

pinion (*n*)   The smaller of two mating gears.

pitch circle (*n*)   An imaginary circle corresponding to the circumference of the friction gear from which the spur gear was derived.

plane (*v*)   To remove material by means of the *planer,* §12.11.

planish (*v*)   To impart a planished surface to sheet metal by hammering with a smooth-surfaced hammer.

plate (*v*)   To coat a metal piece with another metal, such as chrome or nickel, by electrochemical methods.

polish (*v*)   To produce a highly finished or polished surface by friction, using a very fine abrasive.

profile (*v*)   To cut any desired outline by moving a small rotating cutter, usually with a master template as a guide.

punch (*v*)   To cut an opening of a desired shape with a rigid tool having the same shape, by pressing the tool through the work.

quench (*v*)   To immerse a heated piece of metal in water or oil in order to harden it.

rack (*n*)   A flat bar with gear teeth in a straight line to engage with teeth in a gear.

ream (*v*)   To enlarge a finished hole slightly to give it greater accuracy, with a *reamer,* §12.16.

relief (*n*)   An offset of surfaces to provide clearance for machining.

RELIEF

rib (*n*)   A relatively thin flat member acting as a brace or support.

RIB

rivet (*v*)   To connect with rivets or to clench over the end of a pin by hammering, §15.36.

round (*n*)   An exterior rounded intersection of two surfaces, §§7.34 and 12.5.

SAE   Society of Automotive Engineers.

sandblast (*v*)   To blow sand at high velocity with compressed air against castings or forgings to clean them.

scleroscope (*n*)   An instrument for measuring hardness of metals.

scrape (*v*)   To remove metal by scraping with a hand scraper, usually to fit a bearing.

shape (*v*)   To remove metal from a piece with a *shaper,* §12.10.

shear (*v*)   To cut metal by means of shearing with two blades in sliding contact.

sherardize (*v*)   To galvanize a piece with a coating of zinc by heating it in a drum with zinc powder, to a temperature of 575–850 °F.

shim (*n*)   A thin piece of metal or other material used as a spacer in adjusting two parts.

solder (*v*)   To join with solder, usually composed of lead and tin.

spin (*v*)   To form a rotating piece of sheet metal into a desired shape by pressing it with a smooth tool against a rotating form.

spline (*n*)   A keyway, usually one of a series cut around a shaft or hole.

SPLINED HOLE

spotface (*v*)   To produce a round *spot* or bearing surface around a hole, usually with a *spotfacer.* The spotface may be on top of a boss or it may be sunk into the surface, §§7.33 and 12.16.

SPOTFACE

sprue (*n*) A hole in the sand leading to the *gate* which leads to the mold, through which the metal enters, §12.3.

steel casting (*n*) Like cast-iron casting except that in the furnace scrap steel has been added to the casting.

swage (*v*) To hammer metal into shape while it is held over a *swage*, or die, which fits in a hole in the *swage block*, or anvil.

sweat (*v*) To fasten metal together by the use of solder between the pieces and by the application of heat and pressure.

tap (*v*) To cut relatively small internal threads with a *tap*. §12.16.

tape (*n*) Conical form given to a shaft or a hole. Also refers to the slope of a plane surface, §13.33.

taper pin (*n*) A small tapered pin for fastening, usually to prevent a collar or hub from rotating on a shaft.

TAPER PIN

taper reamer (*n*) A tapered reamer for producing accurate tapered holes, as for a taper pin. §§13.33 and 15.35.

temper (*v*) To reheat hardened steel to bring it to a desired degree of hardness, §12.22.

template or *templet* (*n*) A guide or pattern used to mark out the work, guide the tool in cutting it, or check the finished product.

tin (*n*) A silvery metal used in alloys and for coating other metals, such as tin plate.

tolerance (*n*) Total amount of variation permitted in limit dimension of a part, §14.2.

trepan (*v*) To cut a circular groove in the flat surface at one end of a hole.

tumble (*v*) To clean rough castings or forgings in a revolving drum filled with scrap metal.

turn (*v*) To produce, on a lathe, a cylindrical surface parallel to the center line, §12.7.

twist drill (*n*) A drill for use in a drill press, §12.16.

undercut (*n*) A recessed cut or a cut with inwardly sloping sides.

UNDERCUT

upset (*v*) To form a head or enlarged end on a bar or rod by pressure or by hammering between dies.

web (*n*) A thin flat part joining larger parts. Also known as a *rib*.

weld (*v*) Uniting metal pieces by pressure or fusion welding processes, §12.19.

Woodruff key (*n*) A semicircular flat key, §15.34.

WOODRUFF KEYS

wrought iron (*n*) Iron of low carbon content useful because of its toughness, ductility, and malleability.

3 CAD/CAM Glossary*

access time (or disk access time) One measure of system response. The time interval between the instant that data is called for from storage and the instant that delivery is completed—i.e., read time. See also *response time*.

alphanumeric (or alphameric) A term that encompasses letters, digits, and special characters that are machine-processable.

alphanumeric display (or alphameric display) A

work-station device consisting of a CRT on which text can be viewed. An alphanumeric display is capable of showing a fixed set of letters, digits, and special characters. It allows the designer to observe entered commands and to receive messages from the system.

alphanumeric keyboard (or alphameric keyboard) A work-station device consisting of a typewriter-like keyboard that allows the designer to communicate with the system using an English-like command language.

American Standard Code for Information Interchange (ASCII) An industry-standard character code widely used for information interchange

*Extracted from *The CAD/CAM Glossary,* 1983 edition, published by the Computervision Corporation, Bedford, MA 01730; reproduced with permission of the publisher.

among data processing systems, communications systems, and associated equipment.

analog Applied to an electrical or computer system, this denotes the capability to represent data in continuously varying physical quantities.

annotation Process of inserting text or a special note or identification (such as a flag) on a drawing, map, or diagram constructed on a CAD/CAM system. The text can be generated and positioned on the drawing using the system.

application program (or package) A computer program or collection of programs to perform a task or tasks specific to a particular user's need or class of needs.

archival storage Refers to memory (on magnetic tape, disks, printouts, or drums) used to store data on completed designs or elements outside of main memory.

array (*v*) To create automatically on a CAD system an arrangement of identical elements or components. The designer defines the element once, then indicates the starting location and spacing for automatic generation of the array. (*n*) An arrangement created in the above manner. A series of elements or sets of elements arranged in a pattern—i.e., matrix.

ASCII See *American National Standard Code for Information Exchange.*

assembler A computer program that converts (i.e., translates) programmer-written symbolic instructions, usually in mnemonic form, into machine-executable (computer or binary-coded) instructions. This conversion is typically one-to-one (one symbolic instruction converts to one machine-executable instruction). A software programming aid.

associative dimensioning A CAD capabililty that links dimension entities to geometric entities being dimensioned. This allows the value of a dimension to be automatically updated as the geometry changes.

attribute A nongraphic characteristic of a part, component, or entity under design on a CAD system. Examples include: dimension entities associated with geometry, text with text nodes, and nodal lines with connect nodes. Changing one entity in an association can produce automatic changes by the system in the associated entity; e.g., moving one entity can cause moving or stretching of the other entity.

automatic dimensioning A CAD capability that computes the dimensions in a displayed design, or in a designated section, and automatically places dimensions, dimensional lines, and arrowheads where required. In the case of mapping, this capability labels the linear feature with length and azimuth.

auxiliary storage Storage that supplements main memory devices such as disk or drum storage. Contrast with *archival storage.*

benchmark The program(s) used to test, compare, and evaluate in real time the performance of various CAD/CAM systems prior to selection and purchase. A *synthetic* benchmark has preestablished parameters designed to exercise a set of system features and resources. A *live* benchmark is drawn from the prospective user's workload as a model of the entire workload.

bit The smallest unit of information that can be stored and processed by a digital computer. A bit may assume only one of two values: 0 or 1 (i.e., ON/OFF or YES/NO). Bits are organized into larger units called *words* for access by computer instructions.

Computers are often categorized by word size in bits, i.e., the maximum word size that can be processed as a unit during an instruction cycle (e.g., 16-bit computers or 32-bit computers). The number of bits in a word is an indication of the processing power of the system, especially for calculations or for high-precision data.

bit rate The speed at which bits are transmitted, usually expressed in bits per second.

bits per inch (bpi) The number of bits that can be stored per inch of a magnetic tape. A measure of the data storage capacity of a magnetic tape.

blank A CAD command that causes a predefined entity to go temporarily blank on the CRT. The reversing command is *unblank.*

blinking A CAD design aid that makes a predefined graphic entity blink on the CRT to attract the attention of the designer.

boot up Start up (a system).

B-spline A sequence of parametric polynomial curves (typically quadratic or cubic polynomials) forming a smooth fit between a sequence of points in 3D space. The piece-wise defined curve maintains a level of mathematical continuity dependent upon the polynomial degree chosen. It is used extensively in mechanical design applications in the automotive and aerospace industries.

bug A flaw in the design or implementation of a software program or hardware design that causes erroneous results or malfunctions.

bulk memory A memory device for storing a large amount of data, e.g., disk, drum, or magnetic tape. It is not randomly accessible as main memory is.

byte A sequence of adjacent bits, usually eight, representing a character that is operated on as a unit. Usually shorter than a word. A measure of the memory capacity of a system, or of an individual storage unit (as a 300-million-byte disk).

CAD See *computer-aided design*.

CAD/CAM See *computer-aided design/computer-aided manufacturing*.

CADDS® Computervision Corporation's registered trademark for its prerecorded software programs.

CAE See *computer-aided engineering*.

CAM See *computer-aided manufacturing*.

CAMACS™ (CAM Asynchronous Communications Software) Computervision's communications link, which enables users to exchange machine control data with other automation systems and devices, or to interact directly with local or remote manufacturing systems and machines. CAMACS tailors CAD/CAM data automatically for a wide range of machine tools, robots, coordinate measurement systems, and off-line storage-devices.

cathode ray tube (CRT) The principal component in a CAD display device. A CRT displays graphic representations of geometric entities and designs and can be of various types: storage tube, raster scan, or refresh. These tubes create images by means of a controllable beam of electrons striking a screen. The term *CRT* is often used to denote the entire display device.

central processing unit (CPU) The computer brain of a CAD/CAM system that controls the retrieval, decoding, and processing of information, as well as the interpretation and execution of operating instructions—the building blocks of application and other computer programs. A CPU comprises arithmetic, control, and logic elements.

character An alphabetical, numerical, or special graphic symbol used as part of the organization, control, or representation of CAD/CAM data.

characters per second (cps) A measure of the speed with which an alphanumeric terminal can process data.

chip See *integrated circuit*.

code A set of specific symbols and rules for representing data (usually instructions) so that the data can be understood and executed by a computer. A code can be in binary (machine) language, assembly language, or a high-level language. Frequently refers to an industry-standard code such as ANSI, ASCII, IPC, or Standard Code for Information Exchange. Many application codes for CAD/CAM are written in FORTRAN.

color display A CAD/CAM display device. Color raster-scan displays offer a variety of user-selectable, contrasting colors to make it easier to discriminate among various groups of design elements on different layers of a large, complex design. Color speeds up the recognition of specific areas and subassemblies, helps the designer interpret complex surfaces, and highlights interference problems. Color displays can be of the penetration type, in which various phosphor layers give off different colors (refresh display) or the TV-type with red, blue, and green electron guns (raster-scan display).

command A control signal or instruction to a CPU or graphics processor, commonly initiated by means of a menu/tablet and electronic pen or by an alphanumeric keyboard.

command language A language for communicating with a CAD/CAM system in order to perform specific functions or tasks.

communication link The physical means, such as a telephone line, for connecting one system module or peripheral to another in a different location in order to transmit and receive data. See also *data link*.

compatibility The ability of a particular hardware module or software program, code, or language to be used in a CAD/CAM system without prior modification or special interfaces. *Upward compatible* denotes the ability of a system to interface with new hardware or software modules or enhancements (i.e., the system vendor provides with each new module a reasonable means of transferring data, programs, and operator skills from the user's present system to the new enhancements).

compiler A computer program that converts or translates a high-level, user-written language (e.g., PASCAL, COBOL, VARPRO, or FORTRAN) or source, into a language that a computer can understand. The conversion is typically one to many (i.e., one user instruction to many machine-executable instructions). A software programming aid, the compiler allows the designer to write programs in an English-like language with relatively few statements, thus saving program development time.

component A physical entity, or a symbol used in

CAD to denote such an entity. Depending on the application, a component might refer to an IC or part of a wiring circuit (e.g., a resistor), or a valve, elbow, or vee in a plant layout, or a substation or cable in a utility map. Also applies to a subassembly or part that goes into higher level assemblies.

computer-aided design (CAD)   A process that uses a computer system to assist in the creation, modification, and display of a design.

computer-aided design/computer-aided manufacturing (CAD/CAM)   Refers to the integration of computers into the entire design-to-fabrication cycle of a product or plant.

computer-aided engineering (CAE)   Analysis of a design for basic error checking, or to optimize manufacturability, performance, and economy (for example, by comparing various possible materials or designs). Information drawn from the CAD/CAM design data base is used to analyze the functional characteristics of a part, product, or system under design and to simulate its performance under various conditions. In electronic design, CAE enables users of the Computervision Designer system to detect and correct potentially costly design flaws. CAE permits the execution of complex circuit loading analyses and simulation during the circuit definition stage. CAE can be used to determine section properties, moments of inertia, shear and bending moments, weight, volume, surface area, and center of gravity. CAE can precisely determine loads, vibration, noise, and service life early in the design cycle so that components can be optimized to meet those criteria. Perhaps the most powerful CAE technique is finite element modeling. See also *kinematics*.

computer-aided manufacturing (CAM)   The use of computer and digital technology to generate manufacturing-oriented data. Data drawn from a CAD/CAM data base can assist in or control a portion or all of a manufacturing process, including numerically controlled machines, computer-assisted parts programming, computer-assisted process planning, robotics, and programmable logic controllers, CAM can involve production programming, manufacturing engineering, industrial engineering, facilities engineering, and reliability engineering (quality control). CAM techniques can be used to produce process plans for fabricating a complete assembly, to program robots, and to coordinate plant operation.

computer graphics   A general term encompassing any discipline or activity that uses computers to generate, process and display graphic images. The essential technology of CAD/CAM systems. See also *computer-aided design*.

computer network   An interconnected complex (arrangement, or configuration) of two or more systems. See also *network*.

computer program   A specific set of software commands in a form acceptable to a computer and used to achieve a desired result. Often called a *software program* or *package*.

configuration   A particular combination of a computer, software and hardware modules, and peripherals at a single installation and interconnected in such a way as to support certain application(s).

connector   A termination point for a signal entering or leaving a PC board or a cabling system.

convention   Standardized methodology or accepted procedure for executing a computer program. In CAD, the term denotes a standard rule or mode of execution undertaken to provide consistency. For example, a drafting convention might require all dimensions to be in metric units.

core (core memory)   A largely obsolete term for *main storage*.

CPU   See *central processing unit*.

CRT   See *cathode ray tube*.

cursor   A visual tracking symbol, usually an underline or cross hairs, for indicating a location or entity selection on the CRT display. A text cursor indicates the alphanumeric input; a graphics cursor indicates the next geometric input. A cursor is guided by an electronic or light pen, joystick, keyboard, etc., and follows every movement of the input device.

cycle   A preset sequence of events (hardware or software) initiated by a single command.

data base   A comprehensive collection of interrelated information stored on some kind of mass data storage device, usually a disk. Generally consists of information organized into a number of fixed-format record types with logical links between associated records. Typically includes operating systems instructions, standard parts libraries, completed designs and documentation, source code, graphic and application programs, as well as current user tasks in progress.

data communication   The transmission of data (usually digital) from one point (such as a CAD/CAM workstation or CPU) to another point via communication channels such as telephone lines.

data link The communication line(s), related controls, and interface(s) for the transmission of data between two or more computer systems. Can include modems, telephone lines, or dedicated transmission media such as cable or optical fiber.

data tablet A CAD/CAM input device that allows the designer to communicate with the system by placing an electronic pen or stylus on the tablet surface. There is a direct correspondence between positions on the tablet and addressable points on the display surface of the CRT. Typically used for indicating positions on the CRT, for digitizing input of drawings, or for menu selection. See also *graphic tablet*.

debug To detect, locate, and correct any bugs in a system's software or hardware.

dedicated Designed or intended for a single function or use. For example, a dedicated work station might be used exclusively for engineering calculations or plotting.

default The predetermined value of a parameter required in a CAD/CAM task or operation. It is automatically supplied by the system whenever that value (e.g., text, height, or grid size) is not specified.

density (1) A measure of the complexity of an electronic design. For example, IC density can be measured by the number of gates or transistors per unit area or by the number of square inches per component. (2) Magnetic tape storage capacity. High capacity might be 1600 bits/inch; low, 800 bits/inch.

device A system hardware module external to the CPU and designed to perform a specific function—i.e., a CRT, plotter, printer, hard-copy unit, etc. See also *peripheral*.

diagnostics Computer programs designed to test the status of a system or its key components and to detect and isolate malfunctions.

dial up To initiate station-to-station communication with a computer via a dial telephone, usually from a workstation to a computer.

digital Applied to an electrical or computer system, this denotes the capability to represent data in the form of digits.

digitize (1) General description: to convert a drawing into digital form (i.e., coordinate locations) so that it can be entered into the data base for later processing. A digitizer, available with many CAD systems, implements the conversion process. This is one of the primary ways of entering existing drawings, crude graphics, lines, and shapes into the system. (2) Computervision usage: to specify a coordinate location or entity using an electronic pen or other device; or a single coordinate value or entity pointer generated by a digitizing operation.

digitizer A CAD input device consisting of a data tablet on which is mounted the drawing or design to be digitized into the system. The designer moves a puck or electronic pen to selected points on the drawing and enters coordinate data for lines and shapes by simply pressing down the digitize button with the puck or pen.

dimensioning, automatic A CAD capability that will automatically compute and insert the dimensions of a design or drawing, or a designated section of it.

direct access (linkage) Retrieval or storage of data in the system by reference to its location on a tape, disk, or cartridge, without the need for processing on a CPU.

direct-view storage tube (DVST) One of the most widely used graphics display devices, DVST generates a long-lasting, flicker-tree image with high resolution and no refreshing. It handles an almost unlimited amount of data. However, display dynamics are limited since DVSTs do not permit selective erase. The image is not as bright as with refresh or raster. Also called *storage tube*.

directory A named space on the disk or other mass storage device in which are stored the names of files and some summary information about them.

discrete components Components with a single functional capability per package—for example, transistors and diodes.

disk (storage) A device on which large amounts of information can be stored in the data base. Synonymous with *magnetic disk storage* or *magnetic disk memory*.

display A CAD/CAM work station device for rapidly presenting a graphic image so that the designer can react to it, making changes interactively in real time. Usually refers to a CRT.

dot-matrix plotter A CAD peripheral device for generating graphic plots. Consists of a combination of wire nibs (styli) spaced 100 to 200 styli per inch, which place dots where needed to generate a drawing. Because of its high speed, it is typically used in electronic design applications. Accuracy and resolution are not as great as with pen plotters. Also known as *electrostatic plotter*.

drum plotter An electromechanical pen plotter that draws an image on paper or film mounted on a rotatable drum. In this CAD peripheral device a combination of plotting-head movement and drum rotation provides the motion.

dynamic (motion) Simulation of movement using CAD software, so that the designer can see on the CRT screen 3D representations of the parts in a piece of machinery as they interact dynamically. Thus, any collision or interference problems are revealed at a glance.

dynamic menuing This feature of Computervision's Instaview terminal allows a particular function or command to be initiated by touching an electronic pen to the appropriate key word displayed in the status text area on the screen.

dynamics The capability of a CAD system to zoom, scroll, and rotate.

edit To modify, refine, or update an emerging design or text on a CAD system. This can be done on-line interactively.

electrostatic plotter See *dot-matrix plotter*.

element The basic design entity in computer-aided design whose logical, positional, electrical, or mechanical function is identifiable.

enhancements Software or hardware improvements, additions, or updates to a CAD/CAM system.

entity A geometric primitive—the fundamental building block used in constructing a design or drawing, such as an arc, circle, line, text, point, spline, figure, or nodal line. Or a group of primitives processed as an identifiable unit. Thus, a square may be defined as a discrete entity consisting of four primitives (vectors), although each side of the square could be defined as an entity in its own right. See also *primitive*.

feedback (1) The ability of a system to respond to an operator command in real time either visually or with a message on the alphanumeric display or CRT. This message registers the command, indicates any possible errors, and simultaneously displays the updated design on the CRT. (2) The signal or data fed back to a commanding unit from a controlled machine or process to denote its response to a command. (3) The signal representing the difference between actual response and desired response and used by the commanding unit to improve performance of the controlled machine or process. See also *prompt*.

figure A symbol or a part that may contain primitive entities, other figures, nongraphic properties, and associations. A figure can be incorporated into other parts or figures.

file A collection of related information in the system that may be accessed by a unique name. May be stored on a disk, tape, or other mass storage media.

file protection A technique for preventing access to or accidental erasure of data within a file on the system.

firmware Computer programs, instructions, or functions implemented in user-modifiable hardware, i.e., a microprocessor with read-only memory. Such programs or instructions, stored permanently in programmable read-only memories, constitute a fundamental part of system hardware. The advantage is that a frequently used program or routine can be invoked by a single command instead of multiple commands as in a software program.

flatbed plotter A CAD/CAM peripheral device that draws an image on paper, glass, or film mounted on a flat table. The plotting head provides all the motion.

flat-pattern generation A CAD/CAM capability for automatically unfolding a 3D design of a sheet metal part into its corresponding flat-pattern design. Calculations for material bending and stretching are performed automatically for any specified material. The reverse flat-pattern generation package automatically folds a flat-pattern design into its 3D version. Flat-pattern generation eliminates major bottlenecks for sheet metal fabricators.

flicker An undesired visual effect on a CRT when the refresh rate is low.

font, line Repetitive pattern used in CAD to give a displayed line appearance characteristics that make it more easily distinguishable, such as a solid, dashed, or dotted line. A line font can be applied to graphic images in order to provide meaning, either graphic (e.g., hidden lines) or functional (roads, tracks, wires, pipes, etc.). It can help a designer to identify and define specific graphic representations of entities that are view-dependent. For example, a line may be solid when drawn in the top view of an object but, when a line font is used, becomes dotted in the side view where it is not normally visible.

font, text Sets of type faces of various styles and sizes. In CAD, fonts are used to create text for drawings, special characters such as Greek letters, and mathematical symbols.

FORTRAN *FOR*mula *TRAN*slation, a high-level programming language used primarily for scientific or engineering applications.

fracturing The division of IC graphics by CAD into simple trapezoidal or rectangular areas for pattern-generation purposes.

function key A specific square on a data tablet, or a key on a function key box, used by the designer to enter a particular command or other input. See also *data tablet.*

function keyboard An input device located at a CAD/CAM workstation and containing a number of function keys.

gap The gap between two entities on a computer-aided design is the length of the shortest line segment that can be drawn from the boundary of one entity to the other without intersecting the boundary of the other. CAD/CAM design-rules checking programs can automatically perform gap checks.

graphic tablet A CAD/CAM input device that enables graphic and location instruments to be entered into the system using an electronic pen on the tablet. See also *data tablet.*

gray scales In CAD systems with a monochromatic display, variations in brightness level (gray scale) are employed to enhance the contrast among various design elements. This feature is very useful in helping the designer discriminate among complex entities on different layers displayed concurrently on the CRT.

grid A network of uniformly spaced points or cros-shatch optionally displayed on the CRT and used for exactly locating and digitizing a position, inputting components to assist in the creation of a design layout, or constructing precise angles. For example, the coordinate data supplied by digitizers is automatically calculated by the CPU from the closest grid point. The grid determines the minimum accuracy with which design entities are described or connected. In the mapping environment, a grid is used to describe the distribution network of utility resources.

hard copy A copy on paper of an image displayed on the CRT—for example, a drawing, printed report, plot, listing, or summary. Most CAD/CAM systems can automatically generate hard copy through an on-line printer or plotter.

hardware The physical components, modules, and peripherals comprising a system—computer disk, magnetic tape, CRT terminal(s), plotter(s), etc.

hard-wired link A technique of physically connecting two systems by fixed circuit interconnections using digital signals.

high-level language A problem-oriented programming language using words, symbols, and command statements that closely resemble English-language statements. Each statement typically represents a series of computer instructions. Relatively easy to learn and use, a high-level language permits the execution of a number of subroutines through a simple command. Examples are BASIC, FORTRAN, PL/I, PASCAL, and COBOL.

A high-level language must be translated or compiled into machine language before it can be understood and processed by a computer. See also *assembler; low-level language.*

host computer The primary or controlling computer in a multicomputer network. Large-scale host computers typically are equipped with mass memory and a variety of peripheral devices, including magnetic tape, line printers, card readers, and possibly hard-copy devices. Host computers may be used to support, with their own memory and processing capabilities, not only graphics programs running on a CAD/CAM system but also related engineering analysis.

host-satellite system A CAD/CAM system configuration characterized by a graphic workstation with its own computer (typically holding the display file) that is connected to another, usually larger, computer for more extensive computation or data manipulation. The computer local to the display is a satellite to the larger host computer, and the two comprise a host-satellite system.

IC See *integrated circuit.*

IGES See *Initial Graphics Exchange Specification.*

inches per second (ips) Measure of the speed of a device (i.e., the number of inches of magnetic tape that can be processed per second, or the speed of a pen plotter).

Initial Graphics Exchange Specification (IGES) An interim CAD/CAM data base specification until the American National Standards Institute develops its own specification. IGES attempts to standardize communication of drawing and geometric product information between computer systems.

initialize To set counters, switches, and addresses on a computer to zero or to other starting values at the beginning of, or at predetermined stages in, a program or routine.

input (data) (1) The data supplied to a computer program for processing by the system. (2) The process of entering such data into the system.

input devices A variety of devices (such as data tablets or keyboard devices) that allow the user to communicate with the CAD/CAM system, for example, to pick a function from many presented, to enter text and/or numerical data, to modify the picture shown on the CRT, or to construct the desired design.

input/output (I/O) A term used to describe a CAD/CAM communications device as well as the process by which communications take place in a CAD/CAM system. An I/O device is one that makes possible communications between a device and a workstation operator or between devices on the system (such as workstations or controllers). By extension, input/output also denotes the process by which communications takes place. Input refers to the data transmitted to the processor for manipulation, and output refers to the data transmitted from the processor to the workstation operator or to another device (i.e., the results). Contrast with the other major parts of a CAD/CAM system: the CPU or central processing unit, which performs arithmetic and logical operations, and data storage devices (such as memories, disks, or tapes).

insert To create and place entities, figures, or information on a CRT or into an emerging design on the display.

instruction set (1) All the commands to which a CAD/CAM computer will respond. (2) The repertoire of functions the computer can perform.

integrated circuit (IC) A tiny complex of electronic components and interconnections comprising a circuit that may vary in functional complexity from a simple logic gate to a microprocessor. An IC is usually packaged in a single substrate such as a slice of silicon. The complexity of most IC designs and the many repetitive elements have made computer-aided design an economic necessity. Also called a *chip.*

integrated system A CAD/CAM system that integrates the entire product development cycle—analysis, design, and fabrication—so that all processes flow smoothly from concept to production.

intelligent work station/terminal A workstation in a system that can perform certain data processing functions in a stand-alone mode, independent of another computer. Contains a built-in computer, usually a microprocessor or minicomputer, and dedicated memory.

interactive Denotes two-way communications between a CAD/CAM system or workstation and its operators. An operator can modify or terminate a program and receive feedback from the system for guidance and verification. See also *feedback.*

interactive graphics system (IGS) or interactive computer graphics (ICG) A CAD/CAM system in which the workstations are used interactively for computer-aided design and/or drafting, as well as for CAM, all under full operator control, and possibly also for text-processing, generation of charts and graphs, or computer-aided engineering. The designer (operator) can intervene to enter data and direct the course of any program, receiving immediate visual feedback via the CRT. Bilateral communication is provided between the system and the designer(s). Often used synonymously with *CAD.*

interface (n) (1) A hardware and/or software link that enables two systems, or a system and its peripherals, to operate as a single, integrated system (2) The input devices and visual feedback capabilities that allow bilateral communication between the designer and the system. The interface to a large computer can be a communications link (hardware) or a combination of software and hard-wired connections. An interface might be a portion of storage accessed by two or more programs or a link between two subroutines in a program.

I/O See *input/output.*

ips See *inches per second.*

jaggies A CAD jargon term used to refer to straight or curved lines that appear to be jagged or sawtoothed on the CRT screen.

joystick A CAD data-entering device employing a hand-controlled lever to manually enter the coordinates of various points on a design being digitized into the system.

key file A disk file that provides user-defined definitions for a tablet menu. See *menu.*

kinematics A computer-aided engineering (CAE) process for plotting or animating the motion of parts in a machine or a structure under design on the system. CAE simulation programs allow the motion of mechanisms to be studied for interference, acceleration, and force determinations while still in the design stage.

layering A method of logically organizing data in a CAD/CAM data base. Functionally different classes of data (e.g., various graphic/geometric entities) are segregated on separate layers, each of which can be displayed individually or in any desired combination. Layering helps the designer distinguish among different kinds of data in creating a complex product such as a multilayered PC board or IC.

layers User-defined logical subdivisions of data in a CAD/CAM data base that may be viewed on the CRT individually or overlaid and viewed in groups.

learning curve A concept that projects the expected

improvement in operator productivity over a period of time. Usually applied in the first 1 to $1\frac{1}{2}$ years of a new CAD/CAM facility as part of a cost-justification study, or when new operators are introduced. An accepted tool of management for predicting manpower requirements and evaluating training programs.

library, graphics (or parts library) A collection of standard, often-used symbols, components, shapes, or parts stored in the CAD data base as templates or building blocks to speed up future design work on the system. Generally an organization of files under a common library name.

light pen A hand-held photosensitive CAD input device used on a refreshed CRT screen for identifying display elements, or for designating a location on the screen where an action is to take place.

line font See *font, line.*

line printer A CAD/CAM peripheral device used for rapid printing of data.

line smoothing An automated mapping capability for the interpolation and insertion of additional points along a linear entity yielding a series of shorter linear segments to generate a smooth curved appearance to the original linear component. The additional points or segments are created only for display purposes and are interpolated from a relatively small set of stored representative points. Thus, data storage space is minimized.

low-level language A programming language in which statements translate on a one-for-one basis. See also *machine language.*

machine A computer, CPU, or other processor.

machine instruction An instruction that a machine (computer) can recognize and execute.

machine language The complete set of command instructions understandable to and used directly by a computer when it performs operations.

macro (1) A sequence of computer instructions executable as a single command. A frequently used, multistep operation can be organized into a macro, given a new name, and remain in the system for easy use, thus shortening program development time. (2) In Computer-vision's IC design system, macro refers to macroexpansion of a cell. This system capability enables the designer to replicate the contents of a cell as primitives without the original cell grouping.

magnetic disk A flat circular plate with a magnetic surface on which information can be stored by selective magnetization of portions of the flat surface.

Commonly used for temporary working storage during computer-aided design. See also *disk.*

magnetic tape A tape with a magnetic surface on which information can be stored by selective polarization of portions of the surface. Commonly used in CAD/CAM for off-line storage of completed design files and other archival material.

mainframe (computer) A large central computer facility.

main memory/storage The computer's general-purpose storage from which instructions may be executed and data loaded directly into operating registers.

mass storage Auxiliary large-capacity memory for storing large amounts of data readily accessible by the computer. Commonly a disk or magnetic tape.

matrix A 2D or 3D rectangular array (arrangement) of identical geometric or symbolic entities. A matrix can be generated automatically on a CAD system by specifying the building block entity and the desired locations. This process is used extensively in computer-aided electrical/electronic design.

memory Any form of data storage where information can be read and written. Standard memories include RAM, ROM, and PROM. See also *programmable read-only memory; random access memory; read-only memory; storage.*

menu A common CAD/CAM input device consisting of a checkerboard pattern of squares printed on a sheet of paper or plastic placed over a data tablet. These squares have been preprogrammed to represent a part of a command, a command, or a series of commands. Each square, when touched by an electronic pen, initiates the particular function or command indicated on that square. See also *data tablet, dynamic menuing.*

merge To combine two or more sets of related data into one, usually in a specified sequence. This can be done automatically on a CAD/CAM system to generate lists and reports.

microcomputer A smaller, lower-cost equivalent of a full-scale minicomputer. Includes a microprocessor (CPU), memory, and necessary interface circuits. Consists of one or more ICs (chips) comprising a chip set.

microprocessor The central control element of a microcomputer, implemented in a single integrated circuit. It performs instruction sequencing and processing, as well as all required computations. It requires additional circuits to function as a microcomputer. See *microcomputer.*

minicomputer A general-purpose, single processor computer of limited flexibility and memory performance.

mirroring A CAD design aid that automatically creates a mirror image of a graphic entity on the CRT by flipping the entity or drawing on its *x* or *y* axis.

mnemonic symbol An easily remembered symbol that assists the designer in communicating with the system (e.g., an abbreviation such as MPY for *multiply*).

model, geometric A complete, geometrically accurate 3D or 2D representation of a shape, a part, a geographic area, a plant or any part of it, designed on a CAD system and stored in the data base. A mathematical or analytic model of a physical system used to determine the response of that system to a stimulus or load. See *modeling, geometric.*

modeling, geometric Constructing a mathematical or analytic model of a physical object or system for the purpose of determining the response of that object or system to a stimulus or load. First, the designer describes the shape under design using a geometric model constructed on the system. The computer then converts this pictorial representation on the CRT into a mathematical model later used for other CAD functions such as design optimization.

modeling, solid A type of 3D modeling in which the solid characteristics of an object under design are built into the data base so that complex internal structures and external shapes can be realistically represented. This makes computer-aided design and analysis of solid objects easier, clearer, and more accurate than with wire-frame graphics.

modem *MO*dulator-*DEM*odulator, a device that converts digital signals to analog signals, and vice versa, for long-distance transmission over communications circuits such as telephone lines, dedicated wires, optical fiber, or microwave.

module A separate and distinct unit of hardware or software that is part of a system.

mouse A hand-held data-entering device used to position a cursor on a data tablet. See *cursor.*

multiprocessor A computer whose architecture consists of more than one processing unit. See *central processing unit; microcomputer.*

network An arrangement of two or more interconnected computer systems to facilitate the exchange of information in order to perform a specific function. For example, a CAD/CAM system might be connected to a mainframe computer to off-load heavy analytic tasks. Also refers to a piping network in computer-aided plant design.

numerical control (NC) A technique of operating machine tools or similar equipment in which motion is developed in response to numerically coded commands. These commands may be generated by a CAD/CAM system on punched tapes or other communications media. Also, the processes involved in generating the data or tapes necessary to guide a machine tool in the manufacture of a part.

off-line Refers to peripheral devices not currently connected to and under the direct control of the system's computer.

on-line Refers to peripheral devices connected to and under the direct control of the system's computer, so that operator-system interaction, feedback, and output are all in real time.

operating system A structured set of software programs that control the operation of the computer and associated peripheral devices in a CAD/CAM system, as well as the execution of computer programs and data flow to and from peripheral devices. May provide support for activities and programs such as scheduling, debugging, input/output control, accounting, editing, assembly, compilation, storage assignment, data management, and diagnostics. An operating system may assign task priority levels, support a file system, provide drives for I/O devices, support standard system commands or utilities for on-line programming, process commands, and support both networking and diagnostics.

output The end result of a particular CAD/CAM process or series of processes. The output of a CAD cycle can be artwork and hard-copy lists and reports. The output of a total design-to-manufacturing CAD/CAM system can also include numerical control tapes for manufacturing.

overlay A segment of code or data to be brought into the memory of a computer to replace existing code or data.

paint To fill in a bounded graphic figure on a raster display using a combination of repetitive patterns or line fonts to add meaning or clarity. See *font, line.*

paper-tape punch/reader A peripheral device that can read as well as punch a perforated paper tape generated by a CAD/CAM system. These tapes are the principal means of supplying/data to an NC machine.

parallel processing Executing more than one element of a single process concurrently on multiple processors in a computer system.

password protection A security feature of certain CAD/CAM systems that prevents access to the system or to files within the system without first entering a password, i.e., a special sequence of characters.

PC board See *printed circuit board.*

pen plotter An electromechanical CAD output device that generates hard copy of displayed graphic data by means of a ballpoint pen or liquid ink. Used when a very accurate final drawing is required. Provides exceptional uniformity and density of lines, precise positional accuracy, as well as various user-selectable colors.

peripheral (device) Any device, distinct from the basic system modules, that provides input to and/ or output from the CPU. May include printers, keyboards, plotters, graphics display terminals, paper-tape reader/punches, analog-to-digital converters, disks, and tape drives.

permanent storage A method or device for storing the results of a completed program outside the CPU—usually in the form of magnetic tape or punched cards.

photo plotter A CAD output device that generates high-precision artwork masters photographically for PC board design and IC masks.

pixel The smallest portion of a CRT screen that can be individually referenced. An individual dot on a display image. Typically, pixels are evenly spaced, horizontally and vertically, on the display.

plotter A CAD peripheral device used to output for external use the image stored in the data base. Generally makes large, accurate drawings substantially better than what is displayed. Plotter types include pen, drum, electrostatic, and flatbed.

postprocessor A software program or procedure that formats graphic or other data processed on the system for some other purpose. For example, a postprocessor might format cutter centerline data into a form that a machine controller can interpret.

precision The degree of accuracy. Generally refers to the number of significant digits of information to the right of the decimal point for data represented within a computer system. Thus, the term denotes the degree of discrimination with which a design or design element can be described in the data base.

preplaced line (or bus) A run (or line) between a set of points on a PC board layout that has been predefined by the designer and must be avoided by a CAD automatic routing program.

preprocessor A computer program that takes a specific set of instructions from an external source and translates it into the format required by the system.

primitive A design element at the lowest stage of complexity. A fundamental graphic entity. It can be a vector, a point, or a text string. The smallest definable object in a display processor's instruction set.

printed circuit (PC) board A baseboard made of insulating materials and an etched copper-foil circuit pattern on which are mounted ICs and other components required to implement one or more electronic functions. PC boards plug into a rack or subassembly of electronic equipment to provide the brains or logic to control the operation of a computer, or a communications system, instrumentation, or other electronic systems. The name derives from the fact that the circuitry is connected not by wires but by copper-foil lines, paths, or traces actually etched onto the board surface. CAD/CAM is used extensively in PC board design, testing, and manufacture.

process simulation A program utilizing a mathematical model created on the system to try out numerous process design iterations with real-time visual and numerical feedback. Designers can see on the CRT what is taking place at every stage in the manufacturing process. They can therefore optimize a process and correct problems that could affect the actual manufacturing process downstream.

processor In CAD/CAM system hardware, any device that performs a specific function. Most often used to refer to the CPU. In software, it refers to a complex set of instructions to perform a general function. See also *central processing unit.*

productivity ratio A widely accepted means of measuring CAD/CAM productivity (throughput per hour) by comparing the productivity of a design/ engineering group before and after installation of the system or relative to some standard norm or potential maximum. The most common way of recording productivity is Actual Manual Hours/Actual CAD Hours, expressed as 4:1, 6:1, etc.

program (n) A precise sequential set of instructions that direct a computer to perform a particular task or action or to solve a problem. A complete program includes plans for the transcription of data, coding for the computer, and plans for the absorption of the results into the system. (v) To develop a program. See also *computer program*.

Programmable Read-Only Memory (PROM) A memory that, once programmed with permanent data or instructions, becomes a ROM. See *read-only memory*.

PROM See *programmable read-only memory*.

prompt A message or symbol generated automatically by the system, and appearing on the CRT, to inform the user of (a) a procedural error or incorrect input to the program being executed or (b) the next expected action, option(s), or input. See also *tutorial*.

puck A hand-held, manually controlled input device that allows coordinate data to be digitized into the system from a drawing placed on the data tablet or digitizer surface. A puck has a transparent window containing cross hairs.

RAM See *random access memory*.

random access memory (RAM) A main memory read/write storage unit that provides the CAD/CAM operator direct access to the stored information. The time required to access any word stored in the memory is the same as for any other word.

raster display A CAD workstation display in which the entire CRT surface is scanned at a constant refresh rate. The bright, flicker-free image can be selectively written and erased. Also called a digital TV display.

raster scan (video) Currently, the dominant technology in CAD graphic displays. Similar to conventional television, it involves a line-by-line sweep across the entire CRT surface to generate the image. Raster-scan features include good brightness, accuracy, selective erase, dynamic motion capabilities, and the opportunity for unlimited color. The device can display a large amount of information without flicker, although resolution is not as good as with storage-tube displays.

read-only memory (ROM) A memory that cannot be modified or reprogrammed. Typically used for control and execute programs. See also *programmable read-only memory*.

real time Refers to tasks or functions executed so rapidly by a CAD/CAM system that the feedback at various stages in the process can be used to guide the designer in completing the task. Immediate visual feedback through the CRT makes possible real time, interactive operation of a CAD/CAM system.

rectangular array Insertion of the same entity at multiple locations on a CRT using the system's ability to copy design elements and place them at user-specified intervals to create a rectangular arrangement or matrix. A feature of PC and IC design systems.

refresh (or vector refresh) A CAD display technology that involves frequent redrawing of an image displayed on the CRT to keep it bright, crisp, and clear. Refresh permits a high degree of movement in the displayed image as well as high resolution. Selective erase or editing is possible at any time without erasing and repainting the entire image. Although substantial amounts of high-speed memory are required, large, complex images may flicker.

refresh rate The rate at which the graphic image on a CRT is redrawn in a refresh display, i.e., the time needed for one refresh of the displayed image.

registration The degree of accuracy in the positioning of one layer or overlay in a CAD display or artwork, relative to another layer, as reflected by the clarity and sharpness of the resulting image.

repaint A CAD feature that automatically redraws a design displayed on the CRT.

resolution The smallest spacing between two display elements that will allow the elements to be distinguished visually on the CRT. The ability to define very minute detail. For example, the resolution of Computervision's IC design system is one part in 33.5 million. As applied to an electrostatic plotter, resolution means the number of dots per square inch.

response time The elapsed time from initiation of an operation at a workstation to the receipt of the results at that workstation. Includes transmission of data to the CPU, processing, file access, and transmission of results back to the initiating workstation.

restart To resume a computer program interrupted by operator intervention.

restore To bring back to its original state a design currently being worked on in a CAD/CAM system after editing or modification that the designer now wants to cancel or rescind.

resume A feature of some application programs that allows the designer to suspend the data-processing operation at some logical break point and restart it later from that point.

reticle The photographic plate used to create an IC mask. See also *photo plotter*.

rotate To turn a displayed 2D or 3D construction about an axis through a predefined angle relative to the original position.

robotics The use of computer-controlled manipulators or arms to automate a variety of manufacturing processes such as welding, material handling, painting and assembly.

ROM See *read-only memory*.

routine A computer program, or a subroutine in the main program. The smallest separately compilable source code unit. See *computer program; source*.

rubber banding A CAD capability that allows a component to be tracked (dragged) across the CRT screen, by means of an electronic pen, to a desired location, while simultaneously stretching all related interconnections to maintain signal continuity. During tracking the interconnections associated with the component stretch and bend, providing an excellent visual guide for optimizing the location of a component to best fit into the flow of the PC board, or other entity, minimizing total interconnect length and avoiding areas of congestion.

satellite A remote system connected to another, usually larger, host system. A satellite differs from a remote intelligent work station in that it contains a full set of processors, memory, and mass storage resources to operate independently of the host. See *host-satellite system*.

scale (v) To enlarge or diminish the size of a displayed entity without changing its shape, i.e., to bring it into a user-specified ratio to its original dimensions. Scaling can be done automatically by a CAD system. (n) Denotes the coordinate system for representing an object.

scissoring The automatic erasing of all portions of a design on the CRT that lie outside user-specified boundaries.

scroll To automatically roll up, as on a spool, a design or text message on a CRT to permit the sequential viewing of a message or drawing too large to be displayed all at once on the screen. New data appear on the CRT at one edge as other data disappear at the opposite edge. Graphics can be scrolled up, down, left, or right.

selective erase A CAD feature for deleting portions of a display without affecting the remainder or having to repaint the entire CRT display.

shape fill The automatic painting-in of an area, defined by user-specified boundaries, on an IC or PC board layout, for example, the area to be filled by copper when the PC board is manufactured. Can be done on-line by CAD.

smoothing Fitting together curves and surfaces so that a smooth, continuous geometry results.

software The collection of executable computer programs including application programs, operating systems, and languages.

source A text file written in a high-level language and containing a computer program. It is easily read and understood by people but must be compiled or assembled to generate machine-recognizable instructions. Also known as *source code*. See also *high-level language*.

source language A symbolic language comprised of statements and formulas used in computer processing. It is translated into object language (object code) by an assembler or compiler for execution by a computer.

spline A subset of a B-spline where in a sequence of curves is restricted to a plane. An interpolation routine executed on a CAD/CAM system automatically adjusts a curve by design iteration until the curvature is continuous over the length of the curve. See also *B-spline*.

storage The physical repository of all information relating to products designed on a CAD/CAM system. It is typically in the form of a magnetic tape or disk. Also called *memory*.

storage tube A common type of CRT that retains an image continuously for a considerable period of time without redrawing (refreshing). The image will not flicker regardless of the amount of information displayed. However, the display tends to be slow relative to raster scan, the image is rather dim, and no single element by itself can be modified or deleted without redrawing. See also *direct view storage tube*.

stretch A CAD design/editing aid that enables the designer to automatically expand a displayed entity beyond its original dimensions.

string A linear sequence of entities, such as characters or physical elements, in a computer-aided design.

stylus A hand-held pen used in conjunction with a data table to enter commands and coordinate input into the system. Also called an *electronic pen*.

subfigure A part or a design element that may be extracted from a CAD library and inserted intact into another part displayed on the CRT.

surface machining Automatic generation of NC tool paths to cut 3D shapes. Both the tool paths and the shapes may be constructed using the mechanical design capabilities of a CAD/CAM system.

symbol Any recognizable sign, mark, shape or pattern used as a building block for designing meaningful structures. A set of primitive graphic entities (line, point, arc, circle, text, etc.) that form a construction that can be expressed as one unit and assigned a meaning. Symbols may be combined or nested to form larger symbols and/or drawings. They can be as complex as an entire PC board or as simple as a single element, such as a pad. Symbols are commonly used to represent physical things. For example, a particular graphic shape may be used to represent a complete device or a certain kind of electrical component in a schematic. To simplify the preparation of drawings of piping systems and flow diagrams, standard symbols are used to represent various types of fittings and components in common use. Symbols are also basic units in a language. The recognizable sequence of characters END may inform a compiler that the routine it is compiling is completed. In computer-aided mapping, a symbol can be a diagram, design, letter, character, or abbreviation placed on maps and charts, that, by convention or reference to a legend, is understood to stand for or represent a specific characteristic or feature. In a CAD environment, symbol libraries contribute to the quick maintenance, placement, and interpretation of symbols.

syntax (1) A set of rules describing the structure of statements allowed in a computer language. To make grammatical sense, commands and routines must be written in conformity to these rules. (2) The structure of a computer command language, i.e., the English-sentence structure of a CAD/CAM command language, e.g., verb, noun, modifiers.

system An arrangement of CAD/CAM dataprocessing, memory, display, and plotting modules—coupled with appropriate software—to achieve specific objectives. The term CAD/CAM system implies both hardware and software. See also *operating system* (a purely software term).

tablet An input device on which a designer can digitize coordinate data or enter commands into a CAD/CAM system by means of an electronic pen. See also *data tablet*.

task (1) A specific project that can be executed by a CAD/CAM software program. (2) A specific portion of memory assigned to the user for executing that project.

template the pattern of a standard, commonly used component or part that serves as a design aid. Once created, it can be subsequently traced instead of redrawn whenever needed. The CAD equivalent of a designer's template might be a standard part in the data-base library that can be retrieved and inserted intact into an emerging drawing on the CRT.

temporary storage Memory locations for storing immediate and partial results obtained during the execution of a program on the system.

terminal See *workstation*.

text editor An operating system program used to create and modify text files on the system.

text file A file stored in the system in text format that can be printed and edited on-line as required.

throughput The number of units of work performed by a CAD/CAM system or a work station during a given period of time. A quantitative measure of system productivity.

time-sharing The use of a common CPU memory and processing capabilities by two or more CAD/CAM terminals to execute different tasks simultaneously.

tool path Centerline of the tip of an NC cutting tool as it moves over a part produced on a CAD/CAM system. Tool paths can be created and displayed interactively or automatically by a CAD/CAM system, and reformatted into NC tapes, by means of postprocessor, to guide or control machining equipment. See also *surface machining*.

track ball A CAD graphics input device consisting of a ball recessed into a surface. The designer can rotate it in any direction to control the position of the cursor used for entering coordinate data into the system.

tracking Moving a predefined (tracking) symbol across the surface of the CRT with a light pen or an electronic pen.

transform To change an image displayed on the CRT by, for example, scaling, rotating, translating, or mirroring.

transformation The process of transforming a CAD display image. Also the matrix representation of a geometric space.

translate (1) To convert CAD/CAM output from one language to another, for example, by means of a postprocessor such as Computervision's IPC-to-Numerics Translator program. (2) Also, by an editing command, to move a CAD display entity a specified distance in a specified direction.

trap The area that is searched around each digitize to find a hit on a graphics entity to be edited. See also *digitize*.

turnaround time The elapsed time between the moment a task or project is input into the CAD/CAM system and the moment the required output is obtained.

turnkey A CAD/CAM system for which the supplier/vendor assumes total responsibility for building, installing, and testing both hardware and software, and the training of user personnel. Also, loosely, a system that comes equipped with all the hardware and software required for a specific application or applications. Usually implies a commitment by the vendor to make the system work and to provide preventive and remedial maintenance of both hardware and software. Sometimes used interchangeably with stand-alone, although stand-alone applies more to system architecture than to terms of purchase.

tutorial A characteristic of CAD/CAM systems. If the user is not sure how to execute a task, the system will show how. A message is displayed to provide information and guidance.

utilities Another term for system capabilities and/or features that enable the user to perform certain processes.

vector A quantity that has magnitude and direction and that, in CAD, is commonly represented by a directed line segment.

verification (1) A system-generated message to a work station acknowledging that a valid instruction or input has been received. (2) The process of checking the accuracy, viability, and/or manufacturability of an emerging design on the system.

view port A user-selected, rectangular view of a part, assembly, etc., that presents the contents of a window on the CRT. See also *window*.

window A temporary, usually rectangular, bounded area on the CRT that is user-specified to include particular entities for modification, editing, or deletion.

wire-frame graphics A computer-aided design technique for displaying a 3D object on the CRT screen as a series of lines outlining its surface.

wiring diagram (1) Graphic representation of all circuits and device elements of an electrical system and its associated apparatus or any clearly defined

functional portion of that system. A wiring diagram may contain not only wiring system components and wires but also nongraphic information such as wire number, wire size, color, function, component label, and pin number. (2) Illustration of device elements and their interconnectivity as distinguished from their physical arrangement. (3) Drawing that shows how to hook things up.

Wiring diagrams can be constructed, annotated, and documented on a CAD system.

word A set of bits (typically 16 to 32) that occupies a single storage location and is treated by the computer as a unit. See also *bit*.

working storage That part of the system's internal storage reserved for intermediate results (i.e., while a computer program is still in progress). Also called *temporary storage*.

workstation The work area and equipment used for CAD/CAM operations. It is where the designer interacts (communicates) with the computer. Frequently consists of a CRT display and an input device as well as, possibly, a digitizer and a hard-copy device. In a distributed processing system, a work station would have local processing and mass storage capabilities. Also called a *terminal* or *design terminal*.

write To transfer information from CPU main memory to a peripheral device, such as a mass storage device.

write-protect A security feature in a CAD/CAM data storage device that prevents new data from being written over existing data.

zero The origin of all coordinate dimensions defined in an absolute system as the intersection of the baselines of the x, y, and z axes.

zero offset On an NC unit, this features allows the zero point on an axis to be relocated anywhere within a specified range, thus temporarily redefining the coordinate frame of reference.

zoom A CAD capability that proportionately enlarges or reduces a figure displayed on a CRT screen.

4 Abbreviations for Use on Drawings and in Text– American National Standard

[Selected from ANSI Y1.1–1972 (R1984)]

A

absolute	ABS
accelerate	ACCEL
accessory	ACCESS.
account	ACCT
accumulate	ACCUM
actual	ACT.
adapter	ADPT
addendum	ADD.
addition	ADD.
adjust	ADJ
advance	ADV
after	AFT.
aggregate	AGGR
air condition	AIR COND
airplane	APL
allowance	ALLOW
alloy	ALY
alteration	ALT
alternate	ALT
alternating current	AC
altitude	ALT
aluminum	AL
American National Standard	AMER NATL STD
American wire gage	AWG
amount	AMT
ampere	AMP
amplifier	AMPL
anneal	ANL
antenna	ANT.
apartment	APT.
apparatus	APP
appendix	APPX
approved	APPD
approximate	APPROX
arc weld	ARC/W
area	A
armature	ARM.
armor plate	ARM-PL
army navy	AN
arrange	ARR.
artificial	ART.
asbestos	ASB

asphalt	ASPH
assemble	ASSEM
assembly	ASSY
assistant	ASST
associate	ASSOC
association	ASSN
atomic	AT
audible	AUD
audio frequency	AF
authorized	AUTH
automatic	AUTO
auto-transformer	AUTO TR
auxiliary	AUX
avenue	AVE
average	AVG
aviation	AVI
azimuth	AZ

B

Babbitt	BAB
back feed	BF
back pressure	BP
back to back	B to B
backface	BF
balance	BAL
ball bearing	BB
barometer	BAR
base line	BL
base plate	BP
bearing	BRG
bench mark	BM
bending moment	M
bent	BT
bessemer	BESS
between	BET.
between centers	BC
between perpendiculars	BP
bevel	BEV
bill of material	B/M
Birmingham wire gage	BWG
blank	BLK
block	BLK
blueprint	BP

board	BD
boiler	BLR
boiler feed	BF
boiler horsepower	BHP
boiling point	BP
bolt circle	BC
both faces	BF
both sides	BS
both ways	BW
bottom	BOT
bottom chord	BC
bottom face	BF
bracket	BRKT
brake	BK
brake horsepower	BHP
brass	BRS
brazing	BRZG
break	BRK
Brinell hardness	BH
British Standard	BR STD
British thermal unit	BTU
broach	BRO
bronze	BRZ
Brown & Sharpe (wire gage, same as AWG)	B&S
building	BLDG
bulkhead	BHD
burnish	BNH
bushing	BUSH.
button	BUT.

C

cabinet	CAB.
calculate	CALC
calibrate	CAL
cap screw	CAP SCR
capacity	CAP
carburetor	CARB
carburize	CARB
carriage	CRG
case harden	CH
cast iron	CI
cast steel	CS
casting	CSTG

castle nut	CAS NUT	counter clockwise	CCW	eccentric	ECC
catalogue	CAT.	counterdrill	CDRILL	effective	EFF
cement	CEM	counterpunch	CPUNCH	elbow	ELL
center	CTR	countersink	CSK	electric	ELEC
center line	CL	coupling	CPLG	elementary	ELEM
center of gravity	CG	cover	COV	elevate	ELEV
center of pressure	CP	cross section	XSECT	elevation	EL
center to center	C to C	cubic	CU	engine	ENG
centering	CTR	cubic foot	CU FT	engineer	ENGR
chamfer	CHAM	cubic inch	CU IN.	engineering	ENGRG
change	CHG	current	CUR	entrance	ENT
channel	CHAN	customer	CUST	equal	EQ
check	CHK	cyanide	CYN	equation	EQ
check valve	CV			equipment	EQUIP
chord	CHD	**D**		equivalent	EQUIV
circle	CIR			estimate	EST
circular	CIR	decimal	DEC	exchange	EXCH
circular pitch	CP	dedendum	DED	exhaust	EXH
circumference	CIRC	deflect	DEFL	existing	EXIST.
clear	CLR	degree	(°) DEG	exterior	EXT
clearance	CL	density	D	extra heavy	X HVY
clockwise	CW	department	DEPT	extra strong	X STR
coated	CTD	design	DSGN	extrude	EXTR
cold drawn	CD	detail	DET		
cold-drawn steel	CDS	develop	DEV	**F**	
cold finish	CF	diagonal	DIAG		
cold punched	CP	diagram	DIAG	fabricate	FAB
cold rolled	CR	diameter	DIA	face to face	F to F
cold-rolled steel	CRS	diametral pitch	DP	Fahrenheit	F
combination	COMB.	dimension	DIM.	far side	FS
combustion	COMB	discharge	DISCH	federal	FED.
commercial	COML	distance	DIST	feed	FD
company	CO	division	DIV	feet	(') FT
complete	COMPL	double	DBL	figure	FIG.
compress	COMP	dovetail	DVTL	fillet	FIL
concentric	CONC	dowel	DWL	fillister	FIL
concrete	CONC	down	DN	finish	FIN.
condition	COND	dozen	DOZ	finish all over	FAO
connect	CONN	drafting	DFTG	flange	FLG
constant	CONST	drawing	DWG	flat	F
construction	CONST	drill or drill rod	DR	flat head	FH
contact	CONT	drive	DR	floor	FL
continue	CONT	drive fit	DF	fluid	FL
copper	COP.	drop	D	focus	FOC
corner	COR	drop forge	DF	foot	(') FT
corporation	CORP	duplicate	DUP	force	F
correct	CORR			forged steel	FST
corrugate	CORR	**E**		forging	FORG
cotter	COT			forward	FWD
counter	CTR	each	EA	foundry	FDRY
counterbore	CBORE	east	E	frequency	FREQ

front	FR	inclosure	INCL
furnish	FURN	include	INCL
		inside diameter	ID
G		instrument	INST
		interior	INT
gage or gauge	GA	internal	INT
gallon	GAL	intersect	INT
galvanize	GALV	iron	I
galvanized iron	GI	irregular	IREG
galvanized steel	GS		
gasket	GSKT	**J**	
general	GEN		
glass	GL	joint	JT
government	GOVT	joint army-navy	JAN
governor	GOV	journal	JNL
grade	GR	junction	JCT
graduation	GRAD		
graphite	GPH	**K**	
grind	GRD		
groove	GRV	key	K
ground	GRD	keyseat	KST
		Keyway	KWY

H		**L**	
half-round	½RD	laboratory	LAB
handle	HDL	laminate	LAM
hanger	HGR	lateral	LAT
hard	H	left	L
harden	HDN	left hand	LH
hardware	HDW	length	LG
head	HD	length over all	LOA
headless	HDLS	letter	LTR
heat	HT	light	LT
heat-treat	HT TR	line	L
heavy	HVY	locate	LOC
hexagon	HEX	logarithm	LOG.
high-pressure	HP	long	LG
high-speed	HS	lubricate	LUB
horizontal	HOR	lumber	LBR
horsepower	HP		
hot rolled	HR	**M**	
hot-rolled steel	HRS		
hour	HR	machine	MACH
housing	HSG	machine steel	MS
hydraulic	HYD	maintenance	MAINT
		malleable	MALL
I		malleable iron	MI
illustrate	ILLUS	manual	MAN.
inboard	INBD	manufacture	MFR
inch	(″) IN.	manufactured	MFD
inches per second	IPS	manufacturing	MFG

material	MATL	obsolete	OBS
maximum	MAX	octagon	OCT
mechanical	MECH	office	OFF.
mechanism	MECH	on center	OC
median	MED	opposite	OPP
metal	MET.	optical	OPT
meter	M	original	ORIG
miles	MI	outlet	OUT.
miles per hour	MPH	outside diameter	OD
millimeter	MM	outside face	OF
minimum	MIN	outside radius	OR
minute	(′) MIN	overall	OA
miscellaneous	MISC		
month	MO		
Morse taper	MOR T		
motor	MOT		
mounted	MTD		
mounting	MTG		
multiple	MULT		
music wire gage	MWG		

N

national	NATL
natural	NAT
near face	NF
near side	NS
negative	NEG
neutral	NEUT
nominal	NOM
normal	NOR
north	N
not to scale	NTS
number	NO.

O

P

pack	PK
packing	PKG
page	P
paragraph	PAR.
part	PT
patent	PAT.
pattern	PATT
permanent	PERM
perpendicular	PERP
piece	PC
piece mark	PC MK
pint	PT
pitch	P
pitch circle	PC
pitch diameter	PD
plastic	PLSTC
plate	PL
plumbing	PLMB
point	PT
point of curve	PC
point of intersection	PI
point of tangent	PT
polish	POL
position	POS
potential	POT.
pound	LB
pounds per square inch	PSI
power	PWR
prefabricated	PREFAB
preferred	PFD
prepare	PREP
pressure	PRESS.
process	PROC
production	PROD
profile	PF
propeller	PROP
publication	PUB
push button	PB

Q

quadrant	QUAD
quality	QUAL
quarter	QTR

R

radial	RAD
radius	R
railroad	RR
ream	RM
received	RECD
record	REC
rectangle	RECT
reduce	RED.
reference line	REF L
reinforce	REINF
release	REL
relief	REL
remove	REM
require	REQ
required	REQD
return	RET.
reverse	REV
revolution	REV
revolutions per minute	RPM
right	R
right hand	RH
rivet	RIV
Rockwell hardness	RH
roller bearing	RB
room	RM
root diameter	RD
root mean square	RMS
rough	RGH
round	RD

S

schedule	SCH
schematic	SCHEM
scleroscope hardness	SH
screw	SCR
second	SEC
section	SECT
semi-steel	SS
separate	SEP
set screw	SS
shaft	SFT
sheet	SH
shoulder	SHLD
side	S
single	S
sketch	SK
sleeve	SLV
slide	SL
slotted	SLOT.
small	SM
socket	SOC
space	SP
special	SPL
specific	SP
spot faced	SF
spring	SPG
square	SQ
standard	STD
station	STA
stationary	STA
steel	STL
stock	STK
straight	STR
street	ST
structural	STR
substitute	SUB
summary	SUM.
support	SUP.
surface	SUR
symbol	SYM
system	SYS

T

tangent	TAN.
taper	TPR
technical	TECH
template	TEMP
tension	TENS.
terminal	TERM.
thick	THK
thousand	M
thread	THD
threads per inch	TPI
through	THRU
time	T
tolerance	TOL
tongue & groove	T & G
tool steel	TS
tooth	T
total	TOT
transfer	TRANS
typical	TYP

U

ultimate	ULT
unit	U
universal	UNIV

V

vacuum	VAC
valve	V
variable	VAR

versus	VS	watt	W	wrought	WRT	
vertical	VERT	week	WK	wrought iron	WI	
volt	V	weight	WT			
volume	VOL	west	W	*Y*		
		width	W			
W		wood	WD	yard	YD	
		Woodruff	WDF	year	YR	
wall	W	working point	WP			
washer	WASH.	working pressure	WP			

5   Running and Sliding Fits[a]—American National Standard

RC 1   *Close sliding fits* are intended for the accurate location of parts which must assemble without perceptible play.

RC 2   *Sliding fits* are intended for accurate location, but with greater maximum clearance than class RC 1. Parts made to this fit move and turn easily but are not intended to run freely, and in the larger sizes may seize with small temperature changes.

RC 3   *Precision running fits* are about the closest fits which can be expected to run freely, and are intended for precision work at slow speeds and light journal pressures, but are not suitable where appreciable temperature differences are likely to be encountered.

RC 4   *Close running fits* are intended chiefly for running fits on accurate machinery with moderate surface speeds and journal pressures, where accurate location and minimum play are desired.

Basic hole system. Limits are in thousandths of an inch. See §14.8.
Limits for hole and shaft are applied algebraically to the basic size to obtain the limits of size for the parts.
Data in **boldface** are in accordance with ABC agreements.
Symbols H5, g5, etc., are hole and shaft designations used in ABC System.

Nominal Size Range, inches Over To	Class RC 1			Class RC 2			Class RC 3			Class RC 4		
	Limits of Clearance	Standard Limits		Limits of Clearance	Standard Limits		Limits of Clearance	Standard Limits		Limits of Clearance	Standard Limits	
		Hole H5	Shaft g4		Hole H6	Shaft g5		Hole H7	Shaft f6		Hole H8	Shaft f7
0 − 0.12	0.1 0.45	+0.2 −0	−0.1 −0.25	0.1 0.55	+0.25 −0	−0.1 −0.3	0.3 0.95	+0.4 −0	−0.3 −0.55	0.3 1.3	+0.6 −0	−0.3 −0.7
0.12− 0.24	0.15 0.5	+0.2 −0	−0.15 −0.3	0.15 0.65	+0.3 −0	−0.15 −0.35	0.4 1.12	+0.5 −0	−0.4 −0.7	0.4 1.6	+0.7 −0	−0.4 −0.9
0.24− 0.40	0.2 0.6	+0.25 −0	−0.2 −0.35	0.2 0.85	+0.4 −0	−0.2 −0.45	0.5 1.5	+0.6 −0	−0.5 −0.9	0.5 2.0	+0.9 −0	−0.5 −1.1
0.40− 0.71	0.25 0.75	+0.3 −0	−0.25 −0.45	0.25 0.95	+0.4 −0	−0.25 −0.55	0.6 1.7	+0.7 −0	−0.6 −1.0	0.6 2.3	+1.0 −0	−0.6 −1.3
0.71− 1.19	0.3 0.95	+0.4 −0	−0.3 −0.55	0.3 1.2	+0.5 −0	−0.3 −0.7	0.8 2.1	+0.8 −0	−0.8 −1.3	0.8 2.8	+1.2 −0	−0.8 −1.6
1.19− 1.97	0.4 1.1	+0.4 −0	−0.4 −0.7	0.4 1.4	+0.6 −0	−0.4 −0.8	1.0 2.6	+1.0 −0	−1.0 −1.6	1.0 3.6	+1.6 −0	−1.0 −2.0
1.97− 3.15	0.4 1.2	+0.5 −0	−0.4 −0.7	0.4 1.6	+0.7 −0	−0.4 −0.9	1.2 3.1	+1.2 −0	−1.2 −1.9	1.2 4.2	+1.8 −0	−1.2 −2.4
3.15− 4.73	0.5 1.5	+0.6 −0	−0.5 −0.9	0.5 2.0	+0.9 −0	−0.5 −1.1	1.4 3.7	+1.4 −0	−1.4 −2.3	1.4 5.0	+2.2 −0	−1.4 −2.8
4.73− 7.09	0.6 1.8	+0.7 −0	−0.6 −1.1	0.6 2.3	+1.0 −0	−0.6 −1.3	1.6 4.2	+1.6 −0	−1.6 −2.6	1.6 5.7	+2.5 −0	−1.6 −3.2
7.09− 9.85	0.6 2.0	+0.8 −0	−0.6 −1.2	0.6 2.6	+1.2 −0	−0.6 −1.4	2.0 5.0	+1.8 −0	−2.0 −3.2	2.0 6.6	+2.8 −0	−2.0 −3.8
9.85−12.41	0.8 2.3	+0.9 −0	−0.8 −1.4	0.8 2.9	+1.2 −0	−0.8 −1.7	2.5 5.7	+2.0 −0	−2.5 −3.7	2.5 7.5	+3.0 −0	−2.5 −4.5
12.41−15.75	1.0 2.7	+1.0 −0	−1.0 −1.7	1.0 3.4	+1.4 −0	−1.0 −2.0	3.0 6.6	+2.2 −0	−3.0 −4.4	3.0 8.7	+3.5 −0	−3.0 −5.2

[a]From ANSI B4.1—1967 (R1987). For larger diameters, see the standard.

5 Running and Sliding Fits[a]—American National Standard
(continued)

RC 5)
RC 6) *Medium running fits* are intended for higher running speeds, or heavy journal pressures, or both.

RC 7 *Free running fits* are intended for use where accuracy is not essential, or where large temperature variations are likely to be encountered, or under both these conditions.

RC 8)
RC 9) *Loose running fits* are intended for use where wide commercial tolerances may be necessary, together with an allowance, on the external member.

(Clearance)

Nominal Size Range, inches (Over – To)	Class RC 5 Limits of Clearance	Class RC 5 Standard Limits Hole H8	Class RC 5 Standard Limits Shaft e7	Class RC 6 Limits of Clearance	Class RC 6 Standard Limits Hole H9	Class RC 6 Standard Limits Shaft e8	Class RC 7 Limits of Clearance	Class RC 7 Standard Limits Hole H9	Class RC 7 Standard Limits Shaft d8	Class RC 8 Limits of Clearance	Class RC 8 Standard Limits Hole H10	Class RC 8 Standard Limits Shaft c9	Class RC 9 Limits of Clearance	Class RC 9 Standard Limits Hole H11	Class RC 9 Standard Limits Shaft
0 – 0.12	0.6 / 1.6	+0.6 / −0	−0.6 / −1.0	0.6 / 2.2	+1.0 / −0	−0.6 / −1.2	1.0 / 2.6	+1.0 / −0	−1.0 / −1.6	2.5 / 5.1	+1.6 / −0	−2.5 / −3.5	4.0 / 8.1	+2.5 / −0	−4.0 / −5.6
0.12– 0.24	0.8 / 2.0	+0.7 / −0	−0.8 / −1.3	0.8 / 2.7	+1.2 / −0	−0.8 / −1.5	1.2 / 3.1	+1.2 / −0	−1.2 / −1.9	2.8 / 5.8	+1.8 / −0	−2.8 / −4.0	4.5 / 9.0	+3.0 / −0	−4.5 / −6.0
0.24– 0.40	1.0 / 2.5	+0.9 / −0	−1.0 / −1.6	1.0 / 3.3	+1.4 / −0	−1.0 / −1.9	1.6 / 3.9	+1.4 / −0	−1.6 / −2.5	3.0 / 6.6	+2.2 / −0	−3.0 / −4.4	5.0 / 10.7	+3.5 / −0	−5.0 / −7.2
0.40– 0.71	1.2 / 2.9	+1.0 / −0	−1.2 / −1.9	1.2 / 3.8	+1.6 / −0	−1.2 / −2.2	2.0 / 4.6	+1.6 / −0	−2.0 / −3.0	3.5 / 7.9	+2.8 / −0	−3.5 / −5.1	6.0 / 12.8	+4.0 / −0	−6.0 / −8.8
0.71– 1.19	1.6 / 3.6	+1.2 / −0	−1.6 / −2.4	1.6 / 4.8	+2.0 / −0	−1.6 / −2.8	2.5 / 5.7	+2.0 / −0	−2.5 / −3.7	4.5 / 10.0	+3.5 / −0	−4.5 / −6.5	7.0 / 15.5	+5.0 / −0	−7.0 / −10.5
1.19– 1.97	2.0 / 4.6	+1.6 / −0	−2.0 / −3.0	2.0 / 6.1	+2.5 / −0	−2.0 / −3.6	3.0 / 7.1	+2.5 / −0	−3.0 / −4.6	5.0 / 11.5	+4.0 / −0	−5.0 / −7.5	8.0 / 18.0	+6.0 / −0	−8.0 / −12.0
1.97– 3.15	2.5 / 5.5	+1.8 / −0	−2.5 / −3.7	2.5 / 7.3	+3.0 / −0	−2.5 / −4.3	4.0 / 8.8	+3.0 / −0	−4.0 / −5.8	6.0 / 13.5	+4.5 / −0	−6.0 / −9.0	9.0 / 20.5	+7.0 / −0	−9.0 / −13.5
3.15– 4.73	3.0 / 6.6	+2.2 / −0	−3.0 / −4.4	3.0 / 8.7	+3.5 / −0	−3.0 / −5.2	5.0 / 10.7	+3.5 / −0	−5.0 / −7.2	7.0 / 15.5	+5.0 / −0	−7.0 / −10.5	10.0 / 24.0	+9.0 / −0	−10.0 / −15.0
4.73– 7.09	3.5 / 7.6	+2.5 / −0	−3.5 / −5.1	3.5 / 10.0	+4.0 / −0	−3.5 / −6.0	6.0 / 12.5	+4.0 / −0	−6.0 / −8.5	8.0 / 18.0	+6.0 / −0	−8.0 / −12.0	12.0 / 28.0	+10.0 / −0	−12.0 / −18.0
7.09– 9.85	4.0 / 8.6	+2.8 / −0	−4.0 / −5.8	4.0 / 11.3	+4.5 / −0	−4.0 / −6.8	7.0 / 14.3	+4.5 / −0	−7.0 / −9.8	10.0 / 21.5	+7.0 / −0	−10.0 / −14.5	15.0 / 34.0	+12.0 / −0	−15.0 / −22.0
9.85–12.41	5.0 / 10.0	+3.0 / −0	−5.0 / −7.0	5.0 / 13.0	+5.0 / −0	−5.0 / −8.0	8.0 / 16.0	+5.0 / −0	−8.0 / −11.0	12.0 / 25.0	+8.0 / −0	−12.0 / −17.0	18.0 / 38.0	+12.0 / −0	−18.0 / −26.0
12.41–15.75	6.0 / 11.7	+3.5 / −0	−6.0 / −8.2	6.0 / 15.5	+6.0 / −0	−6.0 / −9.5	10.0 / 19.5	+6.0 / −0	−10.0 / −13.5	14.0 / 29.0	+9.0 / −0	−14.0 / −20.0	22.0 / 45.0	+14.0 / −0	−22.0 / −31.0

[a]From ANSI B4.1—1967 (R1987). For larger diameters, see the standard.

6 Clearance Locational Fits[a]—American National Standard

LC Locational clearance fits are intended for parts which are normally stationary, but which can be freely assembled or disassembled. They run from snug fits for parts requiring accuracy of location, through the medium clearance fits for parts such as spigots, to the looser fastener fits where freedom of assembly is of prime importance.

Basic hole system. Limits are in thousandths of an inch. See §14.8.

Limits for hole and shaft are applied algebraically to the basic size to obtain the limits of size for the parts.

Data in **boldface** are in accordance with ABC agreements.

Symbols H6, h5, etc., are hole and shaft designations used in ABC System.

Nominal Size Range, inches Over – To	Class LC 1 Limits of Clearance	Hole H6	Shaft h5	Class LC 2 Limits of Clearance	Hole H7	Shaft h6	Class LC 3 Limits of Clearance	Hole H8	Shaft h7	Class LC 4 Limits of Clearance	Hole H10	Shaft h9	Class LC 5 Limits of Clearance	Hole H7	Shaft g6
0 – 0.12	0 / 0.45	+0.25 / -0	+0 / -0.2	0 / 0.65	+0.4 / -0	+0 / -0.25	0 / 1	+0.6 / -0	+0 / -0.4	0 / 2.6	+1.6 / -0	+0 / -1.0	0.1 / 0.75	+0.4 / -0	-0.1 / -0.35
0.12– 0.24	0 / 0.5	+0.3 / -0	+0 / -0.2	0 / 0.8	+0.5 / -0	+0 / -0.3	0 / 1.2	+0.7 / -0	+0 / -0.5	0 / 3.0	+1.8 / -0	+0 / -1.2	0.15 / 0.95	+0.5 / -0	-0.15 / -0.45
0.24– 0.40	0 / 0.65	+0.4 / -0	+0 / -0.25	0 / 1.0	+0.6 / -0	+0 / -0.4	0 / 1.5	+0.9 / -0	+0 / -0.6	0 / 3.6	+2.2 / -0	+0 / -1.4	0.2 / 1.2	+0.6 / -0	-0.2 / -0.6
0.40– 0.71	0 / 0.7	+0.4 / -0	+0 / -0.3	0 / 1.1	+0.7 / -0	+0 / -0.4	0 / 1.7	+1.0 / -0	+0 / -0.7	0 / 4.4	+2.8 / -0	+0 / -1.6	0.25 / 1.35	+0.7 / -0	-0.25 / -0.65
0.71– 1.19	0 / 0.9	+0.5 / -0	+0 / -0.4	0 / 1.3	+0.8 / -0	+0 / -0.5	0 / 2	+1.2 / -0	+0 / -0.8	0 / 5.5	+3.5 / -0	+0 / -2.0	0.3 / 1.6	+0.8 / -0	-0.3 / -0.8
1.19– 1.97	0 / 1.0	+0.6 / -0	+0 / -0.4	0 / 1.6	+1.0 / -0	+0 / -0.6	0 / 2.6	+1.6 / -0	+0 / -1	0 / 6.5	+4.0 / -0	+0 / -2.5	0.4 / 2.0	+1.0 / -0	-0.4 / -1.0
1.97– 3.15	0 / 1.2	+0.7 / -0	+0 / -0.5	0 / 1.9	+1.2 / -0	+0 / -0.7	0 / 3	+1.8 / -0	+0 / -1.2	0 / 7.5	+4.5 / -0	+0 / -3	0.4 / 2.3	+1.2 / -0	-0.4 / -1.1
3.15– 4.73	0 / 1.5	+0.9 / -0	+0 / -0.6	0 / 2.3	+1.4 / -0	+0 / -0.9	0 / 3.6	+2.2 / -0	+0 / -1.4	0 / 8.5	+5.0 / -0	+0 / -3.5	0.5 / 2.8	+1.4 / -0	-0.5 / -1.4
4.73– 7.09	0 / 1.7	+1.0 / -0	+0 / -0.7	0 / 2.6	+1.6 / -0	+0 / -1.0	0 / 4.1	+2.5 / -0	+0 / -1.6	0 / 10	+6.0 / -0	+0 / -4	0.6 / 3.2	+1.6 / -0	-0.6 / -1.6
7.09– 9.85	0 / 2.0	+1.2 / -0	+0 / -0.8	0 / 3.0	+1.8 / -0	+0 / -1.2	0 / 4.6	+2.8 / -0	+0 / -1.8	0 / 11.5	+7.0 / -0	+0 / -4.5	0.6 / 3.6	+1.8 / -0	-0.6 / -1.8
9.85–12.41	0 / 2.1	+1.2 / -0	+0 / -0.9	0 / 3.2	+2.0 / -0	+0 / -1.2	0 / 5	+3.0 / -0	+0 / -2.0	0 / 13	+8.0 / -0	+0 / -5	0.7 / 3.9	+2.0 / -0	-0.7 / -1.9
12.41–15.75	0 / 2.4	+1.4 / -0	+0 / -1.0	0 / 3.6	+2.2 / -0	+0 / -1.4	0 / 5.7	+3.5 / -0	+0 / -2.2	0 / 15	+9.0 / -0	+0 / -6	0.7 / 4.3	+2.2 / -0	-0.7 / -2.1

[a]From ANSI B4.1—1967 (R1987). For larger diameters, see the standard.

6 Clearance Locational Fits[a] — American National Standard (continued)

Nominal Size Range, inches Over–To	LC 6 Limits of Clearance	LC 6 Hole H9	LC 6 Shaft f8	LC 7 Limits of Clearance	LC 7 Hole H10	LC 7 Shaft e9	LC 8 Limits of Clearance	LC 8 Hole H10	LC 8 Shaft d9	LC 9 Limits of Clearance	LC 9 Hole H11	LC 9 Shaft c10	LC 10 Limits of Clearance	LC 10 Hole H12	LC 10 Shaft	LC 11 Limits of Clearance	LC 11 Hole H13	LC 11 Shaft
0 – 0.12	0.3 / 1.9	+1.0 / −0	−0.3 / −0.9	0.6 / 3.2	+1.6 / −0	−0.6 / −1.6	1.0 / 3.6	+1.6 / −0	−1.0 / −2.0	2.5 / 6.6	+2.5 / −0	−2.5 / −4.1	4 / 12	+4 / −0	−4 / −8	5 / 17	+6 / −0	−5 / −11
0.12 – 0.24	0.4 / 2.3	+1.2 / −0	−0.4 / −1.1	0.8 / 3.8	+1.8 / −0	−0.8 / −2.0	1.2 / 4.2	+1.8 / −0	−1.2 / −2.4	2.8 / 7.6	+3.0 / −0	−2.8 / −4.6	4.5 / 14.5	+5 / −0	−4.5 / −9.5	6 / 20	+7 / −0	−6 / −13
0.24 – 0.40	0.5 / 2.8	+1.4 / −0	−0.5 / −1.4	1.0 / 4.6	+2.2 / −0	−1.0 / −2.4	1.6 / 5.2	+2.2 / −0	−1.6 / −3.0	3.0 / 8.7	+3.5 / −0	−3.0 / −5.2	5 / 17	+6 / −0	−5 / −11	7 / 25	+9 / −0	−7 / −16
0.40 – 0.71	0.6 / 3.2	+1.6 / −0	−0.6 / −1.6	1.2 / 5.6	+2.8 / −0	−1.2 / −2.8	2.0 / 6.4	+2.8 / −0	−2.0 / −3.6	3.5 / 10.3	+4.0 / −0	−3.5 / −6.3	6 / 20	+7 / −0	−6 / −13	8 / 28	+10 / −0	−8 / −18
0.71 – 1.19	0.8 / 4.0	+2.0 / −0	−0.8 / −2.0	1.6 / 7.1	+3.5 / −0	−1.6 / −3.6	2.5 / 8.0	+3.5 / −0	−2.5 / −4.5	4.5 / 13.0	+5.0 / −0	−4.5 / −8.0	7 / 23	+8 / −0	−7 / −15	10 / 34	+12 / −0	−10 / −22
1.19 – 1.97	1.0 / 5.1	+2.5 / −0	−1.0 / −2.6	2.0 / 8.5	+4.0 / −0	−2.0 / −4.5	3.0 / 9.5	+4.0 / −0	−3.0 / −5.5	5 / 15	+6 / −0	−5 / −9	8 / 28	+10 / −0	−8 / −18	12 / 44	+16 / −0	−12 / −28
1.97 – 3.15	1.2 / 6.0	+3.0 / −0	−1.2 / −3.0	2.5 / 10.0	+4.5 / −0	−2.5 / −5.5	4.0 / 11.5	+4.5 / −0	−4.0 / −7.0	6 / 17.5	+7 / −0	−6 / −10.5	10 / 34	+12 / −0	−10 / −22	14 / 50	+18 / −0	−14 / −32
3.15 – 4.73	1.4 / 7.1	+3.5 / −0	−1.4 / −3.6	3.0 / 11.5	+5.0 / −0	−3.0 / −6.5	5.0 / 13.5	+5.0 / −0	−5.0 / −8.5	7 / 21	+9 / −0	−7 / −12	11 / 39	+14 / −0	−11 / −25	16 / 60	+22 / −0	−16 / −38
4.73 – 7.09	1.6 / 8.1	+4.0 / −0	−1.6 / −4.1	3.5 / 13.5	+6.0 / −0	−3.5 / −7.5	6 / 16	+6 / −0	−6 / −10	8 / 24	+10 / −0	−8 / −14	12 / 44	+16 / −0	−12 / −28	18 / 68	+25 / −0	−18 / −43
7.09 – 9.85	2.0 / 9.3	+4.5 / −0	−2.0 / −4.8	4.0 / 15.5	+7.0 / −0	−4.0 / −8.5	7 / 18.5	+7 / −0	−7 / −11.5	10 / 29	+12 / −0	−10 / −17	16 / 52	+18 / −0	−16 / −34	22 / 78	+28 / −0	−22 / −50
9.85 – 12.41	2.2 / 10.2	+5.0 / −0	−2.2 / −5.2	4.5 / 17.5	+8.0 / −0	−4.5 / −9.5	7 / 20	+8 / −0	−7 / −12	12 / 32	+12 / −0	−12 / −20	20 / 60	+20 / −0	−20 / −40	28 / 88	+30 / −0	−28 / −58
12.41 – 15.75	2.5 / 12.0	+6.0 / −0	−2.5 / −6.0	5.0 / 20.0	+9.0 / −0	−5 / −11	8 / 23	+9 / −0	−8 / −14	14 / 37	+14 / −0	−14 / −23	22 / 66	+22 / −0	−22 / −44	30 / 100	+35 / −0	−30 / −65

[a] From ANSI B4.1–1967 (R1987). For larger diameters, see the standard.

7 Transition Locational Fits[a]—American National Standard

LT Transition fits are a compromise between clearance and interference fits, for application where accuracy of location is important, but either a small amount of clearance or interference is permissible.

Basic hole system. Limits are in thousandths of an inch. See §14.8.

Limits for hole and shaft are applied algebraically to the basic size to obtain the limits of size for the mating parts.

Data in **boldface** are in accordance with ABC agreements.

"Fit" represents the maximum interference (minus values) and the maximum clearance (plus values).

Symbols H7, js6, etc., are hole and shaft designations used in ABC System.

Nominal Size Range, inches Over–To	Class LT 1 Fit	Hole H7	Shaft js6	Class LT 2 Fit	Hole H8	Shaft js7	Class LT 3 Fit	Hole H7	Shaft k6	Class LT 4 Fit	Hole H8	Shaft k7	Class LT 5 Fit	Hole H7	Shaft n6	Class LT 6 Fit	Hole H7	Shaft n7
0 – 0.12	−0.10 / +0.50	+0.4 / −0	+0.10 / −0.10	−0.2 / +0.8	+0.6 / −0	+0.2 / −0.2							−0.5 / +0.15	+0.4 / −0	+0.5 / +0.25	−0.65 / +0.15	+0.4 / −0	+0.65 / +0.25
0.12– 0.24	−0.15 / +0.65	+0.5 / −0	+0.15 / −0.15	−0.25 / +0.95	+0.7 / −0	+0.25 / −0.25							−0.6 / +0.2	+0.5 / −0	+0.6 / +0.3	−0.8 / +0.2	+0.5 / −0	+0.8 / +0.3
0.24– 0.40	−0.2 / +0.8	+0.6 / −0	+0.2 / −0.2	−0.3 / +1.2	+0.9 / −0	+0.3 / −0.3	−0.5 / +0.5	+0.6 / −0	+0.5 / +0.1	−0.7 / +0.8	+0.9 / −0	+0.7 / +0.1	−0.8 / +0.2	+0.6 / −0	+0.8 / +0.4	−1.0 / +0.2	+0.6 / −0	+1.0 / +0.4
0.40– 0.71	−0.2 / +0.9	+0.7 / −0	+0.2 / −0.2	−0.35 / +1.35	+1.0 / −0	+0.35 / −0.35	−0.5 / +0.6	+0.7 / −0	+0.5 / +0.1	−0.8 / +0.9	+1.0 / −0	+0.8 / +0.1	−0.9 / +0.2	+0.7 / −0	+0.9 / +0.5	−1.2 / +0.2	+0.7 / −0	+1.2 / +0.5
0.71– 1.19	−0.25 / +1.05	+0.8 / −0	+0.25 / −0.25	−0.4 / +1.6	+1.2 / −0	+0.4 / −0.4	−0.6 / +0.7	+0.8 / −0	+0.6 / +0.1	−0.9 / +1.1	+1.2 / −0	+0.9 / +0.1	−1.1 / +0.2	+0.8 / −0	+1.1 / +0.6	−1.4 / +0.2	+0.8 / −0	+1.4 / +0.6
1.19– 1.97	−0.3 / +1.3	+1.0 / −0	+0.3 / −0.3	−0.5 / +2.1	+1.6 / −0	+0.5 / −0.5	−0.7 / +0.9	+1.0 / −0	+0.7 / +0.1	−1.1 / +1.5	+1.6 / −0	+1.1 / +0.1	−1.3 / +0.3	+1.0 / −0	+1.3 / +0.7	−1.7 / +0.3	+1.0 / −0	+1.7 / +0.7
1.97– 3.15	−0.3 / +1.5	+1.2 / −0	+0.3 / −0.3	−0.6 / +2.4	+1.8 / −0	+0.6 / −0.6	−0.8 / +1.1	+1.2 / −0	+0.8 / +0.1	−1.3 / +1.7	+1.8 / −0	+1.3 / +0.1	−1.5 / +0.4	+1.2 / −0	+1.5 / +0.8	−2.0 / +0.4	+1.2 / −0	+2.0 / +0.8
3.15– 4.73	−0.4 / +1.8	+1.4 / −0	+0.4 / −0.4	−0.7 / +2.9	+2.2 / −0	+0.7 / −0.7	−1.0 / +1.3	+1.4 / −0	+1.0 / +0.1	−1.5 / +2.1	+2.2 / −0	+1.5 / +0.1	−1.9 / +0.4	+1.4 / −0	+1.9 / +1.0	−2.4 / +0.4	+1.4 / −0	+2.4 / +1.0
4.73– 7.09	−0.5 / +2.1	+1.6 / −0	+0.5 / −0.5	−0.8 / +3.3	+2.5 / −0	+0.8 / −0.8	−1.1 / +1.5	+1.6 / −0	+1.1 / +0.1	−1.7 / +2.4	+2.5 / −0	+1.7 / +0.1	−2.2 / +0.4	+1.6 / −0	+2.2 / +1.2	−2.8 / +0.4	+1.6 / −0	+2.8 / +1.2
7.09– 9.85	−0.6 / +2.4	+1.8 / −0	+0.6 / −0.6	−0.9 / +3.7	+2.8 / −0	+0.9 / −0.9	−1.4 / +1.6	+1.8 / −0	+1.4 / +0.2	−2.0 / +2.6	+2.8 / −0	+2.0 / +0.2	−2.6 / +0.4	+1.8 / −0	+2.6 / +1.4	−3.2 / +0.4	+1.8 / −0	+3.2 / +1.4
9.85–12.41	−0.6 / +2.6	+2.0 / −0	+0.6 / −0.6	−1.0 / +4.0	+3.0 / −0	+1.0 / −1.0	−1.4 / +1.8	+2.0 / −0	+1.4 / +0.2	−2.2 / +2.8	+3.0 / −0	+2.2 / +0.2	−2.6 / +0.6	+2.0 / −0	+2.6 / +1.4	−3.4 / +0.6	+2.0 / −0	+3.4 / +1.4
12.41–15.75	−0.7 / +2.9	+2.2 / −0	+0.7 / −0.7	−1.0 / +4.5	+3.5 / −0	+1.0 / −1.0	−1.6 / +2.0	+2.2 / −0	+1.6 / +0.2	−2.4 / +3.3	+3.5 / −0	+2.4 / +0.2	−3.0 / +0.6	+2.2 / −0	+3.0 / +1.6	−3.8 / +0.6	+2.2 / −0	+3.8 / +1.6

[a]From ANSI B4.1—1967 (R1987). For larger diameters, see the standard.

8 Interference Locational Fits[a]—American National Standard

LN *Locational interference fits* are used where accuracy of location is of prime importance, and for parts requiring rigidity and alignment with no special requirements for bore pressure. Such fits are not intended for parts designed to transmit frictional loads from one part to another by virtue of the tightness of fit, as these conditions are covered by force fits.

Basic hole system. Limits are in thousandths of an inch. See §14.8.
Limits for hole and shaft are applied algebraically to the basic size to obtain the limits of size for the parts.
Data in **boldface** are in accordance with ABC agreements.
Symbols H7, p6, etc., are hole and shaft designations used in ABC System.

| Nominal Size Range, inches Over To | Class LN 1 | | | Class LN 2 | | | Class LN 3 | | |
| | Limits of Interference | Standard Limits | | Limits of Interference | Standard Limits | | Limits of Interference | Standard Limits | |
		Hole H6	Shaft n5		Hole H7	Shaft p6		Hole H7	Shaft r6
0 – 0.12	0 0.45	+0.25 −0	+0.45 +0.25	0 0.65	+0.4 −0	+0.65 +0.4	0.1 0.75	+0.4 −0	+0.75 +0.5
0.12– 0.24	0 0.5	+0.3 −0	+0.5 +0.3	0 0.8	+0.5 −0	+0.8 +0.5	0.1 0.9	+0.5 0	+0.9 +0.6
0.24– 0.40	0 0.65	+0.4 −0	+0.65 +0.4	0 1.0	+0.6 −0	+1.0 +0.6	0.2 1.2	+0.6 −0	+1.2 +0.8
0.40– 0.71	0 0.8	+0.4 −0	+0.8 +0.4	0 1.1	+0.7 −0	+1.1 +0.7	0.3 1.4	+0.7 −0	+1.4 +1.0
0.71– 1.19	0 1.0	+0.5 −0	+1.0 +0.5	0 1.3	+0.8 −0	+1.3 +0.8	0.4 1.7	+0.8 −0	+1.7 +1.2
1.19– 1.97	0 1.1	+0.6 −0	+1.1 +0.6	0 1.6	+1.0 −0	+1.6 +1.0	0.4 2.0	+1.0 −0	+2.0 +1.4
1.97– 3.15	0.1 1.3	+0.7 −0	+1.3 +0.7	0.2 2.1	+1.2 −0	+2.1 +1.4	0.4 2.3	+1.2 −0	+2.3 +1.6
3.15– 4.73	0.1 1.6	+0.9 −0	+1.6 +1.0	0.2 2.5	+1.4 −0	+2.5 +1.6	0.6 2.9	+1.4 −0	+2.9 +2.0
4.73– 7.09	0.2 1.9	+1.0 −0	+1.9 +1.2	0.2 2.8	+1.6 −0	+2.8 +1.8	0.9 3.5	+1.6 −0	+3.5 +2.5
7.09– 9.85	0.2 2.2	+1.2 −0	+2.2 +1.4	0.2 3.2	+1.8 −0	+3.2 +2.0	1.2 4.2	+1.8 −0	+4.2 +3.0
9.85–12.41	0.2 2.3	+1.2 −0	+2.3 +1.4	0.2 3.4	+2.0 −0	+3.4 +2.2	1.5 4.7	+2.0 −0	+4.7 +3.5

[a]From ANSI B4.1–1967 (R1987). For larger diameters, see the standard.

9 Force and Shrink Fits[a]—American National Standard

FN 1 — Light *drive fits* are those requiring light assembly pressures, and produce more or less permanent assemblies. They are suitable for thin sections or long fits, or in cast-iron external members.

FN 2 — Medium *drive fits* are suitable for ordinary steel parts, or for shrink fits on light sections. They are about the tightest fits that can be used with high-grade cast-iron external members.

FN 3 — Heavy *drive fits* are suitable for heavier steel parts or for shrink fits in medium sections.

FN 4 }
FN 5 } — *Force fits* are suitable for parts which can be highly stressed, or for shrink fits where the heavy pressing forces required are impractical.

Basic hole system. Limits are in thousandths of an inch. See §14.8.
Limits for hole and shaft are applied algebraically to the basic size to obtain the limits of size for the parts.
Data in **boldface** are in accordance with ABC agreements.
Symbols H7, s6, etc., are hole and shaft designations used in ABC System.

(interference)

| Nominal Size Range, inches | | Class FN 1 | | | Class FN 2 | | | Class FN 3 | | | Class FN 4 | | | Class FN 5 | | |
| | | Limits of Interference | Standard Limits | | Limits of Interference | Standard Limits | | Limits of Interference | Standard Limits | | Limits of Interference | Standard Limits | | Limits of Interference | Standard Limits | |
Over	To		Hole H6	Shaft		Hole H7	Shaft s6		Hole H7	Shaft t6		Hole H7	Shaft u6		Hole H8	Shaft x7
0	0.12	-0.05 / -0.5	+0.25 / -0	+0.5 / +0.3	**-0.2 / -0.85**	**+0.4 / -0**	**+0.85 / +0.6**				**0.3 / 0.95**	**+0.4 / -0**	**+0.95 / +0.7**	0.3 / 1.3	+0.6 / -0	+1.3 / +0.9
0.12	0.24	-0.1 / -0.6	+0.3 / -0	+0.6 / +0.4	-0.2 / -1.0	+0.5 / -0	+1.0 / +0.7				0.4 / 1.2	+0.5 / -0	+1.2 / +0.9	0.5 / 1.7	+0.7 / -0	+1.7 / +1.2
0.24	0.40	-0.1 / -0.75	+0.4 / -0	+0.75 / +0.5	-0.4 / -1.4	+0.6 / -0	+1.4 / +1.0				0.6 / 1.6	+0.6 / -0	+1.6 / +1.2	0.5 / 2.0	+0.9 / -0	+2.0 / +1.4
0.40	0.56	-0.1 / -0.8	+0.4 / -0	+0.8 / +0.5	-0.5 / -1.6	+0.7 / -0	+1.6 / +1.2				0.7 / 1.8	+0.7 / -0	+1.8 / +1.4	0.6 / 2.3	+1.0 / -0	+2.3 / +1.6
0.56	0.71	-0.2 / -0.9	+0.4 / -0	+0.9 / +0.6	-0.5 / -1.6	+0.7 / -0	+1.6 / +1.2				0.7 / 1.8	+0.7 / -0	+1.8 / +1.4	0.8 / 2.5	+1.0 / -0	+2.5 / +1.8
0.71	0.95	-0.2 / -1.1	+0.5 / -0	+1.1 / +0.7	-0.6 / -1.9	+0.8 / -0	+1.9 / +1.4				0.8 / 2.1	+0.8 / -0	+2.1 / +1.6	1.0 / 3.0	+1.2 / -0	+3.0 / +2.2
0.95	1.19	-0.3 / -1.2	+0.5 / -0	+1.2 / +0.8	-0.6 / -1.9	+0.8 / -0	+1.9 / +1.4	-0.8 / -2.1	+0.8 / -0	+2.1 / +1.6	1.0 / 2.3	+0.8 / -0	+2.3 / +1.8	1.3 / 3.3	+1.2 / -0	+3.3 / +2.5
1.19	1.58	-0.3 / -1.3	+0.6 / -0	+1.3 / +0.9	-0.8 / -2.4	+1.0 / -0	+2.4 / +1.8	-1.0 / -2.6	+1.0 / -0	+2.6 / +2.0	1.5 / 3.1	+1.0 / -0	+3.1 / +2.5	1.4 / 4.0	+1.6 / -0	+4.0 / +3.0

[a]ANSI B4.1-1967 (R1987).

9 Force and Shrink Fits[a]—American National Standard (continued)

Limits are in thousandths of an inch. Each cell shows the upper limit and the lower limit (upper / lower).

Nominal Size Range, inches (Over–To)	Class FN 1 Limits of Interference	Class FN 1 Hole H6	Class FN 1 Shaft	Class FN 2 Limits of Interference	Class FN 2 Hole H7	Class FN 2 Shaft s6	Class FN 3 Limits of Interference	Class FN 3 Hole H7	Class FN 3 Shaft t6	Class FN 4 Limits of Interference	Class FN 4 Hole H7	Class FN 4 Shaft u6	Class FN 5 Limits of Interference	Class FN 5 Hole H8	Class FN 5 Shaft x7
1.58–1.97	0.4 / 1.4	+0.6 / –0	+1.4 / +1.0	0.8 / 2.4	+1.0 / –0	+2.4 / +1.8	1.2 / 2.8	+1.0 / –0	+2.8 / +2.2	1.8 / 3.4	+1.0 / –0	+3.4 / +2.8	2.4 / 5.0	+1.6 / –0	+5.0 / +4.0
1.97–2.56	0.6 / 1.8	+0.7 / –0	+1.8 / +1.3	0.8 / 2.7	+1.2 / –0	+2.7 / +2.0	1.3 / 3.2	+1.2 / –0	+3.2 / +2.5	2.3 / 4.2	+1.2 / –0	+4.2 / +3.5	3.2 / 6.2	+1.8 / –0	+6.2 / +5.0
2.56–3.15	0.7 / 1.9	+0.7 / –0	+1.9 / +1.4	1.0 / 2.9	+1.2 / –0	+2.9 / +2.2	1.8 / 3.7	+1.2 / –0	+3.7 / +3.0	2.8 / 4.7	+1.2 / –0	+4.7 / +4.0	4.2 / 7.2	+1.8 / –0	+7.2 / +6.0
3.15–3.94	0.9 / 2.4	+0.9 / –0	+2.4 / +1.8	1.4 / 3.7	+1.4 / –0	+3.7 / +2.8	2.1 / 4.4	+1.4 / –0	+4.4 / +3.5	3.6 / 5.9	+1.4 / –0	+5.9 / +5.0	4.8 / 8.4	+2.2 / –0	+8.4 / +7.0
3.94–4.73	1.1 / 2.6	+0.9 / –0	+2.6 / +2.0	1.6 / 3.9	+1.4 / –0	+3.9 / +3.0	2.6 / 4.9	+1.4 / –0	+4.9 / +4.0	4.6 / 6.9	+1.4 / –0	+6.9 / +6.0	5.8 / 9.4	+2.2 / –0	+9.4 / +8.0
4.73–5.52	1.2 / 2.9	+1.0 / –0	+2.9 / +2.2	1.9 / 4.5	+1.6 / –0	+4.5 / +3.5	3.4 / 6.0	+1.6 / –0	+6.0 / +5.0	5.4 / 8.0	+1.6 / –0	+8.0 / +7.0	7.5 / 11.6	+2.5 / –0	+11.6 / +10.0
5.52–6.30	1.5 / 3.2	+1.0 / –0	+3.2 / +2.5	2.4 / 5.0	+1.6 / –0	+5.0 / +4.0	3.4 / 6.0	+1.6 / –0	+6.0 / +5.0	5.4 / 8.0	+1.6 / –0	+8.0 / +7.0	9.5 / 13.6	+2.5 / –0	+13.6 / +12.0
6.30–7.09	1.8 / 3.5	+1.0 / –0	+3.5 / +2.8	2.9 / 5.5	+1.6 / –0	+5.5 / +4.5	4.4 / 7.0	+1.6 / –0	+7.0 / +6.0	6.4 / 9.0	+1.6 / –0	+9.0 / +8.0	9.5 / 13.6	+2.5 / –0	+13.6 / +12.0
7.09–7.88	1.8 / 3.8	+1.2 / –0	+3.8 / +3.0	3.2 / 6.2	+1.8 / –0	+6.2 / +5.0	5.2 / 8.2	+1.8 / –0	+8.2 / +7.0	7.2 / 10.2	+1.8 / –0	+10.2 / +9.0	11.2 / 15.8	+2.8 / –0	+15.8 / +14.0
7.88–8.86	2.3 / 4.3	+1.2 / –0	+4.3 / +3.5	3.2 / 6.2	+1.8 / –0	+6.2 / +5.0	5.2 / 8.2	+1.8 / –0	+8.2 / +7.0	8.2 / 11.2	+1.8 / –0	+11.2 / +10.0	13.2 / 17.8	+2.8 / –0	+17.8 / +16.0
8.86–9.85	2.3 / 4.3	+1.2 / –0	+4.3 / +3.5	4.2 / 7.2	+1.8 / –0	+7.2 / +6.0	6.2 / 9.2	+1.8 / –0	+9.2 / +8.0	10.2 / 13.2	+1.8 / –0	+13.2 / +12.0	13.2 / 17.8	+2.8 / –0	+17.8 / +16.0
9.85–11.03	2.8 / 4.9	+1.2 / –0	+4.9 / +4.0	4.0 / 7.2	+2.0 / –0	+7.2 / +6.0	7.0 / 10.2	+2.0 / –0	+10.2 / +9.0	10.0 / 13.2	+2.0 / –0	+13.2 / +12.0	15.0 / 20.0	+3.0 / –0	+20.0 / +18.0
11.03–12.41	2.8 / 4.9	+1.2 / –0	+4.9 / +4.0	5.0 / 8.2	+2.0 / –0	+8.2 / +7.0	7.0 / 10.2	+2.0 / –0	+10.2 / +9.0	12.0 / 15.2	+2.0 / –0	+15.2 / +14.0	17.0 / 22.0	+3.0 / –0	+22.0 / +20.0
12.41–13.98	3.1 / 5.5	+1.4 / –0	+5.5 / +4.5	5.8 / 9.4	+2.2 / –0	+9.4 / +8.0	7.8 / 11.4	+2.2 / –0	+11.4 / +10.0	13.8 / 17.4	+2.2 / –0	+17.4 / +16.0	18.5 / 24.2	+3.5 / +0	+24.2 / +22.0

[a] From ANSI B4.1–1967 (R1987). For larger diameters, see the standard.

Dimensions are in millimeters.

10 International Tolerance Grades[a]

Basic sizes			Tolerance grades[b]																	
Over	Up to and Including	IT01	IT0	IT1	IT2	IT3	IT4	IT5	IT6	IT7	IT8	IT9	IT10	IT11	IT12	IT13	IT14	IT15	IT16
0	3	0.0003	0.0005	0.0008	0.0012	0.002	0.003	0.004	0.006	0.010	0.014	0.025	0.040	0.060	0.100	0.140	0.250	0.400	0.600
3	6	0.0004	0.0006	0.001	0.0015	0.0025	0.004	0.005	0.008	0.012	0.018	0.030	0.048	0.075	0.120	0.180	0.300	0.480	0.750
6	10	0.0004	0.0006	0.001	0.0015	0.0025	0.004	0.006	0.009	0.015	0.022	0.036	0.058	0.090	0.150	0.220	0.360	0.580	0.900
10	18	0.0005	0.0008	0.0012	0.002	0.003	0.005	0.008	0.011	0.018	0.027	0.043	0.070	0.110	0.180	0.270	0.430	0.700	1.100
18	30	0.0006	0.001	0.0015	0.0025	0.004	0.006	0.009	0.013	0.021	0.033	0.052	0.084	0.130	0.210	0.330	0.520	0.840	1.300
30	50	0.0006	0.001	0.0015	0.0025	0.004	0.007	0.011	0.016	0.025	0.039	0.062	0.100	0.160	0.250	0.390	0.620	1.000	1.600
50	80	0.0008	0.0012	0.002	0.003	0.005	0.008	0.013	0.019	0.030	0.046	0.074	0.120	0.190	0.300	0.460	0.740	1.200	1.900
80	120	0.001	0.0015	0.0025	0.004	0.006	0.010	0.015	0.022	0.035	0.054	0.087	0.140	0.220	0.350	0.540	0.870	1.400	2.200
120	180	0.0012	0.002	0.0035	0.005	0.008	0.012	0.018	0.025	0.040	0.063	0.100	0.160	0.250	0.400	0.630	1.000	1.600	2.500
180	250	0.002	0.003	0.0045	0.007	0.010	0.014	0.020	0.029	0.046	0.072	0.115	0.185	0.290	0.460	0.720	1.150	1.850	2.900
250	315	0.0025	0.004	0.006	0.008	0.012	0.016	0.023	0.032	0.052	0.081	0.130	0.210	0.320	0.520	0.810	1.300	2.100	3.200
315	400	0.003	0.005	0.007	0.009	0.013	0.018	0.025	0.036	0.057	0.089	0.140	0.230	0.360	0.570	0.890	1.400	2.300	3.600
400	500	0.004	0.006	0.008	0.010	0.015	0.020	0.027	0.040	0.063	0.097	0.155	0.250	0.400	0.630	0.970	1.550	2.500	4.000
500	630	0.0045	0.006	0.009	0.011	0.016	0.022	0.030	0.044	0.070	0.110	0.175	0.280	0.440	0.700	1.100	1.750	2.800	4.400
630	800	0.005	0.007	0.010	0.013	0.018	0.025	0.035	0.050	0.080	0.125	0.200	0.320	0.500	0.800	1.250	2.000	3.200	5.000
800	1000	0.0055	0.008	0.011	0.015	0.021	0.029	0.040	0.056	0.090	0.140	0.230	0.360	0.560	0.900	1.400	2.300	3.600	5.600
1000	1250	0.0065	0.009	0.013	0.018	0.024	0.034	0.046	0.066	0.105	0.165	0.260	0.420	0.660	1.050	1.650	2.600	4.200	6.600
1250	1600	0.008	0.011	0.015	0.021	0.029	0.040	0.054	0.078	0.125	0.195	0.310	0.500	0.780	1.250	1.950	3.100	5.000	7.800
1600	2000	0.009	0.013	0.018	0.025	0.035	0.048	0.065	0.092	0.150	0.230	0.370	0.600	0.920	1.500	2.300	3.700	6.000	9.200
2000	2500	0.011	0.015	0.022	0.030	0.041	0.057	0.077	0.110	0.175	0.280	0.440	0.700	1.100	1.750	2.800	4.400	7.000	11.000
2500	3150	0.013	0.018	0.026	0.036	0.050	0.069	0.093	0.135	0.210	0.330	0.540	0.860	1.350	2.100	3.300	5.400	8.600	13.500

[a]From ANSI B4.2−1978 (R1984).

[b]IT Values for tolerance grades larger than IT16 can be calculated by using the formulas: IT17 = IT × 10, IT18 = IT13 × 10, etc.

a37

11 Preferred Metric Hole Basis Clearance Fits[a]— American National Standard

Dimensions are in millimeters.

Basic Size		Loose Running			Free Running			Close Running			Sliding			Locational Clearance		
		Hole H11	Shaft c11	Fit	Hole H9	Shaft d9	Fit	Hole H8	f7	Fit	Hole H7	Shaft g6	Fit	Hole H7	Shaft h6	Fit
1	Max	1.060	0.940	0.180	1.025	0.980	0.070	1.014	0.994	0.030	1.010	0.998	0.018	1.010	1.000	0.016
	Min	1.060	0.880	0.060	1.000	0.955	0.020	1.000	0.984	0.006	1.000	0.992	0.002	1.000	0.994	0.000
1.2	Max	1.260	1.140	0.180	1.225	1.180	0.070	1.214	1.194	0.030	1.210	1.198	0.018	1.210	1.200	0.016
	Min	1.200	1.080	0.060	1.200	1.155	0.020	1.200	1.184	0.006	1.200	1.192	0.002	1.200	1.194	0.000
1.6	Max	1.660	1.540	0.180	1.625	1.580	0.070	1.614	1.594	0.030	1.610	1.598	0.018	1.610	1.600	0.016
	Min	1.600	1.480	0.060	1.600	1.555	0.020	1.600	1.584	0.006	1.600	1.592	0.002	1.600	1.594	0.000
2	Max	2.060	1.940	0.180	2.025	1.980	0.070	2.014	1.994	0.030	2.010	1.998	0.018	2.010	2.000	0.016
	Min	2.000	1.880	0.060	2.000	1.955	0.020	2.000	1.984	0.006	2.000	1.992	0.002	2.000	1.994	0.000
2.5	Max	2.560	2.440	0.180	2.525	2.480	0.070	2.514	2.494	0.030	2.510	2.498	0.018	2.510	2.500	0.016
	Min	2.500	2.380	0.060	2.500	2.455	0.020	2.500	2.484	0.006	2.500	2.492	0.002	2.500	2.494	0.000
3	Max	3.060	2.940	0.180	3.025	2.980	0.070	3.014	2.994	0.030	3.010	2.998	0.018	3.010	3.000	0.016
	Min	3.000	2.880	0.060	3.000	2.955	0.020	3.000	2.984	0.006	3.000	2.992	0.002	3.000	2.994	0.000
4	Max	4.075	3.930	0.220	4.030	3.970	0.090	4.018	3.990	0.040	4.012	3.996	0.024	4.012	4.000	0.020
	Min	4.000	3.855	0.070	4.000	3.940	0.030	4.000	3.978	0.010	4.000	3.988	0.004	4.000	3.992	0.000
5	Max	5.075	4.930	0.220	5.030	4.970	0.090	5.018	4.990	0.040	5.012	4.996	0.024	5.012	5.000	0.020
	Min	5.000	4.855	0.070	5.000	4.940	0.030	5.000	4.978	0.010	5.000	4.988	0.004	5.000	4.992	0.000
6	Max	6.075	5.930	0.220	6.030	5.970	0.090	6.018	5.990	0.040	6.012	5.996	0.024	6.012	6.000	0.020
	Min	6.000	5.855	0.070	6.000	5.940	0.030	6.000	5.978	0.010	6.000	5.988	0.004	6.000	5.992	0.000
8	Max	8.090	7.920	0.260	8.036	7.960	0.112	8.022	7.987	0.050	8.015	7.995	0.029	8.015	8.000	0.024
	Min	8.000	7.830	0.080	8.000	7.924	0.040	8.000	7.972	0.013	8.000	7.986	0.005	8.000	7.991	0.000
10	Max	10.090	9.920	0.260	10.036	9.960	0.112	10.022	9.987	0.050	10.015	9.995	0.029	10.015	10.000	0.024
	Min	10.000	9.830	0.080	10.000	9.924	0.040	10.000	9.972	0.013	10.000	9.986	0.005	10.000	9.991	0.000
12	Max	12.110	11.905	0.315	12.043	11.950	0.136	12.027	11.984	0.061	12.018	11.994	0.035	12.018	12.000	0.029
	Min	12.000	11.795	0.095	12.000	11.907	0.050	12.000	11.966	0.016	12.000	11.983	0.006	12.000	11.989	0.000
16	Max	16.110	15.905	0.315	16.043	15.950	0.136	16.027	15.984	0.061	16.018	15.994	0.035	16.018	16.000	0.029
	Min	16.000	15.795	0.095	16.000	15.907	0.050	16.000	15.966	0.016	16.000	15.983	0.006	16.000	15.989	0.000
20	Max	20.130	19.890	0.370	20.052	19.935	0.169	20.033	19.980	0.074	20.021	19.993	0.041	20.021	20.000	0.034
	Min	20.000	19.760	0.110	20.000	19.883	0.065	20.000	19.959	0.020	20.000	19.980	0.007	20.000	19.987	0.000
25	Max	25.130	24.890	0.370	25.052	24.935	0.169	25.033	24.980	0.074	25.021	24.993	0.041	25.021	25.000	0.034
	Min	25.000	24.760	0.110	25.000	24.883	0.065	25.000	24.959	0.020	25.000	24.980	0.007	25.000	24.987	0.000
30	Max	30.130	29.890	0.370	30.052	29.935	0.169	30.033	29.980	0.074	30.021	29.993	0.041	30.021	30.000	0.034
	Min	30.000	29.760	0.110	30.000	29.883	0.065	30.000	29.959	0.020	30.000	29.980	0.007	30.000	29.987	0.000

[a]From ANSI B4.2—1978 (R1984). For description of preferred fits, see Table 14.2.

11 Preferred Metric Hole Basis Clearance Fits[a]— American National Standard (continued)

Dimensions are in millimeters.

Basic Size		Loose Running Hole H11	Loose Running Shaft c11	Loose Running Fit	Free Running Hole H9	Free Running Shaft d9	Free Running Fit	Close Running Hole H8	Close Running Shaft f7	Close Running Fit	Sliding Hole H7	Sliding Shaft g6	Sliding Fit	Locational Clearance Hole H7	Locational Clearance Shaft h6	Locational Clearance Fit
40	Max	40.160	39.880	0.440	40.062	39.920	0.204	40.039	39.975	0.089	40.025	39.991	0.050	40.025	40.000	0.041
	Min	40.000	39.720	0.120	40.000	39.858	0.080	40.000	39.950	0.025	40.000	39.975	0.009	40.000	39.984	0.000
50	Max	50.160	49.870	0.450	50.062	49.920	0.204	50.039	49.975	0.089	50.025	49.991	0.050	50.025	50.000	0.041
	Min	50.000	49.710	0.130	50.000	49.858	0.080	50.000	49.950	0.025	50.000	49.975	0.009	50.000	49.984	0.000
60	Max	60.190	59.860	0.520	60.074	59.900	0.248	60.046	59.970	0.106	60.030	59.990	0.059	60.030	60.000	0.049
	Min	60.000	59.670	0.140	60.000	59.826	0.100	60.000	59.940	0.030	60.000	59.971	0.010	60.000	59.981	0.000
80	Max	80.190	79.950	0.530	80.074	79.900	0.248	80.046	79.970	0.106	80.030	79.990	0.059	80.030	80.000	0.049
	Min	80.000	79.660	0.150	80.000	79.826	0.100	80.000	79.940	0.030	80.000	79.971	0.010	80.000	79.981	0.000
100	Max	100.220	99.830	0.610	100.087	99.880	0.294	100.054	99.964	0.125	100.035	99.988	0.069	100.035	100.000	0.057
	Min	100.000	99.610	0.170	100.000	99.793	0.120	100.000	99.929	0.036	100.000	99.966	0.012	100.000	99.978	0.000
120	Max	120.220	119.820	0.620	120.087	119.880	0.294	120.054	119.964	0.125	120.035	119.988	0.069	120.035	120.000	0.057
	Min	120.000	119.600	0.180	120.000	119.793	0.120	120.000	119.929	0.036	120.000	119.966	0.012	120.000	119.978	0.000
160	Max	160.250	159.790	0.710	160.100	159.855	0.345	160.063	159.957	0.146	160.040	159.986	0.079	160.040	160.000	0.065
	Min	160.000	159.540	0.210	160.000	159.755	0.145	160.000	159.917	0.043	160.000	159.961	0.014	160.000	159.975	0.000
200	Max	200.290	199.760	0.820	200.115	199.830	0.400	200.072	199.950	0.168	200.046	199.985	0.090	200.046	200.000	0.075
	Min	200.000	199.470	0.240	200.000	199.715	0.170	200.000	199.904	0.050	200.000	199.956	0.015	200.000	199.971	0.000
250	Max	250.290	249.720	0.860	250.115	249.830	0.400	250.072	249.950	0.168	250.046	249.985	0.090	250.046	250.000	0.075
	Min	250.000	249.430	0.280	250.000	249.715	0.170	250.000	249.904	0.050	250.000	249.956	0.015	250.000	249.971	0.000
300	Max	300.320	299.670	0.970	300.130	299.810	0.450	300.081	299.944	0.189	300.052	299.983	0.101	300.052	300.000	0.084
	Min	300.000	299.350	0.330	300.000	299.680	0.190	300.000	299.892	0.056	300.000	299.951	0.017	300.000	299.968	0.000
400	Max	400.360	399.600	1.120	400.140	399.790	0.490	400.089	399.938	0.208	400.057	399.982	0.111	400.057	400.000	0.093
	Min	400.000	399.240	0.400	400.000	399.650	0.210	400.000	399.881	0.062	400.000	399.946	0.018	400.000	399.964	0.000
500	Max	500.400	499.520	1.280	500.155	499.770	0.540	500.097	499.932	0.228	500.063	499.980	0.123	500.063	500.000	0.103
	Min	500.000	499.120	0.480	500.000	499.615	0.230	500.000	499.869	0.068	500.000	499.940	0.020	500.000	499.960	0.000

[a]From ANSI B4.2−1978 (R1984). For description of preferred fits, see Table 14.2.

a39

12 Preferred Metric Hole Basis Transition and Interference Fits[a]— American National Standard

Dimensions are in millimeters.

Basic Size		Locational Transn. Hole H7	Locational Transn. Shaft k6	Locational Transn. Fit	Locational Transn. Hole H7	Locational Transn. Shaft n6	Locational Transn. Fit	Locational Interf. Hole H7	Locational Interf. Shaft p6	Locational Interf. Fit	Medium Drive Hole H7	Medium Drive Shaft s6	Medium Drive Fit	Force Hole H7	Force Shaft u6	Force Fit
1	Max	1.010	1.006	0.010	1.010	1.010	0.006	1.010	1.012	0.004	1.010	1.020	−0.004	1.010	1.024	−0.008
	Min	1.000	1.000	−0.006	1.000	1.004	−0.010	1.000	1.006	−0.012	1.000	1.014	−0.020	1.000	1.018	−0.024
1.2	Max	1.210	1.206	0.010	1.210	1.210	0.006	1.210	1.212	0.004	1.210	1.220	−0.004	1.210	1.224	−0.008
	Min	1.200	1.200	−0.006	1.200	1.204	−0.010	1.200	1.206	−0.012	1.200	1.214	−0.020	1.200	1.218	−0.024
1.6	Max	1.610	1.606	0.010	1.610	1.610	0.006	1.610	1.612	0.004	1.610	1.620	−0.004	1.610	1.624	−0.008
	Min	1.600	1.600	−0.006	1.600	1.604	−0.010	1.600	1.606	−0.012	1.600	1.614	−0.020	1.600	1.618	−0.024
2	Max	2.010	2.006	0.010	2.010	2.010	0.006	2.010	2.012	0.004	2.010	2.020	−0.004	2.010	2.024	−0.008
	Min	2.000	2.000	−0.006	2.000	2.004	−0.010	2.000	2.006	−0.012	2.000	2.014	−0.020	2.000	2.018	−0.024
2.5	Max	2.510	2.506	0.010	2.510	2.510	0.006	2.510	2.512	0.004	2.510	2.520	−0.004	2.510	2.524	−0.008
	Min	2.500	2.500	−0.006	2.500	2.504	−0.010	2.500	2.506	−0.012	2.500	2.514	−0.020	2.500	2.518	−0.024
3	Max	3.010	3.006	0.010	3.010	3.010	0.006	3.010	3.012	0.004	3.010	3.020	−0.004	3.010	3.024	−0.008
	Min	3.000	3.000	−0.006	3.000	3.004	−0.010	3.000	3.006	−0.012	3.000	3.014	−0.020	3.000	3.018	−0.024
4	Max	4.012	4.009	0.011	4.012	4.016	0.004	4.012	4.020	0.000	4.012	4.027	−0.007	4.012	4.031	−0.011
	Min	4.000	4.001	−0.009	4.000	4.008	−0.016	4.000	4.012	−0.020	4.000	4.019	−0.027	4.000	4.023	−0.031
5	Max	5.012	5.009	0.011	5.012	5.016	0.004	5.012	5.020	0.000	5.012	5.027	−0.007	5.012	5.031	−0.011
	Min	5.000	5.001	−0.009	5.000	5.008	−0.016	5.000	5.012	−0.020	5.000	5.019	−0.027	5.000	5.023	−0.031
6	Max	6.012	6.009	0.011	6.012	6.016	0.004	6.012	6.020	0.000	6.012	6.027	−0.007	6.012	6.031	−0.011
	Min	6.000	6.001	−0.009	6.000	6.008	−0.016	6.000	6.012	−0.020	6.000	6.019	−0.027	6.000	6.023	−0.031
8	Max	8.015	8.010	0.014	8.015	8.019	0.005	8.015	8.024	0.000	8.015	8.032	−0.008	8.015	8.037	−0.013
	Min	8.000	8.001	−0.010	8.000	8.010	−0.019	8.000	8.015	−0.024	8.000	8.023	−0.032	8.000	8.028	−0.037
10	Max	10.015	10.010	0.014	10.015	10.019	0.005	10.015	10.024	0.000	10.015	10.032	−0.008	10.015	10.037	−0.013
	Min	10.000	10.001	−0.010	10.000	10.010	−0.019	10.000	10.015	−0.024	10.000	10.023	−0.032	10.000	10.028	−0.037
12	Max	12.018	12.012	0.017	12.018	12.023	0.006	12.018	12.029	0.000	12.018	12.039	−0.010	12.018	12.044	−0.015
	Min	12.000	12.001	−0.012	12.000	12.012	−0.023	12.000	12.018	−0.029	12.000	12.028	−0.039	12.000	12.033	−0.044
16	Max	16.018	16.012	0.017	16.018	16.023	0.006	16.018	16.029	0.000	16.018	16.039	−0.010	16.018	16.044	−0.015
	Min	16.000	16.001	−0.012	16.000	16.012	−0.023	16.000	16.018	−0.029	16.000	16.028	−0.039	16.000	16.033	−0.044
20	Max	20.081	20.015	0.019	20.021	20.028	0.006	20.021	20.035	−0.001	20.021	20.048	−0.014	20.021	20.054	−0.020
	Min	20.000	20.002	−0.015	20.000	20.015	−0.028	20.000	20.022	−0.035	20.000	20.035	−0.048	20.000	20.041	−0.054
25	Max	25.021	25.015	0.019	25.021	25.028	0.006	25.021	25.035	−0.001	25.021	25.048	−0.014	25.021	25.061	−0.027
	Min	25.000	25.002	−0.015	25.000	25.015	−0.028	25.000	25.022	−0.035	25.000	25.035	−0.048	25.000	25.048	−0.061
30	Max	30.021	30.015	0.019	30.021	30.028	0.006	30.021	30.035	−0.001	30.021	30.048	−0.014	30.021	30.061	−0.027
	Min	30.000	30.002	−0.015	30.000	30.015	−0.028	30.000	30.022	−0.035	30.000	30.035	−0.048	30.000	30.048	−0.061

[a]From ANSI B4.2–1978. For description of preferred fits, see Fig. 12.18.

12 Preferred Metric Hole Basis Transition and Interference Fits[a]— American National Standard (continued)

Dimensions are in millimeters.

Basic Size		Locational Transn. Hole H7	Locational Transn. Shaft k6	Locational Transn. Fit	Locational Transn. Hole H7	Locational Transn. Shaft n6	Locational Transn. Fit	Locational Interf. Hole H7	Locational Interf. Shaft p6	Locational Interf. Fit	Medium Drive Hole H7	Medium Drive Shaft s6	Medium Drive Fit	Force Hole H7	Force Shaft u6	Force Fit
40	Max	40.025	40.018	0.023	40.025	40.033	0.008	40.025	40.042	−0.001	40.025	40.059	−0.018	40.025	40.076	−0.035
	Min	40.000	40.002	−0.018	40.000	40.017	−0.033	40.000	40.026	−0.042	40.000	40.043	−0.059	40.000	40.060	−0.076
50	Max	50.025	50.018	0.023	50.025	50.033	0.008	50.025	50.042	−0.001	50.025	50.059	−0.018	50.025	50.086	−0.045
	Min	50.000	50.002	−0.018	50.000	50.017	−0.033	50.000	50.026	−0.042	50.000	50.043	−0.059	50.000	50.070	−0.086
60	Max	60.030	60.021	0.028	60.030	60.039	0.010	60.030	60.051	−0.002	60.030	60.072	−0.023	60.030	60.106	−0.057
	Min	60.000	60.002	−0.021	60.000	60.020	−0.039	60.000	60.032	−0.051	60.000	60.053	−0.072	60.000	60.087	−0.106
80	Max	80.030	80.021	0.028	80.030	80.039	0.010	80.030	80.051	−0.002	80.030	80.078	−0.029	80.030	80.121	−0.072
	Min	80.000	80.002	−0.021	80.000	80.020	−0.039	80.000	80.032	−0.051	80.000	80.059	−0.078	80.000	80.102	−0.121
100	Max	100.035	100.025	0.032	100.035	100.045	0.012	100.035	100.059	−0.002	100.035	100.093	−0.036	100.035	100.146	−0.089
	Min	100.000	100.003	−0.025	100.000	100.023	−0.045	100.000	100.037	−0.059	100.000	100.071	−0.093	100.000	100.124	−0.146
120	Max	120.035	120.025	0.032	120.035	120.045	0.012	120.035	120.059	−0.002	120.035	120.101	−0.044	120.035	120.166	−0.109
	Min	120.000	120.003	−0.025	120.000	120.023	−0.045	120.000	120.037	−0.059	120.000	120.079	−0.101	120.000	120.144	−0.166
160	Max	160.040	160.028	0.037	160.040	160.052	0.013	160.040	160.068	−0.003	160.040	160.125	−0.060	160.040	160.215	−0.150
	Min	160.000	160.003	−0.028	160.000	160.027	−0.052	160.000	160.043	−0.068	160.000	160.100	−0.125	160.000	160.190	−0.215
200	Max	200.046	200.033	0.042	200.046	200.060	0.015	200.046	200.079	−0.004	200.046	200.151	−0.076	200.046	200.265	−0.190
	Min	200.000	200.004	−0.033	200.000	200.031	−0.060	200.000	200.050	−0.079	200.000	200.122	−0.151	200.000	200.236	−0.265
250	Max	250.046	250.033	0.042	250.046	250.060	0.015	250.046	250.079	−0.004	250.046	250.169	−0.094	250.046	250.313	−0.238
	Min	250.000	250.004	−0.033	250.000	250.031	−0.060	250.000	250.050	−0.079	250.000	250.140	−0.169	250.000	250.284	−0.313
300	Max	300.052	300.036	0.048	300.052	300.066	0.018	300.052	300.088	−0.004	300.052	300.202	−0.118	300.052	300.382	−0.298
	Min	300.000	300.004	−0.036	300.000	300.034	−0.066	300.000	300.056	−0.088	300.000	300.170	−0.202	300.000	300.350	−0.382
400	Max	400.057	400.040	0.053	400.057	400.073	0.020	400.057	400.098	−0.005	400.057	400.244	−0.151	400.057	400.471	−0.378
	Min	400.000	400.004	−0.040	400.000	400.037	−0.073	400.000	400.062	−0.098	400.000	400.208	−0.244	400.000	400.435	−0.471
500	Max	500.063	500.045	0.058	500.063	500.080	0.023	500.063	500.108	−0.005	500.063	500.292	−0.189	500.063	500.580	−0.477
	Min	500.000	500.005	−0.045	500.000	500.040	−0.080	500.000	500.068	−0.108	500.000	500.252	−0.292	500.000	500.540	−0.580

[a]From ANSI B4.2-1978. For description of preferred fits, see Fig. 12.18.

13 Preferred Metric Shaft Basis Clearance Fits[a]—American National Standard

Dimensions are in millimeters.

Basic Size		Loose Running Hole C11	Loose Running Shaft h11	Loose Running Fit	Free Running Hole D9	Free Running Shaft h9	Free Running Fit	Close Running Hole F8	Close Running Shaft h7	Close Running Fit	Sliding Hole G7	Sliding Shaft h6	Sliding Fit	Locational Clearance Hole H7	Locational Clearance Shaft h6	Locational Clearance Fit
1	Max	1.120	1.000	0.180	1.045	1.000	0.070	1.020	1.000	0.030	1.012	1.000	0.018	1.010	1.000	0.016
	Min	1.060	0.940	0.060	1.020	0.975	0.020	1.006	0.990	0.006	1.002	0.994	0.002	1.000	0.994	0.000
1.2	Max	1.320	1.200	0.180	1.245	1.200	0.070	1.220	1.200	0.030	1.212	1.200	0.018	1.210	1.200	0.016
	Min	1.260	1.140	0.060	1.220	1.175	0.020	1.206	1.190	0.006	1.202	1.194	0.002	1.200	1.194	0.000
1.6	Max	1.720	1.600	0.180	1.645	1.600	0.070	1.620	1.600	0.030	1.612	1.600	0.018	1.610	1.600	0.016
	Min	1.660	1.540	0.060	1.620	1.575	0.020	1.606	1.590	0.006	1.602	1.594	0.002	1.600	1.594	0.000
2	Max	2.120	2.000	0.180	2.045	2.000	0.070	2.020	2.000	0.030	2.012	2.000	0.018	2.010	2.000	0.016
	Min	2.060	1.940	0.060	2.020	1.975	0.020	2.006	1.990	0.006	2.002	1.994	0.002	2.000	1.994	0.000
2.5	Max	2.620	2.500	0.180	2.545	2.500	0.070	2.520	2.500	0.030	2.512	2.500	0.018	2.510	2.500	0.016
	Min	2.560	2.440	0.060	2.520	2.475	0.020	2.506	2.490	0.006	2.502	2.494	0.002	2.500	2.494	0.000
3	Max	3.120	3.000	0.180	3.045	3.000	0.070	3.020	3.000	0.030	3.012	3.000	0.018	3.010	3.000	0.016
	Min	3.060	2.940	0.060	3.020	2.975	0.020	3.006	2.990	0.006	3.002	2.994	0.002	3.000	2.994	0.000
4	Max	4.145	4.000	0.220	4.060	4.000	0.090	4.028	4.000	0.040	4.016	4.000	0.024	4.012	4.000	0.020
	Min	4.070	3.925	0.070	4.030	3.970	0.030	4.010	3.988	0.010	4.004	3.992	0.004	4.000	3.992	0.000
5	Max	5.145	5.000	0.220	5.060	5.000	0.090	5.028	5.000	0.040	5.016	5.000	0.024	5.012	5.000	0.020
	Min	5.070	4.925	0.070	5.030	4.970	0.030	5.010	4.988	0.010	5.004	4.992	0.004	5.000	4.992	0.000
6	Max	6.145	6.000	0.220	6.060	6.000	0.090	6.028	6.000	0.040	6.016	6.000	0.024	6.012	6.000	0.020
	Min	6.070	5.925	0.070	6.030	5.970	0.030	6.010	5.988	0.010	6.004	5.992	0.004	6.000	5.992	0.000
8	Max	8.170	8.000	0.260	8.076	8.000	0.112	8.035	8.000	0.050	8.020	8.000	0.029	8.015	8.000	0.024
	Min	8.080	7.910	0.080	8.040	7.964	0.040	8.013	7.985	0.013	8.005	7.991	0.005	8.000	7.991	0.000
10	Max	10.170	10.000	0.260	10.076	10.000	0.112	10.035	10.000	0.050	10.020	10.000	0.029	10.015	10.000	0.024
	Min	10.080	9.910	0.080	10.040	9.964	0.040	10.013	9.985	0.013	10.005	9.991	0.005	10.000	9.991	0.000
12	Max	12.205	12.000	0.315	12.093	12.000	0.136	12.043	12.000	0.061	12.024	12.000	0.035	12.018	12.000	0.029
	Min	12.095	11.890	0.095	12.050	11.957	0.050	12.016	11.982	0.016	12.006	11.989	0.006	12.000	11.989	0.000
16	Max	16.205	16.000	0.315	16.093	16.000	0.136	16.043	16.000	0.061	16.024	16.000	0.035	16.018	16.000	0.029
	Min	16.095	15.890	0.095	16.050	15.957	0.050	16.016	15.982	0.016	16.006	15.989	0.006	16.000	15.989	0.000
20	Max	20.240	20.000	0.370	20.117	20.000	0.169	20.053	20.000	0.074	20.028	20.000	0.041	20.021	20.000	0.034
	Min	20.110	19.870	0.110	20.065	19.948	0.065	20.020	19.979	0.020	20.007	19.987	0.007	20.000	19.987	0.000
25	Max	25.240	25.000	0.370	25.117	25.000	0.169	25.053	25.000	0.074	25.028	25.000	0.041	25.021	25.000	0.034
	Min	25.110	24.870	0.110	25.065	24.948	0.065	25.020	24.979	0.020	25.007	24.987	0.007	25.000	24.987	0.000
30	Max	30.240	30.000	0.370	30.117	30.000	0.169	30.053	30.000	0.074	30.028	30.000	0.041	30.021	30.000	0.034
	Min	30.110	29.870	0.110	30.065	29.948	0.065	30.020	29.979	0.020	30.007	29.987	0.007	30.000	29.987	0.000

[a]From ANSI B4.2–1978. For description of preferred fits, see Fig. 12.18.

13 Preferred Metric Shaft Basis Clearance Fits[a]— American National Standard (continued)

Dimensions are in millimeters.

Basic Size		Loose Running Hole C11	Shaft h11	Fit	Free Running Hole D9	Shaft h9	Fit	Close Running Hole F8	Shaft h7	Fit	Sliding Hole G7	Shaft h6	Fit	Locational Clearance Hole H7	Shaft h6	Fit
40	Max	40.280	40.000	0.440	40.142	40.000	0.204	40.064	40.000	0.089	40.034	40.000	0.050	40.025	40.000	0.041
	Min	40.120	39.840	0.120	40.080	39.938	0.080	40.025	39.975	0.025	40.009	39.984	0.009	40.000	39.984	0.000
50	Max	50.290	50.000	0.450	50.142	50.000	0.204	50.064	50.000	0.089	50.034	50.000	0.050	50.025	50.000	0.041
	Min	50.130	49.840	0.130	50.080	49.938	0.080	50.025	49.975	0.025	50.009	49.984	0.009	50.000	49.984	0.000
60	Max	60.330	60.000	0.520	60.174	60.000	0.248	60.076	60.000	0.106	60.040	60.000	0.059	60.030	60.000	0.049
	Min	60.140	59.810	0.140	60.100	59.926	0.100	60.030	59.970	0.030	60.010	59.981	0.010	60.000	59.981	0.000
80	Max	80.340	80.000	0.530	80.174	80.000	0.248	80.076	80.000	0.106	80.040	80.000	0.059	80.030	80.000	0.049
	Min	80.150	79.810	0.150	80.100	79.926	0.100	80.030	79.970	0.030	80.010	79.981	0.010	80.000	79.981	0.000
100	Max	100.390	100.000	0.610	100.207	100.000	0.294	100.090	100.000	0.125	100.047	100.000	0.069	100.035	100.000	0.057
	Min	100.170	99.780	0.170	100.120	99.913	0.120	100.036	99.965	0.036	100.012	99.978	0.012	100.000	99.978	0.000
120	Max	120.400	120.000	0.620	120.207	120.000	0.294	120.090	120.000	0.125	120.047	120.000	0.069	120.035	120.000	0.057
	Min	120.180	119.780	0.180	120.120	119.913	0.120	120.036	119.965	0.036	120.012	119.978	0.012	120.000	119.978	0.000
160	Max	160.460	160.000	0.710	160.245	160.000	0.345	160.106	160.000	0.146	160.054	160.000	0.079	160.040	160.000	0.065
	Min	160.210	159.750	0.210	160.145	159.900	0.145	160.043	159.960	0.043	160.014	159.975	0.014	160.000	159.975	0.000
200	Max	200.530	200.000	0.820	200.285	200.000	0.400	200.122	200.000	0.168	200.061	200.000	0.090	200.046	200.000	0.075
	Min	200.240	199.710	0.240	200.170	199.885	0.170	200.050	199.954	0.050	200.015	199.971	0.015	200.000	199.971	0.000
250	Max	250.570	250.000	0.860	250.285	250.000	0.400	250.122	250.000	0.168	250.061	250.000	0.090	250.046	250.000	0.075
	Min	250.280	249.710	0.280	250.170	249.885	0.170	250.050	249.954	0.050	250.015	249.971	0.015	250.000	249.971	0.000
300	Max	300.650	300.000	0.970	300.320	300.000	0.450	300.137	300.000	0.189	300.069	300.000	0.101	300.052	300.000	0.084
	Min	300.330	299.680	0.330	300.190	299.870	0.190	300.056	299.948	0.056	300.017	299.968	0.017	300.000	299.968	0.000
400	Max	400.760	400.000	1.120	400.350	400.000	0.490	400.151	400.000	0.208	400.075	400.000	0.111	400.057	400.000	0.093
	Min	400.400	399.640	0.400	400.210	399.860	0.210	400.062	399.943	0.062	400.018	399.964	0.018	400.000	399.964	0.000
500	Max	500.880	500.000	1.280	500.385	500.000	0.540	500.165	500.000	0.228	500.083	500.000	0.123	500.063	500.000	0.103
	Min	500.480	499.600	0.480	500.230	499.845	0.230	500.068	499.937	0.068	500.020	499.960	5.020	500.000	499.960	0.000

[a]From ANSI B.42-1978. For description of preferred fits, see Fig. 12.8.

Dimensions are in millimeters.

Basic Size		Locational Transn. Hole K7	Locational Transn. Shaft h6	Locational Transn. Fit	Locational Transn. Hole N7	Locational Transn. Shaft h6	Locational Transn. Fit	Locational Interf. Hole P7	Locational Interf. Shaft h6	Locational Interf. Fit	Medium Drive Hole S7	Medium Drive Shaft h6	Medium Drive Fit	Force Hole U7	Force Shaft h6	Force Fit
1	Max	1.000	1.000	0.006	0.996	1.000	0.002	0.994	1.000	0.000	0.986	1.000	−0.008	0.982	1.000	−0.012
	Min	0.990	0.994	−0.010	0.986	0.994	−0.014	0.984	0.994	−0.016	0.976	0.994	−0.024	0.972	0.994	−0.028
1.2	Max	1.200	1.200	0.006	1.196	1.200	0.002	1.194	1.200	0.000	1.186	1.200	−0.008	1.182	1.200	−0.012
	Min	1.190	1.194	−0.010	1.186	1.194	−0.014	1.184	1.194	−0.016	1.176	1.194	−0.024	1.172	1.194	−0.028
1.6	Max	1.600	1.600	0.006	1.596	1.600	0.002	1.594	1.600	0.000	1.586	1.600	−0.008	1.582	1.600	−0.012
	Min	1.590	1.594	−0.010	1.586	1.594	−0.014	1.584	1.594	−0.016	1.576	1.594	−0.024	1.572	1.594	−0.028
2	Max	2.000	2.000	0.006	1.996	2.000	0.002	1.994	2.000	0.000	1.986	2.000	−0.008	1.982	2.000	−0.012
	Min	1.990	1.994	−0.010	1.986	1.994	−0.014	1.984	1.994	−0.016	1.976	1.994	−0.024	1.972	1.994	−0.028
2.5	Max	2.500	2.500	0.006	2.496	2.500	0.002	2.494	2.500	0.000	2.486	2.500	−0.008	2.482	2.500	−0.012
	Min	2.490	2.494	−0.010	2.486	2.494	−0.014	2.484	2.494	−0.016	2.476	2.494	−0.024	2.472	2.494	−0.028
3	Max	3.000	3.000	0.006	2.996	3.000	0.002	2.994	3.000	0.000	2.986	3.000	−0.008	2.982	3.000	−0.012
	Min	2.990	2.994	−0.010	2.986	2.994	−0.014	2.984	2.994	−0.016	2.976	2.994	−0.024	2.972	2.994	−0.028
4	Max	4.003	4.000	0.011	3.996	4.000	0.004	3.992	4.000	0.000	3.985	4.000	−0.007	3.981	4.000	−0.011
	Min	3.991	3.992	−0.009	3.984	3.992	−0.016	3.980	3.992	−0.020	3.973	3.992	−0.027	3.969	3.992	−0.031
5	Max	5.003	5.000	0.011	4.996	5.000	0.004	4.992	5.000	0.000	4.985	5.000	−0.007	4.981	5.000	−0.011
	Min	4.991	4.992	−0.009	4.984	4.992	−0.016	4.980	4.992	−0.020	4.973	4.992	−0.027	4.969	4.992	−0.031
6	Max	6.003	6.000	0.011	5.996	6.000	0.004	5.992	6.000	0.000	5.985	6.000	−0.007	5.981	6.000	−0.011
	Min	5.991	5.992	−0.009	5.984	5.992	−0.016	5.980	5.992	−0.020	5.973	5.992	−0.027	5.969	5.992	−0.031
8	Max	8.005	8.000	0.014	7.996	8.000	0.005	7.991	8.000	0.000	7.983	8.000	−0.008	7.978	8.000	−0.013
	Min	7.990	7.991	−0.010	7.981	7.991	−0.019	7.976	7.991	−0.024	7.968	7.991	−0.032	7.963	7.991	−0.037
10	Max	10.005	10.000	0.014	9.996	10.000	0.005	9.991	10.000	0.000	9.983	10.000	−0.008	9.978	10.000	−0.013
	Min	9.990	9.991	−0.010	9.981	9.991	−0.019	9.976	9.991	−0.024	9.968	9.991	−0.032	9.963	9.991	−0.037
12	Max	12.006	12.000	0.017	11.995	12.000	0.006	11.989	12.000	0.000	11.979	12.000	−0.010	11.974	12.000	−0.015
	Min	11.988	11.989	−0.012	11.977	11.989	−0.023	11.971	11.989	−0.029	11.961	11.989	−0.039	11.956	11.989	−0.044
16	Max	16.006	16.000	0.017	15.995	16.000	0.006	15.989	16.000	0.000	15.979	16.000	−0.010	15.974	16.000	−0.015
	Min	15.988	15.989	−0.012	15.977	15.989	−0.023	15.971	15.989	−0.029	15.961	15.989	−0.039	15.956	15.989	−0.044
20	Max	20.006	20.000	0.019	19.993	20.000	0.006	19.986	20.000	−0.001	19.973	20.000	−0.014	19.967	20.000	−0.020
	Min	19.985	19.987	−0.015	19.972	19.987	−0.028	19.965	19.987	−0.035	19.952	19.987	−0.048	19.946	19.987	−0.054
25	Max	25.006	25.000	0.019	24.993	25.000	0.006	24.986	25.000	−0.001	24.973	25.000	−0.014	24.960	25.000	−0.027
	Min	24.985	24.987	−0.015	24.972	24.987	−0.028	24.965	24.987	−0.035	24.952	24.987	−0.048	24.939	24.987	−0.061
30	Max	30.006	30.000	0.019	29.993	30.000	0.006	29.986	30.000	−0.001	29.973	30.000	−0.014	29.960	30.000	−0.027
	Min	29.985	29.987	−0.015	29.972	29.987	−0.028	29.965	29.987	−0.035	29.952	29.987	−0.048	29.939	29.987	−0.061

[a] From ANSI B4.2–1978 (R1984). For description of preferred fits, see Table 14.2.

14 Preferred Metric Basis Transition and Interference Fits[a]— American National Standard (continued)

Dimensions are in millimeters.

Basic Size		Locational Transn. Hole K7	Shaft h6	Fit	Locational Transn. Hole N7	Shaft h6	Fit	Locational Interf. Hole P7	Shaft h6	Fit	Medium Drive Hole S7	Shaft h6	Fit	Force Hole U7	Shaft h6	Fit
40	Max	40.007	40.000	0.023	39.992	40.000	0.008	39.983	40.000	-0.001	39.966	40.000	-0.018	39.949	40.000	-0.035
	Min	39.982	39.984	-0.018	39.967	39.984	-0.033	39.958	39.984	-0.042	39.941	39.984	-0.059	39.924	39.984	-0.076
50	Max	50.007	50.000	0.023	49.992	50.000	0.008	49.983	50.000	-0.001	49.966	50.000	-0.018	49.939	50.000	-0.045
	Min	49.982	49.984	-0.018	49.967	49.984	-0.033	49.958	49.984	-0.042	49.941	49.984	-0.059	49.914	49.984	-0.086
60	Max	60.009	60.000	0.028	59.991	60.000	0.010	59.979	60.000	-0.002	59.958	60.000	-0.023	59.924	60.000	-0.057
	Min	59.979	59.981	-0.021	59.961	59.981	-0.039	59.949	59.981	-0.051	59.928	59.981	-0.072	59.894	59.981	-0.106
80	Max	80.009	80.000	0.028	79.991	80.000	0.010	79.979	80.000	-0.002	79.952	80.000	-0.029	79.909	80.000	-0.072
	Min	79.979	79.981	-0.021	79.961	79.981	-0.039	79.949	79.981	-0.051	79.922	79.981	-0.078	79.879	79.981	-0.121
100	Max	100.010	100.000	0.032	99.990	100.000	0.012	99.976	100.000	-0.002	99.942	100.000	-0.036	99.889	100.000	-0.089
	Min	99.975	99.978	-0.025	99.955	99.978	-0.045	99.941	99.978	-0.059	99.907	99.978	-0.093	99.854	99.978	-0.146
120	Max	120.010	120.000	0.032	119.990	120.000	0.012	119.976	120.000	-0.002	119.934	120.000	-0.044	119.869	120.000	-0.109
	Min	119.975	119.978	-0.025	119.955	119.978	-0.045	119.941	119.978	-0.059	119.899	119.978	-0.101	119.834	119.978	-0.166
160	Max	160.012	160.000	0.037	159.988	160.000	0.013	159.972	160.000	-0.003	159.915	160.000	-0.060	159.825	160.000	-0.150
	Min	159.972	159.975	-0.028	159.948	159.975	-0.052	159.932	159.975	-0.068	159.875	159.975	-0.125	159.785	159.975	-0.215
200	Max	200.013	200.000	0.042	199.986	200.000	0.015	199.967	200.000	-0.004	199.895	200.000	-0.076	199.781	200.000	-0.190
	Min	199.967	199.971	-0.033	199.940	199.971	-0.060	199.921	199.971	-0.079	199.849	199.971	-0.151	199.735	199.971	-0.265
250	Max	250.013	250.000	0.042	249.986	250.000	0.015	249.967	250.000	-0.004	249.877	250.000	-0.094	249.733	250.000	-0.238
	Min	249.967	249.971	-0.033	249.940	249.971	-0.060	249.921	249.971	-0.079	249.831	249.971	-0.169	249.687	249.971	-0.313
300	Max	300.016	300.000	0.048	299.986	300.000	0.018	299.964	300.000	-0.004	299.850	300.000	-0.118	299.670	300.000	-0.298
	Min	299.964	299.968	-0.036	299.934	299.968	-0.066	299.912	299.968	-0.088	299.798	299.968	-0.202	299.618	299.968	-0.382
400	Max	400.017	400.000	0.053	399.984	400.000	0.020	399.959	400.000	-0.005	399.813	400.000	-0.151	399.586	400.000	-0.378
	Min	399.960	399.964	-0.040	399.927	399.964	-0.073	399.902	399.964	-0.098	399.756	399.964	-0.244	399.529	399.964	-0.471
500	Max	500.018	500.000	0.058	499.983	500.000	0.023	499.955	500.000	-0.005	499.771	500.000	-0.189	499.483	500.000	-0.477
	Min	499.955	499.960	-0.045	499.920	499.960	-0.080	499.892	499.960	-0.108	499.708	499.960	-0.292	499.420	499.960	-0.580

[a]From ANSI B4.2—1978 (R1984). For description of preferred fits, see Table 14.2.

15 Screw Threads, American National, Unified, and Metric

AMERICAN NATIONAL STANDARD UNIFIED AND AMERICAN NATIONAL SCREW THREADS[a]

Nominal Diameter	Coarse[b] NC UNC Thds. per Inch	Tap Drill[d]	Fine[b] NF UNF Thds. per Inch	Tap Drill[d]	Extra Fine[c] NEF UNEF Thds. per Inch	Tap Drill[d]	Nominal Diameter	Coarse[b] NC UNC Thds. per Inch	Tap Drill[d]	Fine[b] NF UNF Thds. per Inch	Tap Drill[d]	Extra Fine[c] NEF UNEF Thds. per Inch	Tap Drill[d]
0 (.060)			80	3/64			1	8	7/8	12	59/64	20	61/64
1 (.073)	64	No. 53	72	No. 53	....	....	1 1/16	....	....	....	....	18	1
2 (.086)	56	No. 50	64	No. 50	....	....	1 1/8	7	63/64	12	1 3/64	18	1 5/64
3 (.099)	48	No. 47	56	No. 45	....	....	1 3/16	....	....	....	....	18	1 9/64
4 (.112)	40	No. 43	48	No. 42	....	....	1 1/4	7	1 7/64	12	1 11/64	18	1 3/16
5 (.125)	40	No. 38	44	No. 37	....	....	1 5/16	....	....	....	....	18	1 17/64
6 (.138)	32	No. 36	40	No. 33	....	....	1 3/8	6	1 7/32	12	1 19/64	18	1 5/16
8 (.164)	32	No. 29	36	No. 29	....	....	1 7/16	....	....	....	....	18	1 3/8
10 (.190)	24	No. 25	32	No. 21	....	....	1 1/2	6	1 11/32	12	1 27/64	18	1 7/16
12 (.216)	24	No. 16	28	No. 14	32	No. 13	1 9/16	....	....	....	....	18	1 1/2
1/4	20	No. 7	28	No. 3	32	7/32	1 5/8	....	....	....	....	18	1 9/16
5/16	18	F	24	I	32	9/32	1 11/16	....	....	....	....	18	1 5/8
3/8	16	5/16	24	Q	32	11/32	1 3/4	5	1 9/16	....	....	....	....
7/16	14	U	20	25/64	28	13/32	2	4 1/2	1 25/32	....	....	....	....
1/2	13	27/64	20	29/64	28	15/32	2 1/4	4 1/2	2 1/32	....	....	....	....
9/16	12	31/64	18	33/64	24	33/64	2 1/2	4	2 1/4	....	....	....	....
5/8	11	17/32	18	37/64	24	37/64	2 3/4	4	2 1/2	....	....	....	....
11/16	....	....	....	....	24	41/64	3	4	2 3/4	....	....	....	....
3/4	10	21/32	16	11/16	20	45/64	3 1/4	4	....	....	....	....	....
13/16	....	....	....	....	20	49/64	3 1/2	4	....	....	....	....	....
7/8	9	49/64	14	13/16	20	53/64	3 3/4	4	....	....	....	....	....
15/16	....	....	....	....	20	57/64	4	4	....	....	....	....	....

[a]ANSI/ASME B1.1–1989. For 8-, 12-, and 16-pitch thread series, see next page.
[b]Classes 1A, 2A, 3A, 1B, 2B, 3B, 2, and 3.
[c]Classes 2A, 2B, 2, and 3.
[d]For approximate 75% full depth of thread. For decimal sizes of numbered and lettered drills, see Appendix 16.

15   Screw Threads, American National, Unified, and Metric (continued)

AMERICAN NATIONAL STANDARD UNIFIED AND AMERICAN NATIONAL SCREW THREADS[a] (continued)

Nominal Diameter	8-Pitch[b] Series 8N and 8UN		12-Pitch[b] Series 12N and 12UN		16-Pitch[b] Series 16N and 16UN		Nominal Diameter	8-Pitch[b] Series 8N and 8UN		12-Pitch[b] Series 12N and 12UN		16-Pitch[b] Series 16N and 16UN	
	Thds. per Inch	Tap Drill[c]	Thds. per Inch	Tap Drill[c]	Thds. per Inch	Tap Drill[c]		Thds. per Inch	Tap Drill[c]	Thds. per Inch	Tap Drill[c]	Thds. per Inch	Tap Drill[c]
$\frac{1}{2}$	....	....	12	$\frac{27}{64}$	....	....	$2\frac{1}{16}$	....	....	....	....	**16**	2
$\frac{9}{16}$	....	....	12[e]	$\frac{31}{64}$	....	....	$2\frac{1}{8}$	....	....	12	$2\frac{3}{64}$	16	$2\frac{1}{16}$
$\frac{5}{8}$	....	....	12	$\frac{35}{64}$	....	....	$2\frac{3}{16}$	....	....	....	....	**16**	$2\frac{1}{8}$
$\frac{11}{16}$	....	....	12	$\frac{39}{64}$	....	....	$2\frac{1}{4}$	8	$2\frac{1}{8}$	12	$2\frac{11}{64}$	16	$2\frac{3}{16}$
$\frac{3}{4}$	....	....	12	$\frac{43}{64}$	16[e]	$\frac{11}{16}$	$2\frac{5}{16}$	....	....	....	....	**16**	$2\frac{1}{4}$
$\frac{13}{16}$	....	....	12	$\frac{47}{64}$	16	$\frac{3}{4}$	$2\frac{3}{8}$	....	....	12	$2\frac{19}{64}$	16	$2\frac{5}{16}$
$\frac{7}{8}$	....	....	12	$\frac{51}{64}$	16	$\frac{13}{16}$	$2\frac{7}{16}$	....	....	....	....	**16**	$2\frac{3}{8}$
$\frac{15}{16}$	....	....	12	$\frac{55}{64}$	16	$\frac{7}{8}$	$2\frac{1}{2}$	8	$2\frac{3}{8}$	12	$2\frac{27}{64}$	16	$2\frac{7}{16}$
1	8[e]	$\frac{7}{8}$	12	$\frac{59}{64}$	16	$\frac{15}{16}$	$2\frac{5}{8}$	....	....	12	$2\frac{35}{64}$	16	$2\frac{9}{16}$
$1\frac{1}{16}$	....	....	12	$\frac{63}{64}$	16	1	$2\frac{3}{4}$	8	$2\frac{5}{8}$	12	$2\frac{43}{64}$	16	$2\frac{11}{16}$
$1\frac{1}{8}$	8	1	12[e]	$1\frac{3}{64}$	16	$1\frac{1}{16}$	$2\frac{7}{8}$	....	....	12	....	16	....
$1\frac{3}{16}$	....	....	12	$1\frac{7}{64}$	16	$1\frac{1}{8}$	3	8	$2\frac{7}{8}$	12	....	16	....
$1\frac{1}{4}$	8	$1\frac{1}{8}$	12	$1\frac{11}{64}$	16	$1\frac{3}{16}$	$3\frac{1}{8}$	....	....	12	....	16	....
$1\frac{5}{16}$	....	....	12	$1\frac{15}{64}$	16	$1\frac{1}{4}$	$3\frac{1}{4}$	8	....	12	....	16	....
$1\frac{3}{8}$	8	$1\frac{1}{4}$	12[e]	$1\frac{19}{64}$	16	$1\frac{5}{16}$	$3\frac{3}{8}$	....	....	12	....	16	....
$1\frac{7}{16}$	....	....	12	$1\frac{23}{64}$	16	$1\frac{3}{8}$	$3\frac{1}{2}$	8	....	12	....	16	....
$1\frac{1}{2}$	8	$1\frac{3}{8}$	12[e]	$1\frac{27}{64}$	16	$1\frac{7}{16}$	$3\frac{5}{8}$	....	....	12	....	16	....
$1\frac{9}{16}$	....	....	....	....	16	$1\frac{1}{2}$	$3\frac{3}{4}$	8	....	12	....	16	....
$1\frac{5}{8}$	8	$1\frac{1}{2}$	12	$1\frac{35}{64}$	16	$1\frac{9}{16}$	$3\frac{7}{8}$	....	....	12	....	16	....
$1\frac{11}{16}$	....	....	....	....	16	$1\frac{5}{8}$	4	8	....	12	....	16	....
$1\frac{3}{4}$	8	$1\frac{5}{8}$	12	$1\frac{43}{64}$	16[e]	$1\frac{11}{16}$	$4\frac{1}{4}$	8	....	12	....	16	....
$1\frac{13}{16}$	....	....	....	....	16	$1\frac{3}{4}$	$4\frac{1}{2}$	8	....	12	....	16	....
$1\frac{7}{8}$	8	$1\frac{3}{4}$	12	$1\frac{51}{64}$	16	$1\frac{13}{16}$	$4\frac{3}{4}$	8	....	12	....	16	....
$1\frac{15}{16}$	....	....	....	....	16	$1\frac{7}{8}$	5	8	....	12	....	16	....
2	8	$1\frac{7}{8}$	12	$1\frac{59}{64}$	16[e]	$1\frac{15}{16}$	$5\frac{1}{4}$	8	....	12	....	16	....

[a]ANSI/ASME B1.1–1989.
[b]Classes 2A, 3A, 2B, 3B, 2, and 3.
[c]For approximate 75% full depth of thread.
[d]Boldface type indicates American National threads only.
[e]This is a standard size of the Unified or American National threads of the coarse, fine, or extra fine series. See preceding page.

15   Screw Threads, American National, Unified, and Metric (continued)

METRIC SCREW THREADS[a]

Preferred sizes for commercial threads and fasteners are shown in **boldface** type.

Coarse (general purpose)		Fine	
Nominal Size & Thd Pitch	Tap Drill Diameter, mm	Nominal Size & Thd Pitch	Tap Drill Diameter, mm
M1.6 × 0.35	1.25	—	—
M1.8 × 0.35	1.45	—	—
M2 × 0.4	1.6	—	—
M2.2 × 0.45	1.75	—	—
M2.5 × 0.45	2.05	—	—
M3 × 0.5	2.5	—	—
M3.5 × 0.6	2.9	—	—
M4 × 0.7	3.3	—	—
M4.5 × 0.75	3.75	—	—
M5 × 0.8	4.2	—	—
M6 × 1	5.0	—	—
M7 × 1	6.0	—	—
M8 × 1.25	6.8	**M8 × 1**	7.0
M9 × 1.25	7.75	—	—
M10 × 1.5	8.5	**M10 × 1.25**	8.75
M11 × 1.5	9.50	—	—
M12 × 1.75	10.30	**M12 × 1.25**	10.5
M14 × 2	12.00	**M14 × 1.5**	12.5
M16 × 2	14.00	**M16 × 1.5**	14.5
M18 × 2.5	15.50	**M18 × 1.5**	16.5
M20 × 2.5	17.5	**M20 × 1.5**	18.5
M22 × 2.5[b]	19.5	**M22 × 1.5**	20.5
M24 × 3	21.0	**M24 × 2**	22.0
M27 × 3[b]	24.0	**M27 × 2**	25.0
M30 × 3.5	26.5	**M30 × 2**	28.0
M33 × 3.5	29.5	**M30 × 2**	31.0
M36 × 4	32.0	**M36 × 2**	33.0
M39 × 4	35.0	M39 × 2	36.0
M42 × 4.5	37.5	**M42 × 2**	39.0
M45 × 4.5	40.5	M45 × 1.5	42.0
M48 × 5	43.0	**M48 × 2**	45.0
M52 × 5	47.0	M52 × 2	49.0
M56 × 5.5	50.5	**M56 × 2**	52.0
M60 × 5.5	54.5	M60 × 1.5	56.0
M64 × 6	58.0	**M64 × 2**	60.0
M68 × 6	62.0	M68 × 2	64.0
M72 × 6	66.0	**M72 × 2**	68.0
M80 × 6	74.0	**M80 × 2**	76.0
M90 × 6	84.0	**M90 × 2**	86.0
M100 × 6	94.0	**M100 × 2**	96.0

[a]Metric Fasteners Standard, IFI-500 (1983) and ANSI/ASME B1.13M–1983 (R1989).
[b]Only for high strength structural steel fasteners.

16 Twist Drill Sizes—American National Standard and Metric

AMERICAN NATIONAL STANDARD DRILL SIZES[a]

All dimensions are in inches.

Drills designated in common fractions are available in diameters 1/64″ to 1¾″ in 1/64″ increments, 1¾″ to 2¼″ in 1/32″ increments. 2¼″ to 3″ in 1/16″ increments and 3″ to 3½″ in 1/8″ increments. Drills larger than 3½″ are seldom used, and are regarded as special drills.

Size	Drill Diameter	Size	Drill Diameter	Size	Drill Diameter	Size	Drill Diameter	Size	Drill Diameter	Size	Drill Diameter
1	.2280	17	.1730	33	.1130	49	.0730	65	.0350	81	.0130
2	.2210	18	.1695	34	.1110	50	.0700	66	.0330	82	.0125
3	.2130	19	.1660	35	.1100	51	.0670	67	.0320	83	.0120
4	.2090	20	.1610	36	.1065	52	.0635	68	.0310	84	.0115
5	.2055	21	.1590	37	.1040	53	.0595	69	.0292	85	.0110
6	.2040	22	.1570	38	.1015	54	.0550	70	.0280	86	.0105
7	.2010	23	.1540	39	.0995	55	.0520	71	.0260	87	.0100
8	.1990	24	.1520	40	.0980	56	.0465	72	.0250	88	.0095
9	.1960	25	.1495	41	.0960	57	.0430	73	.0240	89	.0091
10	.1935	26	.1470	42	.0935	58	.0420	74	.0225	90	.0087
11	.1910	27	.1440	43	.0890	59	.0410	75	.0210	91	.0083
12	.1890	28	.1405	44	.0860	60	.0400	76	.0200	92	.0079
13	.1850	29	.1360	45	.0820	61	.0390	77	.0180	93	.0075
14	.1820	30	.1285	46	.0810	62	.0380	78	.0160	94	.0071
15	.1800	31	.1200	47	.0785	63	.0370	79	.0145	95	.0067
16	.1770	32	.1160	48	.0760	64	.0360	80	.0135	96	.0063
										97	.0059

LETTER SIZES

Letter	Diameter	Letter	Diameter	Letter	Diameter	Letter	Diameter	Letter	Diameter
A	.234	G	.261	L	.290	Q	.332	V	.377
B	.238	H	.266	M	.295	R	.339	W	.386
C	.242	I	.272	N	.302	S	.348	X	.397
D	.246	J	.277	O	.316	T	.358	Y	.404
E	.250	K	.281	P	.323	U	.368	Z	.413
F	.257								

[a]ANSI B94.11M—1979 (R1987).

16 Twist Drill Sizes—American National Standard and Metric (continued)

METRIC DRILL SIZES
Decimal-inch equivalents are for reference only.

Drill Diameter		Drill Diameter		Drill Diameter		Drill Diameter		Drill Diameter		Drill Diameter	
mm	in.	mm	in.	mm	in.	mm	in.	mm	in.	mm	in.
0.40	.0157	1.95	.0768	4.70	.1850	8.00	.3150	13.20	.5197	25.50	1.0039
0.42	.0165	2.00	.0787	4.80	.1890	8.10	.3189	13.50	.5315	26.00	1.0236
0.45	.0177	2.05	.0807	4.90	.1929	8.20	.3228	13.80	.5433	26.50	1.0433
0.48	.0189	2.10	.0827	5.00	.1969	8.30	.3268	14.00	.5512	27.00	1.0630
0.50	.0197	2.15	.0846	5.10	.2008	8.40	.3307	14.25	.5610	27.50	1.0827
0.55	.0217	2.20	.0866	5.20	.2047	8.50	.3346	14.50	.5709	28.00	1.1024
0.60	.0236	2.25	.0886	5.30	.2087	8.60	.3386	14.75	.5807	28.50	1.1220
0.65	.0256	2.30	.0906	5.40	.2126	8.70	.3425	15.00	.5906	29.00	1.1417
0.70	.0276	2.35	.0925	5.50	.2165	8.80	.3465	15.25	.6004	29.50	1.1614
0.75	.0295	2.40	.0945	5.60	.2205	8.90	.3504	15.50	.6102	30.00	1.1811
0.80	.0315	2.45	.0965	5.70	.2244	9.00	.3543	15.75	.6201	30.50	1.2008
0.85	.0335	2.50	.0984	5.80	.2283	9.10	.3583	16.00	.6299	31.00	1.2205
0.90	.0354	2.60	.1024	5.90	.2323	9.20	.3622	16.25	.6398	31.50	1.2402
0.95	.0374	2.70	.1063	6.00	.2362	9.30	.3661	16.50	.6496	32.00	1.2598
1.00	.0394	2.80	.1102	6.10	.2402	9.40	.3701	16.75	.6594	32.50	1.2795
1.05	.0413	2.90	.1142	6.20	.2441	9.50	.3740	17.00	.6693	33.00	1.2992
1.10	.0433	3.00	.1181	6.30	.2480	9.60	.3780	17.25	.6791	33.50	1.3189
1.15	.0453	3.10	.1220	6.40	.2520	9.70	.3819	17.50	.6890	34.00	1.3386
1.20	.0472	3.20	.1260	6.50	.2559	9.80	.3858	18.00	.7087	34.50	1.3583
1.25	.0492	3.30	.1299	6.60	.2598	9.90	.3898	18.50	.7283	35.00	1.3780
1.30	.0512	3.40	.1339	6.70	.2638	10.00	.3937	19.00	.7480	35.50	1.3976
1.35	.0531	3.50	.1378	6.80	.2677	10.20	.4016	19.50	.7677	36.00	1.4173
1.40	.0551	3.60	.1417	6.90	.2717	10.50	.4134	20.00	.7874	36.50	1.4370
1.45	.0571	3.70	.1457	7.00	.2756	10.80	.4252	20.50	.8071	37.00	1.4567
1.50	.0591	3.80	.1496	7.10	.2795	11.00	.4331	21.00	.8268	37.50	1.4764
1.55	.0610	3.90	.1535	7.20	.2835	11.20	.4409	21.50	.8465	38.00	1.4961
1.60	.0630	4.00	.1575	7.30	.2874	11.50	.4528	22.00	.8661	40.00	1.5748
1.65	.0650	4.10	.1614	7.40	.2913	11.80	.4646	22.50	.8858	42.00	1.6535
1.70	.0669	4.20	.1654	7.50	.2953	12.00	.4724	23.00	.9055	44.00	1.7323
1.75	.0689	4.30	.1693	7.60	.2992	12.20	.4803	23.50	.9252	46.00	1.8110
1.80	.0709	4.40	.1732	7.70	.3031	12.50	.4921	24.00	.9449	48.00	1.8898
1.85	.0728	4.50	.1772	7.80	.3071	12.50	.5039	24.50	.9646	50.00	1.9685
1.90	.0748	4.60	.1811	7.90	.3110	13.00	.5118	25.00	.9843		

17 Acme Threads, General-Purpose[a]

Size	Threads per Inch	Size	Threads per Inch	Size	Threads per Inch	Size	Threads per Inch
¼	16	¾	6	1½	4	3	2
⁵⁄₁₆	14	⅞	6	1¾	4	3½	2
⅜	12	1	5	2	4	4	2
⁷⁄₁₆	12	1⅛	5	2¼	3	4½	2
½	10	1¼	5	2½	3	5	2
⅝	8	1⅜	4	2¾	3	. . .	. .

[a]ANSI/ASME B1.5—1988.

18 Bolts, Nuts, and Cap Screws—Square and Hexagon—American National Standard and Metric

AMERICAN NATIONAL STANDARD SQUARE AND HEXAGON BOLTS[a] AND NUTS[b] AND HEXAGON CAP SCREWS[c]

Boldface type indicates product features unified dimensionally with British and Canadian standards.
All dimensions are in inches.
For thread series, minimum thread lengths, and bolt lengths, see §15.25.

Nominal Size D Body Diameter of Bolt		Regular Bolts					Heavy Bolts		
		Width Across Flats W		Height H			Width Across Flats W	Height H	
		Sq.	Hex.	Sq. (Unfin.)	Hex. (Unfin.)	Hex. Cap Scr.[c] (Fin.)		Hex. (Unfin.)	Hex. Screw (Fin.)
1/4	**0.2500**	**3/8**	**7/16**	**11/64**	**11/64**	**5/32**	....	....	....
5/16	**0.3125**	**1/2**	**1/2**	**13/64**	**7/32**	**13/64**	....	....	....
3/8	**0.3750**	**9/16**	**9/16**	**1/4**	**1/4**	**15/64**	....	....	....
7/16	**0.4375**	**5/8**	**5/8**	**19/64**	**19/64**	**9/32**	....	....	....
1/2	**0.5000**	**3/4**	**3/4**	**21/64**	**11/32**	**5/16**	**7/8**	11/32	5/16
9/16	**0.5625**	....	**13/16**	....	....	**23/64**	....	....	....
5/8	**0.6250**	**15/16**	**15/16**	**27/64**	**27/64**	**25/64**	**1 1/16**	27/64	25/64
3/4	**0.7500**	**1 1/8**	**1 1/8**	**1/2**	**1/2**	**15/32**	**1 1/4**	1/2	15/32
7/8	**0.8750**	**1 5/16**	**1 5/16**	**19/32**	**37/64**	**35/64**	**1 7/16**	37/64	35/64
1	**1.000**	**1 1/2**	**1 1/2**	**21/32**	**43/64**	**39/64**	**1 5/8**	43/64	39/64
1 1/8	**1.1250**	**1 11/16**	**1 11/16**	**3/4**	**3/4**	**11/16**	**1 13/16**	3/4	11/16
1 1/4	**1.2500**	**1 7/8**	**1 7/8**	**27/32**	**27/32**	**25/32**	**2**	27/32	25/32
1 3/8	**1.3750**	**2 1/16**	**2 1/16**	**29/32**	**29/32**	**27/32**	**2 3/16**	29/32	27/32
1 1/2	**1.5000**	**2 1/4**	**2 1/4**	**1**	**1**	**15/16**	**2 3/8**	1	15/16
1 3/4	**1.7500**	....	**2 5/8**	....	**1 5/32**	**1 3/32**	**2 3/4**	1 5/32	1 3/32
2	**2.0000**	....	**3**	....	**1 11/32**	**1 7/32**	**3 1/8**	1 11/32	1 7/32
2 1/4	2.2500	....	3 3/8	....	1 1/2	1 3/8	3 1/2	1 1/2	1 3/8
2 1/2	2.5000	....	3 3/4	....	1 21/32	1 17/32	3 7/8	1 21/32	1 17/32
2 3/4	2.7500	....	4 1/8	....	1 13/16	1 11/16	4 1/4	1 13/16	1 11/16
3	3.0000	....	4 1/2	....	2	1 7/8	4 5/8	2	1 7/8
3 1/4	3.2500	....	4 7/8	....	2 3/16	....	....	....	....
3 1/2	3.5000	....	5 1/4	....	2 5/16	....	....	....	....
3 3/4	3.7500	....	5 5/8	....	2 1/2	....	....	....	....
4	4.0000	....	6	....	2 11/16	....	....	....	....

[a]ANSI B18.2.1—1981.
[b]ANSI B18.2.2—1987.
[c]Hexagon cap screws and finished hexagon bolts are combined as a single product.

18 Bolts, Nuts, and Cap Screws—Square and Hexagon—American National Standard and Metric (continued)

AMERICAN NATIONAL STANDARD SQUARE AND HEXAGON BOLTS AND NUTS AND HEXAGON CAP SCREWS (continued)

See ANSI B18.2.2 for jam nuts, slotted nuts, thick nuts, thick slotted nuts, and castle nuts.
For methods of drawing bolts and nuts and hexagon-head cap screws, see Figs. 15.29, 15.30, and 15.32.

Nominal Size D Body Diameter of Bolt	Regular Nuts					Heavy Nuts			
	Width Across Flats W		Thickness T			Width Across Flats W	Thickness T		
	Sq.	Hex.	Sq. (Unfin.)	Hex. Flat (Unfin.)	Hex. (Fin.)		Sq. (Unfin.)	Hex. Flat (Unfin.)	Hex. (Fin.)
1/4 0.2500	7/16	7/16	7/32	7/32	7/32	1/2	1/4	15/64	15/64
5/16 0.3125	9/16	1/2	17/64	17/64	17/64	9/16	5/16	19/64	19/64
3/8 0.3750	5/8	9/16	21/64	21/64	21/64	11/16	3/8	23/64	23/64
7/16 0.4375	3/4	11/16	3/8	3/8	3/8	3/4	7/16	27/64	27/64
1/2 0.5000	13/16	3/4	7/16	7/16	7/16	7/8[a]	1/2	31/64	31/64
9/16 0.5625	....	7/8	....	31/64	31/64	15/16	....	35/64	35/64
5/8 0.6250	1	15/16	35/64	35/64	35/64	1 1/16[a]	5/8	39/64	39/64
3/4 0.7500	1 1/8	1 1/8	21/32	41/64	41/64	1 1/4[a]	3/4	47/64	47/64
7/8 0.8750	1 5/16	1 5/16	49/64	3/4	3/4	1 7/16[a]	7/8	55/64	55/64
1 1.0000	1 1/2	1 1/2	7/8	55/64	55/64	1 5/8[a]	1	63/64	63/64
1 1/8 1.1250	1 11/16	1 11/16	1	1	31/32	1 13/16[a]	1 1/8	1 1/8	1 7/64
1 1/4 1.2500	1 7/8	1 7/8	1 3/32	1 3/32	1 1/16	2[a]	1 1/4	1 1/4	1 7/32
1 3/8 1.3750	2 1/16	2 1/16	1 13/64	1 13/64	1 11/64	2 3/16[a]	1 3/8	1 3/8	1 11/32
1 1/2 1.5000	2 1/4	2 1/4	1 5/16	1 5/16	1 9/32	2 3/8[a]	1 1/2	1 1/2	1 15/32
1 5/8 1.6250	....	....	....	....	....	2 9/16	....	....	1 19/32
1 3/4 1.7500	....	....	....	....	....	2 3/4	....	1 3/4	1 23/32
1 7/8 1.8750	....	....	....	....	....	2 15/16	....	....	1 27/32
2 2.0000	....	....	....	....	....	3 1/8	....	2	1 31/32
2 1/4 2.2500	....	....	....	....	....	3 1/2	....	2 1/4	2 13/64
2 1/2 2.5000	....	....	....	....	....	3 7/8	....	2 1/2	2 29/64
2 3/4 2.7500	....	....	....	....	....	4 1/4	....	2 3/4	2 45/64
3 3.0000	....	....	....	....	....	4 5/8	....	3	2 61/64
3 1/4 3.2500	....	....	....	....	....	5	....	3 1/4	3 3/16
3 1/2 3.5000	....	....	....	....	....	5 3/8	....	3 1/2	3 7/16
3 3/4 3.7500	....	....	....	....	....	5 3/4	....	3 3/4	3 11/16
4 4.0000	....	....	....	....	....	6 1/8	....	4	3 15/16

[a] Product feature not unified for heavy square nut.

18 Bolts, Nuts, and Cap Screws—Square and Hexagon— American National Standard and Metric (continued)

METRIC HEXAGON BOLTS, HEXAGON CAP SCREWS, HEXAGON STRUCTURAL BOLTS, AND HEXAGON NUTS

Nominal Size D, mm	Width Across Flats W (max)		Thickness T (max)			
Body Dia and Thd Pitch	Bolts,[a] Cap Screws,[b] and Nuts[c]	Heavy Hex & Hex Structural Bolts[a] & Nuts[c]	Bolts (Unfin.)	Cap Screw (Fin.)	Nut (Fin. or Unfin.)	
					Style 1	Style 2
M5 × 0.8	8.0		3.88	3.65	4.7	5.1
M6 × 1	10.0		4.38	4.47	5.2	5.7
M8 × 1.25	13.0		5.68	5.50	6.8	7.5
M10 × 1.5	16.0		6.85	6.63	8.4	9.3
M12 × 1.75	18.0	21.0	7.95	7.76	10.8	12.0
M14 × 2	21.0	24.0	9.25	9.09	12.8	14.1
M16 × 2	24.0	27.0	10.75	10.32	14.8	16.4
M20 × 2.5	30.0	34.0	13.40	12.88	18.0	20.3
M24 × 3	36.0	41.0	15.90	15.44	21.5	23.9
M30 × 3.5	46.0	50.0	19.75	19.48	25.6	28.6
M36 × 4	55.0	60.0	23.55	23.38	31.0	34.7
M42 × 4.5	65.0		27.05	26.97	....	....
M48 × 5	75.0		31.07	31.07	....	....
M56 × 5.5	85.0		36.20	36.20	....	....
M64 × 6	95.0		41.32	41.32	....	....
M72 × 6	105.0		46.45	46.45	....	....
M80 × 6	115.0		51.58	51.58	....	....
M90 × 6	130.0		57.74	57.74	....	....
M100 × 6	145.0		63.90	63.90	....	....

HIGH STRENGTH STRUCTURAL HEXAGON BOLTS[a] (Fin.) AND HEXAGON NUTS[c]

M16 × 2	27.0	....	10.75	....	....	17.1
M20 × 2.5	34.0	....	13.40	....	....	20.7
M22 × 2.5	36.0	....	14.9	....	....	23.6
M24 × 3	41.0	....	15.9	....	....	24.2
M27 × 3	46.0	....	17.9	....	....	27.6
M30 × 3.5	50.0	....	19.75	....	....	31.7
M36 × 4	60.0	....	23.55	....	....	36.6

[a] ANSI B18.2.3.5M–1979 (R1989), B18.2.3.6M–1979 (R1989), B18.2.3.7M–1979 (R1989).
[b] ANSI B18.2.3.1M–1979 (R1989).
[c] ANSI B18.2.4.1M–1979 (R1989), B18.2.4.2M–1979 (R1989).

19 Cap Screws, Slotted[a] and Socket Head[b]—American National Standard and Metric

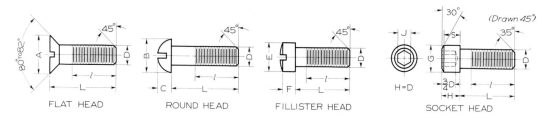

FLAT HEAD   ROUND HEAD   FILLISTER HEAD   SOCKET HEAD

For methods of drawing cap screws, screw lengths, and thread data, see Fig. 15.32.

Nominal Size D	Flat Head[a]	Round Head[a]		Fillister Head[a]		Socket Head[b]		
	A	B	C	E	F	G	J	S
0 (.060)	....	....	....	....	....	.096	.05	.054
1 (.073)	....	....	....	....	....	.118	1/16	.066
2 (.086)	....	....	....	....	....	.140	5/64	.077
3 (.099)	....	....	....	....	....	.161	5/64	.089
4 (.112)	....	....	....	....	....	.183	3/32	.101
5 (.125)	....	....	....	....	....	.205	3/32	.112
6 (.138)	....	....	....	....	....	.226	7/64	.124
8 (.164)	....	....	....	....	....	.270	9/64	.148
10 (.190)	....	....	....	....	....	.312	5/32	.171
1/4	.500	.437	.191	.375	.172	.375	3/16	.225
5/16	.625	.562	.245	.437	.203	.469	1/4	.281
3/8	.750	.675	.273	.562	.250	.562	5/16	.337
7/16	.812	.750	.328	.625	.297	.656	3/8	.394
1/2	.875	.812	.354	.750	.328	.750	3/8	.450
9/16	1.000	.937	.409	.812	.375	....	....	....
5/8	1.125	1.000	.437	.875	.422	.938	1/2	.562
3/4	1.375	1.250	.546	1.000	.500	1.125	5/8	.675
7/8	1.625	....	....	1.125	.594	1.312	3/4	.787
1	1.875	....	....	1.312	.656	1.500	3/4	.900
1 1/8	2.062	....	....	....	....	1.688	7/8	1.012
1 1/4	2.312	....	....	....	....	1.875	7/8	1.125
1 3/8	2.562	....	....	....	....	2.062	1	1.237
1 1/2	2.812	....	....	....	....	2.250	1	1.350

[a]ANSI B18.6.2–1972 (R1983).
[b]ANSI/ASME B18.3—1986. For hexagon-head screws, see §15.29 and Appendix 18.

19 Cap Screws, Slotted[a] and Socket Head[b]— American National Standard and Metric (continued)

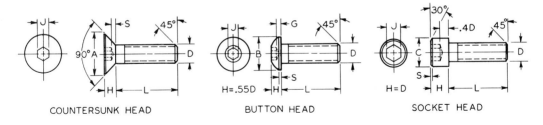

COUNTERSUNK HEAD   BUTTON HEAD   SOCKET HEAD

For methods of drawing cap screws, screw lengths, and thread data, see Fig. 15.32.

Nominal Size D	Countersunk Head[a]			Button Head[b]			Socket Head[c]		Hex Socket Size
	A (max)	H	S	B	S	G	C	S	J
M1.6 × 0.35	....	....	....	....	....	....	3.0	0.16	1.5
M2 × 0.4	....	....	....	....	....	....	3.8	0.2	1.5
M2.5 × 0.45	....	....	....	....	....	....	4.5	0.25	2.0
M3 × 0.5	6.72	1.86	0.25	5.70	0.38	0.2	5.5	0.3	2.5
M4 × 0.7	8.96	2.48	0.45	7.6	0.38	0.3	7.0	0.4	3.0
M5 × 0.8	11.2	3.1	0.66	9.5	0.5	0.38	8.5	0.5	4.0
M6 × 1	13.44	3.72	0.7	10.5	0.8	0.74	10.0	0.6	5.0
M8 × 1.25	17.92	4.96	1.16	14.0	0.8	1.05	13.0	0.8	6.0
M10 × 1.5	22.4	6.2	1.62	17.5	0.8	1.45	16.0	1.0	8.0
M12 × 1.75	26.88	7.44	1.8	21.0	0.8	1.63	18.0	1.2	10.0
M14 × 2	30.24	8.12	2.0	....	....	....	21.0	1.4	12.0
M16 × 2	33.6	8.8	2.2	28.0	1.5	2.25	24.0	1.6	14.0
M20 × 2.5	19.67	10.16	2.2	....	....	....	30.0	2.0	17.0
M24 × 3	....	....	....	....	....	....	36.0	2.4	19.0
M30 × 3.5	....	....	....	....	....	....	45.0	3.0	22.0
M36 × 4	....	....	....	....	....	....	54.0	3.6	27.0
M42 × 4.5	....	....	....	....	....	....	63.0	4.2	32.0
M48 · × 5	....	....	....	....	....	....	72.0	4.8	36.0

Table title: METRIC SOCKET HEAD CAP SCREWS

[a]IFI 535–1982.
[b]ANSI/ASME B18.3.4M–1986.
[c]ANSI/ASME B18.3.1M–1986.

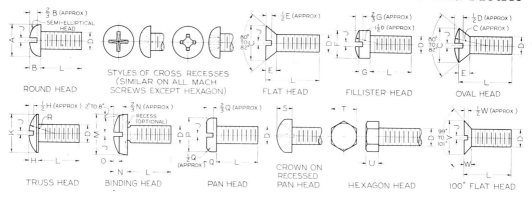

ROUND HEAD · STYLES OF CROSS RECESSES (SIMILAR ON ALL MACH. SCREWS EXCEPT HEXAGON) · FLAT HEAD · FILLISTER HEAD · OVAL HEAD

TRUSS HEAD · BINDING HEAD · PAN HEAD · CROWN ON RECESSED PAN HEAD · HEXAGON HEAD · 100° FLAT HEAD

AMERICAN NATIONAL STANDARD MACHINE SCREWS[a]

Length of Thread: On screws 2″ long and shorter, the threads extend to within two threads of the head and closer if practicable; longer screws have minimum thread length of 1¾″.

Points: Machine screws are regularly made with plain sheared ends, not chamfered.

Threads: Either Coarse or Fine Thread Series, Class 2 fit.

Recessed Heads: Two styles of cross recesses are available on all screws except hexagon head.

Nominal Size	Max. Diameter D	Round Head		Flat Heads & Oval Head		Fillister Head		Truss Head			Slot Width
		A	B	C	E	F	G	K	H	R	J
0	0.060	0.113	0.053	0.119	0.035	0.096	0.045	0.131	0.037	0.087	0.023
1	0.073	0.138	0.061	0.146	0.043	0.118	0.053	0.164	0.045	0.107	0.026
2	0.086	0.162	0.069	0.172	0.051	0.140	0.062	0.194	0.053	0.129	0.031
3	0.099	0.187	0.078	0.199	0.059	0.161	0.070	0.226	0.061	0.151	0.035
4	0.112	0.211	0.086	0.225	0.067	0.183	0.079	0.257	0.069	0.169	0.039
5	0.125	0.236	0.095	0.252	0.075	0.205	0.088	0.289	0.078	0.191	0.043
6	0.138	0.260	0.103	0.279	0.083	0.226	0.096	0.321	0.086	0.211	0.048
8	0.164	0.309	0.120	0.332	0.100	0.270	0.113	0.384	0.102	0.254	0.054
10	0.190	0.359	0.137	0.385	0.116	0.313	0.130	0.448	0.118	0.283	0.060
12	0.216	0.408	0.153	0.438	0.132	0.357	0.148	0.511	0.134	0.336	0.067
¼	0.250	0.472	0.175	0.507	0.153	0.414	0.170	0.573	0.150	0.375	0.075
⁵⁄₁₆	0.3125	0.590	0.216	0.635	0.191	0.518	0.211	0.698	0.183	0.457	0.084
⅜	0.375	0.708	0.256	0.762	0.230	0.622	0.253	0.823	0.215	0.538	0.094
⁷⁄₁₆	0.4375	0.750	0.328	0.812	0.223	0.625	0.265	0.948	0.248	0.619	0.094
½	0.500	0.813	0.355	0.875	0.223	0.750	0.297	1.073	0.280	0.701	0.106
⁹⁄₁₆	0.5625	0.938	0.410	1.000	0.260	0.812	0.336	1.198	0.312	0.783	0.118
⅝	0.625	1.000	0.438	1.125	0.298	0.875	0.375	1.323	0.345	0.863	0.133
¾	0.750	1.250	0.547	1.375	0.372	1.000	0.441	1.573	0.410	1.024	0.149

Nominal Size	Max. Diameter D	Binding Head			Pan Head			Hexagon Head		100° Flat Head		Slot Width
		M	N	O	P	Q	S	T	U	V	W	J
2	0.086	0.181	0.050	0.018	0.167	0.053	0.062	0.125	0.050	. . . .	. . . .	0.031
3	0.099	0.208	0.059	0.022	0.193	0.060	0.071	0.187	0.055	. . . .	. . . .	0.035
4	0.112	0.235	0.068	0.025	0.219	0.068	0.080	0.187	0.060	0.225	0.049	0.039
5	0.125	0.263	0.078	0.029	0.245	0.075	0.089	0.187	0.070	. . . .	. . . .	0.043
6	0.138	0.290	0.087	0.032	0.270	0.082	0.097	0.250	0.080	0.279	0.060	0.048
8	0.164	0.344	0.105	0.039	0.322	0.096	0.115	0.250	0.110	0.332	0.072	0.054
10	0.190	0.399	0.123	0.045	0.373	0.110	0.133	0.312	0.120	0.385	0.083	0.060
12	0.216	0.454	0.141	0.052	0.425	0.125	0.151	0.312	0.155	. . . .	. . . .	0.067
¼	0.250	0.513	0.165	0.061	0.492	0.144	0.175	0.375	0.190	0.507	0.110	0.075
⁵⁄₁₆	0.3125	0.641	0.209	0.077	0.615	0.178	0.218	0.500	0.230	0.635	0.138	0.084
⅜	0.375	0.769	0.253	0.094	0.740	0.212	0.261	0.562	0.295	0.762	0.165	0.094
⁷⁄₁₆	.4375	. . . .	. . . .	. . . .	.865	.247	.305	. . . .	. . . .	. . . .	. . . .	.094
½	.500	. . . .	. . . .	. . . .	.987	.281	.348	. . . .	. . . .	. . . .	. . . .	.106
⁹⁄₁₆	.5625	. . . .	. . . .	. . . .	1.041	.315	.391	. . . .	. . . .	. . . .	. . . .	.118
⅝	.625	. . . .	. . . .	. . . .	1.172	.350	.434	. . . .	. . . .	. . . .	. . . .	.133
¾	.750	. . . .	. . . .	. . . .	1.435	.419	.521	. . . .	. . . .	. . . .	. . . .	.149

20 Machine Screws—American National Standard and Metric (continued)

METRIC MACHINE SCREWS[b]

Length of Thread: On screws 36 mm long or shorter, the threads extend to within one thread of the head: on longer screws the thread extends to within two threads of the head.

Points: Machine screws are regularly made with sheared ends, not chamfered.

Threads: Coarse (general-purpose) threads series are given.

Recessed Heads: Two styles of cross-recesses are available on all screws except hexagon head.

Nominal Size & Thd Pitch	Max. Dia. D, mm	Flat Heads & Oval Head		Pan Heads			Hex Head		Slot Width
		C	E	P	Q	S	T	U	J
M2 × 0.4	2.0	3.5	1.2	4.0	1.3	1.6	3.2	1.6	0.7
M2.5 × 0.45	2.5	4.4	1.5	5.0	1.5	2.1	4.0	2.1	0.8
M3 × 0.5	3.0	5.2	1.7	5.6	1.8	2.4	5.0	2.3	1.0
M3.5 × 0.6	3.5	6.9	2.3	7.0	2.1	2.6	5.5	2.6	1.2
M4 × 0.7	4.0	8.0	2.7	8.0	2.4	3.1	7.0	3.0	1.5
M5 × 0.8	5.0	8.9	2.7	9.5	3.0	3.7	8.0	3.8	1.5
M6 × 1	6.0	10.9	3.3	12.0	3.6	4.6	10.0	4.7	1.9
M8 × 1.25	8.0	15.14	4.6	16.0	4.8	6.0	13.0	6.0	2.3
M10 × 1.5	10.0	17.8	5.0	20.0	6.0	7.5	15.0	7.5	2.8
M12 × 1.75	12.0	....	....	....	....	....	18.0	9.0	....

Nominal Size	Metric Machine Screw Lengths—L[c]																			
	2.5	3	4	5	6	8	10	13	16	20	25	30	35	40	45	50	55	60	65	70	80	90
M2 × 0.4	PH	A	A	A	A	A	A	A	A											
M2.5 × 0.45		PH	A	A	A	A	A	A	A	A										
M3 × 0.5			PH	A	A	A	A	A	A	A	A	A								
M3.5 × 0.6				PH	A	A	A	A	A	A	A	A	A							
M4 × 0.7				PH	A	A	A	A	A	A	A	A	A							
M5 × 0.8					PH	A	A	A	A	A	A	A	A	A	A	A				
M6 × 1						A	A	A	A	A	A	A	A	A	A	A	A	A		
M8 × 1.25						A	A	A	A	A	A	A	A	A	A	A	A	A	A	A
M10 × 1.5							A	A	A	A	A	A	A	A	A	A	A	A	A	A	A	A
M12 × 1.75							A	A	A	A	A	A	A	A	A	A	A	A	A	A	A	A

Min. Thd Length—28 mm

Min. Thd Length—38 mm

[b]Metric Fasteners Standard. IFI-513 (1982).

[c]PH = recommended lengths for only pan and hex head metric screws;

   A = recommended lengths for all metric screw head-styles.

21 Keys—Square, Flat, Plain Taper,[a] and Gib Head

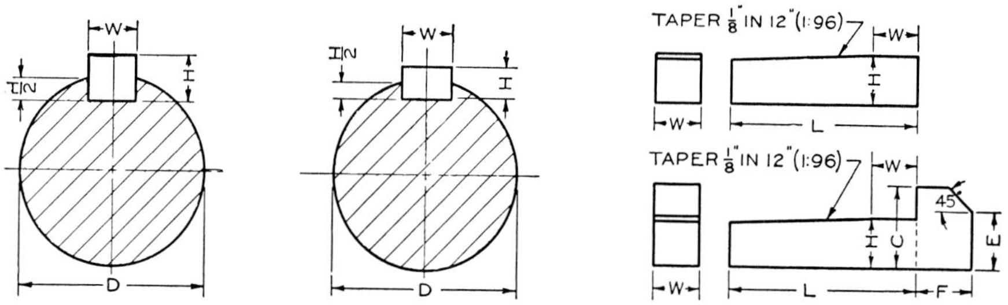

Shaft Diameters	Square Stock Key	Flat Stock Key	Gib Head Taper Stock Key					
			Square			*Flat*		
			Height	Length	Height to Chamfer	Height	Length	Height to Chamfer
D	W = H	W × H	C	F	E	C	F	E
1/2 to 9/16	1/8	1/8 × 3/32	1/4	7/32	5/32	3/16	1/8	1/8
5/8 to 7/8	3/16	3/16 × 1/8	5/16	9/32	7/32	1/4	3/16	5/32
15/16 to 1 1/4	1/4	1/4 × 3/16	7/16	11/32	11/32	5/16	1/4	3/16
1 5/16 to 1 3/8	5/16	5/16 × 1/4	9/16	13/32	13/32	3/8	5/16	1/4
1 7/16 to 1 3/4	3/8	3/8 × 1/4	11/16	15/32	15/32	7/16	3/8	5/16
1 13/16 to 2 1/4	1/2	1/2 × 3/8	7/8	19/32	5/8	5/8	1/2	7/16
2 5/16 to 2 3/4	5/8	5/8 × 7/16	1 1/16	23/32	3/4	3/4	5/8	1/2
2 7/8 to 3 1/4	3/4	3/4 × 1/2	1 1/4	7/8	7/8	7/8	3/4	5/8
3 3/8 to 3 3/4	7/8	7/8 × 5/8	1 1/2	1	1	1 1/16	7/8	3/4
3 7/8 to 4 1/2	1	1 × 3/4	1 3/4	1 3/16	1 3/16	1 1/4	1	13/16
4 3/4 to 5 1/2	1 1/4	1 1/4 × 7/8	2	1 7/16	1 7/16	1 1/2	1 1/4	1
5 3/4 to 6	1 1/2	1 1/2 × 1	2 1/2	1 3/4	1 3/4	1 3/4	1 1/2	1 1/4

[a] Plain taper square and flat keys have the same dimensions as the plain parallel stock keys, with the addition of the taper on top. Gib head taper square and flat keys have the same dimensions as the plain taper keys, with the addition of the gib head.

Stock lengths for plain taper and gib head taper keys: The minimum stock length equals 4W, and the maximum equals 16W. The increments of increase of length equal 2W.

22 Screw Threads,[a] Square and Acme

Size	Threads per Inch	Size	Threads per Inch	Size	Threads per Inch	Size	Threads per Inch
3/8	12	7/8	5	2	2 1/2	3 1/2	1 1/3
7/16	10	1	5	2 1/4	2	3 3/4	1 1/3
1/2	10	1 1/8	4	2 1/2	2	4	1 1/3
9/16	8	1 1/4	4	2 3/4	2	4 1/4	1 1/3
5/8	8	1 1/2	3	3	1 1/2	4 1/2	1
3/4	6	1 3/4	2 1/2	3 1/4	1 1/2	over 4 1/2	1

[a] See Appendix 17 for General-Purpose Acme Threads.

23   Woodruff Keys[a]—American National Standard

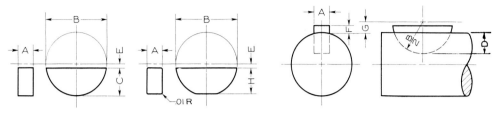

Key No.[b]	Nominal Sizes				Maximum Sizes			Key No.[b]	Nominal Sizes				Maximum Sizes		
	A × B	E	F	G	H	D	C		A × B	E	F	G	H	D	C
204	1/16 × 1/2	3/64	1/32	5/64	.194	.1718	.203	808	1/4 × 1	1/16	1/8	3/16	.428	.3130	.438
304	3/32 × 1/2	3/64	3/64	3/32	.194	.1561	.203	809	1/4 × 1 1/8	5/64	1/8	13/64	.475	.3590	.484
305	3/32 × 5/8	1/16	3/64	7/64	.240	.2031	.250	810	1/4 × 1 1/4	5/64	1/8	13/64	.537	.4220	.547
404	1/8 × 1/2	3/64	1/16	7/64	.194	.1405	.203	811	1/4 × 1 3/8	3/32	1/8	7/32	.584	.4690	.594
405	1/8 × 5/8	1/16	1/16	1/8	.240	.1875	.250	812	1/4 × 1 1/2	7/64	1/8	15/64	.631	.5160	.641
406	1/8 × 3/4	1/16	1/16	1/8	.303	.2505	.313	1008	5/16 × 1	1/16	5/32	7/32	.428	.2818	.438
505	5/32 × 5/8	1/16	5/64	9/64	.240	.1719	.250	1009	5/16 × 1 1/8	5/64	5/32	15/64	.475	.3278	.484
506	5/32 × 3/4	1/16	5/64	9/64	.303	.2349	.313	1010	5/16 × 1 1/4	5/64	5/32	15/64	.537	.3908	.547
507	5/32 × 7/8	1/16	5/64	9/64	.365	.2969	.375	1011	5/16 × 1 3/8	3/32	5/32	8/32	.584	.4378	.594
606	3/16 × 3/4	1/16	3/32	5/32	.303	.2193	.313	1012	5/16 × 1 1/2	7/64	5/32	17/64	.631	.4848	.641
607	3/16 × 7/8	1/16	3/32	5/32	.365	.2813	.375	1210	3/8 × 1 1/4	5/64	3/16	17/64	.537	.3595	.547
608	3/16 × 1	1/16	3/32	5/32	.428	.3443	.438	1211	3/8 × 1 3/8	3/32	3/16	9/32	.584	.4065	.594
609	3/16 × 1 1/8	5/64	3/32	11/64	.475	.3903	.484	1212	3/8 × 1 1/2	7/64	3/16	19/64	.631	.4535	.641
807	1/4 × 7/8	1/16	1/8	3/16	.365	.2500	.375	. . . .	. . . .	. .	. .	. .	. . . .	. . . .	. . . .

[a]ANSI B17.2–1967 (R1978).
[b]Key numbers indicate nominal key dimensions. The last two digits give the nominal diameter B in eighths of an inch, and the digits before the last two give the nominal width A in thirty-seconds of an inch.

24   Woodruff Key Sizes for Different Shaft Diameters[a]

Shaft Diameter	5/16 to 3/8	7/16 to 1/2	9/16 to 3/4	13/16 to 15/16	1 to 1 3/16	1 1/4 to 1 7/16	1 1/2 to 1 3/4	1 13/16 to 2 1/8	2 3/16 to 2 1/2
Key Numbers	204	304 305	404 405 406	505 506 507	606 607 608 609	807 808 809	810 811 812	1011 1012	1211 1212

[a]Suggested sizes; not standard.

25   Pratt and Whitney Round-End Keys

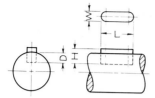

KEYS MADE WITH ROUND
ENDS AND KEYWAYS CUT
IN SPLINE MILLER

Maximum length of slot is 4″ + W. Note that key is sunk two-thirds into shaft in all cases.

Key No.	L^a	W or D	H	Key No.	L^a	W or D	H
1	1/2	1/16	3/32	22	1 3/8	1/4	3/8
2	1/2	3/32	9/64	23	1 3/8	5/16	15/32
3	1/2	1/8	3/16	F	1 3/8	3/8	9/16
4	5/8	3/32	9/64	24	1 1/2	1/4	3/8
5	5/8	1/8	3/16	25	1 1/2	5/16	15/32
6	5/8	5/32	15/64	G	1 1/2	3/8	9/16
7	3/4	1/8	3/16	51	1 3/4	1/4	3/8
8	3/4	5/32	15/64	52	1 3/4	5/16	15/32
9	3/4	3/16	9/32	53	1 3/4	3/8	9/16
10	7/8	5/32	15/64	26	2	3/16	9/32
11	7/8	3/16	9/32	27	2	1/4	3/8
12	7/8	7/32	21/64	28	2	5/16	15/32
A	7/8	1/4	3/8	29	2	3/8	9/16
13	1	3/16	9/32	54	2 1/4	1/4	3/8
14	1	7/32	21/64	55	2 1/4	5/16	15/32
15	1	1/4	3/8	56	2 1/4	3/8	9/16
B	1	5/16	15/32	57	2 1/4	7/16	21/32
16	1 1/8	3/16	9/32	58	2 1/2	5/16	15/32
17	1 1/8	7/32	21/64	59	2 1/2	3/8	9/16
18	1 1/8	1/4	3/8	60	2 1/2	7/16	21/32
C	1 1/8	5/16	15/32	61	2 1/2	1/2	3/4
19	1 1/4	3/16	9/32	30	3	3/8	9/16
20	1 1/4	7/32	21/64	31	3	7/16	21/32
21	1 1/4	1/4	3/8	32	3	1/2	3/4
D	1 1/4	5/16	15/32	33	3	9/16	27/32
E	1 1/4	3/8	9/16	34	3	5/8	15/16

^{a}The length L may vary from the table, but equals at least 2W.

26 Washers,[a] Plain—American National Standard

For parts lists, etc., give inside diameter, outside diameter, and the thickness; for example, .344 × .688 × .065 TYPE A PLAIN WASHER.

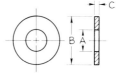

PREFERRED SIZES OF TYPE A PLAIN WASHERS[b]

Nominal Washer Size[c]			Inside Diameter	Outside Diameter	Nominal Thickness
			A	B	C
. . . .	. . . .		0.078	0.188	0.020
. . . .	. . . .		0.094	0.250	0.020
. . . .	. . . .		0.125	0.312	0.032
No. 6	0.138		0.156	0.375	0.049
No. 8	0.164		0.188	0.438	0.049
No. 10	0.190		0.219	0.500	0.049
3/16	0.188		0.250	0.562	0.049
No. 12	0.216		0.250	0.562	0.065
1/4	0.250	N	0.281	0.625	0.065
1/4	0.250	W	0.312	0.734	0.065
5/16	0.312	N	0.344	0.688	0.065
5/16	0.312	W	0.375	0.875	0.083
3/8	0.375	N	0.406	0.812	0.065
3/8	0.375	W	0.438	1.000	0.083
7/16	0.438	N	0.469	0.922	0.065
7/16	0.438	W	0.500	1.250	0.083
1/2	0.500	N	0.531	1.062	0.095
1/2	0.500	W	0.562	1.375	0.109
9/16	0.562	N	0.594	1.156	0.095
9/16	0.562	W	0.625	1.469	0.109
5/8	0.625	N	0.656	1.312	0.095
5/8	0.625	W	0.688	1.750	0.134
3/4	0.750	N	0.812	1.469	0.134
3/4	0.750	W	0.812	2.000	0.148
7/8	0.875	N	0.938	1.750	0.134
7/8	0.875	W	0.938	2.250	0.165
1	1.000	N	1.062	2.000	0.134
1	1.000	W	1.062	2.500	0.165
1 1/8	1.125	N	1.250	2.250	0.134
1 1/8	1.125	W	1.250	2.750	0.165
1 1/4	1.250	N	1.375	2.500	0.165
1 1/4	1.250	W	1.375	3.000	0.165
1 3/8	1.375	N	1.500	2.750	0.165
1 3/8	1.375	W	1.500	3.250	0.180
1 1/2	1.500	N	1.625	3.000	0.165
1 1/2	1.500	W	1.625	3.500	0.180
1 5/8	1.625		1.750	3.750	0.180
1 3/4	1.750		1.875	4.000	0.180
1 7/8	1.875		2.000	4.250	0.180
2	2.000		2.125	4.500	0.180
2 1/4	2.250		2.375	4.750	0.220
2 1/2	2.500		2.625	5.000	0.238
2 3/4	2.750		2.875	5.250	0.259
3	3.000		3.125	5.500	0.284

[a]From ANSI B18.22.1–1965 (R1981). For complete listings, see the standard.
[b]Preferred sizes are for the most part from series previously designated "Standard Plate" and "SAE." Where common sizes existed in the two series, the SAE size is designated "N" (narrow) and the Standard Plate "W" (wide).
[c]Nominal washer sizes are intended for use with comparable nominal screw or bolt sizes.

27   Washers,[a] Lock—American National Standard

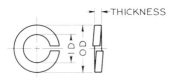

For parts lists, etc., give nominal size and series; for example, ¼ REGULAR LOCK WASHER

PREFERRED SERIES

Nominal Washer Size[b]		Inside Diameter, Min.	Regular		Extra Duty		Hi-Collar	
			Outside Diameter, Max.	Thickness, Min.	Outside Diameter, Max.	Thickness, Min.	Outside Diameter, Max.	Thickness, Min.
No. 2	0.086	0.088	0.172	0.020	0.208	0.027	. . . .	. . . .
No. 3	0.099	0.101	0.195	0.025	0.239	0.034	. . . .	. . . .
No. 4	0.112	0.115	0.209	0.025	0.253	0.034	0.173	0.022
No. 5	0.125	0.128	0.236	0.031	0.300	0.045	0.202	0.030
No. 6	0.138	0.141	0.250	0.031	0.314	0.045	0.216	0.030
No. 8	0.164	0.168	0.293	0.040	0.375	0.057	0.267	0.047
No. 10	0.190	0.194	0.334	0.047	0.434	0.068	0.294	0.047
No. 12	0.216	0.221	0.377	0.056	0.497	0.080	. . . .	. . . .
¼	0.250	0.255	0.489	0.062	0.535	0.084	0.365	0.078
⁵⁄₁₆	0.312	0.318	0.586	0.078	0.622	0.108	0.460	0.093
⅜	0.375	0.382	0.683	0.094	0.741	0.123	0.553	0.125
⁷⁄₁₆	0.438	0.446	0.779	0.109	0.839	0.143	0.647	0.140
½	0.500	0.509	0.873	0.125	0.939	0.162	0.737	0.172
⁹⁄₁₆	0.562	0.572	0.971	0.141	1.041	0.182	. . . .	. . . .
⅝	0.625	0.636	1.079	0.156	1.157	0.202	0.923	0.203
¹¹⁄₁₆	0.688	0.700	1.176	0.172	1.258	0.221	. . .	. . . .
¾	0.750	0.763	1.271	0.188	1.361	0.241	1.111	0.218
¹³⁄₁₆	0.812	0.826	1.367	0.203	1.463	0.261	. . . .	. . . .
⅞	0.875	0.890	1.464	0.219	1.576	0.285	1.296	0.234
¹⁵⁄₁₆	0.938	0.954	1.560	0.234	1.688	0.308	. . . .	. . . .
1	1.000	1.017	1.661	0.250	1.799	0.330	1.483	0.250
1¹⁄₁₆	1.062	1.080	1.756	0.266	1.910	0.352	. . . .	. . . .
1⅛	1.125	1.144	1.853	0.281	2.019	0.375	1.669	0.313
1³⁄₁₆	1.188	1.208	1.950	0.297	2.124	0.396	. . . .	. . . .
1¼	1.250	1.271	2.045	0.312	2.231	0.417	1.799	0.313
1⁵⁄₁₆	1.312	1.334	2.141	0.328	2.335	0.438	. . . .	. . . .
1⅜	1.375	1.398	2.239	0.344	2.439	0.458	2.041	0.375
1⁷⁄₁₆	1.438	1.462	2.334	0.359	2.540	0.478	. . . .	. . . .
1½	1.500	1.525	2.430	0.375	2.638	0.496	2.170	0.375

[a]From ANSI B18.21.1–1972 (R1983). For complete listing, see the standard.
[b]Nominal washer sizes are intended for use with comparable nominal screw or bolt sizes.

28 Wire Gage Standards[a]

Dimensions of sizes in decimal parts of an inch.[b]

No. of Wire	American or Brown & Sharpe for Non-ferrous Metals	Birming-ham, or Stubs' Iron Wire[c]	American S. & W. Co.'s (Washburn & Moen) Std. Steel Wire	American S. & W. Co.'s Music Wire	Imperial Wire	Stubs' Steel Wire[c]	Steel Manu-facturers' Sheet Gage[b]	No. of Wire
7-0's	.651354	. . . .	.4900	. . . .	.500	. . . .	. . . . .	7-0's
6-0's	.580049	. . . .	.4615	.004	.464	. . . .	. . . . .	6-0's
5-0's	.516549	.500	.4305	.005	.432	. . . .	. . . . .	5-0's
4-0's	.460	.454	.3938	.006	.400	. . . .	. . . . .	4-0's
000	.40964	.425	.3625	.007	.372	. . . .	. . . . .	000
00	.3648	.380	.3310	.008	.348	. . . .	. . . . .	00
0	.32486	.340	.3065	.009	.324	. . . .	. . . . .	0
1	.2893	.300	.2830	.010	.300	.227	. . . . .	1
2	.25763	.284	.2625	.011	.276	.219	. . . . .	2
3	.22942	.259	.2437	.012	.252	.212	.2391	3
4	.20431	.238	.2253	.013	.232	.207	.2242	4
5	.18194	.220	.2070	.014	.212	.204	.2092	5
6	.16202	.203	.1920	.016	.192	.201	.1943	6
7	.14428	.180	.1770	.018	.176	.199	.1793	7
8	.12849	.165	.1620	.020	.160	.197	.1644	8
9	.11443	.148	.1483	.022	.144	.194	.1495	9
10	.10189	.134	.1350	.024	.128	.191	.1345	10
11	.090742	.120	.1205	.026	.116	.188	.1196	11
12	.080808	.109	.1055	.029	.104	.185	.1046	12
13	.071961	.095	.0915	.031	.092	.182	.0897	13
14	.064084	.083	.0800	.033	.080	.180	.0747	14
15	.057068	.072	.0720	.035	.072	.178	.0763	15
16	.05082	.065	.0625	.037	.064	.175	.0598	16
17	.045257	.058	.0540	.039	.056	.172	.0538	17
18	.040303	.049	.0475	.041	.048	.168	.0478	18
19	.03589	.042	.0410	.043	.040	.164	.0418	19
20	.031961	.035	.0348	.045	.036	.161	.0359	20
21	.028462	.032	.0317	.047	.032	.157	.0329	21
22	.025347	.028	.0286	.049	.028	.155	.0299	22
23	.022571	.025	.0258	.051	.024	.153	.0269	23
24	.0201	.022	.0230	.055	.022	.151	.0239	24
25	.0179	.020	.0204	.059	.020	.148	.0209	25
26	.01594	.018	.0181	.063	.018	.146	.0179	26
27	.014195	.016	.0173	.067	.0164	.143	.0164	27
28	.012641	.014	.0162	.071	.0149	.139	.0149	28
29	.011257	.013	.0150	.075	.0136	.134	.0135	29
30	.010025	.012	.0140	.080	.0124	.127	.0120	30
31	.008928	.010	.0132	.085	.0116	.120	.0105	31
32	.00795	.009	.0128	.090	.0108	.115	.0097	32
33	.00708	.008	.0118	.095	.0100	.112	.0090	33
34	.006304	.007	.0104	. . . .	.0092	.110	.0082	34
35	.005614	.005	.0095	. . . .	.0084	.108	.0075	35
36	.005	.004	.0090	. . . .	.0076	.106	.0067	36
37	.004453	. . . .	.0085	. . . .	.0068	.103	.0064	37
38	.003965	. . . .	.0080	. . . .	.0060	.101	.0060	38
39	.003531	. . . .	.0075	. . . .	.0052	.099	. . . . .	39
40	.003144	. . . .	.0070	. . . .	.0048	.097	. . . . .	40

[a]Courtesy Brown & Sharpe Mfg. Co.

[b]Now used by steel manufacturers in place of old U.S. Standard Gage.

[c]The difference between the Stubs' Iron Wire Gage and the Stubs' Steel Wire Gage should be noted, the first being commonly known as the English Standard Wire, or Birmingham Gage, which designates the Stubs' soft wire sizes and the second being used in measuring drawn steel wire or drill rods of Stubs' make.

29 Taper Pins[a]—American National Standard

To find small diameter of pin, multiply the length by .02083 and subtract the result from the larger diameter.
All dimensions are given in inches.
Standard reamers are available for pins given above the heavy line.

	7/0	6/0	5/0	4/0	3/0	2/0	0	1	2	3	4	5	6	7	8
Number	7/0	6/0	5/0	4/0	3/0	2/0	0	1	2	3	4	5	6	7	8
Size (Large End)	.0625	.0780	.0940	.1090	.1250	.1410	.1560	.1720	.1930	.2190	.2500	.2890	.3410	.4090	.4920
Shaft Diameter (Approx)[b]		7/32	1/4	5/16	3/8	7/16	1/2	9/16	5/8	3/4	13/16	7/8	1	1 1/4	1 1/2
Drill Size (Before Reamer)[b]	.0312	.0312	.0625	.0625	.0781	.0938	.0938	.1094	.1250	.1250	.1562	.1562	.2188	.2344	.3125
Length L															
.250	X	X	X	X	X										
.375	X	X	X	X	X	X									
.500	X	X	X	X	X	X	X								
.625	X	X	X	X	X	X	X								
.750	X	X	X	X	X	X	X	X	X						
.875	X	X	X	X	X	X	X	X	X	X	X				
1.000		X	X	X	X	X	X	X	X	X	X	X			
1.250		X	X	X	X	X	X	X	X	X	X	X			
1.500				X	X	X	X	X	X	X	X	X			X
1.750				X	X	X	X	X	X	X	X	X	X	X	X
2.000				X	X	X	X	X	X	X	X	X	X	X	X
2.250					X	X	X	X	X	X	X	X	X	X	X
2.500						X	X	X	X	X	X	X	X	X	X
2.750							X	X	X	X	X	X	X	X	X
3.000							X	X	X	X	X	X	X	X	X
3.250									X	X	X	X	X	X	X
3.500										X	X	X	X	X	X
3.750										X	X	X	X	X	X
4.000											X	X	X	X	X
4.250												X	X	X	X
4.500												X	X	X	X

[a]ANSI B18.8.2—1978 (R1983). For Nos. 9 and 10, see the standard. Pins Nos 11 (size .8600), 12 (size 1.032), 13 (size 1.241), and 14 (size 1.523) are special sizes; hence their lengths are special.

[b]Suggested sizes; not American National Standard.

30 Cotter Pins[a]—American National Standard

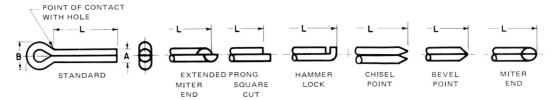

POINT OF CONTACT WITH HOLE

STANDARD

EXTENDED MITER END   PRONG SQUARE CUT   HAMMER LOCK   CHISEL POINT   BEVEL POINT   MITER END

PREFERRED POINT TYPES

All dimensions are given in inches.

Nominal Size or Pin Diameter	Diameter A		Outside Eye Diameter B Min.	Extended Prong Length Min.	Hole Sizes Recommended
	Max.	Min.			
1/32 .031	.032	.028	.06	.01	.047
3/64 .047	.048	.044	.09	.02	.062
1/16 .062	.060	.056	.12	.03	.078
5/64 .078	.076	.072	.16	.04	.094
3/32 .094	.090	.086	.19	.04	.109
7/64 .109	.104	.100	.22	.05	.125
1/8 .125	.120	.116	.25	.06	.141
9/64 .141	.134	.130	.28	.06	.156
5/32 .156	.150	.146	.31	.07	.172
3/16 .188	.176	.172	.38	.09	.203
7/32 .219	.207	.202	.44	.10	.234
1/4 .250	.225	.220	.50	.11	.266
5/16 .312	.280	.275	.62	.14	.312
3/8 .375	.335	.329	.75	.16	.375
7/16 .438	.406	.400	.88	.20	.438
1/2 .500	.473	.467	1.00	.23	.500
5/8 .625	.598	.590	1.25	.30	.625
3/4 .750	.723	.715	1.50	.36	.750

[a]ANSI B18.8.1—1972 (R1983).

31 Metric Equivalents

Length

U.S. to Metric	Metric to U.S.
1 inch = 2.540 centimeters	1 millimeter = .039 inch
1 foot = .305 meter	1 centimeter = .394 inch
1 yard = .914 meter	1 meter = 3.281 feet or 1.094 yards
1 mile = 1.609 kilometers	1 kilometer = .621 mile

Area

1 inch^2 = 6.451 centimeter^2	1 millimeter^2 = .00155 inch^2
1 foot^2 = .093 meter^2	1 centimeter^2 = .155 inch^2
1 yard^2 = .836 meter^2	1 meter^2 = 10.764 foot^2 or 1.196 yard^2
1 acre^2 = 4,046.873 meter^2	1 kilometer^2 = .386 mile^2 or 247.04 acre^2

Volume

1 inch^3 = 16.387 centimeter^3	1 centimeter^3 = .061 inch^3
1 foot^3 = .028 meter^3	1 meter^3 = 35.314 foot^3 or 1.308 yard^3
1 yard^3 = .764 meter^3	1 liter = .2642 gallons
1 quart = .946 liter	1 liter = 1.057 quarts
1 gallon = .003785 meter^3	1 meter^3 = 264.02 gallons

Weight

1 ounce = 28.349 grams	1 gram = .035 ounce
1 pound = .454 kilogram	1 kilogram = 2.205 pounds
1 ton = .907 metric ton	1 metric ton = 1.102 tons

Velocity

1 foot/second = .305 meter/second	1 meter/second = 3.281 feet/second
1 mile/hour = .447 meter/second	1 kilometer/hour = .621 mile/second

Acceleration

1 inch/second^2 = .0254 meter/second^2	1 meter/second^2 = 3.278 feet/second^2
1 foot/second^2 = .305 meter/second^2	

Force

N (newton) = basic unit of force, kg-m/s^2. A mass of one kilogram (1 kg) exerts a gravitational force of 9.8 N (theoretically 9.80665 N) at mean sea level.

32 Welding Symbols and Processes— American Welding Society Standard[a]

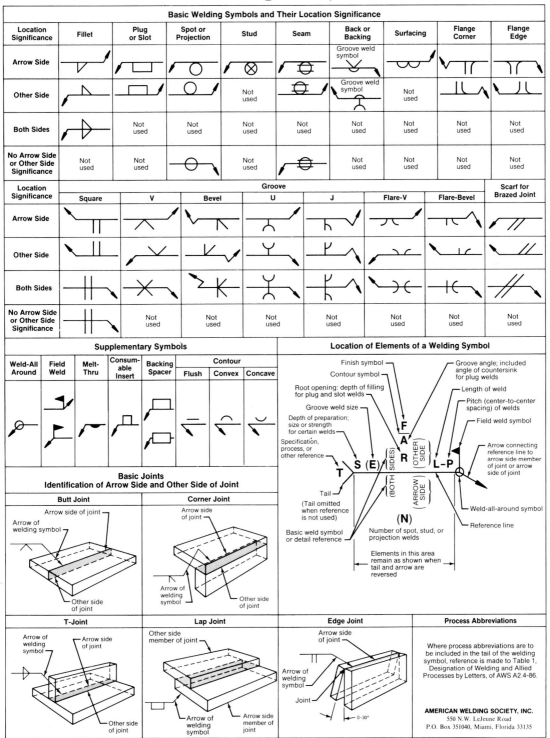

[a]ANSI/AWS A2.4–1986. It should be understood that these charts are intended only as shop aids. The only complete and official presentation of the standard welding symbols is in A2.4.

a67

Typical Welding Symbols

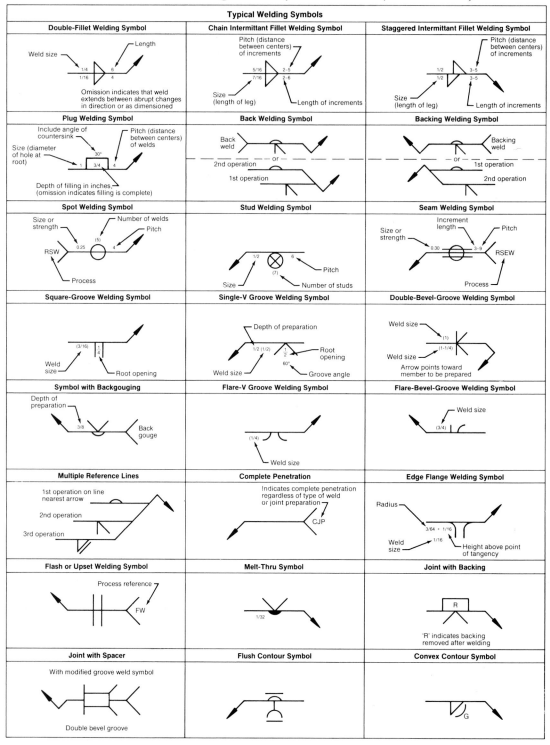

Double-Fillet Welding Symbol	Chain Intermittant Fillet Welding Symbol	Staggered Intermittant Fillet Welding Symbol
Plug Welding Symbol	Back Welding Symbol	Backing Welding Symbol
Spot Welding Symbol	Stud Welding Symbol	Seam Welding Symbol
Square-Groove Welding Symbol	Single-V Groove Welding Symbol	Double-Bevel-Groove Welding Symbol
Symbol with Backgouging	Flare-V Groove Welding Symbol	Flare-Bevel-Groove Welding Symbol
Multiple Reference Lines	Complete Penetration	Edge Flange Welding Symbol
Flash or Upset Welding Symbol	Melt-Thru Symbol	Joint with Backing
Joint with Spacer	Flush Contour Symbol	Convex Contour Symbol

32 Welding Symbols and Processes—
American Welding Society Standard (continued)

MASTER CHART OF WELDING AND ALLIED PROCESSES[c]

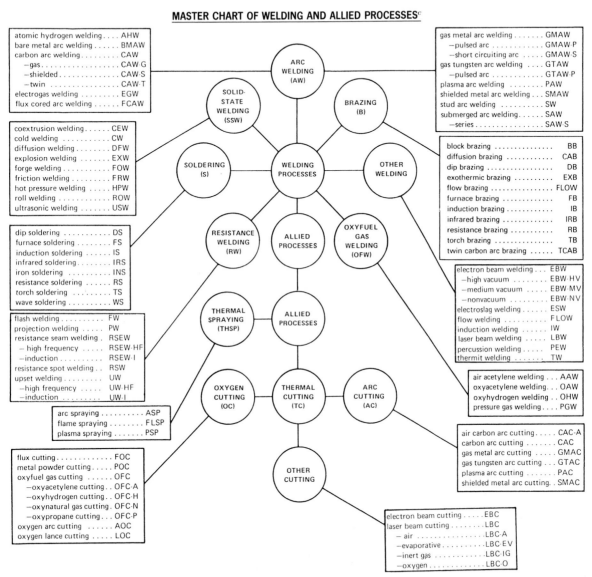

```
atomic hydrogen welding .... AHW
bare metal arc welding ...... BMAW
carbon arc welding ......... CAW
  −gas ................... CAW-G
  −shielded ............. CAW-S
  −twin ................. CAW-T
electrogas welding ........ EGW
flux cored arc welding ...... FCAW
```

```
coextrusion welding ...... CEW
cold welding .......... CW
diffusion welding ....... DFW
explosion welding ...... EXW
forge welding ......... FOW
friction welding ........ FRW
hot pressure welding ..... HPW
roll welding ........... ROW
ultrasonic welding ...... USW
```

```
dip soldering .......... DS
furnace soldering ....... FS
induction soldering ...... IS
infrared soldering ....... IRS
iron soldering ......... INS
resistance soldering ..... RS
torch soldering ........ TS
wave soldering ........ WS
```

```
flash welding ........ FW
projection welding ..... PW
resistance seam welding . RSEW
  − high frequency ..... RSEW-HF
  −induction ......... RSEW-I
resistance spot welding .. RSW
upset welding ........ UW
  −high frequency ..... UW-HF
  −induction ......... UW-I
```

```
arc spraying ......... ASP
flame spraying ........ FLSP
plasma spraying ....... PSP
```

```
flux cutting ........... FOC
metal powder cutting ..... POC
oxyfuel gas cutting ...... OFC
  −oxyacetylene cutting .. OFC-A
  −oxyhydrogen cutting .. OFC-H
  −oxynatural gas cutting . OFC-N
  −oxypropane cutting ... OFC-P
oxygen arc cutting ...... AOC
oxygen lance cutting ..... LOC
```

```
gas metal arc welding ....... GMAW
  −pulsed arc ........... GMAW-P
  −short circuiting arc ..... GMAW-S
gas tungsten arc welding .... GTAW
  −pulsed arc ........... GTAW-P
plasma arc welding ........ PAW
shielded metal arc welding ... SMAW
stud arc welding ......... SW
submerged arc welding ..... SAW
  −series ............... SAW-S
```

```
block brazing ............... BB
diffusion brazing ........... CAB
dip brazing ................ DB
exothermic brazing .......... EXB
flow brazing ............... FLOW
furnace brazing ............ FB
induction brazing ........... IB
infrared brazing ............ IRB
resistance brazing .......... RB
torch brazing .............. TB
twin carbon arc brazing ...... TCAB
```

```
electron beam welding ... EBW
  −high vacuum ....... EBW-HV
  −medium vacuum ..... EBW-MV
  −nonvacuum ........ EBW-NV
electroslag welding ...... ESW
flow welding .......... FLOW
induction welding ...... IW
laser beam welding ..... LBW
percussion welding .... PEW
thermit welding ....... TW
```

```
air acetylene welding ... AAW
oxyacetylene welding... OAW
oxyhydrogen welding .. OHW
pressure gas welding .... PGW
```

```
air carbon arc cutting ..... CAC-A
carbon arc cutting ....... CAC
gas metal arc cutting ..... GMAC
gas tungsten arc cutting ... GTAC
plasma arc cutting ....... PAC
shielded metal arc cutting.. SMAC
```

```
electron beam cutting ..... EBC
laser beam cutting ........ LBC
  − air ................. LBC-A
  −evaporative .......... LBC-EV
  −inert gas ............ LBC-IG
  −oxygen .............. LBC-O
```

[c]ANSI/AWS A3.0−89.

33 Topographic Symbols

Symbol	Name	Symbol	Name
	Highway		National or State Line
	Railroad		County Line
	Highway Bridge		Township or District Line
	Railroad Bridge		City or Village Line
	Drawbridges		Triangulation Station
	Suspension Bridge	BM X 1232	Bench Mark and Elevation
	Dam		Any Location Station (WITH EXPLANATORY NOTE)
	Telegraph or Telephone Line		Streams in General
	Power-Transmission Line		Lake or Pond
	Buildings in General		Falls and Rapids
			Contours
	Capital		Hachures
	County Seat		Sand and Sand Dunes
	Other Towns		Marsh
	Barbed Wire Fence		Woodland of Any Kind
	Smooth Wire Fence		Orchard
	Hedge		Grassland in General
	Oil or Gas Wells		Cultivated Fields
	Windmill		
	Tanks		Commercial or Municipal Field
	Canal or Ditch		Airplane Landing Field Marked or Emergency
	Canal Lock		Mooring Mast
	Canal Lock (POINT UPSTREAM)		Airway Light Beacon (ARROWS INDICATE COURSE LIGHTS)
	Aqueduct or Water Pipe		Auxiliary Airway Light Beacon, Flashing

34 Piping Symbols[a]—American National Standard

	FLANGED	SCREWED	BELL & SPIGOT	WELDED	SOLDERED
1. Joint					
2. Elbow—90°					
3. Elbow—45°					
4. Elbow—Turned Up					
5. Elbow—Turned Down					
6. Elbow—Long Radius					
7. Reducing Elbow					
8. Tee					
9. Tee—Outlet Up					
10. Tee—Outlet Down					
11. Side Outlet Tee—Outlet Up					
12. Cross					
13. Reducer—Concentric					
14. Reducer—Eccentric					
15. Lateral					
16. Gate Valve—Elev.					
17. Globe Valve—Elev.					
18. Check Valve					
19. Stop Cock					
20. Safety Valve					
21. Expansion Joint					
22. Union					
23. Sleeve					
24. Bushing					

[a]ANSI Z32.2.3–1949 (R1953).

35 Heating, Ventilating, and Ductwork Symbols[a]— American National Standard

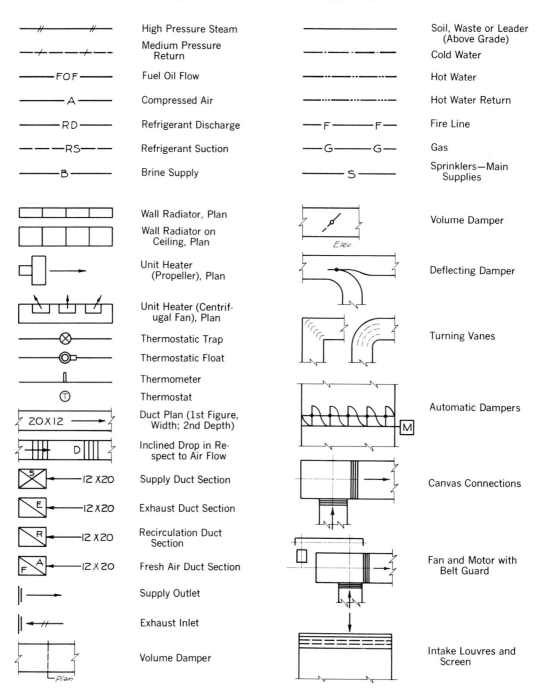

High Pressure Steam

Medium Pressure Return

Fuel Oil Flow

Compressed Air

Refrigerant Discharge

Refrigerant Suction

Brine Supply

Wall Radiator, Plan

Wall Radiator on Ceiling, Plan

Unit Heater (Propeller), Plan

Unit Heater (Centrifugal Fan), Plan

Thermostatic Trap

Thermostatic Float

Thermometer

Thermostat

Duct Plan (1st Figure, Width; 2nd Depth)

Inclined Drop in Respect to Air Flow

Supply Duct Section

Exhaust Duct Section

Recirculation Duct Section

Fresh Air Duct Section

Supply Outlet

Exhaust Inlet

Volume Damper

Soil, Waste or Leader (Above Grade)

Cold Water

Hot Water

Hot Water Return

Fire Line

Gas

Sprinklers—Main Supplies

Volume Damper

Deflecting Damper

Turning Vanes

Automatic Dampers

Canvas Connections

Fan and Motor with Belt Guard

Intake Louvres and Screen

[a]ANSI Z32.2.3–1949 (R1953) and ANSI Y32.2.4–1949 (R1984).

36 American National Standard Graphical Symbols for Electronic Diagrams[a]

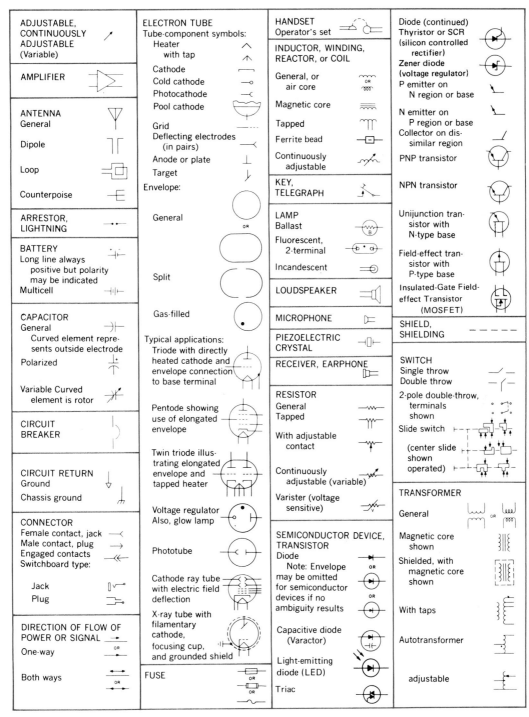

[a]ANSI IEEE 315–1975 (R1989).

37 Form and Proportion of Geometric Tolerancing Symbols[a]

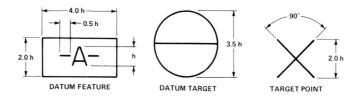

DATUM FEATURE   DATUM TARGET   TARGET POINT

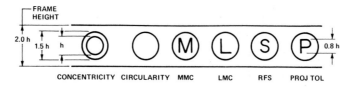

CONCENTRICITY   CIRCULARITY   MMC   LMC   RFS   PROJ TOL

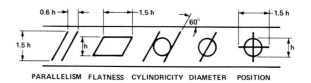

PARALLELISM   FLATNESS   CYLINDRICITY   DIAMETER   POSITION

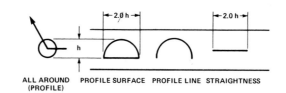

ALL AROUND (PROFILE)   PROFILE SURFACE   PROFILE LINE   STRAIGHTNESS

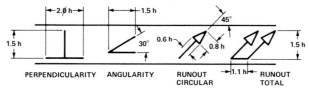

PERPENDICULARITY   ANGULARITY   RUNOUT CIRCULAR   RUNOUT TOTAL

[a]ANSI Y14.5–1982 (R1988).

38   Wrought Steel Pipe[a] and Taper Pipe Threads[b]— American National Standard

All dimensions are in inches except those in last two columns.

Nominal Pipe Size	D Outside Diameter of Pipe	Threads per Inch	L₁[c] Normal Engagement by Hand Between External and Internal Threads	L₂[c] Length of Effective Thread	Sched. 10	Sched. 20[d]	Sched. 30[d]	Sched. 40[d]	Sched. 60[e]	Sched. 80[e]	Sched. 100	Sched. 120	Sched. 140	Sched. 160	Length of Pipe, Feet, per Square Foot External Surface[f]	Length of Standard-Weight Pipe, Feet, Containing 1 cu. ft.[f]
⅛	.405	27	.1615	.2639				.068		.095					9.431	2,533.8
¼	.540	18	.2278	.4018				.088		.119					7.073	1,383.8
⅜	.675	18	.240	.4078				.091		.126					5.658	754.36
½	.840	14	.320	.5337				.109		.147				.188	4.547	473.91
¾	1.050	14	.339	.5457				.113		.154				.219	3.637	270.03
1	1.315	11.5	.400	.6828				.133		.179				.250	2.904	166.62
1¼	1.660	11.5	.420	.7068				.140		.191				.250	2.301	96.275
1½	1.900	11.5	.420	.7235				.145		.200				.281	2.010	70.733
2	2.375	11.5	.436	.7565				.154		.218				.344	1.608	42.913
2½	2.875	8	.682	1.1375				.203		.276				.375	1.328	30.077
3	3.500	8	.766	1.2000				.216		.300				.438	1.091	19.479
3½	4.000	8	.821	1.2500				.226		.318					.954	14.565
4	4.500	8	.844	1.3000				.237		.337		.438		.531	.848	11.312
5	5.563	8	.937	1.4063				.258		.375		.500		.625	.686	7.199
6	6.625	8	.958	1.5125				.280		.432		.562		.719	.576	4.984
8	8.625	8	1.063	1.7125		.250	.277	.322	.406	.500	.594	.719	.812	.906	.443	2.878
10	10.750	8	1.210	1.9250		.250	.307	.365	.500	.594	.719	.844	1.000	1.125	.355	1.826
12	12.750	8	1.360	2.1250		.250	.330	.406	.562	.688	.844	1.000	1.125	1.312	.299	1.273
14 OD	14.000	8	1.562	2.2500	.250	.312	.375	.438	.594	.750	.938	1.094	1.250	1.406	.273	1.065
16 OD	16.000	8	1.812	2.4500	.250	.312	.375	.500	.656	.844	1.031	1.219	1.438	1.594	.239	.815
18 OD	18.000	8	2.000	2.6500	.250	.312	.438	.562	.750	.938	1.156	1.375	1.562	1.781	.212	.644
20 OD	20.000	8	2.125	2.8500	.250	.375	.500	.594	.812	1.031	1.281	1.500	1.750	1.969	.191	.518
24 OD	24.000	8	2.375	3.2500	.250	.375	.562	.688	.969	1.219	1.531	1.812	2.062	2.344	.159	.358

Nominal Wall Thickness

[a] ANSI/ASME B36.10M–1985.
[b] ANSI/ASME B1.20.1–1983.
[c] Refer to §15.22 and Fig. 15.20.
[d] Boldface figures correspond to "standard" pipe.
[e] Boldface figures correspond to "extra strong" pipe.
[f] Calculated values for Schedule 40 pipe.

39 Cast-Iron Pipe, Thicknesses and Weights— American National Standard

Left table

Size, inches	Thickness, inches	Outside Diameter, inches	Avg. per Foot[b]	Per Length
			Weight (lb) Based on	
Class 50: 50 psi Pressure—115 ft Head				
3	.32	3.96	12.4	195
4	.35	4.80	16.5	265
6	.38	6.90	25.9	415
8	.41	9.05	37.0	590
10	.44	11.10	49.1	785
12	.48	13.20	63.7	1,020
14	.48	15.30	74.6	1,195
16	.54	17.40	95.2	1,525
18	.54	19.50	107.6	1,720
20	.57	21.60	125.9	2,015
24	.63	25.80	166.0	2,655
30	.79	32.00	257.6	4,120
36	.87	38.30	340.9	5,455
42	.97	44.50	442.0	7,070
48	1.06	50.80	551.6	8,825
Class 100: 100 psi Pressure—231 ft Head				
3	.32	3.96	12.4	195
4	.35	4.80	16.5	265
6	.38	6.90	25.9	415
8	.41	9.05	37.0	590
10	.44	11.10	49.1	785
12	.48	13.20	63.7	1,020
14	.51	15.30	78.8	1,260
16	.54	17.40	95.2	1,525
18	.58	19.50	114.8	1,835
20	.62	21.60	135.9	2,175
24	.68	25.80	178.1	2,850
30	.79	32.00	257.6	4,120
36	.87	38.30	340.9	5,455
42	.97	44.50	442.0	7,070
48	1.06	50.80	551.6	8,825
Class 150: 150 psi Pressure—346 ft Head				
3	.32	3.96	12.4	195
4	.35	4.80	16.5	265
6	.38	6.90	25.9	415
8	.41	9.05	37.0	590
10	.44	11.10	49.1	785
12	.48	13.20	63.7	1,020
14	.51	15.30	78.8	1,260
16	.54	17.40	95.2	1,525
18	.58	19.50	114.8	1,835
20	.62	21.60	135.9	2,175
24	.73	25.80	190.1	3,040
30	.85	32.00	275.4	4,405
36	.94	38.30	365.9	5,855
42	1.05	44.50	475.3	7,605
48	1.14	50.80	589.6	9,435
Class 200: 200 psi Pressure—462 ft Head				
3	.32	3.96	12.4	195
4	.35	4.80	16.5	265
6	.38	6.90	25.9	415

Right table

Size, inches	Thickness, inches	Outside Diameter, inches	Avg. per Foot[b]	Per Length
			Weight (lb) Based on	
Class 200: 200 psi Pressure—462 ft Head (cont'd)				
8	.41	9.05	37.0	590
10	.44	11.10	49.1	785
12	.48	13.20	63.7	1,020
14	.55	15.30	84.4	1,350
16	.58	17.40	101.6	1,625
18	.63	19.50	123.7	1,980
20	.67	21.60	145.9	2,335
24	.79	25.80	205.6	3,290
30	.92	32.00	297.8	4,765
36	1.02	38.30	397.1	6,355
42	1.13	44.50	512.3	8,195
48	1.23	50.80	637.2	10,195
Class 250: 250 psi Pressure—577 ft Head				
3	.32	3.96	12.4	195
4	.35	4.80	16.5	265
6	.38	6.90	25.9	415
8	.41	9.05	37.0	590
10	.44	11.10	49.1	785
12	.52	13.20	68.5	1,095
14	.59	15.30	90.6	1,450
16	.63	17.40	110.4	1,765
18	.68	19.50	133.4	2,135
20	.72	21.60	156.7	2,505
24	.79	25.80	205.6	3,290
30	.99	32.00	318.4	5,095
36	1.10	38.30	425.5	6,810
42	1.22	44.50	549.5	8,790
48	1.33	50.80	684.5	10,950
Class 300: 300 psi Pressure—693 ft Head				
3	.32	3.96	12.4	195
4	.35	4.80	16.5	265
6	.38	6.90	25.9	415
8	.41	9.05	37.0	590
10	.48	11.10	53.1	850
12	.52	13.20	68.5	1,095
14	.59	15.30	90.6	1,450
16	.68	17.40	118.2	1,890
18	.73	19.50	142.3	2,275
20	.78	21.60	168.5	2,695
24	.85	25.80	219.8	3,515
Class 350: 350 psi Pressure—808 ft Head				
3	.32	3.96	12.4	195
4	.35	4.80	16.5	265
6	.38	6.90	25.9	415
8	.41	9.05	37.0	590
10	.52	11.10	57.4	920
12	.56	13.20	73.8	1,180
14	.64	15.30	97.5	1,605
16	.68	17.40	118.2	1,945
18	.79	19.50	152.9	2,520
20	.84	21.60	180.2	2,970
24	.92	25.80	236.3	3,895

[a]Average weight per foot based on calculated weight of pipe before rounding.

40 Cast-Iron Pipe Screwed Fittings,[a] 125 lb— American National Standard

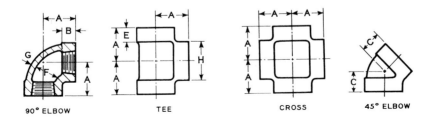

90° ELBOW   TEE   CROSS   45° ELBOW

DIMENSIONS OF 90° AND 45° ELBOWS, TEES, AND CROSSES (STRAIGHT SIZES)

All dimensions given in inches.
Fittings having right- and left-hand threads shall have four or more ribs or the letter "L" cast on the band at end with left-hand thread.

Nominal Pipe Size	Center to End, Elbows, Tees, and Crosses A	Center to End, 45° Elbows C	Length of Thread, Min. B	Width of Band. Min. E	Inside Diameter of Fitting F		Metal Thickness G	Diameter of Band, Min. H
					Max.	Min.		
¼	.81	.73	.32	.38	.58	.54	.11	.93
⅜	.95	.80	.36	.44	.72	.67	.12	1.12
½	1.12	.88	.43	.50	.90	.84	.13	1.34
¾	1.31	.98	.50	.56	1.11	1.05	.15	1.63
1	1.50	1.12	.58	.62	1.38	1.31	.17	1.95
1¼	1.75	1.29	.67	.69	1.73	1.66	.18	2.39
1½	1.94	1.43	.70	.75	1.97	1.90	.20	2.68
2	2.25	1.68	.75	.84	2.44	2.37	.22	3.28
2½	2.70	1.95	.92	.94	2.97	2.87	.24	3.86
3	3.08	2.17	.98	1.00	3.60	3.50	.26	4.62
3½	3.42	2.39	1,03	1.06	4.10	4.00	.28	5.20
4	3.79	2.61	1.08	1.12	4.60	4.50	.31	5.79
5	4.50	3.05	1.18	1.18	5.66	5.56	.38	7.05
6	5.13	3.46	1.28	1.28	6.72	6.62	.43	8.28
8	6.56	4.28	1.47	1.47	8.72	8.62	.55	10.63
10	8.08[b]	5.16	1.68	1.68	10.85	10.75	.69	13.12
12	9.50[b]	5.97	1.88	1.88	12.85	12.75	.80	15.47

[a]From ANSI/ASME B16.4–1985.
[b]This applies to elbows and tees only.

41   Cast-Iron Pipe Screwed Fittings,[a] 250 lb— American National Standard

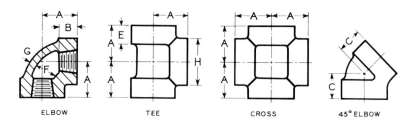

ELBOW   TEE   CROSS   45° ELBOW

DIMENSIONS OF 90° AND 45° ELBOWS, TEES, AND CROSSES (STRAIGHT SIZES)

All dimensions given in inches.
The 250 lb standard for screwed fittings covers only the straight sizes of 90° and 45° elbows, tees, and crosses.

Nominal Pipe Size	Center to End, Elbows, Tees, and Crosses A	Center to End, 45° Elbows C	Length of Thread, Min. B	Width of Band, Min. E	Inside Diameter of Fitting F		Metal Thick- ness G	Outside Diameter of Band, Min. H
					Max.	Min.		
¼	.94	.81	.43	.49	.58	.54	.18	1.17
⅜	1.06	.88	.47	.55	.72	.67	.18	1.36
½	1.25	1.00	.57	.60	.90	.84	.20	1.59
¾	1.44	1.13	.64	.68	1.11	1.05	.23	1.88
1	1.63	1.31	.75	.76	1.38	1.31	.28	2.24
1¼	1.94	1.50	.84	.88	1.73	1.66	.33	2.73
1½	2.13	1.69	.87	.97	1.97	1.90	.35	3.07
2	2.50	2.00	1.00	1.12	2.44	2.37	.39	3.74
2½	2.94	2.25	1.17	1.30	2.97	2.87	.43	4.60
3	3.38	2.50	1.23	1.40	3.60	3.50	.48	5.36
3½	3.75	2.63	1.28	1.49	4.10	4.00	.52	5.98
4	4.13	2.81	1.33	1.57	4.60	4.50	.56	6.61
5	4.88	3.19	1.43	1.74	5.66	5.56	.66	7.92
6	5.63	3.50	1.53	1.91	6.72	6.62	.74	9.24
8	7.00	4.31	1.72	2.24	8.72	8.62	.90	11.73
10	8.63	5.19	1.93	2.58	10.85	10.75	1.08	14.37
12	10.00	6.00	2.13	2.91	12.85	12.75	1.24	16.84

[a]From ANSI/ASME B16.4–1985.

42 Cast-Iron Pipe Flanges and Fittings,[a] 125 lb— American National Standard

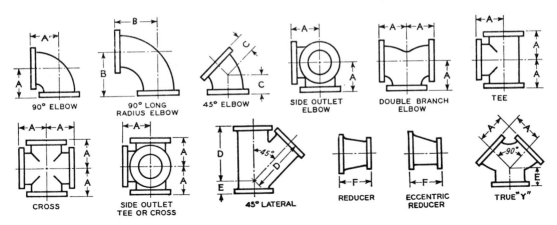

90° ELBOW · 90° LONG RADIUS ELBOW · 45° ELBOW · SIDE OUTLET ELBOW · DOUBLE BRANCH ELBOW · TEE · CROSS · SIDE OUTLET TEE OR CROSS · 45° LATERAL · REDUCER · ECCENTRIC REDUCER · TRUE "Y"

DIMENSIONS OF ELBOWS, DOUBLE BRANCH ELBOWS, TEES, CROSSES, LATERALS, TRUE Y's (STRAIGHT SIZES), AND REDUCERS

All dimensions in inches.

Nominal Pipe Size	Inside Diameter of Fittings	Center to Face 90° Elbow, Tees, Crosses True "Y" and Double Branch Elbow A	Center to Face, 90° Long Radius Elbow B	Center to Face 45° Elbow C	Center to Face Lateral D	Short Center to Face True "Y" and Lateral E	Face to Face Reducer F	Diameter of Flange	Thickness of Flange, Min.	Wall Thickness
1	1.00	3.50	5.00	1.75	5.75	1.75	....	4.25	.44	.31
1¼	1.25	3.75	5.50	2.00	6.25	1.75	....	4.62	.50	.31
1½	1.50	4.00	6.00	2.25	7.00	2.00	....	5.00	.56	.31
2	2.00	4.50	6.50	2.50	8.00	2.50	5.0	6.00	.62	.31
2½	2.50	5.00	7.00	3.00	9.50	2.50	5.5	7.00	.69	.31
3	3.00	5.50	7.75	3.00	10.00	3.00	6.0	7.50	.75	.38
3½	3.50	6.00	8.50	3.50	11.50	3.00	6.5	8.50	.81	.44
4	4.00	6.50	9.00	4.00	12.00	3.00	7.0	9.00	.94	.50
5	5.00	7.50	10.25	4.50	13.50	3.50	8.0	10.00	.94	.50
6	6.00	8.00	11.50	5.00	14.50	3.50	9.0	11.00	1.00	.56
8	8.00	9.00	14.00	5.50	17.50	4.50	11.0	13.50	1.12	.62
10	10.00	11.00	16.50	6.50	20.50	5.00	12.0	16.00	1.19	.75
12	12.00	12.00	19.00	7.50	24.50	5.50	14.0	19.00	1.25	.81
14 OD	14.00	14.00	21.50	7.50	27.00	6.00	16.0	21.00	1.38	.88
16 OD	16.00	15.00	24.00	8.00	30.00	6.50	18.0	23.50	1.44	1.00
18 OD	18.00	16.50	26.50	8.50	32.00	7.00	19.0	25.00	1.56	1.06
20 OD	20.00	18.00	29.00	9.50	35.00	8.00	20.0	27.50	1.69	1.12
24 OD	24.00	22.00	34.00	11.00	40.50	9.00	24.0	32.00	1.88	1.25
30 OD	30.00	25.00	41.50	15.00	49.00	10.00	30.0	38.75	2.12	1.44
36 OD	36.00	28.00	49.00	18.00	....	....	36.0	46.00	2.38	1.62
42 OD	42.00	31.00	56.50	21.00	....	....	42.0	53.00	2.62	1.81
48 OD	48.00	34.00	64.00	24.00	....	....	48.0	59.50	2.75	2.00

[a]ANSI/ASME B16.1–1989.

43 Cast-Iron Pipe Flanges, Drilling for Bolts and Their Lengths,[a] 125 lb—American National Standard

Nominal Pipe Size	Diameter of Flange	Thickness of Flange, Min.	Diameter of Bolt Circle	Number of Bolts	Diameter of Bolts	Diameter of Bolt Holes	Length of Bolts
1	4.25	.44	3.12	4	.50	.62	1.75
1¼	4.62	.50	3.50	4	.50	.62	2.00
1½	5.00	.56	3.88	4	.50	.62	2.00
2	6.00	.62	4.75	4	.62	.75	2.25
2½	7.00	.69	5.50	4	.62	.75	2.50
3	7.50	.75	6.00	4	.62	.75	2.50
3½	8.50	.81	7.00	8	.62	.75	2.75
4	9.00	.94	7.50	8	.62	.75	3.00
5	10.00	.94	8.50	8	.75	.88	3.00
6	11.00	1.00	9.50	8	.75	.88	3.25
8	13.50	1.12	11.75	8	.75	.88	3.50
10	16.00	1.19	14.25	12	.88	1.00	3.75
12	19.00	1.25	17.00	12	.88	1.00	3.75
14 OD	21.00	1.38	18.75	12	1.00	1.12	4.25
16 OD	23.50	1.44	21.25	16	1.00	1.12	4.50
18 OD	25.00	1.56	22.75	16	1.12	1.25	4.75
20 OD	27.50	1.69	25.00	20	1.12	1.25	5.00
24 OD	32.00	1.88	29.50	20	1.25	1.38	5.50
30 OD	38.75	2.12	36.00	28	1.25	1.38	6.25
36 OD	46.00	2.38	42.75	32	1.50	1.62	7.00
42 OD	53.00	2.62	49.50	36	1.50	1.62	7.50
48 OD	59.50	2.75	56.00	44	1.50	1.62	7.75

[a]ANSI B16.1–1989.

44 Shaft Center Sizes

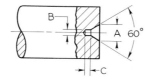

Shaft Diameter D	A	B	C	Shaft Diameter D	A	B	C
³⁄₁₆ to ⁷⁄₃₂	⁵⁄₆₄	³⁄₆₄	¹⁄₁₆	1⅛ to 1¹⁵⁄₃₂	⁵⁄₁₆	⁵⁄₃₂	⁵⁄₃₂
¼ to ¹¹⁄₃₂	³⁄₃₂	³⁄₆₄	¹⁄₁₆	1½ to 1³¹⁄₃₂	⅜	³⁄₃₂	⁵⁄₃₂
⅜ to ¹⁷⁄₃₂	⅛	¹⁄₁₆	⁵⁄₆₄	2 to 2³¹⁄₃₂	⁷⁄₁₆	⁷⁄₃₂	³⁄₁₆
⁹⁄₁₆ to ²⁵⁄₃₂	³⁄₁₆	⁵⁄₆₄	³⁄₃₂	3 to 3³¹⁄₃₂	½	⁷⁄₃₂	⁷⁄₃₂
¹³⁄₁₆ to 1³⁄₃₂	¼	³⁄₃₂	³⁄₃₂	4 and over	⁹⁄₁₆	⁷⁄₃₂	⁷⁄₃₂

45   Cast-Iron Pipe Flanges and Fittings,[a] 250 lb—American National Standard

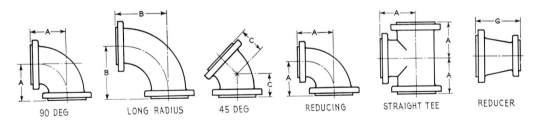

DIMENSIONS OF ELBOWS, TEES, AND REDUCERS

All dimensions are given in inches.

Nominal Pipe Size	Inside Diameter of Fitting, Min.	Wall Thickness of Body	Diameter of Flange	Thickness of Flange Min.	Diameter of Raised Face	Center-to-Face Elbow and Tee A	Center-to-Face Long Radius Elbow B	Center-to-Face 45° Elbow C	Face-to-Face Reducer G
1	1.00	.44	4.88	.69	2.69	4.00	5.00	2.00	....
1¼	1.25	.44	5.25	.75	3.06	4.25	5.50	2.50	....
1½	1.50	.44	6.12	.81	3.56	4.50	6.00	2.75	....
2	2.00	.44	6.50	.88	4.19	5.00	6.50	3.00	5.00
2½	2.50	.50	7.50	1.00	4.94	5.50	7.00	3.50	5.50
3	3.00	.56	8.25	1.12	5.69	6.00	7.75	3.50	6.00
3½	3.50	.56	9.00	1.19	6.31	6.50	8.50	4.00	6.50
4	4.00	.62	10.00	1.25	6.94	7.00	9.00	4.50	7.00
5	5.00	.69	11.00	1.38	8.31	8.00	10.25	5.00	8.00
6	6.00	.75	12.50	1.44	9.69	8.50	11.50	5.50	9.00
8	8.00	.81	15.00	1.62	11.94	10.00	14.00	6.00	11.00
10	10.00	.94	17.50	1.88	14.06	11.50	16.50	7.00	12.00
12	12.00	1.00	20.50	2.00	16.44	13.00	19.00	8.00	14.00
14 OD	13.25	1.12	23.00	2.12	18.94	15.00	21.50	8.50	16.00
16 OD	15.25	1.25	25.50	2.25	21.06	16.50	24.00	9.50	18.00
18 OD	17.00	1.38	28.00	2.38	23.31	18.00	26.50	10.00	19.00
20 OD	19.00	1.50	30.50	2.50	25.56	19.50	29.00	10.50	20.00
24 OD	23.00	1.62	36.00	2.75	30.31	22.50	34.00	12.00	24.00
30 OD	29.00	2.00	43.00	3.00	37.19	27.50	41.50	15.00	30.00

[a]ANSI B161–1989.

46   Cast-Iron Pipe Flanges, Drilling for Bolts and Their Lengths,[a] 250 lb—American National Standard

Nominal Pipe Size	Diameter of Flange	Thickness of Flange, Min.	Diameter of Raised Face	Diameter of Bolt Circle	Diameter of Bolt Holes	Number of Bolts	Size of Bolts	Length of Bolts	Length of Bolt Studs with Two Nuts
1	4.88	.69	2.69	3.50	.75	4	.62	2.50	....
1¼	5.25	.75	3.06	3.88	.75	4	.62	2.50	....
1½	6.12	.81	3.56	4.50	.88	4	.75	2.75	....
2	6.50	.88	4.19	5.00	.75	8	.62	2.75	....
2½	7.50	1.00	4.94	5.88	.88	8	.75	3.25	....
3	8.25	1.12	6.69	6.62	.88	8	.75	3.50	....
3½	9.00	1.19	6.31	7.25	.88	8	.75	3.50	....
4	10.00	1.25	6.94	7.88	.88	8	.75	3.75	....
5	11.00	1.38	8.31	9.25	.88	8	.75	4.00	....
6	12.50	1.44	9.69	10.62	.88	12	.75	4.00	....
8	15.00	1.62	11.94	13.00	1.00	12	.88	4.50	....
10	17.50	1.88	14.06	5.25	1.12	16	1.00	5.25	....
12	20.50	2.00	16.44	17.75	1.25	16	1.12	5.50	....
14 OD	23.00	2.12	18.94	20.25	1.25	20	1.12	6.00	....
16 OD	25.50	2.25	21.06	22.50	1.38	20	1.25	6.25	....
18 OD	28.00	2.38	23.31	24.75	1.38	24	1.25	6.50	....
20 OD	30.50	2.50	25.56	27.00	1.38	24	1.25	6.75	....
24 OD	36.00	2.75	30.31	32.00	1.62	24	1.50	7.50	9.50
30 OD	43.00	3.00	37.19	39.25	2.00	28	1.75	8.50	10.50

[a]ANSI B16.1–1989.

Index

D

Sheet Layouts

A convenient code to identify American National Standard sheet sizes and forms suggested by the authors for title, parts or material list, and revision blocks, for use of instructors in making assignments, is shown here. All dimensions are in inches.

Three **sizes** of sheets are illustrated: **Size A**, Fig. I, **Size B**, Fig. V, and **Size C**, Fig. VI. Metric size sheets are not shown.

Eight **forms** of lettering arrangements are suggested, known as **Forms 1, 2, 3, 4, 5, 6, 7,** and **8,** as shown below and opposite. The total length of **Forms 1, 2, 3,** and **4** may be adjusted to fit **Sizes A4, A3,** and **A2.**

The term **layout** designates a sheet of certain size plus a certain arrangement of lettering. Thus **Layout A–1** is a combination of **Size A,** Fig. I, and **Form 1,** Fig. II. **Layout C–678** is a combination of **Size C,** Fig. VI, and **Forms 6, 7,** and **8,** Figs. IX, X, and XI. **Layout A4–2** (adjusted) is a combination of **Size A4** and **Form 2,** Fig. III, adjusted to fit between the borders. Other combinations may be employed as assigned by the instructor.

Fig. II   Form 1. Title Block

Fig. III   Form 2. Title Block

Fig. IV   Form 3. Title Block

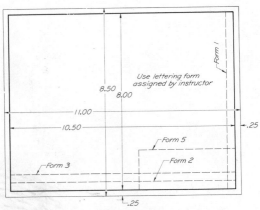

Fig. I   Size A Sheet (8.50″ × 11.00″)

Sheet Sizes

American National Standard

A – 8.50″ × 11.00″
B – 11.00″ × 17.00″
C – 17.00″ × 22.00″
D – 22.00″ × 34.00″
E – 34.00″ × 44.00″

International Standard

A4 – 210 mm × 297 mm
A3 – 297 mm × 420 mm
A2 – 420 mm × 594 mm
A1 – 594 mm × 841 mm
A0 – 841 mm × 1189 mm
      (25.4 mm = 1.00″)

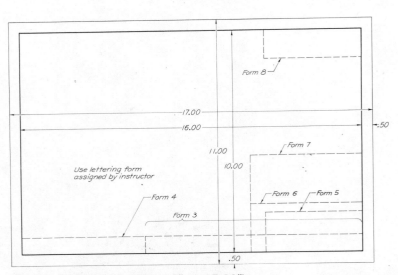

Fig. V   Size B Sheet (11.00″ × 17.00″)